ANNUAL REVIEW OF ASTRONOMY
AND ASTROPHYSICS

ANNUAL REVIEW OF ASTRONOMY AND ASTROPHYSICS

VOLUME 32, 1994

GEOFFREY BURBIDGE, *Editor*
University of California, San Diego

DAVID LAYZER, *Associate Editor*
Harvard College Observatory

ALLAN SANDAGE, *Associate Editor*
Observatories of the Carnegie Institution of Washington

ANNUAL REVIEWS INC   4139 EL CAMINO WAY   P.O. BOX 10139   PALO ALTO, CALIFORNIA 94303-0139

ANNUAL REVIEWS INC.
Palo Alto, California, USA

International Standard Serial Number: 0066-4146
International Standard Book Number: 0-8243-0932-4
Library of Congress Catalog Card Number: 63-8846

Annual Review and publication titles are registered trademarks of Annual Reviews Inc.

Annual Reviews Inc. and the Editors of its publications assume no responsibility for the statements expressed by the contributors to this *Review*.

TYPESET BY TECHBOOKS
PRINTED AND BOUND IN THE UNITED STATES OF AMERICA

PREFACE

This volume was planned at a meeting held on May 9, 1992 in San Francisco, CA. Those who attended the meeting included Geoffrey Burbidge (Editor), David Layzer and Allan Sandage (Associate Editors), John Leibacher, Morton Roberts, Anneila Sargent, Tom Soifer, and Frank Shu (Committee Members). Chris McKee and Stan Woosley came as guests.

In the preface to Volume 31, I pointed out that 25 articles were scheduled for this volume. In fact, there are 14 contained here. At present 29 articles are scheduled for Volume 33 (1995).

Once again, I would like to thank the Production Editor, David Couzens, for doing an excellent job in Palo Alto.

GEOFFREY BURBIDGE

SOME RELATED ARTICLES IN OTHER *ANNUAL REVIEWS*

From the *Annual Review of Earth and Planetary Sciences*, Volume 22 (1994):

Meteorite and Asteroid Reflectance Spectroscopy, C.M. Pieters and L.A. McFadden

Physics of Zodiacal Dust, B.Å.S. Gustafson

From the *Annual Review of Nuclear and Particle Science*, Volume 43 (1993):

The Status of the Solar Neutrino Problem, T. J. Bowles and V. N. Gavrin

The Search for Discrete Astrophysical Sources of Energetic Gamma Radiation, J.W. Cronin, K. G. Gibbs, and T.C. Weekes

Annual Review of Astronomy and Astrophysics
Volume 32, 1994

CONTENTS

WATCHER OF THE SKIES, *E. Margaret Burbidge*   1

ASTEROSEISMOLOGY, *Timothy M. Brown and Ronald L. Gilliland*   37

THE GOLDILOCKS PROBLEM: Climatic Evolution and Long-Term Habitability of Terrestrial Planets, *Michael R. Rampino and Ken Caldeira*   83

PHYSICAL PARAMETERS ALONG THE HUBBLE SEQUENCE, *Morton S. Roberts and Martha P. Haynes*   115

THE R-, S-, AND P-PROCESSES IN NUCLEOSYNTHESIS, *Bradley S. Meyer*   153

ABUNDANCES IN THE INTERSTELLAR MEDIUM, *T.L. Wilson and R.T. Rood*   191

MASSIVE STAR POPULATIONS IN NEARBY GALAXIES, *André Maeder and Peter S. Conti*   227

COOLING FLOWS IN CLUSTERS OF GALAXIES, *A.C. Fabian*   277

ANISOTROPIES IN THE COSMIC MICROWAVE BACKGROUND, *Martin White, Douglas Scott, and Joseph Silk*   319

DYNAMICS OF COSMIC FLOWS, *Avishai Dekel*   371

COSMIC DUSTY PLASMAS, *D.A. Mendis and M. Rosenberg*   419

PRE-MAIN-SEQUENCE BINARY STARS, *Robert D. Mathieu*   465

BARYONIC DARK MATTER, *Bernard Carr*   531

BINARY AND MILLISECOND PULSARS, *E.S. Phinney and S. R. Kulkarni*   591

INDEXES
    Subject Index   641
    Cumulative Index of Contributing Authors, Volumes 22–32   655
    Cumulative Index of Chapter Titles, Volumes 22–32   657

ANNUAL REVIEWS INC. is a nonprofit scientific publisher established to promote the advancement of the sciences. Beginning in 1932 with the *Annual Review of Biochemistry*, the Company has pursued as its principal function the publication of high-quality, reasonably priced *Annual Review* volumes. The volumes are organized by Editors and Editorial Committees who invite qualified authors to contribute critical articles reviewing significant developments within each major discipline. The Editor-in-Chief invites those interested in serving as future Editorial Committee members to communicate directly with him. Annual Reviews Inc. is administered by a Board of Directors, whose members serve without compensation.

For the convenience of readers, a detachable order form/envelope is bound into the back of this volume.

E. Margaret Burbidge

Annu. Rev. Astron. Astrophys. 1994. 32: 1–36

WATCHER OF THE SKIES

E. Margaret Burbidge

Center for Astrophysics and Space Sciences, University of California, San Diego, La Jolla, California 92093-0111

FAMILY BACKGROUND

It is presumptuous to borrow words from a poet for the title of this prosaic account of a lifetime in astronomy, but I do so because of my love of the poetry of Keats. My love of astronomy will, I hope, be clear from this memoir.

My lifetime in astronomy begins with my family; therefore that is how I will begin this account. My father was a chemist—a lecturer at the Manchester School of Technology (MST), and my mother was one of his students. Her determination, her will to succeed, and her interest in all the natural sciences are gifts that she passed on to me.

My mother told me that she had an excellent, supportive teacher in high school, who suggested that, with her ability in mathematics and natural sciences, she should, after leaving school, enroll as an undergraduate at the MST. But my widowed grandfather, typical in his British 19th century view of the traditional path his younger daughter's life should follow, said "no"; she should apply herself to the arts of housekeeping and, since she was very attractive, she could expect to find a suitable husband without difficulty.

Here came the crucial support from my mother's school teacher, who encouraged her to sit for a scholarship exam offered by the MST, and guided her studies in preparation. The exam was taken without my grandfather's knowledge and, not surprisingly, my mother won a scholarship.

It was time for the crucial confrontation: The school teacher made an appointment to talk with my grandfather, and pointed out that it would be civically irresponsible to deny such a promising student, so high in the scholarship exam results, the opportunity to accept the scholarship and attend the college. My grandfather was stern, but basically kind and certainly civically responsible (for example, he was a strong supporter of the well-known Hallé Orchestra of Manchester, and a generous provider of financial support to struggling young musicians). Therefore, he dutifully but reluctantly gave his consent.

1

0066–4146/94/0915–0001$05.00

My mother chose to major in chemistry at MST. She never revealed to me the steps that led to that decision, so I can but guess. Chemistry is fascinating, and its relation to the 19th and early 20th century optimistic view toward improving the quality of life for all, must have played a part. And I cannot rule out the immediate reaction of a sheltered, 18-year-old girl to my father, who was good-looking, a very gifted experimental chemist, and a wonderful teacher, with a strong sense of humor, seventeen years older than herself, and with an eye for a pretty girl. In his class there were only two women students taking chemistry. His twinkling blue eyes must have settled on my mother. She told me that all the students joked about his idiosyncrasies, something that happens only to teachers who have strong personalities and are either excellent or very bad teachers. He was good. He taught inorganic chemical analysis from a small textbook which he had written, although his own natural interests were in organic chemistry. My mother told me that in his classes he described various precipitates resulting from standard inorganic test procedures, in ways that produced giggles from the two women students. A precipitate was described as "lavender-violet in color"— which did he mean? He clearly knew little about colors of women's dresses. But his description of one precipitate (or solution, I do not know which) as "peach-blossom color" set his name for that class forever: Stanley John Peachey was now referred to by the class as a whole as "old Peachblossom."

His name, my maiden name, "Peachey," was reputed to come from Huguenot ancestors, presumably fishermen, who fled from Brittany to England during the persecution of the Huguenots.

The inevitable happened: My mother and father married despite my grandfather's opposition, in 1916 (a signet ring of my father's which I possess bears the inner inscription March 12, 1912, and I never extracted the significance of that date from my mother).

After their marriage, my father obtained lucrative patents for some of his research inventions in rubber chemistry, particularly one which greatly speeded up the vulcanization of rubber. He left his teaching position, they moved to London around 1921, and he set up his own industrial chemical laboratory for further research.

I was born on August 12, 1919, a date which, at age 11 or 12 when I had been given a much Bowdlerised account of the "facts of life" by my mother, struck me as a weird and wonderful coincidence: The beginning of my existence as a few-celled creature must have coincided with the World War I Armistice— November 11, 1918. My excitement in telling my mother this deduction was not greeted with enthusiasm nor with any further explanation.

EARLY YEARS; SCHOOL YEARS, FUN WITH THE STARS, MATHEMATICS, AND SCIENCES

My pathway to astronomy led through what was then a fairly normal route. At age 4, before beginning school, my first view of the beauty of stars in the summer sky during a night-time boat crossing from England to France was the earliest step toward a lifetime love of astronomy. Then I developed an early interest in arithmetic and in numbers (especially large ones with many powers of ten to write out and contemplate); this began in my first years in school. I had learnt to read before going to school, so books were a continuing delight. My parents gave me books written for children on all the natural sciences, and reading these was coupled with both my mother's and father's willingness to show me and tell me about the wonders of the seashore, of flowers, plants, and trees (both my sister and I became passionate tree climbers throughout Hampstead Heath, near which we lived). My love of flowers is lifelong, and has been inherited by my own daughter.

I did not have a telescope; viewing the stars and planets was confined to what I could see with my father's binoculars. My parents subscribed to a weekly publication, *The Children's Newspaper*, which carried a regular feature on an inner page that described currently interesting sights, such as the brightness of Venus in the evening sky, the phases of the moon, close passage of any naked-eye planets by each other or past noticeable stars, and sometimes there were descriptions of constellations. I became fascinated by the changing phase and position of the moon, and it remains an ingrained habit to glance up at the moon, particularly when it is near new or full. Obviously that habit has been deeply embedded in my mind by the observational astronomer's concern with the bright or dark half of the lunar cycle!

In the present-day climate of concern over the quality of education in the U.S., it distresses me to read of people who have never wondered what causes the changing phases of the moon, who have no idea what causes the seasons, and some who do not even know that the Earth moves yearly in orbit around the Sun, much less what causes a lunar eclipse. How many decades of abysmal apathy have led to this state of affairs? We cannot reach out and touch the stars, but it is a revelation to witness the fascination one can arouse by giving a popular talk or slide show to school classes or to groups of teachers.

My father gave me a microscope and a chemistry set. I had great fun with both, although my father was always watching rather anxiously over the latter, exhorting me to be careful and to be clean and tidy in preparing my chemistry experiments. One of the tales my mother told of her undergraduate years was of "Old Peachblossom" reiterating to his class that they should be able to carry out chemistry experiments "on top of a grand piano, wearing evening dress" (although why they should want to do so was a good question). He deplored my mother's lab overall, spotted with acid burns and other spills.

When I was 12 or 13 years old, my grandfather gave me Sir James Jeans' popular books on astronomy. Suddenly, I saw my fascination with the stars, born at age 4, linked to my other delight, large numbers. That the nearest star is 26,000,000,000,000 miles away revived those excitements of my first school years (although falling short of my then favorite contemplation, 1 followed by 36 zeros). I decided then and there that the occupation I most wanted to engage in "when I was grown up" was to determine the distances of the stars. My mother recalled telling me, as I lay on my stomach on the floor reading the wonders described by Jeans, that it was bedtime, and that I pleaded for a little more time: "Mum, it's so exciting!"

At school (Francis Holland School for Girls), the teaching of general science (not much physics, some chemistry, more botany) was not extensive, but mathematics was taught well and the science teacher, Mary Pearson Barter, was excellent; I have never forgotten the debt I owe to her. My last year in school was spent in waiting for my 17th birthday, when the Dean at University College London (UCL) would accept me as an undergraduate. During that last year, I attended only the classes I chose—more mathematics, French and Latin, beginning German, and most enticing, individual lab work in science. Miss Barter gave me permission to work alone in the lab, with some rather primitive physics equipment which was stored in boxes in the greenhouse on the roof. There were magnets, circuits, lenses, mirrors, prisms, and suchlike, all very simple; she recommended an easy first-year undergraduate textbook on experiments that one could set up oneself. All my friends in their final school year were studying for Oxford entrance (history, English literature, languages, etc), or were still studying for what are now called O and A levels. Although discipline at that school was fairly strict, there was no ban on nonconformity and the incipient divergence of our paths through life created no problems—those girls remained my friends.

1936–1947: UNIVERSITY, WORLD WAR II, GRADUATE STUDIES

For my university years, my mother (my father had died after a long illness) chose University College, London, (UCL on Gower Street). It was a wise choice as it turned out, much better than Cambridge would have been. A first year of physics, chemistry, and pure and applied mathematics, was leading toward the choice between chemistry or mathematics as a major, when I discovered that UCL offered a major in astronomy with a math minor. I had thought that my interest in astronomy would have to be pursued as an amateur, but it seemed that there might be career possibilities. So, without worrying about future jobs, I began studies with Christopher Clive Langton Gregory (C. C. L. Gregory, whose son is the well-known expert in visual perception, Richard Gregory) and Elizabeth Williamson.

Experimental work was carried out with some difficulty, considering the frequency of cloudy nights, on two small telescopes in the forecourt of UCL. If the stars were veiled by the clouds over London, we learnt the intricacies of setting up telescopes, determining the errors of alignment, collimation, circle calibration, principles of navigation, and how to measure and compute orbits of binary stars (artificial ones, created by small lights reflecting in a mercury bead!).

In my third year, astronomy students took courses in atomic and molecular physics from Professor Dingle and Dr. Pearse at Imperial College in South Kensington, and, a highlight of the year, we were taken to the University of London Observatory (ULO) in Mill Hill, which seemed at that time to be on the outskirts of London and therefore a "real" observatory. We were shown, without being allowed to use them, the 24-inch reflector, the gift of which Gregory had arranged with a donor (Mr. Wilson) in Ireland, and the twin Radcliffe refractor, an 18-inch visual and a 24-inch photographic instrument, again acquired by Gregory's efforts from Oxford (Gregory 1966), where they had been housed in the Radcliffe building. On learning that the 24-inch was beginning a program of measuring stellar parallaxes, I remembered the James Jeans books of my childhood and, for a few months at least, declared my principal goal in astronomical research to be the determination of the distances of as many stars as possible.

Those three years as an undergraduate passed all too quickly, and created, or rather fostered, an enduring fascination with the physical sciences—the many branches of physics, especially astronomy and astrophysics, and geology. These years also provided an awakening on the social side, involving new male and female friendships. My favorite workplace at the time was a certain desk in the science library, on the ground floor to the left of the imposing steps and portico of UCL. It was, sadly, later demolished by Hitler's bombs during the war; I felt that as a personal loss.

My graduation was marked by no ceremonies; in the summer of 1939 it was obvious that Britain was headed for war with Nazi Germany. What useful work was I fitted for? Gregory provided me with an introduction to a computing firm which worked on the tables for the Nautical Almanac. Today's readers of this Prefatory Chapter should realize that computing, in those far-off years, was done mainly by logarithms and hand-cranked small machines. Helped by my B.Sc. record, and the recommendations of Gregory and Williamson, I was accepted, and left early on a Monday morning for my first day at work. It did not turn out well. A lengthy private luncheon, which I had assumed would be an occasion to welcome me into the firm, with an opportunity to meet other employees, convinced me that I did not want to work there. I made a quick and, to me, important decision: At the end of one day's work, I submitted my resignation.

For a few weeks, I joined my mother who had chosen the ARP—The Air

Raid Wardens—as her war work. The post to which she was assigned was in Hampstead, not far from the flat where we lived. The work was of course unpaid, but as she was in comfortable, though by no means wealthy, circumstances, she did not want to earn wages or salary, she simply wanted to serve her country with useful work to the best of her ability. She hated Hitlerite Germany, and to the end of her life could not forgive even those Germans who had suffered under Hitler.

September 3, 1939, dawned, and England was at war. For months it was referred to as "the phony war," because the air raids which had been expected to start immediately did not happen for months. Thus my "work" in the ARP was as a "gofer"—a runner of errands, a body for first aid learners to practice their coming tasks of rescuing and giving first aid to injured people from bombed or fire-bombed buildings. My mother's tasks in the daytime were clerical, in the ARP post, and at nighttime, checking all houses and flats in the area to make sure all windows were properly blacked-out, and helping those who could not fix their curtains effectively. A census of inhabitants in the area for which her post was responsible was kept up-to-date, so that in the event of direct bomb hits the wardens and police could know how many injured or dead were to be taken care of.

When the Battle of Britain started, I had a good and, I felt, more useful job offered by C. C. L. Gregory. He had been sent to the Admiralty, for classified work concerned with protection against submarine warfare and bombing attacks. At ULO, the astronomer had gone into the Army, and the mechanic/technician had gone into the Airforce for work on repair and maintenance of airplanes. I was offered a job which combined caretaking, maintenance of equipment, seeing about repairs to buildings for shrapnel damage, and the opportunity to use the 24-inch Wilson reflector and a spectrograph for a start on research towards a Ph.D.

Mill Hill, northwest of London, suffered relatively little from the German bombing attacks; it was southeast London that bore the brunt. The 24-inch mirror, deemed replaceable, was left in place, but the lenses of the 18- and 24-inch refractors were removed and stored safely at the bottom of the concrete telescope pier. Starting in 1940, I used the Wilson telescope and its inefficient spectrograph to work at night on Be stars—my Ph.D. thesis project. This ousted parallax work as my abiding interest, and stellar spectroscopy became part of my life. There were two other war-related occupations during the next few years—growing vegetables for the Gregory household in the grounds around the Observatory, and doing subcontract work for the Ministry of Defense on two small individual projects, one assembling a consignment of optical instruments for measuring approach angle and speed of attacking aircraft, and one under a microscope filling two small holes etched in glass plates with a suitable evaporative oil and powdered black enamel, for use in stereoscopic instruments for assessing photographic surveys of allied bombing results on enemy targets.

Those nights, standing or sitting on a ladder in the dome of the Wilson reflector, guiding a star on the slit of the spectrograph, fulfilled my early dreams. I have never tired of the joy of looking through the slit in the darkened dome and watching the stars. The Wilson telescope was so antiquated that it was driven by a sector operated by a hanging weight. One had to remember the times for stopping the drive, setting the sector back, and winding up the weight for its next fall. I often think about the joys of work in an open dome, under the stars, next to the telescope, joys denied to most younger astronomers and students who must sit in a warm console room, facing a television guiding screen and many complex computer interfaces, well removed from the telescope itself. I feel lucky that for 25 years after my career in astronomy really began, my telescope work was always conducted in admittedly ever-improving and more sophisticated ways but ways which my early training had led me to love.

The purpose of my research was to try to understand the physics of the Be stars, using Gamma Cassiopeiae as the prototype. The broad H and He absorption lines indicated fast rotation; double emission lines within the absorptions would come from a rotating ring or disk. The aperiodic variations in relative intensity of the emission doublets had been studied especially by D. B. McLaughlin, Otto Struve, and R. E. Baldwin. The model envisaged involved outflow from the stellar surface, maybe due to equatorial instability, and the appearance of sharp absorptions, strong in metastable He I λ 3889, indicated the formation of an outer shell of gas that was stable for months before eventually dissipating. While sitting on the observing ladder, I used to picture the star, its surface seething with turbulent outflowing gas, driven by radiation pressure and balanced by stellar gravity versus equatorial instability from the rapid rotation. I used to picture those photons traveling through space and time, waiting for me or someone cleverer than me to make sense of the physics of the processes in the outer layers of such stars.

With the end of the European side of WWII, VE Day, the small ULO staff returned to Mill Hill. C. C. L. Gregory, back from the Admiralty, directed the replacement and adjustment of the lenses in the refracting telescopes, stellar parallax and proper motion work resumed, Roger Pring returned from the army to become First Assistant, the mechanic returned from the Air Force, and I became Second Assistant. Observations made at post-sunset and pre-dawn for parallax, and in the middle of the night for proper motion, were resumed, and the limitations of the Wilson Reflector and its spectrograph became increasingly apparent to me. The exchange of astronomical literature recovered after the hiatus caused by the war, and the publications by Otto Struve especially provided the goal and impetus toward my aim to have access to larger telescopes, better instruments, and clear skies. Having read an advertisement in, I think, *The Observatory* for Carnegie Fellowships at Mt. Wilson Observatory, I put together an application; I did not have many publications but my undergraduate record was good and my research plans in stellar spectroscopy would, I felt, give me a reasonable chance for the award of a Carnegie Fellowship.

The letter of denial opened my eyes to a new and somewhat frightening situation: new, because I had never before experienced gender-based discrimination. The turn-down letter simply pointed out that Carnegie Fellowships were available only for men, although the advertisement had not stated that fact. Apparently, women were not allowed to use the Mt. Wilson telescopes, and Carnegie Fellows were guaranteed the right to apply for observing time on those telescopes.

A guiding operational principle in my life was activated: If frustrated in one's endeavor by a stone wall or any kind of blockage, one must find a way around—another route towards one's goal. This is advice I have given to many women facing similar situations. I tell them: Try it, it works.

In the interim, before the prospect of better telescopes and better skies was realized, other important changes in my life occurred.

1947–1951: MARRIAGE, L'OBSERVATOIRE DE HAUTE PROVENCE, PLANNING TOWARD YERKES

In the autumn of 1947, following C. C. L. Gregory's advice, I enrolled in some graduate-level courses at UCL. The college building had suffered such severe bomb damage that the lectures were not conducted in the old physics area, but in Foster Court, where geography had previously been taught; damage there was slight. The courses were given by David Bates, one course straight out of *The Theory of Atomic Spectra,* by Condon and Shortley, and one on the night sky emissions in the upper atmosphere, the application of quantum physics to the excitation and de-excitation of oxygen ions. The small class contained some students who had been pre-war undergraduate contemporaries of mine, and some new graduate students enrolled with Professor H. S. W. Massey. One of these, from Bristol University, sat next to me, and we became friends. His name, Geoffrey Burbidge, is familiar to all readers of this prefatory chapter. We began to talk about many aspects of physics, but also about many other topics—tennis, opera, theatre, the Cotswold countryside, politics, and history. The not-very-long delayed culmination of our friendship was marriage on April 2, 1948.

Geoffrey (hereafter in this article, Geoff) and I rented an apartment created in the Gregorys' house in Mill Hill. Geoff's thesis project was the study of the mesonic Auger effect in cosmic rays; his supervisor was Professor H. S. W. Massey. This work required the calculation of mathematical functions which can nowadays all be found in tabular form. The calculations were done on a Brunsviga hand-cranked calculating machine which we bought, and which we still possess. It played its role later when Geoff used it extensively in the work with Fowler and Hoyle on stellar nucleosynthesis (B^{2}FH—see later). (Willy Fowler called this machine, which he could hear cranking in the office where we sat round the corner from his in the Kellogg Radiation Laboratory at Caltech, our "Babbage machine.")

In Mill Hill, in addition to carrying on his Ph.D. thesis work, Geoff joined in the parallax and proper motion photographic work at the University of London Observatory. The walk of a mile or so between our apartment in the Gregory house to the Observatory involved some interesting encounters with the nighttime police patrols, who, demanding to know what we were doing on the streets at such ungodly hours, found our account of our work puzzling although interesting.

After our marriage in April, 1948, we applied and were accepted as members of the International Astronomical Union, and were permitted to attend the first post-war IAU General Assembly, held in Zurich in the summer of 1948. IAU meetings in that era were so much smaller than the colossal gatherings they are now, and I was able to meet and talk to Otto Struve, then President of IAU Commission 29 on stellar spectra, whose work on stellar spectroscopy was so important to me in guiding my thesis work and my future career. Dr. Struve, hearing about the difficulty I had in getting to the *better telescopes, better instruments, clear skies,* said it should be possible for me to apply for an IAU grant to go to the U.S., and to apply for the new Fulbright travel grants in order to get there.

We did not immediately follow this advice. First, it seemed important to see what we could do in Europe. As a result of contacts made during the Zurich IAU meeting, we applied for permission to travel to L'Observatoire de Haute Provence (OHP), St. Michel, France, to use the 80-cm telescope and spectrograph to continue work on the Be stars. Dr. Ch. Febrenbach gave us some weeks in the summer of 1949 on that telescope, so we applied to the Royal Society, London, for a modest amount of travel funds to get to Haute Provence. Readers will have no trouble in guessing the response from Professor R. Redman: You should not be seeking observational opportunities outside the U.K., use the telescopes we have available here; request denied.

This again activated the principle I have already described: If you meet with a blockage, find a way around it. In this case, the way around was to scrape up all of our available funds (not much!) and pay our own way to OHP.

Once there, in the summer of 1949, our eyes were opened again to the beauty of clear, dark, summer skies, and the advantage of the OHP 80-cm telescope over the Wilson reflector in London. One of the resulting publications was called *Hydrogen and Helium Line Intensities in Some Be Stars* (Burbidge & Burbidge 1951). That summer was wonderful: In addition to the nighttime, there was the OHP dormitory and friendships created there (we became especially attached to Madame Barrachino, who ran the dormitory and diner, and with whom we maintained Christmas-time correspondence for many years). I still remember her wonderful Provencal accent: hard to follow at first, with our school-days education in French. It took about two weeks before we were able, quite suddenly in my case, to take part in the friendly colloquial conversation at the dinner table with the others—Georges Courtès, the Duflot twins, Daguillon, and others whose names I do not recall after nearly half a century.

After these magical weeks in Haute Provence, we were given time at L'Institut d'Astrophysique in Paris to use the Chalonge microphotometer to make tracings of our spectra. This hospitality extended even to providing us, free of charge, a bedroom in the Institute, and initiated a lifelong friendship with Gérard de Vaucouleurs, who taught us how to use the microphotometer. The tracings were done on photographic paper, which then had to be developed in the darkroom; while I ran the microphotometer, Geoff developed the tracings. After many intensive days and nights of this work, he began wearing rubber gloves as protection from the chemicals. It was in this guise that he crept into the back of a lecture hall where a conference on Novae was in progress. Fred Hoyle was a participant in the conference; Fred, like us, was existing on a minimum amount of money, and thus we met at lunchtime in the cheapest nearby restaurant we could find. That was the café where a never-to-be-forgotten incident occurred: Fred, after the meal, looking at the check, saw that the waitress had added the amounts incorrectly, and pointed this out to her. She launched into a spate of rapid Parisian French with which our Provence experience provided no help; Fred, in his best Yorkshire accent, said "I may not understand your lingo, miss, but I *can* do arithmetic!" and pointed to the figures again; she (fairly gracefully) accepted that she had made a mistake.

Thus began our lifelong friendship with Fred Hoyle.

It was now time to follow the suggestions given to us during the IAU meeting in Zurich. Since Otto Struve was the Director of Yerkes Observatory, that seemed the obvious place to try and gain acceptance, and this idea was reinforced during a meeting in London at the Royal Astronomical Society, where we met S. Chandrasekhar, also on the University of Chicago faculty and living at Yerkes Observatory, Williams Bay. It was clear that there were no Mt. Wilson-type restrictions on women using the 40-inch refractor at Williams Bay or the 82-inch telescope at McDonald Observatory in SW Texas. Struve had created McDonald Observatory, which for many years was maintained and operated by the University of Chicago, and he had observed there frequently with various collaborators. Nancy Grace Roman was on the staff at Yerkes, and was a frequent observer at both places.

At the same time, there were Agassiz Fellowships available at Harvard College Observatory, and as we were great admirers of Director Harlow Shapley and the HCO staff, it seemed a good idea for Geoff to apply for an Agassiz

Fellowship. So this is what we did; his application was successful, mine for an IAU grant and both of us for Fulbright funds to travel to the U.S., were also successful, and I was informed that I would be welcome at Yerkes, although there was a disappointment there: Otto Struve had accepted a position at the University of California at Berkeley, so I would not be able to work with him.

During the IAU 1948 meeting in Zurich, we had also made friends with Brad and Bede Wood, at University of Pennsylvania, and—wonderful friends that they were—they said that they would meet us on our arrival on the Queen Mary at New York, to help us through immigration. They invited us to stay with them before we set off for our destination—Geoff to Cambridge, Massachusetts, and I to Williams Bay, Wisconsin.

The last year spent at Mill Hill was devoted, in addition to scientific work, to taking part in helping the rapidly recovering UCL and ULO. Professor Massey, now in charge of astrophysics, took great interest in planning the future of astrophysics and in revitalizing ULO by the addition of new buildings. Before this planning came to fruition, however, and while the planning for our venture to the U.S. was underway, other friendships brought enrichment to our lives.

Two Chinese astronomy scholars, displaced by the 1949 revolution in China, sought refuge in England, and both came to visit ULO with the request that they be permitted to work there, unpaid, just to keep their contacts in astronomy alive. They were T. Kiang, who eventually settled in Armagh, Ireland, and S.-K. Wang, who eventually returned to China to become prominent in the revitalization of China's astronomical re-awakening and repair of the ravages attributed to "the Gang of Four," and to create the radio astronomy program at Beijing University. They were welcomed at ULO.

Two more friends arrived in London—Gérard de Vaucouleurs, who had be-friended us in Paris in 1949, and his wife, Antoinette, whom we had not met previously. Gérard was in London for work with the British Broadcasting Corporation on their French broadcasts, but both he and Antoinette wanted to work as much as possible at ULO. Gérard took part in the Radcliffe refractor programs, and Antoinette measured spectra obtained in the ULO laboratory and with the Wilson reflector. At that time, I had invested in an easy-to-ride motor-bike (top speed: 45 mph, with the starter on the handlebars, and, instead of the usual pedals, it had flat paddles for one's feet). It also had a pillion seat. Gérard used to sit on that pillion while I drove him to the apartment he and Antoinette rented, after a night's observing. There exists a photograph of myself on that bike, C. C. L. Gregory on his massive powerful motor bike, and Anne Carew Robinson, who was hired at ULO during this period.

In 1951, it came time to say *au revoir* to ULO, and set sail for the U.S. and the observatories of Yerkes, McDonald, and Harvard College.

YERKES, McDONALD, AND HARVARD COLLEGE OBSERVATORIES

From the moment I got on the train from New York to Chicago, it was like entering a new world—literally, a time of expanding horizons in all directions, physically, mentally, and spiritually. I cannot remember the journey from Chicago to Williams Bay, but I recollect well the welcome at Yerkes. The Van Biesbroeck family—Georges, his wife and his sister, Marguerite (the librarian at Yerkes)—had a large house on the Yerkes grounds which they ran as a boarding house for students, postdocs, and visitors. Mrs. Van B. ran the house, kitchen, and dining room; all three of them, I believe, worked in the vegetable garden across the road where, European-style, they grew many of the vegetables that we ate in the diner.

At those meals, I met many who have remained friends or of whom I have interesting memories. These included Eberhart Jensen, from Norway, Nancy Grace Roman, Harold Johnson, who was working with Bill Morgan on the famous stellar classification program, and graduate students Don Osterbrock, N. Limber, and Larry Helfer. My grant from the IAU, $1,000, covered board and lodging at the Van Biesbroecks; Geoff's Agassiz Fellowship, $1,600, covered a rented room in the household of a Harvard professor living near the Harvard College Observatory; we had some savings, and planned as soon as possible to try for observing time at McDonald. I started working with Professors W. W. Morgan and W. A. Hiltner. Bill Morgan started me on his observing program with the 40-inch refractor, with which I was privileged to spend nights taking spectra of B stars for Morgan's program of mapping the nearby spiral structure in our Galaxy. How well I remember the meticulous instructions for developing the spectra—metol sulfite developer for the *exact* time, then, after the fixing in hypo, the plates were *not* to be placed in running water for 30 minutes, the usual procedure with which I had been brought up, but should be placed in six successive dishes of water for 5 minutes each. There were other details to this procedure, and Bill Morgan and his assistant, Irene Hansen (later, Irene Osterbrock) would be likely to detect if one had deviated in any way from the instructions. The work under Morgan on spiral structure in our Galaxy initiated my interest in spiral structure in galaxies in general.

Before the cold Yerkes winter set in, Geoff and I prepared a program to submit for McDonald observing time with the Cassegrain spectrograph of Struve/Elvey fame. But the time for submission was past; since we wanted winter time when the December Milky Way was up, we were too late. Here the never-to-be-forgotten kindness of Al Hiltner came to our rescue. He had set me to work on prevention of internal reflections and scattered light in a spectrometer for calibrating coudé plates at McDonald, and he had a month (I believe) scheduled for photometry at McDonald. He said there would be many nonphotometric nights during this period, and if Geoff and I could get ourselves to McDonald with the small stipend allowed to graduate students and postdocs for travel, we could have the nonphotometric nights for spectroscopy, since the Cassegrain photometer and spectrograph were easily interchangeable.

Soon after my arrival at Yerkes, the gorgeous fall colors spread over the woods around Yerkes. Bill Morgan organized what was apparently an annual event—a walk some way along the path around Lake Geneva—a right of way since Indian times. Cameras were encouraged, and later in the fall Bill and Irene organized the annual photo exhibition in the library, for which all entrants must have taken, developed, printed, and mounted their pictures, for the competition. My photos, very much also-rans, were duly displayed; they were mostly pictures of birds, water pools, waves, and rocks taken in the Scilly Isles the summer before we left England. I do not remember who the prize-winners were, but the event is one of my many happy Yerkes memories.

December came, Geoff traveled from Cambridge, Massachusetts to Chicago (courtesy Mr. Fulbright) and we set off for Pecos, Texas by train coach plus Greyhound bus. We were welcomed by the McDonald secretary at Pecos (mistakenly labeled Pecan by me in my ignorance of US names!) and were driven 90 miles to McDonald and installed in one of the small visitors' cottages.

At McDonald, we made another friend, the Superintendent, Marlyn Krebs, son of the Yerkes Superintendent. He, with, I suppose, Al Hiltner, instructed us in the use of the spectrograph, darkrooms, photographic supplies, and the protocol about whether any particular night was suitable or not for Hiltner's photometry.

I shall never forget the excitement of using that 82-inch telescope. Compared with the Yerkes winter nights, when one only ended observing on a clear night when the temperature dropped below $-5°F$, the McDonald nights in the dome were usually quite comfortable on the Cassegrain floor, even when a Texas "blue norther" had blown in.

After the observing run, it was time for Geoff to return to Harvard, and this time I accompanied him. I looked forward to measuring the spectra we had obtained, but Harvard College Observatory had no measuring machine. When I asked Dr. Harlow Shapley, who was welcoming and kind as always to overseas visitors, about a measuring machine, he said he was intending to acquire one—it was much needed at HCO. Years later, Shapley teased me about my persistence on this topic, and reminded me of a conversation one day in his office, when he had just received news that King George VI of England had died and had been succeeded by Queen Elizabeth II. He had greeted me with the words "Well, you now have a new Queen in England!" He recalled my response as being some moments of surprised silence, and then: "Now, what about that measuring machine?"

Geoff was attending lectures by Bart Bok on galactic structure and stellar distribution functions; I was learning all I could from the work of Cecilia Payne

Gaposchkin on stellar element abundances and on variable stars. HCO was a wonderful place for widening horizons. Among the friends we made there were Arne and Ingrid Wyller, who were planning to drive with Ingrid's mother from Cambridge across the country and into Canada for the 1952 summer meeting of the American Astronomical Society (AAS) in Victoria, B.C. They were looking for a fourth passenger to share expenses on the three-week camping trip, and invited either Geoff or me. We tossed a coin, and I won. Our route went through Yellowstone Park, where several days of sightseeing renewed my earlier undergraduate interest in geology.

We reached Victoria—again, new science and new friends. By the good luck I was encountering everywhere, Harlan and Joan Smith, whom we had got to know at Harvard and who had driven out for the Victoria meeting, were looking for someone to share their drive (again, camping) down the coast into California to visit Lick, Mt. Wilson, and Palomar Observatories.

After this, it was time to return to Cambridge, Massachusetts, and the only affordable transportation was by Greyhound Bus. Not many of today's postdocs or graduate students have needed to make that arduous although fascinating five-day journey (I stopped one night in a YWCA in St. Louis to rest my weary bones); the experience with its encounters, both friendly and adversarial, would be a story in itself. I remember well my relief at being met by Geoff around midnight at the Cambridge bus station.

For the next year, Yerkes Observatory offered both Geoff and me postdoc positions, and we started work there again. With the Yerkes facilities, one could tackle our McDonald spectra of Be stars, measuring wavelengths and line intensities. Chandra's book on Radiative Transfer gave us theoretical background to tackle the physics of the outer atmospheres of these stars, including the effects of rotation, electron scattering, and radiative transfer in the hydrogen emission lines. We measured the Balmer decrement in the emission lines to be slower than the standard "case B" planetary nebulae decrement, and we calculated the effect of radiative transfer in the Balmer emission lines and were able to explain the Balmer decrement in the emission lines in Be stars. We calculated absorption line profiles for various rotational speeds and fitted to the observations. Several papers were published (e.g. Burbidge & Burbidge 1953 a,b).

Meanwhile, I worked for Bill Morgan on observing B star spectra at Yerkes for his work on spiral structure in our Galaxy. Geoff started working with Chandra, who was at that time engaged in his magnum opus on magnetohydrodynamics. In common with the rest of the astronomical community, we both found Chandra's lectures to set a precedent for excellence; each topic, of course, appeared as a classic monograph. Geoff started work on the magnetic stability of stars, and this got me interested in the Ap stars with strong and variable magnetic fields. We both, through Chandra, became interested in white dwarfs and in stars with apparent deficiencies in their metal/hydrogen surface abundances.

Two landmarks stand out. First, an application for observing time at Mc-Donald with the coudé spectrograph in the winter of 1952–1953 to continue work on Be and related stars was denied because the Milky Way time was over-subscribed, but we were given time in the spring of 1953, and thus needed to plan a different observational program on stars of higher galactic latitude. With interests aroused through Geoff's contact with Chandra and Fermi on the role of magnetic fields in astronomy, we selected a list of Apm stars, the most important being α^2CVn. Again, the influence of Struve played a role; he had published an enormous wavelength and identification list of absorption lines in α^2CVn. We planned to determine atmospheric abundances, in particular of those elements abnormally strong (Eu, Si, etc).

The second important event, in the spring of 1953, was a conference on "The Origin of the Elements," hosted at Yerkes by Gérard Kuiper, and organized by Maria Mayer and Harold Urey; a short account of which we wrote up for *Observatory* (Burbidge & Burbidge 1953). We had thought only subliminally about the question of the origin of the chemical elements in the solar system (meteorites, the Sun) and in those few stars on which abundance analyses had been carried out. How did the so-called cosmic abundances come about? Why were some stars apparently different (including the Apm stars, the Ba II stars and S stars, the apparently metal-deficient stars, and white dwarfs)? At the conference, George Gamow was in great form with his ylem theory, with jokes about his leaps past the difficulties at $A = 5$ and 8, and Maria Mayer was describing the Mayer-Teller hypothesis, of a primordial polyneutron origin to explain the neutron-rich isotopes (Mayer & Teller 1949). A bell was rung in my mind—the fascinating RAS talk by Fred Hoyle (Hoyle 1946) on physical processes for building the elements in the interiors of stars at late evolutionary stages, when exhaustion of hydrogen as a fuel for nuclear reactions, followed by internal contraction and external expansion in an inhomogeneous structure explaining the red giants, would be followed by further internal collapse to very high T, ρ conditions sufficient to set up equilibrium among the nuclei and produce most of the more abundant elements. How much of Fred Hoyle's work was ahead of its time, ignored by the stolid, unimaginative, mentally constipated run-of-the-mill astronomers!

The McDonald observing run was successful; our spectra had to be microphotometered in the Yerkes basement before we left Yerkes. Our return journey to England, at the expiration of our two-year visitor visas, was by way of Ann Arbor, Michigan, for the famous 1953 summer school at which George Gamow, Walter Baade, Edwin Salpeter, and George Batchelor were lecturing. Geoff and I were housed in an apartment of a U. Michigan professor, and Allan Sandage (with whom our lasting friendship began during this summer school) was staying with other single postdocs and graduate students in a dormitory where they were educated and entertained by nightly bull sessions with Walter

Baade. The lectures by Gamow, in which he expounded his ylem theory, crystallized our own ideas that the elements were not formed in some primordial series of events at the origin of the universe, but were built up out of hydrogen in successive generations of evolving stars.

For the next year, we had two job offers: first from Professor Z. Kopal, who offered us two junior faculty positions at U. Manchester, and second, one position, for Geoff only, with Martin Ryle's radio astronomy group in Cambridge. Geoff, following his interactions with Chandra and with Fermi in Chicago, was very interested in the strong radio sources, about which there was currently much discussion as to the physical mechanism producing the radiation. I had all the McDonald spectra to analyze, and Cambridge seemed to be an ideal place in which to tackle them.

CAMBRIDGE, 1953–55: SYNTHESIS OF THE ELEMENTS IN STARS

We found an apartment in Cambridge, 8A Botolph Lane, just around the corner from Free School Lane, where the old Cavendish Laboratory housed the radio astronomy group. With permission from Prof. Redman (and payment of a "bench fee" to Cambridge University), I was able to use the measuring equipment at the Observatories in Madingley Road. Wavelength reductions on our spectra were done with our Brunsviga, and I measured equivalent widths on the yards of Yerkes microphotometer paper with a planimeter. Our apartment rental did not include permission to keep a bicycle in the narrow access passage, so the daily walk behind Trinity and Johns Colleges and along Grange and Madingley Roads to the Observatories provided time to think, plan, and contemplate.

Geoff was not having a particularly easy time at the Cavendish. As a theoretician, thought to be contaminated by ideas inspired by Fred Hoyle, and as a proponent of the synchrotron mechanism for producing radio emission from the strong extragalactic sources (the Ryle group at that time favored plasma oscillations), the Ryle group did not willingly share their observational work on counts of radio sources with him.

It was in the autumn of 1954 that we made the acquaintance of Willy Fowler and his family—he was spending a sabbatical year at the Cavendish, where he had hoped to do some experimental nuclear physics, but had found that none of the equipment he needed was available or working. At this time, element abundances were emerging from our curve-of-growth analysis of α^2CVn (Burbidge & Burbidge 1955), and the heavy element anomalies were beginning to suggest that somehow neutrons were involved—an idea whose germ had been planted by the Mayer-Teller and Gamow work, but we believed the processes must take place in stars. While we were well-educated in atomic physics, we knew much less about nuclear physics. Geoff attended a $\Delta^2 V$ lecture by Fowler, and asked him afterwards if we could talk about processes involving neutrons in stars. Fowler, a leader in the experimental low-energy nuclear physics program at Kellogg on the light elements, had recently worked with Salpeter and Hoyle while they were visiting Caltech. He was excited by the prospect of adding neutron processes towards a theory to build all the elements in their cosmic abundances through generations of stars—which through evolution, finally produce Hoyle's iron peak elements (Hoyle 1954), and end as supernovae, exploding and enriching the interstellar medium with heavy elements made from the initial ingredient, hydrogen. Fred Hoyle was in Cambridge and we four worked together during that exciting 1954–1955 year, adding together one piece after another of the puzzle. Willy's wife, Ardy Fowler, in her wonderful hospitable way, made available their rented home in Cambridge and we four divided time between there, the Cavendish, and Botolph Lane until the time came when Geoff and I had to think about jobs for next year. Knowing all too well of the ban on women for Carnegie Fellowships, Willy thought Geoff had a good chance for one, and thought he could give me a postdoc fellowship in the Kellogg Radiation Laboratory at Caltech. "This time," he enjoined us, "you must apply for regular immigration visas, not visiting two-year Fulbright visas." So we did.

One more good memory of Cambridge remains to be recalled. I reviewed, for the radioastronomy journal club, a paper by M. Schwarzschild (1954) on masses and the mass-to-light ratio in galaxies, about which little was then known. This was, to me, a new interest, related both to stellar evolution and, more strongly, to Geoff's work on radio galaxies. I thought of an observing program on measuring rotation curves of spiral galaxies to determine their masses and M/L ratios. Geoff was awarded a Carnegie Fellowship, but it was made clear that work with the 100-inch Hooker telescope on Mt. Wilson on galaxies, which had been carried on by Milton Humason in collaboration with Edwin Hubble, was no longer possible; the night sky was too bright. Humason's multi-night exposures, as he sat at the Cassegrain focus of the 100-inch, would have been impossible; all such work in the future should be carried out at Palomar. However, we set off for Pasadena while the work on B^2FH (Burbidge, Burbidge, Fowler, & Hoyle 1957) was only partly completed, with the primary goal to spend most of the next two years on that major project.

CALTECH, MT. WILSON, B^2FH

In Pasadena, we renewed the friendship begun in Ann Arbor with Allan Sandage, and he began a Chinese water-drip campaign on Dr. I. S. Bowen, Director of Mt. Wilson and Palomar Observatories, to consider Geoff's Carnegie Fellowship observing proposals as coming jointly from both Geoff and me, and to allow me to go up Mt. Wilson with Geoff. To stay in the Monastery was, of course, impossible, as was use of Observatory transport up and down Mt. Wilson. Standard reasons for not allowing women on the telescope included the fact that there were only male-oriented bathroom facilities on the mountain, and that the telescope technicians (then called night assistants) would object to operating under directions from a woman. However, Dr. Bowen relented, under pressure from Caltech as well as from Allan Sandage, and we were allowed to go together and stay in the Kapteyn Cottage (a small summer cottage) on the mountain, as long as we brought our own food and used our own transport—an old, but serviceable, Chevrolet. The night assistant "problem" was solved when Arnie Ratzlaff, at the 60-inch, asked us one night why we did not come to the galley for midnight lunch, but brought our own sandwiches. On being told that we did this on Director's orders, Arnie simply laughed and said that was nonsense. So from then on, we joined in the communal night lunch in the galley.

Down in Pasadena, Geoff and I were working hard on B²FH, dividing time between 813 Santa Barbara Street (his scientific abode) and Kellogg (mine). The eight processes were being worked on: H-burning, He-burning, α-process, the e-process (Hoyle's process), the s-process (slow neutron capture for interiors of S-type red giants), and the r-process (rapid neutron capture in supernovae to produce the displaced peaks alongside the s-process peaks at Maria Mayer's "magic numbers" of closed nucleon shells); there were also the p-process, to account for some low-abundance heavy isotopes, and what we called the x-process, for deuterium, Li, Be, and B. Red giants of type S, we felt, had atmospheres that were too complicated to tackle for abundance determination and the $\alpha^2 CVn$ abundance anomalies, which had seemed so promising originally, did not produce the right heavy-element overabundances. At this point, we needed good spectra of a Ba II star, where the enhanced spectral lines were of just those elements in the s-process chain. A problem arose; the brightest Ba II star, ζ Cap, was in Prof. Jesse Greenstein's observational programs, which covered a wide range of forefront problems in stellar spectroscopy (Greenstein 1984), although its analysis had not yet been undertaken. So we chose HD 46407, the next brightest accessible star from the list in Bidelman and Keenan's important paper (1951); in that paper they concluded that the strength of Ba II λ 4554 in these and in S-type and carbon stars was probably not a consequence of ionization and excitation processes.

Our Mt. Wilson observing program also included various low-metal-abundance stars, and some T Tauri stars; these were linked to the complete picture of stellar evolution which was emerging at that time. We worked on this program throughout the winter of 1955–1956, with Geoff doing the plate-cutting and other darkroom work while I worked at the telescope, at the cassegrain focus using the movable ladder in the 60-inch dome, and at the 100-inch, mostly at the coudé focus. Geoff's expertise at plate-cutting with a diamond (this was before Bill Miller's ruling machines were designed) had been learned in the workshops of his father's building business in Chipping Norton, England.

By April 1956, I was beginning to be unable to disguise pregnancy under the loose layers of warm clothing one used to wear for winter observing out in cold domes, and there came a time when climbing up the movable ladder and heaving it around the dome became more than I could manage. So we terminated our observational program, and concentrated on analyzing the spectra. Meanwhile, the work on B^2FH was proceeding marvelously.

For the light elements, the experimental work organized at Kellogg by Fowler and the Lauritsens was crucial, while some more recent nuclear data on energy levels in iron-peak elements produced an e-process curve slightly improved over the original Hoyle calculation. For the neutron-rich heavy elements, our data on HD 46407 (Burbidge & Burbidge 1957) fitted beautifully with the s-process prediction that the product of neutron-capture cross section times abundance, averaged over the s-chain isotopes, would be constant, and, within a factor ~ 2, for 11 out of 15 elements studied, this proved to be the case, even though measurements of the neutron-capture cross-sections at that time were not particularly accurate and, of course, received subsequent improvement. Geoff had noted that the light curve by Baade of the SN in IC 4182 was linear, indicating the sort of decay produced by radioactive elements. Radioactivity proved correct, but we chose the wrong radioactive element, Cf, instead of $Ni \rightarrow Co \rightarrow Fe$, as later proved correct.

Several papers were written but the principal one, always thereafter known as B^2FH, was a 100-page account of all eight processes (Burbidge et al 1957).

REACHING THE GALAXIES—YERKES AND McDONALD OBSERVATORIES AGAIN

Towards the end of Geoff's second year of the Carnegie Fellowship, and of my research fellowship at Caltech, and after the birth of our daughter, we were offered positions in the University of Chicago's astronomy department, i.e. at Yerkes Observatory. Chicago at that time had a "nepotism rule," so that husband and wife could not both be on the faculty. This applied however distinguished and senior the couple; for example, Joseph Mayer had a professorship and therefore Maria Mayer, future Nobel Laureate, could not be on the Chicago faculty and so was employed by the Argonne Laboratory—her teaching and research at the University thus was "voluntary." So Geoff was offered a faculty position, while a Shirley Farr Fellowship was offered to me. We accepted gladly.

Back at Yerkes, we once again had access to McDonald Observatory, and could set about realizing the goal that originated from our interests in Cambridge. The McDonald 82-inch had a prime focus capability; since the telescope diameter was not large enough to accommodate a cage, access to that focus was obtained from two electrically operated "pulpits," off a bridge which rose up and down within the dome slit. Each pulpit could be manually cranked up, so that with suitable manipulations one could reach in and insert plate holders or other instrumentation directly at the prime focus; telescope guiding was then done from an eyepiece and a paddle on a post which could be moved electrically around the rim of the telescope. At first, the experience up there produced a slightly dizzy sensation, as one looked down upon the primary mirror and felt the awesome responsibility of knowing that it was as much as one's life was worth to fumble and drop anything down onto the mirror.

Horace Babcock, years earlier, had designed and constructed a spectrograph that could be placed right at the prime focus—the "B Spectrograph," with small plateholders that pushed into the side. Direct photography was done by sliding the spectrograph over, leaving the place for standard 5 × 7 inch plateholders to be inserted.

The B spectrograph had been used extensively by Thornton Page for his classical work (Page 1975) on velocities of binary galaxies and their statistical mass determinations, but it had fallen into disuse; the only current prime focus user was Georges Van Biesbroeck, who studied comets, asteroids, and proper motion stars. Our first McDonald observing run after arriving at Yerkes renewed our friendship with Marlyn Krebs; he disinterred the B spectrograph from some dusty storage region, and the three of us cleaned it, remounted the collimator, adjusted the whole optical train, and then set to work using it on a program we had prepared on spiral galaxies. The spectrograph was designed to work with 103a-F film, in the red, using the Hα and [NII] emission lines from ionized gas in the galaxies. The spectrograph slit was long enough that one could obtain velocities along half the major axis of these galaxies, or across them at various angles, in single exposures—this proved enormously more economical of observing time than the earlier work of Babcock, and of Mayall and Aller, on velocities of individual H II regions in M31 and M33.

Essential to our operation was the help of the young telescope operator Johnny Carrasco, as he was then called; now he is the well-known and heavily acknowledged chief telescope operator Juan Carrasco, of Palomar fame. As before, Geoff was in charge of photographic operations. The dexterity he had used in plate-cutting at Mt. Wilson was now essential in cutting the half-inch squares of photographic film by a cutting punch which deposited the tiny square into a drawer, *but*—it could fall film-up or film-down, and it is much harder to tell by feel in the dark which is the emulsion side of photographic film than is the case with plates. Only the delicate touch of Geoff's fingertips could do this unerringly. It was an essential part of our operation—imagine spending three, or four, or even five hours standing, exposed to wind and cold in the pulpit of the bridge, eye glued to the eyepiece, with those precious few photons falling only on the back side of the photographic film!

We have, of course, an ample store of anecdotes about our McDonald days. The two most traumatic were the occasion when the motor that moved the post around the telescope rim jammed and trapped Geoff's arm, nearly breaking it; with great presence of mind he yelled "cut the power," and Johnny and I were able to extract his bruised arm. The second occasion was during an attempt to take a spectrum of NGC 5128 at the cassegrain focus. At that time, NGC 5128 had sometimes been hypothesized to be a planetary nebula, but as a strong radio source, Cen A, it was obviously a galaxy; at declination −43°, it was just accessible from McDonald. To reach it, we had the cassegrain floor raised right up; I had removed the safety chains guarding the inner edges of the two halves of the cassegrain floor, and, in my excitement over setting on this far-southerly faint object, I stepped over the edge of the floor and fell some 10 feet down to the main floor, carrying the control paddle with me. Completely winded, but with no broken bones (thanks to the padding of heavy winter clothing), I could not draw breath to tell Johnny and Geoff that I was alive and conscious; I could hear them fumbling their way down the steps in the dark, dreading what they might find at the bottom. Fortunately, bruises, a cut forehead, and a multitude of aching parts were all I earned, and a spoilt night's observing and a day in bed were all the payment exacted. Not surprisingly, we carried out our spectroscopic work on NGC 5128 at the prime focus, where one could lie on the roof of the coudé room to reach the guiding eyepiece.

Such were the hazards of observing "in the old days." Other astronomers fared worse; Joy fell off the Mt. Wilson high Cassegrain platform and suffered numerous fractured bones. The eagerness of the most active and excited astronomers can still lead to peril, as was sadly demonstrated recently at Kitt Peak; we all mourn Aaronson and know that "there, but for the grace . . . " might we have been.

Near the start of our work on galaxies, I took a picture of M51 to try out the direct imaging capability at the 82-inch prime focus. Seeing that image on the 5 × 7 inch photographic plate in the washing dish in the darkroom produced such euphoria that I felt it was almost sinful to be enjoying astronomy so much, now that it was my job and the source of my livelihood.

The long-slit spectra, with the inclined and curved emission lines of Hα and [NII], now so familiar to all extragalactic astronomers, were new to us, and needed to be measured with a two-coordinate screw machine. The rotational velocities thus obtained were the raw material for obtaining the mass distribution throughout these spiral galaxies. Thus began a wonderful collaboration and friendship with Kevin Prendergast. At each point measured, the rotational velocity is derived from the gravitational balance with the integrated mass interior to that point, thus one needs a model for which the necessary integral equation can be set up. Earlier models had led to equations that were awkward to solve; Kevin's expertise in theoretical analysis and mathematics provided the model we three used—an integral converted to a sum of similar spheroids which could be solved analytically quite easily. Thus followed a series of papers, summarized in a chapter in the Sandage volume on galaxies (Burbidge & Burbidge 1975).

The mass distributions that we obtained ended necessarily at the last measurable point on our spectra. We extrapolated the rotation curves further out, assuming a Keplerian fall-off in velocity, and derived mass-to-light (M/L) ratios interior to our last observed points, and assumed these applied to the whole galaxies. However, some 20 years earlier, Horace W. Babcock had made some remarkable measurements on the outer parts of the Andromeda galaxy, M31 (Babcock 1939). He had built an attachment to the spectrograph on the Crossley reflector at Lick Observatory, which enabled him to observe the absorption lines in M31 out to 95 arc minutes either side of the nucleus. He discovered that the rotation curve was still rising, out to his furthest measured point.

While he realized that his absorption-line measurements were subject to considerable probable errors, he noted that the measures, if valid, implied that the M/L ratio in M31 rose from some 1.6 within 30 arc sec of the center, to 18 at 15 arc min out, 43 at 50 arc min out, and 62 at 80 arc min out. This appears to be the first observation that required the existence of considerable dark, or unseen, mass in the outer parts of a normal spiral galaxy.

This remarkable observation appears to have escaped notice until the increasing precision of 21-cm radio observations, which demonstrated horizontal rotation curves beyond the last measurable optical points. Optical measurements of the ionized gas component—Hα and [NII] emission lines—required the 4-m Kitt Peak telescope, as opposed to the 82-inch telescope, plus a much more sophisticated spectrograph, and the skill and determination of Vera Rubin, to show that ionized hydrogen extended beyond our last measurable points, and that the extended rotation curves were *flat*. This, with the 21-cm data, brought into the collective mind of the astronomical community the postulate of dark, unseen matter extending far beyond the obvious visible luminous extent of spiral galaxies. But by then we had moved on to a new and still-abiding interest: the quasi-stellar radio sources—QSRS, QSOs, quasars.

The first identifications of these, by Matthews & Sandage (1963) produced one object, 3C 48, just barely observable at the 82-inch prime focus. We duly and with some difficulty in guiding (since no offset guiding could be done at the 82-inch prime focus) obtained a spectrum. Some possible fuzzy emission features could be seen in the red, but they were barely visible and did not look like any stellar or extragalactic features with which we were familiar, so we did nothing with them. As I described in my 1984 Russell lecture, 20-20 hindsight and careful printing of these spectra in 1983 showed quite well that they were Hβ, [O III] 5007, 4959, but again in hindsight, how could we have convinced anybody at that time that we were looking at a faint stellar object with a large redshift?

During the years 1957–1962 that we spent at Yerkes, mainly working on galaxies, we also continued the collaboration with Fowler and Hoyle, by spending part of the summers in Pasadena. Much of my time there was spent at Santa Barbara Street, using Rudolph Minkowski's 2-coordinate measuring machine on our galaxy spectra. We spent wonderful hours, both social and scientific, with Walter Baade, R. Minkowski, M. Humason, and Allan Sandage. Our good friend Henrietta Swope, who had befriended us earlier in the two years we spent in Pasadena, and who played the role of virtual godmother to our daughter Sarah, preceding, during, and after her birth, was now as always giving us a warm welcome in Pasadena.

Another important friendship developed. Allan Sandage had introduced us to Mrs. Hubble—Grace, widow of Edwin Hubble—during our previous two years (1955–1957) that we had spent in Pasadena. Now, in the years after 1957, every summer during which we spent time in Pasadena, we used to visit Grace and have tea with her in her beautiful home in San Marino. She told us much about Edwin that, I understand, is now guarded in the archives of the Huntingdon Library. Because of his worldwide scientific prominence, Edwin had expected to be appointed Director of the combined Mt. Wilson and Palomar Observatories, so the choice of I. S. Bowen for this position had come as something of a shock, although Ike Bowen's expertise with optical instruments had made him a suitable candidate for the major responsibility of commissioning the Palomar 5-m telescope, which had been mothballed during World War II.

Grace told us that Edwin had died very suddenly after driving home, while he was getting out of the car; as he lay in the driveway Grace thought he had just fainted and did not accept that he had died until doctors convinced her of the bad news. During our visits with Grace, she told us of the work she was doing at the Huntingdon Library on his papers. She showed us his study on the ground floor of her house, which she had kept exactly as he had left it. Grace was a remarkable woman, an intellectual giant, who had many friends in the literary and artistic world. She told us of her and Edwin's friendship with George Arliss and others famous in the theatrical world, and we met Aldous Huxley and Gerald Heard over tea at her house. She had given us many items that we could not otherwise have afforded, before and after the birth of our daughter Sarah who, as a small child, loved the teatime visits with "Aunt Grace" during our later summer visits. Grace who had had no children of her own was extremely kind to Sarah and enjoyed her visits (this is more than can be said of the Hubble 16-year-old black cat, Nicholas, who always retreated into the cellar when Sarah appeared).

Of the many stories Grace told us about Edwin, one demonstrates a side few are aware of. Apparently, during one dinner party, a conversation developed over the relative housekeeping merits of men and women, the hostess maintaining that on the whole men had no ability in this field—for example, consider how bad they were at dusting. Edwin stood up, went over to the door, and, being so tall, reached up and ran his finger over the top of the door, then showed his dust-covered finger to the assembled company. "It was very naughty of him, but she was being provoking," remarked Grace.

THE BEGINNING OF THE YEARS AT THE UNIVERSITY OF CALIFORNIA

After the publication of B^2FH, all four of us received many invitations to give general or specific colloquia and lectures on "The Synthesis of the Elements." One such visit, at the invitation of Roger Revelle, was to the Scripps Institution of Oceanography. Harmon Craig, whom we had first met at the 1952 Yerkes meeting when he was a student of Harold Urey, contributed, with his wife Valerie, to the warm welcome we received at Scripps. Roger Revelle was at the beginning of his creation of the new University of California campus at La Jolla. We were, as I have described, deeply involved in the physics of galaxies, and during the summer visits to Pasadena, Geoff was maintaining the contacts with Fred Hoyle and Willy Fowler.

Having eventually made his point with the radio astronomers in Cambridge that radio galaxies were powered by synchrotron emission, as first published by theoretician Shklovsky, Geoff had next turned to the problem of the energies needed to produce such huge fluxes of radio radiation. Man-made synchrotrons, after all, require far more input energy then the resulting output of high-energy particles; Geoff's calculations (Burbidge 1956, 1959a,b) of the minimum energies in high-energy electrons and magnetic fields in a number of radio galaxies, including M87 and Cyg A, assumed equipartition between field and particles, but he also considered the greater energy requirements if accelerated protons were produced as well as electrons. Time scales were also important; the directivity of the radio jets, as well as the optical and radio jet in M87, clearly demonstrated that the source of the energy was located in the very nuclei of the galaxies, and the radiating particles, with lifetimes dependent on the frequency of the radiation, required continued energy replenishment and a mechanism of ejection. The summertime interactions with Hoyle and Fowler were spent in tackling this problem, and our summer visits to Cambridge began. After looking at multiple supernovae in compact nuclei and quasi-stellar radio sources as ways of producing the observed properties, Fowler, Hoyle, and Geoff settled on the collapse of matter towards the Schwarzschild radius, and a major paper by the four B^2FH authors was written which really initiated the subsequent activity by many in relativistic astrophysics (Hoyle et al 1964). We also collaborated with Sandage, and published a paper on "Evidence for Violent Events in the Nuclei of Galaxies" (Burbidge et al 1963).

Meanwhile, at Yerkes our observational work on galaxies continued, and we became interested in small groups and clusters, and the masses required to hold them together for a Hubble time. A great surprise showed up from our observation of the radial velocity of the fifth member of Stephan's Quintet, a distorted spiral galaxy, whose velocity differed by some 5500 km s^{-1} from the others. I well remember my surprise on taking the McDonald spectrogram out of the washing dish in the darkroom, peering at it with the magnifying glass, and seeing a clear, extended, Hα emission line which could be at no more than $+1000$ km s^{-1} from rest wavelength. This, with other close groups picked out of Vorontosov Velyaminov's catalog, initiated our interest in discordant redshifts—a subject still unresolved after 30 years.

One of our McDonald observing runs was shared with Nick Mayall, who was working with Bill Morgan on the spectral classification of galaxies. Nick's criticisms of the prime focus "B" spectrograph, especially its lack of an offset guider, and his description of the new 120-inch telescope at Lick Observatory, put a spark to a fuse which had been lying ready for ignition. We had been involved with the political troubles at Yerkes, concerned with Gérard Kuiper's wish to start an infrared program that involved negotiations with the University of Texas and culminated in that institutions's taking over control of McDonald Observatory from Chicago. At the invitation of Nick Mayall, Geoff and I took a quarter's leave from Chicago, and spent it on Mt. Hamilton. The 120-inch was newly commissioned; there was as yet no prime focus spectrograph (it was being designed by Albert Whitford and Nick Mayall and was being constructed in the Lick shops; it bore a remarkable similarity to the B spectrograph although with vastly better capability, e.g. choice of gratings, two cameras, offset guiding), and the 120-inch was large enough to accommodate a cage inside the upper end. This cage had an elliptical shape, rather than circular as at Palomar, and was built so as just to fit around a normal-sized person. Stan Vasilevskis initiated me into the equipment for direct photography at the prime focus; this was all that one could do until the spectrograph was completed; again, everything was easier than it had been at the 82-inch prime focus.

To ride with the telescope in that cage was an experience I wish I could share with today's generation of young astronomers. A canvas hood over the top could be zipped up if it was cold and windy, but otherwise one could look out at the spectacular vision of the heavens during a long spectrographic exposure (for direct photography, of course, one's eye was firmly riveted to the offset guiding eyepiece). But spectroscopy was for the future: Our spring quarter visit on Mt. Hamilton ended, and we returned to Yerkes, but were filled with the ambition to move from the University of Chicago to the University of California. Our acquaintance with Roger Revelle, the prestige associated with our part in "The Synthesis of the Elements" and recommendations provided by Willy Fowler, Harold Urey, and Maria Mayer culminated in the offer of positions at the newly formed San Diego campus. (The Mayers had been recruited at La Jolla; she was no longer an "associate" of Joe Mayer, since U. California nepotism rules did not forbid a husband and wife from both having faculty positions, as long as they were in separate departments.) We accepted gladly, although it was a wrench to leave our friends at Yerkes—Chandra, Bill Morgan, the Hiltners, the Chamberlains. Van Biesbroeck had been denied any further observing privileges at McDonald so, at "80 years young," he left Yerkes to spend ten more years at his beloved astronomy in the University of Arizona, on the new 90-inch Steward Observatory telescope on Kitt Peak.

So the next phase of our lives began.

LA JOLLA; NEW DIRECTIONS; THE CHALLENGE OF THE QSOs

The quasars, or QSOs as I shall henceforth call them, were an observational challenge that could be tackled with the new telescope and prime focus spectrograph at Lick, as they were being tackled with the 2.1-m telescope on Kitt Peak by Roger Lynds. We quickly developed a friendship with the Lynds family, and Roger and I delighted in sharing the excitement of spectroscopy and measurement of redshifts of these enigmatic objects. Once Cyril Hazard and the radio astronomers in Australia had produced the accurate position of 3C 273 from lunar occultations, and had identified it with a 13th magnitude stellar object, the riddle of the spectra of the handful of other stellar radio sources identified by Matthews and Sandage was solved by Maarten Schmidt, to whom Hazard provided the accurate position of 3C 273. The story of Maarten's identification of the Balmer emission lines in 3C 273, and his use of Osterbrock's table of ultraviolet emission lines predicted for planetary nebulae, to identify the emission lines in the existing handful of other identified 3C objects and—to the astonishment of the scientific world—to identify lines in 3C 9 at a redshift $z > 2$, is too well known to repeat here.

These large redshifts, interpreted by Hubble's law as representing the distances of these objects, immediately posed theoretical problems because, to be visible at such enormous distances, and indeed to be strong radio emitters, the QSOs must be emitting an enormous energy output. Variability discovered in 3C 273 set stringent limits on the size of the emitting region, and this combination provided a challenge to the theorists. Geoff, following his theoretical work on the radio galaxies, saw at once that if the redshifts were *not* due to the expansion of the Universe, the objects could be closer and the energy problem could be alleviated. This idea is described in the first monograph on the QSOs, which we wrote and published in 1967, and in subsequent papers.

I am continually surprised by the almost religious fervor with which most astronomers demand a single "Big Bang" act of creation for the Universe.

To question that observing very large redshifts implies observing the universe quite near its origin, (i.e. to question the notion of a singular origin of all that comprises our presently observed universe) has, during the past 30 years, been treated as heresy, and the punishment meted out to doubters and questioners and to those who find observational data that conflict with the standard big bang picture, while it does not involve physical torture and death as during past centuries, is severe indeed—deprivation of the opportunity to carry on one's research, either by denial of access to telescopes or by denial of funds to carry out research. The notion of any form of "steady state" universe was, from 1948 onwards, when Hoyle, Bondi, and Gold first proposed it, an anathema, and young aspiring astronomers continue to fear for their livelihood if they dare to question the standard dogma. But I wish to engage in no further polemical discussion here; this account of my lifetime in astronomy leads to more interesting topics.

POLITICS IN ASTRONOMY

The next few years in La Jolla were a time of expanding facilities, the formation of a group in infrared astronomy, and observational work at Lick and the Kitt Peak National Observatory (KPNO) on galaxies and QSOs. We became increasingly interested in the problem of the source of energy of the QSOs, the mechanisms of ejection of highly directed streams of charged particles in radio galaxies, and the problem of discrepant redshifts in small groups of galaxies. This was a period when the number of QSOs discovered and observed was growing rapidly, and Geoff foresaw that it would soon be timely to collect all the data into a catalog. Also, the first observations of absorption lines in QSOs occurred in this period—in 3C 191, by Roger Lynds and ourselves. Shortly thereafter, the discovery of a doublet of narrow absorptions in the short-wavelength wing of CIV λ 1549 emission in the QSO PHL 938 posed a fascinating problem, because the wavelength separation of the pair did not agree with the separation of CIV $\lambda\lambda$ 1548, 1551. It is interesting that the first person to point the way to identifying this doublet was Fred Hoyle, who was visiting La Jolla; he remarked that, strangely, the ratio of wavelengths of the doublet agreed with the ratio of MgII $\lambda\lambda$ 2796, 2803. Kitt Peak observations by Roger Lynds, and mine at Lick, showed the identifications of absorptions with $z_a \ll z_e$ in Ton 1530, PKS 0237-23, and PHL 938. Were these produced by gas in nearer galaxies along the line of sight to distant QSOs, as predicted by Bahcall and Spitzer, or were they related in some still not understood way to the QSOs themselves? Around this time, the first BAL QSO was discovered by Roger Lynds—PHL 5200. I recall Roger telling me that when he first took his spectrum out of the washing dish in the darkroom at Kitt Peak, and saw the enormous gap in the continuous spectrum shortward of C IV emission, his first thought was that there was a flaw in the photographic emulsion at that place! If bulk ejection of absorbing gas at velocities $\sim 0.1c$ was occurring, might some of the absorptions at $z_a \ll z_e$ also be the result of ejected gas?

During these years, Geoff and I used to spend happy weeks during the summers, both at Leiden Observatory, by the hospitality of Jan and Mieke Oort, who became our very good friends, and at Cambridge, by the hospitality of Churchill College and the Observatories. The Fowlers also spent several weeks of the summer in Cambridge, and the work with Fred Hoyle and Willy Fowler continued there. Around this time, Fred saw the need for an Institute of Theoretical Astronomy adjacent to the Cambridge Observatories, and began the fund-raising effort that resulted in the building of this extremely successful Institute. The architecture and furnishings were mostly the result of Fred Hoyle's vision; the Institute of Theoretical Astronomy (IOTA), later the Institute of Astronomy, is now known as the Hoyle Building.

However, scientific research began to mesh with scientific politics. Geoff was appointed as the U. California scientific representative on the AURA Board, that operated KPNO and the new CTIO, and he was also elected to the AAS Council. Both Geoff and I began to think seriously about the problem of gender discrimination in astronomy, which I have described earlier in this article. Outstanding astronomers such as Annie J. Cannon and Cecilia Payne-Gaposchkin had never been considered for the major awards in astronomy, such as the Bruce Medal and the Henry Norris Russell Lectureship. Women were recognized by a prize resulting from a legacy by Cannon, the Annie Jump Cannon prize, which was available exclusively for women. I was well-enough recognized now by the astronomical community to do something about this discrimination which had first hit me in 1947. Thus, when I was informed by the AAS President and Council that they had decided to award the 1971 Cannon Prize to me, I wrote them a letter (which as can be imagined went through many drafts before I sent it) declining to accept the prize and explaining my reasons: The prize, available only for women, was in itself discriminatory. I treasure among my papers a 1971 letter from Beatrice Tinsley, expressing her strong approval of my action.

This struck a powerful blow in the revolution for the recognition of women who had over decades and indeed centuries achieved major advances in astronomy, and for opening up all opportunities for young women entering the field. Geoff, on the AAS Council, told me of the consternation caused by my letter about the Cannon Prize, particularly on the part of the President and Treasurer, who anticipated various legal problems arising from the terms of Cannon's will. The membership of the AAS was also divided over the issue, as became clear later during the effort to pass the Equal Rights Amendment to the U.S. Constitution. But, young and old, enough fair-minded men have been, and are, dedicated fighters for equal opportunities—the AAS Executive Officer, Peter Boyce, is a shining example.

Other political activity was afoot. An excellent telescope, the Anglo-Australian Telescope, was under construction, and it was clear that something comparable was needed by the U.K. in the northern hemisphere; the excellent work by the U.K. radio astronomers had to depend on the U.S. for the necessary complementary optical observations. The Cyg A breakthrough had come about by identification and spectroscopy by Baade and Minkowski.

So a Northern Hemisphere Review Committee was formed by the U.K. Science Research Council (SRC); its membership included the two Astronomers Royal, R. v. d. R. Woolley and H. Brück, Sir Bernard Lovell, Fred Hoyle, J. M. Cassels (Professor of Physics at Liverpool University), and, as well-respected expatriate British astronomers, Wallace W. L. Sargent and Geoffrey Burbidge. The charge to the committee was to recommend the best way to provide first-class optical observational facilities in the Northern Hemisphere.

The report that resulted from the committee's deliberations produced a majority and a minority statement. The majority favored diminishing the power of the Royal Observatories, Greenwich and Edinburgh, and forming a new center, in Cambridge, Sussex, or Manchester, with the responsibility of searching out a first-class site for one or more well-instrumented major optical telescopes, necessarily out of England, presumably somewhere in Southern Europe, and setting up a management structure. The minority, the Astronomers Royal, were against a third center, and, such was their power in the British establishment, the Northern Hemisphere Committee Report was suppressed by the SRC and has never been published. It does, of course, exist in the archival papers of at least two of the majority members.

At this time, however, the retirement of Sir Richard Woolley from RGO (the Royal Greenwich Observatory) at Herstmonceux was imminent, and the SRC had to look for a new appointment. During the summer of 1971, while Geoff and I were in Cambridge, I was approached by Sir Brian Flowers, head of the SRC, and asked whether I would consider accepting this appointment if it were offered. Despite my foray into the U.S. political scene as a spearhead for abolition of gender discrimination, I knew then, and have always known since, that I do not have the temperament to direct a major scientific establishment. I explained this to Flowers, and pointed out that Geoffrey did possess the right abilities and temperament, and the position should much preferably be offered to him. The weak response was that the consensus of U.K. astronomers wanted "an observational astronomer" in the position. Filled with doubts, I said I would think it over, and let them know my decision. Geoff and I then left Cambridge to spend a few weeks at a summer school in Italy.

Skilled politicians, willing to use methods that verge on the unethical, know how to play upon the loyalties of those unskilled in politics. In Italy, over a very bad phone line, I was informed that the British press had learned that I was a candidate for the job, and there would be write-ups in the newspapers, which would be bad for the future of U.K. astronomy and the Northern Hemisphere Observatory (NHO) if I were to refuse the job. I was also informed that the Crown would split the position of Astronomer Royal from the RGO Directorship (they had been linked for some 300 years), and a male scientist would become Astronomer Royal. An SRC senior appointment would be offered to Geoff, and we could continue our joint work at Herstmonceux.

So, misgivings laid aside but not forgotten, I accepted, visited London and the SRC headquarters in the autumn of 1971, and we went on leave of absence from UCSD in the summer of 1972; I knew then that it would be foolish to resign my faculty position at UCSD.

MORE POLITICS; THE ANGLO-AUSTRALIAN TELESCOPE; SEARCH FOR A NORTHERN HEMISPHERE OBSERVATORY SITE

It quickly became apparent that the RGO Herstmonceux staff was divided into two halves. There were our good friends the Pagels and the younger astronomers, largely recruited by Woolley, whose overriding interest was in astronomical research, together with the astronomy department at Sussex University. And there were what I came to label "the old guard"—the sunspot and solar astronomers, the Nautical Almanac group, the Time Service group and two groups who wielded much power—"Staff Side," and the union. Almost immediately after my arrival, the research astronomers were asking me how soon I would manage to engineer the moving of the Isaac Newton 100-inch telescope (INT) to a good site, out of England, where it could be used for frontier astronomical research. Located as it was, some 300 feet above sea level, in an area named Pevensey Marshes, the years of waiting for this telescope to be built had clearly ended in the disastrous choice of its site, S.E. of London (the prevailing wind is usually NW). Next, the first staff meeting of the senior staff was attended by myself and, by my invitation, Geoff. The kindest of "the old guard" was clearly deputed to come to my office sometime afterwards, and to tell me as gently as he could, that Geoff would not be welcome at any future meetings. This caused me much distress, and clearly was out of the spirit of the implied—but *only* implied, promises of the SRC. We endured the situation for a few weeks, then Geoff and our daughter left Herstmonceux and returned to La Jolla. They came back to England for Christmas, 1972, and during this time went through the INT observing books and totaled up the number of hours of observing actually achieved during its 3 years of operation. The telescope was available to university as well as RGO astronomers, upon acceptance of good observing proposals. It was heartbreaking to read in the observing record comments such as this: "Exposure ended; kept the dome open, but watched the guide star slowly disappear into the murk. Closed the dome." Walking to the dome in the early evening, one would walk through a gentle dewy mist rising through the grass. The total annual observing hours averaged 600–800 per year, in comparison to some 2000 per year at the best sites worldwide.

Efforts to initiate plans for moving the INT, when a good NHO site had been found, met bitter opposition from the old guard staff, and the local Hailsham dignitaries (the right honorable Quintin Hogg had been Member of Parliament for Hailsham). Never one to suffer fools gladly, Geoff wrote what became a famous letter to *Nature* (Burbidge 1972) exposing the situation at Herstmonceux. Brian Flowers summoned us both to his office in London, and a bitter confrontation ensued.

At this point, only two factors kept me at Herstmonceux. One was the fact that I was, by virtue of my position, a member of the U.K. half of the Anglo-Australian Telescope Board (Fred Hoyle was chairman, and Jim Hosie was the SRC representative). Meetings were mostly in Canberrra, with trips to Siding Springs, where the dome was under construction. Details concerning the telescope design and construction, and discussion of the bids to be accepted from contractors, were under way. It was Fred Hoyle who pointed out that the telescope design, largely based on the design of the KPNO 4-m Mayall telescope, should be corrected for a problem with the declination mount and drive which had been found at KPNO and had needed correction.

The other factor that kept me at Herstmonceux was the continued loyalty of the research astronomers and the Sussex astronomers and students, too many to name here, who fulfilled the promise of their abilities when the AAT and NHO came into being. The search for a good site for the NHO proceeded at a snail's pace, but the expertise of Merle Walker (of Lick Observatory) was enlisted for comparing various possible sites in the Canary Islands, the Azores, and southern Spain.

The site eventually chosen, on La Palma, has proven to be very good, and the removal of the INT to that site has enabled astronomers to do forefront research with it. The adjustment to the mounting, necessary because of the difference in latitude, had already been considered in a preliminary way while I was at Herstmonceux, and the full-scale engineering design and construction turned out excellently.

Eventually, after one and a half unhappy years, I set about submitting my resignation as Director of RGO. Some meetings in London ensued; I was asked to delay my resignation so as not to upset AAT and NHO negotiations, so I delayed as long as I felt useful. At two meetings in London, after my resignation was finally accepted, I was informed that the news media would be on my track that evening when I returned to Herstmonceux. The warnings proved correct. The next morning I set off to drive to Lewes to take the train and attend the wedding of my nephew and godson. A horrible automobile accident on the way landed me in the hospital and nursing home for some three weeks, during which the wrong leg was labeled on the x rays as needing traction to heal a cracked hip joint. I walked on the injured leg for some weeks before the incorrect reading of the x rays was discovered. All in all, I can look back upon those weeks before my daughter came over to help me pack up from Herstmonceux for the return to California, as due payment of my "blood money" to get out of the intolerable situation. At the orthopedic surgeon's office in La Jolla, after new x rays were taken, the doctor came into the examination room where I was awaiting his opinion, and said "You might as well throw away those crutches; they haven't been doing you any good!" And he and the nurse showed me the new x rays, with right and left hips clearly marked! I had been getting on and off planes to Australia and California, and moving around in Australia, using crutches to keep my left leg off the ground, having been enjoined to put my weight *"only* on the right leg." The experience had, to say the least, been painful, but seems to have left no residual damage to my right hip joint.

MORE POLITICS: PLANNING FOR A LARGE SPACE TELESCOPE; U.S. CITIZENSHIP

I had served on the Space Science Board, whose mandate is to consider and make recommendations on all ventures in space science. A large space telescope for the UV and optical, as envisioned decades earlier by Lyman Spitzer, was being seriously considered, since the state of technology for such a telescope, with a 3-m primary mirror, seemed feasible. Much political activity was underway; astronomers were working on Congress and the NASA hierarchy; the enthusiasm and support for the project by Nancy Roman within NASA was of fundamental importance in the ultimate agreement to fund and plan a 2.4-m (rather than Spitzer's 3-m) telescope with imaging and spectroscopic capability down to the MgF_2 reflective cutoff in the ultraviolet (1150 Å).

James Fletcher was the NASA Administrator. He was more interested in solar system exploration than in galactic or extragalactic astronomy. Astronomers who, throughout their careers, had done their observational work from the ground, were enthusiastic about the prospect of a large, well-instrumented telescope in orbit, above the UV-absorbing ozone layer in Earth's atmosphere, and with a diffraction-limited mirror that could take advantage of the absence of atmospheric-induced "seeing" distortion of images, so they had volunteered their time to attend a Washington meeting to help make an impressive presentation to Administrator Fletcher. The result was that Fletcher was convinced; NASA was awarded the finances for this Space Telescope, with a planned lifetime of 15 years, but it was to be placed in low Earth orbit because it would, as Fletcher had dictated to the Space Science Board, be launched by the Space

Shuttle which would visit it during the 15-year lifetime, for updating, replacements of instruments, etc and, in fact, space astronomy would in the future be tied to launches by the Shuttle.

I return to more about the Space Telescope later, after a digression on my involvement in the American Astronomical Society, acceptance as a U.S. citizen, and service as President of the American Association for the Advancement of Science.

When I was informed of my election as President of the American Astronomical Society, I realized that I would be involved in political activity at the Federal level. During all the years since 1955, Geoff and I had maintained our U.K. citizenship. With no basic language change, there is less incentive for the British immigrants to become U.S. citizens. "Green cards" ("pink cards" as they now are) enable one to pass freely through the passport control desks at airports, so the main disadvantage for "resident aliens" is that they are disenfranchised. I felt that, if I were to be involved in congressional presentations and discussions with politicians, I should do so as a U.S. citizen; I would be able to talk with and write letters to people for whom I had voted (or not voted, as the case might be). So the lengthy and tedious business of applying for citizenship began; in San Diego, this involved a long waiting period. Eventually, the effort succeeded, and I felt that I could honestly serve the American Astronomical Society as President and represent it in some acrimonious business with the U.S. Post Office, which was attempting to require a footnote stating "this is an advertisement" to all pages of Astrophysical Journal publications for which there were page charges. Fortunately, reason prevailed, even with the then Postmaster General.

Further political activities arose when the Equal Rights Amendment to the U.S. Constitution was before the nation; it was ratified by all but three states, and most scientific societies were banning annual meetings from being held in the negative states. The AAS membership was divided about 6 to 5 on the issue; I received mail that ranged from vituperous condemnation of me as President for letting this issue come before the Society, to very supportive letters from men and women anxious to see the removal of the barriers and inequalities suffered by women in science, and indeed in life in general.

The years of my involvement in the American Association for the Advancement of Science did lead to political activity. In my opinion, the AAAS is a great and extremely valuable organization, and, in recent years, its involvement in a massive effort toward improvement of education in science and mathematics in the U.S. has helped to raise the level of concern throughout the country. The AAAS "Project 2061" was just beginning in my third year on the AAAS Board, as Past President, and is now doing its work throughout the nation's schools.

QSOs, COSMOLOGY, AND THE HUBBLE SPACE TELESCOPE

Throughout these years, work in extragalactic astronomy continued. There was a break in the collaborative work with Geoffrey, when he was offered the Directorship of the Kitt Peak National Observatory which he accepted; he moved to Tucson in 1978, and held this position until 1984 when he returned to the University of California.

The next Space Telescope planning involved the choice of the first-generation instruments. Apart from the obvious guide-star acquisition and tracking instrumentation and an on-axis camera, other instrumentation was open for discussion. European participation in the telescope was eagerly accepted; one of the four instruments would be a camera/spectrograph designed and built by astronomers in Europe. The remaining "slots" were allocated to two US spectrographs and a fast photometric system. Teams were rapidly formed; I was invited to join the Harvard/Smithsonian Faint Object Spectrograph Team. Contractors were chosen, instruments were designed, and proposals to NASA were submitted.

For the U.S. spectrographs, many optical experts realized that detectors were of prime importance, and these had better be photon-counting detectors of proven design and performance. Two teams were chosen, both using the Digicon detector of McIlwain and Beaver (Beaver & McIlwain 1971, Beaver et al 1972). One was for the Goddard High Resolution Spectrograph, the other, a small team headed by Richard Harms of UCSD, was for the Faint Object Spectrograph (FOS). Team members were re-designated by NASA; I found myself on the UCSD FOS team.

The award of such a major NASA contract to UCSD provided a considerable perturbation to the smooth running of the business office of the Physics Department. Management of the contract with Martin Marietta, Denver, was seen by NASA to be inadequate, and our Chancellor McElroy was told in no uncertain terms that if something were not done to improve the management, the contract would be canceled. The Chancellor's remedy was to create a new Organized Research Unit, the Center for Astrophysics and Space Sciences, in a building where the High Energy Astrophysics and the Infrared groups already had their labs and offices. The choice of a director for this ORU posed a problem, because of the perturbation, just mentioned, within Physics. As a Co-Investigator of the FOS, I was asked to take on this job. The usual term for director of an ORU is five years, but I was asked to stay on longer. After the failure of some attempts to recruit well-known astronomers from outside UCSD, which were carried out rather ineptly, Larry Peterson, head of the High Energy Astrophysics group, agreed to take on this rather thankless task, and my service ended in 1988.

After launch of the Space Telescope in 1990, the subsequent discovery of the faulty figuring of the primary 2.4-m mirror is too well known to discuss here.

All I wish to emphasize is that astronomers are used to coping with few photons, atmospheric turbulence, less-than-perfect telescopes and instruments, and, one and all, they rose to the challenge and exciting results have been produced by the Space Telescope, named, perhaps unfortunately for his memory, the Hubble Space Telescope. My first observation with the Faint Object Spectrograph, carried out at Goddard Space Flight Center, did cause me euphoria similar to what I felt when taking my first direct photograph of a galaxy—to be looking for the first time at far-UV light never before detected, coming down from outer space.

The QSOs have continued to provide puzzles. The redshifts, if due to the expansion of the universe, require very large energy releases from "the central engines," and highly relativistic outflow of charged particles. There are also the observations of associations between galaxies and QSOs where the QSOs have redshifts very different from the galaxies (the old problem of "discrepant velocities"); these are well documented by Burbidge et al (1990) and Arp (1987). I find myself very much attracted by the new ideas of Hoyle et al (1993), who have described a Quasi-Steady-State-Cosmology, in which matter is created in successive epochs, rather than in a single event marking the "beginning" of the universe. We observe matter pouring out of the centers of active galactic nuclei, and streams of relativistic particles from the centers of radio galaxies and QSOs; the concept of these outpourings as "creation events" in the presence of very strong gravitational fields appeals to me philosophically and observationally.

CONCLUSION

When I was invited by the ARAA Editorial Board to write this account, one reason given was that most of the younger observational astronomers are not aware of what it was like to use optical telescopes before TV, 2-dimensional photon-counting devices, and computers made possible today's remote observing—either remote in the sense that one is in a warm lighted console room next to the telescope dome, or remote at a terminal in an entirely different location. Thus I have concentrated on my past experience, and the literature cited contains publications of which most astronomers who entered the field less than 10, 15, or even 20 years ago will be unaware. In recalling the past, I became aware of how much I have owed to various friends throughout the years, and, above all, that this is an account, not of one life in astronomy, but of two—Geoff's and my own.

ACKNOWLEDGMENT

I close with particular thanks to Betty Travell for her patience, accuracy, and continued help in the major work of preparing this manuscript.

Literature Cited

Arp H. 1987. *Quasars, Redshifts & Controversies.* Berkeley: Interstellar Media

Babcock HW. 1939. *Bull. Lick Obs.* 19:41

Beaver E, Burbidge M, McIlwain C, Epps H, Strittmatter P. 1972. *Ap. J.* 178:95

Beaver EA, McIlwain CE. 1971. *Rev. Sci. Instr.* 42:1321

Bidelman WP, Keenan PC. 1951. *Ap. J.* 114:473

Burbidge EM, Burbidge GR. 1951. *Ap. J.* 113:84

Burbidge EM, Burbidge GR. 1953. *Observatory* 73:69

Burbidge EM, Burbidge GR. 1957. *Ap. J.* 126:357

Burbidge EM, Burbidge GR. 1975. In *Galaxies and the Universe,* ed. A Sandage, M Sandage, J Kristian, p. 81. Chicago: Univ. Chicago Press

Burbidge EM, Burbidge GR, Fowler WA, Hoyle F. 1957. *Rev. Mod. Phys.* 29:547

Burbidge G. 1972. *Nature* 239:117

Burbidge G, Hewitt A, Narlikar JV, Das Gupta P. 1990. *Ap. J. Suppl.* 74:675

Burbidge GR. 1956. *Ap. J.* 124:416

Burbidge GR. 1959a. *Ap. J.* 129:849

Burbidge GR. 1959b. In *Radio Astronomy IAU Symp. No. 9,* ed. RN Bracewell, p. 541. Stanford: Stanford Univ. Press

Burbidge GR, Burbidge EM. 1953a. *Ap. J.* 117:407

Burbidge GR, Burbidge EM. 1953b. *Ap. J.* 118:252

Burbidge GR, Burbidge EM. 1955. *Ap. J. Suppl.* 1:431

Burbidge GR, Burbidge EM, Sandage AR. 1963. *Rev. Mod. Phys.* 35:947

Greenstein JL. 1984. *Annu. Rev. Astron. Astrophys.* 22:1

Gregory CCL. 1966. *Q. J. R. Astron. Soc.* 7:81 (Obituary by EM Burbidge)

Hoyle F. 1946. *MNRAS* 106:343

Hoyle F. 1954. *Ap. J. Suppl.* 1:121

Hoyle F, Burbidge G, Narlikar JV. 1993. *Ap. J.* 410:437

Hoyle F, Fowler WA, Burbidge GR, Burbidge EM. 1964. *Ap. J.* 139:909

Matthews TA, Sandage AR. 1963. *Ap. J.* 138:30

Mayer MG, Teller E. 1949. *Phys. Rev.* 76:1226

Page T. 1975. In *Galaxies and the Universe,* ed. A Sandage, M Sandage, J Kristian, p. 541 Chicago: Univ. Chicago Press

Schwarzschild M. 1954. *Astron. J.* 59:273

Annu. Rev. Astron. Astrophys. 1994. 32: 37–82

ASTEROSEISMOLOGY

Timothy M. Brown

High Altitude Observatory/National Center for Atmospheric Research,[1]
P.O. Box 3000, Boulder, Colorado 80307

Ronald L. Gilliland

Space Telescope Science Institute, 3700 San Martin Drive, Baltimore,
Maryland 21218

KEY WORDS: pulsating stars, magnetic stars, white dwarfs, δ Scuti stars.

1.   INTRODUCTION

Asteroseismology is commonly understood to mean the study of normal-mode pulsations in stars that, like the Sun, display a large number of simultaneously excited modes. The idea of learning about a physical system by examining its oscillation modes is of course an old one in physics, but it is only fairly recently that data of sufficient quality have become available to apply this technique to stars.

The Sun is (and will likely remain) the outstanding example of the progress that can be made using seismological methods. Seismic studies of the Sun have succeeded in mapping the variation of sound speed with depth in the Sun, and the variation of angular velocity with both depth and latitude; they have been used to measure the depth of the Sun's convective envelope, and they have begun to be used to estimate the helium abundance in the convection zone and to reveal at least some of the subsurface structure of solar activity.

Three properties of the solar pulsations have combined to make this progress possible. First, the Sun manages to excite a very large number of modes simultaneously: Something like 10^7 modes are thought to have amplitudes large enough for observation. Each mode carries information about the solar

[1]The National Center for Atmospheric Research is sponsored by the National Science Foundation.

37

0066–4146/94/0915–0037$05.00

interior that is somewhat different from that of any other mode. Thus, one may use mode characteristics to fit for a large number of parameters describing the solar structure. Second, the acoustic modes of the Sun are fairly weakly damped, having lifetimes that are typically several thousand oscillation cycles. This allows one to measure the mode properties (particularly the frequencies) very accurately, permitting delicate tests to distinguish between models. Last, the amplitudes of individual oscillation modes are very small (for the modes with largest amplitude, relative displacements at the surface of the Sun are less than 10^{-7}, and Mach numbers are roughly 10^{-4}). This property assures that the presence of the modes has only a small influence on the structure of the star, and moreover that linear theory is adequate for most purposes involving the character of the modes themselves.

Many stars besides the Sun may be expected to support pulsations with these same three properties. Stars of roughly solar type should of course behave in ways similar to the Sun, and stars of this sort form a large fraction of the potential targets for asteroseismology. But several other kinds of star (δ Scuti stars, roAp stars, and the pulsating white dwarfs) also have the desired pulsation characteristics. Pulsations in some of these stars are, for various reasons, much easier to observe than in the Sun-like stars; indeed, to date virtually all unambiguous observations of multi-mode pulsators relate to these other categories of stars.

Regardless of the type of star or the mechanism driving its pulsations, we will not in the foreseeable future have as much pulsation information about other stars as we have about the Sun. A very large majority of the 10^7 modes seen in the Sun have horizontal wavelengths that are a small fraction of a solar radius. When averaged over the solar disk (as a distant observer would do), the perturbations due to these modes average to zero, rendering them undetectable. It is only in special circumstances that modes with angular degree greater than about 3 are observable on distant stars; the number of modes that may be observed is therefore likely to be at most a few tens. Nevertheless, since oscillation mode frequencies are arguably the most precise measurements relating to a star that we can make, a few tens of such frequencies may still be of great importance to our understanding of stellar structure and evolution.

2. FUNDAMENTAL IDEAS CONCERNING STELLAR OSCILLATIONS

2.1 *Terminology and Basic Physics*

The physical basis for understanding stellar pulsations has been described with great clarity and detail by many authors (see, for instance, Cox 1980, Unno et al 1989, Christensen-Dalsgaard & Berthomieu 1991). Here we give only the minimal description necessary for our purpose. Pulsations in stars are usually characterized by the nature of the restoring force that is principally responsible for the adiabatic oscillatory behavior. The observed pulsations in the Sun, for instance, are essentially sound waves, termed p-modes, in which pressure-gradient forces provide the largest part of the restoring force. In the modes seen in white dwarfs, on the other hand, buoyancy is the dominant restoring force (both gravity and composition gradients are very important in white dwarfs); pulsations of this sort are termed g-modes. Although useful, these designations may sometimes be unduly restrictive. For instance, many pulsation modes in δ Scuti stars have the character of p-modes in the stellar envelopes, but look more like g-modes in the stellar core (Däppen et al 1988, Däppen 1993). Other complications arise in the roAp stars, where there is clear evidence that large-scale magnetic fields play an important role in the mode behavior, especially near the stellar surface (e.g. Shibahashi 1983).

The stars that display pulsations may be described with reasonable accuracy as spheres. For this reason it is possible and convenient to write the pulsation eigenmodes as the product of a function of radius and a spherical harmonic. The spatial and temporal variation of a perturbation to the star's mean state are then

$$\xi_{nlm}(r, \theta, \phi, t) = \xi_{nl}(r) Y_l^m(\theta, \phi) e^{-i\omega_{nlm}t}. \qquad (1)$$

Here ξ is any scalar perturbation associated with the mode (e.g. the radial displacement); r, θ, ϕ, and t are the radial coordinate, the colatitude, the longitude, and time, respectively. The mode's *radial order n* is usually identified with the number of nodes in the eigenfunction that exist between the center of the star and its surface. Since it deals with the depth structure, n is not accessible to direct observation. The *angular degree l* is the product of the stellar radius R_* and the total horizontal wavenumber of the mode; modes with large values of l display many sign changes across a stellar hemisphere, and hence are usually unobservable on distant stars. The *azimuthal order m* is the projection of l onto the star's equator; it is therefore restricted to be less than or equal to l in absolute value. Note that p-modes may be purely radial ($l = 0$), but that g-modes, since they are driven by buoyancy forces, must involve a variation in the horizontal coordinates, and hence always have $l \geq 1$. The mode frequency ω_{nlm} generally depends on n and l in complicated ways, depending on the restoring forces responsible for the pulsation and on the structure of the star. In particular, there is generally no simple harmonic relation between the frequencies of modes with (for instance) given l and successive values of n. Mode frequencies resulting from theoretical calculations are often expressed as angular frequencies ω_{nlm}, as in Equation (1). The results of observations are more commonly written in terms of the circular frequency $\nu_{nlm} \equiv \omega_{nlm}/2\pi$. For stars that are truly spherically symmetric, mode frequencies depend only upon n and l, and are independent of m. This occurs because m depends on the choice of position for the pole of the coordinate system, which is arbitrary for a spherical configuration. Any condition that breaks the spherical symmetry (such as rotation about an axis, or the presence of magnetic fields) can lift this frequency degeneracy.

Observations of stellar pulsations usually involve either the photometric intensity or the radial velocity (for some purposes it is useful to perform simultaneous or near-simultaneous observations in more than one spectral band, so that one may speak of oscillations in the stellar color.) The perturbations in intensity and in velocity are related, of course. The displacements of the stellar plasma cause Doppler shifts directly; the accompanying compressions or displacements from equilibrium height also cause temperature changes, resulting in perturbations to the observed intensity.

The oscillation parameters that one may observe on stars include not only the mode frequencies, but also amplitudes and linewidths. Frequencies usually carry the most information, because they can be measured accurately and because they can usually be calculated with good accuracy considering only adiabatic effects. Mode amplitudes and linewidths, on the other hand, require explicit and detailed treatment of the energy transfer into and out of the oscillation modes, and hence their interpretation is, at present, much more problematic. Recent work on driving and damping mechanisms for Sun-like stars suggests that this situation may soon improve, however (Goldreich & Kumar 1990, Murray 1993). Because of the relatively poor current understanding of the physics of mode driving and damping, henceforth we shall concentrate mostly on the information that may be inferred from mode frequencies.

Although mode frequencies depend in complicated ways on the stellar structure, there is a useful limit ($n \gg l$) in which simple asymptotic formulae give useful approximations to the true frequency behavior (Vandakurov 1968, Tassoul 1980, Christensen-Dalsgaard 1988a). For p-modes, one finds

$$\nu_{nl} = \Delta\nu_0\left(n + \frac{l}{2} + \epsilon\right) - \frac{AL^2 - \eta}{(n + l/2 + \epsilon)}, \tag{2}$$

where $\Delta\nu_0$, A, ϵ, and η are parameters that depend on the structure of the star, and $L^2 \equiv l(l + 1)$. If the parameters A and η were zero, one would therefore find p-mode frequencies to fall in a regular picket fence pattern with frequency spacing $\Delta\nu_0/2$: Modes with odd l would fall exactly halfway between modes with even l, and modes with different n at a given l would always be separated in frequency by multiples of $\Delta\nu_0$. The parameter $\Delta\nu_0$, termed the *large separation*, is simply related to the sound travel time through the center of the star:

$$\Delta\nu_0 = \left(2\int_0^{R_*} \frac{dr}{c}\right)^{-1}, \tag{3}$$

where c is the local sound speed and R_* is the stellar radius. Consideration of the virial theorem (Cox 1980, Gough 1990) shows that this travel time is related to the mean density of the star, so that

$$\Delta\nu_0 \cong 135\left(\frac{M_*}{R_*^3}\right)^{1/2} \mu Hz, \tag{4}$$

where M_* and R_* are the stellar mass and radius in solar units. Equation (4) holds exactly for homologous families of stars, but it is obeyed quite closely even for stars that are not homologous, such as stars of different mass along the main sequence (Ulrich 1986). The large separation is thus easily interpreted in terms of the stellar structure, and moreover it is likely to be straightforward to observe, even in noisy stellar oscillation data.

Parameters A and ϵ in Equation (2) have to do with the structure near the center of the star and near the surface, respectively. Modes with different degree l penetrate to different depths within the star. Modes with $l = 0$ have substantial amplitude even at the center; those with higher values of l avoid a region in the stellar core that grows in radius as l increases. This difference in the region sampled by modes with different l leads to the second term on the right-hand side of Equation (2), removing the frequency degeneracy between modes that differ by (say)-1 in n and $+2$ in l. This effect is often parameterized in terms of the *small separation*, defined as $\delta_{nl} \equiv \nu_{n+1,l} - \nu_{n,l+2}$. The small separation may be written as an integral analogous to that in Equation (3) (Däppen et al 1988):

$$\delta_{n,l} = \Delta\nu_0 \frac{(l+1)}{2\pi^2\nu_{nl}} \int_0^{R_*} \frac{dc}{dr} \frac{dr}{r}. \tag{5}$$

The small separation is thus sensitive to sound speed gradients, particularly in the stellar core. Since these gradients change as nuclear burning changes the molecular weight distribution in the star's energy-producing region, the small separation contains information about the star's evolutionary state.

The parameter ϵ relates to the phase shift suffered by sound waves upon reflection near the star's surface. It depends upon the details of the thermodynamic (and, in the Sun, magnetic) structure near the surface; it is thus difficult to say much of a general nature about it (but see Christensen-Dalsgaard & Pérez-Hernández 1992, Pérez-Hernández & Christensen-Dalsgaard 1993). One may note, however, that the depth at which upward propagating sound waves reflect depends almost entirely on their frequency, and not on l. For this reason, modes with different ν sample different regions near the stellar surface, and structural changes that occur near the surface tend to cause frequency perturbations of the form $\delta\nu_{nl} = f(\nu_{nl})$.

In the asymptotic limit $n \gg l$, g-mode *periods* (not frequencies) become almost equally spaced:

$$T_{nl} = \frac{T_0(n + l/2 + \delta)}{L}, \tag{6}$$

where T_{nl} is the period of a g-mode, T_0 and δ are parameters that play roles similar to those of $\Delta\nu_0$ and ϵ in Equation (2), and the other symbols have their previous meanings. The asymptotic period T_0 depends upon an integral of the Brunt-Väisälä frequency N throughout the star (Tassoul 1980):

$$T_0 = 2\pi^2 \left[\int_0^{R_*} \frac{N}{r} dr \right]^{-1}. \tag{7}$$

2.2 The Solar Example

The Sun provides the best-developed example of seismological inference, and observations of the Sun as a star at once illustrate the phenomena that are observable and motivate the search for similar phenomena on other stars. It is important to remember, however, that many of the successes of helioseismology rest on observation of modes in the range $5 \le l \le 100$, which will be inaccessible on most distant stars.

Figure 1 shows the power spectrum of solar p-modes as measured with the IPHIR full-disk photometer while *en route* to Mars on the Soviet *Phobos* spacecraft (Toutain & Frölich 1992). The IPHIR instrument measured the brightness in several colors, integrated over the visible disk of the Sun, using silicon diode photometers. Except for their low noise level, these observations are thus closely analogous to normal photometric observations of stars. Figure 1 illustrates several important aspects of the solar p-modes. First, the mode frequencies are very well defined, with typical quality factors Q of several thousand. Second, the mode amplitudes are large only within a restricted frequency range, between roughly 2500 and 4000 μHz. Within that range, the low-degree modes to which IPHIR is sensitive are indeed almost evenly spaced in frequency, and in spite of the compressed frequency scale of this figure, many close pairs of modes (corresponding to $l = 0, 2$ or to $l = 1, 3$) may be seen. The separation between pairs of modes turns out to be roughly 68 μHz$=\Delta\nu_0/2$; the separation between modes making up a given $l = 0, 2$ pair is about 9 μHz. The amplitudes of the pulsations are quite small: The largest peaks near 3000 μHz have power corresponding to amplitudes $\delta I/I$ of only about 3×10^{-6}. A similar power spectrum obtained by measuring the disk-integrated solar velocity (see e.g. Claverie et al 1984) has virtually identical mode structure, somewhat smaller background power relative to the mode power, and peak mode amplitudes of roughly 15 cm s^{-1}.

A great deal of information about the Sun's interior has been obtained from the measured pulsation frequencies, including accurate estimates of the variation of sound speed (Christensen-Dalsgaard et al 1985, Vorontsov 1989) and angular velocity (Duvall et al 1984, Brown 1985, Brown et al 1989, Libbrecht 1989) with depth and latitude, and a precise estimate of the depth of the adiabatically stratified region of the solar convection zone (Christensen-Dalsgaard et al 1991). All of these inferences draw upon observations of modes with a substantial range of l, however; results of useful precision cannot be obtained with disk-integrated

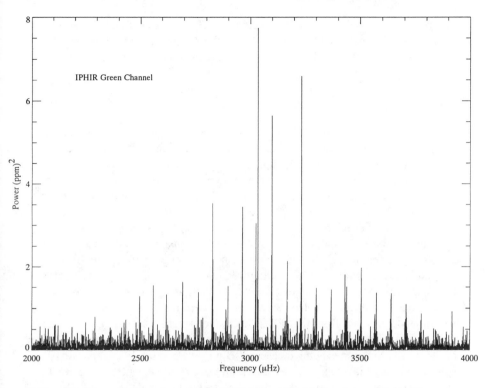

Figure 1 Power spectrum of one month of disk-integrated solar intensity measured by the green channel of the IPHIR experiment. From data provided by C. Frölich (see Toutain & Frolich 1992).

observations alone. Several other inferences do not share the requirement for high-*l* data.

Perhaps most notable among these is the ability to test at least some explanations for the solar neutrino deficit. Since the low-*l* p-modes penetrate close to the solar core, their frequencies (in particular the small frequency separation) may be used to test for the presence of physical effects that might account for the observed paucity of [8]B neutrinos from the Sun. Two astrophysical models to explain the neutrino deficit involve lowering the temperature of the solar core, either by transporting heat by means of Weakly Interacting Massive Particles (WIMPs; see e.g. Steigman et al 1978, Spergel & Press 1985), or by mixing fresh fuel into the Sun's center (Schatzman et al 1981). Models of these processes (Gilliland & Däppen 1988; Christensen-Dalsgaard 1991, 1992; Lebreton et al 1988; Cox et al 1990) yield small frequency separations δ_{02} that are incompatible with the observations.

Another important issue addressable with such data is the frequency splitting of modes with $l > 0$ by the solar rotation (see e.g. Hansen et al 1977, Brown et al 1989, Libbrecht & Morrow 1991). In the simple case in which angular velocity is independent of latitude, the frequencies of modes within a multiplet are given by

$$\nu_{nlm} = \nu_{nl0} + \frac{m\beta_{nl}}{2\pi} \int_0^{R_*} \Omega(r) K_{nl}(r) dr, \qquad (8)$$

where β_{nl} is a correction factor of order unity accounting for Coriolis forces, $\Omega(r)$ is the solar angular velocity, and K_{nl} is a unimodular kernel that is roughly proportional to the local energy density in the mode. Thus, the splitting of low-l multiples (sets of modes with the same n and l but different m) depends somewhat on the angular velocity near the Sun's center, where only eigenmodes with small l have substantial amplitude. An unambiguous measurement of the splitting would reveal much about angular momentum transport within the Sun and similar stars—a process that has been extensively modeled in recent years (Pinsonneault et al 1989, Charbonneau & MacGregor 1993). IPHIR data indicate that the characteristic splitting frequency $\nu_{nl1} - \nu_{nl0}$ is 579 ± 25 nHz for $l = 1$ and 550 ± 23 nHz for $l = 2$, as compared with 462 nHz for the surface equatorial rotation. However, there is room for question regarding the splittings determined from the IPHIR data since the interpretation is difficult because the frequency shifts due to rotation are comparable to the natural widths of the oscillation modes being measured. But if taken at face value, the IPHIR splittings imply that the solar interior within $0.2~R_\odot$ of the center may have an angular velocity almost five times that observed at the Sun's surface (Toutain & Frölich 1992).

Finally, the frequencies of solar p-modes have been observed to change by small amounts (a few parts in 10^5) in response to changing levels of solar activity. The best measurements of this effect rely on observations averaged over a range of l (Libbrecht & Woodard 1990, Bachmann & Brown 1993), but the first such frequency shifts reported came from disk-integrated measurements from the ACRIM radiometer (Woodard & Noyes 1985). Such measurements are in principle possible for other stars, and (especially in the case of stars that are much more active than the Sun) may provide clues to the nature of stellar activity.

As mentioned above, the processes determining the amplitudes and lifetimes of the solar p-modes are only partially understood. It is fairly clear, however, that the source of energy for the Sun's p-modes is acoustic noise generated by high-speed convective motions within a few scale heights of the solar surface (Goldreich & Keeley 1977, Goldreich & Kumar 1990, Kumar & Lu 1991, Cox et al 1991). This implies that all stars with vigorous surface convection zones (which is to say, all stars with spectral types later than roughly F5) should support p-modes that are more or less similar to those observed in the Sun. One attempt to estimate the p-mode amplitudes on other stars (Christensen-Dalsgaard & Frandsen 1983) has been made, giving results that may be parameterized

(Kjeldsen 1993) as

$$\frac{v_*}{v_\odot} = 2\left[\left(\frac{g*}{g_\odot}\right)^{0.6} + \left(\frac{g*}{g_\odot}\right)^{4.5}\right]^{-1},\tag{9}$$

where v_* is the typical rms velocity amplitude of the largest stellar mode, $v_\odot$ is the same quantity for the Sun, and g_* and $g_\odot$ are the stellar and solar surface gravities, respectively. This estimate is very uncertain, and of course says nothing about the likely lifetimes of p-modes on other stars. It is, however, the best analysis currently available. One may reasonably expect surprises, and perhaps considerable advances in the level of theoretical understanding, when p-mode amplitudes are measured for a suitable range of stellar types.

2.3   Estimates of Information Content

As we will show later, useful observations of multi-mode pulsations on other stars require heavy commitments of observing time. It is therefore important to ask whether the results one may obtain from seismology are worth the effort required to get them. What exactly can one learn from oscillation data?

The solar case is an exciting and informative example of the power of seismological methods, but it is by no means a perfect predictor of the usefulness of asteroseismology. First, as noted above, our ability to resolve spatial structures on the Sun leads to measurement opportunities that do not exist for other stars. On the other hand, the Sun is only a single star, and its study (even in great detail) is bound to leave many unanswered questions about other stars with different circumstances or histories. It thus seems likely that studying oscillations in a large sample of stars (even restricting the sample to those of roughly solar type) would allow fundamentally new insights. Finally, there are a great many stars that do not resemble the Sun at all, either in their structure or in their modes of oscillation. The study of these stars allows one to probe physical domains that are quite different from those seen in the Sun. In the remainder of this section, we discuss the information contained in oscillation frequencies for the case of stars that are somewhat like the Sun, and for groups of such stars. Stars of this sort have been fairly thoroughly studied from a theoretical point of view, since the solar example is so well understood. In the next major section we discuss other sorts of stars, for which clear-cut observations exist, but for which the theory is less well developed.

The first efforts to estimate the information content of oscillation frequencies for Sun-like stars were made by Ulrich (1986, 1988) and by Christensen-Dalsgaard (1986b). These authors computed the sensitivity of the frequency separations Δv_0 and δ_{nl} to changes in the stellar mass M and age τ; Ulrich (1986) also considered changes in the initial composition parameters Y (the helium abundance), Z (the heavy element abundance), and in the mixing length

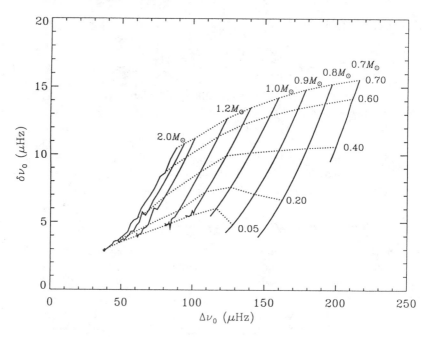

Figure 2   The "asteroseismic H-R diagram" of Christensen-Dalsgaard (1993), showing the variation in large ($\Delta\nu_0$) and small ($\delta\nu_0$) frequency separation with stellar mass and age. Mass is constant
along solid lines; age (parameterized by the central hydrogen abundance) is constant along dotted
lines.

ratio α. They concluded that if the composition were known, then measurement of the two frequency separations would allow the stellar mass and age
to be estimated with useful precision. These results are summarized in the
so-called asteroseismic H-R diagram, shown in Figure 2. If one assumes that
individual mode frequencies may be measured with precision comparable to
the mode linewidth (unknown for other stars, but typically 1 μHz for the Sun),
then Figure 2 suggests that frequency separations could be used to determine
stellar masses to within a few percent, and ages to within perhaps 5% of the
main-sequence lifetime.

Gough (1987) showed that these estimates are too optimistic, since the frequency separations are also quite sensitive to variations in other parameters,
particularly in Z. A realistic uncertainty in Z therefore leads to important uncertainties in mass and age. The generic difficulty Gough illustrated is that
most observable properties of stars (both their oscillation frequencies and more
standard indices such as luminosity and surface temperature) depend to some
extent on all of the parameters of stellar structure. Thus, the two frequency
separations can provide two relations among the five structural parameters

M, Y, Z, τ, α, but without additional information it generally is not possible to arrive at a definite value for any one of them. Other observational constraints are therefore needed, either in the form of different kinds of oscillation data (e.g. frequencies for individual modes, rather than frequency separations only), or in the form of more traditional astronomical measurements (photometry, astrometry, or spectroscopy). A minor complication is that proper interpretation of some photometric and astrometric data requires that the distance to the star be included as another parameter describing its properties. Thus, at least six well-determined observational properties are required to uniquely define the structural parameters of a field star. More may be necessary if it develops that some of the observations are redundant, or if other processes are important in determining the stellar structure, beyond those assumed in this simple 5-parameter description of stars.

Brown et al (1994) performed a more complete treatment of the problem of estimating model parameters from p-mode frequencies. The approach was to perform a least-squares fit of the model parameters to all of the observations that one might reasonably expect to have, including oscillation frequencies or frequency separations. With reasonable errors ascribed to the various observations, several conclusions emerged. Estimates of the age, mixing length, and mass of field stars can be substantially improved by the addition of oscillation frequencies. Indeed, without the frequency data, parameters such as the age are essentially unconstrained, and must be estimated from more general considerations, such as the age of the galaxy. The relative improvement in errors is greatest for distant stars, for which astrometric data are relatively unreliable. The lowest absolute errors, however, occur for nearby stars with high-quality astrometry. It develops that oscillation frequency data is usually unhelpful in constraining the heavy element abundance Z. In the best field star cases, one should be able to reach errors in mass and mixing length of about 3%, and in helium abundance and age of about 12%. Errors of this size would be interesting from the point of view of galactic evolution if they could be obtained for a good sample of stars near the Sun. They are not, however, small enough to allow tests of the physics of stellar structure theory.

A more interesting situation occurs if mode frequencies can be obtained for both stars in a well-observed visual binary (α Cen, for example). In such a case the two stars may be assumed to have the same age, distance, and initial composition, so that the number of parameters required to describe the system is less than twice that for a single star. Moreover, some new observables (the orbital data) provide fundamentally new sorts of information. One result is that parameter errors become smaller for binaries than for field stars. A more important difference is that, with many more observables than model parameters, one may search for inconsistencies between the observations and the best-fit model. If significant inconsistencies are found, then significant

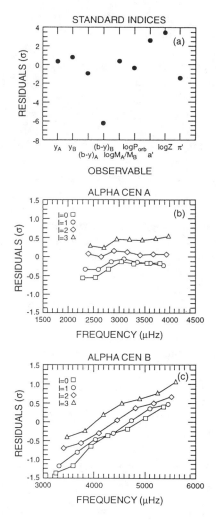

Figure 3 Residual errors between the values of observable quantities for the α Cen A/B visual binary system calculated using LAOL opacities and those calculated from a best-fit model using OPAL opacities. The top panel shows errors (in σ) for standard astronomical indices, including Strömgren magnitudes and colors, orbital data, and spectroscopic metal abundance. The bottom two panels show residual mode frequencies for the two stellar components, as functions of l and frequency. From Brown et al (1994).

errors must exist in the model of the star system. In this way, it may be possible to detect errors in the physics underlying the calculation of stellar structure. Figure 3 shows an example of the sort of discrepancies that might arise. In this case, observed properties of a binary system (chosen to be similar to the α Cen system) were constructed using LAOL opacity tables (Huebner et al 1977), but were fit to a model based on OPAL opacities (Rogers & Iglesias 1992). Since the "true" and assumed models of the system employed different physics, no combination of model parameters can match the constructed observations exactly. Figure 3 shows the residuals between the "true" observations and those implied by a best fit to the model using OPAL opacities. These residuals are generally large enough to be detected in spite of observational errors, and the pattern of discrepancies provides clues to the nature of the error in the assumed model. Not all modifications of the input physics result in changes that are as large as those in this example, and the degree to which different physical effects may produce similar sets of residuals is not yet known. Nonetheless, it seems reasonable that oscillation frequencies, if available, would not only allow measurement of the structural parameters of stars, but would also place constraints on at least some aspects of stellar evolution theory.

Star clusters are the natural extension of the progression from single stars to visual binaries; Gough & Novotny (1993a, b) have begun to investigate the utility of cluster observations by considering a group of very similar Sun-like stars, all with the same age, composition, and distance. They found that assuming the stars to be coeval with identical initial compositions causes a partial cancellation in the errors associated with some observables. In effect, measurements that apply to global properties of the cluster (the age and the composition) have their errors reduced by averaging over all of the observed stars. The errors associated with model properties that are peculiar to each star (mass and mixing length), however, take on values that are essentially independent of the number of stars observed. Further work in this area is needed, especially to extend the treatment so that stars throughout the cluster H-R diagram may be included in the solution.

Though global properties of stars are of great interest, some questions can only be properly answered by measuring the depth dependence of physical conditions within a star, in a way what is relatively model-independent. This is the aim of seismic inverse theory. Investigating its applicability to stars, Gough & Kosovichev (1993) have applied inverse techniques to frequencies arising from a hypothetical set of low-degree ($l \leq 2$) p-modes in a star of 1.1 $M_\odot$. They find that, because of the restricted range of l, little can be said about the star's outer layers. Within 0.4 R_* from the center, however, the mode set that they assume allows them to resolve about four distinct depth samples when inverting for either the sound speed or the angular velocity. This is sufficient information to infer the presence or absence of a convective core, or to detect a region of rapid rotation with a radius of 0.2 R_* or less.

3. SEISMOLOGY OF STARS UNLIKE THE SUN

Throughout the HR diagram, stellar pulsations are found where extension of the classical RR Lyrae and Cepheid instability strips intersect stellar sequences. The rich set of low-amplitude oscillation modes in evidence for some white dwarf, main-sequence, and subgiant stars within the instability strip support asteroseismology on these otherwise disparate stellar classes. In this section we discuss the pulsating white dwarfs, rapidly oscillating Ap stars, and the δ Scuti stars which, taken together, provide nearly all substantive asteroseismology results to date.

The roAp stars have generally been studied with the simplest of techniques: single channel time-series photometry from a single superior location (usually the Sutherland site of the South African Astronomical Observatory where most roAp stars have been detected and studied by Don Kurtz and colleagues; see Kurtz 1990). The pulsating white dwarfs have been most intensively studied using standard two-channel photometers (simultaneously observing program and control stars, allowing for removal of atmospheric transparency variations) on a longitude-distributed network (the Whole Earth Telescope—Nather et al 1990) of several telescopes at average observing sites. The δ Scuti stars with rich oscillation spectra have been most successfully studied using multi-site photometry campaigns by Breger et al (1989, 1990, 1991), Belmonte et al (1991), and Michel et al (1992a). CCD photometry has proved useful in the study of pulsating planetary nebulae nuclei (Bond & Ciardullo 1993) where discrimination against an extended background is required, and for δ Scuti stars in open clusters (Gilliland et al 1991, Kjeldsen & Frandsen 1992, Gilliland & Brown 1992a).

3.1   *Pulsating White Dwarfs*

White dwarf (WD) pulsations were first detected by Landolt (1968); since the first discovery, three (or perhaps four) different classes of pulsating white dwarfs have been identified, most of them oscillating in more than one mode (often many tens of modes are visible). Oscillations in white dwarfs are seen as photometric fluctuations with amplitudes that range from roughly 0.3 magnitude down to the limits of detectability; typical periods fall in the range between 100 s and 2000 s. The combination of fairly large amplitudes, short periods, and many excited modes makes the variable WD stars ideal subjects for asteroseismology. Indeed, the successes of WD pulsation studies provide the best current example of asteroseismological methods, and many of the techniques used by the WD community may be viewed as archetypes for the investigation of other kinds of stars.

At least three varieties of pulsating white dwarfs are presently known. The oscillating planetary nebula nuclei (PNNs) are extremely hot objects ($T_{\text{eff}} \geq 10^5\ K$) characterized by the presence of O VI; the prototypical object is K1-16 (Grauer & Bond 1984). There is disagreement as to whether they should be considered distinct from the DOV (or GW Vir) stars, which are also very hot objects, but are more evolved than the PNN; the prototypical object of this class is PG1159–035 (McGraw et al 1979, Winget et al 1991, Kawaler & Bradley 1993). The DBV stars are of intermediate temperature, and have atmospheres consisting almost entirely of helium. Oscillations in stars of this type have the distinction of having been predicted (Winget et al 1982a) before they were observed (Winget et al 1982b). They occupy an instability strip within the temperature range $21,500 \leq T_{\text{eff}} \leq 24,000$ K (Thejl et al 1991). Finally, the DAV (or ZZ Ceti) stars have atmospheres of almost pure hydrogen, and are found within a narrow temperature range $11,200 \leq T_{\text{eff}} \leq 12,500$ K (Weidemann & Koester 1984). All three sorts of WD pulsators oscillate in g-modes, usually of high radial order n (note that the radial order is usually denoted by k in the white dwarf literature), and the pulsations are thought to be driven by a κ instability arising in an ionization zone near the stellar surface. [It is possible, however, that a hydrogen burning instability may play a role in the DOV stars; see Starrfield et al (1984, 1985), Kawaler et al (1986).] Like Cepheids, pulsating white dwarfs are thought to be otherwise normal stars that simply happen to be passing through an instability strip. There is evidence, however, that not all DA stars within the instability strip are pulsators (Dolez et al 1991, Kepler & Nelan 1993). Models suggest that many WD stars should be unstable to radial pulsations in any of several modes (see e.g. Kawaler 1993), but no such radial pulsations (which would be of very short period) have yet been observed.

In order to understand pulsations in white dwarfs, it is necessary to review a few of the basic facts of their structure. For details beyond the scope of this discussion, see e.g. Winget & Fontaine (1982), Winget (1988), and Kawaler & Hansen (1989). By far the largest part of the mass of the cooler (DA and DB) pulsating white dwarfs is thought to reside in an electron-degenerate core, which probably consists predominantly of carbon and oxygen. Although essentially isothermal (because of the high conductivity of the electrons and the lack of internal heat sources), the core is density stratified in such a way as to make it neutrally stable to overturning motions (e.g. Brassard et al. 1991). Overlying the core, one finds a thin ($10^{-2}M_*$ or less) non- or partially-degenerate envelope, consisting mostly of helium. In the DA stars, the helium envelope is in turn overlain by an even thinner ($< 10^{-4}M_*$) layer of almost pure hydrogen. Because of the very high surface gravity ($\log g \cong 8$), the various constituents of the envelope should be separated by gravitational settling of the heavier components, so that the envelope consists of layers of almost pure H or He, separated by transition layers of somewhat uncertain thickness. This compositional stratification proves to be crucial in understanding the pulsations of white dwarfs. Over the temperature range of interest ($10^5 > T_{\text{eff}} > 10^4$ K, say), thin convection zones and corresponding regions of partial ionization of the underlying species are believed to occur near the C-He and He-H transition zones. At temperatures below 15,000 K or so, the surface hydrogen layer in DA stars becomes partially ionized (and hence connectively unstable) near its top. At yet lower temperatures, the surface convective region thickens, eventually occupying the entire hydrogen layer and (possibly) dredging helium to the surface. The very hot stars like PG1159-035 and PNNs have structures that are

more complicated than those of cooler stars, partly because they are still sufficiently hot for thermal pressure to play a role in their interiors. Qualitatively,
however, the ideas governing pulsations in DA and DB pulsators also apply to
the hotter DO and PNN stars.

Soon after the discovery of variable white dwarfs, overstable g-modes became
the accepted explanation for the phenomenon (e.g. Osaki & Hansen 1972).
There were (and are) compelling reasons for this identification: The observed
periods are reasonable for g-modes, but are two orders of magnitude too long for
p-modes, and the color dependence of the intensity variation is inconsistent with
significant radius variations (Robinson et al 1982). Problems remained with the
g-mode identification, however. The most serious was the very dense spectrum
of available g-modes. For each mode that was seen to be excited, there were
often many modes with similar characteristics and nearly identical frequencies
that were not seen at all (Robinson 1979). Where were all the other modes? A
related problem was that it was not at first possible to identify which oscillation
modes were being observed; even the degree l of the modes being seen was uncertain. Resolving these problems required first, a deeper understanding of the
importance of the layered stratification as it applied to g-modes (the notion of
"mode trapping"), and second, a means for obtaining drastically improved observations (the Whole Earth Telescope.) We shall discuss these topics next, and
then move on to the inferences about white dwarfs that have resulted from them.

Strong surface gravity on WD stars discriminates against p-mode pulsations
(which involve vertical motions), and favors g-modes (in which the motions are
predominantly horizontal). But g-modes may propagate only in regions with
stable density stratifications. In white dwarfs, g-modes are therefore confined
to the stellar envelope, above the degenerate core. The eigenfunctions may be
trapped in an even stronger sense, however, by the rapid jumps in fluid properties that occur at boundaries between different composition layers (Kawaler
& Weiss 1990, Brassard et al 1992). This *mode trapping* occurs if one or more
vertical wavelengths of the eigenfunction fit exactly (or nearly exactly) into
a layer of uniform composition. In such a case, the wavefunction matching
conditions at the layer boundaries force the eigenfunction to have a small amplitude outside the layer in question. This confinement of the eigenfunction
causes modifications to the simple expression for mode frequency spacings
given in Equation (6). Although Equation (6) continues to describe the mean
period spacing, the period separation between modes with consecutive values
of n shows variations near trapped modes. As we shall see below, the detailed
variation of period spacing with mode period leads to powerful diagnostics of
the near-surface structure.

The other extremely important consequence of mode trapping is its connection with mode growth rates. Energy to drive the pulsation modes comes from
the modulation of the radiative flux by changing opacity in the partial ioniza-

tion zones of abundant elements (the κ mechanism; see e.g. Cox 1980 for a detailed discussion). For this mechanism to overpower the damping forces acting on a mode, the relevant inonization zone must lie in a region of the star where the thermal time constant of the overlying material is comparable to the pulsation period. This condition constrains the depth of the ionization zone; the effective temperature regimes where the condition is met for hydrogen and helium outer envelopes correspond to the instability strips for DA and DB white dwarfs, respectively. The situation for DOV and PNN stars is (as usual) a little more complicated, but it is plausible that pulsations in these stars are driven by ionization of heavier elements, probably carbon or oxygen (Starrfield et al 1984, 1985). A general property of modes being driven by the κ mechanism is that the growth rate of the instability is inversely proportional to the energy associated with the mode (when normalized to unit surface amplitude). This is important because modes trapped in a surface layer may involve motions of only a tiny fraction of the envelope mass. For equal surface amplitudes, a trapped mode may therefore have a mode energy that is many orders of magnitude smaller (and a growth rate correspondingly larger) than that of similar untrapped modes. Of course, differences in growth rate need not result in corresponding differences in steady-state amplitude; the end result of such processes can only be inferred from fully nonlinear models of the driving and damping processes. Such models for white dwarfs are not now available, and are not expected soon. Nonetheless, it is highly plausible that mode trapping provides the filter that causes the selective excitation of a few modes, while driving the rest at much smaller amplitude.

Observations of pulsating white dwarfs have always been difficult because of the faintness of the WD pulsators and the necessity to provide precise, time-resolved photometric measurements. The first revolution in WD observing methods, which resolved many basic questions (and exploded many myths) concerning the basic character of WD variability, was the use of two-star photoelectric photometers by McGraw & Robinson (1976). These instruments allowed simultaneous observation of a reference star along with the WD target, so that small fluctuations in sky transparency would not be confused with true variability (Winget 1988). It soon became apparent, however, that the pulsation spectra of WD stars were much more complicated than had originally been thought. There proved to be so much fine structure in the spectra that data spanning only a single night of observation were completely inadequate to resolve it. Even many nights of data from a single site left ambiguities in the interpretation, since inevitable daytime gaps in the temporal coverage resulted in the appearance of spurious frequency components in the computed power spectra.

The spurious frequencies are a straightforward result of incompletely sampled time series. Their form is described in terms of the observational *window function*, which is taken to be unity when observations are available, and zero when they are not. The problem of poorly formed window functions is common to all types of astronomical seismology. Since the problem has to do with information that is genuinely missing, attempts to solve it by clever data analysis do not fare well. In fact, the only really effective cure for a bad window function is to get a better one. In helioseismology the outcome of this realization is the GONG network (Harvey et al 1988, 1993); in white dwarf seismology it is the second revolution in WD observing methods, the Whole Earth Telescope (Nather 1989, Nather et al 1990, Winget 1993).

The Whole Earth Telescope (henceforth WET) is not so much a telescope as a world-wide network of telescopes, augmented during observing campaigns by a central facility for collecting and analyzing data and for organizing the observing effort. The objective is to obtain nearly continuous observations of a few objects by distributing observing sites in longitude. The number of independent sites participating in WET runs has increased from 6 in the first (1988) run, to 12 in 1991, involving telescopes ranging in aperture from 0.4 m to 3.6 m (Nather et al 1990, Winget 1991). The instruments used for WET observations are usually standard two-channel photometers, though some observations have also been taken with three-channel systems. Because of the significant organizational and scheduling burden involved, the WET assembles itself for only about two campaigns per year. The WET has succeeded in obtaining nearly continuous observing runs of up to about 12 days duration (the runs typically include a few scattered observations before and after the main body of the data, so meaningful run lengths are hard to define). The resulting frequency response functions are extremely clean (Winget 1991), and allow very detailed and reliable analyses of the temporal power spectra of the target stars.

One of the most straightforward and yet informative applications of WD seismology has been the direct measurement of WD evolutionary timescales, using observations of period changes in WD pulsators. Some pulsating white dwarfs have power spectra that are dominated by one or a few pulsation modes, and in many such cases the frequencies of these modes have proved to be extremely stable. Changes in g-mode periods in evolving white dwarfs may be thought of as resulting from a combination of changes in the temperature and changes in the radius:

$$\frac{\dot{\Pi}}{\Pi} = -a\frac{\dot{T}}{T} + b\frac{\dot{R}}{R}, \qquad (10)$$

where Π is the period, T and R are the temperature and radius, and a and b are positive constants of order unity (Winget et al 1983). For DAV stars, one expects contraction to be unimportant, so that the cooling term in Equation (10) predominates, causing g-mode periods to increase with time. The relevant time scale in this case should be the characteristic cooling time for WD stars with $T_{\mathrm{eff}} = 10^4$ K, i.e. roughly 10^9 y.

The best characterized case of period change is the prototypical DOV star PG1159-035. As illustrated in Figure 4, this star oscillates in many modes (125 at last count; Winget et al 1991), but a few of these have substantially larger amplitudes than the others. One of these, with a period near 516 s, has been found to have nearly constant amplitude and consistent variation of phase with time over the interval 1979–1989. The rate of period change is quite well determined in this case: $\dot{\Pi} = (-2.49 \pm 0.06) \times 10^{-11}$ corresponding to an evolutionary timescale of about 0.7×10^6 y, with the period decreasing with time. The observed timescale is about what stellar models predict. Unfortunately, however, evolutionary models of DOV stars show that the periods of typical g-modes should *increase* with time, as cooling by radiation and neutrino losses dominates the effects of global contraction in Equation (10) (Kawaler et al 1986). The explanation for this contradiction is probably that the 516 s mode is not typical; it is a trapped mode, for which changes in the stratification dominate effects of changing overall structure. Kawaler (1993) finds that detailed models of PG1159–035 give trapped modes with very nearly the observed period; these modes display negative values of $\dot{\Pi}$, although no model gives negative period changes that are as large in magnitude as those observed.

Interpretation of the power spectrum shown in Figure 4 has yielded a tremendous amount of information concerning PG1159–035 (see Winget et al 1991, Kawaler 1993, and references therein). Among the most firmly established and interesting results are the following: 1. Virtually all of the peaks seen in the power spectrum may be identified as members either of triplets (with $l = 1$) or quintuplets (with $l = 2$). Conclusive evidence for the l identification is provided by the ratio of period spacings for the two kinds of multiplets. Models show that this ratio should be within a few percent of its asymptotic value of $\sqrt{3} \cong 1.73$ while the observed ratio is 1.72. 2. The source of the frequency splitting within multiplets is apparently rotation of the star. This identification is consistent with the presence of $2l + 1$ components in each multiplet (as in Equation 8), and moreover the observed frequency splittings for triplets and quintuplets are consistent with the calculated values of β_{nl} from Equation (8). The inferred rotation period for the star (assuming uniform rotation) is 1.38 ± 0.01 days. 3. The presence of a significant global magnetic field would induce a component of the frequency splitting proportional to m^2. No such symmetric component is observed; the observed limit on its magnitude implies a maximum global field strength of about 6000 G. 4. The mean period spacing between multiplets is 21.5 s for $l = 1$ and 12.7 s for $l = 2$. For stars like PG1159–035, these mean period spacings depend mainly on the stellar mass, with little dependence on other parameters (Kawaler 1986, 1988). The observed period spacings thus allow an accurate estimate of the stellar mass; in this case the seismic mass is $0.59 \pm 0.01 M_\odot$ (Kawaler 1993). 5. Departures from uniform period spacings can be interpreted in terms of the chemical strat-

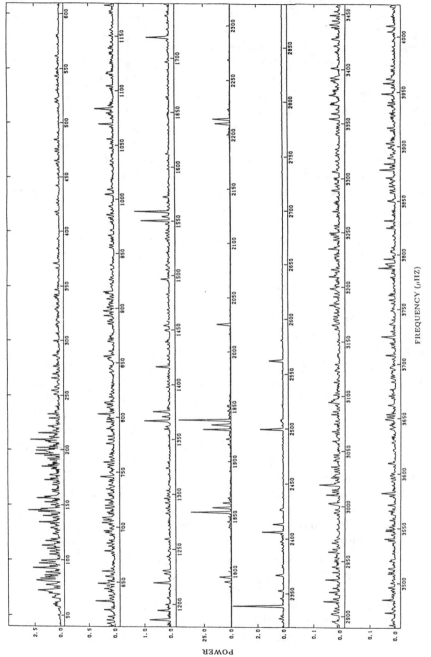

Figure 4 Power spectrum of the intensity pulsations of PG1159–035, as observed by the Whole Earth Telescope. Note the vertical scale differs from panel to panel, to accommodate the large dynamic range. From Winget et al (1991).

ification of the WD envelope. The fairly large $\pm 10\%$ observed variations from the mean spacing can be modeled surprisingly well using either evolutionary models starting from post-AGB progenitors, or using less restrictive structural models, as shown in Figure 5 (Kawaler 1993). The model-fitting process gives what appear to be secure values (listed in the figure) for the mass, the effective temperature, the mass of the helium-enriched surface layer, and the surface helium abundance.

All the conclusions concerning PG1159–035 derive from considerations that are essentially linear in the mode amplitudes. A different sort of phenomenon may be seen in the temporal power spectrum of the DBV star GD358 (Winget 1993; Winget et al 1991, 1993). This spectrum is shown in Figure 6. As with the spectrum of PG1159–035, one may identify many rotationally split triplets, and the frequency splitting and mean period spacing may be used to infer the mass and rotation rate of GD358. The new feature in this spectrum, however, is the presence of a very large number of peaks that correspond to sum-and difference frequencies of the actual modes. These combination frequencies are evidently the result of nonlinear coupling between the modes, though it is not clear whether the nonlinearity arises in the region where the modes are driven (Brickhill 1992) or as an atmospheric response to essentially sinusoidal driving from below (Brassard et al 1992). The spectrum of GD358 displays other unique features (Winget et al 1993). The rotational splitting of the $l = 1$ multiplets is observed to be an increasing function of the radial overtone number n. Since modes with higher n preferentially sample the outer part of the stellar envelope, the most likely explanation for this behavior is that GD358 rotates differentially, with the outer envelope rotating some 1.8 times faster than the core. Besides depending upon n, the frequency dependence of modes within the $l = 1$ multiplets is not linear in m; there is a significant dependence proportional to m^2. This quadratic splitting is consistent with the presence of a global magnetic field, with surface values estimated to be 1300 ± 300 G. Bradley et al (1993) discuss the issues related to seismology of DBV stars; for details specific to GD358, see Bradley & Winget (1994).

Finally, we note that there remain a great many unsettled questions regarding the pulsating white dwarfs. The nature and cause of temporal variability in WD mode amplitudes is obscure. For instance, there is evidence that the minor components of the multiplet containing the 516 s mode of PG1159–035 have increased significantly in amplitude since 1987 (Winget et al 1991). The mechanisms that might lead to such rearrangement of power in the spectrum are not known. Indeed, the distribution of power among the different m-values in the quintuplets of PG1159–035 is somewhat mysterious; it is at best partially explained by assuming the stellar rotation axis to be inclined at 60° to the line of sight. It is not clear that the driving mechanism for DOV stars is understood. Though mode trapping is fairly successful in explaining why some modes are

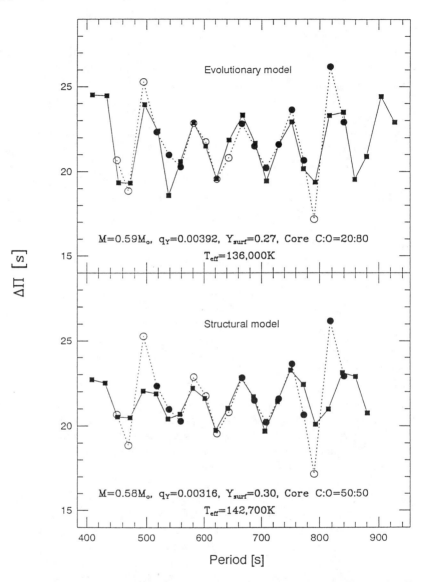

Figure 5 Computed (*squares*) and observed (*circles*) period spacings for the DOV star PG1159–035, from Kawaler & Bradley (1993). Open circles indicate relatively insecure period spacings. Model parameters used to obtain these fits are given in the corresponding panels.

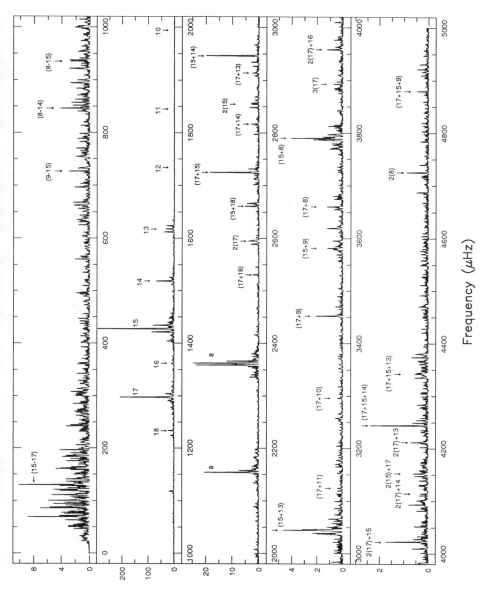

Figure 6 Power spectrum of the WET light curve of the DBV star GD358, showing the identified combination frequencies. From Winget et al (1993).

excited more than others, it develops that mode amplitude alone is not a reliable indicator of trapping.

To sum up, theory and observations of white dwarfs have come together in a way that allows most of the important processes to be understood at least qualitatively, with some notable quantitative successes. But many details still elude explanation; these puzzles will provide the focus for much work in the next few years.

3.2   Rapidly Oscillating Ap Stars

Since the rapidly oscillating Ap (roAp) stars have been reviewed recently by Kurtz (1990) in this series, we will give only a short overview and update on progress of work on these fascinating stars. The Ap stars (Wolff 1983) show rotationally modulated spectral peculiarities suggesting overabundances of Si, Sr, Cr, Eu and other rare earths. In extreme cases like Przybylski's (1961) star, HD 101065, the spectrum is dominated by the rare earth elements with Holmium being the strongest. The Ap stars show (modulated) magnetic fields of 400 to 2200 G; Hd 101065 (Wolff & Hagan 1976) defines the upper range. Accompanying the spectrum and magnetic field modulations are small amplitude photometric variations of 0.01 to 0.10 mag (Catalano et al 1993) with periods of typically 1–10 days. The unifying model to explain Ap stars is the *oblique rotator model* (Babcock 1949, Stibbs 1950) in which the magnetic field is inclined at an angle to the rotation axis; radiative diffusion (Michaud et al 1981, Babel & Michaud 1991, and references therein) influenced by the presence of strong magnetic dipole fields is the likely source of the extreme surface abundance anomalies.

The Ap stars are located along the main sequence in the HR diagram with temperatures spanning a domain that starts just below the cool edge of the classical (δ Scuti) instability strip and extending well beyond the blue edge to hotter temperatures. Many of the Ap stars are located in the instability strip. It was, however, expected that radiative diffusion would deplete He from the region of He II ionization thought to be the source of κ mechanism driving for δ Scuti oscillations, and that Ap stars would therefore be stable. Observations of HD 101065 by Kurtz & Wegner (1979) undertaken to demonstrate the expected stability of Ap stars showed quite the opposite as a regular pulsation of 6 millimag with a 12.14-min period was detected. Kurtz (1982) reported in a seminal work the characteristics of five rapidly oscillating Ap stars, thus establishing this new type of variable. The oscillations have amplitudes $\leq$ 0.008 mag (Johnson B), periods of 6–15 min, and are thought to represent nonradial p-modes of low degree ($l \leq 3$) and large radial overtone ($n \approx 10$–80). Radial velocity variations have been detected in two of the roAp stars (HD 24712, Matthews et al 1988; HD 201601, Libbrecht 1988). Although the roAp stars all fall within the classical instability strip (arguing for the κ

mechanism) the driving mechanism responsible for the oscillations is not yet securely determined. Shibahashi (1983) has proposed a mechanism in which magnetic field tension supplies a restoring force.

At the time of Kurtz's 1990 review, 14 roAp stars had been detected. As of this writing, 25 roAp stars are known with 10 of these following from the Cape Rapidly Oscillating Ap Star Survey (Martinez 1993). Eleven of the roAp stars possess rich enough oscillation spectra and have been observed extensively enough to yield estimates of $\Delta \nu_0$ (Martinez 1993). In general the values of $\Delta \nu_0$ are consistent with Equation (4). Published A star models and $\Delta \nu_0$ values (Heller & Kawaler 1988) and values of T_{eff} inferred from Strömgren photometry (Moon & Dworetsky 1985) allow a prediction of luminosity. For the one star with a reliable parallax—γ Equ (HD 201601)—the asteroseismological luminosity is in excellent agreement with direct determination. *HIPPARCOS* parallaxes for the other roAp stars will allow testing of these asteroseismological luminosities. Tentative values of $\delta_{n,l}$ (Equation 5) are available for HD 24712 (Kurtz et al 1989), HD 60435 (Matthews et al 1987), HD 101065 (Martinez & Kurtz 1990), and HD 119027 (Martinez et al 1993). These may be placed into an asteroseismological HR diagram and compared against eigenvalues computed by Audard & Provost (1993) (see Figure 2), but residual uncertainties about possible effects of the magnetic field on $\delta_{n,l}$ make interpretation premature. Given the extreme abundance anomalies of Ap stars, classic methods of temperature determination yield insecure results; accurate parallaxes coupled with asteroseismology may more reliably constrain the fundamental parameters of the roAp stars.

Frequency changes of the oscillation modes could result from orbital motion, stellar evolution (Heller & Kawaler 1988 show that $|\dot{\Pi}| \sim 10^{-12}-10^{-13}$ are typically expected and could be detectable with ~ 10 y of observations), and possibly from magnetic-cycle-induced changes as are seen for the Sun. In the best case for stellar-induced frequency changes Kurtz et al (1994) review data from HR 3831 over a 12 y time base and show that the primary mode frequency varies smoothly by about 0.1 μHz on a two year timescale. Stellar evolution and orbital motion can be eliminated as causes; interpreted as a magnetic cycle effect the observed frequency changes might impose constraints on magnetic variations in HR 3831 at much lower relative amplitude than could be measured by classical means.

3.3 δ *Scuti Stars*

δ Scuti stars are objects with spectral types in the range $A–F$, occupying positions on the HR diagram either on or somewhat above the main sequence. The effective temperature range occupied by the δ Scutis corresponds well with the extension of the Cepheid and RR Lyrae instability strip to the main sequence. In recent years, they have been the subject of several readable reviews (Breger

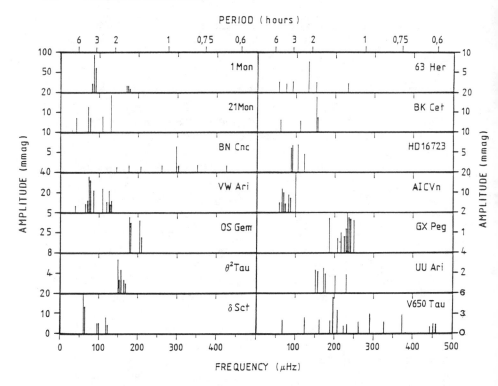

Figure 7   Schematic power spectra of most of the known δ Scuti stars with more than four identified periods. From Belmonte et al. (1993b).

1979, Dziembowski 1990, Däppen 1993, Matthews 1993). These stars have pulsation periods ranging between roughly half an hour and half a day. Their photometric amplitudes range from a few milli-magnitudes to several tenths of a magnitude; low-amplitude pulsators are probably underrepresented in lists of δ Scuti stars, simply because they are difficult to detect. Many δ Scutis are rapid rotators, with $V \sin i \geq 100$ km s^{-1}. Berger (1991) notes that the slowly rotating δ Scutis tend to have large oscillation amplitudes and simple pulsation spectra (only one or two modes excited), whereas the rapidly rotating stars tend to display smaller amplitudes and more complicated pulsation spectra. Some idea of the complication and variety of δ Scuti pulsation behavior may be obtained from Figure 7, taken from Belmonte et al (1993b), which shows the pulsation spectra of most of the known objects that oscillate with more than four periods. For references to specific objects shown in this Figure, see Belmonte et al. Rich δ Scuti p-mode spectra have also been documented (Gilliland & Brown 1992a) for two blue stragglers in M67; asteroseismology may allow important inferences to be reached concerning the origin of this enigmatic class of stars.

In the simplest cases, the pulsations in δ Scuti stars appear to be mostly low-order p-modes. As with the classical pulsating variables, the source of energy for the pulsation is thought to be an instability driven by the κ mechanism operating mostly in the first ionization zone of helium. In the case of the large-amplitude pulsators (sometimes termed "dwarf Cepheids", the modes most often excited are the fundamental radial mode, the first radial overtone, or both. When both the fundamental and first overtone are present, their period ratio can be used to test models of the structure of δ Scuti stars. For many years, this comparison between theory and observation was an embarrassment, both for Cepheids and for δ Scutis: the computed period ratios for the Cepheids were too large, while those for the δ Scutis were too small (see e.g. Christensen-Dalsgaard 1993). Recent improvements in the calculation of the opacity of stellar material (Rogers & Iglesias 1992, 1993; Mihalas et al 1990; Seaton 1993) have drastically reduced the disagreement between observation and theory, and moreover have solved other outstanding problems (see Simon 1982, Osaki 1993). It is worth noting that this fundamental improvement in understanding the microphysics of stellar interiors was, in part, both motivated and validated by the simplest sort of asteroseismological data.

Once δ Scuti stars have undergone significant chemical evolution, the structure of their pulsations becomes more complicated than that of simple p-modes. With increasing helium content in the core, an important gradient of molecular weight develops in the inner part of the star, causing a sharp increase in the buoyancy frequency N in those regions. This "N mountain" eventually dominates the pulsational response of the stellar core, so that global pulsations can acquire a dual character—p-modes in the envelope, and g-modes in the core (Däppen et al 1988). The importance of buoyancy forces in these modes leads to a very dense and complicated spectrum of possible modes. Thus, for somewhat evolved δ Scuti stars, the central problem posed by observations is that of understanding why only a few modes are excited to visible amplitudes, when so many other modes exist. In this respect, δ Scuti pulsations resemble those of the white dwarfs, described in Section 3.1.

To date, two mechanisms have been proposed to explain mode selection in δ Scuti stars. The first relies on mode trapping in the stellar envelope, in a manner analogous to that observed in the pulsating white dwarfs (Dziembowski 1990). Although such trapping certainly might occur, invoking it to explain the observed mode excitation meets with several difficulties. First, the ratio between envelope mass and core mass is probably not so small in δ Scuti stars as the ratio between (say) hydrogen and helium shell masses in the DAV stars. Thus, the ratio of mode energies between trapped and untrapped modes is not so small, and the growth rates should be more nearly equal. Second, model computations (Dziembowski & Królikowska 1990) suggest that only modes with $l = 1$ suffer significant trapping, while modes with other l values have probably been observed, at least on some stars. Finally, one must remember a warning already mentioned in connection with the white dwarfs, namely that growth rates are not necessarily a reliable guide to limiting amplitudes.

The principal alternative to trapping for explaining mode selection in δ Scutis is that of *parametric resonance* (Dziembowski & Królikowska 1985). For this mechanism to operate, it is necessary that one or a few of the low-order modes be unstable because of the opacity mechanism, but that the other pulsation modes be linearly stable. If the unstable modes grow to substantial amplitude, it may then be possible to transfer energy from them to other modes by non-linear processes. The most direct such process involves triples of modes (of which an excited mode is one) satisfying certain selection rules regarding their frequencies and spatial structures. The rather strict demands that this mechanism places on frequency and wavenumber differences between the coupled modes assures that only a small fraction of the available modes may be excited. A likely implication of driving by parametric resonance is that the indirectly excited modes may couple to yet other modes, and those to others, and so on. In this case, one would expect a cascade of energy from the driven mode down through an increasing number of modes with decreasing amplitudes. The result would presumably be a wide spectrum of mode amplitudes for each pulsating star—a prediction that is testable, given observations of sufficient precision.

Whether δ Scuti pulsations are selected for excitation by mode trapping or by parametric resonance, a clear understanding of the process will evidently require correct identification of the modes that are present. Ideally, this would involve correctly typing each mode as p- or g-mode, and specifying the three spatial indices n, l, m. Because of the dense mode spectrum and uncertainties regarding stellar parameters and applicable physics, theory offers only limited help in identifying modes from their frequencies alone. For this reason, methods for inferring l and/or m (and perhaps the mode type) directly from observations are particularly important for δ Scuti stars. There are two such methods available, one depending on photometric data, the other on spectrographic line profiles.

The photometric method (Dziembowski 1977, Balona & Stobie 1979, Watson 1988) involves measurement of the relative phase $\Delta\psi$ and amplitude of pulsations in the flux and in the color. This method of mode identification has the advantage that the necessary data can be obtained relatively easily, even for faint stars. The more useful of the two diagnostics is usually the phase shift $\Delta\psi$. Photometric mode identification is generally most useful for modes with $l \leq 3$.

The spectroscopic method provides a more powerful (but observationally more demanding) approach to mode identification. In the simplest case, one considers a slowly rotating star (i.e. one for which $V \sin i$ is small compared to the intrinsic spectral linewidth), pulsating in low-degree modes. In this case, the pulsations cause not only a mean Doppler shift, but also various distortions of the line profile, the detailed nature of the distortion depending on l, m, and the ratio of horizontal to vertical velocities in the mode in question. As described by Campos & Smith (1980) and Smith (1982), and elaborated by Balona (1986 a, b), the relationships between amplitude and phase of various moments of the line profile allow one to distinguish among modes with different spatial scale and character. More recently, the same principle has been applied to rapidly rotating stars, i.e. stars for which $V \sin i$ is much greater than the intrinsic linewidth. In the spectra of such stars, position within the line profile maps approximately onto longitude on the stellar disk; the number of separate longitudes that can be resolved is roughly equal to the ratio of the rotational velocity to the intrinsic Doppler width of the line. This ratio is 10 or so for typical rotation speeds, $V \sin i = 100$ km s^{-1}. Given this effective spatial resolution, it becomes possible to detect oscillation modes with $l \gg 4$. Walker et al (1987) found that four rapidly rotating δ Scuti stars (21 Mon, κ^2 Boo, vUMa and o^1 Eri) showed narrow moving features within their rotationally-broadened line profiles. These features were shown to be consistent with fairly high-degree nonradial pulsations ($8 \leq l \leq 16$), and inconsistent with other explanations such as starspots (Kennelly et al 1991). When attempting to analyze the observed profiles, the large amount of structure apparent within them has encouraged the use of Fourier transforms in space and time (Kennelly et al 1993a, b; Matthews 1993; Merryfield & Kennelly 1993), instead of the moment analysis of Balona (1986a, b). How best to interpret the resulting m-v power spectra is a question that requires more work, but there is little doubt that this approach to δ Scuti observations will yield more spatial information about the modes than has hitherto been available.

When observing δ Scuti pulsations, one also encounters the problems with temporal sampling that afflict all branches of asteroseismology. Gaps in data coverage are perhaps even more damaging for δ Scutis than for some other stars, because of the fairly common appearance of harmonics, sum and difference frequencies, and other evidence of nonlinear behavior in δ Scuti temporal spectra (Matthews 1993, Mantegazza et al 1993). Unsurprisingly, a solution to these problems has been sought in the form of international networks of telescopes. Though not yet claiming so many sites as the WET network, recent efforts to obtain near-continuous observations of δ Scutis have helped to resolve several puzzling issues, as well as to create new mysteries. Breger et al (1987, 1989, 1990) have reported results of multisite campaigns to observe the δ Scuti variables θ^2 Tau and 4 CVn, finding at least 4 (and probably more) distinct periodicities in each star. Observations from two widely separated sites (Belmonte et al 1991) yielded a list of at least 6 modes in 63 Her, and an extended form of the same network (STEllar PHotometry International, or STEPHI) has yielded long runs with high duty cycle on several stars (GX Peg: Michel et al 1992a, V650 Tau: Michel et al 1992b, BN and BU CnC: Belmonte et al 1993a). In addition to these networks built around two- and three-star photometers, there are plans to organize a network of small telescopes equipped with CCD cameras for the purpose of observing δ Scutis (Frandsen 1993).

In spite of the observational and theoretical progress of recent years, δ Scuti stars remain something of a mystery. Results from some of the network observing runs mentioned above serve to illustrate the outstanding puzzles. The Hyades member θ^2 Tau is observed (Breger et al 1989) to pulsate in 4 (or quite possibly 5) closely spaced frequencies. The frequency ratios and flux-color relationships are such that rotational splitting or sequences of modes with similar l cannot reproduce the observations; presently the only viable mode identification is with mode sets involving rather arbitrary combinations of l and m. Why are these modes excited, while many others are not? The frequencies observed in 4 CVn (Breger et al 1990) are reasonably well matched by rotationally-split p-modes with l of 2 or 3. The pulsations in this star, however, apparently changed in amplitude (first increasing, then decreasing) by a factor of more than 2 between 1966 and 1984. Is this amplitude variation real, and if so, how does it come about? Most multiperiodic δ Scutis concentrate their excited modes within a fairly small range of frequencies, as seen in Figure 6. Some, however (63 Her: Belmonte et al 1991, Mangeney et al 1991; V650 Tau, Michel et al 1993), show a considerable spread in their excited frequency spectrum. Is the same excitation mechanism operating in both sorts of stars? If so, what other stellar parameter accounts for the difference in behavior? Questions of this sort simply confirm what no one would wish to dispute: that δ Scuti stars (and particularly their driving mechanisms) are poorly understood. Advances in observational methods and increased efforts in the theoretical direction do, however, give reason to hope that some of these questions may soon be answered.

4.   PULSATIONS IN SUN-LIKE STARS

There is no cause to believe that the Sun is an extraordinary star of its type, so it is reasonable to expect that other stars like the Sun pulsate in much the same way the Sun does. So far, however, observational efforts have yielded results that are equivocal. In this section we discuss Sun-like stars mostly from an observational point of view, putting most emphasis on the practical difficulties encountered, ways of dealing with them, and prospects for success. In spite of the observational limitations, considerable progress has been made in predicting the oscillation properties of likely target stars for asteroseismic observations (see e.g. Demarque & Guenther 1990, Edmonds et al 1992, Guenther & Demarque 1993).

4.1   *Expected Characteristics of the Pulsations*

The principal impediment to the observation of pulsations in stars like the Sun is their small amplitudes. As mentioned in Section 2.2, the strongest modes in the solar p-mode spectrum show amplitudes of only about 3×10^{-6} in relative intensity or 15 cm s^{-1} in velocity. Unlike the stellar pulsations discussed in Section 3, all of the oscillation modes in the Sun are believed to be intrinsically stable. The reason that pulsations are excited at all (and the reason that the excitation is so undiscriminating, exciting millions of modes to similar amplitudes) is the presence of the solar convection. Motions near the top of the convection zone can reach Mach numbers of order unity (e.g. Nordlund & Stein 1990, Bogdan et al 1993, Rast et al 1993), providing a source of acoustic noise that is loosely coupled to the part of the Sun in which the p-modes propagate. In the Sun, damping of individual p-modes arises partly from radiative losses, but is most likely dominated by scattering from convectively generated inhomogeneities in the outer envelope (Goldreich & Kumar 1991). Expected amplitudes therefore depend in a complicated way on the behavior of the stellar convection and on its coupling to the cavity in which the oscillation energy resides. Christensen-Dalsgaard & Frandsen (1983) estimated oscillation amplitudes as a function of T_{eff} and surface gravity, assuming radiative damping as the dominant energy sink and a now-obsolete form of the theory of convective mode excitation (Goldreich & Keeley 1977). The results suggests a weak dependence of mode surface amplitude on surface gravity (cf Equation 9). These results are highly uncertain, but remain the only ones available. The best hope for more accurate theoretical estimates of mode properties in other stars may lie with numerical simulation of the convection-pulsation interaction (Rast & Toomre 1993, Bogdan et al 1993), but adequate models with the required resolution are probably some years away.

A rough estimate can be made of the frequency range within which oscillations would be seen on distant stars. In the WKB approximation, the reflection of sound waves as they propagate upward through a stellar envelope is governed by the behavior of the *acoustic cutoff frequency*, ω_{ac}. Waves propagate upward until the local value of ω_{ac} becomes greater than the wave frequency, and then they reflect. In the simplest (isothermal layer) approximation, which is adequate for our purposes, one may write

$$\omega_{ac} = \frac{c}{2H} \propto g T^{-1/2}, \tag{11}$$

where c is the local sound speed, H is the pressure scale height, g is the gravitational acceleration, and T is the temperature. In the stellar atmospheres, ω_{ac} reaches a maximum, ω_{ac0}, in the photosphere, where the temperature is minimum. Waves with frequencies above ω_{ac0} never reflect, but rather continue propagating into the tenuous outer parts of the stellar atmosphere. As a result, p-modes with frequencies above ω_{ac0} are not expected to attain significant amplitudes. On the other hand, modes with frequencies much smaller than ω_{ac0} reflect deep in the stellar envelope. This reduces the surface amplitude for a given mode energy, and moreover reduces the coupling between the mode and the near-surface convective driving source. These considerations suggest that maximum p-mode amplitudes should be found at frequencies that are a modest fraction (roughly 0.6, in the Sun) of ω_{ac0}. From Equation (11), it follows that the expected frequency of maximum p-mode amplitude should scale as $g T_{\text{eff}}^{-1/2}$. If one adopts the scaling appropriate to the Sun, this implies cyclic frequencies ranging from about 1 mHz (for F-type subgiants) to about 10 mHz (for M dwarfs). Goldreich & Kumar (1990) have developed a theory explaining the shape of the power envelope of the solar p-modes. This approach offers hope that observations of stellar p-mode amplitudes might provide diagnostics relating to stellar convection and to g.

The foregoing suggests that the most attractive targets for stellar oscillation searches should be stars that have lower surface gravity and are more luminous than the Sun, since such stars should have larger amplitudes (cf Equation 9) and should pulsate with longer periods (simplifying many observational problems). While this conclusion is probably true, the expected frequency separation of p-modes in these stars can work in the other direction. From Equation (4), $\Delta\nu_0$ for p-modes ranges from roughly 30 μHz (for subgiants) to 400 μHz (for M dwarfs). The frequency spacing one would actually see is $\Delta\nu_0/2$, because of the way in which modes with odd and even values of l interleave. Thus, detecting discrete modes in subgiants requires frequency resolution better than 15 μHz, which is a factor of about 2 better than can be attained from one site in a single night. Simulations show that even if one combines multiple nights of single-site observations, the sidelobes that result from the diurnal cycle probably lead to fatal ambiguities when attempting to interpret pulsation time series from such luminous stars (Gilliland & Brown 1992b). This is essentially the same problem faced by the solar and white dwarf communities in obtaining observations adequate for their purposes, and the solutions are the same: One must observe from a network of sites or from space.

4.2   *Photometric Techniques*

Photometric techniques for the detection of solar-like pulsations have a few advantages when compared to the Doppler techniques to be discussed in Section 4.3. They can be applied to fainter stars than are feasible for Doppler observations; moreover photometric detectability does not require sharp stellar spectrum lines, hence is independent of stellar rotation rates. Unfortunately, the easily achieved limits to photometric precision are rather high, when compared to expected oscillation amplitudes in Sun-like stars. The fundamental limiting noise source for ground-based photometry is atmospheric scintillation:

$$\delta I/I = 0.09 D^{-2/3} X^{1.75} \exp(-h/h_0)/(2t_{\text{int}})^{1/2}, \tag{12}$$

where $\delta I/I$ is the rms noise in measured relative intensity, D is the telescope aperture in cm, X is the airmass, h is altitude, h_0 is the atmospheric scale height (typically 8000 m), and t_{int} is the exposure time in seconds (A. T. Young 1967, and 1992 personal communication). Averaged over an 8-hour observing window using 60 s integrations on a 4-m telescope at 2000-m elevation, $\delta I/I$ is about 215 μmag. For a 10-m telescope at 4000 m, the corresponding value is about 120 μmag. In both cases, assuming a broad band filter and an efficient detector, noise from counting statistics is small relative to scintillation for stars brighter than $m_B=13$. Thus, in terms of photometric precision obtainable from the ground with large telescopes, all stars with $m_B \leq 13$ are equally amenable to observation.

What do these examples imply for the pursuit of asteroseismology on Sun-like stars? With many thousands of data points, the highest noise peaks in a power spectrum occur at about 4 times the mean noise level. To obtain a 4σ detection of a single coherent oscillation of amplitude X, given a time series with noise level ϵ, requires a number of observations N_{obs} given by $N_{obs} = 16(\epsilon/X)^2$)(Scargle 1982). Assuming solar amplitude (3 μmag) and $m_B = 13$ (so that scintillation noise and Poisson noise are about equal), the 4-m and 10-m examples given above require continuous observing periods of 111 days and 35 days, respectively. Clearly, even with optimistic assumptions, using ground-based photometry to detect p-modes in close solar analogues is not a likely prospect. Fortunately the situation may be better for stars of higher luminosity than the Sun, since these may (see Equation 9) have significantly larger amplitudes. Because of the quadratic dependence of detection time on oscillation amplitude, pulsations at five times the solar amplitude should be detectable in 30 nights of observing time. Thus, five nights would be needed on each of six 4-m class telescopes. Although still a difficult project, this at least suggests that detecting stellar oscillations on Sun-like stars is feasible with existing telescopes, if the scientific justification can be made sufficiently compelling. Note also that single-site observations would probably not succeed in this application. In keeping with the discussion at the end of Section 4.1, stars with amplitudes as large as assumed probably have large frequency separations $\Delta\nu_0$ that are too small to resolve with observations from a single site. The frequency separation vs amplitude tradeoff for Sun-like stars implies that for stars with large enough amplitudes to be detectable from the ground, a longitude-distributed network campaign will be required; stars with frequency separations large enough to be resolved with single-site observations probably have amplitudes that are too small to detect.

Kjeldsen & Frandsen (1992) provide a thorough review of high-precision time-resolved photometry as might be applied to stellar oscillation detection. A basic measure of precision for a given photometric data set is provided by the mean noise amplitude near 3 mHz resulting from a power spectrum analysis. As of 1992 the lowest noise level reached was about 20 μmag using single-channel photoelectric photometry for some 580 hours on 1-m class telescopes

(Kurtz et al 1989) for a study of the roAp star HR 1217. Belmonte et al (1990) and Mangeney et al (1991) reached a noise level of about 30 μmag using three channel (object, control star, and sky) photometry on a pair of 1.5-m telescopes for 184 hours. In the latter case the control star HD 155543 (F2 V) was analyzed for possible solar-like oscillations. This photometry supports detection of oscillations with amplitudes of about 100 μmag, still a factor of several larger than expected. Analysis of the data does produce evidence suggestive of oscillations (Belmonte et al 1990), but the results are of low significance.

Experiments by Gilliland & Brown (1992b) and Kjeldsen & Frandsen (1992) have shown that CCD ensemble photometry, in which the ensemble mean is constrained to be constant (to control low frequency noise from atmospheric transparency variations), can provide nearly atmospheric scintillation limited results on 2-m class telescopes even in conditions of bright moon and varying transparency that would be disastrous for photoelectric photometry. Ideal stellar fields for the control of observational noise have many sources in a small area of sky (to maintain coherence of transparency fluctuations), but these must be well separated from each other (to allow precise individual intensity estimates). Such fields of $m_B \leq$ 13th mag stars with 10 or so stars within a CCD field of view are not common, but do occur in a few clusters. The stars in some such fields (see Gilliland & Brown 1992b for the specific example of M67) are also of primary interest from a theoretical perspective since they share common properties of equal age, metallicity, and initial hydrogen abundance. As noted in Section 2.3, in such an ensemble of stars the addition of the frequency splittings associated with p-mode oscillations to classical astronomical knowledge may allow testing for consistency of stellar structure and evolution theory, and for fine estimation of stellar parameters (see Gough & Novotny 1993a, b).

In an early attempt at cluster seismology, Gilliland et al (1991) obtained a noise level of about 45 μmag for several giants and subgiants in the cluster M67, using 70 hours of CCD ensemble photometry on 0.9-m class telescopes. Building on that experience, Gilliland et al (1993) conducted the most ambitious CCD photometry campaign to date for the detection of oscillations in Sun-like stars. It involved the collaborative use of 34 allocated nights on 4-m class telescopes within a one-week period. These allocations yielded good time-series data on 20 nights for a total of 156 hours. The duty cycle (the fraction of the 6.0 day time base when at least one site was acquiring time-series data) was 64%. In a technical sense the project was quite successful: CCD ensemble photometry was able to approach quite closely the theoretical noise limits appropriate to 4-m class telescopes. Good results were maintained even with a diverse set of CCDs, many of which had never before been challenged to deliver ultra-high precision, and in spite of observing conditions that were often far from "photometric." The two best stars in the ensemble had noise levels in the amplitude spectrum of about 7 μmag, a factor of three better than the lowest noise figures previously attained. For the best two ensemble stars Gilliland et al (1993) derive unambiguous upper limits for multi-mode oscillations of 27 and 30 μmag; that is, any oscillations this large would have been easily detected and quantified. These limits are larger by about 20–30% than the amplitudes predicted from Equation (9). Although no definite oscillation detections could be claimed, about half of the ensemble stars showed suggestive evidence of multi-mode oscillations. In the best cases, significance reached about 2σ as calibrated by Monte Carlo trials using many cases with realistic noise. The best case (star no. 16 of Gilliland et al 1991; cross references: S984, Sanders 1977; F134, Fagerholm 1906) showed evidence for oscillations with amplitude $\sim 20\,\mu$mag, $\Delta\nu_0/2 \sim 19\,\mu$Hz, and peak frequency near 1 mHz; all these values are quite close to theoretically expected ones.

Gilliland et al (1993) also show that a network could be organized with much greater sensitivity to oscillations with mode separations of 19 μHz. The realized network was very well represented in the U.S. Southwest, but weak 8 hours to the east. Detailed simulations show that augmenting the realized 20 observing nights with an assumed ten clear nights from the CTIO 4-m, WHT 4.2-m, and Russian 6-m telescopes would have improved the sensitivity for stars with mode separations near 20 μHz by a full factor of 2. For main-sequence stars with expected frequency separations of about 40 μHz the simulated gain was about 30%, with most of the improvement following from simple $N_{\mathrm{obs}}^{1/2}$ considerations.

The Gilliland et al (1993) experiment provided for the first time a good chance of detecting oscillations on a few Sun-like stars via photometric observations. Although no unambiguous detections were forthcoming, and thus significant astrophysical results were not obtained, the technique of CCD ensemble photometry on large telescopes was well validated. If the same stellar ensemble could be observed again in a few years using the ten best telescopes possible (including Keck 10-m and MMT 6.5-m), each with a seven-night time allocation, theoretical mode amplitudes suggest that unambiguous detections would follow for five stars in the ensemble, with good chances of success for an additional five stars. Robust quantitative results on at least five stars would provide the potential for a real challenge to stellar structure and evolution theory.

Since ground-based observations are so hampered by atmospheric scintillation, what are the prospects for pursuing photometric detection of stellar oscillations from space? There can be no doubt from a technical standpoint that space-based observations are the best approach—IPHIR (see Figure 1) has amply demonstrated this for the Sun observed as a star. Several recent studies (Hudson et al 1986, Appourchaux et al 1993, Baglin et al 1993, Jones et al 1993) have shown that an excellent asteroseismology experiment could be mounted using a 1-m class spaceborne telescope and available detector technology. Moreover, it may be possible to use data from experiments with other aims (e.g. FRESIP—a proposed photometric search for inner planets circling other stars; Borucki et al 1993) for seismological purposes. The principal requirements for a successful photometric seismology mission are: adequate aperture, a detector with adequate dynamic range, a cycle time short enough to sample pulsations with periods of a few minutes, and an orbit that allows a suitable observing window function. A lesser but still important consideration is the field of view, which determines the number of stars that can be observed simultaneously. Assuming CCDs with high efficiency, a 3000 Å wide filter, and a 1-m telescope, the count rate on an $m_B = 10.0$ star would be roughly 10^8 per minute, giving a Poisson noise of 100 ppm. At this precision, pulsations of solar amplitude could be marginally detected in about 12 days. Lower noise could be achieved on brighter stars, to the extent allowed by the maximum detector counting rate; at $m_B=7$, a marginal detection would take about 1 day. Robust frequency determinations could be made using data with a signal-to-noise ratio three times better than that required for a simple marginal detection. This would take roughly a factor of 10 more observing time. Oscillation detection on early K dwarfs would require measurement of amplitudes a factor of 5 lower (Christensen-Dalsgaard & Frandsen 1983) than solar, and thus require 25 times as long for a basic detection. Continuous monitoring (possible from high orbits) would provide the perfect window function essential for robust interpretation of the data, but which is so hard to come by with ground-based observations. Even in the absence of scintillation, a long time base is evidently essential. Observing the same targets for months or years would, however, support very precise frequency determinations, and would allow testing for frequency changes that may arise from stellar activity cycles. Following about 100 stars each of F, G, and K spectral types appears technically feasible, and would provide fundamentally important information for the understanding of p-mode driving and for stellar structure and evolution theory.

4.3   Doppler-Shift Techniques

In principle, Doppler measurements of solar-like pulsations have several advantages compared to photometric methods. The Doppler shift measurement process is in essence a differential one, conveying some advantages in the detection of small signals. Also, the contrast between the pulsation signal and the background of stellar convective noise is larger in the velocity than in the intensity signal (Harvey 1988). One pays a price for these advantages, however. Because of the tiny wavelength shifts that are associated with expected pulsation signals (15 cm s^{-1} $\Rightarrow$ $\delta\lambda/\lambda = 5 \times 10^{-10}$), successful measurements require both high spectral resolution and very low noise. If one defines "Sun-like" to mean stars of luminosity class IV or V, with spectral types cooler than F5, then there are only three candidates of roughly first magnitude: α Cen

A (G0V) and B (G3V), and Procyon (F5IV). The next brightest star of interest is β Hyi (G2IV), which is 2.3 stellar magnitudes fainter than Procyon. To date, almost all Doppler shift searches for Sun-like pulsations have therefore been aimed at one of these stars.

The most important problem to be solved in pulsation searches to date is that of attaining sufficient Doppler precision to detect the pulsation modes. To see what is involved, let us assume that the largest p-modes may have Doppler amplitudes of 15 cm s^{-1}, and that in one night one can obtain 400 observations with exposure times of roughly 60 s each (a rapid cadence is required to sample the pulsation time scale). To obtain a 4σ detection of the largest modes, one therefore requires the noise associated with each observation to be roughly 75 cm s^{-1}. The fundamental limit to the achievable precision results from photon counting statistics (Connes 1985, Brown 1990). To sufficient accuracy, it may be written as

$$\delta v_{rms} = \frac{cw}{\lambda d (N_{pix} N_{lines} I_c)^{1/2}},\tag{13}$$

where c is the speed of light, w is the width of the spectrum line (including both instrumental and stellar line broadening processes), λ is the center wavelength of the line, d is the fractional line depth, N_{pix} is the number of wavelength samples obtained across the line width, N_{lines} is the number of spectral lines observed, and I_c is the continuum intensity in the measurement, expressed as the number of detected photons. Putting in plausible values for an echelle spectrograph at a 2-m telescope, with an exposure time of 60 s, one finds δv_{rms} for a single spectrum line of moderate strength to be roughly 10 m s^{-1}. To obtain observations suitable for pulsation studies, this noise level must evidently be improved by more than an order of magnitude. This may be done by (a) improving the instrumental resolution, (b) observing many lines, or (c) getting more light through the system. In practice, it is relatively easy to make the instrumental contribution to the line width smaller than the stellar contribution; beyond this point further improvements are of no use. Increasing N_{line} means using a wider bandwidth. This is a practical strategy; with cross-dispersed echelle spectrographs, hundreds or thousands of lines may be measured simultaneously. Finally, increasing I_c requires larger telescopes, more efficient optical systems, or both.

In addition to the limit set by photon statistics, one must also deal with noise from instrumental sources. These problems are not trivial; attaining relative wavelength precision of a few parts in 10^{10} requires maintaining (or monitoring) some dimension within the instrument with similar precision. Several techniques for doing this have been devised, each with its own advantages and drawbacks.

The first serious attempts at seismology applied to Sun-like stars used atomic resonance cells to measure the intensity in one or both wings of strong spectrum lines (the sodium D lines near 595 nm; Gelly et al 1986, 1988). This approach has the advantage that wavelength stability is fairly easy to assure, because of the intrinsic stability of the atomic scattering process being exploited. Moreover, the transmission of such a system can be made very high by comparison with that of competing schemes, so that relatively few photons are wasted. On the other hand, the resonance-line technique is necessarily limited to observation of a single line (or perhaps a multiplet, as in the case of Na D). This limitation in bandwidth ultimately sets a limit of a few m s $^{-1}$ to the precision that can be attained in a nominal 60-s exposure time, even when using apertures as large as 3.6 m (see Gelly et al 1986).

A single-line detection scheme using an interferometric approach has been used by Butcher and colleagues (Butcher & Hicks 1988; Pottasch et al 1992, 1993). These authors used a stabilized Fabry-Perot interferometer combined with an order isolation filter to give a narrow-band (0.012 nm FWHM) tunable filter, with its bandpass located in the vicinity of the FeI 557.6 nm absorption line. In spite of its single-line restriction, this instrument (combined with the 3.6-m ESA telescope at La Silla) has yielded the most convincing evidence yet for pulsations in α Cen (see below). An advantage of interferometry as opposed to atomic resonance methods is the greater choice of spectrum lines.

In the long run, echelle spectroscopy provides the most promising detection method for oscillations in the Doppler shift. Its outstanding advantage is that measurements can be made on many hundreds of lines simultaneously, yielding large potential improvements in the attainable photon-limited Doppler precision. This desirable feature is balanced by a serious difficulty: Spectrographs of the required size are essentially impossible to stabilize to the necessary precision (at least on a reasonable budget—see Brown 1990). The solution to this problem is to measure the important instabilities in a continuous fashion, using as fiducials the spectrum lines provided by any of several stable wavelength sources. One approach to combining the stellar and calibration spectra is to pass the starlight through a component (usually a molecular absorption cell) that impresses absorption features on the spectrum at stable wavelengths. A suitable fit to the observed spectrum can then yield the wavelength shift of the stellar features relative to the reference ones (Campbell & Walker 1979, Marcy & Butler 1992). One limitation of this technique is that the materials suitable for use in absorption cells provide usable spectrum lines only within fairly limited wavelength ranges. Iodine vapor, for instance, is useful only within the range 500 nm $\leq \lambda \leq$ 620 nm, while the best region in terms of stellar line density and stability of the terrestrial atmosphere is 380 nm $\leq \lambda \leq$ 500 nm. An alternate approach is to provide a separate path into the spectrograph (usually using optical fibers) that lies near the path followed by the starlight. This path may be used to introduce a reference spectrum that is recorded on the detector simultaneously with and immediately adjacent to the stellar spectrum. This method allows one to use any desired wavelength range, but suffers because the light paths and spectrograph illumination for the two beams cannot be exactly the same (Brown 1990, Brown et al 1991).

Results from Doppler shift asteroseismology (even negative ones) are available only for three stars: α Cen A, α CMi (Procyon A), and β Hyi.

Gelly et al (1986) were the first to search for pulsations in α Cen A (G2 V, m_v = 0.0), using a sodium resonance cell Doppler analyzer. They believed that they found pulsations with maximum amplitudes of about 1.5 m s^{-1}, and showing a periodicity in the temporal spectrum indicating $\Delta v_0/2 = 82.7$ μHz. This value of $\Delta v_0/2$ is much larger than the $\sim 55\mu$Hz one expects based on Equation (4), given the inferred mass and radius of α Cen A. Brown & Gilliland (1990), using an echelle spectrograph fiber-fed from the CTIO 1.5-m telescope, failed to reproduce the results of Gelly et al (1986). They reported no clear evidence for pulsations, with limits of 70 cm s^{-1} in velocity and roughly 4×10^{-5} in the ratio of line core to continuum intensities. These limits are roughly 5 and 10 times the amplitudes that would be expected from the Sun. Pottasch et al (1992, 1993) used a Fabry-Perot system at the 3.6-m ESO telescope to obtain six consecutive nights of observations of α Cen A. They found evidence for a regular series of modes in the frequency range between 2.2 and 3.4 mHz; the mode amplitudes inferred were between 75 and 120 cm s^{-1}, with $\Delta v_0/2 = 55.3 \pm 3.6$ μHz. This value for the large splitting is consistent with expectations, though it is unclear why the amplitudes should be so much larger than the corresponding solar values, or the limits found by Brown & Gilliland (1990). Finally, Edmonds (1993), using echelle data obtained at the 4-m AAT, found evidence for frequency separation similar to that found by Pottasch et al (1992), but found mode amplitudes that were considerably smaller (≤ 35 cm s^{-1}). In view of the conflicting nature of the evidence about α Cen A, the most one can confidently say is that the p-mode amplitudes are probably not more than 3–5 time the solar values, and that other mode properties are not yet determined.

The first oscillation observations of Procyon (F5IV-V, $m_v = 0.4$) were also reported by Gelly et al (1986, 1988). They reported evidence for pulsations in the frequency range $0.6 \leq v \leq 1.7$ mHz, with amplitudes of about 70 cm s^{-1} per mode. Innis et al (1991) reported a limit of 4 m s^{-1} (with 3σ confidence) based on resonance-cell observations. The most convincing (though still equivocal) results on Procyon were obtained by Brown et al (1991) with an echelle spectrograph at the KPNO 2.1-m telescope. A fiber "double scrambler" (Brown 1990) prevented seeing fluctuations and resulting changes in collimator illumination from compromising the Doppler stability. Using data from six consecutive nights, Brown et al (1991) detected an excess of power in the temporal spectrum of Procyon's Doppler shift, in the frequency band below about 1.4 mHz (Figure 8). The total power in this bandpass corresponds to an rms velocity of about 2.5 m s^{-1}; if this power is interpreted as pulsation modes with

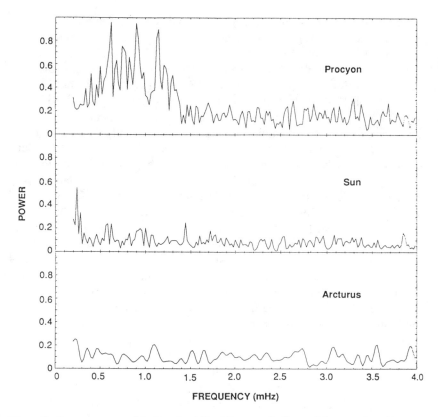

Figure 8   Power spectrum of the Doppler shifts of Procyon, the Sun, and Arcturus, from Brown et al (1991). These spectra are averages of six single-night power spectra. All are on the same scale; unit power indicates a velocity amplitude of roughly 1 m s^{-1}.

the frequency separation expected for Procyon, the individual mode amplitudes turn out to be 50–70 cm s^{-1}. Unfortunately, the small ($\cong 35.5\mu$Hz) magnitude expected for $\Delta\nu_0/2$ in Procyon's spectrum causes severe difficulties in identifying modes in data taken from a single site (Gilliland & Brown 1992b). Brown et al (1991) were unable to identify a clear mode structure in the excess power that they observed; they found some evidence for $\Delta\nu_0/2 = 35\mu$Hz, but other values of the large frequency separation fit the data about equally well. In view of the absence of a definite p-mode signature in the power spectrum, it is possible that the excess power below 1.4 mHz arises not from pulsations but from some other process (such as convection). Further observations, probably from a longitude-distributed network, will be required to resolve this issue.

Observations of β Hyi (G2 IV, $m_v = 2.8$) are so far rather limited. Frandsen (1987) searched for periodic changes in spectrum line core-to-continuum ratios, finding no signals as large as a few times 10^{-5} (roughly ten times the signal expected in the solar case). More recently Edmonds (1993) has searched for Doppler oscillations in β Hyi. No conclusive evidence for pulsations was found at a limiting amplitude of about 1 m s^{-1}. However, the power spectrum of the power spectrum did show evidence for regular peak spacings at various harmonics of 37.1 μHz. This value for $\Delta\nu_0/2$ is in reasonably good agreement with expectations for this star, though on balance Edmonds (1993) found the evidence too weak to support a claim of unambiguous detection of p-modes.

Although Doppler measurements have not yet revealed clear pulsations in any Sun-like star, prospects for doing so in the fairly near future are good. The two major problems impeding progress are the simple shortage of photons, and the necessity (in the case of subgiant stars) for extended observations from worldwide networks of observing sites. Both of these are becoming less serious with the advent of efficient echelle spectrographs at in increasing number of the world's large telescopes. Examples include a new fiber-fed spectrograph at the Hale 5-m telescope (Peri & Libbrecht 1993), and the Advanced Fiber Optic Echelle (Noyes et al 1993a, b). The most impressive recent addition to the available spectrograph arsenal is undoubtedly the HIRES spectrograph at the Keck 10-m telescope (Vogt 1992). Although not designed specifically for Doppler measurements (it is not fiber fed, for instance), this spectrograph can be fitted with an iodine absorption cell for continuous calibration; because of its very large throughput, it should be capable of Doppler precision significantly better than that of any other existing system.

5.   CONCLUSIONS

Asteroseismology is evidently becoming a powerful diagnostic of stellar parameters, providing a valuable probe of unconventional physical effects, and a testing ground of the theory of stellar structure and evolution. But the field is still, to a large extent, in the process of defining itself. The level of maturity reached so far depends strongly on the type of star under consideration. White dwarf studies appear to be the furthest along: The observational methods and theoretical background for these stars suffice to treat at least the most important processes seen to be operating, though there is still room for unexpected phenomena. Sun-like stars are arguably in the least satisfactory state, since we have not yet even solved the problem of detecting their pulsations. One may reasonably hope that the solar example will facilitate interpretation of pulsation data regarding these stars (once they are obtained), but surprises probably lurk in even this well-trodden field. The roAp and δ Scuti stars are somewhere in the middle, in that they can be studied with current observing methods, but the physics governing their behavior is relatively poorly understood. The resulting overall picture of the field is exciting but not completely satisfying; the emphasis one sees in the literature is often on technical progress rather than on physical results about stars. What advances, observational or theoretical, are needed to make asteroseismology a more productive component of stellar astronomy's tool kit?

The first requirement is an increased commitment to global observing efforts. For all types of stars discussed in this review, serious data gaps almost always compromise physical interpretation. Thus, the importance of global networks that can make coordinated observations can scarcely be overstated. Several such networks now exist; others should be expected and encouraged. Particularly in the case of Sun-like stars, for which large apertures are required, this mode of operation will involve substantial amounts of scarce telescope time. Observing campaigns for these purposes will have to be correspondingly well planned, organized, and justified.

Space-borne telescopes provide an alternative to ground-based network operation. This possibility is particularly attractive for photometric observations, since the attainable photometric precision for very many interesting objects is limited by atmospheric scintillation. Several studies of seismology missions have been performed or are in progress (PRISMA, STARS), and one small-aperture experiment (EVRIS) is scheduled to fly in the near future. For Sun-like stars, such space missions may provide important capabilities that are difficult or impossible to achieve from the ground.

Progress on δ Scuti and roAp stars depends first on better methods of mode identification (involving coordinated multicolor photometry and possibly spectroscopy, all from longitude-distributed networks), and second on advances in the theories of mode driving and selection mechanisms for δ Scutis, with the addition of magnetic fields for roAp stars.

Finally, it is important to understand that asteroseismic data carry unique and detailed information about stars, but that these data augment, not replace, other kinds of observation. Seismology encourages one to ask sharper questions than are common in stellar astronomy, but unambiguous answers to these questions depend not just on mode frequencies, but also on accurate photometry, astrometry, and spectroscopy. We therefore expect that seismic investigations of stars should and will be accompanied by coordinated observations of more traditional sorts. Only in this way will the full potential of asteroseismology be achieved.

ACKNOWLEDGMENTS

We thank Paul Charbonneau and Carl Hansen for their useful comments and suggestions concerning this review.

Literature Cited

Appourchaux T, et al. 1993. *PRISMA: Report on the Phase-A Study* ESA SCI (93)3
Audard N, Provost J. 1993. *Astron. Astrophys.* In press
Babcock HW. 1949. *Observatory* 69:191
Babel J, Michaud G. 1991. *Astron. Astrophys.* 248:155
Bachmann KT, Brown TM. 1993. *Ap. J. Lett.* 411:L45
Baglin A. Weiss W, Bisnovatyi-Kogan G. 1993. See Weiss & Baglin 1993, p. 758
Balona LA. 1986a. *MNRAS* 219:111
Balona LA. 1986b. *MNRAS* 220:647
Balona LA, Stobie RS. 1979. *MNRAS* 189:649
Belmonte JA, Chevreton M, Mangeney A, Praderie F, Saint-Pe O, et al. 1991. *Astron. Astrophys.* 246:71
Belmonte JA, Michel E, Alvarez M, Jiang SY, Chevreton M, et al. 1993a. *Astron. Astrophys.* In press
Belmonte JA, Pérez Hernández F, Roca Cortés T. 1990. *Astron. Astrophys.* 231:383
Belmonte JA, Roca Cortés T, Vidal I, Schmider FX, Michel E, et al. 1993b. See Weiss & Baglin 1993, p. 739
Bogdan TJ, Cattaneo F, Malagoli A. 1993. *Ap. J.* 407:316
Bond HE, Ciardullo R. 1993. In *White Dwarfs: Advances in Observations and Theory*, ed. MA Barstow, p. 491. Dordrecht: Kluwer
Borucki WJ, Cochrane WD, Dunham EW, Koch DG, Reitsema H. 1993. *Frequency of Earth-Sized Inner Planets: FRESIP.* NASA study document
Bradley PA, Winget DE. 1994. *Ap. J.* Submitted
Bradley PA, Winget DE, Wood MA. 1993. *Ap. J.* 406:661
Brassard P, Fontaine G, Wesemael F, Hansen CJ. 1992. *Ap. J. Suppl.* 80:369
Brassard P, Fontaine G, Wesemael F, Kawaler SD, Tassoul M. 1991. *Ap. J.* 367:601
Brassard P, Wesemael F, Fontaine G, Talon A. 1992. In *8th European Workshop on White Dwarfs*, ed. MA Barstow, p. 485. Dordrecht: Kluwer
Breger M. 1979. *Publ. Astron. Soc. Pac.* 91:5
Breger M. 1991. *Astron. Astrophys.* 250:107
Breger M, Balona LA, Grothues H-G. 1991. *Astron. Astrophys.* 243:160
Breger M, Garrido R, Huang L, Jiang S-y, Guo Z-h, et al. 1989. *Astrophys.* 214:209
Breger M, Huang L, Jiang S-y, Guo Z-h, Antonello E, Mantegazza L. 1987. *Astron. Astrophys.* 175:117
Breger M, McNamara BJ, Kerschbaum F,

Huang L, Jiang S-y, et al. 1990. *Astron. Astrophys.* 231:56
Brickhill A J. 1992. *MNRAS* 259:529
Brown TM. 1985. *Nature* 317:591
Brown TM. 1990. In *CCDs in Astronomy*, ed. GH Jacoby, ASP Conf. Ser. 8:335. San Francisco: Astron. Soc. Pac.
Brown TM, ed. 1993. *GONG. 1992: Seismic Investigation of the Sun and Stars*, ASP Conf. Ser. 42. San Francisco: Astron. Soc. Pac.
Brown, TM, Christensen-Dalsgaard J, Dziembowski W, Goode P, Gough DO, Morrow CA. 1989. *Ap. J.* 343:526
Brown TM, Christensen-Dalsgaard J, Weibel-Mihalas B, Gilliland RL. 1994. *Ap. J.* In press
Brown TM, Gilliland RL. 1990. *Ap. J.* 350:839
Brown TM, Gilliland, RL Noyes RW, Ramsey LW. 1991. *Ap. J.* 368:599
Butcher HR, Hicks TR. 1988. In *Seismology of the Sun and the Distant Stars* ed. DO Gough, p. 347. Dordrecht: Reidel
Campbell B, Walker GAH. 1979. *Publ. Astron. Soc. Pac.* 91:540
Campos AJ, Smith MA. 1980. *Ap. J.* 238:667
Catalano FA, Renson P, Leone F. 1993. *Astron. Astrophys. Suppl.* 98:269
Charbonneau P, MacGregor KB. 1993 *Ap. J.* 417: In press
Christensen-Dalsgaard J. 1988a. See Christensen-Dalsgaard & Frandsen 1988, p. 3
Christensen-Dalsgaard J. 1988b. See Christensen-Dalsgaard & Frandsen 1988, p. 295
Christensen-Dalsgaard J. 1991. In *Challenges to Theories of the Structure of Moderate-Mass Stars,* ed. D. Gough, J. Toomre, p. 11. Berlin:Springer-Verlag
Christensen-Dalsgaard J. 1992. *Ap. J.* 385–354
Christensen-Dalsgaard J. 1993. See Weiss & Baglin 1993, p. 483
Christensen-Dalsgaard J, Berthomieu G. 1991. In *Solar Interior and Atmosphere*, ed. AN Cox, WC Livingston, MS Matthews, p. 401. Tucson: Univ. Ariz. Press
Christensen-Dalsgaard J, Duvall TL Jr, Gough DO, Harvey JW, Rhodes EJ. 1985. *Nature* 315:378
Christensen-Dalsgaard J, Frandsen S. 1983. *Solar Phys.* 82:469
Christensen-Dalsgaard J, Frandsen S, eds. 1988. *Advances in Helio-and Asteroseismology, IAU Symp No. 123.* Dordrecht:Reide
Christensen-Dalsgaard J, Gough DO, Thompson MJ. 1991. *Ap. J.* 378:413
Christensen-Dalsgaard J, Pérez-Hernández F. 1992. *MNRAS* 257:62

Claverie, et al. 1984. *Mem. Soc. Astron. Ital.* 55:63

Connes P. 1985. *Astrophys. Space Sci.* 110:211

Cox AN, Chitre SM, Frandsen S, Kumar P. 1991. In *Solar Interior and Atmosphere*, ed. AN Cox, WC Livingston, MS Matthews, p. 618. Tucson: Univ. Ariz. Press

Cox AN, Guzik JA, Raby S. 1990. *Ap. J.* 353:698

Cox JP. 1980. *Theory of Stellar Pulsation.* Princeton: Princeton Univ. Press

Däppen W. 1993. See Brown 1993, p. 316

Däppen W, Dziembowski WA, Sienkiewicz R. 1988. See Christensen-Dalsgaard & Frandsen 1988, p. 233

Demarque P, Guenther DB. 1990. In *Progress of Seismology of the Sun and Stars*, ed. Y Osaki, H. Shibahashi, p. 405 Berlin:Springer-Verlag

Dolez N, Vauclair G, Koester D. 1991. In *Proc. 7th European Workshop on White Dwarfs*, ed. G. Vauclair, EM Sion, p. 361. Dordrecht:Reidel

Duvall, TL Jr, Dziembowski W, Goode PR, Gough DO, Harvey JW, Leibacher JW. 1984. *Nature* 310:22

Dziembowski W. 1977. *Acta Astron.* 27:203

Dziembowski W. 1990. In *Progress of Seismology of the Sun and Stars*, ed. Y. Osaki, H. Shibahashi, p. 359. Berlin:Springer-Verlag

Dziembowski W, Królikowska M. 1985. *Acta Astron.* 35:5

Dziembowski W, Królikowska M. 1990. *Acta Astron.* 40:19

Edmonds PD. 1993. Doctoral dissertation. Univ. Sydney

Edmonds P, Cram L, Demarque P, Guenther DB, Pinsonneault MH 1992. *Ap. J.* 394:313

Fagerholm E. 1906. Inaugural Dissertation. Uppsala

Frandsen S. 1987. *Astron. Astrophys.* 181:289

Frandsen S. 1993. See Weiss & Baglin 1993, p. 679

Gelly B, Grec G, Fossat E. 1986. *Astron. Astrophys.* 164:383

Gelly B, Grec G, Fossat E. 1988. See Christensen-Dalsgaard & Frandsen 1988, p. 249

Gilliland RL, Brown TM. 1992a. *Astron. J.* 103:1945

Gilliland RL, Brown TM. 1992b. *Publ. Astron. Soc. Pac.* 104:582

Gilliland RL, Brown TM, Duncan DK, Suntzeff NB, Lockwood GW, et al. 1991. *Astron. J.* 101:541

Gilliland RL, Brown TM, Kjeldsen H, McCarthy JK, Peri ML, et al. 1993. *Astron. J.* In press

Gilliland RL, Däppen W. 1988. *Ap. J.* 324:1153

Goldreich P, Keeley DA. 1977. *Ap. J.* 211:934

Goldreich P, Kumar P. 1990. *Ap. J.* 363:694

Goldreich P, Kumar P. 1991. *Ap. J.* 363:694

Gough DO. 1987. *Nature* 326:257

Gough DO. 1990. In *Astrophysics—Recent Progress and Future Possibilities*, ed. B.

Gustaffson, PE Nissen, R. Danish Acad. Sci. Lett. *Mat. Fys. Medd.* 42:13

Gough DO, Kosovichev AG. 1993. See Brown 1993, p. 351

Gough DO, Novotny E. 1993a. See Weiss & Baglin 1993, p. 550

Gough DO, Novotny E. 1993b. See Brown 1993, p. 355

Grauer AD, Bond HE. 1984. *Ap. J.* 277:211

Guenther DB, Demarque P. 1993. *Ap. J.* 405:298

Hansen CJ, Cox JP, Van Horn HM. 1977. *Ap. J.* 217:151

Harvey JW. 1988. See Christensen-Dalsgaard & Frandsen 1988, p. 497

Harvey JW & the GONG Instrument Development Team 1988. In *Seismology of the Sun and Sun-Like Stars*, ed. EJ Rolfe, p. 203. ESA SP-286

Harvey J, Hill F, Kennedy J, Leibacher J. 1993. See Brown 1993, p. 397

Heller CH, Kawaler SD. 1988. *Ap. J. Lett.* 329:43

Hudson H, Brown TM, Christensen-Dalsgaard J, Cox AN, Demarque P, et al. 1986. *A Concept Study for an Asteroseismology Explorer.* Proposal submitted to NASA

Huebner WF, Merts AL, Magee NH, Argo MF. 1977. *Astrophysical Opacity Library.* Los Alamos Sci. Library Rep. LA-6760-M

Innis JL, Isaak GR, Speake CC, Brazier RI, Williams HK. 1991. *MNRAS* 249:643

Jones A, et al. 1993. *STARS: An Investigation of Stellar Structure and Evolution.* Proposal for the ESA M3 mission

Kawaler SD. 1986. Ph.D thesis. Univ. Texas

Kawaler SD. 1988. See Christensen-Dalsgaard & Frandsen 1988, p. 329

Kawaler SD. 1993. *Ap. J.* 404:294

Kawaler SD, Bradley PA. 1993. *Ap. J.* In press

Kawaler SD, Hansen CJ. 1989. In *White Dwarfs*, ed. G. Wegner, p. 97. Berlin:Springer-Verlag

Kawaler SD, Weiss P. 1990. In *Progress of Seismology of the Sun and Stars*, ed. Y. Osaki, H. Shibahashi, p. 431. Berlin:Springer-Verlag

Kawaler SD, Winget DE, Iben I Jr, Hansen CJ. 1986. *Ap. J. Lett.* 306:L41

Kennelly EJ, Matthews JM, Walker GAH. 1993a. See Weiss & Baglin 1993, p. 743

Kennelly EJ, Walker GAH, Matthews JM, Merryfield WJ. 1993b. See Brown 1993, p. 359

Kennelly EJ, Walker GAH, Yang S. Hubeny I. 1991. *Publ. Astron. Soc. Pac.* 103:1250

Kepler SO, Nelan EP. 1993. *Astron. J.* 105:608

Kjeldsen H. 1993. Doctoral dissertation. Univ. Aarhus

Kjeldsen H, Frandsen S. 1992. *Publ. Astron. Soc. Pac.* 104:413

Kumar P, Lu E. 1991. *Ap. J. Lett.* 375:L35

Kurtz DW. 1982. *MNRAS* 200:807

Kurtz DW. 1990. *Annu. Rev. Astron. Astrophys.* 28:607

Kurtz DW, Matthews JM, Martinez P, Seeman J, Cropper M, et al. 1989. *MNRAS* 240:881

Kurtz DW, et al. 1994. In preparation
Kurtz DW, Wegner G. 1979. *Ap. J.* 232:510
Landolt AU. 1968. *Ap. J.* 153:151
Lebreton Y, Berthomieu G, Provost J. 1988. See Christensen-Dalsgaard & Frandsen 1988, p. 95
Libbrecht KG. 1988. *Ap. J. Lett.* 330:51
Libbrecht KG. 1989. *Ap. J.* 336:1092
Libbrecht KG, Morrow CA. 1991. In *Solar Interior and Atmosphere*, ed. AN Cox, WC Livingston, MS Matthews, p. 479. Tucson: Univ. Ariz. Press
Libbrecht KG, Woodard MF. 1990. *Nature* 345:779
Mangeney A, Däppen W, Praderie F, Belmonte JA. 1991. *Astron. Astrophys.* 244:351
Mantegazza L, Poretti E, Antonello E, Riboni E. 1993. See Weiss & Baglin 1993, p. 733
Marcy GW, Butler RP. 1992. *Publ. Astron. Soc. Pac.* 104:240
Martinez P. 1993. Doctoral dissertation. Univ. Cape Town
Martinez P, Kurtz DW. 1990. *MNRAS* 242:636
Martinez P, Kurtz DW, Meintjes PJ. 1993. *MNRAS* 260:9
Matthews JM. 1993. See Brown 1993, p. 303
Matthews JM, Kurtz DW, Wehlau WH. 1987. *Ap. J.* 313:782
Matthews JM, Wehlau WH, Walker GAH, Yang S. 1988. *Ap. J.* 324:1099
McGraw JT, Robinson EL. 1976. *Ap. J. Lett.* 205:L155
McGraw JT, Starrfield SG, Liebert J, Green RF. 1979. In *White Dwarfs and Variable Degenerate Stars, IAU Colloq. No. 53*, ed. HM Van Horn, V. Weidemann, p. 377. Rochester:Univ. Rochester
Merryfield WJ, Kennelly EJ. 1993. See Brown 1993, p. 363
Michaud G, Mégessier C, Charland Y. 1981. *Astron. Astrophys.* 103:244
Michel E, Belmonte JA, Alvarez M, Jiang SY, Chevreton M, et al. 1992. *Astron. Astrophys.* 255:139
Michel E, Goupil MJ, Lebreton Y. 1993. See Weiss & Baglin 1993, p. 547
Mihalas D, Hummer DG, Mihalas BW, Däppen W. 1990. *Ap. J.* 350:300
Moon TT, Dworetsky MM. 1985. *MNRAS* 217:305
Murray N. 1993. See Brown 1993, p. 3
Nather RE. 1989. In *White Dwarfs*, ed. G. Wegner, p. 109. Berlin:Springer-Verlag
Nather RE, Winget DE, Clemens JC, Hansen CJ, Hine BP. 1990. *Ap. J.* 361:309
Nordlund Å, Stein RF. 1990. *Comput. Phys. Commun.* 59:119
Noyes RW, Brown TM, Horner S, Korzennik S, Nisenson P. 1993a. See Weiss & Baglin 1993, p. 752
Noyes RW, Brown TM, Horner S, Korzennik S, Nisenson P. 1993b. See Brown 1993, p. 485
Osaki Y. 1993. See Weiss & Baglin 1993, p. 512

Osaki Y, Hansen CJ. 1972. *Astrophys J.* 185:277
Pérez-Hernández F, Christensen-Dalsgaard J. 1993. See Brown 1993, p. 343
Peri ML, Libbrecht KG. 1993. See Brown 1993, p. 489
Pinsonneault MH, Kawaler SD, Sofia S, Demarque P. 1989. *Ap. J.* 338:424
Pottash EM, Butcher HR, van Hoesel FHJ. 1992. *Astron. Astrophys.* 264:13
Pottasch EM, Butcher HR, van Hoesel FHJ. 1993. See Weiss & Baglin 1993, p. 717
Przybylski A. 1961. *Nature* 189:73
Rast MP, Nordlund Å, Stein RF, Toomre J. 1993. See Brown 1993, p. 57
Rast MP, Toomre J. 1993. See Brown 1993, p. 41
Robinson EL. 1979. In *White Dwarfs and Variable Degenerate Stars, IAU Colloq. No. 53*, ed. HM Van Horn, V Weidemann, p. 343. Rochester:Univ. Rochester
Robinson EL, Kepler SO, Nather RE. 1982. *Ap. J.* 259:219
Rogers FJ, Iglesias CA. 1992. *Ap. J. Suppl.* 79:507
Rogers FJ, Iglesias CA. 1993. See Brown 1993, p. 155
Sanders WL. 1977. *Astron. Astrophys. Suppl.* 27:89
Scargle JD. 1982. *Ap. J.* 263:835
Schatzman E, Maeder A, Angrand F, Glowinski R. 1981. *Astron. Astrophys.* 96:1
Seaton MJ. 1993. See Weiss & Baglin 1993, p. 222
Shibahashi H. 1983. *Ap. J. Lett.* 275:5
Simon N. 1982. *Ap. J. Lett.* 260:L87
Smith MA. 1982. *Ap. J.* 254:242
Spergel DN, Press WH. 1985. *Astrophys. J.* 294:663
Starrfield SM, Cox AN, Kidman RB, Pesnell WD. 1984. *Ap. J.* 281:800
Starrfield SM, Cox AN, Kidman RB, Pesnell WD. 1985. *Ap. J. Lett.* 293:23
Steigman G, Sarazin CL, Quintana H, Faulkner J. 1978. *Astron. J.* 83:1050
Stibbs DWN. 1950. *MNRAS* 110:395
Tassoul M. 1980. *Ap. J. Suppl.* 43:469
Thejl P, Vennes S, Shipman HL. 1991. *Ap. J.* 370:355
Toutain T, Frölich C. 1992. *Astron. Astrophys.* 257:287
Ulrich RK. 1986. *Ap. J. Lett.* 306:L37
Ulrich RK. 1988. See Christensen-Dalsgaard & Frandsen 1988, p. 299
Unno W, Osaki Y, Ando H, Saio H, Shibahashi H. 1989. *Nonradial Oscillations of Stars.* Tokyo: Univ. Tokyo Press. 2nd ed.
Vandakurov YV. 1968. *Sov. Astron.* 11:630
Vogt SS. 1992. In *ESO Workshop on High Resolution Spectroscopy with the VLT*, ed. M-H Ulrich, p. 223. Garching: ESO
Vorontsov SV. 1989. *Sov. Astron. Lett.* 15:21
Walker GAH, Yang S, Fahlman GG. 1987. *Ap. J. Lett.* 320:L139

Watson RD. 1988. *Astrophys. Space Sci.* 140: 255

Weidemann V, Koester D. 1984. *Astron. Astrophys.* 132:195

Weiss W, Baglin A, eds. 1993. *Inside the Stars,* ASP Conf. Ser. 40. San Francisco: Astron. Soc. Pac.

Winget DE. 1988. See Christensen-Dalsgaard & Frandsen 1988, p. 305

Winget DE. 1991. In *White Dwarfs,* ed. G. Vauclair, E. Sion, p. 129. Dordrecht: Kluwer

Winget DE. 1993. See Brown 1993, p. 33

Winget DE, Fontaine G. 1982. In *Pulsations in Classical and Cataclysmic Variable Stars*, ed. JP Cox, CJ Hansen, p. 46. Boulder:Joint Inst. Lab. Astrophys.

Winget DE, Hansen CJ, Van Horn HM. 1983. *Nature* 303:781

Winget DE, Nather RE, Clemens JC, Provencal J, Kleinman SJ, et al. 1991. *Ap. J.* 378:326

Winget DE, Nather RE, Clemens JC, Provencal J, Kleinman SJ, et al. 1993. *Ap. J.* Submitted

Winget DE, Robinson EL, Nather RE, Fontaine G. 1982b. *Ap. J. Lett.* 262:L11

Winget DE, Van Horn HM, Tassoul M, Hansen CJ, Fontaine G, Carroll BW. 1982a. *Ap. J. Lett.* 252:L65

Wolff SC. 1983. *The A-Type Stars: Problems and Perspectives.* NASA Spec. Publ. 463

Wolff SC, Hagan W. 1976. *Publ. Astron. Soc. Pac.* 88:119

Woodard MF, Noyes RW. 1985. *Nature* 318:449

Young AT. 1967. *Astron. J.* 72:747

Annu. Rev. Astron. Astrophys. 1994. 32: 83–114

THE GOLDILOCKS PROBLEM: Climatic Evolution and Long-Term Habitability of Terrestrial Planets

Michael R. Rampino

Department of Earth System Science, New York University, New York, New York 10003

Ken Caldeira

Global Climate Research Division, Lawrence Livermore National Laboratory, Livermore, California 94551

KEY WORDS: atmospheric evolution, greenhouse effect, Mars, Venus

INTRODUCTION

Why is Venus too hot, Mars too cold, and Earth "just right" for life? (The allusion to the fairy tale involves the three bowls of porridge belonging to Papa Bear, Mama Bear, and Baby Bear—one too hot, one too cold, and one just right—tested by a hungry Goldilocks.) A simplistic answer might be that a planet's surface temperature is to a large extent a function of its distance from the Sun, and Earth just happens to be at the "right" distance for comfortable temperatures and liquid water. However, this is far from the whole story.

The Goldilocks Problem involves the early history of the planets and the evolution of their atmospheres. Its solution must also take into consideration the long-term evolution of the Sun, and hence the so-called faint young Sun problem, that is, the fact that the early Earth was apparently warm enough for liquid water despite the 25–30% lower luminosity of the early Sun (Newman & Rood 1977; Gough 1981). Had Earth been too cold initially for liquid water to exist on its surface, the resulting icy planet would have had a high albedo or reflectivity, lowering temperatures further, and might have become irreversibly ice-covered—the "white Earth catastrophe" (Caldeira & Kasting 1992a). Yet

83

0066–4146/94/0915–0083$05.00

evidence exists that liquid water has been abundant on Earth for at least the last 3.8 billion years.

The white Earth catastrophe might be averted through geologic activity that provides continued outgassing of CO_2, thereby warming the planet, and eventually melting the ice. But could too much CO_2 produce surface conditions too hot for liquid water, arresting the rock weathering reactions that act to remove CO_2 from the atmosphere, and creating a dense, hot CO_2-rich atmosphere, such as present on Venus today?

Many scientists have stressed the importance of the origin and evolution of life on Earth in biogeochemical cycling of carbon and in causing important changes in atmospheric composition over the last 4 billion years. Proponents of the Gaia hypothesis (Lovelock & Margulis 1974; Lovelock 1979, 1989) go further in claiming that life itself has managed to maintain surface conditions on Earth within a fairly narrow window through a series of negative feedbacks involving greenhouse gases, cloud albedo, and other factors.

Thus, despite its trivial sounding name, the Goldilocks Problem touches on major questions in stellar evolution, the origin of planets, geologic activity, the origin and evolution of life, biogeochemical cycles, climate modeling, the search for extraterrestrial intelligence, and on the place of life and human existence in the Cosmos. Recent reviews have emphasized various aspects of the problem (e.g. Prinn & Fegley 1987, Kasting 1989, Priem 1990, Pollack 1991, McKay 1991, Pepin 1991, Hunten 1993); this review seeks to clarify the Goldilocks Problem within the larger questions of stellar and planetary evolution and planetary habitability, and to discuss some of the most recent work on the problem.

PLANETARY SURFACE TEMPERATURES

The effective temperature (T_e) of a planet's surface, determined by the planet's distance from the Sun and its surface albedo, is defined as:

$$\sigma T_e^4 = \frac{S}{4}(1 - A)$$

where S is the amount of solar insolation at the planet's distance from the Sun, A is the planet's albedo, and σ is the Stefan-Boltzmann constant.

Earth, Mars, and Venus have atmospheres that contain the important greenhouse gases, H_2O vapor and CO_2, which absorb outgoing long-wave radiation, and thus warm the planet's surface (Table 1).

Therefore, the surface temperatures (T_s) of the these planets may be defined as the sum of their effective temperatures (T_e) plus the greenhouse effect (δT) provided by the greenhouse gases in the atmosphere ($T_s = T_e + \delta T$) (Goody & Walker 1972, Pollack 1991). Table 2 shows the values of T_e and T_s for Venus, Earth, and Mars.

Table 1   Main constituents of terrestrial planetary atmospheres

Planet	Atmospheric pressure (bars)	Major constituents[a]
Venus	90	$\mathbf{CO_2}$ (0.96), N_2 (0.035), $\mathbf{H_2O}$ (4×10^{-5})
Earth	1	N_2 (0.77), O_2 (0.21), $\mathbf{H_2O}$ (~0.01), Ar (0.009), $\mathbf{CO_2}$ (3.5×10^{-4})
Mars	0.006	$\mathbf{CO_2}$ (0.95), N_2 (0.027), Ar (0.016), $\mathbf{H_2O}$ (3×10^{-4})

[a]Numbers in parentheses show the fractional abundance, by number, of the major atmospheric constituents. Boldface indicates greenhouse gases. Other gases can contribute to greenhouse warming through pressure broadening of the absorption bands of greenhouse gases.

From Table 2, it can be seen that our original conjecture that the Goldilocks Problem could be solved merely by taking into account the differences in the planetary distances from the Sun does not work—T_e for *all three* planets is below the freezing point of water. On Earth, the comfortable conditions for life are made possible by the 33 K of greenhouse warming supplied by H_2O and CO_2 gases in the planet's atmosphere. Note also that although Venus is the closest to the Sun of the three planets, its higher surface temperature is not a direct result of increased solar insolation—cloud-covered Venus absorbs only slightly more solar radiation than barren Mars as a result of the great differences in their present albedos. The greenhouse effect of Venus' dense (90 bar) CO_2 atmosphere adds 521°C of warming. The effectiveness of greenhouse heating on Venus is increased by the fact that greenhouse gases can absorb thermal radiation in their weaker transitions as total atmospheric mass increases, thus the fraction of thermal IR absorbed increases with a planet's surface pressure (Pollack 1991).

Table 2   Variation of effective and surface temperature on three planets showing the influence of greenhouse gases[a]

Planet	Atmosphere pressure (atm)	Greenhouse gases	Orbit (AU)	Solar constant (W m^{-2})	Albedo (%)	Effective temp (K)	Surface temp (K)	Greenhouse warming (°C)
Venus	90	CO_2	0.723	2620	76	229	750	521
Earth	1	H_2O, CO_2	1.00	1368	30	255	288	33
Mars	0.006	CO_2	1.524	589	25	210	218	8

[a]Note that the relationship among the solar insolation values (S/S_0) taking Earth = 1.00, are Venus = 1.91, and Mars = 0.43.

It is important at this point to note that the amount of CO_2 at the surface of Earth is also very high, equivalent to ~60 bar, but on Earth the CO_2 is tied up in carbonate rocks as solid $CaCO_3$ (Ronov & Yaroshevsky 1976, Holland 1978), and additional CO_2 is present in the upper mantle. This suggests that by whatever method CO_2 accumulates in a planet's atmosphere (e.g. outgassing from the interior, delivery by impactors during the planetary accretion process), Venus and Earth may have received about the same complement of the gas. The difference is that on Earth the CO_2 became locked up in surface rocks, whereas on Venus the CO_2 has remained in the atmosphere, creating the extreme greenhouse conditions on the planet today. The question is: Why did things turn out so badly for our sister planet?

TOO HOT: THE RUNAWAY GREENHOUSE

Venus today is an uninhabitable inferno with a dense carbon dioxide atmosphere and a surface temperature of 750 K (477°C)—hot enough to melt lead. The atmosphere is also exceedingly dry—Donahue & Hodges (1992) recently provided evidence that the H_2O mixing ratio is only ~3×10^{-5} in the Venus atmosphere. The high temperatures and lack of water on Venus are commonly attributed to a "runaway greenhouse," in which H_2O was lost, and CO_2 built up in the atmosphere to the high levels seen today.

The idea of a runaway greenhouse was suggested by Hoyle (1955), and has been further developed by a number of workers including Sagan (1960), Gold (1964), Dayhoff et al (1967), Ingersoll (1969), Rasool & DeBergh (1970), Pollack (1971), Goody & Walker (1972), Walker (1975), Watson et al (1981), Matsui & Abe (1986a,b), Abe & Matsui (1988), Kasting (1988, 1989, 1991a), Durham & Chamberlain (1989), and Tajika & Matsui (1992). Early ideas are well summarized in Walker (1977), Henderson-Sellers (1978), and Henderson-Sellers & Cogley (1982). These early considerations of the runaway greenhouse are exemplified in Figure 1 from Rasool & DeBergh (1970) (see Goody & Walker 1972).

The three curves in the figure represent the evolution of the surface temperatures of the terrestrial planets (starting with no atmosphere and an albedo like that of present-day Mars) as water vapor, either with or without CO_2, is released into the atmosphere by planetary volcanic outgassing. The curves show how surface temperatures increase as a result of the greenhouse effect, as water vapor accumulates in the atmospheres of Venus, Earth, and Mars. On Mars and on Earth, the increase is halted when the water vapor pressure is equal to the saturation vapor pressure, and freezing or condensation occurs. The Martian temperatures are so low that gases do not accumulate in the atmosphere for very long before the pressure reaches the saturation pressure of ice. This

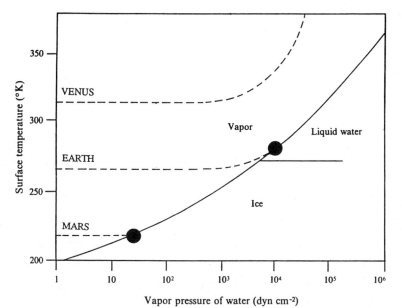

Figure 1 An early interpretation of the runaway greenhouse effect. The dashed curves show how surface temperatures increase, due to the greenhouse effect, as water vapor accumulates in the atmospheres of Venus, Earth, and Mars. On Mars and Earth, the increase is halted when the water vapor pressure is equal to the saturation vapor pressure, and either freezing or condensation occurs. Temperatures on Venus are higher because Venus is closer to the Sun, and saturation is never achieved. Therefore, temperature runs away. Note that the temperatures on the left-hand axis are not the same for Earth and Venus as the effective temperatures in Table 1, as different planetary albedoes were used. (After Goody & Walker 1972.)

produces a thin atmosphere with a small greenhouse effect associated with low atmospheric H_2O vapor pressure.

Because Venus is closer to the Sun, initial surface temperatures were higher. According to this scenario, outgassing of H_2O and CO_2 produced a greenhouse effect that increased the surface temperature, which, in turn, increased the saturation vapor pressure of water in the atmosphere. The greenhouse effect of the increased water vapor is so great that saturation was never achieved, and condensation never occurred. As more water vapor is released, a strong positive feedback develops, temperatures increase further, the saturation vapor pressure of water increases further, etc, and the temperature eventually runs away. In this picture, oceans never formed on Venus; the water has always existed as water vapor in the atmosphere.

The presence of H_2O in the upper atmosphere inevitably leads to loss of water by photodissociation (Goody & Walker 1972). The disappearance of liquid water prevents the so-called Urey silicate-rock weathering reactions, e.g.:

$$CaSiO_3 + CO_2 \Rightarrow CaCO_3 + SiO_2$$

which act to remove CO_2 from planetary atmospheres, from taking place (Walker et al 1981). These reactions occur in the fluid phase, or at the fluid/solid interface, and so require liquid water. Therefore, the continued outgassing of CO_2 will further increase the planet's surface temperature in a positive feedback loop, leading eventually to an extremely high CO_2 atmosphere, such as seen on Venus today (Kasting 1989). In size, Venus is Earth's close twin (equatorial diameter of 12,104 km vs 12,756 km for Earth), and the planet's bulk should have efficiently sealed in the heat produced by accretionary impacts and radioactive decay. Venus is apparently volcanically active, although the presence of some process resembling terrestrial plate tectonics remains debatable (Head et al 1992).

It is now apparent that the early analyses of the runaway greenhouse had two problems:

1. They neglected convection in the lower atmosphere, which would change the vertical profile of the mixing ratio of water vapor in the planet's atmosphere, and reduce the lapse rate (the rate of temperature decrease with altitude). This greatly reduces the magnitude of the greenhouse effect. The elevated temperatures at high altitudes mean that H_2O mixing ratios are relatively high in the upper atmosphere; and

2. Although the low abundance of non-radiogenic rare gases in the atmospheres of the three planets indicates that they did not acquire a primordial atmosphere directly from the solar nebula, the current models of rapid planet formation suggest that planets start out with dense atmospheres from degassing of impactors during accretion, and/or the addition of volatile-rich cometary material in the latter stages of accretion (e.g. Matsui & Abe 1986a,b; Abe & Matsui 1988; Chyba 1990).

Another way of looking at the runaway greenhouse is to consider the critical value of the solar flux incident at the top of a planet's atmosphere above which liquid water cannot exist at the surface (Komabayashi 1967, Ingersoll 1969, Hart 1978, Kasting 1991a, Nakajima et al 1992). For example, if Earth was by some means pushed closer and closer to the Sun (or the Sun's luminosity increased), we would anticipate that at some point, the oceans would be vaporized, and the planet would be enveloped in a dense steam atmosphere. The amount of water in Earth's oceans is 1.4×10^{24} g, which means that the surface pressure of this atmosphere would be $\sim$270 bar, $\sim$50 bar greater than the pressure at the critical point of water.

Nearly all solar models indicate that the Sun has been getting more luminous with time, by $\sim$30% since the Solar System formed (but see Gilliland 1989), because as the Sun converts hydrogen to helium the Sun's core becomes denser and hotter, increasing the rate of thermonuclear fusion (Newman & Rood 1977). The luminosity (S) of the Sun on the main sequence as measured at Earth using a standard solar model (Sackman et al 1990) can be approximated by

$$S(t) = \left(1 - \frac{0.38t}{\tau_0}\right)^{-1} S_0$$

for the interval -4.5 billion years $< t < 4.77$ billion years (Caldeira & Kasting 1992b). Here t is time expressed as years from the present, $\tau_0 = 4.55$ billion years, and the subscript 0 refers to present day values ($S_0 = 1368\,\mathrm{Wm}^{-2}$).

Thus, for the Goldilocks Problem of planetary habitability, especially with changing solar output over time, the most important aspect of a runaway greenhouse is the critical solar flux required to trigger it. How much additional solar insolation would be required to turn Earth into Venus? Or alternately, how much closer to the Sun would Earth need to have been to put it on the path to a runaway greenhouse? How lucky were we?

This critical question has been discussed in a number of papers (e.g. Hart 1978, Rasool & DeBergh 1970, Kasting 1988). Rasool & DeBergh (1970) calculated that the water in the oceans would never have condensed if Earth were formed 4 to 7% (6 to 10×10^6 km) closer to the Sun, and Hart (1978) calculated about 5% closer (0.95 AU). More recently, using a radiative-convective climate model and treating solar flux as a variable, Kasting (1988) made detailed estimates of this energy threshold for a fully saturated, cloud-free atmosphere, where F_{IR} = outgoing IR, and F_s = incoming solar radiation:

$$F_{IR} = \frac{S}{S_0}(F_s).$$

Assuming a surface temperature and vertical temperature profile, and calculating F_{IR} and F_s, the critical solar flux in his model was $1.4\,S_0$. Abe & Matsui (1988) performed a similar calculation for a case in which part of the energy required to trigger a runaway greenhouse was derived from infalling planetesimals during planetary accretion, and the results agreed well with those of Kasting (1988). These findings are in agreement with the inferred history of Venus, where, because of the Sun's lower early luminosity, the solar flux (now $1.91\,S_0$) would have been ~ 1.3 to $1.4\,S_0$ early in the history of the Solar System.

AN ALTERNATIVE: THE MOIST GREENHOUSE

Kasting and co-workers (1984, 1988) proposed an alternative model to the runaway greenhouse, the "moist greenhouse," in which Venus could have lost its water while maintaining liquid oceans. The concept comes from studies suggesting that the vertical distribution of water vapor in a planet's atmosphere should be strongly correlated with its mixing ratio near the surface (Ingersoll

1969). When water vapor is a minor constituent of the lower atmosphere, the concentration declines rapidly with altitude throughout the convective region as a consequence of condensation and rainout (Kasting 1988). This takes place in the present terrestrial atmosphere, where the mixing ratio for water vapor drops from ~ 0.01 near the surface to $\sim 3 \times 10^{-6}$ in the lower stratosphere. The upper troposphere provides a so-called cold trap which prevents much water vapor from entering the stratosphere. Ingersoll (1969) showed that in a convecting atmosphere with a water vapor mixing ratio > 0.1, the release of latent heat by the condensing water keeps the lapse rate at a small value so that the atmospheric temperature decreases very slowly with height. The almost constant mixing ratio means that water can remain at significant levels even at high altitudes, where it can be effectively dissociated by solar ultraviolet (UV) with subsequent loss of hydrogen to space. As long as sufficient solar extreme ultraviolet (EUV, with $l \leq 100$ nm) energy is available to cause the escape, H should escape at about the diffusion-limited rate (Hunten 1973).

The work of Kasting et al (1984) and Kasting (1988) suggested that hydrogen escape becomes very rapid for incident solar fluxes exceeding 1.1 S_0. The calculations indicate that Venus could have lost most of its water without ever experiencing true runaway greenhouse conditions. In a plot of $S/S_0(S_{eff})$ against surface temperature, it seems that early Venus may have been just at the transition region between runaway and moist greenhouse situations (Figure 2). The solar flux at the orbit of Venus early in the history of the Solar

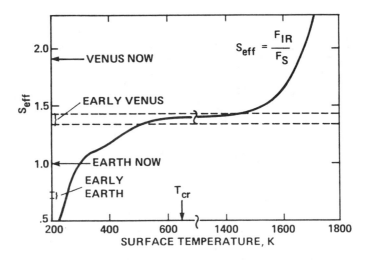

Figure 2   Effective solar constant S_{eff} vs surface temperature. The horizontal dashed lines represent estimates of the solar flux at the orbit of Venus ~ 4.6 billion years ago. Note that there is a break in the horizontal scale between 700 and 1300 K. (After Kasting 1988.)

System ($\sim$1.3 to 1.4 S_0) is so close to the minimum calculated for a runaway greenhouse (1.4 S_0), that Kasting (1988) suggested that the effect of increased clouds, which would reflect more solar radiation back to space, would have tipped the balance in favor of the moist greenhouse. These modeling efforts suggested that Earth could experience a moist greenhouse catastrophe when solar insolation has increased by $\sim$10% (although this may be prevented by negative cloud feedback)

WATER LOSS FROM VENUS

The present Venus atmosphere has only 10^{-5} the water in Earth's oceans. Either Venus was originally water poor, or the planet has lost much of its water. Lewis (1982) suggested that, according to the equilibrium condensation model of planet formation, Venus would have originally been rather dry. This is because the material that condensed out of the solar nebula at Venus' distance from the Sun is predicted to have had much less of the hydrous silicate phases than the condensate at greater distances from the Sun. However, two problems exist with this model: 1. It is apparently kinetically impossible to form these hydrous phases at the densities of the solar nebula in the relatively short time (10^8 yr) in which the planets formed (Prinn & Fegley 1987); the water should have condensed out of the solar nebula primarily as water ice; and 2. Wetherill (1986) has made a good case that the planets did not form where they are now. His calculations suggest that planetary eccentricities were pumped up to high values during the latter stages of accretion, and that an exchange of material from different parts of the solar nebula took place. The cumulative crater density on the inner planets, and the modern ideas on the formation of the Moon by the collision of a large (Mars-sized) body with Earth support this picture (Wetherill 1986), and if this is correct, then Venus could have initially received a significant amount of water, possibly as much water as Earth.

One measure of the amount of water lost from the atmosphere of Venus is the D/H ratio of the present water vapor, which is $\sim$2.4 $\times$ 10^{-2} (Donahue et al 1982, McElroy 1982, DeBergh et al 1991, Donahue & Hartle 1992). This is $\sim$150 times Earth's D/H ratio of 1.6 $\times$ 10^{-4}. Donahue et al (1982) originally interpreted this difference as indicating that Venus once had $\sim$100 times as much water as it does at present. Since the present water abundance on Venus is only 0.0014% that of Earth, the conclusion is that Venus started out with only about 0.1% of Earth's present water budget. However, if some deuterium was lost along with hydrogen during the escape process, then the amount of water that was lost could have been orders of magnitude greater, and Venus could have started out with as much water as Earth (Donahue et al 1982, Kasting & Pollack 1983).

Taking the case of Rayleigh fractionation—escape with no resupply—the time constant τ for the evolution of the D/H ratio in the Venus atmosphere is

$$\tau \sim R/(f\phi),$$

where R is the vertical column abundance of water vapor in the atmosphere, ϕ is the hydrogen escape rate, and f is the D/H fractionation factor (the relative efficiency of D escape compared to H escape):

$$f = \frac{\phi_D/N_D}{\phi_H/N_H}$$

where ϕ is the escape rate, and N is the column abundance in the atmosphere (Grinspoon 1987).

Recently, Gurwell & Yung (1992) estimated a value of f of 0.125, which is ten times greater than the canonical value of 0.013. This higher fractionation factor takes into account reactions between H and hot Oxygen (O^0). Calculations using the new fractionation factor (Grinspoon 1993) give $R_O/R_t = 300$, equivalent to $\sim 0.3\%$ of a terrestrial ocean, i.e. Venus could have originally had ~ 10 meter layer of planetary water. The escape time for water τ_{esc} from the planet could have been ~ 190–240 million years (a minimum value, since f should have been higher early in Venus' history when hydrodynamic escape was possible).

The picture of an early wet Venus was earlier challenged by Grinspoon (1987) and Grinspoon & Lewis (1988), who pointed out that the present D/H enrichment on Venus might be explained if the water abundance in the atmosphere were in a steady state, with loss of water by photodissociation and hydrogen escape balanced by continued influx of water-rich comets. However, it has been suggested that the steady-state picture cannot explain the present D/H ratio of the Venus atmosphere (J. F. Kasting, personal communication 1993).

Venus could have outgassed H_2O that was more enriched in deuterium than that outgassed on Earth. However, the homogenization of planetesimals during accretion (Wetherill 1986) and the fact that the D/H ratio of chondritic meteorites and of Halley's Comet (Eberhardt et al 1987) are similar to the terrestrial value, suggests a single Solar System value. Another possibility is that Venus might have had an ocean at one time that was enriched in D/H by hydrodynamic escape. If Venus had some process like plate tectonics, some of this water could have gone into hydrated seafloor and would have been subducted. Venus could now be outgassing this D-rich water, which is further enriched in D by present-day nonthermal escape mechanisms (J. F. Kasting, personal communication 1993). It is also worth noting that the surface temperature of a planet may affect convection patterns and the chemical structure of the mantle, providing some feedback between outgassed CO_2 and H_2O, planetary surface temperatures, and geologic activity that have yet to be fully explored (Ogawa 1993).

Getting rid of the last few bars of water presents a problem in both the runaway greenhouse and moist greenhouse models (Kasting 1988), but if plate tectonics, or some process like it, exists or once existed on Venus, the remaining water may have gone into the hydrated crust, which was then subducted and returned to the planet's mantle. Volcanic outgassing apparently continued on Venus, increasing atmospheric pCO_2 and the planet's greenhouse effect. Outgassed SO_2 accumulated in the atmosphere as clouds of sulfuric acid. The end result is the present hot and arid conditions on Venus.

Durham & Chamberlain (1989) recently asked the reverse question: What levels of CO_2 are required to produce the climatic conditions thought to have existed on the three planets $\sim$4.0 billion years ago? From a comparative analysis of the planets, making a number of assumptions about the early outgassing history and using a one-dimensional radiative/convective model, they estimated that the Venus atmosphere became unstable at $\geq$ 11 bars of CO_2, whereas early Earth and Mars had stable atmospheres that would not have gone into a moist or runaway greenhouse even if the pCO_2 were equal to 100 bars on Earth. The surface pCO_2 on early Mars could have been $\sim$1.3 to 2.1 bars, whereas pCO_2 on Earth may have been as great as 14 bars, although the results must be considered as only suggestive. One interesting finding was that Rayleigh scattering significantly increases the albedo of atmospheres high in CO_2. However, Durham & Chamberlain (1989) did not take into account CO_2 condensation, and the effect of the resulting high-albedo CO_2 clouds on the early climate of the planets (see below, Kasting 1991b).

EARLY HOT, H_2O-RICH ATMOSPHERES

In a series of papers, Matsui & Abe (1986a,b; Abe & Matsui 1988), and more recently Tajika & Matsui (1992), explored the question of the radiative effects of dense steam atmospheres that may have been present during accretion of the terrestrial planets, and subsequent atmospheric conditions. The source of the water would have been infalling planetesimals (inner Solar System objects and/or late accreting material from the outer Solar System) that released their volatiles when impacting at speeds exceeding 2 to 3 km/sec. Partial release of volatiles by impacting planetesimals may have begun when Venus and Earth were only $\sim$1% of their present masses, and complete release would have occurred when the planets were $\sim$10% of their final masses. For Mars, by contrast, devolatization of impactors would have begun later in the accretionary history, and a significant fraction of the volatiles that went into the formation of Mars was probably not released during accretion (Pollack 1991).

The water and CO_2 released during accretion may have stayed in the atmospheres of Earth and Venus leading to a runaway greenhouse powered by solar and accretionary heating (Matsui & Abe 1986a,b; Zahnle et al 1988). If accretion took $\sim$10^8 years, then enough energy would have been released to create steam atmospheres. Eventually the surface temperatures could have reached the melting point temperatures of surface rocks ($\sim$1500 K) when solution of the water in surface melt rocks would have helped to stabilize the atmosphere. Subsequent to the steam atmosphere phase, stabilization of Earth's CO_2-rich atmosphere could have been established through the carbon cycle (Tajika & Matsui 1992).

TOO COLD: THE ICEHOUSE OF MARS

Mars is a frozen wasteland. The average temperature is 218 K, or $-55°$C (Table 2), and even with the large seasonal and diurnal fluctuations that occur, liquid water does not exist. The low atmospheric pressure on Mars (7 mb) further prevents the condensation of liquid water (Kahn 1985). The low surface temperatures on Mars also mean that there can be very little water vapor in the atmosphere. Appreciable amounts of water are apparently stored in permanent water ice caps and as permafrost beneath the surface (Pollack 1991). Geologic evidence has been interpreted as suggesting at least 440 m of water, present as ice, groundwater, and ground ice. (Overlying the water ice caps at both poles are "seasonal" ice caps of solid carbon dioxide which condense from the atmosphere in winter and sublime to the atmosphere in summer.) Given this deep freeze, water vapor feedback on Mars today is so weak that even with twenty times the surface pressure of CO_2 as Earth, Mars produces less than half the greenhouse warming, and most of the 33 degrees of greenhouse warming on Earth comes from water vapor.

Yet, considerable evidence exists for a warmer, wetter past for Mars. Images of the Martian surface returned by *Mariner* and *Viking* spacecraft show evidence of ancient riverbeds—channels cut into the surface presumably by running water—with dendritic drainage patterns, and meandering paths. Valley networks occur in the ancient cratered terrain of Mars, indicating liquid water on the surface of the planet $\sim$3.8 billion years ago (McKay & Davis 1991). The valleys and channels may have emptied into standing bodies of water—large lakes [possibly ice-covered (McKay 1991)] or small oceans (Parker et al 1993, Baker et al 1991). Liquid water on the planet's surface at much later periods of time is also considered a possibility by some workers (McKay 1991). This evidence suggests that Mars was once warm enough to sustain liquid water ($T > 273$ K). The most plausible explanation of this prior warmth and greater atmospheric pressure was a strong greenhouse warming from an early, much denser CO_2 atmosphere.

Although impact degassing may have been ineffective on Mars, carbon dioxide and water vapor were probably outgassed early in the planet's history. The amount of outgassing, however, is uncertain. McElroy et al (1976, 1977) estimated 8 to 133 m of outgassed water. From evidence of volcanism, Greeley (1987) estimated a global layer 46 m deep released in the first 2 billion years.

There are some indications that very little water remains in the upper mantle of Mars: Dreibus & Wanke (1987) estimated 18–36 ppm in Mars' mantle, and SNC meteorites (apparently originating on Mars) have low water contents of 16–52 ppm.

Strangely enough, the presence of an early massive CO_2 atmosphere on Mars, permitting liquid water to occur on the surface, could be part of the reason Mars became such an inhospitable place. If Mars had a 1 bar CO_2 atmosphere, weathering of silicates could have entirely converted the CO_2 to carbonate rocks in $\sim$10 million years (Pollack 1991). Without replenishment of CO_2, the warm, greenhouse conditions would have been lost quite quickly. On Earth, CO_2 precipitated as carbonate rock is recycled to the atmosphere by a combination of chemical reactions and plate tectonics (Figure 3). Replenishment of atmospheric CO_2 comes about through plate tectonics and resulting volcanism. The heating of carbonate rocks as ocean plates descend into the Earth's mantle causes decomposition of $CaCO_3$, and the resulting CO_2 is released to the atmosphere through volcanoes (Walker et al 1981). Mars shows no signs of a history of plate tectonic activity, but a kind of recycling might have taken place on early Mars, where large-scale lava flows could have buried carbonates and caused release of CO_2. Higher early heat flow ($\sim$3 to 5 times present values) might have made such a process possible (Pollack 1991).

Impact bombardment might have provided a more effective early source of heat on Mars (Carr 1989), although impact heating may not have occurred fast enough to prevent the loss of considerable atmospheric CO_2 through weathering reactions, and impact erosion may have actually removed volatiles from the Martian atmosphere (e.g. Ahrens 1993, Hunten 1993). Pollack (1991) went further in proposing a possible feedback in which the weathering rate would have adjusted to the resupply of CO_2, since the fraction of time liquid water was available (when ground temperatures hovered around 273 K) could have been controlled by the weathering rate needed to balance the CO_2 outgassing rate. This situation would have ended as the global heat flow decreased, bombardment slowed, and global volcanism died down. Carr noted that the early valley drainage networks approximately coincided with the time of heavy bombardment. The decrease in atmospheric pCO_2 to its low present value may have occurred as CO_2 was adsorbed onto regolith, or possibly by dry carbonate formation from CO_2 and water vapor (Pollack 1991). Carbonates have been detected on the surface of Mars by remote sensing (Pollack 1991), and in SNC meteorites of probable Martian origin.

However, all of these scenarios may be moot—a recent simulation of the climate of early Mars (Kasting 1991b) predicted that the low temperatures would permit the formation of CO_2 ice clouds in the upper troposphere. These clouds would further cool the surface by reducing the tropospheric lapse rate (because of the release of latent heat), and by reflecting additional sunlight back

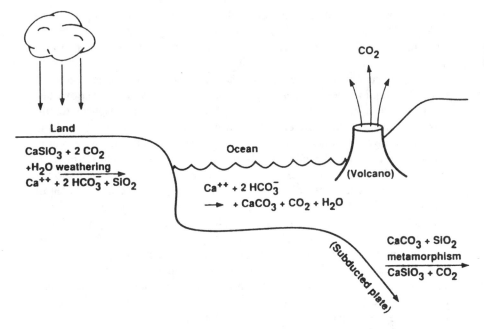

Figure 3 Schematic diagram of the biogeochemical cycle of carbon on Earth (courtesy of J. F. Kasting).

to space. Because CO_2 clouds are poor IR absorbers, their contribution to the greenhouse effect should be relatively small; their primary influence would be to cool the planet. Thus, Kasting's work suggests that a CO_2/H_2O greenhouse for early Mars does not work. This opens up the possibility that other greenhouse gases (CH_4, NH_3?) may have provided a warm climate.

Mars apparently represents a "runaway glaciation" model of planetary development (Hart 1978). Could this have happened to Earth? In early work, Schneider & Gal-Chen (1973) suggested that a decrease of 1.6% in the solar constant (equivalent to Earth being 0.8% farther from the Sun at present) would lead to runaway glaciation, whereas Wetherald & Manabe (1975) found that Global Climate Model (GCM) experiments would not produce a runaway glaciation on Earth even if the solar constant was decreased by 4%. In Hart's (1978) modeling effort, using an early atmosphere averaging $\sim$1.2 atm, composed of CO_2 followed by increases in CH_4 and other reduced gases, if the Earth-Sun distance was increased by 0.01 AU, then Earth became terminally glaciated about 2 billion years ago. Clearly, however, the predictions for runaway glaciation of Earth depend upon assumptions about the composition and density of the early atmosphere.

In the history of Mars, we also see the effect that a planet's size can have on its evolutionary path. Information from the study of the Moon showed that the once volcanically active body was geologically dead by about 3 billion years ago. The reason is most likely that the Moon is too small a planetary body (equatorial diameter of 3,476 km; only ∼1/4 the diameter of Earth, with about 1/80th of Earth's mass) to retain the internal heat trapped during accretion and generated by radioactive elements. Geologic activity, volcanism, and outgassing shut down after less than 1.5 billion years, and with only 1/6th the Earth's gravitational force, the Moon quickly lost to space whatever atmosphere it may have had.

The situation for Mars was apparently similar, but not so drastic. Mars is apparently too small (equatorial diameter = 6,794 km; ∼1/2 Earth's diameter) to have retained its internal heat for the entire 4.6 billion year history of the Solar System, and geologic activity, carbon cycling (if it existed), and volcanism apparently wound down and ended some 0.5 to 1 billion years ago (Hansson 1991). As the Martian interior cooled, degassing of CO_2 would have slowed, and carbon dioxide would have been drawn out of the atmosphere through silicate-rock weathering reactions and converted to calcium carbonate rocks in the crust—a process that could continue only so long as the temperature remained above the freezing point of water. This scenario may explain why Mars' atmospheric pressure is so close to the triple point (the point where solid, liquid, and vapor phases of water coexist).

JUST RIGHT: MOTHER EARTH

Evidently, Earth averted both the deep freeze of Mars and the greenhouse hell of Venus. But how? Pollack (1991) asks the critical question, "Was it just blind luck that enabled the Earth's mean temperature to vary relatively little over almost its entire history so that oceans and life could have persisted over such an extended period, despite a varying solar luminosity?" The alternative to blind luck is that some important feedbacks exist that have controlled the Earth's climate history. Such feedback processes may be geochemical and/ or biogeochemical.

Earth is clearly set apart from the other terrestrial planets by its wetness. The amount of liquid water present near or at the surface of Earth is 1.4×10^9 km^3, more than 97% of which is in the oceans (2% in polar ice and glaciers). This is equal to a layer of water 2.7 km thick covering the entire planet.

The interior of Earth also seems to contain significant amounts of water, although the exact amount is not well known. Estimates range from ∼60 to > 200 ppm in the upper mantle (Dreibus & Wanke 1987, Jambon & Zimmerman 1990, Carr & Wanke 1992). Thus, Earth is apparently wet inside and out. Most of the Earth's water may have been derived from a late volatile-rich veneer, with any earlier water reacting with iron to form FeO and H_2, the latter of which would have been lost (Dreibus & Wanke 1987). The wet interior may be the result of melting of the Earth's surface during accretion, as a result of the development of a steam atmosphere, allowing impact devolatized water at the surface to dissolve into the molten rock (Abe & Matsui 1988). Because of Mars' smaller size and greater distance from the Sun, the Martian surface may not have melted to the same degree, preventing the uptake of water. Or possibly Mars acquired a late volatile-rich layer, which was not folded into the interior (as with the more geologically active Earth) and instead remained as a surface veneer (Carr & Wanke 1992).

An examination of the evolution of surface conditions on Earth quickly leads to what has been termed "the faint young Sun paradox": Calculations showed that the early Earth, with an atmosphere of current composition, would be too cold for liquid water, and hence life, as recently as 2 billion years ago; however, the geologic record clearly shows evidence for life as far back as 3.5 billion years ago, and liquid water back to $\sim$3.8 billion years ago (Figure 4).

Sagan & Mullen (1972) suggested that reduced gases (mainly NH_3 and CH_4) in Earth's early atmosphere could have counteracted the faint young Sun problem by providing an additional greenhouse effect. In particular, NH_3 mixing ratios of only 10^{-5} to 10^{-4} were calculated to produce enough additional greenhouse heating to counteract the faint young Sun. However, these reduced gases

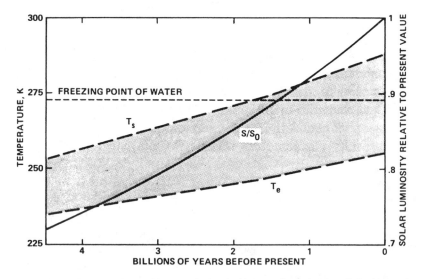

Figure 4 The faint young Sun problem as calculated with a one-dimensional, radiative/convective climate model. The solid curve is Gough's (1981) solar-luminosity parameterization. The dashed curves represent the effective radiating temperature T_e and the surface T_s. The shaded area shows the magnitude of the greenhouse effect. (After Kasting & Grinspoon 1991.)

were found to be too unstable with respect to photodissociation to provide a long-term source of greenhouse heating (Kuhn & Atreya 1979, Owen et al 1979). Owen et al (1979) suggested that a CO_2-rich early atmosphere could have provided the additional greenhouse warming to offset the faint young Sun problem.

The silicate/carbonate rock cycle is a geochemical conveyor belt driven by heat released by the mantle (Berner et al 1983). At present, it takes $\sim$100 million years for new sea floor to upwell at mid-ocean ridges, spread to subduction zones, and, finally, to plunge into the mantle—long enough for atmospheric CO_2 to have cycled through the rock shell of the Earth many times over. Over the 4.6 billion years of planetary evolution there has been a gradual slowing of this conveyor belt, and probably a reduction in the amount of volcanic carbon dioxide released to the Earth's atmosphere.

However, although the greenhouse effect of an early atmosphere much richer in CO_2 is invoked to explain the warm conditions of the early Earth, this leads to the question of how the "thermostat" was regulated subsequently in the face of increasing solar luminosity. A stabilizing climatic feedback process is strongly indicated.

Walker et al (1981) were the first to propose that, on geological time scales, climate is stabilized by factors affecting the rate at which calcium silicate rocks are geochemically weathered. A key point is that the rate at which CO_2 is removed from the atmosphere by rock weathering increases as temperature increases. The two major reasons for this are that: 1. rainfall and runoff increase, exposing more bedrock to erosion; and 2. respiration of soil organisms increases. In the long run, a higher CO_2 level in soil increases weathering rates, and like increased weathering from increased rainfall, this is a negative feedback.

If, for some reason, surface temperature were to fall, the temperature-weathering feedback would cause the removal rate of CO_2 from the atmosphere to fall, while carbon dioxide continued to outgas. The net result would be more atmospheric carbon dioxide, and a greenhouse warming partly compensating for the initial cooling. Conversely, were surface temperature to rise, the increased weathering would drive atmospheric CO_2 levels down, creating a cooling opposed to the initial warming. The long-term effect of weathering is therefore to stabilize global temperature.

The gradual reduction in the CO_2 content of the atmosphere as Earth cooled was probably partly the effect of decreasing geologic activity as radioactive elements decreased and as Earth's internal heat dissipated. Pulses of activity seem to have occurred, as evidenced by variations in ocean-floor spreading rates (Larson 1991). There is also evidence from variations in carbon isotopes and organic-carbon burial rates that during the last 600 million years (the Phanerozoic Eon), atmospheric CO_2 rose and fell by significant factors (Berner 1993).

For example, ~100 million years ago (the mid-Cretaceous Period) was a time of great warmth. This period correlates with a marked increase in the rates of sea-floor creation, and therefore subduction and volcanism. Earth's climate was 10–15°C hotter and the poles probably ice-free, quite likely in response to greenhouse warming from an atmosphere 4–10 times richer in carbon dioxide (Caldeira & Rampino 1991).

Biogeochemical cycling of carbon dioxide helped Earth to avoid the greenhouse that entrapped Venus, but how did we escape becoming a frozen planet like Mars? Early work by Budyko (1969) and Sellers (1969) suggested that a 2 to 5% decrease in the solar constant would lead to an ice-covered Earth. In their models, a small reduction in solar luminosity resulted in the advance of ice and snow cover toward the equator; this additional ice and snow reflected more solar radiation back to space and further cooled the planet. If solar luminosity is reduced beyond some critical value in these models, this ice-albedo feedback results in the catastrophic, irreversible freezing of the entire Earth's surface—the "white Earth catastrophe."

The temperature dependence of the rate of CO_2 uptake by silicate weathering and the negative feedback that it creates may have stabilized Earth against the ice-albedo catastrophe (Walker et al 1981). Walker et al suggested that, given the low early solar luminosities, high atmospheric CO_2 concentrations would be necessary to produce sufficient CO_2 consumption by silicate weathering to balance sources of atmospheric CO_2. Berner, Lasaga & Garrels (1983) incorporated this feedback in a carbonate-silicate cycle model of the global carbon cycle.

However, work by Marshall et al (1988) suggested that the silicate-weathering feedback could produce cold global climates if the continents were clustered near the equator. Exposed silicate rock would then be bathed in a relatively warm and wet environment, conducive to CO_2 consumption by weathering. Thus, for example, lower solar luminosity and the presence of an equatorial, Late Proterozoic (1 billion to 600 million years ago) supercontinent may both help to explain the widespread low-latitude glaciation (Marshall et al 1988, Hambrey & Harland 1985, Walter 1979).

The calculations of Marshall et al (1988) suggested that the silicate-weathering/CO_2 feedback could stabilize the global ice line at low latitudes. However, reanalysis of their model has indicated that this is not possible (Caldeira & Kasting 1992a). Because atmospheric pCO_2 responds much more slowly ($\sim 10^5$ yr) than does sea ice and snow cover (< 1 yr), the negative feedback proposed by Marshall et al (1988) does not apply to rapid fluctuations in the ice line. The silicate-weathering feedback could not act rapidly enough to buffer the Earth against a catastrophic ice advance. Indeed, during the Pleistocene (the last 2 million years) the correlation between glacial ice mass and atmospheric pCO_2 is negative, not positive (Barnola et al 1987). A perturbation analysis (Cahalan

& North 1979) of the Marshall et al (1988) energy-balance model shows that there is no stable ice line equatorward of about 30° latitude, regardless of atmospheric pCO_2 level. Hence, in their model, a low-latitude glaciation should run away to the globally ice-covered state. Could this have happened during the Late Proterozoic?

To study this question, Caldeira & Kasting (1992a) developed a zonally averaged energy-balance climate model based on a one-dimensional radiative-convective climate model (Kasting & Ackerman 1986). At today's solar flux and CO_2 level, the model exhibits four possible steady states (Figure 5): 1. ice free ($x_s = 1$), 2. stable partial ice cover ($x_s = 0.95$), 3. unstable partial ice cover ($x_s = 0.28$), and 4. ice covered ($x_s = 0$).

If Earth was initially in the stable, partially ice-covered state, and was then subjected to a rapid perturbation (relative to the 10^5 yr response time of the silicate weathering feedback) that would either temporarily lower the effective solar flux to about 0.9, or produce transient glaciation equatorward of $x_s = 0.28$, the planet would fall into the ice-covered state ($x_s = 0$). Without any change in atmospheric pCO_2, an increase in solar flux by about 27% would be needed to melt the equatorial ice (Figure 5). At solar fluxes higher than this value, a reverse ice-albedo feedback would apparently completely deglaciate Earth.

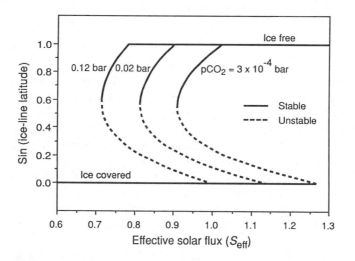

Figure 5 Steady-state ice lines (x_s) as a function of effective solar luminosity, S_{eff}, for three values of atmospheric pCO_2. If a perturbation were to shift Earth from its present state ($S_{eff} = 1$, $pCO_2 = 3 \times 10^{-4}$ bar, $x_s = 0.95$) to an ice-covered state today ($S_{eff} = 1$, $pCO_2 = 3 \times 10^{-4}$ bar, $x_s = 0$), then sufficient CO_2 (~0.12 bar) would accumulate in the atmosphere within ~30 million years to make the ice-covered state unstable. The model would then shift to the ice-free state ($S_{eff} = 1$, $pCO_2 = 0.12$ bar, $x_s = 1$) and silicate-rock weathering would begin to remove the excess CO_2 from the atmosphere. (After Caldeira & Kasting 1992a.)

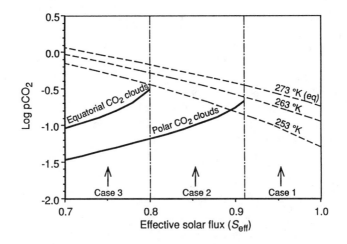

Figure 6   Solid curves indicate the onset of polar and global CO_2-cloud cover; dashed curves show equatorial surface temperature as a function of effective solar luminosity and atmospheric pCO_2.   Case 1: Starting from an ice-covered state with low atmospheric pCO_2, volcanic CO_2 would accumulate in the atmosphere and initiate ice-melting prior to the formation of CO_2 clouds.   Case 2: Polar CO_2 clouds would form prior to the onset of equatorial ice-melting.   Case 3: CO_2 clouds would form globally prior to the onset of ice-melting, in which case CO_2-global warming would likely be incapable of melting the ice.   (After Caldeira & Kasting 1992a.)

But this is not what would actually happen were the present Earth to freeze. In the ice-covered state, little or no silicate rock would be exposed to weathering, so CO_2 from metamorphic and mantle sources could accumulate in the atmosphere at a rate of about $8 \times 10^{12}\,\text{mol}\,\text{yr}^{-1}$ (Holland 1978). In less than ~30 million years, atmospheric pCO_2 would build up to nearly 0.12 bar, and equatorial ice would become unstable. This evolution corresponds to Case 1 in Figure 6.

However, if the change from stable partial ice cover to global glaciation occurred earlier in Earth's history, when solar luminosity was lower, the results might then be quite different. The reason is the early formation of clouds of CO_2 ice, wrapping the planet in a high-albedo blanket (Kasting 1991b). The points where CO_2 clouds begin to form in the Caldeira & Kasting model are indicated by the solid curves in Figure 6. The earlier phase of Earth's history is divided into two parts, labeled "Case 2" and "Case 3." When S_{eff} is less than about 0.92, i.e. earlier than about 1 billion years ago (Gough 1981), the model predicts that CO_2 clouds would start to form at the poles (Case 2). When S_{eff} is less than about 0.8, i.e. earlier than about 3 billion years ago (Gough 1981), CO_2 cloud cover would extend all the way to the equator.

If the albedo of such a CO_2 cloud-covered planet were as high as 0.8, the model predicts that Earth would not be able to emerge from that state, even at the present solar luminosity. Thus, if Earth had experienced runaway glacia-tion prior to about 3 billion years ago, the situation might have been effectively irreversible. These considerations imply that our planet might now be uninhab-itable had it not been warm during the early part of its history.

THE GAIA HYPOTHESIS

The most controversial idea related to the Goldilocks Problem may be the Gaia Hypothesis (Lovelock & Margulis 1974; Lovelock 1979, 1989), which holds that living organisms on Earth actively regulate atmospheric composition and climate in the face of challenges such as the increasing luminosity of the Sun. Versions of the Gaia Hypothesis range from "strong Gaia," which proposes that life maintains planetary conditions at some "optimum" for living things, to "weak Gaia," which maintains that biological processes and feedbacks affect global climate (see Schneider & Boston 1991). Although a compelling case has by no means been made for the "strong"—planetary homeostasis—version of Gaia, most earth scientists would accept a "weak" version—that biological processes and feedbacks are important (see Schneider & Boston 1991). It is worth noting that Krumbein & Schellnhuber (1990) estimated that the total mass of all organisms that have ever lived on Earth as ranging from 2.5×10^{26} to 2.5×10^{33} g, or from 10% to 10^5 times the mass of Earth itself. This in itself suggests that the biosphere cannot be ignored as a major player in cycling energy and matter during the history of the planet.

Although many geochemists believe carbon dioxide variations over geologic time can be explained primarily through abiotic interactions such as rock weath-ering (e.g. Holland et al 1986), such explanations are increasingly challenged by advocates of Gaia, or Gaian-type feedback mechanisms (Lovelock 1979, 1989; see also Schneider & Boston 1991). For example, the weathering rate feedback, in which CO_2 is removed as it builds up and causes planetary warm-ing, is now understood to be accelerated by organisms (Volk 1987). Important changes apparently took place ~3 billion years ago when the first soil-forming microorganisms created an environment in which CO_2 concentrations in contact with weathering rocks could be greatly elevated (Schwartzman & Volk 1989), and ~100 million years ago when angiosperms replaced gymnosperms as the dominant plant group, possibly causing an increase in weathering rates through higher rates of root respiration and increased levels of soil pCO_2 (Volk 1989a).

At the same time, other forms of life may be accelerating the degassing of CO_2 from carbonate rock, and, hence, increasing the CO_2 supply to the atmosphere. For example, prior to the Cretaceous (~140 million years ago), most carbonate accumulated in shallow-water environments and open ocean carbonate accumulation was a minor component in the global carbon cycle (Boss & Wilkinson 1991). The evolution of calcareous plankton and their spread in the Cretaceous changed that situation. Since that time, there has been a gradual shift from shallow-water to deep-water carbonate accumulation (Opdyke & Wilkinson 1988). Deep-water carbonate is transported rapidly to subduction zones; hence, there may already be enhanced CO_2 degassing due to the subduction-zone metamorphism of the tiny shells of open ocean organisms (Volk 1989b; Caldeira 1991, 1992).

Other Gaian mechanisms that have been invoked to produce a planetary homeostasis include the emission and uptake of greenhouse gases such as CO_2 and methane by life in the ocean and in soils, and the release of dimethyl sulfide gas (DMS), which provides cloud condensation nuclei for clouds (Charlson et al 1987), by ocean plankton. An increase in cloud condensation nuclei can cause clouds to become more reflective, thus increasing the planetary albedo and cooling the Earth. Cloud effects on solar UV have also been noted. Mass extinctions of ocean plankton, such as occurred at major geological boundaries, should decrease DMS and cloud condensation nuclei, causing drastic climate warming (Rampino & Volk 1987), and some evidence of this has been found in the record.

Gaian mechanisms might go further and affect the inner workings of Earth, and thus impact the carbon cycle by changing the major driving force behind volcanic outgassing and the return of volatiles to the mantle (so-called ingassing). For example, an increase in volatiles (H_2O, CO_2) in the upper mantle lowers the kinematic viscosity, allowing more rapid convection (McGovern & Schubert 1989). Periods when volatiles were more efficiently ingassed into the mantle should be followed by increased convection, and presumably faster sea-floor spreading rates. Such a situation may have arisen in the Cretaceous Period of Earth history (140 to 65 million years ago). As we mentioned, the evolution and diversification of calcareous plankton in the Early Cretaceous caused a significant increase in carbonate deposited on ocean crust. This carbonate was rapidly recycled by subduction, with a fraction going back into the mantle as CO_2 (in addition to water from hydrated ocean crust and sediments).

Such an increase in mantle volatiles is predicted to decrease mantle viscosity, possibly leading to a period of increased spreading rates on ocean ridges, increased subduction, and volcanic outgassing of CO_2. Mid-Cretaceous (120 to 80 million years ago) spreading rates were exceptionally high (Larson 1991). The increased convection would act to cool the mantle, causing an increase in viscosity, and hence a slowdown in convection and sea-floor spreading. The pulse of activity could be self-limiting, and might run in cycles tens to hundreds of millions of years long.

Another possibility is that since the viscosity of the mantle in the vicinity of the subduction zones would be most dramatically decreased with the arrival of the subducted carbonate, the subducting slabs may begin to sink more rapidly. If slab pull is the determining force for plate motion (Kearey & Vine 1990), ocean crust production rates might increase quite rapidly. Thus, strangely enough, the evolution of calcareous plankton may have led to increases in ocean crustal production rates.

CONTINUOUSLY HABITABLE ZONES

The habitable zone (HZ) around stars has been defined as that region in which planetary temperatures are neither too high nor too low for life to develop, whereas the continuously habitable zone (CHZ) around a main sequence star is defined as the region around the star within which planetary temperatures remain within the temperature constraints for habitability, taking into consideration the evolution of the star's luminosity (Hart 1978, 1979). (Others have defined habitability in human terms, suggesting that habitable planets are those on which large numbers of people can live comfortably, e.g. Dole & Asimov 1964). Using computer simulations of Earth's atmospheric composition and surface temperatures, Hart (1978, 1979) considered solar luminosity changes, variations in Earth's albedo, atmospheric pressure and composition, the greenhouse effect, variation in biomass, and a variety of geochemical processes, in order to calculate CHZs around main sequence stars.

Hart's finding of a relatively narrow CHZ around the Sun (0.95 AU to 1.01 AU) during its lifetime, and even narrower CHZs around other stars, prompted responses from planetary scientists and climate modelers (Owen et al 1979, Schneider & Thompson 1980). Hart seemed to be reducing an important variable in the Drake Equation to a very low number, and hence his results had important repercussions for the fledgling field of SETI (Search for Extraterrestrial Intelligence). Hart's modeling has been criticized on the basis of a number of his assumptions that govern the sensitivity of his climate model (e.g. the composition of the early atmosphere, that the fraction of the planet covered by clouds is proportional to the total mass of water vapor in the atmosphere, and that a strong negative feedback exists between the amount of clouds and surface temperatures), as well as for his particular scenarios of planetary and atmospheric evolution.

Recently, the question of habitable zones around main sequence stars has been taken up by Kasting et al (1993). The inner edge of the HZ is determined in their model by the loss of water through photolysis and hydrogen escape. The outer edge of the HZ is determined by the formation of CO_2 clouds that cool a planet's surface by increasing its albedo and by lowering the convective lapse rate. Conservative estimates in the Solar System for these distances are 0.95 and 1.37 AU. The width of the HZ is slightly greater for planets that are larger than Earth and which have higher N_2 partial pressures in their atmospheres. Climate stability between these two limits is controlled by the weathering rate/temperature feedback mechanism, in which higher temperatures lead to higher weathering rates of silicate rock, which lowers atmospheric CO_2 and hence lowers surface temperature and weathering rate, and vice versa.

During the evolution of the Sun, the HZ evolves outward (Figure 7) as the Sun's luminosity increases. Kasting et al (1993) conservatively estimated that the CHZ over 4.6 billion years ranges from 0.95 to 1.15 AU (without cold starts—that is, a planet that starts out cold, with clouds of frozen CO_2, probably cannot escape global glaciation), or a width of 0.2 AU, which is much greater than Hart's (1978 and 1979) estimates of 0.06 and 0.046 AU, respectively.

Kasting et al (1993) have looked at CHZs around stars from 0.5 to $1.5M_\odot$ (stars more massive than $1.5M_\odot$ have lifetimes too short to be interesting from a habitability standpoint), and found that all of the calculated CHZs are 4 to 20 times wider than those calculated by Hart (1979). Planets around late K and M type stars may not be habitable as they can become trapped in synchronous rotation as a result of tidal damping, although mid to early K stars provide another possible location for habitable planets in addition to G stars like the Sun. This has important ramifications for SETI.

THE FATE OF THE EARTH

When will solar luminosity be so high that liquid water can no longer exist on Earth's surface? To answer this question, some estimate needs to be made regarding the future atmospheric content of greenhouse gases. Lovelock & Whitfield (1982) proposed that atmospheric CO_2 content will tend toward zero in approximately 10^8 yr from now, due to a CO_2-weathering feedback of the type proposed by Walker, Hays & Kasting (1981).

Photosynthetic organisms, at the base of the food chain, extract carbon from the atmosphere and from dissolved CO_2. These carbon reducers, dependent on an adequate atmospheric pCO_2 for their survival, are the source of carbon for the rest of the biosphere. Hence, atmospheric pCO_2 reductions brought about by increased solar luminosity could effectively cut off the carbon flux to the biosphere. At present-day geologic CO_2 degassing rates, this could happen in less than 1 billion years, unless organisms evolve to more efficiently exchange CO_2 with the atmosphere (Caldeira & Kasting 1992b).

The shift of carbonate deposition to the deep oceans with the appearance of calcareous plankton about 140 million years ago has already been mentioned. If subducting deep-ocean carbonate efficiently degasses its CO_2 to the atmosphere or ocean, then a complete transfer of the sedimentary carbonate mass to the pelagic realm could increase the CO_2 supply and chemical weathering rates by an order of magnitude. This might mean that thermal limits and/or loss of water, and not CO_2-starvation may ultimately limit the lifespan of the biosphere.

However, even without this biologically mediated accelerated CO_2-degassing, atmospheric CO_2 content would probably not go all the way to zero. Model cal-

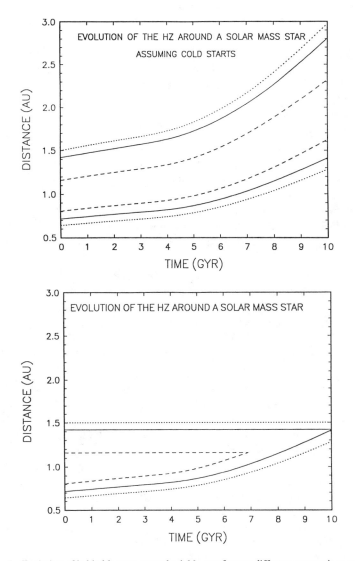

Figure 7   Evolution of habitable zone around a 1 $M_\odot$ star for two different assumptions concerning the possibility of "cold starts": (*top*) cold starts permitted, (*bottom*) no cold starts allowed. The term cold start refers to whether or not a planet will warm up once the stellar luminosity increases to the appropriate critical values (see text). The three pairs of curves correspond to the different habitability estimates discussed in the text: long dashes—water loss and first CO_2 cloud condensation limits (most conservative); solid curves—runaway and maximum greenhouse limits; dotted curves—recent Venus and early Mars limits (most optimistic). (After Kasting et al 1993.)

culations by Caldeira & Kasting (1992b) indicated that when the Earth's mean temperature becomes warm (> 300 K), silicate dissolution may proceed more rapidly than CO_2 and H_2SO_4 are supplied to the atmosphere by volcanism. This would leave an excess of Ca^{2+} and Mg^{2+} cations that could not be precipitated as carbonates or sulfates, which would probably precipitate in the oceans in a variety of silicate phases. Geologic carbon inputs to the ocean and atmosphere must be largely balanced by carbonate sedimentation, suggesting that the oceans would continue to be saturated with respect to some carbonate phase.

If future ocean waters are in approximate chemical equilibrium with silicates and carbonates, ocean pH would be buffered close to its present value. If ocean pH increases by no more than half a pH unit and the oceanic Ca^{2+} concentration changes by no more than an order of magnitude, then application of carbonate equilibria indicates that atmospheric pCO_2 would be reduced no more than two orders of magnitude from today's value. Even if ocean chemistry is not this well buffered, Henry's law implies that more CO_2 will be partitioned into the atmosphere on a warmer Earth, so the atmosphere may still contain up to a few ppm of CO_2.

According to the calculations, if the atmospheric pCO_2 is taken to be 1 ppm in the distant future, Earth's mean global surface temperature approaches 80°C in about 1.5 billion years. If there is more CO_2 in the atmosphere, this temperature could be reached sooner. As the surface temperature approaches $\sim$80°C, the stratospheric H_2O mixing ratio reaches $\sim$2.5%. Above this mixing ratio, the loss of hydrogen to space is limited by the solar EUV heating rate to $\sim 6 \times 10^{11}$ atoms $H\,cm^{-2}\,s^{-1}$ (Watson, Donahue & Walker 1981), giving a time scale for ocean loss of $\sim$1 billion years. Hence, liquid water on Earth could be completely eliminated $\sim$2.5 billion years from now. This would bring terrestrial life to its final close. After Earth's water is lost, silicate-rock weathering will cease; hence, volcanic CO_2 should accumulate in the atmosphere, creating a climate much like that of Venus.

ANCIENT GLACIAL DEPOSITS AND EARLY CLIMATE

Much of the climate and carbon-cycle modeling related to the long-term evolution of the Earth's climate uses paleoclimate information as constraints (e.g. Kasting 1989). A major constraint has been the occurrence of ancient glaciations, including an Early Proterozoic (Huronian) Glaciation at $\sim$2.3 billion years ago, and the widespread Late Proterozoic glaciations between about 1 billion years and 600 million years ago (Hambrey & Harland 1981). During glacial epochs, T_s for Earth could not have been more than 20°C (293 K) (Kasting 1989). However, these glacial periods occur at times when many models predict high atmospheric pCO_2, and thus warm planetary temperatures (Kasting 1989). The Huronian glaciation has been a particular problem, as most models of atmospheric composition at 2.3 billion years ago predict large amounts of atmospheric CO_2, making such an early glacial epoch difficult to produce (Kasting 1989). Evidence for the Late Proterozoic glaciations occurs in many parts of the world, some at sites that are reconstructed at low latitudes, suggesting a worldwide glaciation (Hambrey & Harland 1981).

All of this, however, rests on the correct identification of these periods as times of glaciation. Recent studies suggest that some of the sedimentary deposits that represent the primary evidence for glaciation may have been misidentified (Rampino 1992, 1994; Oberbeck et al 1993). The evidence of past glaciations is primarily in the form of so-called tillites—typically poorly sorted sedimentary rocks, with boulders in a fine-grained matrix, interpreted as analogues of the glacial deposits (till) of the more recent ice ages. Diagnostic criteria include faceted and striated stones, striated rock pavements, and laminated siltstones containing large "ice-rafted" clasts. Late Proterozoic tillites in Australia are up to 6,000 m thick (Hambrey & Harland 1981).

The first clue that something might be amiss was the fact that essentially all studies based on modern sedimentological and stratigraphic analysis have concluded that many recognized "tillites" were not deposited passively by melting glaciers, but are in reality debris-flow and related mass-flow deposits, with a significant component from fallout of coarse material [some deposits have been called "debris-rain" sediments (Rampino 1994)].

Based on these analyses, tillites have generally been reinterpreted as debris-flow deposits formed where glaciers deposited material into the sea or large lakes. The current models of sedimentation envisions these debris-flow sequences as generated directly by rapid dumping of debris at a marine ice front, or by deposition of unstable sedimentary accumulations on the shallow sea floor by iceberg rainout, with later re-sedimentation in submarine debris flows. Melting icebergs are inferred to have provided additional rainout of coarse material.

Re-examination of the characteristics of these deposits, however, shows that they apparently are similar in distribution and textures to the expected deposits of ballistic ejecta of large impacts (Oberbeck et al 1993, Rampino 1994). In ballistic sedimentation, debris ejected at high velocity from an impact site, and material produced by spall from the ground as the shock waves move outward, sweep rapidly away from the impact site(s). This ballistic debris strikes the ground at greater and greater velocity farther away from the impact, imparting an outward momentum to a mixture of target debris and entrained local material, which moves rapidly outward as a ground-hugging debris flow. Local sediments are entrained within the fast-moving flows. For a 100-km crater on the moon (with no atmosphere) the horizontal component of velocity of the debris flows could be as great as ~ 300 m s^{-1}. On Earth, the largest ejecta would not be significantly affected by atmospheric drag. Known ejecta deposits on Earth are composed of poorly sorted debris containing abundant striated and faceted stones, resting in some places, on eroded and striated bedrock surfaces (Rampino 1994).

A number of climatic enigmas, such as the seeming conflict between an early CO_2-rich atmosphere and the Huronian glaciation, and the apparent worldwide extent of Late Proterozoic glaciations, might be solved if the tillite deposits associated with these ice ages, or at least some of them, are in reality impact debris (Oberbeck et al 1993, Rampino 1994).

TERRAFORMING

Terraforming, or planetary engineering—the transformation of non-habitable planets into habitable places—has received considerable scientific attention in recent years (e.g. Fogg 1989, McKay & Haynes 1990, McKay et al 1991). If Earth eventually becomes uninhabitable, then terraforming other planets may represent an alternative for life. Mars is the most likely candidate for terraforming, and a number of possible schemes for re-engineering that planet's atmosphere and climate have been suggested (McKay et al 1991). An early NASA study (prior to analysis of *Viking* mission data) suggested that the greenhouse warming potential of the CO_2 in the Martian polar caps could raise planetary temperatures above freezing (Averner & MacElroy 1976). It was proposed, for example, that a reduction in the albedo of the polar regions—by covering them with low-albedo dust or dark plants—could release significant amounts of CO_2 into the atmosphere. However, we now know that the NASA terraformers overestimated the CO_2 mass of these caps available for greenhouse warming—the "dry ice" caps at Mars' poles are almost entirely volatilized and recondensed each Martian year (Tillman et al 1993).

McKay et al (1991) have made the most extensive study of terraforming Mars. They suggest that if a large amount of volatiles exist on Mars in subsurface reservoirs, planetary engineering schemes could exploit runaway effects involving release of water and CO_2 from the polar caps and regolith. McKay et al (1991) estimated that an amount of CO_2 adsorbed onto surface rocks equivalent to ~300 mbar could be released into the Martian atmosphere. However, the amount of carbon dioxide that exists either as adsorbed CO_2, or in carbonate rocks is poorly known, and further exploration of Mars is required to provide additional information on global inventories of volatiles such as CO_2, H_2O, and N_2.

In a speculative study, Lovelock & Allaby (1984) suggested the creation of a more habitable Mars by using chlorofluorocarbons as greenhouse gases. In their scenario, CFCs could be sent from Earth in missiles. However, the authors may have significantly underestimated the mass of CFCs needed, and a better approach might be to manufacture on Mars new greenhouse molecules engineered to strongly and broadly absorb infrared radiation at very low concentrations (McKay et al 1991).

With a long-term commitment, terraforming of planetary atmospheres could become a future goal of humanity (e.g. Savage 1993). As McKay et al (1991) note "If . . . investigations indicate that it is feasible to make Mars habitable, the motivation to do so may depend on the potential for life (human and non-human) on Mars, the economic payoff, and the nature of other large-scale efforts that provide habitats in space."

CONCLUSIONS

The Goldilocks Problem touches on many of the latest developments relating to the evolution of planets and their atmospheres. Recent work has stressed the early evolution of the atmospheres of Venus, Earth, and Mars, and the light that is shed on the probability of habitable planets around main sequence stars.

The two key factors relating to the habitability of a planet seem to be the size of the planet and its distance from the Sun, although the conclusions are not straightforward. Venus is too close to the Sun. The planet underwent a runaway greenhouse, or moist greenhouse, early in its history. In the runaway greenhouse scenario, the relatively high temperatures on early Venus would have prevented water from condensing, creating a dense H_2O atmosphere and intense greenhouse effect. The lack of liquid water prevented the silicate-rock weathering reactions that act to remove CO_2 gas from the atmosphere and store it as solid calcium carbonate on the planet's surface, allowing CO_2 to build up to present high levels. Water was lost as the high Venus temperatures allowed H_2O vapor to reach high altitudes where the H_2O molecules were broken by solar UV, and escaped from the atmosphere driven by solar extreme UV radiation, sealing Venus' fate.

In the moist greenhouse scenario, liquid water was able to condense on early Venus, but the low lapse rate of the warm Venus atmosphere allowed water vapor mixing ratios to remain high at great altitudes, facilitating the breakup and escape of H_2O by solar UV. As in the runaway greenhouse model, the loss of water allowed the buildup of the present massive CO_2-rich Venus atmosphere.

Mars is too small. Although apparently experiencing warm temperatures and the presence of liquid water early in its history, as the small planet lost its internal heat, its internal geologic engines shut down. Mars lost the ability (if it ever had it) to recycle carbon and release volcanic CO_2 to provide a replenishment of atmospheric CO_2, and thus maintain a significant greenhouse effect. The planet became a frozen world.

Earth was apparently lucky, but more than blind luck was involved. The relatively large size of the planet has kept its internal heat from leaking away too rapidly, and plate-tectonic activity has maintained the recycling of carbon dioxide. Earth is far enough away from the Sun so that early temperatures, probably maintained by a CO_2-rich atmosphere, remained within the range of liquid water. The water-mediated silicate rock weathering reactions have been effective in removing CO_2 from the atmosphere, with a built-in temperature-dependence feedback, so that as the Sun's luminosity increased, the resulting warmer surface temperatures on Earth caused weathering rates to increase, thereby removing more of the atmospheric CO_2, and cooling the planet's surface. Thus, a balance has been maintained, probably with the help of additional feedbacks related to life. However, as the Sun's luminosity continues to increase, biological and abiological mechanisms for homeostasis may fail, and recent model calculations suggest a hot, dry, uninhabitable Earth within about a billion years.

Recent estimates of the continuously habitable zones around stars like the Sun suggest a width of ~0.2 AU (from 0.95 to 1.15 AU), considerably more optimistic than previous estimates of a continuously habitable zone only about 25% of that width. This has positive implications for the distribution of life in the Cosmos, and the search for extraterrestrial intelligence.

ACKNOWLEDGMENTS

We thank M. I. Hoffert and T. Volk for discussions and information, and J. F. Kasting for a critical review.

Literature Cited

Abe Y, Matsui T. 1988. *J. Atmos. Sci.* 45:3081–101

Ahrens TJ. 1993. *Annu. Rev. Earth Planet. Sci.* 21:525–55

Averner MM, MacElroy RR, eds. 1976. *On the Habitability of Mars. NASA SP414.* Washington DC:US Govt. Print. Off.

Baker VR, Strom RG, Gulick VC, Kargel JS, Komatsu G, Kale VS. 1991. *Nature* 352:589–94

Barnola JM, Raynaud D, Korotkevich YS, Lorius C. 1987. *Nature* 329:408–14

Berner RA. 1993. *Science* 261:68–70

Berner RA, Lasaga AC, Garrels RM. 1983. *Am. J. Sci.* 283:641–683

Boss SK, Wilkinson BH. 1991. *J. Geol.* 99:497–513

Budyko MI. 1969. *Tellus* 21:611–19

Cahalan RF, North GR. 1979. *J. Atmos. Sci.* 36:1178–88

Caldeira K. 1991. *Geology* 19:204–6

Caldeira K. 1992. *Nature* 357:578–81

Caldeira K, Kasting JF. 1992a. *Nature* 359:226–28

Caldeira K, Kasting JF. 1992b. *Nature* 360:721–23

Caldeira K, Rampino MR. 1991. *Geophys. Res. Lett.* 18:987–90

Carr MH. 1989. *Icarus* 79:311–27

Carr MH, Wanke H. 1992. *Icarus* 98:61–71

Charlson RJ, Lovelock J, Andreae MO, Warren SG. 1987. *Nature* 326:655–61

Chyba CF. 1990. *Nature* 343:129–33

Dayhoff MO, Eck R, Lippincott ER, Sagan C. 1967. *Science* 155:556–57

DeBergh C, Bezard B, Owen T, Crisp D, Maillard J-P, Lutz BL. 1991. *Science* 251:547–49

Dole SH, Asimov I. 1964. *Planets for Man.* New York:Random House

Donahue TM, Hartle RE. 1992. *Geophys. Res. Lett.* 19:2449–52

Donahue TM, Hodges RR Jr. 1992. *J. Geophys. Res.* 97:6083–90

Donahue TM, Hoffman JH, Hodges RR Jr, Watson AJ. 1982. *Science* 216:630–33

Dreibus G, Wanke H. 1987. *Icarus* 71:225–40

Durham R, Chamberlain JW. 1989. *Icarus* 77:59–66

Eberhardt P, Hodges RR, Krankowsky D, Berthelier JJ, Schultz W, et al. 1987. *Lunar Planet. Sci.* XVIII:252–53

Fogg MJ. 1989. *J. Brit. Interplanet. Soc.* 42:577–82

Gilliland RL. 1989. *Palaeogeogr. Palaeoclimatol. Palaeoecol. (Global and Planet. Change Sect.)* 75:35–55

Gough DO. 1981. *Sol. Phys.* 74:21–34

Gold T. 1964. In *The Origin and Evolution of Atmospheres and Oceans,* ed. PJ Brancazio, AGW Cameron, pp. 249–56. New York:Wiley

Goody RM, Walker JCG. 1972. *Atmospheres.* Englewood Cliffs, NJ:Prentice-Hall

Greeley R. 1987. *Science* 236:1653–54

Grinspoon DH. 1987. *Science* 238:1702–4

Grinspoon DH. 1993. *Nature* 363:428–31

Grinspoon DH, Lewis JS. 1988. *Icarus* 74:430–36

Gurwell MA, Yung YL. 1992. *Planet. Space Sci.* 41:91–104

Hambrey MJ, Harland WB, eds. 1981. *Earth's Pre-Pleistocene Glacial Record.* Cambridge:Cambridge Univ. Press

Hambrey MJ, Harland WB. 1985. *Palaeogeogr. Palaeoclimatol. Palaeoecol.* 51:255–72

Hansson A. 1991. *Mars and the Development of Life.* New York:Ellis Horwood

Hart MH. 1978. *Icarus* 33:23–39

Hart MH. 1979. *Icarus* 37:351–57

Head JW, Crumpler LS, Aubele JC, Guest JE, Saunders RS. 1992. *J. Geophys. Res.* 97:13,153–97

Henderson-Sellers A. 1978. *The Origin and Evolution of Planetary Atmospheres.* London:Adam Hilger

Henderson-Sellers A, Cogley JG. 1982. *Nature* 298:832–35

Holland HD. 1978. *The Chemistry of the Atmosphere and Oceans.* New York:Wiley Intersci.

Holland HD, Lazar B, McCaffrey M. 1986. *Nature* 320:27–33

Hoyle F. 1955. *Frontiers in Astronomy.* London:Heinemann

Hunten DM. 1973. *J. Atmos. Sci.* 30:1481–94

Hunten DM. 1993. *Science* 259:915–20

Ingersoll AP. 1969. *J. Atmos. Sci.* 26:1191–98

Jambon A, Zimmerman JL. 1990. *Earth Planet. Sci. Lett.* 101:323–31

Kahn R. 1985. *Icarus* 62:175–90

Kasting JF. 1988. *Icarus* 74:472–94

Kasting JF. 1989. *Palaeogeogr. Palaeoclimatol. Palaeoecol. (Global and Planet. Change Sect.)* 75:83–95

Kasting JF. 1991a. In *Planetary Sciences, American and Soviet Research,* ed. TM Donahue, KK Trivers, DM Abramson, pp. 234–45 Washington, DC:Natl. Acad. Press

Kasting JF. 1991b. *Icarus* 94:1–13

Kasting JF, Ackerman TP. 1986. *Science* 234:1383–85

Kasting JF, Grinspoon DH. 1991. In *The Sun in Time,* ed. CP Sonett, MS Giampapa, MS Matthews, pp. 447–62 Tucson:Univ. Ariz. Press

Kasting JF, Pollack JB. 1983. *Icarus* 53:479–508

Kasting JF, Pollack JB, Ackerman TP. 1984. *Icarus* 57:335–55

Kasting JF, Toon OB, Pollack JB. 1988. *Sci. Am.* 258(2):48–53

Kasting JF, Whitmire DP, Reynolds RT. 1993. *Icarus* 101:108–28

Kearey P, Vine FJ. 1990. *Global Tectonics.* Oxford:Blackwell

Komabayashi M. 1967. *J. Meteorol. Soc. Jpn.* 45:137–39

Krumbein WE, Schellnhuber H-J. 1990. In *Facets of Modern Biogeochemistry,* ed. V Ittekkot, S Kempe, W Michaelis, A Spitzy, pp. 5–22. Berlin:Springer-Verlag

Kuhn WR, Atreya SK. 1979. *Icarus* 37:207–13

Larson RI. 1991. *Geology* 19:963–66

Lewis J. 1982. *Icarus* 16:241–52

Lovelock JE. 1979. *Gaia.* Oxford:Oxford Univ. Press

Lovelock JE. 1989. *The Ages of Gaia.* Oxford:Oxford Univ. Press

Lovelock JE, Allaby M. 1984. *The Greening of Mars.* New York:Warner

Lovelock JE, Margulis L. 1974. *Tellus* 26:2–9

Lovelock JE, Whitfield M. 1982. *Nature* 296:561–63

Marshall HG, Walker JCG, Kuhn WR. 1988. *J. Geophys. Res.* 93:791–801

Matsui T, Abe Y. 1986a. *Nature* 319:303–5

Matsui T, Abe Y. 1986b. *Nature* 322:526–28

McElroy MB, Prather MJ, Rodriquez JM. 1982. *Science* 215:1614

McElroy, MB Kung T-Y, Yung YL. 1977. *J. Geophys. Res.* 82:4379–88

McElroy MB, Yung YL, Nier AO. 1976. *Science* 194:70–72

McGovern PJ, Schubert G. 1989. *Earth Planet. Sci. Lett.* 96:27–37

McKay CP. 1991. *Icarus* 91:93–100

McKay CP, Davis WL. 1991. *Icarus* 90:214–21

McKay CP, Haynes RH. 1990. *Sci. Am.* Dec:109–16

McKay CP, Toon OB, Kasting JF. 1991. *Nature* 352:489–96

Nakajima S, Hayashi Y-Y, Abe Y. 1992. *J. Atmos. Sci.* 49:2256–66

Newman MJ, Rood RT. 1977. *Science* 198:1035–37

Oberbeck VR, Marshall JR, Aggarwal H. 1993. *J. Geol.* 101:1–19

Ogawa M. 1993. *Icarus* 102:40–61

Opdyke BN, Wilkinson BH. 1988. *Paleoceanogr.* 3:685–703

Owen T, Cess RD, Ramanathan V. 1979. *Nature* 277:640–42

Parker TJ, Gorsline DS, Saunders RS, Pieri DC, Schneeberger DM. 1993. *J. Geophys. Res.* 98:11,061–78

Pepin RO. 1991. *Icarus* 92:2–79

Pollack JB. 1971. *Icarus* 14:295–306

Pollack JB. 1991. *Icarus* 91:173–98

Priem HNA. 1990. *Geol. Mijnbouw* 69:391–406

Prinn RG, Fegley B Jr. 1987. *Annu. Rev. Earth Planet. Sci.* 15:171–212

Rampino MR. 1992. *Eos, Trans. Am. Geophys. Union (Suppl)* 73:99

Rampino MR. 1994. *J. Geol.* In press

Rampino MR, Volk T. 1987. *Nature* 332:63–65

Rasool SI, DeBergh C. 1970. *Nature* 226:1037–39

Ronov AB, Yaroshevsky AA. 1976. *Geochem. Int.* 12:89–121

Sackman I-J, Boothroyd AI, Fowler WA. 1990. *Astrophys. J.* 360:727–36

Sagan C. 1960. *JPL Tech. Rep. No. 32-34*

Sagan C, Mullen G. 1972. *Science* 177:52–56

Savage MT. 1993. *The Millennial Project.* Denver:Empyrean

Schneider SH, Boston P, eds. 1991. *The Science of Gaia.* Cambridge:MIT Press

Schneider SH, Gal-Chen T. 1973. *J. Geophys. Res.* 78:6182–94

Schneider SH, Thompson SL. 1980. *Icarus* 41:456–69

Schwartzman D, Volk T. 1989. *Nature* 340:457–60

Sellers WD. 1969. *J. Appl. Meteorol.* 8:392–400

Tajika E, Matsui T. 1992. *Earth Planet. Sci. Lett.* 113:251–66

Tillman JE, Johnson NC, Guttorp P, Percival DB. 1993. *J. Geophys. Res.* 98:10,963–71

Volk T. 1987. *Am. J. Sci.* 287:763–79

Volk T. 1989a. *Geology* 17:107–10

Volk T. 1989b. *Nature* 337:637–40

Walker JCG. 1975. *J. Atmos. Sci.* 32:1248–56

Walker JCG. 1977. *Evolution of the Atmosphere.* New York:McMillan

Walker JCG, Hays PB, Kasting JF. 1981. *J. Geophys. Res.* 86:9776–82

Walter M. 1979. *Am. Sci.* 67:142

Watson AJ, Donahue TM, Walker JCG. 1981. *Icarus* 48:150–66

Wetherald RT, Manabe S. 1975. *J. Atmos. Sci.* 32:2044–59

Wetherill GW. 1986. In *Origin of the Moon,* ed. WK Hartmann, RJ Phillips, GJ Taylor, pp. 519–50. Houston:Lunar Planet. Inst.

Zahnle K, Kasting JF, Pollack JB. 1988. *Icarus* 74:62–97

Annu. Rev. Astron. Astrophys. 1994. 32: 115–52

PHYSICAL PARAMETERS ALONG THE HUBBLE SEQUENCE

Morton S. Roberts

National Radio Astronomy Observatory,[1] Charlottesville, Virginia 22903

Martha P. Haynes

Center for Radiophysics and Space Research and National Astronomy and Ionosphere Center,[2] Cornell University, Ithaca, New York 14853, and National Radio Astronomy Observatory, Green Bank, West Virginia 24944

KEY WORDS: galaxies, physical properties, morphological type, mass, star formation

1. INTRODUCTION

One of the remarkable aspects of galaxies is that they can be classified into relatively few categories. The well-ordered sequence of galaxy types appears to offer a clue to possible formation and evolutionary processes. It is thus not surprising that morphology is so frequently an underlying theme in the study of galaxies, and serves as the principal subject of this review. Excellent discussions of recent classification systems are given by Buta (1992a,b). For other reviews with historical references, see de Vaucouleurs (1959) and Sandage (1975).

Hubble (1926, 1936) introduced an early scheme to categorize galaxies; its concepts are still in use. In its simplest form, three basic types are recognized: ellipticals, spirals, and irregulars. Most modern schemes try to employ multiple classification criteria. There are two systems in common use today, both similar in application and notation, and both derived from Hubble's original classification scheme. One is the Hubble system as detailed by Sandage (1961)

---

[1] The National Radio Astronomy Observatory is operated by Associated Universities, Inc., under a cooperative agreement with the National Science Foundation.

[2] The National Astronomy and Ionosphere Center is operated by Cornell University under a cooperative agreement with the National Science Foundation.

115

0066–4146/94/0915–0115$05.00

(Sandage & Tammann 1987, Sandage & Bedke 1993). The other system, developed by de Vaucouleurs (1959), adds more descriptive details to the notation and extends Hubble's original spiral sequence beyond Sc. Because the application of this system to over 20,000 galaxies in the Third Reference Catalog of Bright Galaxies (RC3, de Vaucoulers et al 1991) has given it wide usage, it is adopted here. A few percent of all galaxies are unclassifiable. Many of these have unusual morphology because they are interacting systems. For the current purpose of looking for trends among average galaxies, we exclude these peculiar objects from our discussion.

Although the criteria for a type assignment are well recognized, the process is in reality subjective. Rather, we seek to replace qualitative measures with quantitative ones and ultimately to uncover the physics underlying galactic structure. As (we hope) will become evident, various trends do exist, but regardless of the parameters, the dispersion within each type is always large, much more so than errors of measurement. One of these trends, that of color with type, has been long recognized (Hubble 1936). Others, e.g. the H_2 content, are only now being evaluated.

Most recently, numerous authors have attempted to classify galaxies using multivariate analysis of available quantitative measures (Whitmore 1984, Watanabe et al 1985). Such quantitative studies show the existence of two principal categories of galaxy parameters: those that measure the absolute scale (size, luminosity, mass) and those that describe more its form (morphology). Because he undertook his classification scheme at a time when distance estimates were available for only a handful of galaxies, Hubble could not discriminate the scale dimension: that intrinsically bright galaxies are bigger and more massive than faint ones of the same morphology.

With notable exceptions and for obvious reasons the study of galaxies is directed primarily to those listed in catalogs, e.g. the New General Catalog (NGC), the Shapley-Ames Survey and its update (RSA), the Uppsala General Catalogue (UGC), and the various editions of the Reference Catalogues (RC). They are all flux- or diameter-limited. They all suffer from Malmquist bias and they are all deficient in low surface brightness systems. They list the brighter galaxies though generally not the most or least luminous systems. And they mostly contain nearby galaxies, i.e. $z \ll 0.1$. These are systems for which we have the most data, and it is such "catalog galaxies" that we discuss here. We also construct an approximately volume-limited sample; mean values of intrinsic properties, e.g. linear diameter, absolute magnitude, so derived are generally smaller because of Malmquist bias, but show trends with type similar to those for the biased samples.

The ability to measure properties is not equal for all types. Thus HI is rare in elliptical systems, and we lack any meaningful HI parameters for these systems. Similarly, total mass estimates for Es, when available, are based on different approaches and assumptions than for spirals. For the latter, rotational velocities are derivable from 21 cm HI observations, while the most common kinematic parameter measured for ellipticals is the central velocity dispersion (see Whitmore et al 1985 for a catalog.) Comparison of total mass estimates between Es and spirals are accordingly uncertain and are not made. Other properties such as CO and its derivative, H_2 content, or X-ray luminosity are available only for relatively small samples and suffer accordingly. For X rays, the best that can be done at present is to contrast data for ellipticals with those for the overall spiral category; we note interesting differences.

The most obvious omission is the lack of any distinction between regular and barred spirals. Here we have been guided by Holmberg's (1958) remarks in his classic paper on the photometry of galaxies. He notes:

> ... that the majority of spiral nebulae exhibit a more or less pronounced bar; the bar may not always be recognizable on the blue plate, but is usually visible on the photovisual exposure. It seems quite possible that a bar is a structural detail common to all, or most, spiral nebulae and that the observed differences are of a quantitative rather than qualitative nature.

This is strikingly illustrated for M51 (Zwicky 1957, Figure 41) by means of a composite of yellow and blue sensitive images. In this composite M51, a classic "regular" spiral shows a small but pronounced bar in its central region.

Some caveats must be emphasized. There are many type-dependent trends to be found in the literature, based on widely ranging sample sizes of various levels of confidence. We are unable to discuss all or even a significant fraction of these. Since our focus is on type-dependencies, other and possibly related issues are treated only briefly. We apologize at the onset if your favorite relationship is omitted and hope that the references will guide the reader to more extensive discussions on these topics.

2. OVERVIEW: THE MORPHOLOGICAL DEPENDENCE OF FUNDAMENTAL PROPERTIES

In an early study of the integral properties of galaxies, Roberts (1969) analyzed 98 spiral and irregular galaxies for which total mass, neutral hydrogen content, luminosity, color, and radius were available. Over the past several decades, the database of observed quantities for extragalactic objects has expanded at an enormous rate. The availability today of large catalogs of galaxies and compilations of data in digital form makes statistical and graphical analysis possible as it has not been before. In preparing this review, we have drawn upon such catalogs to explore several of the morphological-dependence issues.

In this section, we present the results of our own analysis which we discuss in comparison with the findings of others in later sections.

Construction of Samples for Analysis

For the current purpose, we make use of two primary compilations: first, the RC3 and second, a private catalog maintained by R. Giovanelli and M. Haynes that we refer to by its familiar name, the Arecibo General Catalog (AGC). The latter catalog primarily adds a significant body of HI line data including upper limits for non-detected objects, a variety of measurements of the 21 cm line width, and qualitative indicators of profile shape.

Currently, redshift surveys extend relatively deeper in the north than in the south (as visible in Figure 2 of Giovanelli & Haynes 1991). Because of the northern hemisphere bias in redshift survey depth, we use as the prime deep sample the compilation of objects that are included both in the RC3 and in the *Uppsala General Catalogue* (Nilson 1973). We refer to the sample of objects common to both catalogs as the "RC3-UGC sample." It should be noted that, because the RC3 is intended to be complete only for objects of high surface brightness, the lowest surface brightness objects are found only in the UGC, and are underrepresented in the current analysis. Likewise, the UGC, being angular-diameter limited, is biased against high surface brightness, compact objects and becomes incomplete for early-type galaxies especially at the larger distances.

When one selects galaxies of fixed flux, the volume element containing the more distant, intrinsically brighter objects is larger than that occupied by the nearer, intrinsically fainter population. This "Malmquist bias" affects all galaxy catalogs that are flux-limited. In order to examine (and counteract) the effects of Malmquist bias, we have also constructed a nearby volume-limited one that should be complete but has relatively fewer galaxies. Since the volume occupied by the Local Supercluster has been well-studied by most available multiwavelength techniques, we have identified 4972 RC3 objects with redshifts implying membership in the Local Supercluster, i.e. with $V_{LG} < 3000$ km s^{-1}. This subset is referred to as the "RC3-LSc sample." Note that it contains galaxies that are not in the RC3-UGC sample.

Since the backbone of our compilation is the RC3, the reader is referred to its first volume for an explanation of its contents. Our general philosophy has been to use all of the corrected parameters directly from the RC3 when available since its authors have gone to considerable length to reduce parameters obtained from different sources to a standard system. For the present comparative purposes, the consistency of approach is perhaps more critical than absolute prescription. Most parameters as detailed below have been taken directly from the RC3. Additional radial velocities, 21 cm parameters, and far infrared data from *IRAS* come from the AGC.

DISTANCES   In order to convert velocities to distances and to further calculate intrinsic parameters, it is necessary to adopt a value of the Hubble constant and a model of the local velocity field. Heliocentric velocities $V_\odot$ are taken from the AGC preferentially if a good quality 21 cm spectrum is available; otherwise, the available optical velocity is used. The velocity with respect to the Local Group V_{LG} was calculated by applying the standard correction to $V_\odot$ given in the RC3: $300 \sin l \cos b$. For objects in the Local Supercluster, a nonlinear infall model was used to calculate the distance to an object with the observed V_{LG}. The model adopted follows the outline of Schechter (1980) with the assumptions of a distance of 20.0 Mpc and an overdensity of 2 for Virgo and an infall velocity at the Local Group of 300 km s^{-1}. The Local Supercluster boundary is taken to be at $V_{LG} = 3000$ km s^{-1}. For more distant objects, distance is computed merely from the Hubble ratio using V_{LG} and is not referenced to any other frame. The assumptions that we have made are not intended to be an endorsement of any particular solution but are chosen for convenience. Most important is our emphasis on consistency. Throughout this paper, we adopt a Hubble constant H_o of 50 km s^{-1} Mpc^{-1}.

OPTICAL SIZE, LUMINOSITY, AND SURFACE MAGNITUDE   The linear size follows from the RC3 and the calculated distance. Likewise, the prescriptions outlined in the RC3 for correcting magnitudes are adopted along with a value of the solar absolute magnitude M$_B$ of $+5.48$. The surface magnitude Σ_B used here is defined simply as $\Sigma_B = B_T^0 + 2.5 \log ab$, with a equal to D$_{25}$ and b is the corresponding minor axis obtained from the RC3 axial ratio. Note that the area term here is different from that used below for surface densities.

NEUTRAL HYDROGEN MASS AND SURFACE DENSITY   The total neutral hydrogen mass M_{HI} in solar units is calculated from the integrated 21 cm line emission $M_{HI} = 2.36 \times 10^5 \, D^2 \int S dV$, where D is the distance in Mpc and $\int S dV$ is the HI line flux in Jy km s^{-1}. For objects for which only a value of the rms noise per velocity interval in the emission spectrum is available, the upper limit to M_{HI} is calculated assuming the emission is rectangular, of amplitude 1.5 times the rms noise and width equal to that expected for an Sa–Sb galaxy of similar luminosity, properly corrected for inclination. The latter relationship was derived from the detected objects. Objects showing emission confused with other sources or HI in absorption cannot be used properly in the analysis and have been ignored. The HI surface density, σ_{HI}, has been calculated as $M_{HI}/\pi R^2$ where R is the optical linear radius. Although the use of the optical area makes σ_{HI} a hybrid quantity, Hewitt et al (1983) have shown that on average, the HI and optical sizes scale linearly. Most authors use the quantity σ_{HI} or some variant thereof as the indicator of HI content in comparative studies.

FAR INFRARED LUMINOSITY AND SURFACE DENSITY   The far infrared luminosity is derived from the fluxes measured by *IRAS* in the 60 and 100 micron bands

$F_{FIR}(Jy) = 2.58\, F_{60\mu} + F_{100\mu}$, as $L_{FIR}(L_\odot) = 3.86 \times 10^5\, D^2\, F_{FIR}$. Similar to σ_{HI}, a hybrid far infrared luminosity surface density σ_{FIR} is calculated as $L_{FIR}/\pi R^2$.

TOTAL MASS AND SURFACE DENSITY FOR SPIRALS   For the galaxies for which 21 cm line emission is detected, profile widths are available to provide an estimate of the circular rotation velocity. Since widths are often measured using different algorithms, we have selected a subset of the available data that meet the following criteria: 1. the level at which the width was measured must be either at 20% of one or more peaks or 50% of the mean intensity; 2. the detection must be a good one, that is, not poor or confused; and 3. the inclination must be greater than $40°$ for the width to be corrected to edge-on. While these restrictions cut down the number of galaxies for which corrected 21 cm line widths are available, they insure greater certainty of the resultant correlations. The total mass M_T determined in this way is available only for non E-type systems, and is calculated according to $M_T(< R)\,(M_\odot) = 2.325 \times 10^5 R V_{rot}^2$. As a practical application, we use the corrected 21 cm line width as the measure of $2V_{rot}$ and D_{25} as the indicator of $2R$. Note that we have not applied a correction for turbulent velocity. The total mass surface density σ_T is likewise calculated as $M_T/\pi R^2$.

The Limitations of Galaxy Catalogs

All catalogs have limitations because of the adopted inclusion criteria and their degree of completeness, and conclusions drawn from the examination of any catalog (or data set) must consider those limitations. Of particular importance, as noted by many authors, are biases against low surface brightness objects or the lack of homogenous sky coverage or survey depth. All of these issues are relevant to the current analysis. Given the catalogs from which we have drawn the sample analyzed herein, even before beginning, we want to emphasize the following failings of our analysis:

1. Low surface brightness galaxies, including both the low mass, low luminosity dwarfs and the low surface brightness giants such as Malin 1, are not included in our results.

2. Likewise, compact high surface brightness objects are also excluded. These include both compact BCDs and the smaller galaxies of intermediate luminosity in clusters.

3. Malmquist bias certainly is present in these data. Nearby samples include a predominance of low luminosity, low mass objects that are absent in samples at larger distances. Non-detections of HI and *IRAS* flux provide meaningful upper limits only for nearby samples. This is discussed further below.

4. Types earlier than Sa suffer from a large fraction of non-detections in the 21 cm line measurements. While estimates of the HI mass and surface density can incorporate upper limits to the detected flux (as discussed below), total mass calculations that use the 21 cm line width as a measurement of the rotational velocity cannot be made for non-detections. Hence all discussions of such properties as total mass, mass surface density, and mass-to-light ratio are limited to S0/a–Im. For the Im objects, only the brighter, higher surface brightness members of the class are included; for the S0s only those with HI detections are included.

MALMQUIST BIAS   The RC3 is to some degree magnitude-limited, and in most instances, the Malmquist bias seriously affects the availability of data for the present analysis. Figure 1 demonstrates that the Malmquist bias seriously affects the sampling of optical luminosities L_B. The upper panel shows clearly that the low luminosity galaxies are only sampled nearby, becoming increasing absent in samples at larger distances. The figure also demonstrates that there is a maximum luminosity near $\sim 10^{11.5}\ L_{\odot}$.

The lower panel examines the effect of Malmquist bias on the derived total mass to luminosity ratio M_T/L_B, in logarithmic units. Although this ratio is dependent linearly on distance, there is essentially no distance bias in this ratio.

In the following section, we present the analysis of both the RC3-UGC and RC3-LSc samples separately in order that we might keep the effect of Malmquist bias in perspective in interpreting our results.

Summary of Results

Given the available data, we are able to calculate the following properties: linear radius R_{lin}, blue luminosity L_B, far infrared luminosity L_{FIR}, total mass M_T, neutral hydrogen mass M_{HI}, the ratios M_T/L_B, M_{HI}/L_B and M_{HI}/M_T, the blue surface magnitude Σ_B, and the surface densities Σ_T, σ_{HI}, and σ_{FIR}. Figures 2–4 summarize our examination of morphological dependence in the fundamental properties for both the RC3-UGC (circles) and RC3-LSc (squares) samples as described above. In each panel, the median (filled symbol) and mean (open symbol) values for each morphological class are plotted along with an indication of the interquartile range (vertical bar extending from the values at the 25th and 75th percentiles). Although in most cases the median and mean values are not significantly different, many of the distributions are significantly non-Gaussian and sufficiently broad and skewed that the mean value is not a useful indicator. In settling on this presentation, we have also examined histograms of each distribution. Note that, although a logarithmic scale is used in the display, all quantities have been calculated using linear variables where logarithms are not already involved (e.g. B-V). Where necessary, survival analysis has been applied in the calculation of mean values but it should be

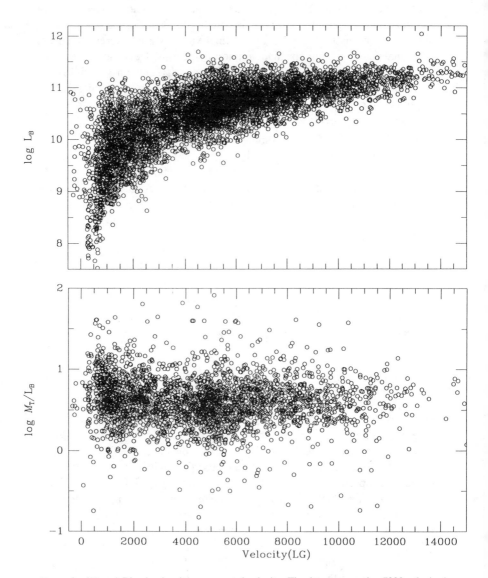

Figure 1 (*Upper*) Blue luminosity vs corrected velocity. The data representing 7930 galaxies in the RC3-UGC sample clearly illustrate the Malmquist bias. (*Lower*) The total mass-to-luminosity ratio for 2864 galaxies in the RC3-UGC sample vs velocity. This ratio has a distance term but the data show essentially no distance or Malmquist bias.

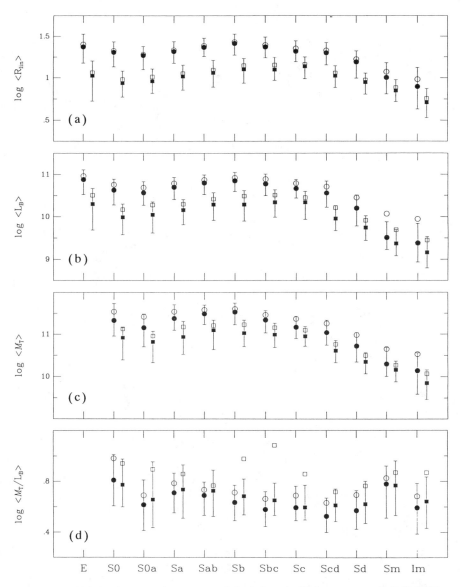

Figure 2 Global galaxy parameters vs morphological type. Circles represent the RC3-UGC sample; squares the RC3-LSc sample. Filled symbols are medians; open ones are mean values. The lower bar is the 25th percentile; the upper the 75th percentile. Their range measures half the sample. The sample size is given in Table 1. (*a*) log linear radius R_{lin}(kpc) to an isophote of 25 B mag/arcsec^2 , (*b*) log blue luminosity L_B in solar units, (*c*) log total mass M_T in solar units, (*d*) log total mass-to-luminosity ratio M_T/L_B.

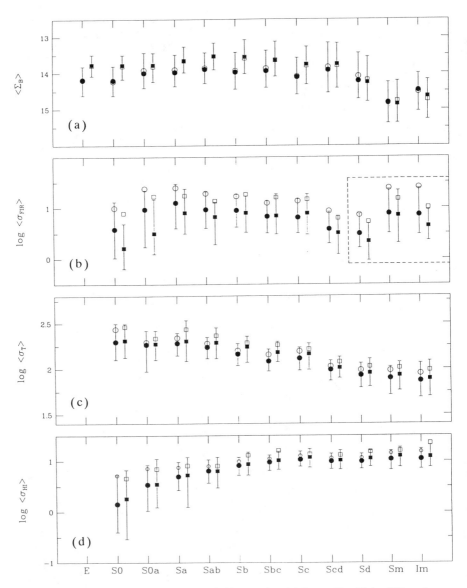

Figure 3 Same as Figure 2, for (*a*) optical (blue) surface brightness Σ_B, (*b*) log FIR surface density σ_{FIR}, (*c*) log total mass surface density Σ_T, (*d*) log HI surface density σ_{HI}. Dashed lines delineate the types with significantly fewer data.

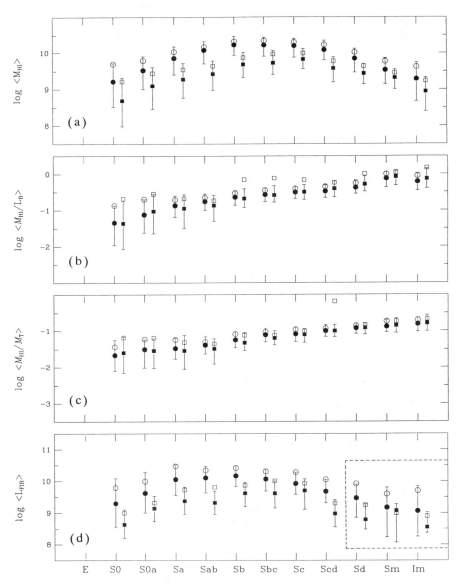

Figure 4   Same as Figure 2, for (*a*) log total HI mass M_{HI}, (*b*) log HI mass-to-blue luminosity ratio M_{HI}/L_B, (*c*) log HI mass fraction M_{HI}/M_T, (*d*) log FIR luminosity L_{FIR}. The dashed lines indicate significantly fewer data for these types.

noted that medians and percentiles treat non-detections as detections. Median and quartile values, along with the number of galaxies in each subsample, are also presented in Table 1 (see the end of this chapter). Since we note that for many properties there is little variation over adjacent types, pairs of morphological classes are combined in producing the table in order to shorten the presentation.

It is convenient to adopt Binggeli's (1993) nomenclature for giant "classical" and dwarf systems. He refers to spirals of type Sa, Sb, and some Scs as "classical spirals" with the term "dwarf irregulars" applying to types Sd, Sm, and Im as well as late Scs. This is not only useful for descriptive purposes but, as evident in the figures and Table 1, marks an important dividing line in certain of the global properties of galaxies.

In the next sections, we use these graphical and tabular summaries in the discussion of the properties of the normal galaxy population.

3.   DISCUSSION OF GALAXY PROPERTIES

As mentioned in the introduction, the classification of galaxies by morphological appearance and the variation in scale properties within a single class are the primary discriminants among galaxies. Here we review the individual properties available from large surveys and attempt to identify the variations that relate more to form than to scale.

Optical Colors

A classic study of galaxies was done by Holmberg (1958) who compiled and analyzed photometric data (integrated magnitudes, colors, and diameters) for 300 galaxies. One of his main conclusions was the dependence of color on morphological type. Objects of different morphological classes show clear differences in their optical colors as measured by the color indices $(U - B)$ and $(B - V)$. Figure 5 shows the well-established trend between morphology and mean color. The E and S0 galaxies are clearly redder than their spiral counterparts, and the trend from redder to bluer is nearly monotonic. At the same time, the range of colors among the Sa galaxies overlaps that of Sc galaxies: some Scs are as red as some Sas while some Sas are as blue as some Scs. It is unlikely that this overlap results from misclassification or observational errors, but rather that the scatter reflects true variations in the colors, and presumably the current star-formation rates, in individual objects.

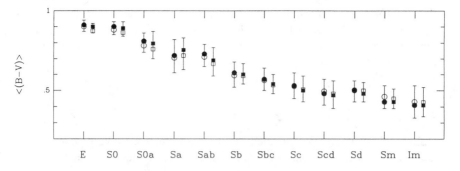

Figure 5   (B − V) color vs morphological type. (Same symbols as in Figure 2.)

Optical Linear Size

Both the median and mean values of linear diameter show subtle differences along the Hubble sequence as evident in Figure 2*a*, with the most distinguishing feature being the "smallness" of the latest types. Within the RC3-LSc sample, the classical spirals show a small systematic increase in size toward the later types. Such a trend is less obvious in the flux-limited RC3-UGC sample. The largest early-type galaxies, the cDs, are underrepresented in the current sample since they are too rare to be found nearby. Their location in regions of highest local density suggests that their large sizes are related to their spatial locations in the deepest potential wells.

It should be noted that the Malmquist bias also affects diameter-limited catalogs in the sense that objects at larger distances have characteristically larger linear diameters. The measurement of optical size enters into the debate concerning the degree of extinction internal to a spiral disk and into the surface brightness level to which a diameter measurement refers (Valentijn 1991, Burstein et al 1991, Giovanelli et al 1994).

Optical Luminosity

The optical luminosity L_B is a parameter of scale. Like the linear size, the range of median L_B values characteristic of classical galaxies varies only slightly, until the latest types, where the distinctiveness of the dwarfs becomes evident (Figure 2*b*). The ellipticals here are slightly brighter than spirals.

LUMINOSITY FUNCTION   Binggeli et al (1988) have carefully reviewed what is currently known about the luminosity function $\Phi(L)$. In their study of the Virgo cluster (Binggeli et al 1985), they derive the luminosity function $\Phi(L, T)$ for each morphological type separately. The range of luminosities representative of the classical galaxies is seen to be similar, as in Figure 2*b*, with the later spirals and dwarfs showing a characteristic decrease and being clearly separate. Binggeli et al find that while the brightest galaxies are ellipticals, the most common galaxies are the dEs. Because of morphological segregation, they conclude that $\Phi(L, T)$ cannot be universal. We discuss the effects of morphological segregation in Section 4.

Optical Surface Brightness

Holmberg's survey (1958) included photometric parameters for both magnitude and diameter with which he investigated the surface brightnesses of galaxies and studied the effects of internal extinction. Because internal extinction causes a sytematic change in the observed surface brightness (and color) as a function of inclination, we adopt a definition of surface magnitude that attempts to account for extinction. As mentioned above, the issues of internal extinction are still under significant debate (Disney et al 1989, Byun 1993).

Figure 3a shows the global surface brightness within D_{25}, Σ_B. The distribution of Σ_B is nearly constant for the E, S0, and classical spiral galaxies, but distinguishes clearly the late dwarf categories Sm and Im. We emphasize again that, as Disney & Phillipps (1983) discussed, current catalogs are biased against systems of low surface brightness, and some of the relative constancy found here across most of the classical galaxy sequence may be due to catalog selection.

DISK SURFACE BRIGHTNESS AND SCALE LENGTH   By examining the available surface photometry, Freeman (1970) found that the face-on central surface brightnesses of most spiral disks are nearly constant, with small scatter: 21.67 $\pm$ 0.30 B-mag arcsec^{-2}. Deviations occur at the ends of the spiral sequence among the S0s on the one side and dwarfs on the other. Surprisingly, most Es also seem to have constant central surface brightness. With the caveat that current catalogs are indeed biased against the low surface brightness systems, most E, S0, and classical spiral galaxies have the same scale length as a function of luminosity regardless of morphology. A thorough discussion of the details of this issue, including the effects of selection bias, is presented in Gilmore et al (1990).

Far Infrared Emission

The *IRAS* all-sky survey has provided measurements, or their upper limits, of the far infrared (FIR) flux in the 12, 25, 60, and 100μ bands. FIR flux measurements of normal galaxies show a clear distinction between elliptical and spiral galaxies. Ellipticals have a much poorer detection rate and when detected are generally lower in both FIR luminosity and in the ratio L_{FIR}/L_B (de Jong et al 1984, Bothun et al 1989, Sauvage & Thuan 1993). Exceptions are found in those early-type systems experiencing starbursts, e.g. NGC 1275 and NGC 1316.

The type dependences of FIR surface density σ_{FIR} and luminosity L_{FIR} are illustrated in Figs 3b and 4d. Many of the discussions in the literature focus on FIR-bright systems. These tend to be prominent starburst galaxies (e.g. M82), often peculiar in their morphological appearance and/or clearly involved in interaction with another system (Sanders et al 1987, but see Haynes & Herter 1988; Kennicutt 1990). In this review we avoid discussion of such galaxies.

Three possible origins of this FIR radiation are commonly identified:

1. Dust heated by nearby young, massive stars and reradiating in the infrared. The dust-molecular cloud complexes in the plane of the Milky Way are an example.

2. Dust reradiating the heating by the general interstellar radiation field, e.g. the Galactic cirrus cloud population.

3. Thermal and/or nonthermal radiation from active galactic nuclear regions, e.g. Seyfert galaxies.

Reviews of IR and FIR radiation are given by Soifer et al (1987), Telesco (1988), Cox & Mezger (1989), Rowan-Robinson (1990).

If the massive-star population, i.e. O-B stars, are the dominant dust-heating source then the FIR is a good measure of star formation (Devereux & Young 1991, 1992). The cirrus clouds in our Galaxy as well as the obvious presence of a general stellar radiation field have given rise to a two-component model to describe a galaxy's FIR radiation (e.g. Lonsdale Persson & Helou 1987, Buat & Deharveng 1988) with proponents supporting the importance of one model over the other.

In a study of the luminosity ratio $L_{FIR}/L_{H\alpha}$, Sauvage & Thuan (1992) find a systematic decrease from early- to late-type spirals and propose that the cirrus fraction responsible for the FIR luminosity decreases from ~86% for Sas to ~3% for Sdms—a result which would require a large, systematic, type-related correction to the use of L_{FIR} as a measure of star formation. They call attention to an alternative explanation for the $L_{FIR}/L_{H\alpha}$ type correlation: that the initial mass function (IMF) changes with type. They are reluctant to accept this possibility because of the proposal of a "universal IMF" (Scalo 1986), although there are many instances in our own Galaxy of differing IMFs (Gilmore & Roberts 1988). As noted in the discussion on HII regions, there is a strong dependence of number and luminosity of HII regions with galaxy type. This increase in both number and luminosity of HII regions with later type implies a type dependence of the IMF (Kennicutt 1988, 1989; Kennicutt et al 1989) similiar to that suggested above.

The strong correlation of FIR and radio radiation for spirals is difficult to explain in a model that does not invoke massive stars as a heating mechanism (Xu 1990; but see Devereux & Eales 1989). This interpretation does not appear to hold for normal elliptical galaxies where no FIR-radio correlation is found (Bregman et al 1992). Though visible patches of dust are frequently found in elliptical galaxies, here again there is no correlation with FIR radiation, and the evaluation of the amount of dust in elliptical galaxies from their (weak) FIR luminosity is at best uncertain.

The type dependence of the 1. FIR detection rate, 2. FIR luminosity, and 3. the luminosity ratio FIR/Radio (Fabbiano et al 1988, Condon et al 1991) is most impressive in separating ellipticals from spirals, with S0s somewhat intermediate in these various quantities. As illustrated in Figure 3b, the distinction in the FIR surface density within the spiral classes Sa–Sc, if present at all, is only slight (Bothun et al 1989, Sauvage & Thuan 1993).

Radio Continuum Emission

The radio radiation from galaxies generally has two spatial components: 1. that within a nuclear or central region where there may also be various levels of substructure, e.g. jets and knots, and 2. a more extended region, where again there may be structure. One or the other component may be very weak or absent (to the current levels of detection). The literature on radio radiation from normal galaxies is extensive; for reviews see Condon (1992) and Hummel (1990).

There are two basic mechanisms responsible for the radio continuum radiation in normal galaxies: synchrotron radiation arising from relativistic electrons accelerated by supernova remnants, and free-free radiation primarily from HII regions. A third origin due to dust heated by starlight is significant only at wavelengths < 1 mm. Because this review is intended to discuss typical galaxies, we omit consideration of "radio galaxies," i.e. radio-loud systems having radio luminosities > 10^{33} watts (e.g. Virgo A, Cygnus A).

Most normal galaxies are, at best, weak radio sources, and the statistics concerning their radio properties can be correspondingly poor. Thus we note that many of the nearer radio-loud galaxies are of early type, although most early-type galaxies are of very low radio luminosity. Frequently, only upper limits to the radio flux are available.

The early-type galaxies, the Es and S0s, clearly differ from the later-type systems in that compact core sources are common in the former. Extended sources comparable to the visible disk are much more common in the later types. When extended sources are found in the early-type galaxies, they are narrow and suggest jet-shaped sources (Hummel et al 1984, Condon & Broderick 1988, Condon et al 1991).

Another strong distinction between ellipticals and later-type systems is found in the comparison of radio emission with FIR. These two quantities correlate remarkably well for spirals but not for elliptical galaxies and only poorly for S0s (Bregman et al 1992). This holds over a radio frequency range from at least 151 MHz (Fitt et al 1988) to 4.85 GHz (Condon et al 1991) for galaxy types Sa–Im

(Sauvage & Thuan 1993). The proposal here is that the infrared measures primarily the reradiation from dust located near sites of active star formation while the radio measures primarily synchrotron radiation from electrons accelerated by supernova remnants resulting from this active massive star formation.

X-Ray Emission

Most of our understanding of the X-ray properties of galaxies comes from the results of *Einstein Observatory* measurements in the 0.2–3.5 keV X-ray band. Fabbiano (1989) provides a detailed review of X rays from normal galaxies; Sarazin (1992) gives a briefer summary. Catalogs include those by Fabbiano et al (1992) and Roberts et al (1991). *ROSAT* data and reports are appearing as this review is being written. In the near future we can expect important extensions to our knowledge of X-ray sources.

X rays have been detected from all galaxy types except the low luminosity dE and dS0 classes. *ROSAT* detections of X rays in the direction of such dwarfs appear to be background sources (e.g. Gizis et al 1993). There is a correspondence between optical and X-ray luminosities such that only the nearest Im-type systems, those in the Local Group, have thus far been seen in X rays. The optical (blue)-X-ray luminosity ratio differs with galaxy type. Es and S0s have, with large dispersion, an X-ray luminosity that varies approximately as the square of their optical luminosity (Canizares et al 1987, Bregman et al 1992). In contrast, the late-type galaxies show, again with large dispersion, a linear dependence between these two quantities (Fabbiano et al 1988).

Here we limit our discussion to normal galaxies, which have X-ray luminosities in the range 10^{38}–10^{42} ergs s^{-1}. The X-ray spectrum differs with galaxy type, and is harder for spirals. There are two distinctly different origins of the (non-nuclear) radiation: (*a*) a component which is basically stellar in origin: supernova remnants and X-ray binaries, and (*b*) diffuse emission from hot $(10^6$–10^7 K) gas.

In spirals, the stellar constituent can have both Population I and II components: for the former, supernova remnants and high mass X-ray binaries; for Population II, the so-called low-mass X-ray binaries. In our Galaxy, low-mass binaries are found primarily in globular clusters and in the central bulge region, although some have also been identified in the disk. The million degree gas is the principal source of X rays in the luminous early-type galaxies. This thermal plasma has its origin in the mass lost by evolving stars, stars which in the bulge component of early-type systems have random motions of typically a few hundred km s^{-1}. Ejecta from these evolving stars collide at velocities corresponding to the observed X-ray temperature. Such a source for the X rays so prominently emitted by ellipticals solves the long-recognized dilemma of locating the mass lost during the evolution of these stellar-rich systems.

In summary, X rays are seen from all galaxy types with the properties of radiation dependent on the fraction of the Population types in each galaxy. This variation is exemplified by bright ellipticals with relatively high luminosity soft X rays and by the bright, late-type spirals with lower luminosity, harder X radiation. The Sas with prominent bulges as well as star-forming disks have X-ray characteristics between those of Es and the later spirals.

Neutral Hydrogen Mass and Content

Because of the sensitivity available at centimeter wavelengths, the 21 cm line has been a valuable tool in measuring the redshifts of galaxies and serves as a general indicator of HI content. Since HI line fluxes, or upper limits, are available for some 15,000 galaxies skywide, the HI content surpasses nearly all other quantitative indicators of the potential for star formation. The HI content of galaxies has been the subject of recent reviews by Haynes et al (1984) and Giovanelli & Haynes (1990). Here we focus only on the morphological dependence among normal objects.

Several quantities are generally used in analyzing the total HI content of galaxies: the total HI mass M_{HI}, the hydrogen mass to luminosity ratio M_{HI}/L_B, and the HI surface density, σ_{HI}. While both M_{HI}/L_B and σ_{HI} have been used as the comparative measure of HI content, a residual dependence of the former ratio on L_B exists, in the sense that higher luminosity galaxies have systematically lower values of M_{HI}/L_B. Numerous authors (e.g. Bottinelli & Gouguenheim 1974) have shown that caution is necessary in using M_{HI}/L_B if the Malmquist bias might play a role. Furthermore, since L_B includes contributions from both disk and bulge, and the HI is a disk property, the morphological dependence on M_{HI}/L_B is complicated. Finally, the fraction of the total mass in the form of HI can be examined via the ratio M_{HI}/M_T. Figures 4a–c show the results of our analysis of HI properties.

The total HI mass M_{HI} is a scale parameter that is seen to vary over at least 2 orders of magnitude in the interquartile ranges and 4 orders of magnitude in the extremes. It is well know that early-type systems—E and S0 galaxies—contain proportionately lower HI masses, and in fact show a much larger range in all measures of HI content, both in M_{HI} and σ_{HI} relative to the later spirals. While some Es and S0s have HI contents similar to those of Sb–Sc spirals, others contain several orders of magnitude less HI. For this reason and because the HI within S0s is often located in an annulus exterior to the optical disk, van Driel & van Woerden (1991) and others have suggested that the HI gas has an external origin—the result of a tidal interaction or the infall of a dwarf companion. Bregman et al (1992) propose that almost no true Es have detectable HI gas except for those few instances where, through the HI kinematics and distribution, infall is indicated. As evident in Figure 4c, the fractional HI mass increases systematically from E/S0 to Im.

Among the later-type spirals, σ_{HI} is useful as an indirect probe of the effect of local environment on star-formation potential. In the study of the HI deficiency of cluster galaxies relative to their counterparts in low density regions, Haynes et al (1984) have defined the measure of the depletion of the HI content as the difference between the observed HI mass (in logarithmic units) and that expected for a galaxy of the same linear diameter and morphological type for isolated objects. Specifically, the HI deficiency parameter is defined as

$$< DEF > = \log[M_{\mathrm{HI}}(T, D)]_o - \log[M_{\mathrm{HI}}(T, D)_{\mathrm{obs}}].$$

The use of the HI deficiency parameter has led numerous authors to conclude that spiral galaxies that pass through a hot X-ray intracluster medium are stripped of their HI gas. Haynes & Giovanelli (1986) found a one-to-one correspondence between high HI deficiency and shrunken HI size. Warmels (1988a,b) and Cayatte et al (1990) also find that the HI disks of Virgo core galaxies are indeed shrunken with respect to their field counterparts or objects outside the core. Although ram-pressure stripping is the favored explanation for the HI deficiency, analysis of the the current data do not discriminate among the alternatives of ram pressure, conductive heat transport, turbulent viscosity, or tidal effects (Magri et al 1988). In other instances, slow but close, prograde tidal encounters can similarly remove the majority of a galaxy's interstellar HI. The effect of environment is discussed further in Section 4.

Carbon Monoxide

Verter (1985, 1990) and Young & Scoville (1991) have summarized extragalactic CO observations. More recent survey results are given by Braine et al (1993), Sage (1993), and Ohta et al (1993). There is a strong correlation between FIR and CO fluxes. However, a significant fraction of the data in the literature refer to galaxies chosen by some IR criterion which in turn reflects active star formation, frequently triggered by interactions with companion galaxies. For such interacting systems Braine & Combes (1993) and Sage (1993) find a higher mass fraction of molecular hydrogen, $M_{\mathrm{H_2}}/M_{\mathrm{T}}$, than for more isolated galaxies. Sage derives a mean increase of about a factor of two. Thus some of the statistical studies have samples rich in CO because active star-forming galaxies were preferentially chosen. Reviews and conference proceedings include those by Young & Scoville (1991), Combes (1992), and Combes & Casoli (1991).

Working with a volume-limited sample Sage (1993) found that for galaxies of type Sa–Sc, $M_{\mathrm{H_2}}$ is approximately 2% of the total mass within the region where the H_2 is measured. H_2 appears to be a smaller fraction in Sd-type galaxies, significantly so if the upper limits on $M_{\mathrm{H_2}}$ in his data set are considered. To derive $M_{\mathrm{H_2}}$ Sage assumes that the conversion factor for CO to H_2 is the same in all galaxies and equal to the Galactic value. Essentially all values of $M_{\mathrm{H_2}}$ in the literature make this assumption. This appears to be a reasonable approach for M31 and M33 but not so in later-type galaxies (Cohen et al 1988, Rubio et al 1991, Ohta et al 1993). What is evident for these late-type systems is the general faintness of the CO radiation, even though current star formation is ongoing. We must conclude that values of the molecular hydrogen content in late-type systems (Sd, Sm, Ir) derived in this manner are uncertain and possibly too low by up to an order of magnitude.

Uncertainty of a different sort also exists for CO results in early-type systems. For ellipticals there is the well recognized problem of classification uncertainty (Buta 1992a,b; Roberts et al 1991; Bregman et al 1992; Hogg et al 1993). If we wish to speak of the cool gas content in different sorts of early-type galaxies, e.g. E or S0, it is important that we minimize the classification uncertainty. Unfortunately, this is difficult and the literature is correspondingly confusing. As an example, Buta (1992a) calls attention to a well recognized Sb spiral, NGC 3928, which because of its compactness is typed as an E0 on the Palomar Sky Survey prints. This "elliptical" stands out in the *IRAS* lists and was successfully searched for CO (Gordon 1990). This is an extreme case. Generally the disagreement exists among the similar-in-appearance types of E, E/S0, S0, and Sa. But it is just in this range of galaxy type that the question is raised of bifurcation or continuity in various properties. Until quantitive measurement of type becomes possible, uncertainty will remain.

Weak CO emission has been detected in only two of over a dozen of the RSA catalog ellipticals that have been searched in depth (Sofue & Wakamatsu 1993, Roberts et al 1991, Bregman et al 1992). Only 3 (out of 64 searched) isolated RSA ellipticals have HI detections. Two of these appear to be examples of capture and the third, NGC 2974, shows spiral features on deep imaging and therefore is an Sa with low surface brightness arms. Thus cool gas, CO and HI, is a rarity in the earliest of galaxies, the ellipticals, while hot X-ray emitting gas is quite common in these systems. Cool gas is more frequently found in the early galaxies of type S0, S0/Sa, and Sa—the frequency of detection increasing in this sense. There are also instances of stringent upper limits for these galaxy types for CO and for HI. Thus, unless survival analysis which evaluates detections together with upper limits for non-detections (Feigelson & Nelson 1985) is used to derive average values of M_{H_2} and M_{HI}, trends regarding these quantities for the early-type categories will be too high.

It is both of these effects—the conversion factor for CO to H_2 for late-type galaxies and inclusion of upper limits for early-type systems—that could well alter any trend with morphological type. With this caution in mind we consider the various relationships for the molecular gas content with type [described in the review by Young & Scoville (1991)]. Those appropriate to our discussion are:

1. M_{H_2}/L_B. This ratio is essentially constant for types Sa–Sc and then decreases for later types, Scd-Sdm; the scatter is large. Using a sample selected differently, Sage (1993) found a similar trend in terms of fractional

mass content, M_{H_2}/M_T. This near constancy over Sa–Sc followed by a decrease for later types is also seen in the intrinsic quantities of radius, blue luminosity, and total mass. However, the decrease in M_{H_2}/L_B could also reflect a changing CO to H_2 conversion factor. We conclude that M_{H_2}/L_B and M_{H_2}/M_T are nearly constant over types Sa, Sb, and Sc. All we can say for later types is that the appropriately normalized CO luminosity is less.

2. M_{H_2}/M_{HI}. This ratio decreases with type S0/Sa–Sd/Sm; the scatter is large. Sage (1993) found the same trend for a different sample. He attributes it to the long recognized increase of M_{HI}/M_T with type since M_{H_2}/M_T is essentially constant (for Sa–Sc). The range and details of the M_{H_2}/M_T variation with type are obviously dependent on the conversion factor and the statistical treatment of the upper limits. He further notes, as do others (e.g. Devereux & Young 1990), the spatial anticorrelation of these two forms of hydrogen in galaxies, including our own; H_2 is more centrally concentrated than HI. We conclude that the global decrease in M_{H_2}/M_{HI} is in the expected sense and reflects the increase in M_{HI}/M_T with type.

3. $(M_{H_2} + M_{HI})$/Area. This quantity, the total cool gas surface density (note that the area is based on the optical diameter D_{25}), increases with later type. This is the sense of the trend for σ_{HI}, as well as for σ_{H_2} (Young & Knezek 1989). For the latter, constancy of σ_{H_2} over a limited mid-type range is also permissible from the data since there is the usual uncertainty at the extreme types. We conclude that the surface density of cool gas increases with later type. This is an important systematic trend, similar to that displayed by HII regions within galaxies of different type.

HII Regions

Regions of ionized hydrogen are one of the distinguishing features of galactic morphology. The first quantitative description relating galaxy type and HII regions is given by Sersic (1960) who studied their sizes as measured on broadband photographic plates. More recently, major progress has been made by Kennicutt and his colleagues using $H\alpha$ imaging in their study of nearly 100 galaxies. In a series of seminal papers (Kennicutt 1988, 1989; Kennicutt et al 1989) a census of HII region properties is given for a galaxy type range of Sa–Im. Summarized here are those that are related to morphological type. Figure 6 (taken from Kennicutt et al 1989) which shows the cumulative HII region luminosity function for galaxies of different type illustrates some of these points:

1. The brightest HII regions in late-type systems are on average ~50 times more luminous than the brightest in early-type spirals. Galaxies intermediate in type have their brightest HII regions intermediate in value, again, on average;

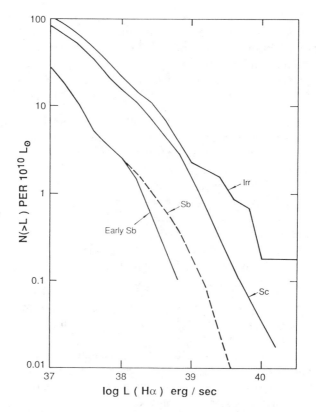

Figure 6   The cumulative HII region luminosity function for different galaxy types from Kennicutt et al (1989).

the dispersion is large. The number of bright HII regions per unit galaxy luminosity or the number expressed as a surface density varies systematically with type. Both measures are smallest for early-type spirals (Sab–Sb) and largest for Sm–Im. The difference in the mean is close to an order of magnitude. These results are for HII regions of Hα luminosity $> 10^{38}$ erg s^{-1} (and can extend to $10^{40.8}$ ergs s^{-1}). For comparison, Orion has $L(\text{H}\alpha) \sim 10^{37}$ ergs s^{-1}.

2. The diameters of the first-rank HII regions vary in a similar sense: The larger ones are found in the later-type systems for galaxies of similar absolute magnitude. This luminosity normalization allows for any possible size-number effect.

3. The slope of the luminosity functions for HII regions in different galaxies shows a weak but systematic trend with type, with later-type systems hav-

ing a shallower slope. The bright end of the luminosity function shows a systematic truncation in the sense that earlier type galaxies are lacking in bright HII regions. Kennicutt et al (1989) illustrate this point by noting that 30 Dor in the LMC is ~20 times brighter than any HII region in M31, a much larger galaxy of type Sb. Furthermore the ten brightest HII regions in the LMC are all brighter than M31's brightest.

All of these trends indicate a massive star-formation rate determined by the galaxy type in which it occurs. Related to this is the systematic occurrence of the brightest stellar associations in the later-type galaxies (Wray & de Vaucouleurs 1980).

Kennicutt (1989) examined the relationship between the (disk-averaged) Hα surface brightness and the HI and CO surface density and found a threshold value for the latter, $\sim 3 M_\odot$ pc^{-2}, below which star formation is rare. He notes that Toomre's (1964) model for the stability of a simple single-fluid disk is consistent with these results. Hogg et al (1993) find a similar threshold value for the onset of star formation within a sample of S0s and Sas.

These results are encouraging in supporting an intuitive expectation that star formation proceeds where there is "enough" material to do so. An understanding of the next link as to why the bright end of the initial mass function varies with galaxy type will offer an important insight into massive star formation.

Chemical Abundances

Chemical abundance determinations are now of a quality and number to enable relationships with other galaxian parameters to be described. This has been done for such quantities as luminosity (Skillman et al 1989), mass (Garnett & Shields 1987, Vila-Costas & Edmunds 1992), and type. Figure 7 displays values of O/H as derived from emission line strengths measured in HII regions, as a function of type (upper panel) and of luminosity (lower panel). The trend is in the sense that the fainter, less massive, and later-type galaxies are lower in O/H.

Using absorption lines and colors, a trend for Fe/H as a function of luminosity is well described for Es and S0s, in the sense of a lower Fe value in fainter galaxies (Pagel & Edmunds 1981, Da Costa 1992). With a reasonable assumption regarding the O/H and Fe/H abundances, Skillman et al (1989) show that the spheroidals and dwarf elliptical galaxies form a smooth extension of the faint end of the abundance-luminosity relation found for spirals and irregular-type galaxies.

Intercomparison between abundances based on emission lines from HII regions and on absorption lines and colors from a stellar population is uncertain. The approaches and assumptions differ significantly (e.g. Pagel & Edmunds 1981). Atomic processes guide the emission-line analysis, while the absorption

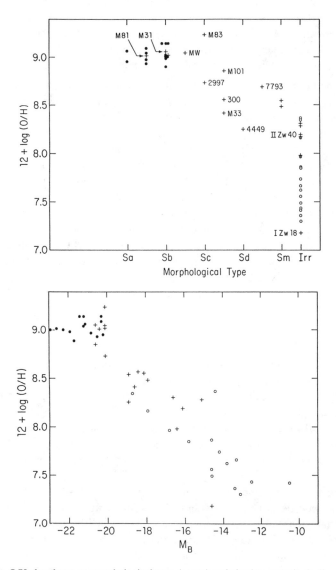

Figure 7   O/H abundance vs morphological type (*upper*) and absolute magnitude (*lower*). Different symbols represent different sources: filled circles—Oey & Kennicutt (1993), plus signs—Garnett & Shields (1987), open circles—Skillman et al (1989). The absolute magnitudes are as given in these references.

lines must be evaluated in terms of both the composite stellar population and metallicity using one of two methods: evolutionary synthesis or stellar population synthesis (e.g. Rocca-Volmerange 1992, Chiosi et al 1992, Bruzual 1992).

Even within the spiral category, where there is a systematic trend of HII region size and luminosity, the emission-line analysis can be type dependent. For early spirals, the widely used temperature indicator [O III]λ4363 line is often too weak to use. Lacking such data, an empirical method is applied in which the abundance is derived from ([OII]λ3727+ [OIII]λ5007)/Hβ and its calibration versus [O/H], a procedure which may be uncertain in the high-abundance regime. These problems and the derivation of abundances for early-type galaxies are discussed by Diaz (1992) and by Oey & Kennicutt (1993).

Although different techniques and often large uncertainties are encountered, the following consistent picture emerges:

1. The least luminous and least massive of galaxies, Ims and dEs, have the lowest abundances (O/H or Fe/H), 10^{-1} to 10^{-2} solar. The most massive and most luminous of galaxies, both ellipticals and (the central regions of) spirals are overabundant in the "heavy" elements by factors of up to a few (Strom & Strom 1978, Aaronson & Mould 1985, Worthey et al 1992). Gorgas et al (1990) suggest that, in the mean for their sample of Es and S0s, the value is near solar.

2. When measurable, most galaxies including our own show an abundance gradient in the sense of a lower metallicity at the outskirts, i.e. at larger galactocentric radii (Diaz 1992, Vila-Costas & Edmunds 1992). There appears to be a type dependence for these gradients. Early-type spirals have shallower [O/H] gradients than do late-type systems (Vila-Costas & Edmunds 1992, Oey & Kennicutt 1993). Gradients within Es and S0s show a wide range and appear not to be correlated with any galaxian parameter (Gorgas et al 1990).

3. An abundance-type effect is present among spirals and irregulars in the sense that earlier galaxies have a higher abundance. This trend reaches an approximately constant value among the early spirals. However, the entire effect may only reflect a luminosity or mass distribution in the highly selected sample of galaxies thus far studied for abundances (Garnett & Shields 1987, Skillman et al 1989, Oey & Kennicutt 1993).

There are exceptions to these various findings. Bertola et al (1993) described a metal-poor luminous early-type galaxy NGC 5018. They found that the inner region has a relatively blue color together with a weak Mg$_2$ index—both measures of low metallicity. Shields et al (1991) described the first sample of cluster galaxies studied for abundances. They found higher interstellar abundances in

Virgo cluster Scs than for field galaxies of similar type, which they attribute to the absence of low abundance infalling material.

A type-abundance dependence in field galaxies is present only in the broadest of terms. More fundamental may be the abundance trend with luminosity and with mass over the range from dwarf galaxy to giant (classical) galaxy.

Total Masses and Related Quantities

The study of total masses has been approached both through the measurement of global HI profile widths and through the detailed study of rotation curves. In an early work, Brosche (1971) noted that the maximum rotational velocity decreased along the spiral sequence from Sa to Im as if the morphological sequence could be understood as a sequence of angular momentum at constant mass. Today the issue appears more complicated. The masses and mass-to-light ratios of galaxies were the subject of a review by Faber & Gallagher (1979) to which the reader is referred for a general discussion.

Similarly to Faber & Gallagher, we have examined the total mass within the optical radius by using the 21 cm line width as the indicator of rotational velocity. While their analysis made use of 121 disk galaxies, the present sample contains mass estimates for over 3000 objects. Figures 2 and 3 include the distribution of properties relevant to mass: M_T, the total mass surface density σ_T, and the total mass-to-light ratio M_T/L_B.

As with luminosity and size we find, in Figure 2c, a small systematic variation of total mass among the classical spirals followed by a pronounced decrease for the dwarf systems. It is important to note that the three parameters L, R, and M are strongly coupled through the Tully-Fisher relations (Tully & Fisher 1977), i.e. both R and V vary with L. Unfortunately our understanding of the Tully-Fisher relations is too poor to point to the fundamental parameter(s).

TOTAL MASS-TO-LIGHT RATIO   As seen in Figure 2d, median values of M_T/L_B are essentially constant at a value near five over the entire type range S0–Im. Although the trends of L_B, R_{lin}, and M_T are each in the same sense over this range, it still comes as a surprise to these authors to find neither a type dependence nor a distance dependence (Figure 1) in M_T/L_B. In particular, a type dependence has been found frequently in analysis of much smaller samples, though the range in slopes found in these determinations includes both positive and negative values, as well as a zero slope (Meisels 1983).

ROTATION CURVES   Recent studies of the mass distribution within galaxies have employed the detailed rotation curves obtained either from the Hα or HI 21 cm line. Rubin (1991) presents a summary of the results, particularly those of the optical studies that look for environmental effects. Broeils (1992) has examined the HI rotation curves and optical surface photometry of 23 spiral and irregular galaxies to look for the morphological dependences in the distribution of mass and light. An extensive bibliography of published rotation curves is available in Corradi & Capaccioli (1991).

In studying the dark matter distribution, HI rotation curves are more suitable since they generally extend far beyond the point where $H\alpha$ is detected, and thus better constrain the halo distribution. Most recently, Broeils (1992) has combined new observations with those available in the literature to look at morphologial dependences in the mass and light distributions and the applicability of dark matter and alternative models. He sees a clear indication that the dark matter component increases for the latest types, but along the spiral sequence evidence for significant morphological trends is lacking, partly due to the still-small number of objects mapped.

With the discovery of two galaxies with declining rotation curves, Casertano & van Gorkom (1991) have found correlations between the peak circular velocity of a galaxy, its central surface brightness, and the slope of the outer rotation curve (see also Persic & Salucci 1991). As in Broeils' sample, they do not sample the full range of the spiral sequence. Kent (1987) has pointed out that Sa galaxies may require no dark matter within the optical disk, but few Sa galaxies have been mapped in HI, primarily because their lower characteristic HI surface densities make synthesis observations more difficult. The current observations suggest two conclusions:

1. Most late-type dwarf galaxies show rotation curves that are still rising at their last measured point.

2. Classical spirals show rotation curves whose outer portions are rising, flat, or falling.

MASS SURFACE DENSITY   As evident in Figure 3c, the total mass surface density σ_T shows a clear decrease along the spiral sequence. The trend is slow but monotonic from S0/a to Sc with a steepening toward the dwarfs.

4.   VARIATIONS IN GALAXY PROPERTIES

Some of the systematics in the properties of galaxies point to a common thread—that of star formation. In this section, we investigate trends along the Hubble sequence and consider also the critical parameter of local environment.

Global Properties

We have discussed 18 measures of the properties of galaxies. For five of these there is a clear monotonic variation with morphological type over the entire spiral-irregular range. A sixth, that of the luminosity function of HII regions, also shows such a monotonic type-dependent variation, though here the data are far fewer. All six are normalized quantities, i.e. the "size" of the system is allowed for: σ_T, σ_{HI}, M_{HI}/L_B, M_{HI}/M_T, $(B - V)$, and HII region luminosity functions. However not all of the normalized (nor absolute) quantites vary systematically and monotonically, e.g. M_T/L_B, Σ_B, σ_{FIR}. Five of the six can be identified with star-formation activity: past, present, or future potential. The sixth, the mass surface density, is also likely related to such activity as a triggering mechanism. Thus, $(B - V)$ measures past and current formation activity, HII regions relate current activity, and the cool gas content measures the reservoir for future star formation. In addition, those parameters for which the data are too few to describe any detailed type dependence other than elliptical versus spiral—radio, X-ray, and chemical abundance—are also star formation related. What is surprisingly lacking from this overall list is the FIR surface density. It does have the elliptical vs spiral dependence of others but shows no variation within the classical spiral category (and only an uncertain trend in the later types due to fewer data). This lends support to the two-component model for the origin of the FIR described previously. Clearly, the morphological sequence of normal, isolated galaxies measures star-formation activity.

The Effect of Environment

Perhaps the strongest differences among galaxies of different morphologies are seen in their clustering tendencies. In individual cases, the importance of mergers, tidal interactions, and sweeping within clusters can readily be demonstrated, yet the overall imprint of recent processes on galaxy morphology remains unclear. Below, we review the evidence for morphological segregation and the importance of interactions between galaxies and their surroundings.

MORPHOLOGICAL SEGREGATION   As early as the 1880s, Wolf noticed that the distribution of nebulae was not uniform in the sense that more elliptical nebulae were concentrated in the Virgo direction than elsewhere. By the 1930s, the morphological differences between field and cluster galaxies were well established (Hubble & Humason 1931). Using his survey of 55 rich clusters, Dressler (1980) has quantified the concept of morphological segregation, showing the steady decrease in spiral fraction and the corresponding increase in the E/S0 population with local galaxy density, a variation in population fraction that is slow but monotonic. Extending Dressler's study, Postman & Geller (1984) have shown that the morphology-density relation holds over six orders of magnitude in space density to regimes where the dynamical timescale approaches the Hubble time.

THE CLUSTER ENVIRONMENT   In the highest density environments, the possible morphology-altering mechanisms are many; galaxy-galaxy, galaxy-cluster, and galaxy-intracluster medium interactions can all lead to significant changes in morphology and star-formation potential. Indeed, the occurrence of pathological and disturbed objects in high density regions is well-recognized. Dressler (1984) reviews the models for morphological alteration in clusters according to the relative importance of initial conditions or late evolution. Whitmore (1990) gives a recent summary of the various galaxy properties that are seen to vary significantly between cluster and field galaxies.

In addition to the obvious variation in morphological make-up, various authors have attempted to identify density dependences in the fundamental properties under consideration in this review. It has already been mentioned in Section 3 that spiral galaxies passing through the center of rich X-ray clusters appear to lose up to 90% of their interstellar HI. At the same time, their molecular consitutents, as measured by their CO content and distribution, remain relatively unaffected (Kenney & Young 1989). Evidence for the stripping of spirals in clusters and constraints on the responsible processes are reviewed in Haynes (1990).

Most recently, studies have addressed the possibility of environmental variations in the distribution of mass within galaxies, and the results are conflicting. The detailed studies of rotation curves of spiral galaxies in clusters by Rubin et al (1988) and Whitmore et al (1988) suggest that, in inner cluster members, the halo is either partially stripped or not allowed to form—a conclusion based on the observation of falling rotation curves in centrally located galaxies. However, Distefano et al (1990), using $H\alpha$ rotation curves, and Guhathakurta et al (1988), using ones derived from HI synthesis maps, do not see such environmental effects in their respective studies of Virgo members.

THE GROUP ENVIRONMENT   In loose groups, where the velocity dispersion is low, slow close prograde tidal encounters can remove significant fractions of a galaxy's interstellar material. A classic, graphical discussion of the tidal phenomenon is given by Toomre & Toomre (1972), and numerous examples of the success of these models in reproducing tidal bridges and tails are now available. Both radio emission and far infrared emission are strongly enhanced in the instance of tidal interactions. The importance of interactions in a wide range of phenomena from the formation of shells and polar rings to the driving of spiral stucture and starburst phenomena have been discussed by numerous authors, most recently Barnes & Hernquist (1992).

GALAXIES AT HIGH REDSHIFT   Evidence is now accumulating that significant evolution of the cluster population has occured between the present time and the epoch corresponding to a redshift $z \simeq 0.4$ (for a review see Koo & Kron 1992). Clusters in the range $0.4 \leq z \leq 1$ show a higher fraction of blue galaxies than do their low redshift counterparts, the so-called Butcher-Oemler effect. Gunn (1990) reviews the evidence for the Butcher-Oemler effect including the increase in emission-line and Seyfert objects and the presence of the "E + A" population. Recent high resolution imaging of the blue cluster members confirms their spiral nature (Lavery et al 1992, Dressler & Gunn 1992). The relationship of the present-day population to the distant cluster galaxies and their field counterparts is critical to our understanding of the process of galaxy evolution and the development of the morphological characteristics evident today.

The S0 Problem

For the most part in this review, we have not emphasized the importance of the S0 class. If a spiral loses 90% of its interstellar HI, its potential for future star formation must be greatly diminished and thus its spiral structure should fade (Larson et al 1980). The resultant object would look like an S0. As pointed out by Giovanelli & Haynes (1985), the S0 fraction at a given density in Dressler's (1980) survey is higher in X-ray luminous clusters, while the corresponding spiral fraction is lower. At the same time though, fundamental differences in the bulge to disk ration and in surface magnitude still distinguish the S0 galaxies from present-day spirals (Dressler 1980). Whereas one fifth of the S0s in the RSA contain significant amounts of HI, others of similar properties are lacking in HI to much lower limits (Bregman et al 1992). The HI detection rate is so much higher than that for ellipticals that a capture hypothesis is unlikely. Pogge & Eskridge (1993) found a similar dichotomy in the presence of HII regions in S0 disks, but concluded that the S0s merely represent the early-type end of a continous variation in star-formation activity along the spiral sequence. Because of the large dispersion in properties, the S0 class still remains enigmatic.

Classical Galaxies vs Dwarfs

We concur with, among others Sandage et al (1985) and Binggeli (1993), that there are fundamental differences between bright ("classical") and faint ("dwarf") galaxies. The RC3 contains too few dEs for an interpretation of their characteristics in our analysis, but repeatedly in Figures 2 and 3 and Table 1, clear distinctions are seen in the latest types. Sandage (1990) summarizes the results of the Las Campanas surveys of the Virgo and Fornax clusters and selected loose groups conducted by him and his collaborators. In particular, these studies have explored the relative distributions of dE and Im dwarfs. They conclude that:

1. Dwarf galaxies have the same general space distribution as giant galaxies. The dEs are predominantly found in dense groups whereas the Ims are more widely dispersed. The dEs at least seem to form only in the vicinity of more massive galaxies.

2. The faint end slope of the luminosity function for dEs in the field is flatter than that found for similar galaxies in Virgo or Fornax.

3. The dwarf to classical ("giant") galaxy ratio decreases in low richness regions relative to ratios found in clusters.

5. SUMMARY

In this review we have discussed quantifiable properties of galaxies and their dependence on morphological type. Why the present-day galaxy population follows these trends is a major challenge of theories of galaxy formation and evolution. Here, we conclude with the following summary of the behavior of the "typical" galaxies found in today's large catalogs:

1. There are *three* general categories of galaxies: elliptical, classical spiral, and dwarf. This last category includes Sd and Sm. The nature of S0s remains a matter of debate, perhaps because they are indeed a transition type.

2. There is a pronounced spatial segregation of morphologies in the sense that early-type galaxies are more clustered.

3. For a volume-limited sample, we find that the classical spirals—Sa, Sb, Sc— show a small systematic variation in median size and in median luminosity, becoming larger and brighter as they become later. There is a suggestion of a similar trend for total mass. We do not find these dependences in the flux-limited sample.

4. The late-type dwarfs are indeed different from the classical galaxies. The dwarfs become smaller, fainter, and less massive over the range Scd–Im.

5. The mass-to-luminosity ratio is essentially constant over the entire sequence S0–Im. Although this ratio is distance dependent, it shows essentially no Malmquist bias.

6. The total mass surface density σ_T decreases systematically over the sequence S0/a–Im.

7. There are a number of systematic trends with type in these data that are clearly identifiable with star-formation processes. Thus global color is a measure of past and present star-formation activity, as are X-ray and radio radiation. The HII region parameters relate to current activity; the cool gas components measure the potential for future star formation. The trend of abundances is also consistent with the above picture but FIR measurements are not.

8. We must stress the wide range to be found for any parameter within any type. The range is always larger than even the most pessimistic error estimates. Much of this wide range is intrinsic. The only instance where overlap does not occur at least over the interquartile interval is for Ims vs classical spirals for the three basic parameters of R_{lin}, L_B, and M_T.

9. The wide range in any of the parameters described above must not be interpreted to mean that parameter space is unbounded. There is an upper limit to L_B as displayed in Figure 1. But, similarly, there are lower bounds to at least the classical spirals. In particular, there are no very faint or very small Sb galaxies.

In such a summary it seems worthwhile to restate the morphological distinctions for the sequence elliptical–classical spiral–dwarf. The sequence is one of bulge-to-disk ratio, largest for the ellipticals and zero for the dwarfs. It also follows the development of spiral arms, ending with what might be thought of as a limiting case, an Im as just an arm.

ACKNOWLEDGMENTS

We thank Riccardo Giovanelli for his collaboration in developing the database used herein and both he and David Hogg for many discussions relevant to our review. Thanks are also given to Harold Corwin for suppling us with a digital copy of the RC3 and to Fran Verter, Trinh Thuan, Jim Condon, and Joel Bregman for answering inquiries and supplying data. M. P. H. receives support through NSF grants AST-9014850 and AST-9023450.

Table 1 Statistical properties of the morphological classes

	Sample		E,S0	S0a,Sa	Sab,Sb	Sbc,Sc	Scd,Sd	Sm,Im
R_{lin}	UGC	median	21.1	19.8	25.1	22.4	17.7	8.5
(kpc)		25%	14.0	13.9	18.4	16.4	11.8	4.9
		75%	28.8	26.5	32.1	29.8	24.0	14.0
		Number	1363	798	1488	1139	2223	919
	LSc	median	9.0	9.8	12.0	13.2	9.3	6.0
		25%	5.8	6.9	8.3	9.5	6.8	4.1
		75%	13.2	13.6	16.7	17.7	12.2	8.3
		Number	705	342	454	616	1037	958
L_B	UGC	median	52.5	43.6	69.2	52.5	25.7	2.7
(10^9 $L_\odot$)		25%	21.4	21.4	38.0	29.5	9.8	1.0
		75%	93.3	79.4	107.2	95.5	53.7	7.1
		Number	1302	761	1391	1029	1527	416
	LSc	median	11.0	13.2	19.5	21.9	6.8	1.9
		25%	4.2	5.4	8.1	9.3	3.2	0.8
		75%	24.0	23.4	38.0	41.7	13.2	4.1
		Number	665	315	407	563	834	595
M_T	UGC	median		22.6	32.4	19.0	7.9	1.6
(10^{10} $M_\odot$)		25%		8.7	17.0	9.5	3.5	0.5
		75%		49.0	52.5	33.9	16.6	4.0
		Number		292	808	639	1471	490
	LSc	median		7.1	11.0	9.1	2.8	1.0
		25%		3.2	4.9	5.1	1.4	0.4
		75%		17.4	21.4	16.2	5.1	1.8
		Number		119	251	389	701	534
M_T/L_B	UGC	median		4.9	4.4	3.8	3.5	4.2
($M_\odot/L_\odot$)		25%		3.1	3.2	2.9	2.6	2.6
		75%		7.2	6.0	5.4	5.0	6.8
		Number		278	749	567	1025	245
	LSc	median		5.1	5.0	4.2	4.1	5.0
		25%		3.1	3.4	3.2	3.0	3.1
		75%		8.8	6.9	6.0	6.0	7.9
		Number		112	238	362	574	342

Table 1   Statistical properties of the morphological classes (*cont.*)

	Sample		E,S0	S0a,Sa	Sab,Sb	Sbc,Sc	Scd,Sd	Sm,Im
Σ_B	UGC	median	14.20	13.98	13.96	14.00	14:02	14.59
		25%	13.82	13.47	13.44	13.44	13.29	14.08
		75%	14.62	14.39	14.41	14.48	14.63	15.23
		Number	1591	977	1694	1154	1663	457
	LSc	median	13.78	13.73	13.55	13.72	14.02	14.73
		25%	13.51	13.36	13.10	13.23	13.39	14.21
		75%	14.15	14.10	13.99	14.22	14.65	15.35
		Number	682	317	412	571	858	649
σ_T $(M_\odot \mathrm{pc}^{-2})$	UGC	median		188.9	154.7	124.2	91.4	74.5
		25%		133.4	115.2	94.5	67.8	49.2
		75%		249.8	199.9	167.6	119.6	113.2
		Number		292	808	639	1471	490
	LSc	median		200.8	178.2	148.5	95.8	81.4
		25%		121.7	125.4	108.1	69.6	52.1
		75%		323.1	242.8	194.2	131.8	120.0
		Number		119	251	389	701	534
σ_{HI} $(M_\odot \mathrm{pc}^{-2})$	UGC	median	1.31	4.64	7.70	9.83	9.80	10.85
		25%	0.33	2.12	4.62	6.92	6.93	6.86
		75%	4.93	9.12	11.39	14.00	13.82	16.71
		Number	403	535	1200	1053	2106	902
		Detect	217	450	1149	1031	2071	894
	LSc	median	1.40	4.45	7.95	10.84	10.82	12.09
		25%	0.26	1.25	4.48	7.20	7.39	7.51
		75%	5.81	11.11	13.86	16.36	16.04	19.23
		Number	288	212	349	561	967	906
		Detect	146	178	340	557	963	901
σ_{FIR} $(L_\odot \mathrm{pc}^{-2})$	UGC	median	3.77	11.47	9.22	6.73	3.63	7.44
		25%	1.00	3.32	4.23	3.21	1.84	3.07
		75%	13.15	28.45	18.85	14.14	8.07	22.88
		Number	256	271	684	698	887	90
		Detect	127	205	549	517	439	73
	LSc	median	1.41	5.98	7.44	7.54	2.49	4.93
		25%	0.51	1.78	3.06	3.04	1.03	2.26
		75%	4.64	20.21	16.02	18.48	5.83	17.33
		Number	151	108	167	211	283	60
		Detect	58	83	145	191	172	48

Table 1 Statistical properties of the morphological classes (*cont.*)

	Sample		E,S0	S0a,Sa	Sab,Sb	Sbc,Sc	Scd,Sd	Sm,Im
M_{HI}	UGC	median	1.24	5.62	15.14	15.85	9.33	2.40
$(10^9\ M_\odot)$		25%	0.23	1.78	6.92	7.94	4.07	0.74
		75%	5.01	13.49	26.30	26.30	17.78	6.17
		Number	410	537	1204	1058	2121	942
		Detect	222	452	1153	1036	2086	934
	LSc	median	0.52	1.41	4.17	5.89	3.02	1.27
		25%	0.10	0.42	1.67	3.02	1.41	0.43
		75%	1.58	4.36	8.71	12.59	5.76	2.69
		Number	295	213	354	567	987	974
		Detect	153	179	345	563	983	969
M_{HI}/L_B	UGC	median	0.04	0.12	0.21	0.29	0.36	0.66
$(M_\odot/L_\odot)$		25%	0.01	0.04	0.12	0.19	0.24	0.36
		75%	0.12	0.24	0.33	0.43	0.56	1.10
		Number	394	510	1126	955	1444	406
		Detect	207	425	1075	934	1424	399
	LSc	median	0.03	0.10	0.20	0.30	0.47	0.78
		25%	0.01	0.03	0.11	0.18	0.28	0.44
		75%	0.14	0.28	0.34	0.48	0.76	1.32
		Number	283	201	325	513	780	555
		Detect	141	167	316	509	776	551
M_{HI}/M_T	UGC	median		0.03	0.05	0.08	0.11	0.15
		25%		0.02	0.03	0.05	0.08	0.09
		75%		0.06	0.09	0.12	0.15	0.23
		Number		292	808	639	1471	490
	LSc	median		0.03	0.04	0.07	0.11	0.15
		25%		0.10	0.02	0.04	0.07	0.09
		75%		0.07	0.08	0.11	0.16	0.24
		Number		119	251	389	700	534
L_{FIR}	UGC	median	1.71	9.89	14.26	9.87	4.05	1.63
$(10^9\ L_\odot)$		25%	0.33	2.41	5.78	4.35	1.53	0.27
		75%	12.07	25.40	30.15	21.92	9.72	10.63
		Number	263	272	688	701	893	95
		Detect	131	205	553	519	442	78
	LSc	median	0.40	1.96	3.74	4.84	0.78	0.55
		25%	0.15	0.75	1.27	1.51	0.34	0.17
		75%	0.90	3.98	7.88	11.52	2.16	1.83
		Number	155	108	171	214	289	69
		Detect	59	83	149	193	175	53

Table 1 Statistical properties of the morphological classes (*cont.*)

	Sample		E,S0	S0a,Sa	Sab,Sb	Sbc,Sc	Scd,Sd	Sm,Im
(B–V)	UGC	median	0.90	0.78	0.64	0.55	0.48	0.42
		25%	0.86	0.66	0.55	0.47	0.42	0.35
		75%	0.94	0.83	0.73	0.62	0.57	0.53
		Number	484	161	243	320	161	168
	LSc	median	0.89	0.78	0.62	0.52	0.48	0.42
		25%	0.84	0.65	0.55	0.44	0.42	0.35
		75%	0.92	0.84	0.71	0.59	0.56	0.51
		Number	421	156	182	282	180	210

Literature Cited

Aaronson M, Mould J. 1985. *Ap. J.* 290:191
Barbuy B, Renzini A., eds. 1992. *Proc. IAU Symp. 149.* Dordrecht:Kluwer
Barnes JE, Hernquist LE. 1992. *Annu. Rev. Astron. Astrophys.* 30:705
Bertola F, Burstein D, Buson LM. 1993. *Ap. J.* 403:573
Binggeli B. 1993. In *Panchromatic View of Galaxies.* In press
Binggeli B, Sandage A, Tammann GA. 1985. *Astron. J.* 90:1681
Binggeli B, Sandage A, Tammann GA. 1988. *Annu. Rev. Astron. Astrophys.* 26:509
Bothun GD, Lonsdale CJ, Rice W. 1989. *Ap. J.* 341:129
Bottinelli L, Gouguenheim L. 1974. *Astron. Astrophys.* 36:461
Braine J, Combes F. 1993. *Astron. Astrophys.* 269:7
Braine J, Combes F, Casoli F, Dupraz C, Gerin M, et al. 1993. *Astron. Astrophys.*S 97:887
Bregman JN, Hogg DE, Roberts MS. 1992. *Ap. J.* 387:484
Broeils AH. 1992. *Dark and Visible Matter in Spiral Galaxies.* PhD thesis. Univ. Groningen
Brosche P. 1971. *Astron. Astrophys.* 13:293
Bruzual G. 1992. See Barbuy & Renzini 1992, p. 311
Buat V, Deharveng JM. 1988. *Astron. Astrophys.* 195:60
Burstein D, Haynes MP, Faber S. 1991. *Nature* 353:515
Buta R. 1992a. In *Morphological and Physical Classification of Galaxies,* ed. G Longo, M Capaccioli, G Busarello, p. 1. Dordrecht:Kluwer

Buta R. 1992b. See Thuan et al 1992, p. 3
Byun Y. 1993. *Publ. Astron. Soc. Pac.* 105:993
Canizares CR, Fabbiano G, Trinchieri G. 1987. *Ap. J.* 312:503
Casertano S, van Gorkom JH. 1991. *Astron. J.* 101:1231
Cayatte V, van Gorkom JH, Balkowski C, Kotanyi C. 1990. *Astron. J.* 100:604
Chiosi C, Bertelli G, Bressan A. 1992. See Barbuy & Renzini 1992, p. 321
Cohen RS, Dame TM, Garay G Montani J, Rubio M, Thaddeus P. 1988. *Ap. J.L* 331:L95
Combes F. 1992. See Thuan et al 1992, p. 35
Combes F, Casoli F., eds. 1991. *Proc. IAU Symp. 146.* Dordrecht:Kluwer
Condon JJ. 1992 *Annu. Rev. Astron. Astrophys.* 30:575
Condon JJ, Broderick JJ. 1988. *Astron. J.* 96:30
Condon JJ, Frayer DT, Broderick JJ. 1991. *Astron. J.* 101:362
Corradi RLM, Capaccioli M. 1991. *Astron. Astrophys.* 90:121
Cox P, Mezger PG. 1989. *Astron. Astrophys. Rev.* 1:49
Da Costa GS. 1992. See Barbuy & Penzini 1992, p. 191
de Jong T, Clegg PE, Soifer BT, Rowan-Robinson M, et al. 1984. *Ap. J.L* 278:L67
de Vaucouleurs G. 1959. In *Handbuch der Physik,* Vol. 53, ed. S. Flugge, p. 275. Berlin:Springer-Verlag
de Vaucouleurs G, de Vaucouleurs A, Corwin HG, Buta RJ, Paturel G, Fouque P. 1991. *Third Reference Catalogue of Bright Galaxies.* New York:Springer-Verlag
Devereux NA, Eales SA. 1989. *Ap. J.* 340:708

Devereux NA, Young JS. 1990. *Ap. J.* 359:42
Devereux NA, Young JS. 1991. *Ap. J.* 371:515
Devereux NA, Young JS. 1992. *Astron. J.* 103:1536
Diaz AI. 1992. In *Evolutionary Phenomena In Galaixes*, ed. JE Beckman, BEJ Pagel, p. 377. Cambridge:Cambridge Univ. Press
Disney M, Davis J, Phillipps S. 1989. *MNRAS* 239:939
Disney M, Phillips S. 1983. *MNRAS* 205:1253
Distefano A, Rampazzo R, Chincarini G, de Souza R. 1990. *Astron. Astrophys.* S 86:7
Dressler A. 1980. *Ap. J.* 236:351
Dressler A. 1984. *Annu. Rev. Astron. Astrophys.* 22:185
Dressler A, Gunn JE. 1992. *Ap. J.* S 78:1
Fabbiano G. 1989. *Annu. Rev. Astron. Astrophys.* 27:87
Fabbiano G, Gioia IM, Trinchieri G. 1988. *Ap. J.* 324:749
Fabbiano G, Kim D-W., Trinchieri G. 1992. *Ap. J.* S 80:531
Faber SM, Gallagher J. 1979. *Annu. Rev. Astron. Astrophys.* 17:135
Feigelson ED, Nelson PI. 1985. *Ap. J.* 293:192
Fitt AJ, Alexander P, Cox MJ. 1988. *MNRAS* 233:907
Freeman KC. 1970. *Ap. J.* 160:811
Garnett DR, Shields GA. 1987. *Ap. J.* 317:82
Gilmore G, Hodge P, van der Kruit PC. 1990. *The Milky Way as a Galaxy*, Mill Valley:Univ. Sci. Books
Gilmore G, Roberts MS. 1988. *Comments Astrophys.* 12:123
Giovanelli R, Haynes MP. 1985. *Ap. J.* 292:404
Giovanelli R, Haynes MP. 1990. In *Galactic and Extragalactic Radio Astronomy*, ed. GL Verschuur, KI Kellermann, p. 522. Berlin:Springer-Verlag
Giovanelli R, Haynes MP, Salzer JJ, Wegner G, da Costa LN, Freudling W. 1994. *Astron. J.* In press
Giovanelli R, Haynes MP. 1991. *Annu. Rev. Astron. Astrophys.* 29:499
Gizis JE, Mould JR, Djorgovski S. 1993. *Publ. Astron. Soc. Poc.* 105:871
Gordon MA. 1990. *Ap. J.* L 350:L29
Gorgas J, Efstathiou G, Salamanca A. 1990. *MNRAS* 245:217
Guhathakurta P, van Gorkom JH, Kotanyi CG, Balkowski C. 1988. *Astron. J.* 96:851
Gunn JE. 1990. See Oegerle et al 1990, p. 341
Haynes MP. 1990. See Oegerle et al 1990, p. 177
Haynes MP, Giovanelli R. 1986. *Ap. J.* 306:466
Haynes MP, Giovanelli R, Chincarini GL. 1984. *Annu. Rev. Astron. Astrophys.* 22:445
Haynes MP, Herter T. 1988. *Astron. J.* 96:504
Hewitt J, Haynes MP, Giovanelli R. 1983. *Astron. J.* 88:272
Hogg DE, Roberts MS, Sandage A. 1993. *Astron. J.* 106:907
Holmberg E. 1958. *Medd. Lund. Ast. Obs. Ser.* II, No. 136
Hubble E. 1926. *Ap. J.* 64:321
Hubble E. 1936. *The Realm of the Nebulae.* New Haven:Yale Univ. Press
Hubble E, Humason M. 1931. *Ap. J.* 74:43
Hummel E. 1990. In *Windows on Galaxies*, ed. G Fabbiano, JS Gallagher, A Renzini, p. 141. Dordrecht:Kluwer
Hummel E, van der Hulst JM, Dickey JM. 1984. *Astron. Astrophys.* 134:207
Kenney J, Young JS. 1989. *Ap. J.* 344:171
Kennicutt RC. 1988. *Ap. J.* 334:144
Kennicutt RC. 1989. *Ap. J.* 344:685
Kennicutt RC. 1990. In *The Interstellar Medium in Galaxies*, ed. HA Thronson, JM Shull, p. 405. Dordrecht:Kluwer
Kennicutt RC, Edgar BK, Hodge PW. 1989. *Ap. J.* 337:761
Kent S. 1987. *Astron. J.* 93:816
Koo DC, Kron RG. 1992. *Annu. Rev. Astron. Astrophys.* 30:613
Larson RB, Tinsley BM, Caldwell N. 1980. *Ap. J.* 237:692
Lavery RJ, Pierce MJ, McClure RD. 1992. *Astron. J.* 104:2067
Lonsdale Persson CJ, Helou G. 1987. *Ap. J.* 314:513
Magri C, Haynes MP, Forman W, Jones C, Giovanelli R. 1988. *Ap. J.* 333:136
Meisels A. 1983. *Astron. Astrophys.* 118:21
Nilson P. 1973. *Uppsala General Catalogue of Galaxies*, Acta Univ. Ups. Ser. V:A, Vol. 1, Uppsala
Oegerle WR, Fitchett MJ, Danly L, eds. 1990. *Clusters of Galaxies.* New York:Cambridge Univ. Press
Oey MS, Kennicutt RC. 1993. *Ap. J.* 411:137
Ohta K, Tomita A, Saito M, Sasaki M, Nakai N. 1993. *Publ. Astron. Soc. Japan.* 45:L21
Pagel BEJ, Edmunds MG. 1981. *Annu. Rev. Astron. Astrophys.* 19:77
Persic M, Salucci P. 1991. *Ap. J.* 368:60
Pogge RW, Eskridge PB. 1993. *Astron. J.* 106:1405
Postman M, Geller M. 1984. *Ap. J.* 281:95
Roberts MS. 1969. *Astron. J.* 74:859
Roberts MS, Hogg DE, Bregman JN, Forman WR, Jones G. 1991. *Ap. J.* S 75:751
Rocca-Volmerange B. 1992. See Barbuy & Renzini, 1992. p. 357
Rowan-Robinson M. 1990. In *The Interstellar Medium in Galaxies*, ed. HA Thronson, JM Shull, p. 121. Dordrecht:Kluwer
Rubin VC. 1991. In *After the First Three Minutes*, ed. SS Holt, CL Bennett, V Trimble, p. 371. New York:Am. Inst. Phys.
Rubin VC, Whitmore BC, Ford WK Jr. 1988. *Ap. J.* 333:522
Rubio M Garay G, Montani J, Thaddeus P. 1991. *Ap. J.* 368:173
Sage LJ. 1993. *Astron. Astrophys.* 272:123
Sandage A. 1961. *The Hubble Atlas of Galaxies*, Washington:Carnegie Inst. Washington

Sandage A. 1975. In *Stars and Stellar Systems,* Vol. 9, ed. A Sandage, M Sandage, J Kristian, p. 1. Chicago:Univ. Chicago Press

Sandage A. 1990. See Oegerle et al 1990, p. 201

Sandage A, Bedke J. 1993. *Carnegie Atlas of Galaxies,*, Washington:Carnegie Inst. Washington. In press

Sandage A, Binggeli B, Tammann GA. 1985. *Astron. J.* 90:1759

Sandage A, Tammann GA. 1987. *A Revised Shapley-Ames Catalog of Bright Galaxies.* Washington:Carnegie Inst. Washington

Sanders DB, Soifer BT, Neugebauer G, Scoville N, Madore BF, et al. 1987. In *Star Formation in Galaxies,* ed. CJ Lonsdale Persson, p. 411. Washington:NASA

Sarazin CL. 1992. See Thuan et al 1992, p. 51

Sauvage M, Thuan TX. 1992. *Ap. J.*L 396:L69

Sauvage M, Thuan TX. 1993. *Ap. J.* In press

Scalo JM. 1986. *Fund. Cosmic Phys.* 11:1

Schechter P. 1980. *Astron. J.* 85:801

Sersic JL. 1960. *Z. Astrophys.* 50:168

Shields GA, Skillman ED, Kennicutt RC. 1991. *Ap. J.* 371:82

Skillman ED, Kennicutt RC, Hodge PW. 1989. *Ap. J.* 347:875

Sofue Y, Wakamatsu K. 1993. *Publ. Astron. Soc. Jpn.* 45:529

Soifer BT, Houck JR, Neugebauer G. 1987. *Annu. Rev. Astron. Astrophys.* 25:187

Strom KM, Strom SE. 1978. *Astron. J.* 83:73

Telesco CM. 1988. *Annu. Rev. Astron. Astrophys.* 26:343

Thuan TX, Balkowski C, Van JTT, eds. 1992.

Physics of Nearby Galaxies. Gif-sur-Yvette: Frontiers

Toomre A. 1964 *Ap. J.* 139:1217

Toomre A, Toomre J. 1972. *Ap. J.* 178:623

Tully RB, Fisher JR. 1977. *Astron. Astrophys.* 54:661

Valentijn E. 1991, *Nature* 346:153

van Driel W, van Woerden H. 1991. *Astron. Astrophys.* 243:71

Verter F. 1985 *Ap. J.* S 57:261

Verter F. 1990 *Publ. Astron. Soc. Pac.* 102:1281

Vila-Costas MB, Edmunds MG. 1992. *MNRAS* 259:121

Warmels RH. 1988a. *Astron. Astrophys.*S 72:19

Warmels RH. 1988b. *Astron. Astrophys.*S 72:57

Watanabe M, Kodaira K, Okamura S. 1985. *Ap. J.* 292:72

Whitmore, B. 1984. *Ap. J.* 278:61

Whitmore B. 1990. See Oegerle et al 1990, p. 139

Whitmore BC, Forbes DA, Rubin VC. 1988. *Ap. J.* 333:542

Whitmore BC, McElroy DB, Tonry JL. 1985. *Ap. J.* S 59:1

Worthy G, Faber S, Gonzalez J. 1992. *Ap. J.* 398:69

Wray JD, de Vaucouleurs G. 1980 *Astron. J.* 85:1

Xu C. 1990 *Ap. J.* L 365:L47

Young JS, Knezek PM. 1989. *Ap. J.* L 347:L55

Young JS, Scoville NZ. 1991. *Annu. Rev. Astron. Astrophys.* 29:581

Zwicky F. 1957. *Morphological Astronomy.* Berlin:Springer-Verlag

Annu. Rev. Astron. Astrophys. 1994. 32: 153–90

THE r-, s-, AND p-PROCESSES IN NUCLEOSYNTHESIS

Bradley S. Meyer

Department of Physics and Astronomy, Clemson University, Clemson, South Carolina 29634-1911

KEY WORDS: heavy elements, supernovae, AGB stars, abundances

1.   INTRODUCTION

Burbidge et al (1957) and Cameron (1957) laid out the framework for our understanding of the formation of the heavy nuclei (those nuclei with mass number $A \gtrsim 70$). From systematics in the solar system abundance distribution, Burbidge et al determined that the heavy nuclei were formed in three distinct nucleosynthetic processes, which they termed the r-, s-, and p-processes. That we still use these terms today is a credit to the soundness of this work done 37 years ago.

We may understand how Burbidge et al and Cameron arrived at their conclusions from Figure 1. One population of nuclei, the s-nuclei, shows an abundance distribution with peaks near mass numbers 87, 138, and 208. These nuclei are made in a *slow* neutron-capture process, the s-process. A *rapid* neutron-capture process, the r-process, is responsible for the r-nuclei, whose abundance distribution shows peaks at mass numbers 80, 130, and 195. The p-process is responsible for production of the rarer, more proton-rich heavy isotopes (the p-nuclei) that cannot be made by neutron capture.

The first quantitive evaluations of the ideas of Burbidge et al and Cameron came to light in the early 1960s with work on the s-process (Clayton et al 1961, Seeger et al 1965) and the r-process (Seeger et al 1965). These calculations further elucidated the mechanisms for heavy-element formation and showed the plausibility of the framework developed in the 1950s. Subsequent work has focused on determining the astrophysical sites where the r-, s-, and p-processes occurred with the help of improved nuclear details, stellar models, and abundances. A goal of this paper is to review the recent progress astrophysicists,

153

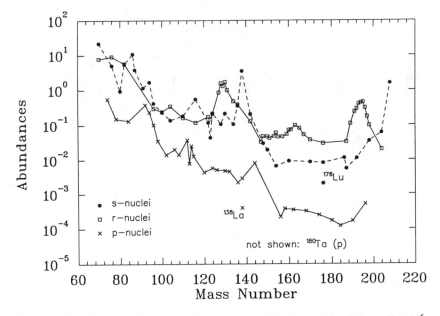

Figure 1 The solar system abundances of r-nuclei, s-nuclei, and p-nuclei, relative to Si $= 10^6$. Only isotopes for which 90% or more of the inferred production comes from a single process are shown. The data are from Anders & Grevesse (1989) and Käppeler et al (1989).

astronomers, and physicists have made in these directions and to point out the problems that remain in our understanding of the formation of the heavy nuclei. Another, perhaps deeper, goal is to to seek some understanding of why there are three major processes available to nature for synthesis of heavy elements.

It is impossible for a single paper to cover all relevant aspects of the r-, s-, and p-processes; therefore, where possible, references to other reviews are given. Readers should turn to these reviews for more details. Nevertheless, it is hoped that the present paper gives some flavor for the rich field of heavy-element synthesis.

2.   GENERAL CONSIDERATIONS

The Master said, "Ssu, I believe you look upon me as one whose aim is simply to learn and retain in mind as many things as possible." He replied, "That is what I thought. Is it not so?" The Master said, "No; I have one thread on which I string them all."

The Analects of Confucius

The r-, s-, and p-processes are distinct nucleosynthetic mechanisms. They occur in different environments and under quite different conditions. Nevertheless, it is useful to seek some unifying concept by which we may understand these disparate processes. We will see that entropy is the concept we need. Careful consideration of entropy in the various nucleosynthetic processes will clarify our discussion and give us insight into how these processes occur and why they occur where they do. With this insight we will see how the r-, s-, and p-processes are each unique answers to the same question: "How does nature produce heavy elements?"

2.1   *Entropy and Equilibrium*

Let us begin our discussion by considering a given thermally isolated system at constant volume. The system has total energy E_0. The entropy of this system is

$$S = k \ln \Gamma, \tag{1}$$

where k is Boltzmann's constant and Γ is the number of energetically allowed macroscopic states available to the system. By macroscopic state we mean a particular distribution of the constituents of the system among their single-particle quantum-mechanical states. In essence, a macroscopic state of the system is one particular way the constituents of the system can share the total energy E_0. Suppose the system does not have all macroscopic states of energy E_0 available to it. In this case, the entropy is less than its maximum. The system will evolve by the Second Law of Thermodynamics and add more macroscopic states to its repertoire. In this evolution, the entropy will thus increase. The system will continue to evolve until all macroscopic states of energy E_0 are available to the system. Once the system reaches this point, it is at maximum entropy and experiences no further evolution. The system has attained equilibrium. We thus see that equilibrium, or maximum entropy, is the evolutionary endpoint of any thermally isolated system.

We can now use these considerations to ask what happens to the nucleons and nuclei in some nucleosynthetic environment. If the system of nucleons and nuclei is thermally isolated and out of equilibrium, it will evolve towards equilibrium. Given enough time, the system will reach equilibrium and attain maximum entropy. At this point, it is in nuclear statistical equilibrium (NSE), and it is a simple matter to compute the abundance of any nuclide (Burbidge et al 1957). We find (e.g. Meyer 1993) that, for a nuclear species of atomic number Z and mass number A, the abundance per baryon $Y(Z, A)$ is

$$Y(Z, A) = G(Z, A)[\zeta(3)^{A-1}\pi^{(1-A)/2}2^{(3A-5)/2}]A^{3/2}\left(\frac{kT}{m_N c^2}\right)^{3(A-1)/2}$$
$$\times \phi^{1-A}Y_p^Z Y_n^{A-Z}e^{[\frac{B(Z,A)}{kT}]}, \tag{2}$$

where $G(Z, A)$ is the nuclear partition function, $\zeta(3)$ is the Riemann zeta function of argument 3, T is the temperature, m_N is the mass of a single baryon, ϕ is the photon-to-baryon ratio, Y_p is the abundance per baryon of protons, Y_n is the abundance per baryon of neutrons, and $B(Z, A)$ is the binding energy of nucleus (Z, A). We note that ϕ is given by

$$\phi = \frac{2}{\pi^2} \frac{1}{(\hbar c)^3} \frac{\zeta(3)(kT)^3}{\rho \mathcal{N}_A}, \tag{3}$$

where $\mathcal{N}_A$ is Avagadro's number and ρ is the baryon mass density. The binding energy of nucleus (Z, A) is

$$B(Z, A) = [Zm_p + Nm_n - m(Z, A)]c^2, \tag{4}$$

where $N = A - Z$ and $m(Z, A)$, m_p, and m_n are the masses of nucleus (Z, A), the proton, and the neutron, respectively.

From Equation (2), we see that the NSE abundance of nuclei is nonzero. This might at first be surprising. If we combine free neutrons and protons into nuclei, we decrease the number of free particles of the system. This would yield fewer ways of sharing the total energy of the system and thus decrease the number of macroscopic states available to the system. The nuclear reactions also release binding energy, however, which increases the number of photons in the system, the energy available to leptons, and the excitation energy in the nuclei. These effects increase the number of ways the system can share the total energy of the system and, hence, can increase the number of macroscopic states available to the system. This increase can more than compensate for the decrease in the number of states due to the loss of free particles and can lead to an increase in the entropy. Once the system has evolved to the point that it experiences no net increase in the number of macroscopic states by changing the abundance of any particular nucleus, the system has reached NSE.

Which nuclei dominate the abundance distribution in NSE? This depends on the photon-to-baryon ratio or, equivalently, the entropy per baryon since this latter quantity scales monotonically with the photon-to-baryon ratio (e.g. Meyer & Walsh 1993). From Equation (2) we see that the abundance of some heavy nucleus (Z, A) depends on ϕ^{1-A}. At some given temperature then, the larger ϕ is, the smaller will be the abundance of the heavy nucleus (Z, A). Fewer heavy nuclei means more light nuclei and free nucleons. The strong dependence of the NSE abundances on ϕ will be crucial for the r-process. This dependence on ϕ is apparent in Figure 2. If we fix the temperature, the larger ϕ is, the more likely nucleons are free or contained in light nuclei.

2.2   Equilibrium Nucleosynthesis

We have seen that the goal of any nucleosynthetic process is equilibrium. How does this goal vary with temperature? The answer is important because nature

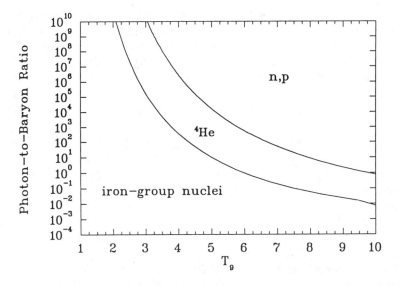

Figure 2 The dominant species in NSE for a gas with electron-to-baryon ratio $Y_e = 0.5$. The different regions of the plot show at what temperatures and photon-to-baryon ratios ϕ the various species dominate the gas. At high temperatures and high photon-to-baryon ratios, free neutrons and protons (labeled n, p) dominate. At lower temperatures and photon-to-baryon ratios, the nucleons are mostly locked up in ^{4}He nuclei. At still lower temperatures and photon-to-baryon ratios, the gas will predominantly be a distribution of iron-group nuclei.

synthesizes nuclei at high temperature and ejects them into the medium between the stars at low temperature. If the nuclei stayed in equilibrium throughout this processing, we could use Equation (2) to determine the abundance of the various species. That the nuclei emerging from some nucleosynthetic process are not in general in an equilibrium distribution gives us important clues about how they were in fact made. We will turn to this point after discussing equilibrium nucleosynthesis in some detail.

Consider a system in NSE at some temperature. If the system expands adiabatically, the temperature will decrease. By Le Chatelier's Principle we know the system will respond to this stress by tending to counter it. How does the system do this? The answer is that nuclear reactions occur and release binding energy. This energy heats up the system and tends to counter the temperature decrease from the expansion. Via these nuclear reactions, the system finds a new equilibrium if the reactions occur faster than the expansion.

As we have noted in Section 2.1, if nuclear reactions assemble free particles and light nuclei into heavier nuclei, there is a loss of macroscopic states and a decrease in the entropy in baryons (free nucleons and nuclei). The binding energy released in the reactions goes into photons and leptons. This increases the numbers of macroscopic states and, hence, the entropy in the photons and leptons. We see that in building up the abundances of heavy nuclei, we transfer entropy from the baryons to the photons and leptons. This is apparent in Figure 3. Suppose a system at an entropy per baryon of $10k$ begins in NSE at $T_9 = 10$ ($T_9 \equiv T/10^9\text{K}$) and expands adiabatically. The system cools. Because the system maintains equilibrium, entropy leaves the baryons as the baryons assemble themselves into heavier nuclei via nuclear reactions. This entropy goes into the photons and leptons (electron-positron pairs).

The entropy in baryons continues to fall until $T_9 \approx 4$. At this point, the system has released essentially all its nuclear binding energy. There is little further transfer of entropy from the baryons to the other constituents of the system. We also note that beginning around $T_9 = 4$ the electron-positron pairs annihilate into two photons. This transfers entropy from the $e^+ - e^-$ pairs into the photons adiabatically. At late times, after the $e^+ - e^-$ pairs have annihilated, the system is left with a residual abundance of electrons to ensure charge neutrality. After the pairs have disappeared, the system is a mix of relativistic photons and nonrelativistic nuclei and electrons. Now for adiabatic expansion, $\rho \propto T^3$ for relativistic particles and $\rho \propto T^{3/2}$ for nonrelativistic particles. The system at late times, because it is a mix of these two types of particles, actually expands such that $\rho \propto T^b$ where b is between 3/2 and 3. This means that entropy now is transferred from the photons to the translational

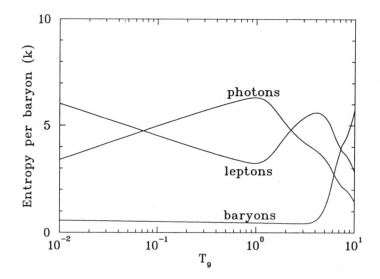

Figure 3 The entropy per baryon in the consitituents of a system expanding adiabatically with total entropy per baryon $10k$ and electron-to-baryon ratio Y_e. Note that the sum of the entropies of the constituents is always $10k$.

entropy of the nonrelativistic particles. This is why the entropy in photons declines after $T_9 = 1$ and that in leptons and baryons rises.

What nuclei are made in such an expansion? Again we note that nuclear reactions tend to occur only if there is a release of binding energy to compensate for the loss of macroscopic states due to the decrease in the number of free particles. As our system expands and cools, the nuclei present in equilibrium will increase in mass. They will continue to increase in mass until they have reached the nuclide with the largest binding energy per nucleon. Once this nuclide dominates the abundance of nuclei, there is no further evolution in the abundances. This is because if the system attempts to arrange the nucleons into a more massive nucleus with a lower binding energy per nucleon, there will be a net decrease in the total binding energy of the system. Reactions giving this rearrangement of nucleons lead to a decrease in the number of free particles and in the energy available to the photons and leptons. There will thus be fewer ways to share the energy of the system and a decrease in the entropy. An expanding and cooling system can maintain NSE by driving the system to a composition dominated by the nuclear species with the largest binding energy per nucleon.

Figure 4 shows the mass fraction $X(Z, A) = AY(Z, A)$ in NSE at temperatures $T_9 = 1$–10 for an entropy per baryon of $10k$ and a net electron-to-baryon ratio $Y_e = 0.5$. If we imagine the system begins at $T_9 = 10$, alpha particles

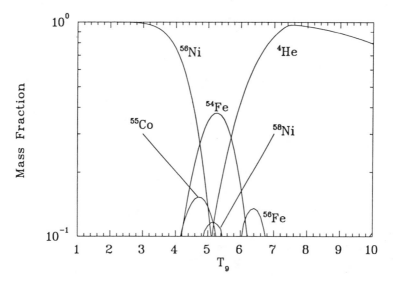

Figure 4 The mass fraction of species in NSE as a function of temperature for a system with entropy per baryon $10k$ and $Y_e = 0.5$.

dominate the mass of the system. As the system expands and cools adiabatically, the abundances of heavier nuclei increase. At late times, however, all nucleons are locked up in ^{56}Ni nuclei. This is because ^{56}Ni has one proton for every two nucleons ($Y_e = 0.5$), and it is the nucleus at $Y_e = 0.5$ with the strongest binding energy per nucleon.

Figure 5 shows what happens if the entropy per baryon is $100k$ instead of $10k$. Because of the larger entropy, the system has more free particles and light nuclei at a given temperature. This is why free neutrons and protons dominate the abundances around $T_9 = 10$ in Figure 5 while alpha particles dominate the abundances in Figure 4. We also see that alpha particle abundances build up only after T_9 drops below 10 for entropy per baryon of $100k$. As in the lower entropy case, however, the nucleons eventually end up all in ^{56}Ni.

What happens now if Y_e is different from 0.5? Figure 6 shows what happens if the entropy per baryon is $10k$ and $Y_e = 0.4$. Here there are six neutrons for every four protons. We see that the system at late times locks up all its nucleons into ^{70}Ni. For $Y_e = 0.4$, this is the nucleus with the largest binding energy per nucleon.

From Figure 6 we see that the final nucleus on which the system converges at low temperature is dependent on Y_e because the final nuclei present must accommodate all of the neutrons and protons initially in the system. This final nucleus, however, is the nucleus with the largest binding energy for that Y_e.

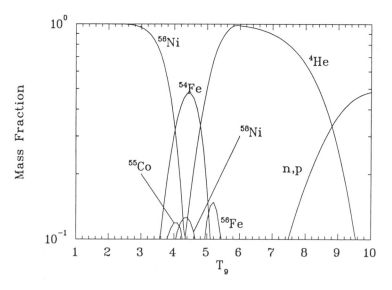

Figure 5 The mass fraction of species in NSE as a function of temperature for a system with entropy per baryon $100k$ and $Y_e = 0.5$.

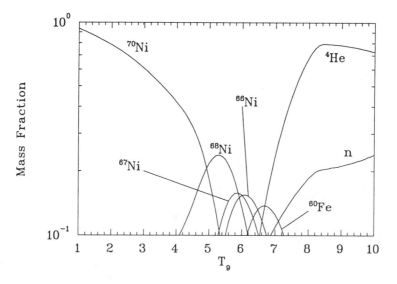

Figure 6   The mass fraction of species in NSE as a function of temperature for a system with entropy per baryon $10k$ and $Y_e = 0.4$.

Figure 7 gives an idea of what that nucleus is for Y_e ranging from 0.35 to 0.6. In all cases, the nuclei are iron-group nuclei ($Z = 26–34$).

2.3   How to Make Heavy Nuclei

We are now in a position to ask the question "How does nature make heavy nuclei?" We have seen in the previous sections that if we eject nuclei into cold interstellar space, and if these nuclei are always in NSE, then these nuclei must be those with the largest binding energy per nucleon for whatever Y_e is appropriate for the environment in which the nuclei find themselves. From Figure 7, however, we see that for a large range of Y_e, these nuclei are simply iron-group nuclei. If NSE always pertains, stars and supernovae can only eject iron-group nuclei. Our observations of uranium on Earth tell us that this is not what happens.

The escape from this dilemma is the fact that nucleosynthetic systems cannot always be in NSE. There are only two ways this can happen. The first possibility is that the system never has time to come into NSE before the star ejects the nucleons into the interstellar medium. In this case the nucleons assemble themselves part of the way up to iron. This is the "falling short of equilibrium" scenario. The second possibility is that the system begins in NSE at high temperature. As the system expands and cools, the equilibrium changes. As the temperature drops, some nuclear reactions slow down. Eventually these

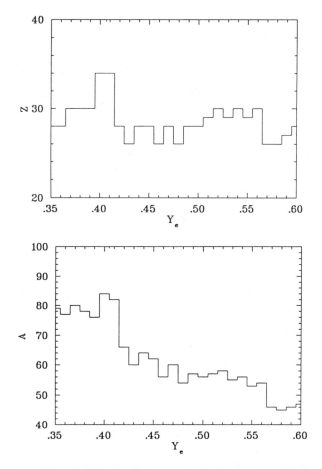

Figure 7 The proton number Z (*top*) and the mass number A (*bottom*) of the species with the largest binding energy per nucleon in each bin in Y_e. Where experimental data are available, they are from Wapstra et al (1988). Where experimental data are unavailable, nuclear masses are from Möller & Nix (1988).

reactions become too slow to allow the system to maintain equilibrium. This is the "freezeout from equilibrium" scenario. These are the only two ways a nucleosynthetic system can be out of equilibrium, and they are the only two options nature has to assemble heavy nuclei.

How in fact does nature make heavy elements in these two scenarios? In the freezeout from equilibrium, a nucleosynthetic system falls out of equilibrium. At this point the composition reflects the last NSE abundance distribution the system attained before it fell out of equilibrium. This composition is a mix of free nucleons, light nuclei, and iron-group nuclei. While some of the reactions necessary to maintain equilibrium become too slow, others such as capture of free nucleons and light nuclei on iron-group nuclei can continue. The iron-group nuclei can serve as "seeds" for the capture of the remaining free nucleons and light nuclei. These captures can produce heavy nuclei. Because the system assembles its own seed nuclei in such a scenario, the system can make heavy elements without any pre-existing seed nuclei. A process that does not require pre-existing seed nuclei to make new nuclei is called primary.

In the other scenario—the falling short of equilibrium scenario—the material never achieves NSE because the timescale to reach equilibrium is always too long compared to the dynamical timescale of the system. The system assembles light nuclei into somewhat heavier nuclei but these are still less massive than the iron-group nuclei characteristic of NSE. An accidental effect of the main nuclear reactions between the light nuclei in this scenario is the liberation of nucleons. If iron-group or heavier nuclei already exist in the system, they can capture these nucleons to produce new heavy nuclei. In this scenario, then, the system must have pre-existing seed nuclei in order to produce heavy elements. We call a process that requires pre-existing seed nuclei to make new elements a secondary process.

These are the only two ways nature can assemble heavy elements. We should not be surprised then that there are two major distributions of heavy nuclei—the r-process and the s-process distributions. We shall see that the r-nuclei in our Solar System likely formed in an environment that experienced a freezeout from equilibrium while the s-nuclei must have formed in an environment that was striving for, but never reached, NSE. The differing character of these scenarios results in the different character of the r- and s-process abundance distributions.

Once an abundance of heavy elements is available, nature may make modifications to it by exposing it to a flux of photons, neutrinos, or nucleons. Such events are probably responsible for the production of the majority of p-nuclei.

3.   THE r-PROCESS

Nature does not endure sudden mutations without great violence.

François Rabelais, *Gargantua*

We turn now to a discussion of the r-process. We seek to understand in some detail how and where the r-process occurs. How the r-process occurs depends on whether it is a primary or secondary process. We will see that the evidence available to us today indicates that the r-process is a primary, freezeout from equilibrium process. As for where the r-process occurs, the rapid timescales associated with the r-process point to violent events such as supernovae or disruptions of neutron stars. Winds from nascent neutron stars are currently believed to be the most probable site. Our discussion will be brief. For more details, the reader should turn to the many excellent reviews in the literature (e.g. Hillebrandt 1978, Mathews & Cowan 1990, Cowan et al 1991).

3.1   The Primary r-Process

Let us return to the notion of a freezeout from equilibrium. As a system in NSE expands and cools, the abundances shift to maintain NSE. We have seen that in doing so the entropy of the system moves from the baryons into the photons and leptons and the abundance of heavier nuclei grows at the expense of free nucleons and light nuclei. Eventually the temperature of the system is too low or the abundance of reactants too small for certain reactions to go fast enough to maintain NSE. These reactions freeze out. The first reactions to freeze out are charged-particle reactions. Neutron-capture reactions can continue, however, because they are not impeded by a nuclear Coulomb barrier. The nuclei present at the time of the freezeout of the charged-particle reactions then eventually capture the remaining neutrons.

If we suppose that the system is quite neutron rich, many free neutrons should exist after freezeout of the charged-particle reactions. One might imagine that as we increase the neutron richness of the system, we simply have more neutron-rich isotopes of the nuclei present in the system. However, there is a limit to how neutron rich the nuclei can get. This is because the nuclei eventually encounter neutron drip, at which point the binding energy of the next neutron captured is negative. Once the nuclei reach neutron drip, they cannot contain any more neutrons. Thus, for sufficiently neutron-rich material, there may be many free neutrons at the time of charged-particle reaction freezeout. The system establishes an equilibrium between the neutron-capture (n, γ) and neutron-disintegration (γ, n) reactions. Beta decays then occur which increase the charge on the nucleus and allow further neutron capture. This phase of the freezeout from equilibrium in which only neutron-capture, neutron-disintegration, and beta-decay reactions occur is the r-process.

How neutron rich must material be to undergo an r-process? We know that the r-process produces uranium ($A = 238$) from seed nuclei ($A \approx 50$–100); therefore, there must be roughly 100 free neutrons per seed nucleus at the time of charged-particle reaction freezeout. The required neutron richness thus depends on the abundance of seed nuclei which in turn depends on the entropy per baryon in the system.

If the entropy per baryon is less than or roughly $10k$, NSE favors iron-group nuclei. These are the seed nuclei for the r-process once the charged-particle reactions have frozen out. Because the r-process needs around 100 free neutrons per seed, a seed nucleus like ^{78}Ni, typical in a low entropy, neutron-rich freezeout, requires a total neutron-to-proton (n/p) ratio of $(50 + 100)/28 \approx 5.4$ or a $Y_e = p/(n + p) \approx 0.16$. This is quite neutron-rich material.

If the entropy per baryon is high ($\gtrsim 100k$), NSE favors ^{4}He at high temperature. As the temperature drops in the expanding material, NSE begins to favor iron-group nuclei. For large entropy per baryon, this occurs late in the expansion. At this late time, the two reaction sequences that begin the assembly of alpha particles into iron-group nuclei, $^4\text{He} + {}^4\text{He} + {}^4\text{He} \rightarrow {}^{12}\text{C}$ and $^4\text{He} + {}^4\text{He} + \text{n} \rightarrow {}^9\text{Be}$ followed by $^9\text{Be} + {}^4\text{He} \rightarrow {}^{12}\text{C} + \text{n}$, may not be operating at a significant level or may even have frozen out. We note that these reaction sequences rely on three-body interactions which are highly sensitive to the density. The higher the entropy per baryon at a given temperature, the larger the photon-to-baryon ratio and the lower the density. A lower density results in slower three-body reaction rates. Thus the higher the entropy per baryon, the higher the temperature at which these three-body reactions freeze out and the lower the abundance of seed nuclei. Unlike the three-body reactions, alpha captures on ^{12}C and heavier nuclei occur rapidly and build up heavy nuclei. As the system falls out of equilibrium, there will be more alpha particles around than there would be in NSE. Some of these alpha particles capture on the iron-group nuclei present to make heavier nuclei ($A \approx 100$). Eventually these reactions freeze out also, and the system is left with an abundance of free neutrons, seed nuclei, and many ^4He nuclei. At this point the free neutrons can capture on the seed nuclei, but not on the ^4He nuclei. In this way an r-process occurs.

The degree of neutron richness necessary for an r-process in a high entropy environment is less than in a low entropy environment. A typical composition for an entropy per baryon of $400k$ and a neutron-to-proton ratio of 1.6 ($Y_e = 0.385$) is 20% free neutrons by mass, 10% seed nuclei, and 70% alpha particles (Woosley & Hoffman 1992). Suppose the seed nucleus is $Z = 35$, $A = 100$ (also typical), then the free neutron-seed nucleus ratio is 200 in the case: more than sufficient for an r-process. We thus see that the degree of neutron richness we need for a primary r-process directly depends on the entropy per baryon.

We have to this point discussed how a primary r-process would occur in the freezeout from NSE in low entropy and high entropy environments. The r-process might also occur in a very low entropy or zero-entropy environment. Matter inside a neutron star is extremely neutron rich and highly-degenerate. If the neutron star is more than a few hours old, it is cold ($T \lesssim 10^9$ K) on a nuclear energy scale (e.g. Baym & Pethick 1979), implying that the entropy per baryon is quite low ($\lesssim 0.5k$). At densities below nuclear density ($\rho \approx 2 \times 10^{14} \text{g cm}^{-3}$) but above neutron-drip density ($\rho \approx 4 \times 10^{11} \text{g cm}^{-3}$), extremely neutron-rich nuclei exist in strong equilibrium with degenerate neutrons and weak equilibrium with the degenerate electrons (Baym et al 1971). Above nuclear density, the material is comprised of free nucleons and electrons. Weak equilibrium forces the Y_e of the material to be of order 0.05 or less (Lattimer et al 1985, Lattimer & Swesty 1991).

Only the strong gravity of the neutron star keeps such matter from exploding apart. If a piece of cold neutron-star matter were to escape from the neutron star into interstellar space, it would decompress. If this escape occurred without too much violence, there would not be a dramatic increase in the entropy. The material would remain cold. The material would consist of neutrons and nuclei, and as the material expanded, an r-process could occur (Lattimer et al 1977, Meyer 1989). In such a system, the only heating that would occur would be from the beta decays and nuclear fissions during the r-process, which are irreversible processes. Notice that such an r-process would not be a freezeout from NSE. The system was not in nuclear statistical equilibrium prior to expansion, and it would not attain NSE later in the expansion unless the beta decays and nuclear fissions that occur could drive the temperature up high enough, something that probably does not happen (Meyer 1989). On the other hand, this would be a freezeout from weak equilibrium. Thus, such an r-process would be primary because it is the formation of the neutron star that creates the seed nuclei and the weak equilibrium would erase the entire previous history of the nucleons. Moreover, material that began at densities above nuclear matter density would experience a phase transition from free nucleons into neutrons and nuclei during the expansion. In this case, the seed nuclei would form during the decompression.

Now that we understand how the r-process occurs in freezeout from equilibrium, we may consider the question of what astrophysical sites could give low, very low, or high entropy r-processes. Let us begin with low entropy sites. The earliest site considered for a low entropy r-process was at the mass cut of a type II supernova (Burbidge et al 1957, Cameron 1957). The mass cut is the boundary between the matter that escapes into space from the supernova and the matter that remains as part of the remnant neutron star. The first serious time-dependent calculations of such an r-process were made by Seeger et al (1965), although this work assumed constant temperature and neutron number density. Later calculations treated the r-process as dynamical, that is, with varying temperature and neutron density (Cameron et al 1970, Schramm 1973, Sato 1974, Kodama & Takahashi 1975, Hillebrandt et al 1976, Hillebrandt 1978). While some of these workers found fairly good fits to the solar system r-process distribution, the models studied provided no natural way of explaining why the particular r-process distribution we see should emerge from a supernova and, more seriously, why a supernova should eject only a small mass of r-process matter (see Section 3.3). It is also important to note that current type II supernova models do not yield Y_e as low as 0.1–0.2 at the mass cut, as required by a low entropy r-process (e.g. Wilson & Mayle 1993, Woosley et al 1994).

Due to the difficulties with the mass-cut site, astrophysicists turned to other supernova scenarios. In particular, workers sought other means of ejecting low entropy, neutron-rich r-process matter. Rotating stellar cores with (LeBlanc & Wilson 1970, Meier et al 1976, Müller & Hillebrandt 1979) and without magnetic fields (Symbalisty 1984, Symbalisty et al 1985) can eject some neutron-matter. It is not clear, however, that the supernova cores can attain the high rotation rates and magnetic fields required for such ejection. Moreover, there is no natural explanation for why the material has just the right conditions to make a solar system r-process distribution.

What about very low entropy r-processes? Lattimer & Schramm (1974, 1976) considered the tidal disruption of a neutron star by a black hole. The ejected neutron-star matter could then undergo a very low entropy r-process (Lattimer et al 1977, Meyer 1989). Neutron star–neutron star collisions could also lead to the ejection of neutron-rich matter (Symbalisty & Schramm 1982, Eichler et al 1989; see also Kochanek 1992 and Colpi et al 1989, 1991, 1993). If this material were not strongly disturbed during the collision, it would undergo a very low entropy r-process. It is likely in such an event, however, that the material would be shocked to entropies of order several k per baryon, so that in fact a low entropy would probably ensue (Evans & Mathews 1988). In any case, these sites suffer the two difficulties of uncertain occurrence rates and the lack of any natural reason a solar system r-process distribution should result.

This leaves the high entropy r-process. A promising site for such an r-process is in the neutrino-driven winds from nascent neutron stars. We consider this site in more detail in Section 3.4.

3.2 The Secondary r-Process

A system that does not achieve equilibrium may also produce r-process elements. As we have seen, as the bulk of the nuclei strive to reach the iron group, the reactions that carry them in this direction may release neutrons that then capture on pre-existing seed nuclei. If the number of neutrons released is sufficiently large, a solar system r-process abundance distribution may result. The nuclear reactions that would be the dominant producers of neutrons are (α, n) reactions, and in particular the reactions $^{13}C(\alpha, n)^{16}O$, $^{22}Ne(\alpha, n)^{25}Mg$, and $^{25}Mg(\alpha, n)^{28}Si$.

A crucial feature of a secondary r-process is that it does not achieve $(n, \gamma) - (\gamma, n)$ equilibrium. Such equilibrium occurs when the flows from (n, γ) and (γ, n) reactions come into balance. The most abundant isotope of some element in such equilibrium is then the one for which the rate for that isotope (Z, A) to capture a neutron is equal to the rate for the resultant isotope $(Z, A + 1)$ to suffer a disintegration (γ, n) reaction. The neutron-separation energy S_n, that is, the binding energy of the least tightly bound neutron, of this nucleus is, upon neglect of the A dependence of nuclear partition functions (e.g. Sato 1974, see also Howard et al 1993, and Meyer 1994),

$$S_n(\text{MeV}) = \frac{T_9}{5.04}\left(34.08 + 1.5\log_{10} T_9 - 1.5\log_{10} n_n\right), \tag{5}$$

where n_n is the neutron number density in units of cm^{-3}. In secondary r-process sites, the neutron number density is typically $\sim 10^{19} \mathrm{cm}^{-3}$ and $T_9 \sim 1$; thus, the dominant nuclei in $(n, \gamma) - (\gamma, n)$ equilibrium have $S_n \approx 3.0\,\mathrm{MeV}$. Figure 8 shows the neutron-separation energies for the isotopes of neodymium $(Z = 60)$. We see that the dominant isotope in $(n, \gamma) - (\gamma, n)$ equilibrium should be the one with neutron number $N \approx 110$. The most neutron-rich beta-stable isotope of neodymium is $^{150}\mathrm{Nd}$ $(N = 90)$. The neutron sources in a secondary r-process would have to supply and maintain some 20 neutrons per seed nucleus in order to establish $(n, \gamma) - (\gamma, n)$ equilibrium. This is something they cannot do.

Let us now consider possible secondary r-process sites. One possible site is the helium shell of an exploding massive star (Truran et al 1978, Thielemann et al 1979; see also Cowan et al 1980, 1983, 1985; Blake et al 1981; Klapdor et al 1981). As the supernova shock wave traverses this shell, it heats up the material to a temperature $T_9 \approx 1$. Although this temperature is not high enough to force the material into NSE, it is high enough to drive material strongly in that direction. Among the main nuclear reactions that occur are (α, n) reactions that liberate neutrons which can then capture on the pre-existing seed nuclei. Truran et al (1978) found that neutron captures could modify an s-process seed abundance distribution into an r-process distribution. Later work indicated that a seed distribution that is enhanced with respect to the solar system heavy-element distribution is required to produce the solar r-nuclei. Such an enhanced distribution could result from s-processing prior to shock passage if protons were to mix down into the helium shell to make $^{13}\mathrm{C}$ from the abundant $^{12}\mathrm{C}$ [via

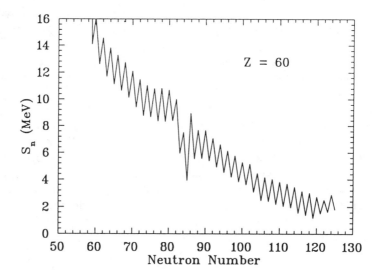

Figure 8 Neutron-separation energies for the isotopes of neodymium. The data are derived from Möller & Nix (1988).

$^{12}C(p, \gamma)^{13}N(\beta^+)^{13}C$]. It is apparent today, however, that the amount of ^{13}C required is unrealistically large (Cowan et al 1985, Cameron et al 1985).

Lee et al (1979) considered supernova shock passage through the carbon shell. Here the $^{22}Ne(\alpha, \gamma)^{25}Mg$ reaction produces the neutrons. ^{25}Mg is also a neutron poison, however; it absorbs many of the liberated neutrons before they have a chance to capture on heavier seed nuclei. Wefel et al (1981) found that some heavy nuclei could be synthesized, but not the bulk of the r-nuclei.

Another intriguing secondary site is again the helium burning shell, but now the effects of neutrino inelastic scattering on ^4He nuclei are included. The neutrinos come from the cooling nascent neutron star resulting from the core-collapse event. These neutrinos spall neutrons from the ^4He (and other light) nuclei. The neutrons can then capture on the seed nuclei and drive an r-process (Epstein et al 1988). A detailed study of this "ν-process" showed that some interesting nucleosynthesis may occur in such an event, but it could not have been a major contributor to the solar system r-process abundances (Woosley et al 1990).

All of the secondary r-process models studied to date have had profound difficulties which have rendered them implausible. Moreover, as shown in the next section, there are other reasons for favoring a primary over a secondary r-process. Nevertheless, the study of secondary r-process models has been valuable because they may have important applications to isotopic anomalies in meteorites (e.g. Clayton 1989, Howard et al 1992, Cameron et al 1993).

3.3   Observational Constraints

We have discussed how the r-process works in both primary and secondary scenarios and possible sites for the r-process. Let us turn now to the "observational" data to evaluate the plausibility of the proposed sites. These data will be astronomical, meteoritical, and nuclear.

We begin with the observations of old stars carried out by Sneden and collaborators (Sneden & Parthasarathy 1983, Sneden & Pilachowski 1985, Gilroy et al 1988; see also Wheeler et al 1989). These workers find that the *elemental* abundances, i.e. abundances as a function of Z, the proton number, of heavy elements in the atmospheres of these old stars match rather well the solar system r-process elemental abundances; s-process elemental abundances do not provide a good fit. One conclusion we may draw from this is that the r-process dominated the s-process at early times in our Galaxy. This would strongly suggest that the r-process is primary because a secondary r-process must produce r-nuclei from an already abundant population of s-nuclei. Another important conclusion is that the r-process mechanism has remained essentially the same throughout the Galaxy's history. A caveat to both of these conclusions is that they are based on *elemental* abundances, whereas a truly definitive identification of the heavy nuclei in old stars as r-process in origin would require isotopic abundances, information that spectroscopy cannot yet generally give us. Nevertheless, the evidence for a primary r-process that has operated in the same fashion throughout the Galaxy's history is compelling.

The next item of evidence we may consider is the question of the timescale for heavy element formation. Mathews & Cowan (1990) and Mathews et al (1992, 1993) have fit chemical evolution models to observations of the elemental abundances of europium (a mostly r-process element) and iron in the atmospheres of old stars. The only models that give good fits to the data are a primary r-process in type II and Ib (that is, core-collapse of a massive star) supernovae or a secondary r-process in which the neutron source itself is primary. An example of a primary neutron source in a secondary r-process event is the ^{13}C produced by mixing of protons down into the helium shell for supernova shock-induced helium or carbon burning. This is effectively a primary source because the star itself constructs the ^{13}C from its initial supply of protons and from ^{12}C made during helium burning. Another example is the neutrino-induced r-process. The neutrons in this source come mostly from inelastic scatterings of neutrinos on ^{4}He nuclei in the helium shell. The star itself produced almost all of the ^{4}He nuclei, so the neutron source is primary. Other proposed sites seemingly ruled out by these timescale arguments are: tidal disruptions of neutron stars by black holes, neutron star–neutron star collisions, a secondary r-process with a secondary neutron source such as ^{22}Ne, neutron-star accretion disks (Hogan & Applegate 1987), core helium flash in low-mass stars (Cowan et al 1982), and classical novae (Hoyle & Clayton 1974).

Experimental nuclear physicists have also provided us with important clues about the r-process. In particular, K.-L. Kratz and collaborators have measured the beta-decay lifetimes of r-process "waiting pointing" nuclei on the $N = A - Z = 50$ and $N = 82$ closed neutron shells (Kratz et al 1988, 1990). These nuclei have particularly strongly bound valence neutrons. Now the total abundance of the element Z is a sum over the abundances of all of the isotopes of that element:

$$Y_Z = \sum_A Y(Z, A). \tag{6}$$

The average rate of beta-decay flow out of element Z is

$$\lambda_Z = \frac{\sum_A \lambda_\beta(Z, A) Y(Z, A)}{Y_Z}, \tag{7}$$

where $\lambda_\beta(Z, A)$ is the beta-decay rate of nucleus (Z, A). The abundance Y_Z decreases during the r-process by beta decay out of Z into $Z + 1$ and increases by beta decay into Z from $Z - 1$; thus,

$$\frac{dY_Z}{dt} = -\lambda_Z Y_Z + \lambda_{Z-1} Y_{Z-1}. \tag{8}$$

Where the r-process path crosses the $N = 50$ and $N = 82$ closed shells, a single isotope dominates the abundances because of the strong neutron binding. For these nuclei, $Y_Z \approx Y(Z, Z + 50)$ for the $N = 50$ closed neutron shell and $Y_Z \approx Y(Z, Z + 82)$ for $N = 82$. If the conditions for $(n, \gamma) - (\gamma, n)$ equilibrium hold for a long enough time, the system will achieve approximate steady beta flow such that $dY_Z/dt \rightarrow 0$. In this case $\lambda_Z Y_Z \rightarrow$ constant. What is particularly significant is that the decay rates found by Kratz and co-workers times the appropriate abundances along the closed neutron shells as inferred directly from the solar system r-process data show that $\lambda_Z Y_Z$ is approximately constant for the $N = 50$ and $N = 82$ waiting-point nuclei. This is strong, albeit circumstantial, evidence that the r-process achieved the conditions of $(n, \gamma) - (\gamma, n)$ equilibrium and steady beta flow. This would indicate that the r-process is primary since a secondary r-process does not achieve $(n, \gamma) - (\gamma, n)$ equilibrium. Another important result found from this work is that the constant $\lambda_Z Y_Z$ for the $N = 50$ closed neutron shell is different from that for the $N = 82$ closed shell. The r-process did not achieve global steady beta flow (Kratz et al 1993). We interpret this to mean that the r-process abundance distribution actually results from a sum of components resulting from a varying set of r-process conditions—no single set of conditions gives rise to the solar system abundance curve. While this point was long appreciated by theoreticians (Seeger et al 1965, Kodama & Takahashi 1975, Hillebrandt et al 1976), the work of Kratz and collaborators establishes this result directly from nuclear physics and meteoritic abundances. The actual r-process site in nature must naturally give rise to a varying set of r-process conditions.

The final observational constraint we consider is the mass of r-process material in the Galaxy. From meteorites and abundances in the Sun we can infer that the mass fraction of r-nuclei in the Galaxy is 2×10^{-7} (from the data in Anders & Grevesse 1989). If the mass of our Galaxy is $1.5 \times 10^{11} M_{\odot}$, there are some $10^4 M_{\odot}$ of r-process material in our Galaxy. Now the rate of supernovae in our Galaxy is between 0.1 yr^{-1} and 0.01 yr^{-1} (e.g. Tammann 1982, Van den Bergh & Tammann 1991) and the Galaxy is of order 10^{10} years old; therefore, there have been some 10^8 to 10^9 supernovae in our Galaxy's history. If each of these supernovae produced r-process material, we expect each supernova to make 10^{-5} to $10^{-4} M_{\odot}$ of r-nuclei. This is a tiny fraction of the total mass ejected in a supernova and its smallness provides an important constraint on the site of the r-process.

From the astronomical and nuclear observations, we have found that the r-process is likely a primary event occurring in core-collapse supernovae. If the r-process comes from low entropy, neutron-rich material ejected from the core in a supernova, then we know from the discussion in Section 2.1 that we need material with $Y_e \lesssim 0.2$ to make actinide nuclei. Such material must come from fairly deep in the core. Hillebrandt et al (1976) found that a supernova needs to eject of order $0.1M_\odot$ of material to get $Y_e \approx 0.1$–0.2 material out of the core. Such a large amount would overproduce the r-nuclei by a factor of 10^3 if all supernovae ejected this much neutron-rich matter. It may be that only rare supernovae, such as those with high magnetic fields and rotation rates (LeBlanc & Wilson 1970, Meier et al 1976, Müller & Hillebrandt 1979, Symbalisty 1984, Symbalisty et al 1985), can eject neutron-rich matter. It remains to be seen, however, whether the high magnetic fields and/or rotation rates in these scenarios are indeed achieved in nature. In conclusion, we have no natural scenario for a low entropy primary r-process that yields $10^{-4}M_\odot$ of r-process material per event. Fortunately, we do have a high entropy r-process that gives this yield of r-nuclei per supernova. We turn to this scenario in the next section.

3.4   The r-Process in Nascent Neutron Star Winds

Woosley & Hoffman (1992) proposed the winds from nascent neutron stars as the site of the r-process. In this section we discuss the physics of these winds. Only detailed models will allow us to determine whether these winds are indeed the site of the r-process. On the other hand, simple arguments will illustrate the basic features of these winds and why they are attractive as an r-process site.

A core-collapse (i.e. type II or Ib) supernova may leave a hot ($kT \approx 10$ MeV) neutron star as a remnant (e.g. Bethe 1990). This neutron star cools by neutrino emission on a timescale of order 10 seconds. Salpeter & Shapiro (1981) were among the first to consider the thermal evolution of such a young neutron star. They were able to show that, although the neutrino luminosity is always sub-Eddington, the photon luminosity can be super-Eddington. Thus, for sufficiently hot nascent neutron stars, there is a neutrino-driven wind.

Duncan et al (1986) repeated the arguments of Salpeter & Shapiro and showed that a spherically symmetric, nascent neutron star of radius 10.6 km and mass $1.4M_\odot$ has a super-Eddington photon luminosity if the neutron star temperature is $\gtrsim 0.4$ MeV. Duncan et al also showed that the star cannot stabilize itself by changing its size or by transporting energy by convection. A wind must therefore blow from the surface of the neutron star. Moreover, Duncan et al showed that the mass loss rate in this wind is of order $10^{-5} M_\odot$ s^{-1}, which gives, if the wind lasts for the early cooling time of the nascent neutron star (~ 10 s), about $10^{-4} M_\odot$. If this material forms r-process nuclei, the wind naturally gives the correct amount of r-process matter per supernova. [For more details on this wind, see Woosley et al (1994) and references therein.]

What about the entropy in the wind? The net heating of a matter element lifting off the surface of the neutron star is governed by the heating due to neutrino interactions with the wind material and cooling due to emission by the matter. The heating goes as $F_\nu \sigma \langle E_\nu \rangle$, where σ is the neutrino-matter interaction cross section, F_ν is the number flux of neutrinos, and $\langle E_\nu \rangle$ is the average neutrino energy. $F_\nu \propto L_\nu \langle E_\nu \rangle^{-1} r^{-2}$ where L_ν is the neutrino luminosity and r is the radial position of the matter element. $\sigma \propto \langle E_\nu \rangle^2$ (e.g. Tubbs & Schramm 1975). The heating thus goes as $L_\nu \langle E_\nu \rangle^2 r^{-2}$. $\langle E_\nu \rangle$ and L_ν fall off on the neutron-star cooling timescale of roughly 10 seconds. The matter elements move out on timescales faster than 10 seconds, so the heating rate falls off roughly as r^{-2}. The cooling goes as T_m^6, where T_m is the local matter temperature. Above the surface of the neutron star, T_m^6 falls off more steeply than r^{-2}. As the mass element lifts off the star, the initial heating is slow because the matter and neutrino temperatures are nearly equal. As the mass element moves out, the matter temperature T_m drops. Once the mass element passes the "gain radius," where the heating and cooling rates are equal, T_m is too low for the material to cool off as fast as it is heated (Bethe & Wilson 1985). As heat is added, the entropy rises. In this way, the entropy can reach values of $100k$ or more before neutrino interactions with the wind material freeze out. These are the values required for a high entropy r-process.

Note that the entropy in mass elements leaving the neutron star at late times will be larger than in mass elements leaving the star at early times. This is because most of the heating occurs fairly near the surface of the neutron star because the neutrino flux and hence the net heating rate falls off as $1/r^2$. As a neutron star ages, it shrinks in radius [from ~ 100 km at a few tenths of a second after core bounce to ~ 10 km at several seconds after bounce in the models of Wilson & Mayle (1993)]. The decrease in the initial r from which the mass elements begin increases the heating rate more than the slow fall off in L_ν and $\langle E_\nu \rangle$ decrease it. The net heating and entropy in the later mass elements is thus larger.

The last question we must consider is the neutron richness of the wind material. Y_e is set in the wind by the reactions $\nu_e + n \rightarrow p + e^-$ and $\bar{\nu}_e + p \rightarrow n + e^+$. If the fluxes and energies of the ν_e and $\bar{\nu}_e$ were equal, Y_e would be slightly larger than 0.5 because the mass of the proton is slightly less than that of the neutron. As a nascent neutron star cools, however, it becomes neutron rich. The opacity for ν_es becomes larger than that for $\bar{\nu}_e$s because of the reaction $\nu_e + n \rightarrow p + e^-$. The $\bar{\nu}_e$s thus have a longer mean free path and originate deeper in the neutron star. This means that they are more energetic than the ν_es. This necessarily drives the material neutron rich. At late times ($t \approx 10$ s after core bounce) $\langle E_{\bar{\nu}_e} \rangle \approx 2 \langle E_{\nu_e} \rangle$ which makes $Y_e \approx 0.33$ (Qian et al 1993). This is certainly neutron rich enough for a high entropy r-process.

As a last point, let us re-emphasize that the entropy and Y_e in a mass element vary according to when that mass element lifts off the neutron star. Each mass element thus undergoes a somewhat different nucleosynthesis. The final r-process abundance distribution, however, is a sum of all of these different components. The r-process in nascent neutron star winds thus naturally satisfies the requirement imposed by the work of Kratz et al (1993). Notice that the wind dynamics and thermodynamics are completely determined by the mass, radius, and temperature of the neutron star. Any given neutron star probably passes through roughly the same sequence of temperatures and radii as a function of time as it cools. We thus expect to get essentially the same r-process out of every supernova. This satisfies our expectations from the observations of the r-process elements in old stars (see Section 3.3).

Does a solar system distribution naturally emerge in such a wind? Meyer et al (1992), using a schematic model based on output from Wilson & Mayle (1993), found the resulting abundances matched the solar system distribution quite well (see also Howard et al 1993 and Takahashi et al 1994). A more detailed model, using mass element trajectories calculated directly in Wilson and Mayle's supernova code produced abundances that also agree well with the solar distribution (Woosley et al 1994). This latter model also gives the correct amount of r-process mass per supernova ($\sim 10^{-4} M_\odot$). Confirmation of nascent neutron star winds as the site for the r-process will require a full survey of the nucleosynthesis in detailed, realistic wind models. Nevertheless, nascent neutron star winds seem extremely promising as the site for the r-process.

4. THE s-PROCESS

> Must we, as Solon advises, always keep the goal in sight?
>
> Aristotle, *Nichomachean Ethics*

The s-process is the other major nucleosynthetic process that assembles heavy elements. We know that the s-process path in the neutron number–proton number plane crosses the neutron closed shells at the valley of beta stability. This tells us that the s-process occurred in an environment with a much lower neutron density than the r-process. Also, the s-process occurred over a much longer time period.

In this section we seek to understand how the s-process occurs. We then turn to the question of s-process sites. Finally we consider constraints on those sites.

4.1   *The s-Process Mechanism*

Because of the neutron densities and timescales inferred for the s-process from the abundance peaks, we can infer that the s-process is not a freeze out from equilibrium. Instead, it is a neutron-capture process that occurs in a system striving to reach equilibrium, but falling short of its goal. The main reactions carrying the bulk of the nuclei towards the iron group can liberate neutrons. Pre-existing seed nuclei capture these neutrons and produce the s-nuclei. The s-process is clearly a secondary process.

The dominant reactions that can liberate neutrons are $^{13}C(\alpha, n)^{16}O$ and $^{22}Ne(\alpha, n)\,^{25}Mg$. In these reactions, the neutron-rich isotopes, ^{13}C and ^{22}Ne give up their excess neutrons to heavier nuclei. At this point, we may ask where these excess neutrons came from in the first place. The answer to this interesting question illustrates an important point about the overall nuclear evolution of the universe.

The abundances that emerge from the Big Bang are roughly 90% by number ^{1}H and 10% ^{4}He (e.g. Walker et al 1991). This yields $Y_e = 0.88$. On the other hand, we may make the observation that ^{1}H and ^{3}He are the only proton-rich (that is, with proton number greater than neutron number) stable isotopes in nature. This means that in order for nature to put the nucleons in the universe into nuclei with the strongest binding energy per nucleon (iron-group nuclei), the Y_e of the universe must decrease.

Most of the decrease in Y_e comes from the weak decays in the p-p chains and the CNO cycle during hydrogen burning. These interactions drop Y_e from 0.88 to 0.5 in material that has completed hydrogen burning. ^{4}He itself does not have any excess neutrons, but some production of excess neutrons occurs in the CNO cycle due to reactions like $^{12}C(p, \gamma)^{13}N(\beta^+)^{13}C$. The net result is the conversion of a free proton into an excess neutron, and a drop in Y_e. The ^{22}Ne production builds up from abundant ^{14}N produced in the CNO cycle. The sequence is $^{14}N(\alpha, \gamma)^{18}F(\beta^+)^{18}O(\alpha, \gamma)^{22}Ne$. Here it is the fact that the only stable isotope of flourine is neutron rich that leads to a decrease in Y_e. We see that the excess neutrons in ^{13}C and ^{22}Ne are a consequence of the overall drive to decrease Y_e in stars. We must keep the goal of the nuclei in sight to understand where the excess neutrons come from that drive the s-process.

The first attempts to understand the details of the s-process led to the classical model. The neutron density is always low in the s-process (compared to the r-process). If a nucleus is unstable to β^- decay following neutron capture in the s-process, it will almost always β^- decay to the first available stable isobar before it can capture another neutron. Thus, it generally suffices in s-process studies to follow only the abundances as a function of mass number, which only change by neutron capture. In this approximation, the rate of change of the abundance N_A of nuclei with mass number A is

$$\frac{dN_A}{dt} = -n_n \langle \sigma v \rangle_A N_A + n_n \langle \sigma v \rangle_{A-1} N_{A-1}, \tag{9}$$

where n_n is the neutron number density and $\langle \sigma v \rangle_A$ is the thermally averaged neutron-capture cross section for the stable isobar of mass number A. We can write $\langle \sigma v \rangle_A$ as $\sigma_A v_T$, where v_T is the thermal velocity of neutrons and σ_A is an average cross section, given in terms of v_T. With the definition of the neutron exposure

$$\tau = \int n_n v_T dt, \tag{10}$$

we find

$$\frac{dN_A}{d\tau} = -\sigma_A N_A + \sigma_{A-1} N_{A-1}. \tag{11}$$

Note that the neutron exposure τ is a fluence. It has units of inverse millibarns (1 barn = 10^{-24} cm^2). Because it is a neutron flux integrated over time, it is an appropriate evolutionary parameter for the s-process. If the s-process achieves a steady state, then $dN_A/d\tau \rightarrow 0$ and $\sigma_A N_A \rightarrow$ constant.

Clayton et al (1961) were able to show that a single neutron exposure τ could not reproduce the solar system's abundance of s-only nuclei. Seeger et al (1965) showed that an exponential distribution of exposures, given by

$$\rho(\tau) = \frac{f N_{56}}{\tau_0} e^{-\tau/\tau_0}, \tag{12}$$

where f is a constant and N_{56} is the initial abundance of ^{56}Fe seed, did reproduce the solar distribution of s-nuclei. For the distribution of exposures given in Equation (12), Clayton & Ward (1974) found that for an exponential average of flows in the s-process

$$\sigma_A N_A = f N_{56} \tau_0 \prod_{A'=56}^{A} [1 + (\sigma_{A'} \tau_0)^{-1}]^{-1}. \tag{13}$$

A fit to the empirical $\sigma_A N_A$ for s-only nuclei then gives the quantities f and τ_0.

A complication to the above classical model is the branching that occurs at certain isotopes. Here it may be that the β^- decay rate is not considerably greater than the neutron-capture rate. In some cases the nucleus may β^- decay before neutron capture and in others it may neutron capture before suffering β^- decay. The assumptions leading to Equation (9) thus break down. Ward et al (1976) developed an analytic treatment of branching in the case of a time-independent neutron flux. For time-dependent neutron fluxes, it is necessary in general to solve a full network of nuclei numerically (e.g. Howard et al 1986). Since the s-process branchings will in general be temperature and neutron density dependent, s-nuclei branchings are important diagnostics of the environment in which the s-process occurred. We will see this in more detail in Section 4.3.

4.2 s-Process Sites

To obtain a good fit of the σN curve to the solar system s-process abundance distribution, three distinct exponential distributions of neutron exposures may be necessary (Clayton & Rassbach 1967, Clayton & Ward 1974). One exposure, with $\tau_0 \approx 0.30$ mb^{-1}, produces most of the nuclei in the mass range $90 < A < 204$. This is the main component. Another exposure, with $\tau_0 \approx 0.06$ mb^{-1} contributes to the $A \lesssim 90$ s-nuclei abundances. This weak component is required in order to explain the σN curve around $A \sim 90$. These two components indicate that two separate sites contributed to the abundance of solar s-nuclei. Finally, a strong component, with $\tau_0 \approx 7.0$ mb^{-1}, may be necessary to explain the abundances of the $A = 204$–209 nuclei. One possible explanation of this component is that the distribution of exposures in the main component is not exactly exponential, but rather is higher than exponential at large τ. There is probably no need for a separate site for the strong component of the s-process.

The weak s-process component likely comes from He burning in the cores of massive stars ($\gtrsim 15M_\odot$) (Truran & Iben 1977, Lamb et al 1977), where the temperature is high enough for the $^{22}Ne(\alpha, n)^{25}Mg$ reaction to produce a substantial amount of neutrons. These stars also have strong winds that eject this material into the interstellar medium. Recent work has confirmed the plausibility of this site (Arnett & Thielemann 1985, Busso & Gallino 1985, Prantzos et al 1987, Langer et al 1989, Raiteri et al 1991a, Baraffe et al 1992). Uncertainties in the $^{22}Ne(\alpha, n)^{25}Mg$ and $^{22}Ne(\alpha, \gamma)^{26}Mg$ reaction rates prevent us from predicting the neutron exposure in these models to high accuracy. Recent results on these rates may indicate that the s-process is somewhat more robust in this site than previously thought (e.g. Baraffe & El Eid 1994). This may complicate the separation of the $A \lesssim 90$ s-nuclei into those coming from the weak and main components.

Some s-processing may also occur in core carbon burning or shell helium burning in massive stars. This has been studied by Arcoragi et al (1991) and Raiteri et al (1991b). The results indicate that this processing does not contribute in a significant way to the weak component.

The main component of the s-process is likely to occur in the helium-burning shell in asymptotic giant branch (AGB) stars (Weigert 1966, Schwarzchild & Härm 1967, Ulrich 1973). The structure of such a star is an inert carbon-oxygen core, on top of which lies a convective helium-burning shell. On top of this helium-burning shell is the hydrogen-rich envelope, which itself is convective. The original idea was that the convective helium shell might reach out far enough into the hydrogen-rich envelope that protons and ^{12}C (the result of helium burning) could mix and produce ^{13}C, as discussed in Section 4.1. The ^{13}C would then be the source of neutrons for the s-process. [The current picture is that convection does not provide the mixing, but that protons reach down into the carbon-rich shell by diffusion or semiconvection (see below).]

An attractive feature of this model is the fact that the helium burning occurs in pulses. Between pulses, hydrogen burns quiescently in a thin shell. Once the supply of helium from the hydrogen burning builds up, a helium-burning pulse occurs. The energy liberated expands the star and shuts off the hydrogen burning. After the pulse has occurred, the star settles down again and begins hydrogen-shell burning anew. Pulses last of order tens of years while the interpulse periods are of order thousands of years. The significance for the s-process is that there is an overlap of mass zones experiencing successive helium-buring pulses. Ulrich (1973) was able to show that the mixing and burning sequence could naturally give rise to an exponential distribution of neutron exposures. Alternating overlap of convection zones can carry the newly produced s-nuclei into the envelope (the so-called "third dredge up"). These nuclei would then find their way into the interstellar medium via winds or by the ejection of the atmosphere in a planetary nebula phase.

This nice model for the s-process suffered a setback when Iben showed that an entropy barrier prohibited mixing of protons into the helium shell (Iben 1975a,b, 1976). It was then proposed instead that $^{22}\text{Ne}(\alpha, n)^{25}\text{Mg}$ be the source (e.g. Iben & Renzini 1983). The helium core grows by accreting the ashes of the hydrogen-burning shell. The products of CNO burning are ^4He and ^{14}N in that shell, which combine to give ^{22}Ne early in helium burning, as discussed in Section 4.1. The $^{22}\text{Ne}(\alpha, n)^{25}\text{Mg}$ reaction then drives the s-process. The pulse and mixing that occurs gives an exponential distribution of neutron exposures. This model has some difficulties, however. Basically, the shell flashes in most AGB stars are not hot enough to liberate most of the ^{22}Ne neutrons, and the massive ABG stars that are hot enough are too rare. This has led workers to consider alternative neutron sources in low-mass AGB stars ($M < 3M_\odot$).

In low-mass AGB stars, the temperature is too low in the helium-burning shell for the $^{22}\text{Ne}(\alpha, n)^{25}\text{Mg}$ reaction to be the major source of neutrons. Iben & Renzini (1982) argued, however, that, despite the entropy barrier to convection, semiconvection or diffusion could cause the mixing of protons with ^{12}C in the interpulse period. This produces pockets of ^{13}C atop the He zones which can liberate neutrons during convective ingestion by the next pulse. Recent work indicates that this is a promising site for the s-process (Gallino et al 1988; Boothroyd & Sackmann 1988a,b,c,d; Hollowell & Iben 1988; Käppeler et al 1990). In particular, these models seem to give a good fit to the main component of the solar σN curve (e.g. Käppeler et al 1990). We must note that these s-process calculations are post-processing calculations, which means that the neutron density is a parameterized quantity. Even more serious is the lack of a demonstrated occurrence of the needed ^{13}C-rich pocket, which is therefore taken on faith at the present time. It remains to be seen whether the good agreement with the solar s-process abundances will hold up when the s-process calculations are directly coupled to complete stellar models. Such coupled calculations may be available in the not-too-distant future. It will also be important to include the effects of energy generation by all the nuclear reactions on the stellar structure (Bazan & Lattanzio 1993).

4.3   Constraints on s-Process Sites

What constraints can help to evaluate the proposed sites discussed in the previous section? s-process branchings are the first important constraints. The likelihood that a beta-unstable nucleus in the s-process beta decays depends on the rate of beta decay compared to the rate of neutron capture. Evidence for branching provides information about these rates. In particular, with knowledge of the beta-decay rate from laboratory experiments, the degree of branching constrains the neutron capture rate $n_n \langle \sigma v \rangle$. Then knowledge of $\langle \sigma v \rangle$ from the laboratory constrains n_n, the neutron number density during the s-process. On the other hand, if the beta-decay rate is temperature sensitive (e.g. Takahashi & Yokoi 1987), branching data yield constraints on the temperature during the s-process. Branching data may also yield constraints on the mass density during the s-process through electron capture rates. Finally, branching data can constrain the duration of the neutron pulses (Ward & Newman 1978). If the pulse period were much shorter than the lifetime of the branching point isotope, there would be no branching. Pulses that were too long in duration would allow too much neutron capture.

What do we find for the s-process in nature? For the main component of the s-process, the isotopes ^{134}Cs, ^{148}Pm, ^{151}Sm, ^{154}Eu, ^{170}Yb, and ^{185}W are branch-point isotopes with potential as diagnostics of the temperatures and neutron densities prevailing during the s-process. Beer et al (1984) used ^{151}Sm, ^{170}Yb, and ^{185}W to find limits on the neutron number density and temperature. Uncertainty in the population of the 137 keV isomeric state in ^{148}Pm during the s-process makes conclusions from this isotope difficult. Uncertainties in cross sections and abundances limit the usefulness of ^{134}Cs and ^{154}Eu.

As for the mass density, Yokoi & Takahashi (1983) noticed that ^{163}Dy could beta decay in stars, even though it is stable on Earth. In stars, the ^{163}Dy atom is ionized so that in fact the daughter atom ^{163}Ho would be at slightly lower mass. ^{163}Ho then could either capture a neutron or electron capture back to ^{163}Dy. The electron capture rate depends on the density of electrons, which in turn depends on the mass density. Beer et al (1985) were able to constrain the mass density in the s-process in this way.

Finally, Beer & Macklin (1988) studied ^{151}Sm in order to determine a lower limit to the duration of the neutron pulse in the s-process. Studies of ^{86}Kr may give an upper limit to the pulse duration (Beer & Macklin 1989). Unfortunately the weak component in this region introduces ambiguities into such an analysis.

The net results of branching studies in the context of the classical model give a temperature for the main component of 2.8–3.9×10^8 K, a neutron density of 2.3–4.5×10^8 cm^{-3}, a mass density of 2.6–13×10^3 g cm^{-3}, and a pulse duration of greater than 3 years (Käppeler et al 1989). These numbers agree reasonably well with those expected from stellar models. A similar analysis for the weak component yields a temperature of 1.8–3.0×10^8 K and a neutron density of 0.8–1.9×10^8 cm^{-3} (Käppeler et al 1989).

The relatively high temperatures found in this analysis for the main component suggest that ^{22}Ne$(\alpha, n)^{25}$Mg is the neutron source for the s-process. Howard et al (1986) studied the s-process nucleosynthesis with this neutron source. They obtained poor fits to the solar σN curve when they used parameters derived from stellar models. In particular, the average neutron density during the pulses was too high to reproduce the correct branchings. Busso et al (1988) have confirmed these results. On the other hand, the high temperature ^{22}Ne source may simply be the last hot part of the neutron burst that was primarily from ^{13}C at lower temperature (see below).

Let us consider now the evidence from observations of stars. It was the observation of technetium in certain red giant stars (Merrill 1952) that showed that stars do indeed synthesize elements and led Cameron (1955) to work out many of the details of the s-process. Since all isotopes of Tc are unstable, any Tc present in the surface of a star must have been synthesized in the interior of the star by the s-process and then dredged up to the surface. Recent observations show that red giant stars in the solar neighborhood that do have s-process abundance enhancements in their atmospheres do not show the accompanying enhancements of ^{25}Mg and ^{26}Mg that one would expect from alpha capture on ^{22}Ne (e.g. Smith & Lambert 1986, McWilliam & Lambert 1988). In addition, observations of Rb and ^{96}Zr constrain s-process branching at ^{85}Kr and ^{95}Zr. Astronomers find that the s-process occurring in the interiors of the stars observed must be happening at low neutron densities ($n_n \lesssim 10^9$cm^{-3}), not the high neutron densities characteristic of the ^{22}Ne$(\alpha, n)^{25}$Mg reaction (e.g. Lambert 1993).

From this evidence, it appears that ^{13}C is more promising as the source of s-process neutrons, indicating that low-mass AGB stars are probably the site of the s-process. Such stars give a low temperature s-process ($\sim 1.5 \times 10^8$K) which would seem to contradict the higher temperatures found from the analysis of the s-process branchings in the classical model ($T = 2.8$–3.9×10^8K) discussed above. In the low-mass AGB star s-process calculations that do show good agreement with solar abundances (e.g. Gallino et al 1988, Käppeler et al 1990), there are two bursts of neutrons per pulse: a strong burst due to the ^{13}C$(\alpha, n)^{16}$C reaction at $T \sim 1.5 \times 10^8$K, and a second, weaker one, due to the ^{22}Ne$(\alpha, n)^{25}$Mg reaction. This weaker burst occurs when the helium shell contracts following the first burst and heats to a temperature of $T \sim 3 \times 10^8$K. It resets the branch-point thermometers to this higher temperature, in agreement with the analysis from the classical model.

More evidence for ^{13}C as the dominant source for neutrons in the s-process comes from studies of galactic abundance evolution. Mathews et al (1992, 1993) studied the evolution of the Ba/Fe ratio in our Galaxy. Ba is predominantly an s-process element and hence must be secondary (i.e. made from initial Fe). Mathews et al found that only an s-process behaving as a primary process fit well the observations of Ba abundances in the atmospheres of old stars. ^{13}C is a primary neutron source, as discussed in Section 3.3. ^{22}Ne is secondary because it must be built up from pre-existing CNO nuclei. The Fe seeds are of course secondary. Clayton (1988a) described how the secondary s-process with the ^{13}C neutron source is able to mimic primary nucleosynthesis. The idea here is that while the galactic abundance of Fe seed for the s-process grows with time, so does the abundance of s-process neutron poisons.

A final point of increasing relevance is the new information from pre-solar SiC grains found in the Murray and Murchison meteorites. These grains are carriers of isotopic anomalies in s-process isotopes (Srinivasan & Anders 1978, Tang & Anders 1988). In addition, these grains are anomalous in their Si and C (Zinner et al 1987, Anders & Zinner 1993). It appears that these grains have condensed in carbon-star atmospheres, which are s-process enriched and have variable ^{13}C-rich compositions (Lambert et al 1986). As surviving stardust, the grains are almost pure endmembers in the "cosmic chemical memory" theory for interpreting isotopic anomalies in solar system samples (Clayton 1978, 1982). From studies of trace s-isotopes in these grains (Ott & Begemann 1990 a,b; Zinner et al 1991; Richter et al 1992; see Anders & Zinner 1993 for a review) the abundance ratios only fit if the grains come from low-mass AGB stars (Gallino et al 1990). A vexing problem with this idea, however, is that such stars cannot explain the anomalous Si isotopes (a major constituent of the grains). One suggested answer is higher mass AGB stars, in which burning of Mg isotopes in late pulses resets the ratio of ^{29}Si/^{30}Si(Brown & Clayton 1992). Galactic abundance evolution of Si isotopes may also hold the key (Clayton 1988b, Gallino et al 1994). Alternatively, some other site may be responsible for these grains (e.g. Arnould & Howard 1993). These tiny, sturdy grains have traveled from afar carrying important messages about the s-process which have yet to be deciphered.

In summary, low-mass AGB stars are at present the most promising site for the main component of the s-process. Confirmation of this site will require continued interplay of nuclear physics, meteoritics, stellar evolution and structure theory, nucleosynthesis theory, galactic abundance evolution theory, and stellar astronomy. Many people will be busy for quite some time to come!

5.   THE p-PROCESS

   ... and the elements shall melt with fervent heat ...

                                                                II Peter 3:10

We turn finally to the p-nuclei. These are the 35 nuclei bypassed by the r- and s-processes. As we see from Figure 1, except for the light p-nuclei (^{92}Mo, ^{94}Mo, ^{96}Ru, ^{98}Ru), the abundances of p-nuclei are considerably less than those of their r- and s-nuclei counterparts. Furthermore, the p-process abundance distribution shows interesting structure with peaks at ^{92}Mo and ^{144}Sm. These are important clues for determining where the p-process occurs.

It is probably wrong to think that the p-process occurs in a single site. We can imagine many astrophysical settings where conditions are right to modify a pre-existing supply of r- and s-nuclei to form p-nuclei. The relevant question is really what site contributes the bulk of the p-nuclei. For more details on the p-process, the reader should consult the excellent review by Lambert (1992).

5.1 General Considerations

It is impossible to produce p-nuclei by neutron capture. How then can nature make these nuclei? The first possibility that suggests itself is proton capture. It may be that in the course of the evolution of some system striving to reach NSE, protons are liberated which can capture on pre-existing seed nuclei to make p-nuclei. Alternatively it may be that in a freezeout from proton-rich NSE, free protons could capture on seed iron-group nuclei. β^+ reactions could allow further capture of protons to higher mass.

We can explore the conditions required for such p-processes with the help of Figures 9 and 10. Figure 9 shows the timescale for capture of a proton [a (p, γ) reaction] by the most proton-rich beta-stable isotope of each element at the fixed temperature of $T_9 = 1$ for different values of the proton mass density ρY_p. The timescales for a proton-disintegration (γ, p) reaction or for a neutron-disintegration (γ, n) reaction for these isotopes are all greater than 10^{10} seconds. Note that for a site with a proton mass density ρY_p of $1 \, \mathrm{g\,cm^{-3}}$, it would take ^{92}Mo $(Z = 42)$ about 10^4 s to capture a proton. Capture of protons on higher-charge isotopes would take even longer. The timescale for proton capture decreases if the setting has a higher density of protons available. For example, for $\rho Y_p = 10^3 \mathrm{g\,cm^{-3}}$, the timescale for capture of protons on ^{92}Mo

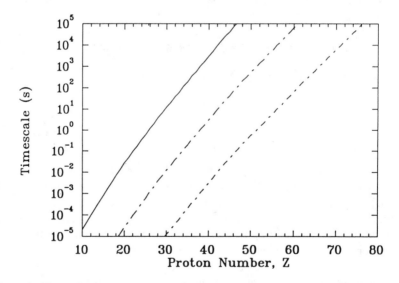

Figure 9 Timescales for proton capture on the most proton-rich isotope of each element at the fixed temperature of $T_9 = 1$. The curves are for mass densities in protons of $\rho Y_p = 1 \, \mathrm{g\,cm^{-3}}$ (*solid curve*), $10^3 \mathrm{g\,cm^{-3}}$ (*long dashed-dotted curve*), and $10^6 \mathrm{g\,cm^{-3}}$ (*short dashed-dotted curve*). The rates are computed from expressions in Woosley et al (1975).

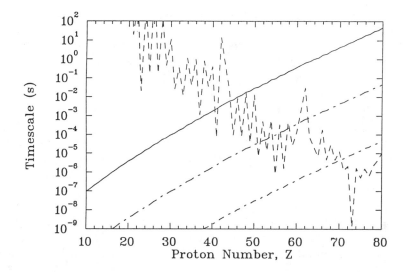

Figure 10 Same as Figure 9 but for $T_9 = 3$. The short-dashed jagged curve shows the timescale for (γ, n) reactions on these nuclei. The (γ, n) rates are computed from neutron-capture cross sections in Woosley & Hoffman or Cowan et al (1991) and neutron-separation energies are derived from Möller & Nix (1988).

would be about 10 s while for $\rho Y_p = 10^6 \mathrm{g\,cm}^{-3}$ it would be about 10^{-2} s. If an astrophysical site could maintain a mass density in protons of $10^6 \mathrm{g\,cm}^{-3}$ for 10^5 s at $T_9 = 1$, the proton-rich isotopes of all elements up to platinum ($Z = 78$) could capture a proton. The question for finding the p-process site is whether such conditions are possible. It is unlikely.

The proton-capture rates also increase with increasing temperature because the reactants have a higher relative kinetic energy compared to the Coulomb barrier than at lower temperature. This leads us to ask what happens to the timescales if we increase the temperature. We see the effect in Figure 10 for $T_9 = 3$. The timescale for proton capture does indeed decrease, but so does the timescale for a (γ, n) reaction. For $\rho Y_p = 1 \mathrm{g\,cm}^{-3}$, the timescale for a (γ, n) reaction is less than that for a (p, γ) reaction at $Z \approx 40$. This means that it is more likely under these conditions for a proton-rich nucleus to suffer a (γ, n) reaction than to capture a proton. Of course this makes sense from our discussion in Section 2. If the system is evolving towards NSE, nuclei more massive than the nucleus with the highest binding energy per nucleon will tend to disintegrate nucleons to increase the number of macroscopic states available to the system.

For higher proton mass densities, the (γ, n) reactions do not dominate the proton captures until higher nuclear charge: $Z \approx 40$–50 for $\rho Y_p = 10^3 \mathrm{g\,cm}^3$ and $Z \approx 70$ for $\rho Y_p = 10^6 \mathrm{g\,cm}^{-3}$. One might imagine that the system could then produce p-nuclei at least up to ytterbium ($Z = 70$) under such high temperature and high density conditions. Such conditions are extremely difficult for nature to achieve, however. The dilemma for making p-nuclei is clear. If nature is to make these nuclei by proton capture at low temperature where the flow will not be impeded by disintegrations, a large supply of protons must be available for a long time. If the proton capture process is to occur at higher temperature where the capture timescales are shorter, disintegration reactions will dominate the flow and prevent capture to higher mass.

The escape from this dilemma is the realization that proton captures need not make most of the p-nuclei. The various disintegration reactions do the job. In particular, if pre-existing r- and s-nuclei are exposed to high temperature, nuclear reactions will occur and tend to drive the abundances toward NSE. The first reactions to occur are the (γ, n) reations which produce quite proton-rich nuclei. Once the nuclei become sufficiently proton rich, they then begin a (γ, p) and (γ, α) cascade. In this way the nuclei "melt" towards iron. If the high temperature drops off quickly enough, the system does not reach NSE and the melting will be incomplete, leaving an abundance of proton-rich heavy nuclei— the p-nuclei. Where the disintegration flow crosses the $N = 50$ and $N = 82$ closed neutron shells, the disintegration timescales become large because of the particularly strong binding energies. This is apparent in Figure 10 as the peaks in the (γ, n) timescales at the closed shell nuclei ^{92}Mo ($Z = 42$) and ^{144}Sm ($Z = 62$). Because of the long disintegration timescales, abundances build up at these nuclei. In this way we can explain the peaks in the solar system's p-process abundance distribution. As a final note, the extremely short (γ, n) timescale for ^{180}Ta ($Z = 73$ in Figure 10) explains why this fragile species is the rarest stable isotope in nature.

5.2   p-Process Sites

The early papers on the p-process considered the proton capture mechanism (Burbidge et al 1957, Ito 1961, Macklin 1970, Truran & Cameron 1972, Audouze & Truran 1975). The site was imagined to be the hydrogen-rich envelope in massive stars undergoing a supernova explosion. The supernova shock passing through this region would heat up the material and proton capture reactions would produce the p-nuclei. However, the densities, temperatures, and timescales required are unrealistic for the hydrogen-rich envelope (see, for example, the discussion in Woosley & Howard 1978).

Arnould (1976) computed the p-process in the hydrostatic oxygen burning phase in stars. The timescales are longer in this site than in the supernova site and would allow for more proton capture. In this site, temperatures were high enough for disintegrations [especially (γ, n) reactions] to be important. A major challenge for this model is to eject the new p-nuclei without significantly modifying their abundances during the subsequent supernova explosion.

Woosley & Howard (1978) computed the p-process in the O/Ne shell in type II, that is, core-collapse, supernovae. The supernova shock heats up this shell and causes the partial melting of the nuclei. In this model, only disintegrations are important, hence the alternative name "gamma-process" (see Rayet et al 1990, Prantzos et al 1990, Rayet et al 1992 for extensions of this model). This model successfully reproduced most of the p-nuclei in their solar system proportions, although it seriously underproduced the light p-nuclei. As in the hydrostatic oxygen burning model, it was necessary to superimpose several abundance distributions to get a realistic distribution of p-nuclei. Type II supernovae should naturally give a distribution of conditions depending on the layers of the proto-SN considered that would naturally give a distribution of abundances. The inner regions of the O/Ne shell will achieve the highest temperatures and thus get closest to NSE. These regions make the lighter p-nuclei. Outer regions produce the heavier p-nuclei because the "melting" is less complete. Prantzos et al (1990) computed the p-process abundance distribution for a specific model of supernova 1987A and found that a distribution of conditions naturally arose and gave a solar system p-process abundance distribution, except for underproduction of light p-nuclei.

The problems with the underproduction of the light p-nuclei in the gamma-process led Howard et al (1991) to consider the production of p-nuclei in the outermost layers of a carbon-oxygen white dwarf star suffering a type Ia supernova explosion. In this model, s-processing prior to the explosion built up the abundances of $A \approx 90$ nuclei. The high density during the explosion then allowed proton capture reactions to produce many of the light p-nuclei while the normal gamma-process made the heavier p-nuclei. Later calculations using the realistic type Ia models of Khoklov (1990) have not been as successful in producing the light p-nuclei (Howard & Meyer 1992). More studies of this promising site are required.

Some p-nuclei may also be produced in spallation reactions. Most notably, neutrinos may spall neutrons from heavy nuclei during type II supernovae to make p-nuclei. Such a process is only likely to produce significant amounts of the rarest p-nuclei such as ^{138}La and ^{180}Ta (Woosley et al 1990).

5.3   Some Constraints on p-Process Models

As mentioned above, the p-process probably occurs in several places in nature. Any astrophysical setting in which high temperatures but sufficiently short timescales lead to incomplete melting of heavy nuclei can produce p-nuclei. What we really seek is the site that produces the bulk of the p-nuclei. We can attempt to analyze this requirement by means of overproduction factors.

The overproduction factor of an isotope in the product material of some nucleosynthetic process is the ratio of its mass fraction in that material to its mass fraction in the solar system. In order for the process to be responsible for the production of the bulk of the solar system's supply of a given isotope, that isotope must have the largest overproduction factor in the product material. If the process is to be responsible for the bulk of the solar system's supply of two or more isotopes, those isotopes must all have comparably large overproduction factors.

Prantzos et al (1990) found an average overproduction factor of 0.96 for p-nuclei in $15 M_\odot$ of ejecta in their model for SN 1987A. The overproduction factor for ^{16}O in the same model was 11.5 (Thielemann et al 1990). Since type II supernovae made most of the solar system's ^{16}O, this would indicate that the type II supernova site could not be responsible for the production of the bulk of the solar system p-process elements. Prantzos et al argue, however, that there are extenuating factors. First, the SN 1987A model used Large Magallenic Cloud metallicities, which are lower than those in our Galaxy. Milky Way metallicities could give enhanced seed abundances and thus higher p-process overproduction factors. Also, it may be that p-nuclei production relative to oxygen could be higher in higher mass stars. Surveys over a range of star masses, such as that of Arnould et al (1992), will be important for understanding the contribution of type II supernovae to the solar system p-process abundances.

As for the type Ia model, Howard et al (1991) noted that besides possibly producing p-nuclei, these supernovae made most of the solar system's ^{56}Fe. The requirements that type Ia supernova models make ^{56}Fe and p-nuclei in solar proportions and that they produce $0.5-1.0 M_\odot$ of ^{56}Fe, and the fact that the typical p-process overproduction factors in this model are $\sim 10^4$, lead to the conclusion that the zones that produce p-nuclei comprise $0.04-0.08 M_\odot$ of the white dwarf. This is in good agreement with the models (e.g. Khoklov 1990). It appears that type Ia supernovae are capable of producing the bulk of the solar system's p-nuclei.

Another important constraint on the p-process is the presence of live ^{146}Sm in the early solar system (Lugmair et al 1983, Prinzhofer et al 1989). From general galactic abundance evolution arguments, Prinzhofer et al inferred from their measurements that the production ratio of $^{146}Sm/^{144}Sm$ should be between 0.07 and 0.5. This ratio causes problems for the gamma-process for which the production ratio for these isotopes is typically ~ 0.02 for the type II model (Woosley & Howard 1978) and ~ 0.05 for the type Ia model (Howard et al 1991). The inferred production ratio is in fairly good agreement with that found from proton-capture models (e.g. Audouze & Truran 1975), which led Prinzhofer et al to favor such models for production of p-process elements. As we have seen, however, such models are not astrophysically realistic, so the measurements present a challenge for gamma-process models. Woosley & Howard (1990) found, however, that if the branching ratios for (γ, n) and (γ, α) reactions on ^{148}Gd are varied within experimental uncertainties, the $^{146}Sm/^{144}Sm$ production ratio could be increased dramatically. It is clearly a worthy goal for experimental nuclear physicists to attempt to make accurate measurements of the disintegration rates of ^{148}Gd. Another important effect that may alleviate the ^{146}Sm/^{144}Sm production problem is galactic infall (Clayton et al 1993).

In summary, plausible models exist that do a good job of satisfying most of the constraints on the p-process. The most vexing puzzle that remains is the underproduction of the light p-nuclei. Perhaps, with more work, we will see that type Ia supernovae can produce these nuclei at the correct levels. On the other hand, it may be that we will need to turn our attention to some other site, such as the α-process in the mass cut in type II supernovae (Woosley & Hoffman 1992) or proton-capture reactions in Thorne-Żytkow objects (Cannon et al 1992), to explain the origin of these isotopes.

6.   CONCLUSION

The terms r-, s-, and p-processes have been successful in clarifying our picture of heavy-element formation. In this paper, we have seen that this is because these different processes reflect the responses a nucleosynthetic system can have to being out of equilibrium. In this way, our understanding of the mechanisms of heavy-element synthesis is firm.

Future work will continue to focus on using all clues at our disposal to resolve the problems surrounding the astrophysical sites for these processes. Important questions we must seek answers to in the near future include: 1. Do nascent neutron star winds have high enough entropies to drive a full r-process?, 2. Do ^{13}C pockets really form in low-mass AGB stars, and do they give the right neutron exposures for the s-process?, and 3. Can type Ia supernovae make the p-nucleus ^{92}Mo? That we can ask such specific questions indicates that we have good ideas about the sites for the r-, s-, and p-processes. That we do not have answers shows that we have much work to do.

ACKNOWLEDGMENTS

The author is grateful to G. M. Fuller, W. M. Howard, G. J. Mathews, and S. E. Woosley for discussions and to N. Luo for assistance. He is especially grateful to M. F. El Eid for discussions and for some of the subroutines necessary to compute the results shown in Figures 3, 4, 5, and 6, and to D. D. Clayton for a critical reading of the manuscript. This work was supported in part by NASA grant NAGW-3480.

Literature Cited

Anders E, Grevesse N. 1989. *Geochim. Cosmochim. Acta* 53:197
Anders E, Zinner E. 1993. *Meteoritics* 28:490
Arcoragi J-P, Langer N, Arnould M. 1991. *Astron. Astrophys.* 249:134
Arnett WD, Thielemann F-K. 1985. *Ap. J.* 295:589
Arnould M. 1976. *Astron. Astrophys.* 46:117
Arnould M, Howard WM. 1993. *Meteoritics* 28:561
Arnould M, Rayet M, Hashimoto M. 1992. In *Unstable Nuclei in Astrophysics,* ed. S Kubono, T Kajino, p. 23. Singapore: World Scientific
Audouze J, Truran JW. 1975. *Ap. J.* 202:204
Baraffe I, El Eid MF. 1994. In *Proc. 7th Ringberg Conf. on Nuclear Astrophysics.* In press
Baraffe I, El Eid MF, Prantzos N. 1992. *Astron. Astrophys.* 258:357
Baym G, Pethick C. 1975. *Annu. Rev. Astron. Astrophys.* 17:415
Baym G, Bethe HA, Pethick C. 1971. *Nucl. Phys. A* 175:225
Bazan G, Lattanzio JC. 1993. *Ap. J.* 409:762
Beer H, Macklin RL. 1988. *Ap. J.* 331:1047
Beer H, Macklin RL. 1989. *Ap. J.* 339:962
Beer H, Walter G, Macklin RL. 1985. In *Capture Gamma-Ray Spectroscopy and Related Topics 1984,* ed. S Raman, p. 778. New York: Am. Inst. Phys.
Beer H, Walter G, Macklin RL, Patchett PJ. 1984. *Astron. Astrophys.* 211:245
Bethe HA. 1990. *Rev. Mod. Phys.* 62:801
Bethe HA, Wilson JR. 1985. *Ap. J.* 295:14
Blake JB, Woosley SE, Weaver TA, Schramm DN. 1981. *Ap. J.* 248:315
Boothroyd AI, Sackmann I-J. 1988a. *Ap. J.* 328:632
Boothroyd AI, Sackmann I-J. 1988b. *Ap. J.* 328:641
Boothroyd AI, Sackmann I-J. 1988c. *Ap. J.* 328:653
Boothroyd AI, Sackmann I-J. 1988d. *Ap. J.* 328:671
Brown LE, Clayton DD. 1992. *Ap. J. Lett.* 392:L79
Burbidge EM, Burbidge GR, Fowler WA, Hoyle F. 1957. *Rev. Mod. Phys.* 29:547
Busso M, Gallino R. 1985. *Astron. Astrophys.* 151:205
Busso M, Picchio G, Gallino R, Chieffi A. 1988. *Ap. J.* 326:196
Cameron AGW. 1955. *Ap. J.* 121:144
Cameron AGW. 1957. Chalk River Rep. CRL-41, Atomic Energy Can. Ltd.
Cameron AGW, Cowan JJ, Truran JW. 1985. In *Nucleosynthesis: Challenges and New Developments,* ed. WD Arnett, JW Truran, p. 190. Chicago: Univ. Chicago Press
Cameron AGW, Delano MD, Truran JW. 1970. *CERN Rep. 70-30* 2:735

Cameron AGW, Thielemann F-K, Cowans JJ. 1993. *Phys. Rep.* 227:283
Cannon RC, Eggleton PP, Podsiadlowski P, Żytkow AN. 1992. *Ap. J.* 386:206
Clayton DD. 1978. *Moon and Planets* 19:109
Clayton DD. 1982. *Q.J. R. Astron. Soc.* 23:174
Clayton DD. 1988a. *MNRAS* 234:1
Clayton DD. 1988b. *Ap. J.* 334:191
Clayton DD. 1989. *Ap. J.* 340:613
Clayton DD, Fowler WA, Hull TE, Zimmerman BA. 1961. *Ann. Phys.* 12:331
Clayton DD, Hartmann DH, Leising MD. 1993. *Ap. J. Lett.* 415:L25
Clayton DD, Rassbach ME. 1967. *Ap. J.* 148:69
Clayton DD, Ward RA. 1974. *Ap. J.* 193:397
Colpi M, Shapiro SL, Teukolsky SA. 1989. *Ap. J.* 339:318
Colpi M, Shapiro SL, Teukolsky SA. 1991. *Ap. J.* 369:422
Colpi M, Shapiro SL, Teukolsky SA. 1993. *Ap. J.* 414:717
Cowan JJ, Cameron AGW, Truran JW. 1980. *Ap. J.* 241:1090
Cowan JJ, Cameron AGW, Truran JW. 1982. *Ap. J.* 252:348
Cowan JJ, Cameron AGW, Truran JW. 1983. *Ap. J.* 265:429
Cowan JJ, Cameron AGW, Truran JW. 1985. *Ap. J.* 294:656
Cowan JJ, Thielemann F-K, Truran JW. 1991. *Phys. Rep.* 208:267
Duncan RC, Shapiro SL, Wasserman I. 1986. *Ap. J.* 309:141
Eichler D, Livio M Piran T, Schramm DN. 1989. *Nature* 340:126
Epstein RI, Colgate SA, Haxton WC. 1988. *Phys. Rev. Lett.* 61:2038
Evans CR, Mathews GJ. 1988. In *Origin and Distribution of the Elements,* ed. GJ Mathews, p. 619. Singapore: World Scientific
Gallino R, Busso M, Picchio G, Raiteri CM, Renzini A. 1988. *Ap. J. Lett.* 334:L45
Gallino R, Busso M, Raiteri CM. 1990. *Nature* 348:298
Gallino R, Raiteri CM, Busso M. 1994. *Ap. J.* Submitted
Gilroy KK, Sneden C, Pilachowski CA, Cowan JJ. 1988. *Ap. J.* 327:298
Hillebrandt W. 1978. *Space Sci. Rev.* 21:639
Hillebrandt W, Takahashi K, Kodama T. 1976. *Astron. Astrophys.* 52:63
Hogan CJ, Applegate JH. 1987. *Nature* 320:236
Hollowell DE, Iben I Jr. 1988. *Ap. J. Lett.* 333:L25
Howard WM, Goriely S, Rayet M, Arnould M. 1993. *Ap. J.* 417:713
Howard WM, Mathews GJ, Takahashi K, Ward RA. 1986. *Ap. J.* 309:633
Howard WM, Meyer BS. 1992. In *Nuclei in the Cosmos,* ed. F Käppeler, K Wisshak, p. 575. Bristol: Inst. Phys.
Howard WM, Meyer BS, Clayton DD. 1992. *Meteoritics* 27:404

Howard WM, Meyer BS, Woosley SE. 1991. *Ap. J. Lett.* 373:L5

Hoyle F, Clayton DD. 1974. *Ap. J.* 191:705

Iben I Jr. 1975a. *Ap. J.* 196:525

Iben I Jr. 1975b. *Ap. J.* 196:549

Iben I Jr. 1976. *Ap. J.* 208:165

Iben I Jr, Renzini A. 1982. *Ap. J.* 260:821

Iben I Jr, Renzini A. 1983. *Annu. Rev. Astron. Astrophys.* 21:271

Ito K. 1961. *Prog. Theor. Phys.* 26:990

Käppeler F, Beer H, Wisshak K. 1989. *Rep. Prog. Phys.* 52:945

Käppeler F, Gallino R, Busso M, Picchio G, Raiteri CM. 1990. *Ap. J.* 354:630

Khoklov A. 1990. *Astron. Astrophys.* 245:114

Klapdor HV, Oda T, Metzinger J, Hillebrandt W, Thielemann F-K. 1981. *Z. Phys.* A 299:213

Kochanek CS. 1992. *Ap. J.* 398:234

Kodama T, Takahashi K. 1975. *Nucl. Phys.* A 239:489

Kratz K-L, Bitouzet J-P, Thielemann F-K, Möller P, Pfeiffer B. 1993. *Ap. J.* 403:216

Kratz K-L, Harms V, Hillebrandt W, Pfeiffer B, Thielemann F-K, Wöhr A. 1990. *Z. Phys.* A 336:357

Kratz K-L, Thielemann F-K, Hillebrandt W, Möller P, Harms V, Wöhr A, Truran JW. 1988. *J. Phys.* G 24:S331

Lamb SA, Howard WM, Truran JW, Iben I Jr. 1977. *Ap. J.* 217:213

Lambert DL. 1992. *Astron. Astrophys. Rev.* 3:201

Lambert DL. 1993. In *Origin and Evolution of the Elements,* ed. N Prantzos, E Vangioni-Flam, M Cassé, p. 257. Cambridge: Cambridge Univ. Press

Lambert DL, Gustafsson B, Eriksson K Hinkle KH. 1986. *Ap. J. Suppl.* 62:373

Langer N, Arcoragi J-P, Arnould M. 1989. *Astron. Astrophys.* 210:187

Lattimer JM, Mackie F, Ravenhall DG, Schramm DN. 1977. *Ap. J.* 213:225

Lattimer JM, Pethick CJ, Ravenhall DG, Lamb DQ. 1985. *Nucl. Phys.* A 432:646

Lattimer JM, Schramm DN. 1974. *Ap. J. Lett.* 192:L145

Lattimer JM, Schramm DN. 1976. *Ap. J.* 210:549

Lattimer JM, Swesty 1991. *Nucl. Phys.* A 535:331

LeBlanc JM, Wilson JR. 1970. *Ap. J.* 161:541

Lee T, Schramm DN, Wefel JP, Blake JB. 1979. *Ap. J.* 232:854

Lugmair GW, Shimamura T, Lewis RS, Anders E. 1983. *Science* 222:1015

Macklin RL. 1970. *Ap. J.* 162:353

Mathews GJ, Bazan G, Cowan JJ. 1992. *Ap. J.* 391:719

Mathews GJ, Bazan G, Cowan JJ, Schramm DN. 1993. *Phys. Rep.* 227:175

Mathews GJ, Cowan JJ. 1990. *Nature* 345:491

McWilliam RA, Lambert DL. 1988. *MNRAS* 230:573

Meier DL, Epstein RI, Arnett WD, Schramm DN. 1976. *Ap. J.* 204:869

Merrill PW. 1952. *Science* 115:484

Meyer BS. 1989. *Ap. J.* 343:254

Meyer BS. 1993. *Phys. Rep.* 227:257

Meyer BS. 1994. *Ap. J.* Submitted

Meyer BS, Mathews GJ, Howard WM, Woosley SE, Hoffmann R. 1992. *Ap. J.* 399:656

Meyer BS, Walsh JH. 1993. In *Nuclear Physics in the Universe,* ed. MW Guidry, MR Strayer, p. 9. Bristol:Inst. Phys.

Möller P, Nix JR. 1988. *At. Data Nucl. Data Tables* 39:213

Müller E, Hillebrandt W. 1979. *Astron. Astrophys.* 88:147

Ott U, Begemann F. 1990a. *Ap. J. Lett.* 353:L57

Ott U, Begemann F. 1990b. *Lunar Planet. Sci.* XXI:920

Prantzos N, Arnould M, Arcoragi J-P. 1987. *Ap. J.* 315:209

Prantzos N, Hashimoto M, Rayet M, Arnould M. 1990. *Astron. Astrophys.* 238:455

Prinzhofer A, Papanastassiou DA, Wasserburg GJ. 1989. *Ap. J. Lett.* 344:L81

Qian Y-Z, Fuller GM, Mathews GJ, Mayle RW, Wilson JR, Woosley SE. 1993. *Phys. Rev. Lett.* 71:1965

Raiteri CM, Busso M, Gallino R, Picchio G. 1991a. *Ap. J.* 371:665

Raiteri CM, Busso M, Gallino R, Pulone L. 1991b. *Ap. J.* 367:228

Rayet M, El Eid M, Arnould M. 1992. In *Nuclei in the Cosmos,* ed. F Käppeler, K Wisshak, p. 613. Bristol:Inst. Phys.

Rayet M, Prantzos N, Arnould M. 1990. *Astron. Astrophys.* 227:271

Richter S, Ott U, Begemann F. 1992. *Lunar Planet Sci.* XXIII:1147

Salpeter EE, Shapiro SL. 1981. *Ap. J.* 251:311

Sato K. 1974. *Prog. Theor. Phys.* 51:726

Schramm DN. 1973. *Ap. J.* 185:293

Schwarzchild M, Härm R. 1967. *Ap. J.* 150:961

Seeger PA, Fowler WA, Clayton DD. 1965. *Ap. J. Suppl.* 11:121

Smith VV, Lambert DL. 1986. *Ap. J.* 311:843

Sneden C, Parthasarathy M. 1983. *Ap. J.* 267:757

Sneden C, Pilachowski CA. 1985. *Ap. J. Lett.* 288:L55

Srinivasan B, Anders E. 1978. *Science* 201:51

Symbalisty EMD. 1984. *Ap. J.* 285:729

Symbalisty EMD, Schramm DN. 1982. *Astrophys. Lett.* 22:143

Symbalisty EMD, Schramm DN, Wilson JR. 1985. *Ap. J. Lett.* 291:L59

Takahashi K, Witti J, Janka H-Th. 1994. *Astron. Astrophys.* In press

Takahashi K, Yokoi K. 1987. *Atom. Data Nucl. Data Tables* 36:375

Tammann GA. 1982. In *Supernovae: A Study of Current Research,* ed. M. Rees, R. Stoneham, p. 371. Dordrecht:Reidel

Tang M, Anders E. 1988. *Geochim. Cosmochim.*

Acta 52:1235

Thielemann F-K, Arnould M, Hillebrandt W. 1979. *Astron. Astrophys.* 74:175

Thielemann F-K, Hashimoto M, Nomoto K. 1990. *Ap. J.* 349:222

Truran JW, Cameron AGW. 1972. *Ap. J.* 171:89

Truran JW, Cowan JJ, Cameron AGW. 1978. *Ap. J. Lett.* 222:L63

Truran JW, Iben I Jr. 1977. *Ap. J.* 216:797

Tubbs DL, Schramm DN. 1975. *Ap. J.* 201:467

Ulrich RK. 1973. In *Explosive Nucleosynthesis,* ed. DN Schramm, WD Arnett, p. 139. Austin: Univ. Texas Press

Van den Bergh S, Tammann GA. 1991. *Annu. Rev. Astron. Astrophys.* 29:363

Walker TP, Steigman G, Schramm DN, Olive KA, Kang H-S. 1991. *Ap. J.* 376:51

Wapstra AH, Audi G, Hoekstra R. 1988. *At. Data Nucl. Data Tables* 39:281

Ward RA, Newman MJ. 1978. *Ap. J.* 219:195

Ward RA, Newman MJ, Clayton DD. 1976. *Ap. J. Suppl.* 31:33

Wefel JP, Schramm DN, Blake JB, Pridmore-Brown D. 1981. *Ap. J. Suppl.* 45:565

Weigert A. 1966. *Z. Astrophys.* 64:395

Wheeler JC, Sneden C, Truran JW. 1989. *Annu. Rev. Astron. Astrophys.* 27:279

Wilson JR, Mayle RW. 1993. *Phys. Rep.* 227:97

Woosley SE, Fowler WA, Holmes JA, Zimmerman BA. 1975. Cal. Inst. Tech., Kellogg Rad. Lab. Preprint No. OAP-422

Woosley SE, Hartmann DH, Hoffman RD, Haxton WC. 1990. *Ap. J.* 356:272

Woosley SE, Hoffman RD. 1992. *Ap. J.* 395:202

Woosley SE, Howard WM. 1978. *Ap. J. Suppl.* 36:285

Woosley SE, Howard WM. 1990. *Ap. J. Lett.* 354:L21

Woosley SE, Mathews GJ, Wilson JR, Hoffman RD, Meyer BS. 1994. *Ap. J.* In press

Yokoi K, Takahashi K. 1981. In *Proc. 4th Int. Conf. on Nuclei Far from Stability,* CERN Rep. 81-09, p. 351

Zinner E, Tang M, Anders E. 1987. *Nature* 330:730

Zinner E, Amari S, Lewis RS. 1991. *Ap. J. Lett.* 382:L47

Annu. Rev. Astron. Astrophys. 1994. 32: 191–226

ABUNDANCES IN THE INTERSTELLAR MEDIUM

T. L. Wilson

Max Planck Institut für Radioastronomie, Auf dem Hügel 69, 53231 Bonn, Germany

R. T. Rood

Department of Astronomy, University of Virginia, Charlottesville, Virginia 22903

KEY WORDS: isotopes, evolution of the Galaxy, cosmology, nucleosynthesis

1. INTRODUCTION

Today, it is believed that all of the deuterium (^{2}H), most of the helium (^{3}He and ^{4}He) and some of the ^{7}Li were produced in the big bang. The simplest version of this scheme is referred to as Standard Big Bang Nucleosynthesis (hereafter SBBN; Boesgaard & Steigman 1985). The comparison between theoretical SBBN and observed abundances has been one of the cornerstones of modern cosmology. Heavier elements, sometimes rather loosely referred to as "metals," are thought to be predominantly produced in stars. The study of this stellar nucleosynthesis has been one of the most fruitful areas of astrophysics since midcentury. While it is tempting to consider that any analysis of elemental and isotopic abundances can be divided into two areas—the first big bang nucleosynthesis and the second stellar nucleosynthesis—the problem is not so simple. The products of SBBN are modified; they are either net produced or destroyed by stellar processing. As the study of cosmological element production has progressed from a simple confirmation of a phase of hot nucleosynthesis to one of the forefront areas of particle physics, ever increasing demands have been placed on the accuracy of the primordial abundances. In parallel developments modern supercomputers have led to more detailed and quantitative predictions of stellar element production (Woosley & Weaver 1993; Meyer 1994). Rather

191

elaborate chemical evolution models dealing with specific elements rather than a generalized heavy element fraction are now possible (Woosley et al 1993).

At the heart of all nucleosynthesis studies are the observed abundances. Observations that show how abundances vary in time and with location in the Milky Way and other galaxies are particularly important. Here we will primarily review radio wavelength abundance determinations from the local interstellar medium (ISM), H II regions, and dense and diffuse clouds throughout the galaxy, concentrating on the light elements. Abundances determined in this way are important for a number of reasons. Comparisons of abundances in the local interstellar medium with Solar System abundances allow a determination of the *evolution* of abundances in the 4.5 Gyr since the formation of the Solar System. At larger distances from the Sun, in the disk of our galaxy, obscuration by dust at optical wavelengths is a limiting factor. Thus, radio observations are the primary technique for exploring abundance *gradients* in the disk of our own galaxy. In the environments available to radio studies it is often easy to observe different isotopes of certain elements. In a few cases (e.g. ^{13}C in red giants) stellar abundances are more a probe of evolutionary processes in stars than an indication of a "cosmic abundance." We note that although radio abundances are not without their difficulties, at least these difficulties are different than those encountered when using other techniques.

Although the main thrust of this review is not directed at SBBN we will briefly address important recent abundance measurements. SBBN theory as summarized by Boesgaard & Steigman (1985) and as compared with the latest data by Walker et al (1991), is also discussed by Pagel & Edmunds (1981), Waddington (1989), Rana (1991), and Wilson & Matteucci (1992, hereafter WM) with emphasis on nucleosynthesis theory and chemical evolution models. (Such models describe the processing of matter in stars and the return of this processed material to the ISM.) Pagel & Edmunds (1981), Edmunds (1989), and Pagel (1989) have reviewed abundance determinations from optical studies; these results are also extensively discussed in AIP Conference 183 on Cosmic Abundances (Waddington 1989). The review of Rana (1991) focused on comparisons with stellar populations in the Solar neighborhood; the data base for such comparisons has been expanded by Edvardsson et al (1993a,b), who have obtained abundances for 189 nearby F and G dwarfs.

The elemental and isotopic abundances for the Solar System, from solar photosphere and C1 chondrite meteorite data, have been compiled by Anders & Grevesse (1989). Since there are some variations in the isotope ratios of meteorites, the Solar System must have been produced from material with different processing histories, and there are a number of lines of evidence which suggest that the Solar System abundances may not be strictly representative of the local ISM 4.5 Gyr ago (Wasserburg 1985; Cameron 1985, 1993; Gies & Lambert 1992; Steigman 1993).

During the 1970s and 1980s there was some reason to wonder whether one should trust any isotope abundances determined from radio astronomy data. However much progress has been made in the past few years, and a summary of ISM abundances with emphasis on radio data seems appropriate.

2.   GENERAL BACKGROUND

A description of the structure and content of our galaxy is contained in the review of Combes (1991). The Solar System is located in the Norma-Scutum spiral arm, 8.5 kpc from the Galactic center. Outside the solar radius, $D_\odot$, interstellar clouds contain a substantial amount of H I; inside $D_\odot$ these consist mostly of H_2. A large concentration of molecular clouds and H II regions is found in the "molecular ring" at galactocentric distance $D_{GC} \simeq 4$ kpc. For $D_{GC} \leq 3$ kpc, there is a lack of molecular material and H II regions; within a few hundred pc from the Galactic center, the number of molecular clouds increases once again. There seems to be little evidence for an exchange of material between the Galactic center and the molecular ring. Since it is likely that the chemical evolution of the Galactic disk and the Galactic center are different, there is no reason that extrapolations of abundances or isotope ratios obtained in the Galactic disk should equal values at the Galactic center. Although it is generally agreed that a larger fraction of the gas in the inner part of the Galaxy is in the form of molecules, it is unclear whether the structure of clouds at $D_{GC} \simeq 4$ kpc differs qualitatively from those at 8.5 kpc. If so, the mass distribution of stars formed, the Initial Mass Function, might be affected. In addition, since the star formation rate must depend on the amount of dense, cold material available (Lada et al 1991), it is clear that the ISM in the inner part of our galaxy has undergone more stellar processing and thus there should be a higher abundance of metals. Although theories of disk galaxy formation exist (see e.g. Larson 1976), observational data about infall, one of the important parameters, are meager. It was hoped that the study of high velocity clouds (HVCs) would aid in limiting the range of infall parameters (see Wakker 1991). In one case, an HVC toward the dwarf galaxy I Zw 18, Kunth et al (1994) have determined that the O/H ratio is 0.1 of the solar value. For four other HVCs, Savage et al (1993) find that the Mg/H ratio is roughly 0.1 of the solar value indicating that there has been some stellar enrichment, and thus, it appears that the HVCs are not composed of primordial material.

3.   ISM ABUNDANCE DETERMINATIONS

Unlike stars, the sources used for radio studies are frequently larger than the telescope beam. The emission line results are averages over the beam; at radio wavelengths, typically 2' beams are used. If the cloud being studied is 7 kpc from the Sun, this corresponds to a linear size of 4 pc. Absorption lines sample those parts of clouds toward continuum sources. In the radio range, the background sources are sometimes extended. A $1'$ source size is typical; if the molecular cloud is 3 kpc from the Sun, the linear size of the region sampled is 1 pc. In the visible and UV range, stellar disk sizes are in the milliarcsecond range; for clouds 100 pc from the Sun observed in absorption against such stars, the sizes sampled are 2×10^{-4} pc or 6×10^{14} cm (or 40 AU). Presumably the larger the region sampled, the more representative the result.

An important assumption made when comparing Galactic chemical evolution models with data is that after a few hundred million years the constituents produced at D_{GC} should become uniformly distributed in a fairly narrow annulus about D_{GC}. This assumption is important for comparisons with Galactic chemical models, which are made on the basis of axisymmetric distributions.

Selected element and isotope abundances for the Solar System and local ISM are given in Tables 1 and 2. The ^{4}He, CNO, and heavier element abundances for the local ISM are taken from measurements of NGC 1976, Orion A, the brightest nearby unobscured H II region. There are some differences in the abundances; the largest is for Ne. However, Baldwin et al (1991) comment that their Ne estimate (the larger value) might be uncertain since this is based on an ionization state that has a small amount of the total population. Although less of an effect than for cold clouds, the abundances in H II regions still may be affected by the condensation of certain elements onto grains—depletion. The data for Orion A are taken to represent the present-day ISM far from the center of our galaxy. The Solar System results are taken to represent a sample for the ISM in this region 4.5 Gyr ago. As will be discussed later, there are differences. It is possible that the solar sample may not have been typical, or that the gas has moved inward relative to the stars (see e.g. Russell & Dopita 1992).

The study of H II regions is a complement to studies of O and B star abundances: Although the temperatures in H II regions are similar to those in O/B stellar atmospheres, the excitation conditions will be rather different. As with stars, the interpretation of H II region abundances may be complicated by different stages of ionization; also, because the line widths in H II regions are larger than $10 \, \text{km s}^{-1}$, the separation of CNO isotopes is not possible. Helium is a special case; details of measurements of ^{3}He are given in Section 4.2.1.

Diffuse clouds consist mostly of atoms and dust grains. Depletion onto grains can be important (see e.g. Spitzer & Jenkins 1975), although van Dishoeck & Black (1986) find the extent of this effect to be smaller than previously thought. In any event, this effect is unimportant for isotopes. For the lighter elements, isotope shifts are larger than typical line widths, so that measurements can be carried out using atomic lines; UV absorption line spectroscopy is especially suited for determining the D/H abundance ratio. Denser and more extincted clouds consist mostly of molecules. Emission lines are well suited for studies of isotopes of C, N, and O. Molecular spectroscopy is needed, since for masses larger than 12, atomic isotope shifts are usually smaller than thermal line widths. However, the interpretation of molecular line data is complicated by molecular excitation and interstellar chemistry. Most essential are comparisons between data sets for different molecular species to eliminate systematic errors.

3.1   Observations of H II Regions

The best studied H II region, Orion A, NGC 1976, is $\sim$ 500 pc from the Sun. Compared to the Solar System, the O/H and N/H abundances are factors of $\sim$ 2 lower (see Table 1). For O, one might suppose that some percentage of the atoms might be bound in dust grains. If so, one would expect more O in the gas phase near the Trapezium, where grains would be destroyed by stellar winds and the intense UV field. However, no O/H gradients have been found in Orion A. Peimbert & Torres-Peimbert (1977), Walter et al (1992), and Peimbert et al (1993) have proposed to eliminate this discrepancy by assuming that fluctuations in the electron temperature T_e are present. The good agreement of determinations in radio (biased toward low values of T_e) and optical (biased toward high values of T_e) ranges does not support this hypothesis (Wilson & Jäger 1987). The O abundance determined for Orion A does not appear to be anomalous, because 18 main sequence B stars in the Orion association have similar O/H ratios (Cunha & Lambert 1992). The N and O abundances are expected to increase with time due to the chemical evolution of the Galaxy. Thus Orion A should have *larger* abundances of O and N than solar. This is not so. While the cause is unclear, this adds to the accumulating evidence that the Solar System abundances may not be representative of the local ISM 4.6 Gyr ago (e.g. Gies & Lambert 1992, Steigman 1993).

The O/H ratio is taken to represent the abundance of metals (see e.g. Pagel

Table 1   Abundances for CNO and heavier elements

Element	Solar System[a]	Orion A[b,c]
(C/H)	3.6×10^{-4}	3.4 to $\sim$ 2.1 $\times 10^{-4}$
(O/H)	8.5×10^{-4}	4.0 to 3.8 $\times 10^{-4}$
(N/H)	1.1×10^{-4}	6.8 to 8.7 $\times 10^{-5}$
(Ne/H)	1.2×10^{-4}	$\sim$ 8 to $\sim$ 40 $\times 10^{-5}$
(Si/H)	3.6×10^{-4}	3.0×10^{-6}
(S/H)	1.8×10^{-5}	8.5 to 13.3 $\times 10^{-6}$
(Ar/H)	3.6×10^{-6}	4.5 to 2.1 $\times 10^{-6}$

[a] Anders & Grevesse (1989).
[b] Rubin et al (1991) for first value listed.
[c] Baldwin et al (1991) for second value listed.

& Edmunds 1981). Large radial gradients, i.e., with distance to the center, in the strength of [O III] to Hβ in spiral galaxies are taken to be gradients in O/H ratios. A direct determination of an O/H gradient for our galaxy is difficult because of extinction. For H II regions more than 3 kpc distant, dust extinction is usually large and studies must be made in the IR or radio wavelength regions. Indirect estimates of element gradients are possible, because a larger abundance of metals leads to lower T_e values due to the enhanced cooling of H II regions (see e.g. Rubin 1985). For our galaxy, the angular resolutions and sensitivities suffice for useful measurements of radio recombination lines from hundreds of individual regions (Wink et al 1983, Shaver et al 1983); for external galaxies, only a few results are available (Anantharamaiah & Goss 1990). Gradients in T_e with D_{GC} have been established using many different radio recombination line transitions; the best results are obtained from short centimeter wavelength data, for which maser effects are not important. The most complete survey at 2 cm gives an electron temperature gradient T_e of $+280 \pm 30$ K kpc^{-1} for $D_{GC} \geq 3$ kpc (Wink et al 1983). It is likely that this gradient is also related to an abundance gradient, but proving this is not simple. A statistical relation between T_e and the total CNO abundances for more distant H II regions was given by Mezger et al (1979). Shaver et al (1983) put this relation on a firmer basis by combining optical measurements of abundances with T_e determinations for 33 southern H II regions. The determinations of T_e from radio recombination lines are simpler and more reliable than optical determinations. Nevertheless, the radio results are in good agreement with optical estimates. Using both radio and optical determinations, element gradients can be obtained for H II regions up to a few kpc from the Sun. These results have been extended by Fich & Silkey (1991), who determined abundances in outer parts of the northern Galaxy. From this, it appears that the N/H abundance gradient is even flatter than predicted from the results of Shaver et al (1983). Data for 20 B stars in 4 young clusters, by Fitzsimmons et al (1990), show no abundance gradients over a range $D_{GC} = 5.5$ to 10.5 kpc. This result might arise because abundances depend on some factor(s) other than D_{GC}. At the least, one expects some scatter at a given D_{GC}. Either some selection effect or just the small number statistics in dealing with 4 clusters could lead to observing no element gradient even though one really exists. The IR measurements of the column densities of O and N measured using IR fine structure lines (see e.g. Lester et al 1987) are not very dependent on T_e, allowing what should be a more secure determination of the relative abundance of O or N ions. If corrections for the different sizes of Strömgren spheres are applied (e.g. Rubin et al 1988), the FIR results give an (N/O) abundance that is larger than solar, as opposed to the optical results which give an abundance lower than solar. However, according to Rubin et al (1988), there is no strong evidence for a gradient with D_{GC}. These authors interpret the results as indicating that N is produced mainly by a primary mechanism. (It is ordinarily thought that N is a product of secondary nucleosynthesis, being produced by CNO processing of ^{12}C and ^{16}O from earlier generations of stars. If N is a primary element produced by CNO processing, the "seed" ^{12}C or ^{16}O must have been produced in earlier helium burning stages of the same star in which the CNO processing takes place.)

The relation between T_e and metal abundance found for the Galactic disk does not hold for the Galactic center region, because the values of T_e are too large in view of the expected high abundance of metals (Wink et al 1983). This discrepancy may indicate that local surroundings affect the value of T_e; a high value of T_e might result from a higher electron density rather than any dependence on metal abundance. However, the calculations of Rubin (1985) show that electron densities are less important than O or N abundances in determining T_e. Even so, it may be that molecular clouds near the Galactic center have larger H_2 densities and T_{kin} values (Bally et al 1987). If the HII regions are embedded in such clouds or are subjected to a high pressure caused by hot ionized gas (Spergel & Blitz 1992), the expansion of the HII regions would be hindered, giving rise to higher electron densities. Examples of the effects of large electron densities on T_e have been observed for nearby HII regions such as NGC 2024 and Orion A, for which there are gradients in T_e on arcminute scales (Wilson et al 1990, Wilson & Jäger 1987).

3.2   *Diffuse Clouds*

Clouds with absorption coefficients in V ranging from $A_V \sim 0.1^m$ to $\sim 1^m$ and local densities $\leq 50\,\text{cm}^{-3}$ consist mostly of HI. These clouds have $T_{kin} \geq 50\,\text{K}$. The column densities can be determined from absorption against hot background stars [see e.g. Kulkarni & Heiles (1988) for a review of 21-cm line measurements, and Savage (1987) for a review of UV data]. These diffuse clouds are not gravitationally bound. The line widths, and therefore the kinetic energies are much larger than needed to balance self gravity. The basic source list for such objects is contained in Tables 1 and 2 of McCullough (1992). In the optical and ultraviolet wavelength range, the most studied cloud is ζ Oph. This cloud is not very diffuse but is favorably positioned in front of a O9.5V star, and is $\leq 160\,\text{pc}$ from the Sun. For element ratios measured in diffuse clouds, accounting for the populations of all ionization states may be complex; for refractory elements, depletion is a large effect.

3.3   *Dense Molecular Clouds*

Giant Molecular Clouds (GMCs) are clumpy, nonrelaxed structures found mostly in spiral arms. GMCs are believed to be in virial equilibrium; there are thousands in our galaxy (see the results presented by Sanders et al 1986,

Dame et al 1987, Solomon & Rivolo 1987, Combes 1991). The cores of GMCs are used for the study of abundances; detailed molecular line investigations have concentrated on fewer than 50 clouds; these are frequently used for isotope studies (see lists in Wannier 1980, Wilson & Walmsley 1989, Langer & Penzias 1990). The densities and extinctions are orders of magnitude larger than for diffuse clouds. These clouds are both chemically and physically very inhomogeneous, with structures on all spatial scales. UV radiation may penetrate deep into these clouds because of their clumpy structure. Lines of CO or isotopes of CO show widths that exceed the sound speed by factors of a few. In the disk, clouds without internal heat sources have the kinetic temperatures T_{kin} of ~ 10 K. These clouds are the birth sites of stars; there is an observed association of high surface brightness H II regions and molecular clouds (Myers et al 1986). Embedded newly born stars give rise to outflows which may cause supersonic line widths; these stars also contribute to the heating and possible disruption of molecular clouds. Even in such warm molecular regions, T_{kin} rarely exceeds a few 100 K, and then only over regions of < 0.1 pc. The H_2 densities range from ~ 50 cm^{-3} over large regions to $> 10^6$ cm^{-3} in regions < 0.1 pc. In the Galactic center region, the average densities and T_{kin} are larger. Isotopic measurements made using emission lines refer to the cores of GMCs, where H_2 densities and column densities are large.

3.3.1 EXCITATION   A summary of our knowledge of molecular line excitation is given by Genzel (1993). Briefly, the ISM is bathed in the diluted hot radiation field of stars, as well as IR fields caused by the absorption and reradiation of starlight by dust, and the 2.7 K background radiation. In molecular clouds, fragile species are shielded by dust, and these molecules sometimes collide with H_2 or dust grains. The temperature of the dust grains is determined by the competition between heating by starlight and the efficiency of grain radiation. T_{kin} is usually lower than the color temperature of the radiation field. These different temperatures tend to drive the populations of molecules away from Local Thermodynamic Equilibrium (LTE). The moderate to large optical depths of many molecular transitions and the velocity fields of molecular clouds complicate any analysis of these processes. To form the ratio of two isotopic species of a given molecule (to which we refer as isotopomers) the total population of each species is needed. Unfortunately, usually only one transition of each isotopomer is measured, so excitation effects can be important. If the molecules are collisionally excited, and the optical depths are similar, the ratio of the total population will be accurately represented by the ratio obtained from a single transition of each isotopomer. This is so for $^{13}C^{32}S$ and $^{12}C^{34}S$, or $H^{12}C^{15}N$ and $H^{13}C^{14}N$. For studies of some ^{12}C and ^{13}C substitutions, optical depths differ by factors of up to 60. This difference can give rise to "photon trapping," in which photons emitted by the more abundant species cannot as freely escape from the cloud as photons of the rarer species, because of resonant absorption. Then, the upper levels of the more abundant species will be more highly populated; in effect, photon trapping gives rise to excitation differences even among isotopomers that are collisionally excited. For $^{12}C^{18}O$ and $^{13}C^{18}O$ only small corrections are needed, since the optical depths of rotational lines of both isotopomers are less than unity. The corrections are larger for isotopes of H_2CO. (For molecular species, where no superscript is given the normal isotopomer is meant.) Measurements of 6-cm wavelength H_2CO K-doublet absorption lines toward intense background sources give better signal-to-noise ratios than emission measurements of the same lines. The determination of isotope ratios using 6-cm lines involves comparisons between H_2CO and $H_2^{13}CO$; the optical depths of corresponding lines of different isotopic species differ by factors of 20 to 100. For H_2CO, the photon trapping effect raises the excitation temperatures of the H_2CO transitions more than in $H_2^{13}CO$, and also raises the populations of more highly excited (and usually unmeasured) energy levels of H_2CO. It is usual to measure the 6-cm K-doublet lines of H_2CO and $H_2^{13}CO$. In a few sources, the corresponding 2-cm K-doublet lines have been measured, in even fewer the 1-cm K-doublet lines. For some regions, information about the spatial structure of H_2CO absorption in the 6-cm line has been provided by high angular resolution maps of Martin-Pintado et al (1985) and by maps in the 6- and 2-cm lines by Dickel & Goss (1990). All these data show that the optical depths of the 6-cm wavelength lines are not very large; however, the optical depths of rotational transitions in the millimeter wavelength range are frequently > 1. When combined with model calculations, these data should give fairly accurate corrections for photon trapping (see e.g. Henkel et al 1980, 1983). For measurements of the isotope ratio using the 6-cm line of H_2CO, such corrections will always raise the observed value of the $^{12}C/^{13}C$ ratio; the trapping corrections can be significant and should be applied to the measured ratios for those sources where the necessary data are available.

If isotope ratios can be more simply determined from measurements of $^{13}C^{18}O$ and $C^{18}O$, one might wonder why H_2CO data are used at all. One problem is obtaining accurate measurements for CO, since thermally excited emission lines with low optical depths have extremely small intensities. The survey of isotope ratios using the $J = 1 - 0$ line of $^{13}C^{18}O$ and $C^{18}O$ (Langer & Penzias 1990, 1993) required years for completion. Also, although these $^{12}C/^{13}C$ isotope ratios have good signal-to-noise ratios, measurements of a number of species are needed to check on the size of systematic effects, which will be discussed in the next two sections.

3.3.2 CHEMICAL FRACTIONATION   Interstellar chemistry is a complex and not fully understood topic involving nonequilibrium, non-LTE processes. For simpler polar molecules such as CO, the formation involves ions, which are produced by the action of cosmic rays, followed by charge exchange (Herbst & Klemperer 1973). In warm very dense clouds, it is likely that some molecular species are produced on dust grain surfaces. Even a comparison of isotopes of a single species might be affected by interstellar chemistry, in the form of chemical fractionation. This effect arises from the small differences in the molecular binding energies, caused by the different zero-point vibration energies, related to the nuclear masses. For D and H, chemical fractionation occurs via an exothermic reaction with H_3^+. This favors the production of HD since the zero-point energy of H_2 is 500 K higher than that for HD. For more complex molecules, such as HCN, H_2D^+ plays a similar role. Thus there should be a enrichment of D in molecules. If the reactions reach equilibrium, the relation of measured to actual ratio is

$$\left(\frac{D}{H}\right)_{measured} = \left(\frac{D}{H}\right)_{actual} \times \exp\left(\frac{\Delta E}{kT_{kin}}\right) \tag{1}$$

For typical kinetic temperatures, chemical fractionation can lead to the enrichment of D in molecules by factors of thousands. For very compact molecular clouds near newly born stars, such as the Orion Hot Core (see e.g. Genzel & Stutzki 1989), dynamical time scales are $< 10^4$ yr. There is evidence that chemical equilibrium may not be reached for times $< 10^4$ yr, so in these situations, the D/H ratio depends on the past history of the cloud (see e.g. Walmsley et al 1987).

Although chemical fractionation is known to have a large effect on the D/H ratios measured using H_2 and HD, reliable ratios can be gotten from measurements of D I and H I in diffuse clouds. If the uncertainty is a factor of ≤ 2, the D/H ratios are very useful for estimating the baryon/photon ratio, $n_B/n_\gamma = \eta$, in the big bang model. In contrast, any interpretation of the measured $^{12}C/^{13}C$ ratios must be more subtle, because ISM values are at most only a factor of two different from the Solar System ratio (see Table 4, Figure 2a). For this reason, systematic uncertainties in the $^{12}C/^{13}C$ ratios must be less than 20%. The effect of chemical fractionation for carbon isotopes is much smaller than for HD, since the difference in zero-point energies is 35 K. Still, this effect may have a significant influence on interpretations (Watson et al 1976). As with HD, the reaction involves ions; for carbon fractionation, the presence of C^+ is needed. In cold molecular clouds, an exchange reaction of $^{13}C^+$ with CO leads to an enrichment of ^{13}C, with the liberation of 0.003 eV (equivalent to $T = 35$ K) of energy. In the fractionation models of Langer et al (1984), ^{13}CO contains most of the ^{13}C. Isotopomers of less abundant species which are not formed from CO should show a deficiency in ^{13}C. One such species is H_2CO. As for Equation (1), in equilibrium, the relation of the measured to actual ratio is

$$\left(\frac{^{13}CO}{^{12}CO}\right)_{measured} = \left(\frac{^{13}CO}{^{12}CO}\right)_{actual} \times \exp\left(\frac{\Delta E}{kT_{kin}}\right) \tag{2}$$

Since C^+ is thought to be mostly in the outer parts of molecular clouds, this reaction may not affect molecules that are in dense cores (see e.g. Langer et al 1984). It is possible that molecular clouds consist of very small dense clumps, with large surface areas, so that the influence of chemical fractionation will depend on the details of cloud structure. The most certain isotope ratios are provided by CH^+ measurements, because CH^+ is formed in an endothermic reaction, and must be produced at high temperatures so that chemical fractionation is not important (Watson et al 1976).

3.3.3 SELECTIVE DISSOCIATION   If the molecules are dissociated by spectral line radiation, and the optical depth of one isotopomer is large, the more abundant species can shield itself against the effects of UV radiation. The most extreme case is the (HD/H_2) ratio, where the enrichment predicted by Equation (1) is nearly balanced by selective dissociation (Watson 1974). Less drastic but still measurable effects have been observed for CO isotopes (Bally & Langer 1982). The calculations of Chu & Watson (1983), Glassgold et al (1985), van Dishoeck & Black (1988), and Viala et al (1988) treat this effect. The results (see e.g. Figure 11 of van Dishoeck & Black 1988) show that the carbon and oxygen isotope ratios in CO are little affected by selective dissociation. If so, this effect will not change isotope ratios in other molecules either.

3.3.4 CARBON ISOTOPE RATIOS: CHEMISTRY OR NUCLEAR PHYSICS?   The most convincing proof that fractionation is not a very large effect is based on comparison with CH^+. The best source for such measurements is ζ Oph. Data taken by Hawkins et al (1985) gave a $^{12}C/^{13}C$ number ratio of 43 ± 4 (all quoted ratios are by number); newer data for ζ Oph give ratios in the range 60 to 70, each with a quoted RMS error of 10% (Stahl et al 1989, Crane et al 1991, Stahl & Wilson 1992, Hawkins et al 1993). The best value at present is $\simeq 70$. The agreement of the average CH^+ ratio for ζ Oph with ratios from radio astronomy for the local ISM (Table 4 and Section 5.1) indicates that fractionation is not a large effect for cloud cores.

3.3.5 THE REAL WORLD OF ISOTOPE RATIOS   Any discussion of the radio astronomical measurements of isotope ratios would not be complete without a brief history of the subject. In regard to radio data, Wannier et al (1976) published the first $^{12}C/^{13}C$ ratios obtained with good signal-to-noise ratios and accurate calibrations. These were made using the $J = 1 - 0$ lines of $C^{18}O$ and ^{13}CO. Two important assumptions were used to obtain the final ratios. First, since it was recognized that peak optical depths of the $J = 1 - 0$ line of ^{13}CO are > 1, isotope ratios were formed using the line wings to avoid optical depth effects. In addition, it was assumed that the $^{16}O/^{18}O$ ratio was constant as a function of D_{GC}. Although both assumptions were open to question, the first turned out to be far more critical since line wings arose from the outer parts of the molecular clouds (see e.g. Penzias 1983), where C^+ was abundant. Consequently the effect of fractionation was large (Section 3.3.2), and this heavily influenced the value of the final ratios. Dickman et al (1979) and Langer et al (1980) measured spatial gradients in ($C^{18}O/^{13}CO$) from the edge to the center of clouds. If C^+ were present in the cloud envelopes, such gradients would be predicted from Equation (2). After these findings became known, confidence in the isotope ratios from radio astronomy sunk to a low, and what seemed simple and definite in 1975 became very uncertain and complex in 1980–1990. Only recently have astronomers begun to accept radio measurements of isotope ratios again. The main reason for this is the recognition that $^{12}C/^{13}C$ ratios from CH^+ agree well with ratios from radio astronomy data for the local ISM. Although this seems reassuring, surprises have happened in the past, and may happen again. The only way to check for the presence of systematic errors is to measure $^{12}C/^{13}C$ using many molecular species.

Although it is accepted that carbon isotope ratios from CH^+ give the actual $^{12}C/^{13}C$ ratios, similar data from other molecules may not. Rather, these illustrate the processes described in the previous two sections. For ζ Oph, ratios have been determined from ($^{12}CO/^{13}CO$) since the optical depths of millimeter wavelength rotational lines of ^{12}CO are not large. The best radio astronomical measurements give a combined lower limit of > 60 (Langer et al 1987, Wilson et al 1992). More significant is the result obtained by Sheffer et al (1992), (see also Lambert et al 1994) using UV absorption lines of ^{13}CO measured with the Hubble Space Telescope, combined with previously measured CO data from another band. Their ratio of 150 ± 27 is much larger than other values. The analysis of Scheffer et al (1992) shows that selective dissociation is the dominant effect in this diffuse cloud, causing a marked underabundance of ^{13}CO. In this case, chemical fractionation seems to play a minor role. Crane & Hegyi (1988) determined a ($^{12}CN/^{13}CN$) ratio of $47.3^{+5.5}_{-4.4}$, while Hawkins et al (1993) find a ratio of 100^{+88}_{-33}. The weighted average ratio from CN is below the actual $^{12}C/^{13}C$ ratio. A comparison of these ratios is complicated by the fact that the CN and CH^+ arise from different regions (Lambert et al 1990). However, assuming that the $^{12}C/^{13}C$ ratio in ζ Oph is ~ 70, CN is enriched in ^{13}C. For clouds other than ζ Oph, absorption line intensities are much lower (see e.g. Crane et al 1990, Centurion & Vladilo 1991, Vladilo et al 1993). Hawkins & Jura (1987) have determined ratios for a number of sources in Taurus and Perseus, finding an average ($^{12}CH^+/^{13}CH^+$) ratio of ~ 40. Toward ζ Per, Hawkins et al (1993) find a ($^{12}CH^+/^{13}CH^+$) ratio of 51 ± 14; at the upper RMS level, this overlaps with the ζ Oph result. Toward ζ Per, Kaiser et al (1991) found a ($^{12}CN/^{13}CN$) ratio of 77^{+27}_{-18}. The larger carbon isotope ratio from CN indicates that the interstellar UV field has a larger effect toward ζ Per than toward ζ Oph.

In the previous discussion, we emphasized the pitfalls in the interpretation of measurements, and pointed to the anomalous cases, to show that care is needed.

Table 2 Selected isotopic ratios for light elements by number

Relative abundance	Present-day solar system[a]	Protosolar	Nearby ISM
(D/H)	3.4×10^{-5}	$(2.6 \pm 1.0) \times 10^{-5}$ [b]	1.6×10^{-5} [c]
(^{4}He/H)	9.8×10^{-2}	0.10 ± 0.01[d]	8.9×10^{-2} [e]
(^{3}He/^{4}He)	1.4×10^{-4}	$(1.5 \pm 0.3) \times 10^{-4}$ [b]	$\leq 10^{-4}$ to $\geq \times 10^{-2}$ [f,g]
(^{7}Li/H)	1.9×10^{-9}	...	...
(^{7}Li/^{6}Li)	12.3	...	6.2[h] to 12.5[i]

[a] Table 3 of Anders & Grevesse (1989).
[b] Geiss (1993).
[c] McCullough (1992), Linsky et al (1993).
[d] Based on solar structure model (Turck-Chièze et al 1988).
[e] Average of result from Baldwin et al (1991) and Osterbrock et al (1992).
[f] Rood et al (1992).
[g] Balser et al (1994).
[h] Meyer et al (1993).
[i] Lemoine et al (1992).

Actual isotope ratios cannot be directly obtained from the data in every case, therefore some interpretation is necessary. In the following paragraphs, we take the point of view that by a proper choice of molecular species, the systematic effects can be minimized, and one can obtain the actual isotope ratios. Even with these data bases, there are some systematic effects, but we believe that the size of such effects is no larger than uncertainties caused by noise.

4.   RESULTS FOR D, He, AND Li

In Figure 1, the calculated abundances of Kernan et al (1994) are plotted as a solid lines versus η, the baryon to photon ratio. In addition to η, there are parameters such as neutron halflife and the number of lepton flavors which affect the calculated abundances by small amounts (see Kernan et al 1994 for details). The temperature of the microwave background is taken to be 2.75 K (Mather et al 1990, Palazzi et al 1990, Gush et al 1990), the number of neutrino families to be 3 (see Figure 3 of Walker et al 1991) and, from the average over many experiments, the mean life of the neutron is found to be 889 sec (see e.g. Walker et al 1991). Then, the SBBN theory and light element abundances determine η (and thus the baryon density of the universe). If the ratio of the baryon density ρ_B to the critical density ρ_c is Ω_B then $\Omega_B h_{50}^2 = 0.015\, T_{2.75}^3 \eta_{10}$, where h_{50} is the value of the Hubble constant (in units of $50\,\mathrm{km\,sec^{-1}\,Mpc^{-1}}$), $T_{2.75}$ is the temperature of the cosmic background radiation in units of 2.75 K, and η_{10} is $10^{10}\,\eta$. As can be seen from Figure 1, the ^{4}He abundance does not have a strong dependence on η, while the ^{3}He and, particularly, the D abundances are sensitive functions of this parameter. One of the strengths of SBBN has been the relatively narrow band in η for which the calculated abundances agreed with those observed. For the survival of the SBBN theory, this concordance must remain as the error bars tighten.

4.1   *Deuterium*

By assumption, deuterium is produced in the big bang and destroyed in stars. Then a measurement of the ISM D/H ratio will allow a lower limit to the primordial D/H ratio (Yang et al 1984). Measurements are difficult, however, because of the small abundance of D and because D can chemically combine with other elements. For example, in dense clouds, an unknown fraction of D is in HD. For the Solar neighborhood, the D/H ratio was first measured using the UV absorption spectroscopy of diffuse clouds with the *Copernicus* and *IUE* satellites (see e.g. York & Rogerson 1976, McCullough 1992). New measurements with the Hubble Space Telescope offer higher sensitivity and wavelength resolution. Along the line of sight toward the double star Capella, Linsky et al (1993) find a D/H ratio of $1.65^{+0.07}_{-0.18} \times 10^{-5}$. Most of the uncertainty is associated with the blending of D I and H I lines, and an accurate estimate of column density of H I from the optically thick Lyman α line. Further observations (Linsky et al 1994) at another orbital phase of Capella allow a direct determination of the the underlying stellar spectrum which must be subtracted to get the interstellar lines. Thus, a source of systematic error is reduced. The preliminary results are consistent with the earlier value. They also have found D/H $= (1.40 \pm 0.10) \times 10^{-5}$ for an additional line of sight toward Procyon. In the *Copernicus* data some sources, such as ϵ Per, were observed to show time variability in the Ly α absorption line of D on time scales of hours. This was attributed to confusion of D with blue-shifted H I outflows (Gry et al 1983). McCullough (1992) has re-analyzed 14 selected sources, observed with the *IUE* and *Copernicus* satellites, and finds a value consistent with $\simeq 1.5 \times 10^{-5}$, with an RMS uncertainty of 10%. The underlying assumption of McCullough (1992) is that there is an average value, and the scatter about this single value should be symmetric, while Gry et al (1983) and Vidal-Madjar et al (1983) have taken the attitude that large systematic errors may be present in some cases, and that the D/H ratios may be a factor of 4 lower than previously reported. The remarkable agreement of the D/H ratios obtained by McCullough (1992) and Linsky et al (1993, 1994) encourages one to believe in a single D/H ratio for the local ISM, but it is still difficult to decide whether these data are free from all systematic errors.

UV line measurements are restricted to clouds within 1 kpc of the Sun. Measurements of more distant clouds can only be carried out using radio or IR lines. In the radio wavelength range, measurements must be made by using the 92-cm hyperfine "spin flip" line of D I. This is a first-order forbidden, magnetic dipole transition with a half life of 6.85×10^8 years. (The analogous transition for H I, at 21-cm wavelength, has only a slightly shorter half life, 1.1×10^7 years, but along any line of sight there are $\sim 10^5$ more H I atoms!) A large column density is required to produce even a weak D I radio line. In the most determined effort so far, Heiles et al (1993) have obtained a limit for the clouds toward Cas A. The column density of D I is reported to be $\leq 2.1 \times 10^{-6}$. However in these moderately dense clouds, significant amounts of D and H might be in molecular form. From a detailed analysis, Heiles et al (1993) concluded that $\leq 14\%$ of D is in atomic form.

In dense, warm molecular clouds, HD is the most abundant deuterated molecule. So far the best upper limit for the (DH)/H_2ratio, observing the $J = 1 - 0$ line at $112\,\mu m$ from the Orion KL nebula is $\leq 4 \times 10^{-5}$ (Watson et al 1987). Since the photon energy is equivalent to 128 K, this line will be found in emission only in warm clouds. Thus, fractionation effects would be diminished, but the required column density of H_2 is probably not known to better than a factor of 2. In the millimeter wavelength range, D/H ratios have been obtained recently from measurements of NH_2D (Walmsley et al 1987), HDO (Henkel et al 1987), CH_3OD (Mauersberger et al 1988, Schulz et al 1991), D_2CO (Turner 1990), CH_2DCN (Gerin et al 1992), and CH_2DOH (Jacq et al 1993); see Wannier (1980) for a summary of previous measurements. All of these species show a very large enrichment in D; for example, NH_2D data for the Hot Core in Orion-KL give a D/H ratio $\simeq 0.06$. One must conclude that the mechanism described in connection with Equation (1) has an important effect on D abundances. Thus, while measurements of deuterated species give unique information about interstellar chemistry, it is difficult to obtain actual ratios. However, there is no other way to estimate D/H ratios in distant regions. It is particularly interesting to determine this ratio for Sgr B2, since this source is taken to be representative for the Galactic center, where there has been a large amount of stellar processing. Since D is destroyed in such processing, the D/H ratio should be much lower than in the ISM near the Sun (see e.g. Audouze et al 1976). New data for Sgr B2 are presented by Walmsley & Jacq (1991); the general conclusion is that the D/H ratio in the Galactic center is $\simeq 0.25$ that of the disk value, in keeping with the results of Penzias (1980). Audouze et al (1976) concluded that even the diminished amount of D probably requires recent infall of primordial gas into the galactic center because the active stellar processing would destroy any D in less than 0.5 Gyr.

In Figure 1 we show the D/H ratio from Linsky et al (1993) and its 2σ errors. The arrows are drawn to the left to indicate that the D abundance serves as an upper limit to η. The vertical dotted line indicates twice the Linsky et al upper limit. The chemical evolution models of Steigman & Tosi (1992) show that the current abundance of D is normally within a factor of 2 of the primordial value. Thus, the dotted line should serve as a somewhat soft lower limit to η.

4.2   *Helium Isotopes*

4.2.1 ^{3}He   In addition to SBBN there are many potential sources of ^{3}He including reprocessed SBBN D (Rood et al 1984, Bania et al 1987, Balser et al 1994). While primordial ^{3}He can be destroyed in massive stars, with reasonable initial mass functions stars are a net source of ^{3}He (Dearborn et al 1986). That the current ^{3}He abundance is almost invariably greater than the primordial abundance is confirmed by the models of Steigman & Tosi (1992), despite the fact that they consider no sources for ^{3}He other than SBBN and reprocessed D. Thus an observed ^{3}He/H ratio (in particular the lowest reliably measured ratio) can provide a lower limit to η, i.e. SBBN should not overproduce ^{3}He.

However, the abundance of ^{3}He is less than 10^{-3} that of ^{4}He, and the isotope shifts of recombination lines are smaller than the line widths. As first suggested by Townes (1957), the hyperfine line of the hydrogen-like ion ^{3}He$^+$ does not occur for ^{4}He$^+$, since this nucleus has no spin. Thus there is no confusion of ^{4}He$^+$ and ^{3}He$^+$ lines, and an unambiguous measurement of the ^{3}He abundance is possible using the ^{3}He$^+$ hyperfine line. Because of the 24 eV ionization potential, ^{3}He$^+$ is present only in H II regions and planetary nebulae (PN). The excitation of the hyperfine line of ^{3}He$^+$ is similar to that for the hyperfine line of H I. Since the radiative lifetime is 2×10^4 years, the hyperfine transition is thermalized by long range collisions with free electrons. Because of the small abundance and large excitation temperature, the hyperfine line is almost certainly optically thin.

The first attempts to measure the hyperfine line of ^{3}He$^+$, made by Predmore et al (1971), gave only upper limits. The measurements are difficult because the ^{3}He$^+$ line has a peak intensity of a few milliKelvin, and arises from compact sources. Thus the largest radio telescopes and most sensitive receivers are needed for these measurements. Even more daunting is the fact that the ^{3}He$^+$ line must usually be observed against a radio continuum source, whose intensity is $\sim 10^4$ times larger than the line intensity. This continuum emission gives rise to instrumental features in the frequency baseline which can mimic spectral line emission. The size of instrumental effects have been determined by comparing the intensity ratios of H and ^{4}He recombination lines of different orders. Observing methods have evolved to minimize the effect of instrumental baselines (see Bania et al 1993b, Balser et al 1994).

Another difficulty arises because the measurements of the hyperfine line of ^{3}He$^+$ yield only column densities (i.e. the line strength is proportional to the electron density n_e). However, determining the abundance ratio also requires knowing the amount of H$^+$ or ^{4}He$^+$, either of which is found using the radio continuum or recombination lines which depend on the emission measure (proportional to n_e^2). Hence models of the density structure of the H II regions are needed to obtain the ^{3}He$^+$/H$^+$ ratios. Using comparisons of models and high resolution measurements Bania et al (1987) estimated that H$^+$ and

^{4}He$^+$ column densities are typically accurate to 20–40%, but can range up to a factor of 2 or more below the "homogeneous sphere" value. A reassessment of source modeling by Balser et al (1994) shows that models made with the heterogeneous high resolution data available to Bania et al may not be reliable. New high resolution data have been obtained and analysis is underway.

The current status of the ^{3}He$^+$ experiment is reported in Rood et al (1993) and Balser et al (1994). They report abundances based only on "homogeneous sphere" models. They find ^{3}He$^+$/H$^+$ $\approx$ 1–2 × 10^{-5} in several H II regions inside the solar circle. Several outer galaxy sources have ^{3}He$^+$/H$^+$ $\approx$ 4 × 10^{-5}. Since this does not fit the expected pattern for overall Galactic enrichment they suggest local enrichment within the outer galaxy H II regions, perhaps from Wolf-Rayet winds.

For limiting η the most important sources are those with the lowest abundance of ^{3}He. The line parameters for the H II region W43 are very well determined and give a homogeneous sphere abundance of ^{3}He$^+$/H$^+$ = 1.13 × 10^{-5}. This and its 2σ errors are indicated on Figure 1. The rightward arrows indicate that this is a lower limit for η. W43 is a complex H II region and the more detailed modeling in progress may yield a higher abundance. Early efforts at such modeling gave abundance corrections typically of $\sim$ 10% and in the case of W3 a factor of two or larger. It seems unlikely that all the low abundance sources have large correction factors. Hence, we adopt twice the W43 value as a conservative upper limit to the primordial ^{3}He. This value is indicated by the upper dot-dash arrow. We note that one detection reported by Balser et al has a very low abundance: for W49 the ^{3}He$^+$/H$^+$ ratio is 0.68 × 10^{-5}. This is indicated by the lower dot-dash arrow.

Another important result is the detection of ^{3}He in NGC 3242, a planetary nebula with a low mass progenitor. The ^{3}He$^+$/H$^+$ ratio is $\sim$ 10^{-3} (Rood et al 1992, Bania et al 1993a). The detection of the ^{3}He$^+$ line in NGC 3242 is the first direct proof that ^{3}He is produced in low mass stars, as proposed by Rood et al (1976). We take this as strong evidence that stellar sources of ^{3}He cannot be ignored.

4.2.2 ^{4}He   On the average, less than 10% of the presently observed ^{4}He should have been made in stars. However, since η is only a weak function of the relative abundance of ^{4}He (see Figure 1), even this is too much to neglect in obtaining an estimate of the baryon density from the primordial ^{4}He. A major goal of observational cosmology is to determine how much of the observed ^{4}He abundance is primordial. Radio recombination line data (Thum et al 1980) have shown that the Strömgren sphere of He$^+$ is significantly smaller than that of H$^+$. This is illustrated by the map of H$^+$ and ^{4}He$^+$ recombination lines for Orion A (Jaffe & Pankonin 1978). Radioastronomical studies have been limited by lower angular resolutions. The different Strömgren sphere sizes (caused by ionization structure) leads to a bias to lower values in the observed

$^4\text{He}^+/\text{H}^+$ ratios, especially for more distant, unresolved sources. Thus, the radio astronomical data may give either lower (ionization structure of H II regions) or higher (stellar production of ^4He) ratios than primordial. For the data of Thum et al (1980) with $40''$ spatial resolution, the average $^4\text{He}^+/\text{H}^+$ ratio is 0.072 ± 0.02, or $Y^+ = 0.238 \pm 0.069$, where Y is the mass fraction of helium as in Figure 1. Most likely, the largest part of the scatter is caused by ionization effects; corrections will raise the Y values. Peimbert et al (1988) also investigated angular resolution effects on radio recombination line results. They find that high resolution radio data are in good agreement with optical data for Orion A and M17.

It may be possible to correct for the presence of neutral He using FIR measurements of fine structure lines of Ar II and Ar III (Herter 1989). At present, single dish radio determinations of the $^4\text{He}/\text{H}$ ratios are not very useful for cosmology. Although radio astronomical data indicate a small gradient in the $^4\text{He}/\text{H}$ ratio with D_{GC}, the ionization effects mentioned above indicate that the evidence for this gradient should be viewed with caution.

In the past ten years there has been more than an order of magnitude improvement in angular resolution. The most extraordinary result is the $^4\text{He}/\text{H}$ ratios for W3A, an H II region in the Galactic anticenter, $\sim 2.4\,\text{kpc}$ from the Sun. Measurements with a $1.9''$ resolution give a $^4\text{He}^+/\text{H}^+$ number ratio of ~ 0.18 (corresponding to a Y^+ of ~ 0.47). In a lower surface brightness region, the $^4\text{He}^+/\text{H}^+$ number ratio rises to ~ 0.4, giving a Y^+ of ~ 0.7 (Roelfsma & Goss 1991, Roelfsma et al 1992). It is unlikely that these high Y values are caused by excitation effects, which raise the intensity of the He^+ but not the H^+ line. Thus, the data show that there are large fluctuations in the $^4\text{He}^+/\text{H}^+$ ratios in regions of $\sim 7 \times 10^{16}\,\text{cm}$ size. In the case of W3, the large amount of ^4He may be caused by local stellar enrichment, perhaps from winds from a cluster of Wolf-Rayet stars. For more distant regions, such local enhancements might not be recognized because of blending.

Because of the better angular resolution and the possibility of investigating a wider range of excitation conditions, optical observations give more reliable estimates of the ^4He abundances. The question is how to relate these to primordial abundances. This is a subject with nearly a 30 year history (see e.g. Hoyle & Tayler 1964, Kunth & Sargent 1983, Peimbert 1986, Pagel et al 1992). The abundance of helium is measured for systems of different metal abundance (usually the O/H abundance), and the primordial value is obtained by extrapolating to zero metallicity. The "traditional value" of the change in helium with respect to metallicity is $\Delta Y / \Delta Z \sim 3$ determined originally from comparing Galactic H II regions to those in the LMC and SMC (Peimbert & Torres-Peimbert 1974, 1976) and eventually to low metallicity compact galaxies (Lequeux et al 1979). More recently, Pagel et al (1992) have argued that $\Delta Y / \Delta Z$ is ~ 4. WM pointed out that such a high value of $\Delta Y / \Delta Z$ is not easy to explain on the basis of the usually accepted stellar nucleosynthesis ideas. Maeder (1992) has recently computed new nucleosynthetic yields for models of differing stellar metallicity. His models include metallicity-dependent winds and explore the effect of varying the lower mass boundary where stars collapse into black holes rather than explode as supernovae. For his "standard" models $\Delta Y / \Delta Z$ is ~ 1; including mass loss can increase the ratio to ~ 2. The slope can be increased to the observed values if stars with $M > 20$–$25\, M_\odot$ produce black holes thus the burying heavy elements they have produced. Giovagnoli & Tosi (1994) have used the Maeder (1992) yields in chemical evolution models, and while they confirm the steeper slopes they find that they cannot satisfy some of the usual constraints applied to such models.

Pagel et al (1992) use a maximum likelihood fit to extrapolate their data to zero O/H and obtain a primordial value of $Y_p = 0.228 \pm 0.005$ (where the error is the standard error of the mean). Using a different database, Baldwin et al (1991) obtained a smaller $\Delta Y / \Delta Z$, leading to a Y_p value of 0.240 ± 0.002.

Two recent developments have suggested ways in which lower $\Delta Y / \Delta Z$ might result. Skillman et al (1993) have noted that the extrapolation to zero metallicity might be affected by the small number of low metallicity galaxies in the previous samples—basically the primordial Y_p was being driven by I Zw 18. They report their first results including newly discovered low metallicity galaxies and new helium emissivities and find $Y_p = 0.238 \pm 0.003$ and a smaller $\Delta Y / \Delta Z$ than Pagel et al (1992). They note that the scatter in the data is large compared to formal error of the fit and hypothesize that systematic effects produce a real scatter. Campbell (1992) has considered the possible origin of such a scatter, arguing that Y_p is best determined by using the lower envelope of observed Y_p values rather than the mean result. By doing so Campbell (1992) finds an increase of 0.005 in Y_p over the Pagel et al (1992) value and also finds that $\Delta Y / \Delta Z$ must be almost a factor of 3 smaller than that obtained by Pagel et al (1992).

Another recent development suggests that the extrapolation to O/H $= 0$ might be avoided altogether. Kunth et al (1994) have measured a very low O/H for I Zw 18. They argue that Y for I Zw 18 is essentially the primordial value. Ultimately the precision of this result is set by the limited accuracy to which the helium in I Zw 18 is known. Based on the data of Davidson et al (1989), Kunth et al (1994) give a value of $Y_p = 0.234 \pm 0.016$.

We note the following results in Figure 1: The systematic effects noted by Skillman et al (1993) and Campbell (1992) all increase Y_p—hence we adopt the Pagel et al (1992) result $Y_p = 0.228$ as a lower limit (left solid mark). Reducing the Pagel et al (1992) result by 1σ gives a more conservative lower limit of $Y_p = 0.223$ (the leftmost dot-dash mark). The Skillman et al (1993) value of 0.238 and that value increased by 0.005 (i.e. to 0.243) for a Campbell-like correction are marked with heavier shorter lines. The upper limit obtained

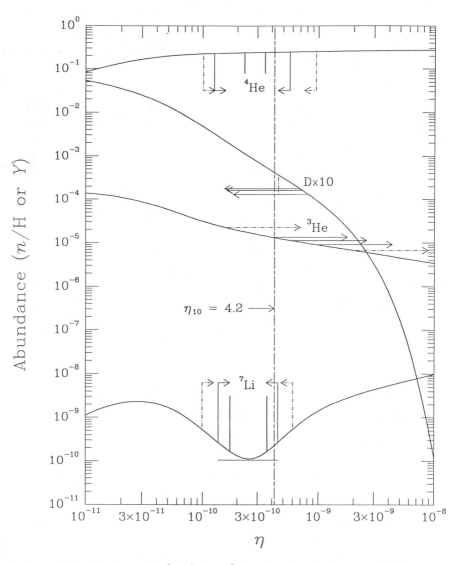

Figure 1 The abundances of D, ^{3}He, ^{4}He, and ^{7}Li as a function of η (Kernan et al 1994). The abundances of D, ^{3}He, and ^{7}Li are number ratios relative to H; for ^{4}He the mass fraction Y is shown. Note that the D abundance has been multiplied by 10 for clarity. The observed abundances are marked with arrows and vertical lines. The leftward arrows for D denote that it serves as an upper limit to η. The rightward arrows for ^{3}He denote that it serves as a lower limit. Both ^{4}He and ^{7}Li restrict η to ranges indicated. See text for further explanation.

from the Skillman result $+0.01$ to account for some systematic error is the solid line to the right of the $\eta_{10} = 4.2$ line. An even more conservation upper limit marked by the dot-dash line is obtained by increasing the Skillman et al (1992) result by 0.01 and a "Campbell shift."

4.3 *Lithium*

^{6}Li is a spallation product, while ^{7}Li is produced in SBBN, in spallation, and in stellar nucleosynthesis (see e.g. Boesgaard & Steigman 1985). The only definite detection of Li in the ISM has been made by Lemoine et al (1993) toward ρ Oph and by Meyer et al (1993) toward ζ Oph and ζ Per. Applying the ionization correction for the (unobserved) Li$^+$, Lemoine et al obtain a ^{7}Li/H ratio of 3.4×10^{-9}. This value agrees well with the Solar System value (Table 2), and the abundance in Pop I stars.

Steigman (1993) points out that the real importance of the ISM Li abundances rests not in the actual abundance which contains large corrections for ionization and absorption onto grains, but in the ^{7}Li/^{6}Li ratio which is $12.5^{+2.2}_{-1.7}$ toward ρ Oph and $6.8^{+1.4}_{-1.7}$ and $5.5^{+1.3}_{-1.1}$ toward ζ Oph and ζ Per, respectively. Steigman (1993) concludes that there has been essentially no stellar production of ^{7}Li since the formation of the Solar System and that ^{6}Li may have even decreased with stellar destruction exceeding production. Since such a conclusion is inconsistent with chemical evolution models he offers the alternate explanation that the Solar System abundances are not a representative sample of the ISM 4.6 Gyr ago.

The primordial abundance of Li has recently been reviewed by Duncan (1993), Vauclair (1993), and Thorburn (1994). The primordial Li abundance is obtained from optical spectra of very low metallicity Pop II stars (Spite & Spite 1982a,b; Rebolo et al 1988; Spite et al 1992). Interpreting Li abundances obtained from stellar spectra is an old and complex problem (Wallerstein & Conti 1969). Because Li is a fragile element its surface abundance can possibly be depleted[1] by nuclear burning at the base the convective envelope. However, nuclear burning rates are not high enough in standard models to even explain the depletion in the Solar photosphere relative to the meteoric value. Agreement between "theory" and the observations can be achieved only by invoking nonconvective mixing processes (diffusion, rotationally driven mixing, turbulence driven by rotation, or the like). There are a number of general trends for Li depletion in Pop I stars: the amount of depletion increases with decreasing stellar effective temperature; the amount of depletion increases gradually with stellar age; large depletion factors are found in a band of near log $T_{\rm eff} \sim 6600$ K known as the "Boesgaard gap" (Boesgaard & Tripico 1986). While one might

---

[1]In discussing stellar Li, "depletion" is used in a different sense than it was used earlier where it referred to grains in the ISM.

have hoped to have understood the first two trends with "extra-mixing" beyond the base of the convective envelope (the depth of which increases with decreasing log T_{eff}), a quantitative understanding of Li depletion proved to be a surprisingly difficult problem.

Unlike the case in Pop I, Spite & Spite found the Li abundance in Pop II stars to be constant over a range of effective temperatures (often referred to as the "Spite plateau"). In addition, the Li abundance appeared to be independent of the stellar metallicity. The natural explanation for these observations is: 1. that the depletion mechanisms affecting Pop I stars are absent in Pop II objects and 2. that the lithium in extreme Pop II stars is primordial. The only problem with this interpretation is that, as discussed by Duncan (1993) and Vauclair (1993), the depletion mechanism(s) in the Sun and other Pop I stars are in some sense not really understood (i.e. without introducing one or more free parameters). Given that, how can we accept *any* stellar Li measurements as indicative of the initial abundance? One can note cynically that this is one of those situations in astrophysics where there are too many high quality observations. We see so many things [such as the Boesgaard gap or, more recently, some Pop II stars in the Spite plateau with almost no Li (Hobbs et al 1991, Thorburn 1992, Spite et al 1993, Thorburn & Beers 1993)] that we become confused.

Still, much progress has been made in the theoretical modeling of Li depletion processes (Pinsonneault et al 1992; Charbonnel et al 1992, 1993; Charbonnel & Vauclair 1992; Deliyannis et al 1993). This modeling predicts that degree of Li depletion should vary (often significantly) with log T_{eff} just as observed in Pop I. If Pop II stars have a constant Li abundance over a range of log T_{eff} the simplest explanation is that in the Pop II stars, Li has not been significantly depleted. Finding a mechanism that produces a large constant depletion over a range of effective temperature in Pop II stars at first appears an unlikely prospect; however, Vauclair (1988) suggests a mechanism that does just that.

Again, newer observations add to complexity; both Spite & Spite (1993) and Thorburn (1994) find a small trend of decreasing Li with decreasing log T_{eff}, which indicates that there is at least mild depletion in Pop II stars. Observations of ^{6}Li (Smith et al 1993), which is much more easily depleted by nuclear burning than ^{7}Li, should lead to tighter constraints on stellar Li depletion mechanisms (Steigman et al 1993). Spite & Spite (1993) and Thorburn (1994) also find an increasing trend in Li with [Fe/H]. This is the first direct observation of Galactic production of Li during the formation of the halo.

Thorburn (1994) suggests $(\text{Li/H})_p = 1.66^{+0.97}_{-0.61} \times 10^{-10}$ based on the hottest Spite plateau stars with lowest [Fe/H]. These are presumably those least depleted and the least enriched. We adopt these values for Figure 1. Because of the minimum in the theoretical curve, a given value ^{7}Li/H can give two (or zero) values of η. Thorburn's central value is marked with the heavier solid lines; the upper limit is marked with lighter solid lines with the arrows indicating that this restricts η_{10} to the range 1.4–4.5; otherwise the Big Bang overproduces ^{7}Li. The observed lower limit lies slightly below the theoretical curve and is denoted by the horizontal line. Vauclair (1993) has cautioned that such limits underestimate the errors that could be present if Li depletion was occurring in Pop II stars. Depletion factors of two or larger could be present and the the observed upper limit could be lower than the primordial value. Following this caution we take twice the Thorburn (1994) observed upper limit as a conservative upper limit to the primordial value and indicate this with dot-dashed lines. This increases the allowed range in η_{10} to 1.0–6.0.

While one can obtain good agreement with the predicted primordial Li abundance in the framework of the SBBN model using the Pop II halo abundances (cf Boesgaard & Steigman 1985), there remains the problem of explaining the factor of 10 growth in Li between Pop I and Pop II. Thorburn (1994) argues that Galactic cosmic ray α–α reactions can produce the growth in Li found in the most metal poor stars, but notes that some other source is necessary to provide the full Pop I Li. Smith & Lambert (1990) provide direct evidence for the stellar production of Li in Magellanic Cloud Li-rich M supergiants. Novae, red giants, asymptotic giant branch stars, and massive stars have been suggested as sources of ^{7}Li. Models for galactic evolution of lithium have been explored by a number of groups (see e.g. Audouze et al 1983, Mathews et al 1990). D'Antona & Matteucci (1991) have investigated the possibility that novae, together with massive asymptotic giant branch stars (5–8 $M_\odot$), can cause the increased Li abundance found for Pop I stars. However, the mechanism for Li production in stars is quite uncertain. Although very unlikely, a net destruction of Li in stars cannot be completely ruled out. In summary, while there may be problems with the details, it seems likely that ^{7}Li is produced in stars and the abundance increases with time.

In addition to Li, Be and B could also have been produced in the SBBN if baryon densities were very large. Even though this is ruled out by the observed value of Y and D, some observations initially suggested a primordial origin for Be and B. The abundance of Be in HD140283 ([Fe/H] $= -2.26$) seemed high and early reports of its B/Be ratio were not consistent with a spallation origin. There are now observations of Be in 20 stars (Gilmore et al 1991, 1992; Ryan et al 1992; Boesgaard & King 1993) and of B in 3 metal-poor stars (Duncan et al 1992). Both Be and B increase linearly with metallicity; the ratio between the two is about 10. The data are consistent with primordial origin for the Spite plateau ^{7}Li and a cosmic-ray spallation origin for Be and B (Olive & Schramm 1992, Walker et al 1993, Prantzos et al 1993).

4.4 Summary and Interpretation of Light Element Abundances

To allow a test of the SBBN models, we have collected the ISM abundance data relevant to cosmology in Table 2. The relation of this data to the SBBN model is given in Table 3. The various abundances discussed above are plotted

Table 3   Cosmological implications of light element abundances

Isotope	Source	Abundance	η_{10}
$(D \times 10^5)/H$	Linsky et al (1993) line of sight to Capella	$1.65^{+0.07}_{-0.18}$	$< 7.23^{-0.17}_{+0.47}$
	chemical evolution upper limit	3.72	4.53
$(^3He \times 10^5)/H$	Balser et al (1994) lowest abundance with high reliability (W43)	1.13 ± 0.11	$> 5.78 \pm 1.05$
	lowest abundance source with detected $^3He^+$ line (W49)	0.68 ± 0.10	$> 19.7 \pm 5.7$
	twice W43 abundance, i.e. assuming a large structure correction to the abundance	2.26	> 1.56
$Y\,(^4He)$	lower limit — Pagel et al (1992)	0.228	> 1.26
	conservative lower limit	0.223	1.01
	Skillman et al (1993)	0.238	2.32
	Skillman et al with an offset as suggested by Campbell (1992)	0.243	3.49
	upper limit — Skillman et al +0.01	0.248	< 5.57
$(^7Li \times 10^{10})/H$	Thorburn (1994) hot Spite plateau — low [Fe/H]	1.66	1.74 or 3.64
	Thorburn (1994) — upper bound	2.69	> 1.37 and < 4.50
	twice Thorburn (1994) upper bound	4.58	> 0.99 and < 6.03

in Figure 1 with arrows indicating the direction of the constraint on η. In a few cases extreme limits are indicated by dot-dash lines. Considering only D, ^4He, and ^7Li we find that η (measured in units of 10^{-10}) is restricted to the range $\eta_{10} = 1.5$–4.4. Even allowing for a doubling of the primordial Li to account for possible depletion increases the range only to 1.0–6.0. Unless more D has been destroyed in stars than in typical chemical evolution models, we are restricted to the upper end of this range.

It has been common to use the sum D + ^3He to constrain η and in particular to rule out models in which large amounts of D have been destroyed. However, given the current situation one can obtain stronger constraints from ^3He alone. To do so we assume no net stellar destruction of ^3He. Given the results of Steigman & Tosi (1992) and the direct evidence for stellar production from the detection of ^3He in the planetary nebula NGC 3242, this seems to be a safe assumption. Considering only the "homogeneous sphere" abundance for W43 together with the earlier constraints restricts η_{10} to a narrow range near 4.2. (Aficionados of cosmological numbers might argue for the usage $\eta_{11} = 42$.) Although one might be tempted to quote $\eta_{10} = 4.2\pm0.3$ the errors are systematic and their true magnitude almost impossible to access. In one direction, realistic structure models could increase the ^{3}He abundance of W43; in the other, there are sources such as W49 for which even doubling the homogeneous sphere abundance of ^{3}He would seem to require $\eta_{10} > 4.2$. As with the D chemical evolution argument, the ^{3}He abundance is another indication that η_{10} lies at the upper end of the range allowed by ^{4}He and ^{7}Li.

The ultimate question is whether the universe will keep expanding, or whether the matter densities are large enough to halt the expansion. All of the light element abundances seem to be consistent with the SBBN model, and indicate that the baryonic density is unable to halt the expansion. From data reviewed by Faber & Gallagher (1979), Peebles (1993) derives $\Omega_{vis} \sim 0.004$, and the light element abundances plus SBBN suggest $\Omega_B \sim (0.02\text{--}0.06)/h_{50}^2$ with evidence leaning toward the upper end of this range. If h_{50} is near the upper end of its range and η_{10} near the lower end, then $\Omega_B \sim 0.005 \sim \Omega_{vis}$. Otherwise there is a sizable amount of dark matter in the form of baryons, but under no circumstances are there nearly enough to close the universe. If the philosophical desire to have a critical density in the universe is to be fulfilled, a great deal of non-baryonic dark matter is needed.

5.   RESULTS FOR CNO ISOTOPES

The measured isotope ratios for selected species are given in Table 4 for a number of locations and sources. For the Galactic center, the ratios are essentially those in Wannier (1980), but the new values are quite different for the "4 kpc molecular ring" and the local ISM.

5.1   *Carbon*

In Figure 2a we plot the isotope ratios of ^{12}C/^{13}C determined from H$_2$CO and CO as a function of D_{GC}. There are three conclusions from these results. First, nearly all the ratios obtained from H$_2$CO lie above those from CO (a very low ratio from NGC 6334 is not used). This is consistent with interstellar chemistry models (Langer et al 1984) which predict that, for a given cloud, the isotope ratio from H$_2$CO gives an upper limit and the ratio from CO gives a lower limit to the actual ratio. Second, there is clearly a decrease in the ratio between clouds near the Sun and regions near 4 kpc from the Galactic center. Third, the H$_2$CO ratios have a larger scatter than ratios obtained from CO data. Results of a least squares unweighted fit to all of the data give

$$(^{12}C/^{13}C) = (7.5 \pm 1.9)D_{GC} + (7.6 \pm 12.9).$$

The uncertainties are RMS values. The slope is slightly steeper than the value 4.7 ± 1.6, obtained by Tosi (1982). Near the Sun, an average for sources, using the data in Figure 2a, gives 76 ± 7 (Table 4), in contrast to the previous

Table 4   Ratios for Galactic center, 4 kpc molecular ring, carbon stars, Solar System, local ISM, and galaxies

Isotope	Galactic center	4 kpc molecular ring	Local ISM[b]	Solar System[c]	Carbon stars[d]	Nuclei of galaxies
$(^{12}C/^{13}C)$	~ 20	53 ± 4^b	77 ± 7^b	89	> 30	$\sim 40^h$
$(^{14}N/^{15}N)$	> 600	375 ± 38^b	450 ± 22^b	270	> 515	...
$(^{16}O/^{18}O)$	250	327 ± 32^b	560 ± 25^b	490	320 to 1260 > 2700	$\sim 200^i$
$(^{18}O/^{17}O)$	3.2 ± 0.2^e	3.2 ± 0.2^e	3.2 ± 0.2^e	5.5	0.6 to 0.9 < 1	8^i
$(^{32}S/^{34}S)$	$\sim 22^f$	$\sim 22^f$	$\sim 22^f$	22	...	...
$(^{29}Si/^{30}Si)$	1.5^g	1.5^g	1.5^g	1.5	...	...

[a] Wannier (1980).
[b] Fits to data shown in Figure 2. The error given is that of the mean.
[c] Anders & Grevesse (1989).
[d] Kahane et al (1992), Johansson et al (1984).
[e] Penzias (1981b).
[f] Frerking et al (1980).
[g] Penzias (1981a).
[h] Henkel et al (1993a,b).
[i] Sage et al (1991); Henkel et al (1993a,b).

value, 52. The newer results give a higher ratio for two reasons: 1. the previous averages included more ratios from CO data, which are lower limits to the actual ratios, and 2. the previous averages used a weighted fit. From Figure 2a, we see that the first factor is more important. There is a large scatter between the data and the fit. Some, perhaps most, of this is due to noise and systematic errors, but there is evidence for actual differences. The best evidence comes from the H_2CO ratios for two groups of clouds toward the intense background source Cas A (Henkel et al 1982). The value for the local cloud is $\simeq 90$, while 3 kpc from the Sun, the ratio is $\simeq 50$. (The local cloud has an A_v of $\sim 1^m$, but this is not expected to cause selective dissociation in H_2CO.) The more distant clouds might have been enriched in ^{13}C by stars near the SNR Cas A.

The $^{12}C/^{13}C$ ratio is a measure of the primary to secondary processing. The ^{12}C is thought to be produced on rapid time scales primarily via He burning in massive stars (see Meyer 1994, this volume). The ^{13}C is thought to be produced primarily via CNO processing of ^{12}C seeds from earlier stellar generations. This occurs on a slower time scale primarily during the red giant phase in low and intermediate mass stars or novae (see WM). As discussed by WM, production time scales can give rise to differences even among secondary products. Isotopes or elements produced from C or O, or with different time scales, might

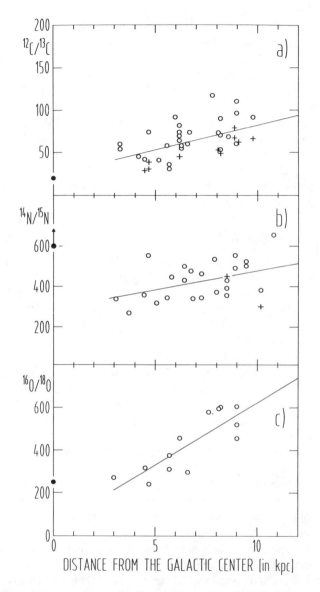

Figure 2   Plots of isotope ratios as a function of distance from the Galactic center, D_{GC}. The distance of the Sun from the center is taken to be 8.5 kpc. The lines are the result of unweighted least-squares fits to the data. (*a*) Data for the $^{12}C/^{13}C$ ratio, from CO (*filled circles*) and H_2CO (*crosses*). The CO survey data were taken from Langer & Penzias (1990, 1993). The H_2CO data were taken from surveys by Gardner & Whiteoak (1979), Henkel et al (1980, 1983, 1985), and Güsten et al (1985). (*b*) $^{14}N/^{15}N$ ratios. The data shown as circles are from the HCN surveys of Wannier et al (1981) and Dahmen et al (1993), while the crosses are from the NH_3 survey of Güsten & Ungerechts (1985). (*c*) $^{16}O/^{18}O$ data from the H_2CO surveys of Gardner & Whiteoak (1981) and Schüller (1985). In each case the value for the Galactic center is shown as a filled circle; the $^{14}N/^{15}N$ value is a lower limit.

show a different dependence on D_{GC}. In previous models of Galactic chemical evolution the delay time in the production of secondary elements in long-lived stars was not taken into account. The newer calculations give the sense of the spatial gradient, although the magnitude is somewhat too small. The Solar System ratio, $^{12}C/^{13}C$ is 89 (see Table 4). As a function of time, ^{13}C is expected to increase with respect to ^{12}C, as with all secondary products. Thus, the ISM $^{12}C/^{13}C$ ratio should be lower than the Solar System value. This is the case, as seen in Table 4. This ratio is only about 20% above the isotope ratio in the local ISM, so over the last 4.5 Gyr, the effect of stars such as IRC+10216 on the average local interstellar ($^{12}C/^{13}C$) ratio is rather small.

That the optical and radio determinations of $^{12}C/^{13}C$ ratios give a consistent picture is important, since these ratios are needed to obtain $^{14}N/^{15}N$ or $^{16}O/^{18}O$ ratios from double isotope measurements.

5.2 Nitrogen Isotopes

There have been two large-scale surveys using $H^{13}CN$ and $HC^{15}N$ (by Wannier et al 1981 for the northern Galaxy, and by Dahmen et al 1993 for southern sources). Additional isotope data for eight carbon stars are given in Wannier et al (1991). To obtain $^{14}N/^{15}N$ ratios from $H^{13}CN$ and $HC^{15}N$ data, $^{12}C/^{13}C$ ratios must be known. We take these from the fit to Figure 2a. Applying these results to the HCN data, we obtain the $^{14}N/^{15}N$ ratios plotted in Figure 2b. In an NH_3 survey, Güsten & Ungerechts (1985) measured ratios for four sources. Although the optical depths of the NH_3 inversion lines are greater than unity, line optical depths can be estimated from the ratio of hyperfine components. For the disk sources, there is good agreement of the HCN and NH_3 isotope ratios. The fit to the combined HCN data gives

$$(^{14}N/^{15}N) = (19.0 \pm 8.5)D_{GC} + (288.6 \pm 65.1)$$

The gradient found from this fit is only slightly lower than the previous value, 31 ± 13 (Tosi 1982). Because of the larger number of sources, especially in the inner part of the Galactic disk, the errors here are much lower.

The Solar System $^{14}N/^{15}N$ ratio is 270. In the ISM near the Sun, the data are consistent with an $\sim 70\%$ increase in this ratio since the formation of the Solar System. The Galactic center region and the carbon star IRC+10216 also have a high metallicity and $^{14}N/^{15}N$ ratios >500 (Table 4). This is consistent with the trend toward a higher ratio in the local ISM, 370 ± 23 (see Table 4) and gives support to the notion that $^{14}N/^{15}N$ ratios are directly correlated with metallicity. However, a corresponding increase of $^{14}N/^{15}N$ ratios is not observed for the inner Galactic disk; there the data show a lower ratio, 263 ± 25, but studies of H II regions indicate a metallicity higher than solar. Thus, the chemical evolution of the inner disk must have been very different from that of the Galactic center, IRC+10216, or the local ISM. WM have argued that models in which novae are the main contributors to ^{15}N could explain these results.

5.3 Oxygen

There are three stable oxygen isotopes. The usual method to obtain the ^{16}O/^{18}O ratio is to form a double ratio, in this case, from $(H_2C^{18}O)/(H_2^{13}CO)$; one must then use the carbon isotope ratios (Figure 2a) to separate the contributions of carbon from oxygen. The isotope values are plotted in Figure 2c. The fit gives

$$(^{16}O/^{18}O) = (58.8 \pm 11.8)D_{GC} + (37.1 \pm 82.6)$$

The fit is surprisingly good, but all data are from H_2CO; ratios from another species would be useful. Compared to the gradient obtained by Tosi (1982), 78 ± 28, this slope is lower but agrees to within the errors.

Ratios of ^{18}O to ^{17}O have been obtained from a CO survey by Penzias (1981b); these ratios show no gradient with D_{GC}. The average value of this ratio in the disk, the Galactic center, and molecular ring is markedly below those of the Solar System, but is larger than the carbon star value. The average ^{16}O/^{18}O ratio in the ISM near the Sun is comparable to that found in carbon stars, while the ratio decreases moving inward in the Galactic disk. The value found for the Galactic center is comparable to that found in the 4 kpc molecular ring.

If the isotopes are formed in stars more massive than $10M_\odot$, then ^{16}O is a primary nucleosynthesis product, while ^{17}O is secondary. The most important production scheme ^{18}O is not clear at present. As noted in the discussion of the C isotopes one expects a radial gradient in a primary/ secondary ratio but not in a secondary/secondary ratio. The observed spatial gradient of ^{16}O/^{18}O and the lack of a gradient in ^{18}O/^{17}O suggest that ^{18}O is secondary. However, there are problems with this conclusion. The observed ^{16}O/^{18}O ratio for the local ISM, 560 ± 25 (Table 4), when compared to the Solar System value of 490 is seen to increase with time, whereas primary/secondary ratios are predicted to decrease with time. Thus, this model cannot reproduce the evolution of either ratio, from a comparison of Solar System and local ISM ratio. To explain both the temporal and spatial behavior of ^{16}O/^{18}O one must assume there is an (unknown) mechanism that has inhibited the ^{18}O production in the last few 10^9 years (Tosi 1982). Since the nucleosynthesis of ^{18}O is still poorly known, this might be possible. As expected for a "purely" secondary/secondary ratio, the ^{18}O/^{17}O ratio with D_{GC} is constant. The observations show a marked decrease in the ^{18}O/^{17}O ratio between the time of the formation of the Solar System and the present; this discrepancy could be due to the fact that, so far, models have been constructed using the assumption that massive stars were the only production sites of ^{17}O. If ^{17}O is produced in low and intermediate mass stars, the $^{18}O/^{17}O$ ratio is a secondary/secondary result, with ^{17}O having a longer production time scale, i.e. more secondary than ^{18}O. Then in the past few billion years, the abundance of ^{17}O should grow faster than that of ^{18}O; this would allow agreement with the decrease of the ratio in time. Another alternative (yet again) is that the Solar System is not a representative sample of the ISM 4.6 Gyr ago.

5.4   Sulfur and Silicon Isotope Ratios

For these isotopes, there is a deficit of data compared to CNO results. Frerking et al (1980) interpreted their results as showing little or no variation in the $^{32}S/^{34}S$ ratio with D_{GC}. The ISM ratio agrees with the Solar System value, so there seems to be no change with time. This is expected on the basis of chemical evolution models, and indicates, to the accuracy of the data and modeling, a good agreement. We take this (optimistically) as evidence for the correctness of both, but more data are needed, especially for the Galactic center region.

The isotopes ^{29}Si and ^{30}Si have been measured in the Galactic center region, and to within the errors, the ratio agrees with that measured for the Solar System. Thus, it would appear that, as with $^{32}S/^{34}S$, there is no gradient in the $^{29}Si/^{30}Si$ ratios with D_{GC} (Penzias 1981a).

5.5   Current Enrichment of the ISM

The study of the circumstellar envelope of the carbon star IRC+10216 has contributed greatly to our understanding of the evolution of isotopic ratios. There has been a sizable decrease in the $^{12}C/^{13}C$ ratio compared to the Solar System value (Johansson et al 1984). Red giant stars have been observed by Wannier & Sahai (1987) using millimeter lines of CO isotopes; these measurements have been repeated by Kahane et al (1992), who also observed the planetary nebula NGC 7027. In their Table 11, Smith & Lambert (1989) give ratios from infrared measurements. All these results give $^{12}C/^{13}C$ ratios significantly below the Solar System value. For the data in Table 4, we have taken limits given by Kahane et al (1992) as values. For the carbon stars, the unweighted average ratio is 40 with an RMS scatter of 15. Only NGC 7027 shows a ratio remotely close to solar; the four other carbon stars have ratios of about 30. According to theory, the lower isotope ratios are related to increases in metallicity, caused by stellar processing.

Measurements of the $^{18}O/^{17}O$ ratios in five carbon-rich stellar envelopes (Table 4) give a ratio well below the Solar System and ISM values. These stars also show average $^{12}C/^{13}C$ ratios between one half to one third of the Solar System value. This carbon isotope ratio is similar to that determined for the Galactic center, but in the Galactic center, the $^{18}O/^{17}O$ isotope ratio is the same as for clouds near the Sun (Penzias 1981b). On this basis, the constancy of the $^{18}O/^{17}O$ ratio in the ISM seems to rule out such stars as major contributors of ^{17}O and ^{18}O in the ISM.

5.6   Summary of CNO Isotope Studies

Wannier (1980) concluded that there was no gradient in the $^{12}C/^{13}C$ ratio with D_{GC}, contrary to Figure 2a. Since that review, Langer & Penzias (1990) have presented more evidence for a gradient, making the opposite conclusion a certainty. Wannier (1980) also concluded that there were no source-to-source variations. Measured source-to-source variations in $^{12}C/^{13}C$ ratios are present; some may be caused by excitation or fractionation effects. However, for a few sources there are differences that exceed both noise and systematic errors. If source-to-source variations occur, then the best relation of $^{12}C/^{13}C$ with D_{GC} is obtained from a least-squares unweighted fit rather than from ratios for individual sources. Given the $^{12}C/^{13}C$ result, there is excellent evidence for a gradient in the $^{16}O/^{18}O$ ratio, and reasonable evidence for a gradient in the $^{14}N/^{15}N$ ratio with D_{GC}. The values of the $^{16}O/^{18}O$ ratio near the Sun, and the $^{14}N/^{15}N$ ratio in the 4 kpc molecular ring cannot be explained with the current Galactic chemical evolution models.

Comparisons can be made with the nuclei of galaxies. Casoli et al (1992) find that ^{12}CO is enhanced relative to ^{13}CO and $C^{18}O$ in NGC 3256. Sage et al (1991) find very large $^{18}O/^{17}O$ ratios for NGC 253, M82, and IC342. For NGC 253, Henkel et al (1993a) have obtained a $^{12}C/^{13}C$ ratio of ~ 40 and $^{16}O/^{18}O$ of ~ 250 from an analysis of a variety of molecules. For NGC 4945, Henkel et al (1993b) find a $^{12}C/^{13}C$ ratio of ~ 50, an $^{16}O/^{18}O$ ratio of ~ 150 and an $^{18}O/^{17}O$ ratio of ≥ 8. These data can be explained by a combination of two effects: 1. the infall of gas deficient in ^{13}C and ^{18}O from the disk to the central region, and 2. the enrichment of this material by the ejecta of massive stars.

6.   SUMMARY AND FUTURE OUTLOOK

For more critical tests of SBBN, it is necessary to determine primordial abundances of D, ^3He, and ^7Li for more sources. In particular, the D/H ratio along more lines of sight in the local diffuse clouds might be obtained with the Hubble Space Telescope. In addition to further observations, more accurate models of H II regions are required for improved ^3He abundances and further progress in the Li depletion phenomenon is required for ^7Li. The *Lyman Far Ultraviolet Spectroscopic Explorer* (*FUSE*), scheduled by NASA for launch in 2000, will determine D/H ratios by measurement of the higher members of the Lyman series in regions of differing stellar processing; for example, the disc and halo of the Galaxy, other galaxies, and intergalactic clouds. Access to the higher members of the Lyman series is important to correct for saturation in the H lines, a limiting factor for Hubble Space Telescope measurements for some objects. This mission may also measure the resonance lines of ^{3}He and ^{4}He (see e.g. Hurwitz & Bowyer 1985). Although there are measurements of ^{3}He abundances in H II regions and PNe, measurements for the local ISM would be an important complement to these results.

Although ^{4}He is the most abundant of the light elements, its abundance is only weakly dependent on the baryon-to-photon ratio η. Thus, estimates of the stellar contribution to ^{4}He, that is, corrections of ΔY for ΔZ, are crucial. Recent values for Y_p include 0.228 ± 0.005 (Pagel et al 1992) and 0.234 ± 0.016 (Kunth et al 1994). Both Skillman et al (1993), by adding new low metallicity galaxies to the sample, and Campbell (1992), in an analysis of the systematic errors, give results that will increase these values of Y_p and decrease $\Delta Y / \Delta Z$. It is interesting to note that while most effort has gone into determining $\Delta Y / \Delta Z$ at low metallicity, its value at solar metallicity and above is also very important, for instance, in determining the origin of the "UV upturn" in elliptical galaxies (Dorman et al 1994).

At present, all of the suitably corrected data give primordial abundances that are in agreement with the SBBN model. Considering only D, ^{4}He, and ^{7}Li we find that $\eta_{10} = 1.5$–4.5. Including ^{3}He [with the realization that the more realistic source modeling now underway by Balser et al (1994) could change the abundances], restricts η_{10} to a narrow range near 4.2—the allowed range appears to be $\sim \pm 0.3$, although its true size is hard to estimate because the errors are systematic. All indications are that the amount of baryonic matter is *far* too small to close the universe—although larger than the amount of luminous matter.

For elements and isotopes of stellar origin, there is now good agreement of radio and optical ^{12}C/^{13}C ratios near the Sun (Table 4), based on data for ζ Oph. A larger number of optical determinations of this ratio would be useful. There are a large number of carbon isotope ratios from radio data; near the Sun, there is good agreement of optical CH$^+$ ratios for the best source, ζ Oph, with the average isotope ratio from radio data. Since in this case the isotope ratios from radio data are representative, we assume that this is true in other parts of the Galaxy. Thus, there is a gradient in the ^{12}C/^{13}C ratio with D_{GC}. There is also a gradient in the ^{16}O/^{18}O and ^{14}N/^{15}N ratios, but again more data would be helpful. This is particularly so for the ^{16}O/^{18}O ratios measured using H_2CO. There is a striking difference between the Solar System and ISM ratio for ^{18}O/^{17}O; this should be checked using a non-CO species. In addition to ISM data, obtaining isotopic ratios for a larger sample of carbon-rich stars should be pursued. Isotope determinations for the nuclei of other galaxies are now possible; these data might allow us to discriminate between the ejecta of high and low mass stars. For carbon isotopes, we seem to understand the interpretation for our ISM, so we can extend these studies to other galaxies. This may not be so simple for oxygen and nitrogen isotopes, where there are still some interpretive difficulties. Some of these problems might be eliminated if the Solar System abundances were not typical of the local ISM at the time of formation.

ACKNOWLEDGMENTS

We thank James Lequeux, Bernard Pagel, Monica Tosi, Allan Sandage, Sylvie Vauclair, and Bill Langer for comments on preliminary drafts and for pointing out important references. We especially thank Terry Walker for supplying the data necessary for constructing Figure 1. TLW's research is partially supported by the Max Planck Forschungpreis of the A. v. Humboldt-Stiftung; RTR is partially supported by NSF grant AST-9121169.

Literature Cited

Anantharamaiah KR, Goss WM. 1990. In *Radio Recombination Lines: 25 Years of Investigation,* ed. MA Gordon, RL Sorochenko, pp. 267–75. Kluwer: Dordrecht
Anders E, Grevesse N. 1989. *Geochim. Cosmochim. Acta* 53:197–214
Audouze J, Boulade O, Malinie G, Poilane Y. 1983. *Astron. Astrophys.* 127:164–68
Audouze J, Lequeux J, Reeves H, Vigroux L. 1976. *Ap. J.* 208:L51–54
Baldwin J, Ferland GJ, Martin PG, Corbin MR, Cota SA, et al. 1991. *Ap. J.* 374:580–609
Bally J, Langer WD. 1982. *Ap. J.* 255:143–48
Bally J, Stark AA, Wilson RW, Henkel C. 1987. *Ap. J. Suppl.* 65:13–82
Balser DS, Bania TM, Brockway CJ, Rood RT, Wilson TL. 1994. *Ap. J.* in press
Bania TM, Rood RT, Wilson TL. 1987. *Ap. J.* 323:30–43
Bania TM, Rood RT, Wilson TL. 1993a. In *The Origin and Evolution of the Elements,* ed. N Prantos, E Vangioni-Flam, M Cassé, pp. 107–11. Cambridge: Cambridge Univ. Press
Bania TM, Rood RT, Wilson TL. 1993b. *MPIfR Tech. Ber.* 75
Boesgaard AM, King J. 1993. *Astron. J.* 106:2309–23
Boesgaard AM, Steigman G. 1985. *Annu. Rev. Astron. Astrophys.* 23:319–78
Boesgaard AM, Tripico MJ. 1986. *Ap. J.* 303:724–39
Cameron AGW. 1985. In *Protostars and Planets II,* ed. DC Black, MS Matthews, pp. 1073–99. Tucson:Univ. Ariz. Press
Cameron AGW. 1993. In *Protostars and Plan-*

ets III, ed. EH Levy, JI Lunine, pp. 47–74. Tucson:Univ. Ariz. Press
Campbell A. 1992. *Ap. J.* 401:157–67
Casoli F, Dupraz C, Combes F. 1992. *Astron. Astrophys.* 264:49–54
Centurión M, Vladilo G. 1991. *Astron. Astrophys.* 251: 245–52
Charbonnel C, Vauclair S. 1992. *Astron. Astrophys.* 265: 55–64
Charbonnel C, Vauclair S, Maeder A, Meynet G, Schaller G. 1993. *Astron. Astrophys.* Submitted
Charbonnel C, Vauclair S, Zahn J-P. 1992. *Astron. Astrophys.* 255: 191–99
Chu Y-H, Watson WD. 1983. *Ap. J.* 267:151–55
Combes F. 1991. *Annu. Rev. Astron. Astrophys.* 29:195–237
Crane P, Hegyi DJ. 1988. *Ap. J.* 326:L35–38
Crane P, Lambert DL, Palazzi E. 1990. *Ap. J.* 363:192–96
Crane P, Lambert DL, Hegyi DJ. 1991. *Ap. J.* 378:181–85
Cunha K, Lambert DL. 1992. *Ap. J.* 399:586–98
D'Antona F, Matteucci F. 1991. *Astron. Astrophys.* 248:62–71
Dahmen G, Wilson TL, Matteucci F. 1993. In prep.
Dame T, Ungerechts H, Cohen RS, deGeus EJ, Grenier JA, et al. 1987. *Ap. J.* 322:706–20
Davidson K, Kinman TD, Friedman SD. 1989. *Astron. J.* 97:1591–99
Dearborn DSP, Schramm DN, Steigman G. 1986. *Ap. J.* 302:35–38
Deliyannis CP, Demarque P, Kawaler SD, Krauss LD, Romanelli P. 1989. *Phys. Rev. Lett.* 62:1583–86

Deliyannis CP, Pinsonneault MH, Duncan DK. 1993. *Ap. J.* 414:740–58

Dickel HR, Goss WM. 1990. *Ap. J.* 351:189–205

Dickman R, McCutcheon WA, Schuter WLH. 1979. *Ap. J.* 234:100–10

Dorman B, Rood RT, O'Connell RW. 1993. *Ap. J.* 419:596–614

Duncan DK. 1993. In *Texas/PASCOS 92: Relativistic Astrophysics and Particle Cosmology,* ed. CW Akerlof, MA Srednicki, pp. 757–62. New York: NY Acad. Sci.

Duncan DK, Lambert DL, Lemke M. 1992. *Ap. J.* 401:584–93

Edmunds MG. 1989. In *Evolutionary Phenomena in Galaxies,* ed. JE Beckman, BEJ Pagel, pp. 356–67. Cambridge: Cambridge Univ. Press

Edvardsson B, Andersen J, Gustafsson B, Lambert DL, Nissen PE, Tomkin J. 1993a. *Astron. Astrophys.* 275:101–52

Edvardsson B, Andersen J, Gustafsson B, Lambert DL, Nissen PE, Tomkin J. 1993b. *Astron. Astrophys. Suppl.* 102:602–5

Faber SM, Gallagher JS. 1979. *Annu. Rev. Astron. Astrophys.* 17:135–87

Fich M, Silkey M. 1991. *Ap. J.* 366:107–14

Fitzsimmons A, Brown PJF, Dufton PL, Lennon DJ. 1990. *Astron. Astrophys.* 232:437–42

Frerking MA, Wilson RW, Linke RA, Wannier PG. 1980. *Ap. J.* 240:65–73

Gardner FF, Whiteoak JB. 1979. *MNRAS* 188:331–41

Gardner FF, Whiteoak JB. 1981. *MNRAS* 194:37–41p

Geiss J. 1993. *Origin and Evolution of the Elements,* ed. N Prantzos, E Vangioni-Flam, M Cassé, pp. 89–106. Cambridge: Cambridge Univ. Press

Genzel R. 1993. In *The Galactic Interstellar Medium, Saas-Fee Advanced Course 21,* ed. WB Burton, BG Elmegreen, R Genzel, pp. 275–391. Heidelberg: Springer-Verlag

Genzel R, Stutzki J. 1989. *Annu. Rev. Astron. Astrophys.* 27:41–85

Gerin M, Combes F, Wlodarczak G, Jacq T, Guélin M, et al. 1992. *Astron. Astrophys.* 259:L35–38

Giovagnoli A, Tosi M. 1994. *MNRAS* Submitted

Gies DR, Lambert DL. 1992. *Ap. J.* 387:673–700

Gilmore G, Edvardsson B, Nisson PE. 1991. *Ap. J.* 378:17–21

Gilmore G, Gustafsson B, Edvardsson B, Nisson PE. 1992. *Nature* 357:379–84

Glassgold AE, Huggins PJ, Langer WD. 1985. *Ap. J.* 290:615–26

Gry C, Laurent C, Vidal-Madjar A. 1983. *Astron. Astrophys.* 124:99–102

Gush HP, Halpern M, Wishnow EH. 1990. *Phys. Rev. Lett.* 65:537–40

Güsten R, Henkel C, Batrla W. 1985. *Astron. Astrophys.* 149:195–98

Güsten R, Ungerechts H. 1985. *Astron. Astrophys.* 145:241–50

Hawkins I, Craig N, Meyer DM. 1993. *Ap. J.* 407:185–97

Hawkins I, Jura M. 1987. *Ap. J.* 317:926–50

Hawkins I, Jura M, Meyer DM. 1985. *Ap. J.* 294:L131–35

Heiles C, McCullough PR, Glassgold AE. 1993. *Ap. J. Suppl.* 89:271–92

Henkel C, Güsten R, Gardner FF. 1985. *Astron. Astrophys.* 143:148–52

Henkel C, Mauersberger R, Wiklind T, Hüttemeister S, Lemme C, Millar TJ. 1993a. *Astron. Astrophys.* 268 L17–20

Henkel C, Mauersberger R, Wilson TL, Snyder LE, Menten KM, Wouterloot JGA. 1987. *Astron. Astrophys.* 182:299–304

Henkel C, Walmsley CM, Wilson TL. 1980. *Ap. J.* 82:41–47

Henkel C, Whiteoak JB, Mauersberger R. 1993b. *Astron. Astrophys.* 274:730–42

Henkel C, Wilson TL, Bieging JH. 1982. *Astron. Astrophys.* 109:344–51

Henkel C, Wilson TL, Walmsley CM, Pauls TA. 1983. *Astron. Astrophys.* 127:388–94

Herbst E, Klemperer W. 1973. *Ap. J.* 185:505–33

Herter T. 1989. In *Infrared Spectroscopy in Astronomy,* ed. BH Kaldeich, pp. 403–14. Paris: Eur. Space Agency

Hobbs LM, Welty DE, Thorburn JA. 1991. *Ap. J.* 373:L47–50

Hoyle F, Tayler RJ. 1964. *Nature* 203:1108–10

Hurwitz M, Bowyer S. 1985. *Publ. Astron. Soc. Pac.* 97:214–18

Jaffe DT, Pankonin V. 1978. *Ap. J.* 226:869–82

Jacq T, Walmsley CM, Mauersberger RM, Anderson T, Herbst E, DeLucia FC. 1993. *Astron. Astrophys.* 271:276–81

Johansson LEB, Andersson C, Elldér J, Friberg P, Hjalmarson A, et al. 1984. *Astron. Astrophys.* 130:227–56

Kahane C, Cernicharo J, Gomez-Gonzales J, Guelin M. 1992. *Astron. Astrophys.* 256:235–43

Kaiser ME, Hawkins I, Wright EL. 1991. *Ap. J.* 379:267–70

Kernan P, Walker TP, Steigman G. 1994. In prep.

Kulkarni SR, Heiles C. 1988. In *Galactic and Extragalactic Radio Astronomy,* ed. GL Verschuur, KI Kellermann, pp. 95–153. Heidelberg: Springer-Verlag

Kunth D, Sargent WLW. 1983. *Ap. J.* 273:81–98

Kunth D, Sargent WLW, Lequeux J, Viallefond F. 1994. *Astron. Astrophys.* 282:709–16

Lada EA, Bally J, Stark AA. 1991. *Ap. J.* 368:432–44

Lambert DL, Sheffer Y, Crane P. 1990. *Ap. J.* 359:L19–22

Lambert DL, Sheffer Y, Gilliland RL, Federman SR. 1994. *Ap. J.* 420:756–71

Langer WD, Goldsmith PF, Carlson ER, Wilson RW. 1980. *Ap. J.* 235:L39–44

Langer WD, Graedel TE, Frerking MA, Armentrout PB. 1984. *Ap. J.* 277:581–604
Langer WD, Glassgold AE, Wilson RW. 1987. *Ap. J.* 322:450–62
Langer WD, Penzias AA. 1990. *Ap. J.* 357:477–92
Langer WD, Penzias AA. 1993. *Ap. J.* 408:539–47
Larson R. 1976. *MNRAS* 176:31–52
Lemoine M, Ferlet R, Vidal-Madjar A, Emerich C, Bertin P. 1993. *Astron. Astrophys.* 269:469–76
Lequeux J, Peimbert M, Rayo SF, Serrano A, Torres-Peimbert S. 1979. *Astron. Astrophys.* 80:155–66
Lester D, Dinerstein HL, Werner MW, Watson DM, Genzel R, Storey JWV. 1987. *Ap. J.* 320:573–85
Linsky JL, Brown A, Gayley K, Diplas A, Savage BD, et al. 1993. *Ap. J.* 402:694–709
Linsky JL, Diplas A, Ayres TR, Wood B, Brown A. 1994. *Bull. Am. Astron. Soc.* 25: 1464
Maeder A. 1992. *Astron. Astrophys.* 264:105–20
Martin-Pintado J, Wilson TL, Johnston KJ, Henkel C. 1985. *Ap. J.* 299:386–404
Mather JC, Cheng ES, Shafer RA, Bennett CL, Boggess NW, et al. 1990. *Ap. J.* 354:L37–40
Mathews GJ, Alcock CR, Fuller GM. 1990. *Ap. J.* 349:449–57
Mauersberger R, Henkel C, Jacq T, Walmsley CM. 1988. *Astron. Astrophys.* 194:L1–4
McCullough PR. 1992. *Ap. J.* 390:213–18
Meyer 1994. *Annu. Rev. Astron. Astrophys.* 32:this vol.
Meyer DM, Hawkins I, Wright EL. 1993. *Ap. J.* 409:L61–64
Mezger PG, Pankonin V, Schmid-Burgk J, Thum C, Wink J. 1979. *Astron. Astrophys.* 80:L3–5
Myers PC, Dame TM, Thaddeus P, Cohen RS, Silverberg RF, Dwek E, Hauser MG. 1986. *Ap. J.* 301:398–422
Osterbrock DE, Tran HD, Veilleuse S. 1992. *Ap. J.* 389:305–24
Olive KA, Schramm DN. 1992. *Nature* 303:439–42
Pagel BEJ. 1989. In *Evolutionary Phenomena in Galaxies*, ed. JE Beckman, BEJ Pagel, pp. 368–76. Cambridge: Cambridge Univ. Press
Pagel BEJ, Edmunds MG. 1981. *Annu. Rev. Astron. Astrophys.* 19:77–113
Pagel BEJ, Simonson EA, Terlevich RJ, Edmunds MG. 1992. *MNRAS* 255:325–45
Palazzi E, Mandolesi N, Crane P, Kutner M, Blades JC, Hegyi DJ. 1990. *Ap. J.* 357:14–22
Peebles PJE. 1993. *Principles of Physical Cosmology*, p. 123. Princeton: Princeton Univ. Press. 716 pp.
Peimbert M. 1986. *Publ. Astron. Soc. Pac.* 98:1057–60
Peimbert M, Storey PJ, Torres-Peimbert S. 1993. *Ap. J.* 414:626–31
Peimbert M, Torres-Peimbert S. 1974. *Ap. J.* 193:327–33
Peimbert M, Torres-Peimbert S. 1976. *Ap. J.* 203:581–86
Peimbert M, Torres-Peimbert S. 1977. *MNRAS* 179:217–34
Peimbert M, Ukita N, Hasegawa T, Jugaku J. 1988. *Publ. Astron. Soc. Jpn.* 40:581–91
Penzias AA. 1980. *Science* 208:663–69
Penzias AA. 1981a. *Ap. J.* 249:513–17
Penzias AA. 1981b. *Ap. J.* 249:518–23
Penzias AA. 1983. *Ap. J.* 273:195–201
Pinsonneault MH, Deliyannis CP, Demarque P. 1992. *Ap. J. Suppl.* 78:179–203
Prantzos N, Cassé M, Vangioni-Flam E. 1993. *Ap. J.* 403:630–43
Predmore CR, Goldwire HC, Walters GK. 1971. *Ap. J.* 168:L125–29
Rana NC. 1991. *Annu. Rev. Astron. Astrophys.* 29:129–62
Rebolo R, Molaro P, Beckman J. 1988. *Astron. Astrophys.* 192:192–205
Rolfsema PR, Goss WM. 1991. *Astron. Astrophys. Suppl.* 87:177–214
Roelfsma PR, Goss WM, Mallik DCV. 1992. *Ap. J.* 394:188–95
Rood RT, Bania TM, Balser DS, Wilson TL. 1993. See Duncan 1993, pp. 751–56
Rood RT, Bania TM, Wilson TL. 1984. In *Proc. IAU Symp. 105, Observational Tests of Stellar Evolution Theory*, ed. A Maeder, A Renzini, pp. 567–70. Dordrecht: Kluwer
Rood RT, Bania TM, Wilson TL. 1992. *Nature* 355:618–20
Rood RT, Steigman G, Tinsley BM. 1976. *Ap. J.* 207:L57–60
Rubin RH. 1985. *Ap. J. Suppl.* 57:349–87
Rubin RH, Simpson JP, Erickson EF, Haas MR. 1988. *Ap. J.* 327:377–88
Rubin RH, Simpson JP, Haas MR, Erickson EF. 1991. *Ap. J.* 374:564–79
Russell SC, Dopita MA. 1992. *Ap. J.* 384:508–22
Ryan SG, Norris JE, Bessel MS, Deliyannis CP. 1992. *Ap. J.* 388:184–89
Sage LJ, Mauersberger R, Henkel C. 1991. *Astron. Astrophys.* 249:31–35
Sanders DB, Clemens DP, Scoville NZ, Solomon PM. 1986. *Ap. J. Suppl.* 60:1–298
Savage BD. 1987. In *Interstellar Processes*, ed. DJ Hollenbach, HA Thronson, pp. 123–41. Dordrecht: Reidel
Savage BD, Lu LM, Bahcall JN, Bergeron J, Boksenberg A, Hartig GF, et al. 1993. *Ap. J.* 413: 116-36
Schüller M. 1985. Diplom thesis. Bonn Univ., Germany
Schulz A, Güsten R, Serabyn E, Walmsley CM. 1991. *Astron. Astrophys.* 246:L55–58
Shaver P, McGee RX, Newton LM, Danks AC, Pottasch SR. 1983. *MNRAS* 204:3–112
Sheffer Y, Federman SR, Lambert DL, Cardelli JA. 1992 *Ap. J.* 397:82–91

Skillman ED, Terlevich RJ, Terlevich E, Kennicutt RC, Garnett DR. 1993. See Duncan 1993, pp. 739–44

Smith VV, Lambert DL. 1989. *Ap. J.* 311:43–63

Smith VV, Lambert DL. 1990. *Ap. J.* 361:69–72

Smith VV, Lambert DL, Nissen PE. 1993. *Ap. J.* 408:262–76

Solomon PM, Rivolo AR. 1987. In *The Galaxy, NATO ASI Vol. 207,* ed. G Gilmore, B Carswell, pp. 105–40. Dordrecht: Reidel

Spergel DN, Blitz L. 1992. *Nature* 357:665–67

Spite F, Spite M. 1982a. *Astron. Astrophys.* 115:357–66

Spite F, Spite M. 1982b. *Nature* 297:483–85

Spite F, Spite M, François P. 1992. In *High Resolution Spectroscopy with the VLT,* ed. MH Ulrich, pp. 159–62. Garching: ESO Publ.

Spite F, Spite M. 1993. *Astron. Astrophys.* 279:L9–12

Spite M, Moloro P, François P, Spite F. 1993 *Astron. Astrophys.* 271:L1–4

Spitzer L, Jenkins EB. 1975. *Annu. Rev. Astron. Astrophys.* 13:133–64

Stahl O, Wilson TL. 1992. *Astron. Astrophys.* 254:327–30

Stahl O, Wilson TL, Henkel C, Appenzeller I. 1989. *Astron. Astrophys.* 221:321–25

Steigman G. 1993. *Ap. J.* 413:L73–76

Steigman G, Fields BD, Olive KA, Schramm DN, Walker TP. 1993. *Ap. J.* 415:L35–38

Steigman G, Tosi M. 1992. *Ap. J.* 401:150–56

Thorburn JA. 1992. *Ap. J. Lett.* 399:L83–86

Thorburn JA. 1994. *Ap. J.* 421:318–343

Thorburn JA, Beers TC. 1993. *Ap. J. Lett.* 404:L13–16

Thum C, Mezger PG, Pankonin V. 1980. *Astron. Astrophys.* 87:269–75

Tosi M. 1982. *Ap. J.* 254:699–707

Townes CH. 1957. In *IAU Symp. No. 4, Radioastronomy,* ed. HC. van de Hulst, pp. 92–103. Cambridge: Cambridge Univ. Press

Turner BE. 1990. *Ap. J.* 362:L29–33

Turck-Chiéze S, Cahen S, Cassé M, Doom C. 1988. *Ap. J.* 345:415–24

van Dishoeck EF, Black JH. 1986. *Ap. J. Suppl.* 62:109–45

van Dishoeck EF, Black JH. 1988. *Ap. J.* 334:771–802

Vauclair S. 1988. *Ap. J.* 335:971–5

Vauclair S. 1993. In *Astroparticle Physics,* ed. J. Tran Vanth Van. In press. Gif-sur-Yvette: Ed. Frontières

Viala YP, Letzelter C, Eidelsberg M, Rostas F. 1988. *Astron. Astrophys.* 193:265–72

Vidal-Madjar A, Laurent C, Gry C, Bruston P, Ferlet R, York DG. 1983. *Astron. Astrophys.* 120:58–62

Vladilo G, Centurión M, Càssola C. 1993. *Astron. Astrophys.* 273:239–46

Waddington CJ, ed. 1989. *AIP Conf. on Abundances of Matter.* New York: Am. Inst. Phys. 426 pp.

Wakker BP. 1991. In *Proc. IAU Symp. 144, The Interstellar Disk-Halo Connection in Galaxies,* ed. H Bloemen, pp. 27–40. Dordrecht: Kluwer

Walker TP, Steigman G, Schramm DN, Olive KA, Kang HS. 1991. *Ap. J.* 376:51–69

Walker TP, Steigman G, Schramm DN, Olive KA, Fields B. 1993. *Ap. J.* 413:562–70

Wallerstein G, Conti P. 1969. *Annu. Rev. Astron. Astrophys.* 7:99–120

Walmsley CM, Hermsen W, Henkel C, Mauersberger R, Wilson TL. 1987. *Astron. Astrophys.* 172:311–15

Walmsley CM, Jacq T. 1991. In *Atoms, Ions, and Molecules,* ed. AD Haschick, PTP Ho, p. 305. San Francisco: Astron. Soc. Pac.

Walter DK, Duffour RJ, Hester JJ. 1992. *Ap. J.* 397:196–213

Walker TP, Steigman G, Schramm DN, Olive KA, Kang HS. 1991. *Ap. J.* 376:51–69

Wannier PG. 1980. *Annu. Rev. Astron. Astrophys.* 18:399–437

Wannier PG, Andersson BG, Olofsson H, Ukita N, Young K. 1991. *Ap. J.* 380:593–605

Wannier PG, Linke RA, Penzias AA. 1981. *Ap. J.* 247:522–29

Wannier PG, Penzias AA, Linke RA, Wilson RW. 1976. *Ap. J.* 204:26–42

Wannier PG, Sahai R. 1987. *Ap. J.* 319:367–82

Wasserburg GJ. 1985. In *Protostars and Planets II,* ed. DC Black, MS Matthews, pp. 703–37. Tucson: Univ. Ariz. Press

Watson DM, Genzel R, Townes CH, Storey JWV. 1987. *Ap. J.* 298:316–27

Watson WD. 1974. *Ap. J.* 181:L129–33

Watson WD, Anichich VG, Huntress WT. 1976. *Ap. J.* 205:L165–68

Wilson TL, Hoang-Binh D, Stark AA, Filges L. 1990. *Astron. Astrophys.* 238:331–36

Wilson TL, Jäger B. 1987. *Ap. J.* 184:291–99

Wilson TL, Matteucci F. 1992. *Astron. Astrophys. Rev.* 4:1–33 (WM)

Wilson TL, Mauersberger R, Langer WD, Glassgold AE, Wilson RW. 1992. *Astron. Astrophys.* 262:248–50

Wilson TL, Walmsley CM. 1989. *Astron. Astrophys. Rev.* 1:141–77

Wink JE, Wilson TL, Bieging J. 1983. *Astron. Astrophys.* 127:211–19

Woosley SE, Timmes FX, Weaver TA. 1993. *J. Phys. G—Nucl. Part. Phys.* 19:S183–96

Woosley SE, Weaver TA. 1993. *Phys. Rep.* 67:65–97

Yang J, Turner MS, Steigman G, Schramm DN, Olive K. 1984. *Ap. J.* 281:493–511

York DG, Rogerson JB. 1976. *Ap. J.* 203:378–85

Annu. Rev. Astron. Astrophys. 1994. 32: 227–75

MASSIVE STAR POPULATIONS IN NEARBY GALAXIES

André Maeder

Geneva Observatory, CH-1290 Sauverny, Switzerland

Peter S. Conti

Joint Institute for Laboratory Astrophysics, University of Colorado, Boulder, Colorado 80309-0440

KEY WORDS: Wolf-Rayet stars, starbursts, supergiants, initial mass function, CNO abundances

1.   INTRODUCTION

Massive stars are among the main drivers of the evolution of galaxies. These O type stars, along with their highly evolved descendants, the even more energetic Wolf-Rayet objects, are major contributors to the UV radiation and power the far-infrared luminosities through the heating of dust. Their stellar winds are important sources of mechanical power. As progenitors of supernovae, massive stars are agents of nucleosynthesis and may be intimately involved in the initiation of new star formation processes. Hence, massive star evolution is a key study in the exploration of the nearby and distant Universe.

The laws of physics are, so far as we know, the same throughout the Universe. Why should we study massive stars in other galaxies, which is certainly more difficult than studying these objects nearby? We do so because the initial compositions of those stars, in particular their modes of star formation and environments, may well differ from place to place. This leads to different evolutionary histories with a number of observable consequences. For example, it has been known for quite some time that the number ratio of blue to red supergiants shows a gradient in the Milky Way and seems to be different from the ratios found in the Magellanic Clouds. Also, the relative frequency of Wolf-Rayet (W-R) stars to their O-type progenitors appears to be much larger in inner Galactic regions compared to some low-metallicity galaxies. Similarly,

227

the ratio of W-R stars of subtypes WN to those of type WC also changes by a factor of about 20 or more between metal-rich and metal-poor environments. Furthermore, the studies of starburst galaxies containing recently born massive stars show the existence of conspicuous differences in their massive star population statistics. Finally, spectroscopic abundance determinations in AGN and QSOs give us evidence of a very different chemical history among their constituent gaseous and stellar content. These few striking examples illustrate that large differences may exist in massive star populations among galaxies. It is thus essential to present a good description of such differences and to have a proper understanding of them.

Until about 20 years ago, it was generally thought that the evolution of massive stars was fully understood. With an internal physics governed by electron scattering opacities and a simple equation of state, the stars were supposed to gently leave the main sequence (MS) and finally explode as red supergiants, giving rise to SN II. More recent years have demonstrated the major role of mass loss and initial metallicity, in addition to the initial mass function (IMF), and the star formation rate (SFR) for shaping massive star evolution and population statistics. The color-magnitude diagrams of young clusters, the stellar abundances of He and CNO elements, and the studies about SN 1987A and its precursor have led to numerous additional investigations on the role of convection and mixing in massive star evolution.

In Section 2 we present some of the statistical properties of massive stars that can be studied individually, and consider what differences have been found between those in three relatively well-known galaxies (the Milky Way and the Magellanic Clouds). We examine the evolution models of OB stars and supergiants in Section 3, and compare them with the observations. We consider the properties of W-R stars, those highly evolved descendents of the most massive stars, in confrontation with the predictions of stellar evolution models in Section 4. Observations and models of even more distant galaxies containing starburst phenomena are considered in Section 5. In these cases, we are usually dealing only with integrated properties of stars in galaxies. We intimate some directions for the future in Section 6.

2.   DISTRIBUTION OF INDIVIDUAL OB STARS

2.1   *Overview*

Massive stars are, for the most part, located in stellar associations born in giant molecular clouds; a powerful enough birth event would be called a "starburst." Initially these stars are surrounded by the dense molecular gas cloud and shrouded by the commonly associated dust. These ensembles will radiate strongly in the IR and radio regions due to the heating of dust and gas excitation, but might be completely hidden optically (e.g. W51 in our Galaxy). After some time, the molecular clouds are dissociated and the dust is dissipated by the radiation and stellar winds from the O stars within, and the region becomes visible as an optical H II or giant H II (GH II) region (e.g. 30 Doradus). The appearance of the spiral arms of galaxies in the visible is primarily determined by the distributions of the H II and GH II regions within them. For the nearer galaxies, the individual stars may be investigated, but for more distant ones, only the integrated properties of the association as it affects the excited gas (and dust) can be studied.

Actual counts of massive stars in associations can be used *directly* to give estimates of the slope of the IMF, along with the related upper and lower mass limits M_{upper} and M_{lower}. We consider these parameters for a group of associations in our Galaxy and the Magellanic Clouds in Section 2.2. In more distant GH II regions and starburst galaxies (Section 5), we turn to *indirect* methods used to confront the questions of "How many?" and "What kinds of massive stars are present?". In such distant galaxies, we make use of the integrated spectra and of global properties, such as their far-infrared (FIR) luminosities, and their optical and UV imaging. For *indirect* methods, we examine the use of 30 Doradus as a fundamental *calibrator.*

2.2   *Direct Star Counts—Census*

Pioneering efforts to elucidate the numbers and types of massive stars in various environments have been made primarily by Massey and associates, using both photometry and spectroscopy. As Massey (1985) has shown, even the unreddened UBV colors for the hottest stars are degenerate, i.e. one cannot distinguish between the hottest and coolest O type stars on the basis of their photometry alone, as is commonly done with luminosity functions. Massey and his associates' homogeneous approach to the determination of the IMF for various associations of the Galaxy and Magellanic Clouds assures us that their comparisons among various stellar groupings ought to be consistent with each other.

2.2.1   PROCEDURE   One first acquires deep CCD UBV frames of the relevant stellar associations. Accurate photometry (to 0.02 mag) must be accomplished, and color-color plots used to estimate the extinction and identify the bluest stars. The UBV colors are used to determine the brightness of the stars but spectra suitable for classification are needed to determine the effective temperature, T_{eff}, of all stars earlier in type than B1V or so. Obtaining spectra is a time-consuming effort requiring large telescopes; the photometry can be done on modest ones.

Distances of the associations by classic spectroscopic parallax methods are obtained for the Galactic clusters; for the Magellanic Clouds, the standard distances are used. M_V and spectral type (or unreddened color) are converted to M_{bol} and T_{eff} using a calibration procedure. These "observed" parameters for the association stars are plotted on a "theoretical" HR diagram with evolutionary tracks. Finally, one counts the numbers of stars in each mass interval along the track, and plots the values as a function of mass. The slope of this relationship is referred to as Γ, defined in the equation

$$f(M) = AM^{\Gamma-1} \tag{1}$$

where $f(M)$ is the fractional number of stars per unit mass interval M, A is a scaling constant, and the *Salpeter* value for Γ is -1.35. By the above procedure, one is essentially measuring the slope of the *Present Day Mass Function* (PDMF). An important assumption is that this number is identical to the slope of the IMF; in other words, stellar deaths can be ignored. This is reasonable for the very youngest O associations, and one can, if necessary, account for the already highly evolved W-R stars that are present in some of the regions studied. A further typical assumption is that the spread in formation time is of the order of, or less than, the evolution time, which seems reasonable for O associations (but see Section 2.2.2). Finally, one ignores the binary membership. Unless the binary fraction, typically considered to be 40% for hot stars (Garmany et al 1980), is different from place to place, this assumption is also not unreasonable in the context of seeking similarities and differences in the numbers and types of massive stars in various environments.

2.2.2 MASSIVE STAR CENSUS IN O-ASSOCATIONS   In Table 1 (adapted from Conti 1994), we summarize the statistics for ten associations in the Galaxy and Magellanic Clouds, along with those for the solar vicinity. They have been grouped by galaxy abundance, thus sampling the composition of the environment out of which these stars formed.

Table 1   Massive star statistics for various regions

Association	$-\Gamma$	$\# > 60 M_\odot$	Galaxy	References
Field $< R_\odot$	1.3		Milky Way	Garmany et al 1982
Field $> R_\odot$	2.1			Garmany et al 1982
Cyg OB2	1.0	7		Massey & Thompson 1991
Car OB1	1.3	7		Massey & Johnson 1993
Ser OB1	1.1	1		Hillenbrand et al 1993
LH9	1.6	0	LMC	Parker et al 1992
LH10	1.1	4		Parker et al 1992
LH58	1.7	0		Garmany et al 1993
LH117	1.8	2		Massey et al 1989a
LH118	1.8	0		Massey et al 1989a
30 Dor	1.4	21		Parker 1991
NGC 346	1.8	3	SMC	Massey et al 1989b

The two entries for the solar vicinity were obtained by a method similar to the others but the models used were older and the numbers not quite comparable; the values are just noted for completeness. The authors of the various papers cited suggest that Γ has been determined to an accuracy of ± 0.2 in each association. If there is a dependence on metal abundance, Z, which, given the uncertainties, is by no means clear, then Γ gets *shallower* as the metal abundance increases. This is in opposition to the prevailing view (Shields & Tinsley 1976) that Γ becomes steeper with increasing abundance.

The entries in Table 1 also give a quantitative indication of M_{upper} for the most massive objects by listing the actual numbers of stars with masses larger than $60 \, M_\odot$ as inferred from the M_{bol} and T_{eff}. We obtained this by inspection of the HR diagrams plotted in the various papers cited. The values, which are more or less proportional to the total numbers of stars in each region, range farther upward in mass to somewhere between 80 to 100 $M_\odot$; this is a reasonable estimate of M_{upper} for starburst modeling purposes (see also Section 3.4). There is no dependence of either the numbers of massive stars or the M_{upper} on the galaxy environment, or on metallicity Z.

It has been suggested from indirect arguments that the M_{lower} limit in some starburst regions might be significantly larger than the canonical 0.1 $M_\odot$ found near the Sun, and more like a few $M_\odot$ (e.g. M 82—Rieke 1991, McLeod et al 1993). Is there any direct evidence for this? Among well-studied energetic GH II regions, 30 Dor would be the place to look. However, Parker's (1991) survey was complete only to an apparent magnitude corresponding to a few $M_\odot$, i.e. just where it begins to get interesting. Despite the (crowding) difficulties, a deeper CCD photometric survey of 30 Dor needs to be made to investigate its M_{lower} limit and this issue in general.

In their study of NGC 6611 = Ser OB1, Hillenbrand et al (1993) have gone sufficiently deep in their CCD survey to be able to say something about the stellar population at 3–7 $M_\odot$. Remarkably, they find that these stars are above the main sequence, in the pre-main sequence phase. Furthermore, the ages of these still contracting stars are a few $\times 10^5$ years, appreciably less than the turn-off time of the upper main sequence, which is a few $\times 10^6$ years! Thus in this association, the presence of (at least) 13 O stars has not inhibited further star formation of lower mass stars. Whether ten or one hundred times that many O stars would inhibit subsequent lower mass star formation remains problematic.

Hillenbrand et al (1993) also call attention to several luminous stars that sit well to the right of the main body of massive stars in NGC 6611. They argue that these stars are indeed cluster members, which must have formed before most of the rest. Eye examination of the rest of the associations referenced above also invariably reveals a few stars in similar, advanced, evolutionary stages. Hillenbrand et al suggest star formation might proceed much like "popcorn" when it is heated; a few kernels "pop" before the main body, and a few more lag behind.

2.2.3 LUMINOSITY FUNCTIONS   Massey (1985) noted that using only photometry, it is difficult to distinguish among the most massive stars employing
luminosity functions alone. In particular, Massey et al (1989b) show that had
they taken only their UBV photometry for the analysis of NGC 346, they would
have found Γ to be -2.5, instead of the -1.8 value listed in Table 1. Hill et
al (1994) have derived Γ for 14 OB associations in the Magellanic Clouds using CCD photometry. They too find no difference in this parameter between
the LMC and SMC, thus no dependence on Z. While their mean values of Γ
are somewhat larger than those listed in Table 1 (and probably for the reason
mentioned above), this distinction is probably not significant. Their uncertainties in Γ are also larger than those found by Massey and associates with their
techniques.

Massey et al (1986) have done CCD photometry of several associations in
M31. Using plots and evolution tracks similar to those discussed above, they
obtained the curious result (shown in their Figures 31 and 32) that there are no
stars in the luminous associations OB78 and OB48 more massive than 40 $M_\odot$,
although both have W-R stars present. This inference is probably faulty as it
is based on photometry only. Thus the use of luminosity functions for massive
stars must be treated with caution, as both the derived Γ and the M_{upper} might
be suspect. For luminous stars of M31 and other Local Group galaxies, spectra
are difficult but not impossible to obtain on the largest telescopes, especially
with multiple slit instrumentation.

3.   STELLAR MODELS AND OBSERVATIONS

3.1   Recent Progress in the Input Physics

Only in the Milky Way and a few galaxies of the Local Group are individual
observations of massive stars feasible. Before interpreting the integrated spectra
of more distant galaxies, we must first proceed to careful tests of the current
models by observing stars in nearby galaxies and checking whether these models
are realistic.

Over recent years many grids of stellar models of massive stars at different
metallicities Z have been produced with various physical assumptions (Brunish
& Truran 1982a,b; Chin & Stothers 1990; Maeder 1990; Arnett 1991; Baraffe
& El Eid 1991; Schaller et al 1992; Schaerer et al 1993a,b; Charbonnel et al
1993; Woosley et al 1993; Alongi et al 1993; Bressan et al 1993; de Loore
& Vanbeveren 1994; Meynet et al 1994). The general input physics for stellar
models has been extensively discussed (Iben 1974, Iben & Renzini 1983, Chiosi
& Maeder 1986, Maeder & Meynet 1989, Chiosi et al 1992a, Schaller et al
1992). We shall limit our review to those points most critical for massive stars.

Evaporation by stellar winds is a dominant feature of massive star evolution
and all model predictions are influenced. Stellar wind models have been de-

veloped by several groups (Abbott 1982, Pauldrach et al 1986, Owocki et al 1988, Kudritzki et al 1987, Schaerer & Schmutz 1994). However, the observed mass loss rates, $\dot{M}$, and the wind momentum in O-type stars are generally still larger than predicted (Lamers & Leitherer 1993). Thus, taking theoretical $\dot{M}$ is probably not yet a comfortable choice and it seems preferable to base the models on the empirical values. Those compiled by de Jager et al (1988) have commonly been adopted. These mainly depend on luminosity and to a smaller extent on T_{eff}; the role of rotation is still uncertain (Nieuwenhuijzen & de Jager 1988, 1990; Howarth & Prinja 1989). These average $\dot{M}$ might be too low (Schaerer & Maeder 1992). Considerable uncertainties remain in the adopted $\dot{M}$, particularly for the red supergiants which lose mass at very high values (de Jager et al 1988, Stencel et al 1989, Jura & Kleinmann 1990). Presently there is no complete theory for the winds of red supergiants. For these, evidence of dust ejection is provided by *IRAS* observations, which show that some of them possess extended circumstellar shells (Stencel et al 1989) potentially leading to OH/IR sources (Cohen 1992). Evidence for strong winds in the previous red supergiant phase of SN 1987A has also been presented by Fransson et al (1989), and for the more recent SN 1993J by Höflich et al (1993).

Different treatments of convection and mixing in stellar interiors have been advocated, giving a major uncertainty in massive star models. We identify the following different assumptions regarding convection and mixing in massive star models:

- Schwarzschild's criterion,

- Schwarzschild's criterion and core overshooting,

- Overshooting below the convective envelope,

- Ledoux criterion,

- Semiconvection or semiconvective diffusion,

- Turbulent diffusion or other forms of rotational mixing.

All of these models are claimed by their authors to fit the observations and the debate has been lively in recent years (Chiosi & Maeder 1986; Maeder & Meynet 1989; Brocato et al 1989; Lattanzio et al 1991; Stothers 1991a,b; Stothers & Chin 1990, 1991, 1992a,b). There is at present no definite theoretical or observational proof in favor of any model. However, a few useful indications on the limits and possibilities of the various models must be mentioned.

Although claims have been made in favor of substantial overshooting from convective cores with respect to what is predicted by Schwarzschild's criterion, it now seems clear that the overshooting distance is limited to about $(0.2–0.4)H_p$ (Maeder & Meynet 1989, Stothers & Chin 1991, Napiwotzki et al

1993, Meynet et al 1993). The main effect is to increase the main sequence width and lifetimes, while the helium burning lifetimes are reduced (due to higher L); the blue loops are also shorter. In contrast to overshooting, which extends convective zones, the Ledoux criterion tends to prevent convective mixing in zones with variable mean molecular weights (Kippenhahn & Weigert 1990). Some recent comparisons with observations seem to favor Ledoux's rather than Schwarzschild's criterion for convection (Stothers & Chin 1992a,b); however, the result may depend on the adopted $\dot{M}$. Other recent theoretical work (Grossman et al 1993) shows that the Ledoux criterion has no bearing at all in stratified stellar layers. Thus, both at the theoretical and observational levels, the convective criteria remain uncertain.

Semiconvection occurs in zones that are convectively unstable according to Schwarzschild's criterion, but not according to Ledoux's. Semiconvection may thus produce some mixing in zones with a gradient of the mean molecular weight. Various treatments of the problem have been made (Chiosi & Maeder 1986, Langer at al 1989, Arnett 1991, Chiosi et al 1992a, Langer 1992, Alongi et al 1993). In a semiconvective zone, the nonadiabatic effects (radiative losses) produce a progressive increase of the amplitudes of oscillations at the Brunt-Väisälä frequency around a stability level (Kippenhahn & Weigert 1990). The growth of amplitudes is generally rapid compared to the evolutionary timescale, so that a situation equivalent to Schwarzschild's criterion is established. However, this might not be true in massive stars, as shown by Langer et al (1985), who discussed the timescales involved in semiconvective mixing. They propose a diffusion treatment that is equivalent to Schwarzschild's criterion when the mixing timescale is short compared to the evolutionary timescale, and to the Ledoux criterion in the opposite case. Models with such diffusion are of special relevance to the discussion about the blue progenitor of SN 1987A (Langer et al 1989; Langer 1991a,c) as well as about the evolutionary status of blue supergiants (cf Section 3.5).

The effects of rotationally induced mixing may be important for massive stars. The radiative viscosity is so large that dissipative processes may have a timescale comparable to the evolutionary timescales of massive stars (Maeder 1987b). Mixing could produce chemically homogeneous or nearly homogeneous evolution on the main sequence and thus lead directly as a result of nuclear burning to the formation of He stars, which would be observed as W-R stars. Models of massive stars losing mass and angular momentum have been calculated by Sreenivasan & Wilson (1985). As Langer (1992b) emphasized, models with semiconvective and rotational mixing may solve several problems: the existence of the WN+WC stars (Langer 1991b; cf Section 4.2), the origin of nitrogen enhancement in OB supergiants, the alleged mass discrepancy for OB main sequence stars (cf Section 3.3), and the nature of the blue progenitor of SN 1987A. Curiously enough, the claims in favor of semiconvection mean *less* mixing in the convective zone with varying mean molecular weight, while the claims in favor of rotational mixing mean *more* mixing from inner material into the outer radiative zone (Langer 1993). The situation is still uncertain, but we think it likely that mass and metallicity are not sufficient to describe massive star evolution and that rotational velocity will be an unavoidable additional parameter, as well as (for some) membership in close binary systems.

3.2 *Metallicity Effects in Massive Stars*

Metallicity, like other effects such as nonconstant star formation rates and peculiar initial mass functions (Section 5), is a key factor influencing massive star populations in galaxies. Metallicity effects can enter evolution through at least four possible doors:

1. Nuclear production. Metallicity Z may influence the nuclear rates; a good example occurs for the CNO cycle. A very slight contraction or expansion to a new equilibrium state may compensate for a change in nuclear rates (Schwarzschild 1958). In massive stars, a lower Z also produces a more active H-burning shell in the post-main sequence evolution and this favors a blue location in a part or the whole of the He-burning phase (Brunish & Truran 1982a,b; Schaller et al 1992). This was one of the initial explanations proposed for the blue precursor of SN 1987A (Truran & Weiss 1987).

2. Opacity effects. In the interiors of massive stars, electron scattering, which is independent of Z, is the main opacity source. Thus, in contrast to the case of low and intermediate mass stars, metallicity has no great direct effect on the inner structure of massive stars.

3. Stellar winds. In the very external layers, Z may strongly influence the opacity and thus the atmospheres and winds. Wind models for O stars by Abbott (1982) suggested a Z-dependence of the mass loss rates $\dot{M}$ of the form $\dot{M} \propto Z^{\alpha}$, with $\alpha = 1.0$. Other models gave a value of α between 0.5 and 0.7 (Kudritzki et al 1987, 1991; Leitherer & Langer 1991; Kudritzki 1994). It is likely that this is the main effect by which Z may influence massive star evolution (Maeder 1991a). For yellow and red supergiants, there are no models (Lafon & Berruyer 1991) nor observations (Jura & Kleinmann 1990) giving reliable $\dot{M}$ vs Z information; thus a major uncertainty in post-MS evolution remains.

4. Helium content. A ratio $\Delta Y / \Delta Z$ greater than 3 between the relative enrichments in helium and heavy elements has been established from low-Z H II regions (Peimbert 1986, Pagel et al 1992). Thus, changes in Z imply large changes in Y, which have a direct effect on the models.

3.3 *Main Sequence Evolution*

3.3.1 HR DIAGRAM, LIFETIMES, MASSES   Let us examine a few of the main properties of the models of massive stars at various metallicities. At low metallicity Z, the zero-age main sequence (ZAMS) is shifted to the blue due to the lower opacity in the external layers. Between the sequences at $Z = 0.001$ and 0.04, the shift in log T_{eff} amounts to $+0.06$ dex at 20 $M_\odot$ (Schaller et al 1992, Schaerer et al 1993b) and a lowering in luminosity by 0.10 dex. The reason is that at low Z the hydrogen content is higher, thus the electron scattering opacity is larger. The width of the MS band is predicted to change considerably according to metallicity. The main feature is a prominent "paunch," which is displaced to lower luminosities for lower mass loss rates and metallicities. Two physical effects are responsible for this paunch (Maeder 1980). First, the large mass fraction of the He core, resulting from the removal of the outer layers, favors the redward extension of the tracks. Second, when the surface hydrogen content becomes lower than $X_s = 0.3$ or 0.4 as a result of mass loss (Chiosi & Maeder 1986), the lowering of the surface opacity moves the star back to the blue. Thus, the paunch appears in the range of masses where mass loss is sufficient to increase the core mass fraction, but not high enough to lower X_s below the critical limit. An increase of overshooting or opacity may enhance the paunch. Models with enhanced opacities may have a MS band covering all the HR diagram (Stothers & Chin 1977, Nasi & Forieri 1990).

The lifetimes in the various nuclear phases change with Z. For the H-burning phase, the lifetimes $t(H)$ are typically longer by 35% for a 20 $M_\odot$ model at $Z = 0.001$ compared to $Z = 0.040$. The reason rests on the lower luminosity and the larger reservoir of hydrogen. The lifetimes $t(He)$ in the He-burning phase are generally longer in models with higher Z, due to the higher mass loss rates which lead to a drastic decrease of the luminosities in this phase. For models of 15 to 120 $M_\odot$, the $t(He)/t(H)$ is typically 9 to 10%; these ratios are between 11 and 19% at $Z = 0.04$ and they may amount to 50% if the mass loss rates are increased by a factor of 2 (Meynet et al 1994). These large factors show how our ignorance of the exact mass loss rates at various Z may affect massive star models.

There is an apparent lack of O stars close to the theoretical zero-age sequence (Garmany et al 1982). This is also quite clear in recent gravity and T_{eff} determinations by Herrero et al (1992). We notice that for massive stars, the accretion timescale of the protostellar cloud is longer than the Kelvin-Helmholtz timescale (Yorke 1986). The consequence is that no massive pre-MS star should be visible (Palla et al 1993)—a fact that could contribute to obscured stars close to the ZAMS. Wood & Churchwell (1989) and Chiosi et al (1992a) suggest that 10% to 20% of the O stars are still embedded in their parent molecular clouds. An alternative explanation is that there is no true ZAMS corresponding to a chemically homogeneous stage for O stars, because nuclear reactions ignite early during the contraction phase (Appenzeller 1980) and may thus make stars inhomogeneous before the end of the contraction phase.

Another potential problem is the so-called mass discrepancy for O stars. Spectroscopic masses derived from gravity and terminal velocity determinations were claimed to be smaller than predicted by stellar models (Bohannan et al 1990, Groenewegen et al 1989, Herrero et al 1992, Kudritzki et al 1992). In other words, spectroscopy suggests that O stars are overluminous for their masses and the discrepancy amounts up to about 50%. Langer (1992) interprets the overluminosity of O stars as a sign of rotational or tidal mixing enlarging the helium core. Apart from the fact that the force multiplier may not be correctly predicted by non-LTE wind models, the reality of the mass discrepancy has been questioned recently by Lamers & Leitherer (1993). They show that large discrepancies exist between theoretical and observed mass loss rates, as is true for the terminal velocities; they also argue that the discrepancies cannot be solved by adopting smaller masses for O stars. According to Schaerer & Schmutz (1994), the use of plane-parallel models for O stars may lead to significant errors for spectroscopic gravities, masses, and helium abundances. It is thus possible that the mass discrepancy is due to the inadequate modeling of stellar atmospheres. This view seems confirmed by the most recent work of Pauldrach et al (1994) and Kudritzki (1994), who do not find the mass discrepancy once additional wind opacity due to iron transitions is taken into account.

3.3.2 ABUNDANCES ON THE MS   The surface abundances in He and CNO elements offer a powerful test of stellar evolution. Evidence of CN processing is provided by He and N enhancements together with C depletion, while O depletion only occurs for advanced stages of processing. The abundances may cover a range from solar values (C/N = 4, O/N = 10) to CNO equilibrium values in the extreme case which is reached in WN stars (C/N = 0.02, O/N = 0.1; Maeder 1983, 1987a). Models with mass loss but no extra-mixing predict He and N enrichment in MS stars only for initial masses larger than about 50 $M_\odot$ depending on the mass loss rates. Models with rotational mixing may lead to a precocious appearance of the products of the CNO cycle (cf Maeder 1987b, Langer 1992). The observations of 25 OB stars by Herrero et al (1992) show that most MS stars have normal He and N abundances. The same is true for MS B-type stars (Gies & Lambert 1992). For example, even the most massive object, Melnick 42 (O3f), appears to show normal abundance ratios (Pauldrach et al 1994; but see Heap et al 1991). However, there are also exceptions for O and B stars. For example, the O4f star ζ Pup presents evidence of an atmosphere with CNO burned material (Bohannan et al 1986, Pauldrach et al 1994). Fast rotators are also an exception and they generally show He and N enhancements (Herrero et al 1992). Another case is the group of ON stars, i.e. O stars with N-enrichments (Walborn 1976, 1988; Howarth & Prinja 1989); this group contains at least 50% short-period binaries (Bolton & Rogers 1978).

An analysis of the association Per OB1 (Maeder 1987b) suggests that there is a bifurcation in stellar evolution: While most stars follow the tracks of inhomogeneous evolution, a fraction of about 15%, mainly composed of fast rotators and binaries, may evolve homogeneously and become ON blue stragglers.

3.4 The Eddington Limit and LBV Stars

The value of the mass of the most massive stars in galaxies has been a much debated subject. Recent photometric and spectroscopic studies suggest stellar masses up to about 100 $M_\odot$ (Section 2.2.2; Table 1; Divan & Burnichon-Prevot 1988, Kudritzki 1988, Heydari-Malayeri & Hutsemékers 1991, Massey & Johnson 1993). Recently Pauldrach et al (1994) have suggested that the most massive star known is Melnick 42 in the LMC, which may have a mass of up to 150 $M_\odot$.

There is an upper luminosity limit to the distribution of stars in the HR diagram. It runs from $\log L/L_\odot = 6.8$ at $T_{\text{eff}} = 40,000$ K to $\log L/L_\odot = 5.8$ at 15,000 K and it stays constant at lower T_{eff} (Humphreys & Davidson 1979; Humphreys 1989, 1992). The theoretical location of the Eddington limit has been examined by Lamers & Fitzpatrick (1988) on the basis of model atmospheres including metal line opacities. The limit was shown to agree with the observed limit in the Milky Way and in the LMC. Subsequent investigations indicate that the Eddington limit rises again at low T_{eff} since the opacities decrease considerably there (Lamers & Noordhoek 1993). Thus, the lowest part of the limit was called the "Eddington trough"; its location in the HR diagram is, of course, higher for stars of lower Z since they have lower opacities in the external layers.

The Eddington limit or "trough" may prevent the redward evolution of very massive stars in the HR diagram (Maeder 1983, Lamers & Noordhoek 1993). The region inside the trough will be empty except for unstable stars during their outbursts. The upper luminosity limit is determined by stars that can just pass under the Eddington trough. Thus, because the location of the trough depends on Z, the upper luminosity of red supergiants may not be an ideal standard candle, contrary to expectations (Humphreys 1983b).

The Luminous Blue Variables (LBVs, Conti 1984), also called hypergiants, S Dor, or Hubble-Sandage Variables, are optically the brightest blue supergiants. They show irregular and violent outbursts, with average mass loss rates up to about $10^{-3} M_\odot \text{yr}^{-1}$ (Davidson 1989, Lamers 1989). The group continues toward lower T_{eff} as the so-called cool hypergiants (Humphreys 1992), which are the most luminous F, G, K, and M stars. These also show evidence of variability, of high mass loss, and extensive circumstellar dust. The He and CNO abundances in LBVs (Davidson et al 1986) are in agreement with products of the CNO cycle at equilibrium, which confirms that LBVs are post-MS supergiants (Maeder 1983). About 30 LBVs have been identified by various authors in nearby galaxies (see list by Humphreys 1989) including the LMC, M31, M33,

NGC 2403, M81, and M101. Among hypergiants, the OH/IR supergiants, revealed by radio and IR observations, are the most extreme M supergiants, likely having optically thick dust shells. About two dozen cool hypergiants are currently known in the Milky Way (Humphreys 1991).

The bolometric luminosities of LBVs are constant (Appenzeller & Wolf 1981) during an outburst. However, the matter ejection, particularly during the outbursts, modifies the photospheric radius and T_{eff}, and as a consequence also the bolometric correction and visual luminosity. During their outbursts, LBVs essentially move back and forth horizontally along the HR diagram. Obscuration by gas and dust may also affect the emitted light (Davidson 1987). The circumstellar environment of these stars is peculiar and may affect the distance estimates (Viotti et al 1993). The evolutionary changes of P Cyg over the past two centuries have been recently discussed by Lamers & de Groot (1992), de Groot & Lamers (1992), and El Eid & Hartman (1993) and have been shown to correspond to recent theoretical estimates. At its minimum visual light, the star is hotter ($T = 20,000-25,000$ K) than at its maximum light (where $T = 9000$ K).

In the past, LBVs have been assigned to all possible evolutionary stages, but they are currently interpreted as a short stage in the evolution of massive stars with initial $M > 40\,M_\odot$. A likely scenario is

$$\text{O star} \rightarrow \text{Of/WN} \rightarrow \text{LBV} \rightarrow \text{Of/WN} \rightarrow \text{LBV} \cdots \rightarrow \text{WN} \rightarrow \text{WC}.$$

After central H-exhaustion, the star undergoes redward evolution in the HR diagram and is likely to reach the Eddington limit or the "trough." Strong mass loss occurs with shell ejection (LBV). As a result, stability and bluer location in the HR diagram are restored (Of/WN). Internal evolution again brings the star to the red in a few centuries—a time that may depend on the stellar mass and amount of ejected mass (cf Maeder 1989, 1992b). The star again moves toward the Eddington limit, and the cycle of evolution between the Of/WN and LBV stages continues until, as a result of mass loss, the surface hydrogen content is low enough ($X_s \leq 0.3$) so that the star definitely settles in the Wolf-Rayet stage. The overall duration of the LBV phase is fixed by the amount of mass ΔM to be lost between the end of the MS phase and the entry in the W-R phase. For a typical $\Delta M = 10\,M_\odot$ and an average $\dot{M}$ of 10^{-3} to $10^{-4}M_\odot\text{yr}^{-1}$, the typical duration would be $\approx 10^4$ to 10^5 yr. This general scenario is consistent with several properties of LBVs: their location in the Hertzsprung-Russell diagram (Humphreys 1989, Massey & Johnson 1993), their $\dot{M}$ rates (Lamers 1989), their high N/C and N/O abundance ratios (Davidson et al 1986), and the existence of transition objects, as discussed below.

Many observational studies have been made of these transition objects, which are often of spectral type Of/WN and present spectral variability. Examples are S Dor (Appenzeller & Wolf 1981, Wolf et al 1988), R 71 (Appenzeller & Wolf 1981, Wolf et al 1981), AG Car (Caputo & Viotti 1970, Viotti et al 1993), R 127

(Stahl et al 1983; Stahl 1986, 1987; Wolf 1989), R 84 (Schmutz et al 1991), and He 3-519 (Davidson et al 1993). It is also possible that after the LBV phase, some stars go to the stage of OH/IR object and then become W-R stars. This different, but not contradictory scenario, could happen to stars with masses low enough to enable them to go below the "trough." The special cases of Var A in M33 (Humphreys 1989) and IRC+104020—an extreme galactic F-supergiant with a very large IR excess from circumstellar dust (Jones et al 1993)—might correspond to such a scenario.

The physical origin of the outbursts in LBVs and hypergiants is still a matter of controversy and several models have been considered (e.g. Stothers & Chin 1983; Doom et al 1986; Appenzeller 1989; de Jager 1992; Maeder 1989, 1992b). The most striking property of these models is the strong density inversion occurring in the outer layers, where a thin gaseous layer floats upon a radiatively supported zone. This zone results from the opacity peak which leads to supra-Eddington luminosities in some layers. The idea of a density inversion has a 40 year history (Underhill 1949, Mihalas 1969, Osmer 1972, Bisnovatyi-Kogan & Nadyozhin 1972, Stothers & Chin 1983). A review of the literature shows that essentially three different kinds of conclusions were drawn: 1. A Rayleigh-Taylor instability occurs as a result of the density inversion, which is therefore washed out by the instability. 2. The supra-Eddington luminosity drives an outward acceleration and mass loss without a density inversion. 3. Strong convection and turbulence develop and the inversion is maintained.

A difficulty with most models is that they look for a hydrostatic solution to the problem. However, the resolution likely lies in the context of hydrodynamical models. Although the second of the above conclusions seems preferable, it is still unclear whether or not the density inversion is maintained. Another noticeable peculiarity in the physics of LBVs is that the thermal timescale in the outer layers is shorter than the dynamical timescale. During an outburst, which is at the dynamical timescale, the ionization front is able to substantially migrate inward (Maeder 1992b), so that some layers of matter may participate in the ejection and produce the observed shells (Hutsemékers 1994).

3.5 *Blue and Red Supergiants*

Conti (1991b) recently reviewed the observations of hot massive stars in galaxies and a complete list of the observations of red supergiants in galaxies has been given by Humphreys (1991). Humphreys (1983b, 1991) has also reviewed the potential role of red supergiants as distance indicators. Amazingly, many problems and controversies remain about supergiants, for which evolution is even more uncertain than for W-R stars! The reason is that W-R stars are dominated by the overwhelming effect of mass loss, which washes out most effects related to uncertainties in convection and mixing. Supergiants are often close to a neutral state between a blue and a red location in the HR diagram

(Tuchmann & Wheeler 1989, 1990); even minor changes in convection and mixing processes may greatly affect their evolution.

3.5.1 CHEMICAL ABUNDANCES   Walborn (1976, 1988) proposed that *ordinary* OB supergiants have an atmospheric composition enriched in helium and nitrogen and depleted in carbon, as a result of CNO processing. According to Walborn, it may just be the small group of the so-called OBC supergiants that have normal cosmic abundances (Howarth & Prinja 1989). Herrero et al (1992) showed that most OB supergiants and Of stars show helium enhancements. As for all rules, there are exceptions: A few B-supergiants do not show He and N excesses (Dufton & Lennon 1989). Herrero et al also show that fast rotators of all luminosities present evidence of CNO processing. Enhancements of nitrogen and helium abundances have also been found for post-MS B type stars by Gies & Lambert (1992), and by Voels et al (1989) in the O9.5 Ia star α Cam. As expected, the so-called OBN stars show evidence of He and N excesses with C depletion (Walborn 1988, Schönberner et al 1988).

Abundance determinations have also been made for B supergiants in the LMC and SMC, particularly interesting in relation to the progenitor of SN 1987A. These supergiants generally show He and N enhancements (Reitermann et al 1990, Kudritzki et al 1990, Lennon et al 1991). A recent high-dispersion study of LMC B-supergiants also confirms such enrichments (Fitzpatrick & Bohannan 1993). Among 62 stars of types B0.7 to B3, only 7 are OBC stars (Fitzpatrick 1991). These authors conclude, in agreement with the Walborn hypothesis, that the "typical" supergiants show contaminated surfaces, and only the rare nitrogen weak stars (OBC) have retained their original main sequence composition. The progenitor of SN 1987A, which was a B-supergiant, had N/C and N/O ratios larger than solar values by 37 and 12, respectively (Fransson et al 1989). From all these results, it is clear that most B-type supergiants in the Galaxy, the LMC, and the SMC generally show evidence of CNO processing on their surfaces.

The above observations place severe constraints on stellar models, which do not usually predict He and N enrichments in blue supergiants at solar composition. At solar Z, blue loops with the associated He and N enrichments (as a result of dredge-up in red supergiants) only occur for $M \leq 15 M_\odot$. This is the case for the models with Schwarzschild's criterion and overshooting (Schaller et al 1992), and with the Ledoux criterion (Stothers & Chin 1992a,b; Brocato & Castellani 1993). Models with semiconvection (Arnett 1991) have the same difficulty: At solar composition, the evolution goes straight to the red supergiant phase and there are no enriched blue supergiants. At lower metallicity, the blue loops are generally more developed and thus blue supergiants are predicted with He and N enrichments. However, even in this case it seems necessary (Langer 1992) to advocate some rotational mixing to account for the observed abundances.

The study of CNO abundances in three A-type supergiants (Venn 1993) reveals N-enrichments larger than predicted by the first convective dredge-up, if these stars have first gone to the red supergiant stage. This supports the idea of additional mixing. Analyses of four F-type supergiants by Luck & Lambert (1985) show material processed by the CN cycle at a level that may be higher than predicted. Analyses of some F and K supergiants in the SMC by Barbuy et al (1991) indicate solar N/Fe and C/Fe ratios, and thus no evidence of CNO processing. A further study by Barbuy et al (1992) of 14 Galactic F-supergiants shows an absence of CNO processed material in stars with low rotational velocities. For F supergiants with high rotation, the derivation of CNO abundances is unfortunately masked by the line broadening. Yellow supergiants also show sodium overabundance by a factor of 3 to 4 (Boyarchuk et al 1988; cf also Lambert 1992). Boyarchuk et al have suggested an increase of this overabundance with initial stellar masses. An interpretation put forward by Denissenkov & Ivanov (1987) and Denissenkov (1988, 1989) rests on proton capture by the isotope Ne^{22}, supposed to be overabundant. However, it is not clear why Ne^{22} should be overabundant, whether it is present initially, or whether it results from N-burning in the helium core.

Red supergiants of type G-K Ib exhibit some sodium overabundances, but less pronounced than in F supergiants (Lambert 1992). For red supergiants, the presence of CNO processed elements, as a result of dredge-up in the deep convective envelopes, is both expected and observed (Lambert et al 1984, Harris & Lambert 1984). Comparisons show a general agreement (Maeder 1987a), with possible indications that some extra mixing may be needed.

3.5.2   THE BLUE HERTZSPRUNG GAP   There are many more stars outside of the MS band than predicted (Meylan & Maeder 1982). The problem is particularly serious in the SMC and LMC. In the Milky Way, excesses of A-type supergiants have also been suggested (Stothers & Chin 1977, Chiosi et al 1978). The observed and theoretical numbers can be brought into agreement if the MS phase would also include the B- and A-type supergiant stages. This discrepancy is related to the problem of the so-called blue Hertzsprung gap (BHG), which is predicted by most stellar models to occur at the end of the MS and is not observed. Instead, the true star distribution appears continuous from the MS to the A-type supergiants (Nasi & Forieri 1990, Fitzpatrick & Garmany 1990, Chiosi et al 1992b).

Various explanations have been proposed for the lack of a BHG. Opacity effects may produce a "paunch" on the MS as discussed above (Section 3.3), but with present opacities (Iglesias et al 1992) and mass loss rates, the paunch occurs at luminosities too high to account for the observations (Schaller et al 1992). Extended atmospheres and the role of binaries (Tuchmann & Wheeler 1989, 1990) have also been advocated as explanations. Mixing reduces, but does not suppress, the gap (Langer 1991c). The temperature scale may also be a problem given the photometric nature of most of the observations of stars. A gap between $T_{eff} = 35,000$ K and 20,000 K corresponds only to a difference of 0.04 in $(B - V)$ color, which is quite small and may be blurred by other effects. The adjustment of individual isochrones on star clusters (Meynet et al 1993), together with a mapping of He and CNO abundances and the use of $(U - B)$ colors, may eventually inform us as to the reality of the gap problem and the exact status of blue supergiants.

3.5.3 THE SUPERGIANT DISTRIBUTION AND THE SN 1987A PROGENITOR   A drop-off in the distribution of LMC supergiants in the HR diagram to the right of an oblique line between $\log T_{eff} = 4.2$ and 3.9 was noted by Fitzpatrick & Garmany (1990), and called a "ledge." A further study of 5050 LMC stars with new calibrations and reddening corrections (Gochermann 1994) shows that the "ledge" might be less significant. In data from the Galaxy it appears marginally (Blaha & Humphreys 1989). Two kinds of models are able to produce high numbers of blue supergiants and to produce a "ledge" or at least a marked decrease in the star distribution in the HR diagram: (*a*) models with low mass loss, (*b*) models with blue loops.

Models with low mass loss (Brunish & Truran 1982a,b; Schaller et al 1992) predict that most of the He core burning phase is spent in the blue supergiant phase directly after the MS. This may give a ledge; however such models do not provide red supergiants, in disagreement with observations in the LMC and SMC. The blue location of models with low mass loss is due to the large intermediate convective zone which homogenizes a part of the star (Stothers & Chin 1979, Maeder 1981). Mass loss, even if small, reduces this zone and favors the redward motion in the HR diagram, and thus the star becomes a red supergiant early during the He phase. However, the uncertainties about mass loss are critical. As an example, 25 $M_\odot$ models at $Z = 0.008$ with typical $\dot{M}$ (Schaerer et al 1993a) spend most of their He-burning phase in the blue with $\log T_{eff}$ between 4.3 and 3.9. An enhancement of $\dot{M}$ by a factor of 2 leads to a red location of the whole He-burning phase (Meynet et al 1994). Thus, as long as the $\dot{M}$ rates are imprecise, it may be difficult to derive conclusions about semiconvection, diffusion, and rotational mixing from the distribution of supergiants.

Models with blue loops also enhance the number of blue supergiants, and are simultaneously able to account for some He and N enrichments in blue supergiants, but often not as much as required by the observations (Section 3.5.1). Models with Schwarzschild's criterion and overshooting at $Z \leq 0.008$ have well-developed blue phases at all masses (Schaerer et al 1993a). This is also the case for models with the Ledoux criterion for $Z < 0.004$ (Brocato & Castellani 1993) and for models with semiconvection by Arnett (1991) at $Z < 0.007$, which well reproduce the numbers of blue and red supergiants in the LMC and lead to a blue location of the supernova progenitor at 20 $M_\odot$, as must be the case for SN 1987A (Arnett 1991, Langer 1991c). The physical connection leading to the blue progenitor is most interesting (Langer 1991a,c) and illustrates a general stellar property. The mild mixing reduces the He content in the He burning shell, which is therefore less efficient. At the end of the He core burning phase, when the CO core contracts, the He shell acts as a weak mirror and produces only a moderate expansion in the intershell region between the He and H shells. Thus, the H shell is not extinct and it keeps very active. It acts as a strong mirror responding to the moderate intershell expansion by a strong contraction of the external envelope; therefore a blue final location results.

In conclusion, uncertainties in mass loss rates may allow many models to fit some features of the supergiant distributions. However, the situation is not fully satisfactory regarding the He and N enhancements, the BHG, and the "ledge." These features are usually not predicted at solar Z. At lower Z, most models predict blue loops and better fit the observations, but even there the agreement does not seem complete for either the BHG, or the He or N abundances. It is essential that the models reproduce the observations at all metallicities and this is not yet achieved.

3.5.4   THE RATIO OF BLUE TO RED SUPERGIANTS (B/R)

The B/R of supergiants was among the first stellar properties to be shown to vary through galaxies (van den Bergh 1968, Humphreys & Davidson 1979, Humphreys 1983a, Meylan & Maeder 1983, Humphreys & McElroy 1984, Brunish et al 1986). Studies have been made in the Milky Way, the LMC, the SMC, and also in M33 (Humphreys & Sandage 1980, Freedman 1985). The B/R depends on the range of luminosities considered, being found to be slightly larger at higher luminosities. The main trend is that B/R increases steeply with Z: for M_{bol} between -7.5 and -8.5, B/R is up to 40 or more in inner Galactic regions and only about 4 in the SMC (Humphreys & McElroy 1984). A difference in B/R by an order of magnitude between the Galaxy and the SMC was also found on the basis of well-selected clusters (Meylan & Maeder 1982). Further studies of the young SMC cluster NGC 330 confirm the high number of red supergiants (Carney et al 1985).

Many star models are able to account for the occurrence of blue supergiants, with predicted B/R more or less in agreement with the observations (Brunish et al 1986). As already mentioned, one reason for this is the flexibility offered by the uncertain mass loss rates. Indeed, B/R may change from infinity in the case of no mass loss to about 0 for high mass loss rates. Thus, we emphasize that the real difficulty is not to account for some average observed B/R in the LMC or the Galaxy, but to account also for its change with metallicity. The models with Schwarzschild's criterion and overshooting (Schaller et al 1992, Alongi et al 1993), the models with the Ledoux criterion (Brocato & Castellani 1993), and the models with semiconvection (Arnett 1991), even if they are able to fit some average B/R, all appear to predict higher B/R at lower Z, in contradiction to the observations. This is a major problem, which is not solved by any of the published models.

We also point out an interesting change is the T_{eff} of red supergiants according to the metallicity of the parent galaxy. Red supergiants are hotter at lower Z, the difference amounting to about 800 K between models at $Z = 0.001$ and at $Z = 0.040$ (Schaller et al 1992). This is consistent with observations (Humphreys 1979, Elias et al 1985) which show that red supergiants in the Milky Way have spectral types between M0 and M5, while in the SMC they are between types K3 and M2. This difference in the range of the spectral types of red supergiants is mainly the result of increased opacities at higher metallicities. We recall that a significant star formation rate is also a necessary condition for the presence of red supergiants in a galaxy. As an example, the paucity of red supergiants in M31 has been assigned to the low star formation rate rather than to the metallicity (Humphreys et al 1988). However, when B/R ratios are considered in galaxies, the effects of possible differences in the SFR and IMF are quite small and mainly result from effects of Z. These various tests show just how essential the studies of galaxies with different metallicities are for stellar evolution.

4. W-R STARS: OBSERVATIONS AND PREDICTIONS

4.1 *Overview*

Recent general reviews on Wolf-Rayet stars have been made by Abbott & Conti (1987), Willis (1987, 1991), Conti & Underhill (1988), Smith (1991a), van der Hucht (1991, 1992), Maeder (1991c), and Massey & Armandroff (1991). W-R stars are nowadays considered as "bare cores" resulting mainly from stellar winds peeling off of single stars initially more massive than about 25 to 40 $M_{\odot}$. Close binaries might also lose their outer layers from Roche lobe overflow (RLOF). The main evidence for the bare core model as reviewed by Lamers et al (1991) are the following:

1. H/He ratios in W-R stars are low or zero.

2. The CNO ratios are typical of nuclear equilibrium (Section 4.2.1) in WN stars.

3. The continuity of the abundances in the sequence of types O, Of, WNL, WNE, WCL, WCE, and WO corresponds nicely to a progression in peeling off the outer material from evolving massive stars.

4. The observed $\dot{M}$ in progenitor O stars and in supergiants are high enough to remove the stellar envelopes within the stellar lifetimes. Also, the average

$\dot{M}$ in W-R stars (Conti 1988) are able to accomplish further significant mass loss.

5. W-R stars have low average masses (between 5 and 10 $M_\odot$; Abbott & Conti 1987); moreover, they fit well the mass-luminosity relation for He stars (Smith & Maeder 1989).

6. W-R stars are present in young clusters and associations with ages smaller than 6 Myr (Humphreys & McElroy 1984, Schild & Maeder 1984).

7. Transition objects Of/WN between Of and W-R stars and between LBV and WN stars exist (Section 3.3).

8. He- and N-rich shells are present around some W-R stars (Esteban & Vilchez 1991).

9. The W-R/O and WN/WC number ratios are consistent with theoretical expectations in galaxies with different Z (Section 4.5).

With their bright emission lines and their high luminosities, W-R stars are observable at large distances and are thus the stars for which we have the best sampling in other galaxies. Their emission lines also can become visible in the integrated spectrum of galaxies with active star formation, which enables us to extend the studies of young massive stars even farther out in the Universe. The discovery of the differences in W-R populations in galaxies has a long history starting with Roberts (1962) in the Milky Way and later Smith (1968a,b,c) for the Magellanic Clouds. Further studies in the LMC and SMC (Azzopardi & Breysacher 1979, 1985; Breysacher 1981) confirmed the variations of W-R populations. These were also noticed (e.g. Kunth & Sargent 1981) in a sample of blue compact galaxies, which are dwarf galaxies with very active star formation, and in the so-called H II or W-R galaxies (Conti 1991a). The observed variations concern mostly the statistics, and in particular, ratios such as W-R/O or WC/WN.

The existence and origin of the variations of W-R populations in galaxies has been extensively debated over the past decade (see Section 4.6). Indeed, properties and statistics of W-R stars depend on many parameters: metallicity Z, star formation rate (SFR), initial mass function (IMF), age and duration of the bursts, binary frequency, etc. It is essential to distinguish between (*a*) regions in galaxies where the assumption of an *average* constant star formation rate over, say, the past 20 Myr is valid, and (*b*) regions or galaxies where a strong recent *burst* has recently occurred so that the assumption of a constant SFR does not apply. The first case concerns selected volumes in the Milky Way and in other galaxies (where individual W-R stars may be counted), where a stationary situation for star formation can be assumed. Within these regions, the effects of metallicity intrinsic to stellar evolution can be assumed to be the dominant factor responsible for the differences in W-R populations. This case is examined here first, as its proper understanding is a prerequisite for the studies of the second case (Section 5), which concerns distant GH II regions, W-R galaxies, and other starbursts.

4.2   Subtypes and Chemical Abundances

The basic physical parameters of W-R stars, i.e. their masses M, luminosities L, mass loss rates $\dot{M}$, radii R, and temperatures T_{eff}, have been discussed by Conti (1988), Abbott & Conti (1987), and van der Hucht (1992). The M are in the range of 5 to 50 $M_\odot$ with an average of about 10 $M_\odot$; the L are between $10^{4.5}$ and $10^6 L_\odot$; the $\dot{M}$ between 10^{-5} and $10^{-4} M_\odot$/yr with an average of $4 \times 10^{-5} M_\odot$/yr; and the observed T_{eff} are between 30,000 K and 90,000 K. There are two main groups of Wolf-Rayet stars: subtypes WN and WC; a small additional subset is labeled WO (Barlow & Hummer 1982).

4.2.1   WN STARS   The WN types are nitrogen-rich helium stars (Smith 1973) showing the "equilibrium" products of the CNO hydrogen burning cycle. The late WN stars of types WN6–WN9, abbreviated WNL (Vanbeveren & Conti 1980, Conti & Massey 1989), generally still contain some hydrogen with H/He ratios between 5 and 1 by number (Conti et al 1983a; Willis 1991; Hamann et al 1991, 1993; Crowther et al 1991). The early WN stars of types WN2–WN6, or WNE, generally show no evidence of hydrogen in their spectra. (There are a few exceptions to this general abundance relation with spectral subtype.) In a smaller sample of objects, Hamann et al (1993) found a correlation of the hydrogen content with the T_{eff} of WN stars, rather than with their subtypes; the coolest WN stars showed hydrogen and the hottest ones had none. The presence or absence of hydrogen certainly should be a determining factor influencing the opacity, temperature, and the structure of the outer layers.

   The observed abundance ratios in mass fraction are C/He = $(0.21–8) \times 10^{-4}$; N/He = $(0.035–1.4) \times 10^{-2}$; and C/N = $(0.6–6.0) \times 10^{-2}$ (cf Willis 1991, Nugis 1991). The corresponding solar ratios are respectively 1.1×10^{-2}, 3.3×10^{-3}, and 3.25 (Grevesse 1991 and references therein). The well-studied WN5 star HD 50896 gives similar results (cf Hillier 1987a,b, 1988). Not much is known concerning the oxygen content of WN stars. Studies of ring nebulae around some WN stars also show strong overabundances of N and He with respect to the Sun (Parker 1978; Esteban & Vilchez 1991, 1992; Esteban et al 1992, 1993). These authors suggest that the ring nebulae have been ejected at the end of the red supergiant phase. In our opinion, it is more likely that a large part of the shell ejection, or even most of it, occurs during the first thousand years after the entry in the WNE and WC stages, which are marked by extreme mass loss, as predicted by the $\dot{M}$ vs M relation for W-R stars (cf Langer 1989b). Among WN stars, eight have inordinately strong CIV lines (Conti & Massey 1989).

Labeled WN/WC stars, these are suggested to be transition objects between WN and WC stars (Section 4.5).

The observed abundance ratios span the range of equilibrium values of the CNO cycle (Maeder 1983, 1991b), with C/N and O/N ratios two orders of magnitude smaller than solar. Interestingly enough, such values are essentially independent of the various model assumptions and mainly reflect the nuclear cross sections and initial composition. The good agreement between the observed and predicted values of CNO equilibrium indicates the general correctness of our understanding of the CNO cycle and of the relevant nuclear data.

The initial CNO content, which depends of the initial Z, determines the amount of nitrogen in WN stars (Maeder 1990, Schaller et al 1992). The N abundance is thus lower for lower initial Z, but equilibrium ratios such as C/N are predicted to be independent of the initial Z. In this connection, one can understand the result by Smith (1991a) who noticed that the ratio of the $\lambda 4686$ He II line to the $\lambda 4640$ N III line is stronger for WNL stars in the LMC compared to the Galaxy.

4.2.2 WC STARS   WC stars contain no hydrogen and, as a result of mass loss, are mainly He, C, and O cores as a result of mass loss (Smith & Hummer 1988, Torres 1988, de Freitas-Pacheco & Machado 1988, Hillier 1989, Willis 1991, Nugis 1991, Eenens & Williams 1992, de Freitas-Pacheco et al 1993). These stars represent objects in which we see at the surface the result of triple-α and other helium burning reactions. A most interesting finding is the one by Smith & Hummer, who showed that the C/He ratio is increasing for earlier WC subtypes. Smith & Maeder (1991) emphasize that a measured (C+O)/He ratio is to be preferred to the C/He and C/O ratios which go up and down during helium processing. They propose the following calibration in (C+O)/He number ratios: WC9, 0.03–0.06; WC8, 0.1; WC7, 0.2; WC6, 0.3; WC5, 0.55; WC4, 0.7–1.0; WO, >1.

The sequence of types WC9 to WO appears as a progression in the exposure of the products of He burning. The rare WO stars (Barlow & Hummer 1982, Kingsburgh & Barlow 1991, Polcaro et al 1992, Kingsburgh et al 1994) simply appear to be the most extreme type in this sequence. Comparisons of observations and model predictions show a generally good agreement (Willis 1991, 1994); however, closer comparisons (Schaerer & Maeder 1992) suggest that $\dot{M}$ in previous evolutionary phases could be higher by a factor of 2 with respect to current values by de Jager et al (1988). The above connection between WC subtypes and the (C+O)/He ratios is the key to understanding the Z-dependence of the distribution of WC stars in galaxies of different metallicities (Section 4.6).

A present uncertainty concerns neon in WC stars. Models predict a substantial abundance of neon (larger than 0.03 in mass fraction)—essentially Ne^{22} at high initial Z and Ne^{20} at low Z (Maeder 1991a). However, the only available data, which come from *IRAS* observation of Ne II at 15.5 microns, indicate an abundance of 0.005 in the WC8 star γ Vel (Barlow et al 1988). Some of the nuclear cross sections in the chain leading to Ne^{22} are still very uncertain and the ashes of N^{14} could possibly be stocked in the form of O^{18} (indistinguishable from O^{16} in W-R stars) rather than in the form of Ne^{22}. The problem is of importance for explaining the role of WC stars as a possible site for s-elements (Prantzos et al 1990), since these elements should be formed by $Ne^{22}(\alpha, n)Mg^{25}$. The role of W-R stars as producers of radioactive Al^{26}, detectable through γ ray observations, does not seem important according to Signore & Dupraz (1990), but according to Meynet (1994a) their role could be as important as that of supernovae.

4.3 Physical Properties

In addition to the usual model ingredients, the W-R star models specifically require special attention on a number of points. Concerning microphysics, the W-R models demand Rosseland opacities for the appropriate He-C-O mixtures (Iglesias & Rogers 1993) and detailed calculations of the ionization balance for heavy elements (Langer et al 1986, Schaller et al 1992). The $\dot{M}$s in stages previous to the W-R phases and their dependence on Z are very critical. Also, the adopted definitions (based on surface abundances) for the transitions from LBV to WNL, WNL to WNE, and WNE to WC are of importance for the comparison of models and observations.

For $\dot{M}$ in the W-R stage, the average observed rates (Abbott et al 1986, Conti 1988) have often been used. However, these rates have led to masses and luminosities that are too high with respect to the observations (Schmutz et al 1989). A number of convergent suggestions have been provided recently in favor of a mass dependence of $\dot{M}$ in WNE and WC stars. In particular, models by Langer (1989b) suggest a relation of the form: $\dot{M}$ (W-R) = (0.6–1.0) $\times 10^{-7}(M/M_\odot)^{2.5}$ $M_\odot$/yr, where the first coefficient applies to WNE and the second to WC stars. Similar mass-dependent $\dot{M}$s have been provided from binaries (Abbott et al 1986, St Louis et al 1988), and from modeling the wind properties (Turolla et al 1988, Bandiera & Turolla 1990, Schaerer & Maeder 1992). The $\dot{M}$ vs M relation generally leads to an enormous mass loss at the entry in the WNE stage, which results in very low final W-R masses. This has a considerable impact on the chemical yields of massive stars (Maeder 1992a), also resulting in an increase of the W-R lifetimes. A question remains as to whether the WN luminosities predicted from models with standard $\dot{M}$ (de Jager et al 1988) are not too high with respect to the observed ones (Howarth & Schmutz 1992). Indeed, the relatively low observed luminosities of some WN stars support larger $\dot{M}$ in previous stages. There are at present no indications of a mass dependence of $\dot{M}$ for WNL stars, although such a relation would not be too surprising.

The maximum mass for the vibrational stability of a He star is about 16 $M_\odot$

(Noels & Maserel 1982, Noels & Magain 1984). Thus, if a star enters the helium configuration with a mass larger than critical (which occurs in current models), it may be expected to be vibrationally unstable with high mass loss as a consequence (Maeder 1985). The fact that regular pulsations have been recently observed in the WN8 star WR 40 (Blecha et al 1992) might give some support to this claim. However, the nature of these pulsations is still under discussion (Kirbiyik 1987) and different pulsation modes have been proposed by Glatzel et al (1993) and Kiriakidis et al (1993). Attempts are also being made to explain the strong W-R winds by multi-scattering and purely radiative processes (Pauldrach et al 1988, Cassinelli 1991), by radiation and turbulence (Blomme et al 1991), or by radiation and Alfven waves (Dos Santos et al 1993). The main difficulty, which is not satisfactorily resolved, is to explain why the wind momentum of W-R stars may be up to 30 times the photon momentum (Barlow et al 1981, Cassinelli 1991; but see Lucy & Abbott 1993).

The problems of the atmospheres of hot stars have been reviewed by Kudritzki & Hummer (1990) and the different definitions of the radii and T_{eff} in extended atmospheres by Bascheck et al (1991). Values of T_{eff} have been given recently by Conti (1988), Schmutz et al (1989, 1993), and Koesterke et al (1992): They range between about 3×10^4 and 10^5 K. In order to compare the observed T_{eff} with data from interior models, a simple correction scheme has been proposed to roughly account for the optically thick winds of W-R stars (de Loore et al 1982, Langer 1989a). More refined procedures have been established by Kato & Iben (1992), by Schaller et al (1992), and in particular by Schaerer & Schmutz (1994). The net result is that from a surface temperature of $1-1.5 \times 10^5$ K (without the wind), the W-R stars are shifted down to $T_{eff} = 3-10 \times 10^4$ K according to their $\dot{M}$ rates, and thus also according to their masses and luminosities since there is a M-L-$\dot{M}$ relation.

Since W-R stars of types WNE and WC are He-C-O cores, they have a rather simple internal structure with little compositional difference between center and surface. W-R properties and the relations between the subtypes are mostly independent of their formation. Evolutionary models predict M-L-$\dot{M}$-T_{eff} relations (Schaerer & Maeder 1992) for W-R stars without hydrogen. The mass-luminosity M-L relation is (Maeder 1983, Langer 1989a, Beech and Mitalas 1992, Schaerer & Maeder 1992):

$$\log L/L_{\odot} = 3.03 + 2.695(\log M/M_{\odot}) - 0.461(\log M/M_{\odot})^2 . \qquad (2)$$

For $M > 10 M_{\odot}$, a linear relation may be appropriate. On the observational side, the M-L relation has been confirmed (Smith & Maeder 1989, Smith et al 1994). The $\dot{M}$ vs M relation is supported by binary observations as discussed above. The M-L-$\dot{M}$-T_{eff} relations also indicate that WNE stars and WC stars should follow well-defined tracks in the HR diagrams. Such alignments seem to be present in the data of Hamann et al (1991, 1993; but see also Maeder & Meynet 1994).

For WNL stars, the luminosities are generally higher than for WNE stars (Conti 1988). Models indicate that WNL luminosities are mainly related to the initial masses. The reason is that the luminosity depends on the size of the He cores, which is determined mainly by the initial mass rather than by the actual mass, as long as the He cores are not themselves peeled off. Models also suggest, in agreement with observations (cf Hamann et al 1993), that the WNL T_{eff} are mainly determined by the remaining hydrogen content.

4.4 Initial Masses, Lifetimes, Formation

Observationally, most W-R stars appear to originate from stars initially more massive than about $40 M_\odot$ (Conti et al 1983b, Conti 1984, Humphreys et al 1985, Tutukov & Yungelson 1985). From the presence of W-R stars in clusters down to type B0, it is clear that a few W-R stars may originate from initial masses down to 20–25 $M_\odot$ (Firmani 1982, Thé et al 1982, Schild & Maeder 1984). The modeling of W-R ring nebulae (Esteban et al 1992) also supports the above values of initial masses. The minimum mass for forming WC stars does not seem significantly higher than that for WN stars.

Maeder & Meynet (1994) have obtained the lifetimes in the W-R stage for two different cases: 1. the standard case with mass loss rates by de Jager et al (1988) in pre-W-R stages and the scaling with $Z^{0.5}$ at other metallicities; and 2. the case with $\dot{M}$ arbitrarily twice as large as in pre-W-R stages. Indeed, several observations, in particular the chemical abundances in WC stars (Section 4.2), the W-R luminosities (Section 4.3), and the number ratios of W-R stars (Section 4.5) clearly support the case of enhanced mass loss, for which the W-R lifetimes are shown in Figure 1. From this figure we note that:

- The minimum mass for forming W-R stars ranges from about 20 to 25 $M_\odot$ at $Z = 0.04$ to about $80 M_\odot$ at $Z = 0.001$. The lower mass value is in agreement with the observations in the Milky Way. The change of the minimum initial mass with Z is a key effect to explaining the W-R statistics in galaxies.

- The lifetimes in the W-R stage go up with M and Z, which is just the opposite of the behavior on the main sequence. The average W-R lifetime weighted by the IMF is about 0.6 Myr at $Z = 0.02$ (for $\dot{M}$ rates twice the standard ones). Figure 1 shows lifetimes up to about 2 Myr in extreme cases.

- A detailed inspection of the model results indicates that the largest initial stellar masses spend most of the W-R phase in the WNL stage (Langer 1987, Maeder 1991a). If the WNL phase is defined from the H abundance at the surface (Section 4.2.1), then the WNL stage can even be entered during the *main sequence phase of the most massive stars.*

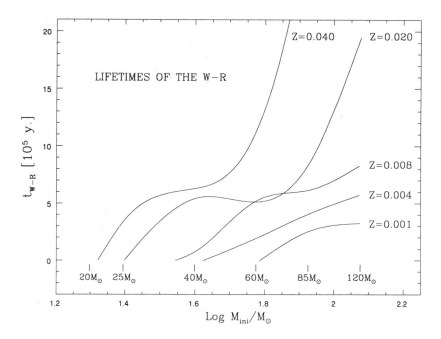

Figure 1 Durations of the W-R phase as a function of initial stellar masses for different Z in the case of enhanced mass loss rates (from Maeder & Meynet 1994).

- The lower initial masses have very short WNL phases, and spend much more time in the WNE and WC phases.

The formation of W-R stars is largely dominated by the overwhelming effects of mass loss, as first proposed in the "Conti" scenario (Conti 1976). Mixing processes due to rotation or tidal distortion in binaries may favor in some cases the formation of W-R stars and increase their lifetimes (Maeder 1981, 1987b). Semiconvection or some mild mixing at the edge of the He core seems necessary to account for the existence of intermediate WN/WC stars (Langer 1991b). These stars represent about 4% of the W-R stars (Conti & Massey 1989), while models without extra mixing only predict 1% or less of WN/WC stars. These stars cannot be explained by binary evolution (Vrancken et al 1991). Some additional mixing is necessary, but at the same time the small observed fraction of WN/WC stars implies that the part of the stellar mass that is actually mixed is quite small, and this puts a limit on the role of mixing at the edge of the He core.

Detailed investigations of W-R binaries have been carried out in the Galaxy by Massey (1981) and in the Magellanic Clouds by Moffat 1988 and Moffat et al 1990. Binary mass transfer by RLOF, which is an extreme case of tidal interaction, may contribute to the formation of WR+O binaries (de Loore 1982; De Greve et al 1988; Vanbeveren 1988, 1991, 1994; Schulte-Ladbeck 1989; De Greve 1991, de Loore & Vanbeveren 1994). The fraction of all stars (single+binaries) undergoing RLOF is estimated to be between 20 and 40% (Podsiadlowski et al 1992). We may note that this percentage also includes binaries that could be mixed by tidal interactions and would thus evolve homogeneously, without large increase of their radius and thus without RLOF. Indeed, the importance of RLOF in W-R+O binaries is still unclear. From the similarity of the relatively large orbital eccentricities in W-R+O and O+O binaries, Massey (1981) concluded that mass transfer probably did not play a *major* role in the formation of W-R+O binaries. We may conjecture that several effects contribute to the formation of W-R stars; it is likely that the relative importance of these effects changes with Z as discussed below.

4.5   W-R Statistics

4.5.1   BASIC DATA AND ITS INTERPRETATION   W-R stars are observed in several galaxies of the Local Group, and provide statistical data on their relative frequencies at various metallicities. In the Milky Way, the catalogs by van der Hucht et al (1988), and by Conti & Vacca (1990) provide rather complete samples up to about 2.5 kpc. These data show that the number density of W-R stars projected onto the Galactic plane is strongly increasing with decreasing galactocentric distance. Deep surveys are extending the sampling (Shara et al 1991). The LMC and SMC catalogs are cornerstones for data at other Z (Azzopardi & Breysacher 1979, 1985; Breysacher 1981, 1986). A few additional W-R stars have also been identified (Morgan & Good 1985, Testor & Schild 1990, Schild et al 1991, Morgan et al 1991). Most (75%) of the W-R stars in the center of the 30 Dor Nebula are WNL stars of types WN6–WN7 (Moffat et al 1987), while in the surroundings the proportion is much lower. The excess of WNL stars in giant H II regions is a common feature, as evidenced by those in M33 (Drissen et al 1990, 1991). The subtype distribution of W-R stars in the Magellanic Clouds has been considered by Smith (1991b). Data on W-R stars in M31 have been obtained by Moffat & Shara (1983, 1987), Massey et al (1986, 1987a), Armandroff & Massey (1991), and Willis et al (1992). For M33, studies have been made by Wray & Corso (1972), Conti & Massey (1981), Massey & Conti (1983), Massey et al (1987a, b), Schild et al (1990), and Armandroff & Massey (1991). In both M31 and M33, the samples are still incomplete. In the two small galaxies NGC 6822 and IC 1613 of the Local Group, many W-R stars were proposed by Armandroff & Massey (1985) and Massey et al (1987a), but further analyses by Azzopardi et al (1988; see also Smith 1988) confirmed only four W-R stars in NGC 6822 and one, which was already found by Davidson & Kinman (1982), in IC 1613. The study of W-R stars in other galaxies is continuing. The galaxy IC 10 at 1.5 Mpc exhibits a very high density of W-R stars with a WC/WN number ratio of about 0.5 (Massey et al 1992). Ten individual W-R stars have been detected in the galaxy NGC 300 at a distance of 1.5 Mpc (Schild & Testor 1991, 1992).

Analyses of W-R star statistics in nearby galaxies have been made by Azzopardi et al (1988), Smith (1988), Massey & Armandroff (1991), Maeder (1991a), and Maeder & Meynet (1994). Table 2 gives the available data on the W-R/O, WC/W-R, and WC/WN ratios for galaxies of the Local Group, when an indication of the metallicity Z is available. The table is adapted from Maeder (1991a) with revisions according to recent data from Conti & Vacca (1990) for the Milky Way; for M31 the W-R/O is from Cananzi (1992); for the SMC the new W-R star found by Morgan et al (1991) is included; the WN/WC ratios are in agreement with those found by Armandroff & Massey (1991). Due to small number statistics, the ratios for NGC 6822 and IC 1613 are not significant.

Such number ratios are to be preferred to surface densities, which would depend not only on stellar evolution but also on the current SFR. For the Galaxy, the statistics for O stars is based on the survey by Garmany & Conti (1982). In other galaxies, the numbers of O stars were estimated by Azzopardi et al (1988) on the basis of UV data from Geneva and Marseille balloon experiments and on the basis of Lequeux's (1986) luminosity function. These indirect estimates lead to larger uncertainties in the numbers of O stars than in those for W-R stars.

The origin of the observed variations of the relative number of W-R stars in different environments was attributed to metallicity by Smith (1973) and Maeder et al (1980), who suggested that high Z favors mass loss, which in turn favors the formation of W-R stars. The variations of W-R subclasses in M31 were also attributed to a metallicity effect by Moffat & Shara (1983). The total dependence on Z was criticized by several authors, who attributed

Table 2   Observed W-R/O, WC/W-R, and WC/WN in galaxies of various metallicities

Galaxy	Z	W-R/O	WC/W-R	WC/WN
M31	0.035	0.24	0.44	0.79
Milky Way				
ring 6–7.5 kpc	0.029	0.205	0.55	1.22
ring 7.5–9 kpc	0.020	0.104	0.48	0.92
ring 9.5–11 kpc	0.013	0.033	0.33	0.49
M33	0.013	0.06	0.52	1.08
LMC	0.006	0.04	0.20	0.26
NGC 6822	0.005	0.02	-	-
SMC	0.002	0.017	0.11	0.13
IC 1613	0.002	0.02	-	-

the differences in the W-R populations mainly to changes in the IMF and SFR (Bertelli & Chiosi 1981, 1982; Garmany et al 1982; Armandroff & Massey 1985; Massey 1985; Massey et al 1986; Massey & Armandroff 1991). As suggested by these authors, it is possible that prominent departures from the assumption of constant star formation, such as is the case in 30 Dor in the LMC (Moffat et al 1987) or in giant HII regions of M33 (Conti & Massey 1981, Drissen et al 1990) where recent bursts of SFR have occurred, may produce peculiar W-R number ratios (Section 5). However, we note that no systematic difference in the IMF slope has been found between the Galaxy, the LMC, and SMC (Humphreys & McElroy, 1984, Mateo 1988, Massey et al 1989a, Parker et al 1992, Section 2). Also, the Galactic gradient of the surface density of W-R stars is much steeper than that of their precursor O stars (Meylan & Maeder 1982, van der Hucht et al 1988)—a fact that is reflected by the changes of the W-R/O in Table 2. Thus, it is likely that the basic effect in stellar evolutionary models is Z, and that effects connected to the SFR and perhaps to the IMF are population parameters that may also influence the relative frequencies of W-R stars in G HII regions and bursts.

4.5.2 PREDICTED W-R/O AND WC/WN VALUES   As illustrated in Table 2, the W-R/O ratio increases with the metallicity of the parent galaxy (Maeder et al 1980; Azzopardi et al 1988; Smith 1988, 1991a). This general trend is also confirmed by studies of the integrated properties of H II or W-R galaxies (Arnault et al 1989; Conti 1991a, b; Smith 1991a; Vacca & Conti 1992; Mas-Hesse & Kunth 1991a, b; Mas-Hesse 1992). In stellar models a growth of the W-R/O with Z is predicted (Maeder 1991a, Maeder & Meynet 1994), resulting from the lowering of the minimum initial mass for forming W-R stars and from the increase of the lifetimes with increasing Z (and mass loss). Figure 2 compares the observations of Table 2 with theoretical values for models with enhanced $\dot{M}$ as defined in Section 4.4 and for stars with a Salpeter IMF. The general agreement is quite good, confirming that metallicity is a key factor in the variations of the W-R/O ratio. Although some scatter appears, it might be due to local departures from the simple assumption of a past constant SFR and to the averaging over some range of Z in large galaxies (Smith 1991a). No satisfactory agreement between observed and predicted W-R/O can be achieved for models with standard $\dot{M}$ (de Jager et al 1988). The differences would be especially large at high Z, while at $Z = 0.002$, the W-R/O values are in both cases of mass loss equal to about 0.005. Such neglible W-R/O values at low Z are in agreement with the low fraction of W-R stars observed in metal deficient galaxies. Also, the study of the integrated spectrum and He II 4686Å feature in dwarf galaxies shows a general absence of W-R contribution for galaxies with very low oxygen content, corresponding to about $Z = 0.002$ (Arnault et al 1989, Smith 1991a).

The observed WC/WN and WC/W-R numbers in Table 2 show a general growth with increasing Z, but it is not monotonic and shows an appreciable

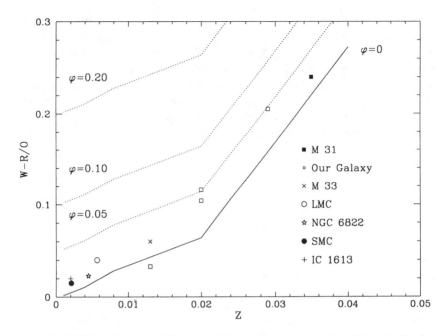

Figure 2   W-R/O as a function of Z in nearby galaxies compared to model predictions by Maeder & Meynet (1994). The solid line represents the predictions of single star models with enhanced mass loss rates (Section 4.4) and Salpeter's IMF. The dotted lines show the same for different values φ, the fraction of O stars undergoing mass transfer in binaries.

scatter. This was noted by Armandroff & Massey (1991) and Massey & Armandroff (1991) as an argument supporting the fact that metallicity is not the only determining factor for explaining the W-R statistics. From models, the change in WC/WN results from the higher $\dot{M}$ which leads to an earlier visibility of the products of He burning (Maeder 1991a, Maeder & Meynet 1994). Interestingly enough, for larger $\dot{M}$ the predicted WC/WN ratios, instead of further increasing as expected, go down again as Z becomes greater than 0.01. This occurs because the WN phase of the most massive stars has already been entered during the main sequence phase and is therefore much longer. This model result accounts for the nonmonotonic behavior of WC/WN found by Armandroff & Massey (1991).

The comparison between observed and theoretical WC/WN or WC/W-R shows, as for the W-R/O, that a better fit is obtained for models with enhanced $\dot{M}$ in previous phases. Nevertheless, the scatter is still there, and reflects departures from the assumption of a constant SFR. Such departures are prominent in some giant H II regions, such as 30 Dor in the LMC, or NGC 592, 595, and 604 in M33, which show evidence of intense star formation (Conti & Massey 1981; Drissen et al 1990, 1991). Armandroff & Massey (1985) and Massey & Armandroff (1991) noticed that regions with metallicity similar to that of the LMC and SMC have different W-R numbers. Smith & Maeder (1991) suggested that in large spirals the W-R populations will be heavily weighted toward properties of high Z values. Thus both the effects of bursts of star formation (Section 5) and of the averaging over Z may contribute to increases in the W-R/O and WC/WN ratios, a situation which may apply particularly to M33.

Close comparisons between models and observations must also account for the various channels of W-R formation and in particular the Roche lobe overflow (RLOF) in close binaries (Vanbeveren 1991, 1994; Vanbeveren & de Loore 1993). The fraction of O stars becoming W-R stars as a result of RLOF was estimated by the latter authors to be about 35% (see also Podsiadlowski et al 1992). A new analysis quoted above (Maeder & Meynet 1994) suggests that the fraction of W-R stars owing their existence to RLOF is highly variable with the metallicity of the parent galaxy; it is nearly 100% at low Z, like in the SMC (Smith 1991b), and lower than 10% in the inner regions of the Milky Way.

The above results are compatible as shown in Figure 2 with a relatively low fraction φ, at most 10%, of the ensemble of the O stars that become W-R as a result of binary mass transfer. Such a low fraction is consistent with by Massey's (1981) result on the distribution of the eccentricities of WR+O binaries. A close investigation of the young cluster Tr 14 (Penny et al 1993) reveals a general absence of close binaries among the brightest O stars. In further support of a low fraction of O stars undergoing RLOF, we remark that at the low Z in the SMC, the W-R/O ratio is about 0.017, while it is 0.21 in inner Galactic regions. Thus, if the fraction of O stars becoming W-R as a result of RLOF is the same in both areas, this fraction would be at most $0.017/0.21 = 8\%$, assuming that all W-R stars in the SMC are formed by RLOF. This fixes the upper limit of the fraction of O stars undergoing RLOF under the mentioned assumption.

Stellar winds produce fewer W-R stars at low Z, and only the binary channel seems efficient in forming WR stars. As an example, in the SMC, eight of the nine observed W-R stars may be binaries; five are confirmed (Moffat 1988, Smith 1991a). These binaries were likely formed by mass transfer (De Greve et al 1988). They are of type WNE, which suggests that the binary channel may mainly lead to WNE stars. Consistent with the SMC observations, the proportion of W-R binaries has been found to be larger toward the anticenter in the Milky Way (van der Hucht et al 1988). The hypothesis that a small fraction of the W-R stars is formed by RLOF also gives a better fit to the observed behavior of the WNE/W-R and WNL/W-R ratios with Z (Maeder & Meynet 1994). We may also note that the average ages of WNL and WNE stars are probably not the same; Moffat et al (1991) suggest that the former are relatively younger than the latter; WNL stars are also more luminous than WNE stars (Conti 1988). These facts are consistent with the result (Section 4.5) that WNL stars mainly originate from the most massive stars.

4.5.3 WC SUBTYPE STATISTICS   The distributions of WC and WO stars in galaxies exhibit a number of distinct properties.

1. It is well known that WC stars are relatively more numerous in inner galactic regions (cf van der Hucht et al 1988, Conti & Vacca 1990). More specifically, the later WC subtypes—WC9 and WC8—are only found in inner galactic regions with higher Z, while outer regions with lower Z contain WC stars of earlier subtypes, mainly WC6–WC4. This is also true in the LMC, where only WC6–WC4 subtypes are found, with very uniform properties (Breysacher 1986, Smith et al 1990, Smith 1991b). In more extreme low-Z dwarf galaxies, the rare WC stars only belong to subtypes WC4 and WO. In M31 as in the Milky Way, late WC stars (WCL) are found in inner galactic regions and early ones (WCE) in outer regions (Moffat & Shara 1987). However, the metal-rich galaxy M31 contains no WC9 and WC8 stars (Massey et al 1987a).

2. The luminosity of earlier WC subtypes is lower than that for later WC subtypes (Lundström & Stenholm 1984, van der Hucht et al 1988, Conti 1988).

3. Stars of a given WC subtype seem brighter in a galaxy with lower Z as suggested by Smith & Maeder (1991), who point out that LMC WC4 stars are brighter than Galactic WC5–WC6 stars.

These various facts can be easily understood on the basis of recent models and of the relation between WC subtypes and the (C+O)/He ratios as shown by Smith & Maeder (1991). The entry points and lifetimes in the WC9 to WO sequence are very dependent on M and Z. At high Z and high M, due to high mass loss the WC stage is entered very early during the He burning phase, so that the surface (C+O)/He ratio is very low, which implies a type WC9 or WC8. As evolution continues, mass and luminosity decline and (C+O)/He decreases and thus the sequence of types WC9 $\rightarrow$ WO is described. The entry point in that sequence occurs at lower L and earlier WC types for lower masses. At lower Z, as in the LMC, the entry in the WC phase occurs at a later stage of central He-burning, i.e. with higher O/C ratios (Smith et al 1990) and also with higher (C+O)/He ratios at the surface, which means earlier WC types, typically WC5–WC4 (Smith & Maeder 1991). Then, the further evolutionary sequence described is short. This behavior also explains the two above-mentioned points concerning the luminosity of WC stars at different Z. That the luminosity for a WC subtype depends on the initial Z may have consequences for the interpretation of WC lines in integrated spectra of galaxies.

From W-R stars in clusters and in galaxies, some relationships between subtypes can be understood (van der Hucht et al 1988; see also Schild & Maeder 1984, Moffat et al 1986, Moffat 1988, Schild et al 1991): For galactocentric radius $R < 8.5$ kpc, we have WNL $\rightarrow$ WCL; while for $R > 6.5$ kpc, we find WNL $\rightarrow$ WCE $\rightarrow$ WO and WNE $\rightarrow$ no WC stars.

These connections are supported by the model results (Maeder 1991a). The first connection is typical of high mass and high Z: A long WNL phase, followed by a negligible WNE phase, emerges on the late and luminous part of the WC sequence, typically at WC9 or WC8. The second connection is typical of large masses with solar or lower metallicity, while the third one corresponds to the lowest part of the mass range able to form W-R stars.

We conclude this section by underlining that the observations and understanding of W-R stars in nearby galaxies has brought the clarification of many problems, which also have far-reaching consequences for the injection of mass, momentum, and energy into the interstellar medium (Leitherer et al 1992a).

5.   MASSIVE STARS IN STARBURSTS

5.1   *Integrated Spectra of Galaxies*

The spectra of galaxies are primarily those of the underlying stellar population (absorption lines) with the addition of nebular emission lines from the GH II and young starburst regions which are present if recent star formation has occurred. We first consider how one might infer the distribution of massive stars from the nebular line analyses, and then discuss spectral synthesis of the stellar features.

5.1.1   NEBULAR LINE ANALYSES   The number of exciting stars of ionized hydrogen regions (NO^*) can readily be estimated from analysis of the emission line spectra (e.g. Kennicutt 1984, Shields 1990). One observes the spatially integrated Hα flux (or the Hβ or radio recombination measure), accounts for the extinction (if necessary), and from simple recombination theory infers the total number of Lyman continuum photons (NLyc) being emitted. This last step contains the nominal nebular analysis assumptions of "Case B"—no dust, and ionization bounded. The NLyc is a direct measure of the number of exciting stars. The total number released is a product of the numbers of hot stars, the slope of the IMF, and the NLyc at each spectral subtype.

Vacca (1991, 1994) has thoroughly reviewed and quantified this procedure. He introduces the parameter η_0 as defined in the equation

$$NO^* = NO7V/\eta_0, \tag{3}$$

where $NO7V$ is the number of "equivalent" O7V type stars, and NO^* is the total number of O stars. The $NO7V$ is simply the *observed* NLyc divided by the number of Lyman continuum photons emitted by an O7V star (1×10^{49} s^{-1}). The η_0 can be calculated as a function of Γ, T_{eff}, and $\log g$ using the Kurucz stellar atmosphere models. Its value, tabulated by Vacca, depends also on M_{upper}, and the M_{lower} for Lyman photon production—roughly the OB star boundary. This latter parameter depends on the metal abundance Z via the stellar structure models. This procedure makes the assumption that all O stars in H II and GH II regions are main sequence; one can separately allow for massive star evolution.

How accurately does this method work? It has been calibrated using 30 Dor in the LMC. Parker (1991, 1993) and Parker & Garmany (1993) have made a detailed census of the hot star population in an $7' \times 7'$ area centered on R136. They find about 400 O stars. Vacca (1991, 1994) has analyzed the nebular spectrum of a trailed, 15-minute spectrophotometric exposure of an $8' \times 8'$ area scan centered on R136 (taken by M. Phillips). Using the Z value for the LMC and Parker's value for $\Gamma(-1.4)$, he finds η_0 to be 0.44 and estimates that about 330 O stars are present, within 30% of the actual count (allowing for the slight difference in areas)! This gives us confidence in the procedure which has been applied to W-R and other emission line galaxies by Vacca & Conti (1992) as we show below (Section 5.1.4).

5.1.2 EMISSION LINE GALAXIES—STARBURSTS   These galaxies with emission line spectra like those of H II regions were noted by Sargent & Searle (1970). Given the often substantial numbers of O-type stars found within these galaxies, they can be understood to be examples of very young "starbursts." Using slit spectroscopy, one may obtain the numbers of exciting stars by a procedure similar to that outlined above for 30 Dor. For galaxies, however, we might have additional complications if the slit width does not include the entire region of interest, if the nebulosity is density bounded, or if dust is present in sufficient quantities to absorb a considerable fraction of the Lyman photons.

5.1.3 STARBURST MODELS   Several models for starburst populations have been produced to simulate various properties of the galaxy spectra such as: the overall stellar and nebular spectrum (Leitherer 1990, 1991; Leitherer et al 1992b; Bernlohr 1992, 1993), the strengths of the Si IV and CIV UV wind features emitted by O stars (Leitherer & Lamers 1991; Mas-Hesse & Kunth 1991a,b), the widths of these UV lines (Robert et al 1993), the far-infrared and radio emission by the dust (Mas-Hesse 1992, Desert 1993), the emission line ratios such as $\lambda4686$ He II/ Hβ or the so-called W-R "bump" $\lambda4650$/ Hβ (Arnault et al 1989, Meynet 1994b), or other line ratios such as $\lambda4686$He II/$\lambda4650$CIII, which is sensitive to the WN/WC ratio (Kruger et al 1992).

The basic physical parameters for a starburst model are the star formation rate—in particular the intensity and duration of the burst, the age after its beginning, the IMF, and Z. The hope of starburst models is to disentangle these various parameters. Let us consider the didactical, but not unrealistic, case of an instantaneous "burst" (formation over a time of up to 1 Myr—small with respect to the massive star evolution time). In this case, an evolved W-R population with its prominent emission features results from only a part of the mass range of the potential W-R progenitors, while in the case of a constant SFR the W-R population results from an equilibrium mixture of the whole potential mass range. We may schematically distinguish four different epochs in the evolution of a burst according to recent models of massive stars and W-R stars (Maeder & Meynet 1994):

1. O-phase: For an age $t \leq 2$ Myr, massive stars are in their O-type phase, giving rise to H II regions without W-R features.

2. WNL phase: From $t = 2$ Myr to about 3 Myr, a large number of W-R stars are present, nearly all of them are of WNL subtype.

3. WC+WNL+WNE phase: After $t = 3$ Myr, the three subtypes may coexist with fractions depending on the mass loss and/or Z. At solar Z and standard $\dot{M}$, WC stars dominate, followed by WNL in the first part of the period and by WNE stars in the second part. The higher the $\dot{M}$ and Z, the more numerous are the WC stars. At lower Z and $\dot{M}$, single star evolution only leads to WNL stars and very few WC stars, but we may also expect (Section 4.5.2) a fraction of W-R stars of WNE type from RLOF binary evolution.

4. O-phase (again): For $t \geq 7$ Myr, the W-R stars have disappeared but at up to 10 Myr there are still O-type stars to produce H II region nebular emission line features. One way to distinguish between this phase and the first O-phase might be to examine the equivalent width of $H\beta$ nebula emission; according to Copetti et al (1986) this parameter steadily decreases in value with age in model H II regions (while the number of Lyc photons remains more or less steady, the starlight at $\lambda4860$ steadily increases).

During these phases of a burst, the various ratios of W-R subtypes to O stars are much larger than in the case of a constant SFR. As an example, at solar Z the average WNL/O ratio is up to six times larger, and the enhancement is even greater at lower Z. The reason is that the duration of the W-R–rich phase is shorter at lower Z, meaning that stars of only a narrow mass range become W-R stars, thus the contrast between the cases of a burst and of constant SFR is much larger. For bursts longer than 1 Myr, the situation is intermediate between the instantaneous burst and constant SFR. We also notice that if an observed H II region consists of a burst plus a region of lower but constant SFR, we have for the 2 Myr after the burst a much lower W-R/O ratio than for constant SFR, since at this time the W-R stars from the burst have not yet appeared. On the whole, then, one must be cautious before making quantitative inferences from number ratios alone.

5.1.4 WOLF-RAYET GALAXIES   These are a subset of emission line galaxies in which, in addition to the nebular line spectrum, one observes broad emission at $\lambda4686$ Å due to the presence of Wolf-Rayet *stars* (Conti 1991a,b, and references therein). The starburst phenomena illustrated by W-R galaxies represent an extreme burst of star formation, in which hundreds to thousands (or more) of massive stars have been born. There are currently about 50 examples of such systems, most of which have been discovered serendipitously. Many are Markarian or Zwicky galaxies and exhibit disturbed morphologies which may be the result of interactions or mergers. Examples of W-R galaxies may be found among Blue Compact Dwarf Galaxies (Sargent 1970), isolated extragalactic H II regions (Sargent & Searle 1970), dwarf irregulars (Dinerstein & Shields 1986), "amorphous" galaxies (Walsh & Roy 1987), spiral galaxies containing knots or GH II regions (Keel 1982), recent galaxy mergers (Rubin et al 1990), or powerful *IRAS* galaxies (Armus et al 1988).

The broad emission features are seen in contrast to the galaxy stellar population continua. The dilution is such that the few hundred Å equivalent width of an emission line of typical single W-R stars is usually only a few Å in W-R galaxies. Examples of the spectra are given in Vacca & Conti (1992) and Conti (1993). Many of these galaxies are metal-weak but this may be a selection effect, given that starburst galaxies with more normal composition are likely to have a brighter underlying stellar population, which could "drown out" the W-R stars even if present.

Vacca & Conti (1992) have recently analyzed optical spectra of ten Wolf-Rayet and four other emission line galaxies. The nebular line ratios indicate that the excitation is caused by stars. The strength of the $\lambda 4686$ Å may be used to infer the numbers of W-R stars present; this uses a calibration of the line flux for single W-R stars in the LMC, dividing that number into the measured line flux in the galaxy. Typically, tens to hundreds of W-R stars are inferred to be present in the starbursts. This procedure has been quantitatively checked by using the area spectrophotometry of 30 Dor in which a broad $\lambda 4686$ He II line is measured; the inferred numbers of W-R stars (20) are similar to the census of Moffat et al (1987). The contribution of the W-R stars to the observed NLyc is subtracted from the observed value for the galaxy, and the remainder is treated, as above, to determine the number of O-type stars present. As it is not possible to derive the slope of an IMF from this procedure, a Γ equivalent to that of 30 Dor was adopted for each galaxy; similarly the M_{upper} was assumed to be the same ($100\,M_{\odot}$). With the usual assumptions that low mass stars have formed along with the high mass ones, typical star formation rates range from 2×10^{-2} to $3\,M_{\odot}$ per year (the former is the value for 30 Dor).

Following the nebular line analysis taken for 30 Dor, spectrophotometric studies have provided quantitative values of WNL/O for W-R galaxies (Vacca & Conti 1992, Conti 1993). Figure 3 shows the comparison of these observations with continuous star formation and newly constructed burst models (Meynet

1994b; see also Maeder & Meynet 1994). The observed WNL/O ratios in W-R galaxies are much higher than the predictions of models with constant SFR, in agreement with previous estimates (Arnault et al 1989). In Figure 3, burst models with two different IMF slopes are shown which bracket the observations nicely. Notice also the importance of Z to the predictions. The observed values are, strictly speaking, the number of WN stars derived from the strength of λ4686 He II, but in most of these W-R galaxies, there is little or no evidence of WC stars. The largest source of uncertainty in the observed values is a possible mismatch between the slit width (1.5″) and the starburst (typically 2 to 3″). This underestimates the number of O stars. Arbitrarily correcting for this would *lower* the ratio by a factor 2 (see a more thorough discussion in Conti 1993).

W-R binaries might enhance the production of W-R stars, but this effect is believed to be small except at the lowest Z. While Masegosa et al (1991) suggest that SN contamination could affect the starburst properties, particularly the "blue bump" at λ4650, this may be unimportant as no other effects (e.g. shocks)

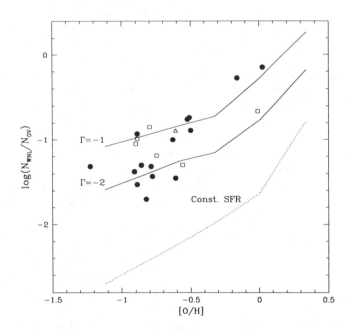

Figure 3 WNL/OV ratio in W-R galaxies, as a function of the oxygen abundance [O/H]; data adapted from Conti (1993); stellar models from Maeder & Meynet (1994). (•) Galaxies of Vacca & Conti (1992) ; (□) galaxies of W. D. Vacca (private communication); (△) 30 Dor (from Vacca 1991). The dotted line denotes the predictions for "continuous" star formation; the solid lines are for values estimated for a "burst" (see text). Γ is the slope of the IMF as defined in Section 2.2.

are seen in the nebular spectra of W-R galaxies (Vacca & Conti 1992). Could there be different values for Γ in various starburst regions? This possibility has been raised in recent years (Scalo 1989; Mas-Hesse & Kunth 1991a,b; Rieke 1991; Joseph 1991; Bernlohr 1993). We have already noted (Section 2) that in regions where the O star populations can be counted *directly*, Γ does appear to differ from region to region by a small, but significant amount. Thus differences in Γ from starburst region to starburst region are possible, but final conclusions are still uncertain.

From Figure 3 we conclude that for W-R galaxies, the W-R/O ratios are well above the predictions for "continuous" star formation, and nicely match the predictions of "burst" models. We infer that the starbursts observed in W-R galaxies are typically going on for only a relatively "brief" interval, typically $\lesssim 10^6$ yr. The energetics are similar to 30 Doradus at the faint end to more than $100\times$ larger. Are these nearby starbursts paradigms for the first phases of massive star evolution in very young galaxies?

Phillips & Conti (1992) discovered evidence for WC9 stars in a GH II region at the edge of the bar in the metal-rich ($+0.5$ dex) galaxy NGC 1365. As their spectrum only covered the yellow region, there is currently no information on the numbers of WN stars in their strong starburst region. However, the presence of late type WC stars in a strong Z environment nicely follows the models predictions (Section 4.5.3).

5.1.5 SPECTRAL SYNTHESIS   Optical spectroscopy of most galaxies shows absorption line spectra from an old evolved population of late type giant stars. In the W-R galaxies studied by Vacca & Conti (1992) the optical absorption spectra are those of late B and A type stars (narrow K line, upper Balmer series in absorption, no other features). In these objects, even though there are large numbers of very hot stars, the optical spectra are dominated by other stars. However, in the UV the O (and W-R) stars become the dominant contributors to the continua. In emission line galaxies containing OB stars, P Cygni spectral features at $\lambda 1400$ Si IV and $\lambda 1550$ C IV are seen; if W-R stars are present, an emission line at $\lambda 1640$ He II is found.

In Figure 4 we show *IUE* spectra of a $3' \times 3'$ area of the GH II region 30 Dor in the LMC and NGC 1741, a W-R galaxy. UV spectra such as these can be used directly to estimate the numbers of hot stars present by spectral synthesis techniques (Leitherer et al 1992a,b; Robert et al 1993). One can also ascertain the age of the starburst and tell whether or not the formation of the massive stars has been "continuous" or a "burst." In work in progress, Vacca et al (1994) have found a reasonable fit for the Si IV, C IV, and He II profiles for ages between 1.5 and 2 Myr for the UV spectrum of 30 Dor illustrated in Figure 4. For younger ages, insufficient W-R stars have been produced to match the $\lambda 1640$ emission line; for older ages, the Si IV and C IV profiles begin to diverge from the observations. It is also possible to estimate the

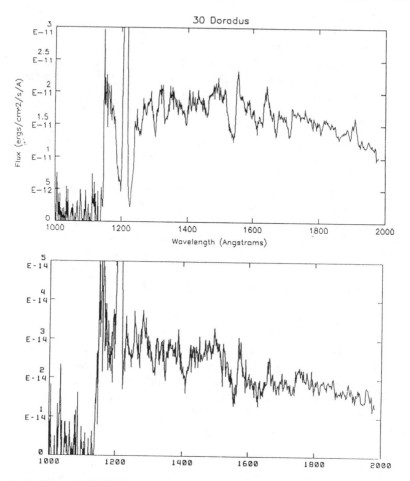

Figure 4   Observed *IUE* SWP spectra of 30 Dor (*top*) and NGC 1741 (*bottom*), a W-R galaxy. Stellar line features at λ1400 Si IV, λ1550 C IV, λ1640 He II, and λ1718 N IV are seen in both objects. The lines shortward of Si IV are primarily interstellar.

number of O stars from the extinction-corrected *IUE* continuum of 30 Dor, as the models predict this quantity. The agreement with the census from Parker (1991, 1993) and Parker & Garmany (1993) is good. Unfortunately, even the brightest starburst galaxies are at the limit of *IUE* sensitivity. The $10'' \times 20''$ slit is typically larger than the starburst, so dilution of the spectral signatures can be a problem. For the number counts, but not for the age estimate, the UV extinction is critical and remains a serious problem for more general application of this method.

5.2   *Global Properties*

5.2.1   FAR-INFRARED LUMINOSITIES   Radiation from dusty H II and GH II regions and galaxies in the far-infrared appears to come from heating of the dust by their stellar populations. Devereux & Young (1991) have argued that the far-infrared (FIR) luminosities relate directly to the content of *hot stars*. They have derived the FIR luminosity of 124 spiral galaxies from the *IRAS* archives, which also have Hα measurements. From the individual *IRAS* colors, they find that this dust is at a temperature of 30 to 40 K, corresponding to heating by *hot* stars; they suggest that ambient heating by stars of all types would be more like 15–20 K. In their Figure 5 they show a correlation between the FIR and the Hα emission extending over two orders of magnitude. Part of the dispersion in their relation could come from the differences in T_{eff} of the exciting stars. Although Devereux & Young claim that one can infer the actual T_{eff} values, we don't believe that the modeling used is adequate.

With considerable caution, one could perhaps use the FIR luminosities of star forming regions to infer the total numbers of hot stars present, but, absent other information, one cannot infer an IMF slope (Γ) or an M_{upper}. The FIR luminosities for normal spiral galaxies range from 10^9 to 10^{11} $L_\odot$ and, according to Devereux & Young (1991), show no dependence on the Hubble type. This seems at odds with the appearance of stellar images on direct plates. It would be nice to confirm that the relationship between FIR luminosity and numbers of hot exciting stars is as orderly as claimed. This might be accomplished with identifications and counts of O type stars in otherwise obscured GH II regions of our Galaxy, such as W51, through classification K band spectroscopy (Conti et al 1993).

5.2.2   IMAGING OF EMISSION LINE AND W-R GALAXIES   Optical CCD photometry of galaxies is a growth industry: Spectral syntheses of the continua can give us valuable information on stellar populations of all but the hottest and most massive stars. As we have already noted, even in W-R galaxies, demonstrably the youngest examples of starburst phenomena, the optical light comes primarily from stars of type A. Thus in this wavelength region we are sampling stars with lifetimes (not necessarily ages) of 50 million years or more.

Conti & Vacca (1994) have used the *HST* with FOC camera and the λ2200 Å filter to obtain a UV image of the W-R galaxy He 2-10; the spatial resolution with image restoration is $\approx 0.1''$. At this wavelength one is primarily sampling the OB stars, which have lifetimes of a few $\times 10^6$ to 10^7 yr. Given the presence of W-R stars in this galaxy, the age is closer to the lower of those numbers. This W-R galaxy has recently been found by Corbin et al (1993), using deep CCD photometry, to be at the center of a faint elliptical galaxy (3 kpc diameter)! There are three starburst regions, the strongest (containing W-R stars) at the center. Figure 5 is a reconstructed *HST* UV image of the central starburst in

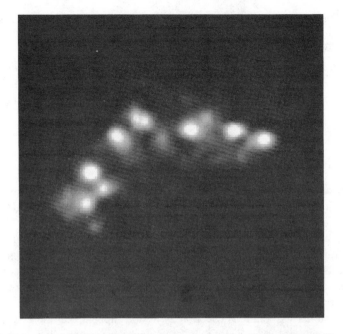

Figure 5   Reconstructed UV image of the central starburst in He2-10 (from Conti & Vacca 1994). Nine individual starburst knots are readily seen. The spatial scale across the figure is $3''$, corresponding to 125 pc. The parent elliptical galaxy (Corbin et al 1993) is about $80''$ in diameter, which is much larger than the scale of this starburst.

He 2-10. One can see ten individual knots of activity, several just at the limit of resolution (10 pc in diameter), with separations of about two to three times this number.

Conti & Vacca (1994) estimate the average $\lambda 2200$ luminosities of the individual starburst knots in He 2-10 to be $\approx 10^{38}$ erg s^{-1} Å^{-1}. This corresponds to M_V of -13.9, and mean total masses (assuming a normal IMF) of a few $\times 10^5 M_\odot$. These masses are similar to those of globular clusters, but in objects with ages of less than 10 Myr! Whitmore et al (1993) have recently identified about 40 blue objects in the galaxy merger remnant NGC 7252 as being globular clusters of age about 500 Myr, on the basis of their optical colors and absolute magnitudes. Their average M_V are about one magnitude fainter than those of He 2-10, which could be an evolution effect since the starburst in NGC 7252 is older.

Conti & Vacca (1994) have noted that at least a dozen other W-R galaxies for which they have *HST* UV imaging also show multiple knots of star forming activity. The luminosities of the knots in several of these galaxies are substantially larger than those of He 2-10. Many of these other galaxies are clear examples of merging or interacting galaxies (the status of He 2-10 is not quite clear on this point). They thus suggest that galaxy merger events may lead to star formation episodes which produce knots of activity and eventual production of what we now recognize as globular clusters. Of course, it is not certain that where large numbers of massive stars have been produced, as in these starburst regions, low mass stars are also born. In nearby galactic associations we do consistently see a lower main sequence, but in highly energetic star formation episodes, this might not always be the case.

6. CONCLUSIONS

Theoretical considerations and extensive modeling suggest that substantial distinctions in the evolutionary history of massive stars will arise because of the importance of mass loss and its strong dependence on metallicity. We have reviewed and discussed the properties, chemical abundances, and populations of individual OB stars, supergiants, and Wolf-Rayet stars in various galaxy environments and find order of magnitude differences from place to place. These are understood to depend on Z, along with the IMF and SFR, but not all these parameters are completely sorted out at the present time. From detailed star counts in stellar associations (Section 2), it does appear that there is no simple dependence of the IMF slope, Γ, on Z, but there may be differences from place to place. There is no evidence that the M_{upper} limit depends on Z; data concerning potential differences in the M_{lower} limit with location is currently lacking. A major unresolved question is that concerning the SFR: What does this parameter depend upon? Clearly one needs sufficient molecular gas; future studies should address the relationship between this parameter and the actual numbers of massive stars in very quantitative terms.

We have lightly touched upon the connection between studies of nearby massive star formation regions where stellar statistics may be accomplished, and measurement and analyses of their integrated properties. These objects, such as 30 Dor in the LMC and NGC 604 in M33, can be used as "stepping stones" in our understanding of similar phenomena in more distant galaxies. It will be important to improve the "calibration" of these types of massive star groupings in various galaxy environments to better understand those even more energetic and less common starbursts found at larger distances.

In starbursts we are dealing nearly exclusively with integrated spectral properties. For those containing massive stars of O and W-R type we have an advantage that the lifetimes are less than 10 Myr, and the formation time scales appear to be only 1 Myr. The modeling can be thus simplified to that of a "burst." Probably SN will not have played much of a role, as yet, in the energetics of the phenomena we are observing; the excitation of the gas will be primarily due to stars. Multiwavelength studies of many galaxies have already been made but there has been little quantitative integration of all these data for the same galaxies and the same starbursts. In addition to the inferences concerning hot stars, one would like to know the cool star population, and the quantity of neutral and molecular gas. This requires observations at IR wavelengths and in the sub-mm and cm regimes. One certain caution is that of aperature; it is critically important that these are matched, independent of the wavelengths, so that the same volume elements are being examined in each case. Even more significant is the probability that as one goes to shorter and shorter wavelengths, differential internal galactic extinction might necessarily shield from our view the "back side" of a starburst episode.

With our new-found knowledge of the properties and evolution of massive stars, we now can begin to study and understand the appearances of ever more distant galaxies, and one would hope, to delineate their past history. "In the beginning" there were undoubtedly many massive stars in newly forming galaxies. Our improved comprehension of newly born local massive stars can help clarify our knowledge of some of the earliest stages in the evolution of our Universe.

ACKNOWLEDGMENTS

The authors are grateful to Lorraine Volsky and the JILA publications office for editorial assistence. PSC appreciates continuous support by the National Science Foundation. AM acknowledges support of the Fonds National Suisse de la Recherche Scientifique.

Literature Cited

Abbott DC 1982. *Ap. J.* 259:282–301
Abbott DC, Bieging JH, Churchwell E, Torres AV. 1986. *Ap. J.* 303:239–61
Abbott DC, Conti PS. 1987. *Annu. Rev. Astron. Astrophys.* 25:113–50
Alongi M, Bertelli G, Bressan A, Chiosi C, Fagotto F, et al. 1993. *Astron. Astrophys. Suppl.* 97:851–71
Appenzeller I. 1980. In *Star Formation, 10th Saas-Fee Course,* ed A Maeder, L Martinet, pp. 3–73. Geneva:Geneva Obs.
Appenzeller I. 1989. See Davidson et al 1989, pp. 195–204
Appenzeller I, Wolf B. 1981. In *The Most Massive Stars,* ed. S d'Odorico, pp. 131–39. Garching:ESO Workshop
Armandroff TE, Massey P. 1985. *Ap. J.* 291:685–92
Armandroff TE, Massey P. 1991. *Astron. J.*

102:927–50
Armus L, Heckman T, Miley G. 1988. *Ap. J. Lett.* 326:45–49
Arnault P, Kunth D, Schild H. 1989. *Astron. Astrophys.* 224:73–85
Arnett D. 1991. *Ap. J.* 383:295–307
Azzopardi M, Breysacher J. 1979. *Astron. Astrophys.* 75:243–46
Azzopardi M, Breysacher J. 1985. *Astron. Astrophys.* 149:213–16
Azzopardi M, Lequeux J, Maeder A. 1988. *Astron. Astrophys.* 189:34–38
Bandiera R, Turolla R. 1990. *Astron. Astrophys.* 231:85–88
Baraffe I, El Eid MF. 1991. *Astron. Astrophys.* 245:548–60
Barbuy B, Medeiros JR, Maeder A. 1992. *Int. Symp. on Nuclear Astrophysics,* Karlsruhe, ed. F Käppeler, K Wisshack, pp. 35–40. Bris-

tol and Philadelphia:Inst. Phys.
Barbuy B, Spite M, Spite F, Milone A. 1991. *Astron. Astrophys.* 247:15–19
Barlow MJ, Hummer DG. 1982. See de Loore & A J Willis 1982, pp. 387–92
Barlow MJ, Roche PF, Aitken DK. 1988. *MNRAS* 232:821–34
Barlow MJ, Smith LJ, Willis AJ. 1981. *MNRAS* 196:101–10
Baschek B, Scholz M, Wehrse R. 1991. *Astron. Astrophys.* 246:374–82
Beech M, Mitalas R. 1992. *Astron. Astrophys.* 262:483–86
Bernlohr K. 1992. *Astron. Astrophys.* 263:54–68
Bernlohr K. 1993. *Astron. Astrophys.* 268:25–34
Bertelli G, Chiosi C. 1981. In *The Most Massive Stars*, ed. S d'Odorico, D Baade, K Kjär, pp. 211–13. Garching:ESO Workshop
Bertelli G, Chiosi C. 1982. See de Loore & Willis 1982, pp. 359–63
Bisnovatyi-Kogan GS, Nadyozhin DK. 1972. *Astrophys. Space Sci.* 15:353–74
Blaha C, Humphreys RM. 1989. *Astron. J.* 98:1598–608
Blecha A, Schaller G, Maeder A. 1992. *Nature* 360:320–21
Blomme R, Vanbeveren D, Van Rensbergen W. 1991. *Astron. Astrophys.* 241:479–87
Bohannan B, Abbott DC, Voels SA, Hummer DG. 1986. *Ap. J.* 308:728–35
Bohannan B, Voels SA, Hummer DG, Abbott DC. 1990. *Ap. J.* 365:729–37
Bolton CT, Rogers GL. 1978. *Ap. J.* 222:234–45
Boyarchuk AA, Gubeny I, Kubat I, Lyubimkov LS, Sakhibullin NA. 1988. *Astrofizika* 28:197–202
Bressan A, Fagotto F, Bertelli G, Chiosi C. 1993. *Astron. Astrophys. Suppl.* 100:647–64
Breysacher J. 1981. *Astron. Astrophys. Suppl.* 43:203–7
Breysacher J. 1986. *Astron. Astrophys.* 160:185–94
Brocato E, Buonanno R, Castellani V, Walker AR. 1989. *Ap. J. Suppl.* 71:25–46
Brocato E, Castellani V. 1993. *Ap. J.* 410:99–109
Brunish WM, Gallagher JS, Truran JW. 1986. *Astron. J.* 91:598–601
Brunish WM, Truran JW. 1982a. *Ap. J.* 256:247–58
Brunish WM, Truran JW. 1982b. *Ap. J. Suppl.* 49:447–68
Cananzi K. 1992. *Astron. Astrophys.* 259:17–24
Caputo F, Viotti R. 1970. *Astron. Astrophys.* 7:266–78
Carney BW, Janes KA, Flower PJ. 1985. *Astron. J.* 90:1196–210
Cassinelli JP. 1991. See van der Hucht & Hidayat 1991, pp. 289–307
Charbonnel C, Meynet G, Maeder A, Schaller G, Schaerer D. 1993. *Astron. Astrophys. Suppl.*

101:415–19
Chin CW, Stothers RB. 1990. *Ap. J. Suppl.* 73:821–40
Chiosi C, Bertelli G, Bressan A. 1992a. *Annu. Rev. Astron. Astrophys.* 30:235–85
Chiosi C, Bertelli G, Bressan A. 1992b. In *Instabilities in Evolved Super- and Hypergiants*, ed. C de Jager, H Nieuwenhuijzen, pp. 145–55. Amsterdam:North-Holland
Chiosi C, Maeder A. 1986. *Annu. Rev. Astron. Astrophys.* 24:329–75
Chiosi C, Nasi E, Sreenivasan SR. 1978. *Astron. Astrophys.* 63:103–24
Cohen RI. 1992. In *Instabilities in Evolved Super- and Hypergiants*, ed. C de Jager, H Nieuwenhuijzen, pp. 55–59. Amsterdam: North-Holland
Conti PS. 1976. *Mem. Soc. R. Sci. Liege* 9:193–212
Conti PS. 1984. In *Observational Tests of the Stellar Evolution Theory, IAU Symp. 105*, ed. A Maeder, A Renzini, pp. 233–54. Dordrecht:Reidel
Conti PS. 1988. In *O-stars and WR stars, NASA SP-497*, ed. PS Conti, AB Underhill, pp. 81–269. Washington:NASA
Conti PS. 1991a. *Ap. J.* 377:115–25
Conti PS. 1991b. See Leitherer et al 1991, pp. 21–43
Conti PS. 1993. In *Massive Stars: Their Lives in the Interstellar Medium*, ed. JP Cassinelli, EB Churchwell. *ASP Conf. Ser.* 35:449–62
Conti PS. 1994. In *Space Sci. Rev.* In press
Conti PS, Block DL, Geballe TR, Hanson MM. 1993. *Ap. J. Lett.* 406:21–23
Conti PS, Garmany CD, de Loore C, Vanbeveren D. 1983b. *Ap. J.* 274:302–12
Conti PS, Leep EM, Perry DN. 1983a. *Ap. J.* 268:228–45
Conti PS, Massey P. 1981. *Ap. J.* 249:471–80
Conti PS, Massey P. 1989. *Ap. J.* 337:251–71
Conti PS, Underhill AB, eds. 1988. *O Stars and WR Stars, NASA SP-497*. Washington, DC: NASA. 428 pp.
Conti PS, Vacca WD. 1990. *Astron. J.* 100:431–44
Conti PS, Vacca WD. 1994. *Ap. J. Lett.* In press
Copetti MVF, Pastoriza MG, Dottori HA. 1986. *Astron. Astrophys.* 156:111–20
Corbin MR, Korista KT, Vacca WD. 1993. *Astron. J.* 105:1313–17
Crowther PA, Smith LJ, Willis AJ. 1991. See van der Hucht & Hidayat, pp. 97
Davidson K. 1987. *Ap. J.* 317:760–64
Davidson K. 1989. See Davidson et al 1989, pp. 101–8
Davidson K, Dufour RJ, Walborn NR, Gull TR. 1986. *Ap. J.* 305:867–79
Davidson K, Humphreys RM, Haijan A, Terzian Y. 1993. *Ap. J.* 411:336–41
Davidson K, Kinman RD. 1982. *Publ. Astron. Soc. Pac.* 94:634–39
Davidson K, Moffatt AJF, Lamers HJGLM, eds.

1989. *Physics of Luminous Blue Variables.* Dordrecht:Kluwer

de Freitas Pacheco JA, Costa RDD, de Araujo FX, Petrini D. 1993. *MNRAS* 260:401–7

de Freitas Pacheco JA, Machado MA. 1988. *Astron. J.* 96:365–70

De Greve JP 1991. See van der Hucht & Hidayat 1991, p. 213

De Greve JP, Hellings P, van den Heuvel EPJ. 1988. *Astron. Astrophys.* 189:74–80

de Groot MJH, Lamers H. 1992. *Nature* 355:422–23

de Jager C. 1992. In *Instabilities in Evolved Super- and Hypergiants,* ed. C de Jager, H Nieuwenhuijzen, pp. 98–103. Amsterdam:North-Holland

de Jager C, Nieuwenhuijzen H, eds. 1992. *Instabilities in Evolved Super- and Hypergiants.* Amsterdam:North-Holland

de Jager C, Nieuwenhuijzen H, van der Hucht KA. 1988. *Astron. Astrophys. Suppl.* 72:259–89

de Loore C. 1982. See de Loore & Willis 1982, pp. 343–58

de Loore C, Hellings P, Lamers HJGLM. 1982. See de Loore & Willis, pp. 53–56

de Loore C, Vanbeveren D. 1994. *Astron. Astrophys. Suppl.* 103:67–82

de Loore C, Willis AJ, eds. 1982. *Wolf-Rayet Stars: Observation, Physics, Evolution, IAU Symp. 99.* Dordrecht:Reidel

Denissenkov PA. 1988. *Sov. Astron. Lett.* 14:435–37

Denissenkov PA. 1989. *Astrofizika* 31:293–308

Denissenkov PA, Ivanov W. 1987. *Sov. Astron. Lett.* 13:214–16

Desert FX. 1993. In *First Light in the Universe,* ed. B Rocca-Volmerange, B Guideroni, M Dennefeld, J Tran Thanh Van, pp. 193–98. Gif-sur-Yvette:Editions Frontieres

Devereux NA, Young JS. 1991. *Ap. J.* 371:515–24

Dinnerstein HL, Shields GA. 1986. *Ap. J.* 311:45–57

Divan L, Burnichon-Prévot ML. 1988. In *O-stars and WR Stars, NASA SP-497,* ed. PS Conti, AB Underhill, pp. 1–78. Washington:NASA

Doom C, de Greve JP, de Loore C. 1986. *Ap. J.* 303:136–45

Dos Santos LC, Jatenco-Pereira V, Opher R. 1993. *Ap. J.* 410:732–39

Drissen L, Moffat AFJ, Shara MM. 1990. *Ap. J.* 364:496–512

Drissen L, Moffat AFJ, Shara MM. 1991. See van der Hucht & Hidayat, pp. 595–600

Dufton PL, Lennon DJ. 1989. *Astron. Astrophys.* 211:397–401

Eenens PRJ, Williams PM. 1992. *MNRAS* 255:227–36

El Eid MF, Hartmann DH. 1993. *Ap. J.* 404:271–75

Elias JH, Frogel JA, Humphreys RM. 1985. *Ap.*

J. Suppl. 57:91–131

Esteban C, Smith LJ, Vilchez JM, Clegg RES. 1993. *Astron. Astrophys.* 272:299–320

Esteban C, Vilchez JM. 1991. See van der Hucht & Hidayat 1991, p. 422

Esteban C, Vilchez JM. 1992. *Ap. J.* 390:536–40

Esteban C, Vilchez JM, Smith LJ, Clegg RES. 1992. *Astron. Astrophys.* 259:629–48

Firmani C. 1982. See de Loore & Willis 1982, pp. 499–513

Fitzpatrick EL. 1991. *Publ. Astron. Soc. Pac.* 103:1123–48

Fitzpatrick EL, Bohannan B. 1993. *Ap. J.* 404:734–38

Fitzpatrick EL, Garmany CD. 1990. *Ap. J.* 363:119–30

Fransson C, Cassetella A, Gilmozzi R, Kirshner RP, Panagia N, et al. 1989. *Ap. J.* 336:429–41

Freedman W. 1985. *Astron. J.* 90:2499–507

Garmany CD, Conti PS. 1982. *Catalogue of Galactic O-type Stars.* Greenbelt, Md.:Goddard Space Flight Center, Astron. Data Center

Garmany CD, Conti PS, Chiosi C. 1982. *Ap. J.* 263:777–90

Garmany CD, Conti PS, Massey P. 1980. *Ap. J.* 242:1063–76

Garmany CD, Massey P, Parker JW. 1993. *Astron. J.* 106:1471–83

Gies DR, Lambert DL. 1992. *Ap. J.* 387:673–700

Glatzel W, Kiriakidis M, Fricke KJ. 1993. *MNRAS* 262:L7–11

Gochermann J. 1994. *Space Sci. Rev.* In press

Grevesse N. 1991. In *Evolution of Stars: The Photospheric Abundance Connection, IAU Symp. No. 145,* ed. G Michaud, A Tutukov, pp. 63–69. Dordrecht:Kluwer

Groenewegen MAT, Lamers HJGLM, Pauldrach AWA. 1989. *Astron. Astrophys.* 221:78–80

Grossman SA, Narayan R, Arnett D. 1993. *Ap. J.* 407:284–315

Hamann WR, Dünnebeil G, Koesterke L, Schmutz W, Wessolowski U. 1991. *Astron. Astrophys.* 249:443–54

Hamann WR, Koesterke L, Wessolowski U. 1993. *Astron. Astrophys.* 274:397–414

Harris MJ, Lambert DL. 1984. *Ap. J.* 281:739–45

Heap SR, Altner B, Ebbets D, Hubeny I, Hutchings JB, et al. 1991. *Ap. J. Lett.* 377:29–32

Herrero A, Kudritzki RP, Vilchez JM, Kunze D, Butler K, Haser S. 1992. *Astron. Astrophys.* 261:209–34

Heydari-Malayeri M, Hutsemékers D. 1991. *Astron. Astrophys.* 243:401–4

Hill RJ, Madore, BF, Freedman, WL. 1994. *Ap. J.* In press

Hillenbrand LA, Massey P, Strom SE, Merrill KM. 1993. *Astron. J.* 106:1906–46

Hillier DJ. 1987a. *Ap. J. Suppl.* 63:947–64

Hillier DJ. 1987b. *Ap. J. Suppl.* 63:965–81

Hillier DJ. 1988. *Ap. J.* 327:822–39
Hillier DJ. 1989. *Ap. J.* 347:392–408
Höflich P, Langer N, Duschinger M. 1993. *Astron. Astrophys.* 275:L25–28
Howarth ID, Prinja RK. 1989. *Ap. J. Suppl.* 69:527–92
Howarth ID, Schmutz W. 1992. *Astron. Astrophys.* 261:503–22
Humphreys RM. 1979. *Ap. J.* 231:384–87
Humphreys RM. 1983a. *Ap. J.* 265:176–93
Humphreys RM. 1983b. *Ap. J.* 269:335–51
Humphreys RM. 1989. See Davidson et al 1989, pp. 3–14
Humphreys RM. 1991. See Leitherer et al 1991, pp. 45–47
Humphreys RM. 1992. See de Jager & Nieuwenhuijzen 1992, pp. 13–17
Humphreys RM, Davidson K. 1979. *Ap. J.* 232:409–20
Humphreys RM, McElroy DB. 1984. *Ap. J.* 284:565–77
Humphreys RM, Nichols M, Massey P. 1985. *Astron. J.* 90:101–8
Humphreys RM, Pennington RL, Jones TJ, Ghigo FD. 1988. *Astron. J.* 96:1884–907
Humphreys RM, Sandage AR. 1980. *Ap. J.* 44:319–81
Hutsemékers D. 1994. *Astron. Astrophys.* 281:L81–84
Iben I. 1974. *Annu. Rev. Astron. Astrophys.* 12:215–77
Iben I, Renzini A. 1983. *Annu. Rev. Astron. Astrophys.* 21:271–342
Iglesias CA, Rogers FJ. 1993. *Ap. J.* 412:752–60
Iglesias CA, Rogers FJ, Wilson BG. 1992. *Ap. J.* 397:717–28
Jones TJ, Humphreys RM, Gehrz RD, Lawrence GF, Zickgraf FJ, et al. 1993. *Ap. J.* 411:323–35
Joseph R. 1991. See Leitherer et al 1991, pp. 259–70
Jura M, Kleinmann SG. 1990. *Ap. J. Suppl.* 73:769–80
Kato M, Iben I. 1992. *Ap. J* 394:305–12
Keel WC. 1982. *Publ. Astron. Soc. Pac.* 94:765–68
Kennicutt RC Jr. 1984. *Ap. J.* 287:116–30
Kingsburgh RL, Barlow MJ. 1991. See van der Hucht & Hidayat, p. 101
Kingsburgh RL, Barlow MJ, Storey PJ. 1994. *Astrophys. Space Rev.* In press
Kippenhahn R, Weigert A. 1990. *Stellar Structure and Evolution.* Berlin, Heildberg:Springer-Verlag
Kirbiyik H. 1987. *Astrophys. Space Sci.* 136:321–30
Kiriakidis M, Fricke KJ, Glatzel W. 1993. *MNRAS* 264:50–62
Koesterke L, Hamann WR, Wessolowski U. 1992. *Astron. Astrophys.* 261:535–43
Kruger H, Fritze v. Alvensleben U, Fricke KJ, Loose H-H. 1992. *Astron. Astrophys.* 259:L73–76
Kudritzki RP. 1988. In *Radiation in Moving Gaseous Media, 18th Saas-Fee Course,* pp. 3–192. Geneva:Geneva Obs.
Kudritzki RP. 1994. *Space Sci. Rev.* In press
Kudritzki RP, Gabler R, Kunze D, Pauldrach AWA, Puls J. 1990. See Leitherer et al 1991, pp. 59–96
Kudritzki RP, Hummer DG. 1990. *Annu. Rev. Astron. Astrophys.* 28:303–45
Kudritzki RP, Hummer DG, Pauldrach AWA, Puls J, Najarro F, Imhoff J. 1992. *Astron. Astrophys.* 257:655–62
Kudritzki RP, Pauldrach A, Puls J. 1987. *Astron. Astrophys.* 173:293–98
Kudritzki RP, Pauldrach A, Puls J, Voels SR. 1991. In *The Magellanic Clouds, IAU Symp. 148,* ed. R Haynes and D Milne, pp. 279–84. Dordrecht:Kluwer
Kunth D, Sargent WLW. 1981. *Astron. Astrophys.* 101:L5–8
Lafon JPJ, Berruyer N. 1991. *Astron. Astrophys. Rev.* 2:249–89
Lambert DL. 1992. See de Jager & Nieuwenhuijzen 1992, pp. 156–70
Lambert DL, Brown JA, Hinkle KH, Johnson HR. 1984. *Ap. J.* 284:223–37
Lamers HJGLM. 1989. See Davidson et al 1989, pp. 135–47
Lamers HJGLM, de Groot MJH. 1992. *Astron. Astrophys.* 257:153–62
Lamers HJGLM, Fitzpatrick EL. 1988. *Ap. J.* 324:279–87
Lamers HJGLM, Leitherer C. 1993. *Ap. J.* 412:771–91
Lamers HJGLM, Maeder A, Schmutz W, Cassinelli JP 1991. *Ap. J.* 368:538–44
Lamers HJGLM, Noordhoek R. 1993. In *Massive Stars and Their Lives in the Interstellar Medium,* ed. JP Cassinelli, E Churchwell. *ASP Conf. Ser.* 35:517–21
Langer N. 1987. *Astron. Astrophys.* 171:L1–4
Langer N. 1989a. *Astron. Astrophys.* 210:93–113
Langer N. 1989b. *Astron. Astrophys.* 220:135–43
Langer N. 1991a. *Astron. Astrophys.* 243:155–59
Langer N. 1991b. *Astron. Astrophys.* 248:531–37
Langer N. 1991c. *Astron. Astrophys.* 252:669–88
Langer N. 1992. *Astron. Astrophys.* 265:L17–20
Langer N. 1993. In *Inside the Stars, IAU Colloq. 137,* ed. W Weiss, A Baglin. *ASP Conf. Ser.* 40:426–36
Langer N, El Eid MF, Baraffe I. 1989. *Astron. Astrophys.* 224:L17–20
Langer N, El Eid MF, Fricke KJ. 1985. *Astron. Astrophys.* 145:179–91
Langer N, El Eid MF, Fricke KJ. 1986. In *Nucleosynthesis and its Implication on Nuclear*

and *Particle Physics,* 20th Moriond Astrophys. Meet., ed. J Audouze, N Mathieu, pp. 177–87. Dordrecht:Reidel

Lattanzio JC, Vallenari A, Bertelli G, Chiosi C. 1991. *Astron. Astrophys.* 250:340–50

Leitherer C. 1990. *Ap. J. Suppl.* 73:1–20

Leitherer C. 1991. See Leitherer et al 1991, pp. 1–19

Leitherer C, Gruenwald R, Schmutz W. 1992b. In *Physics of Nearby Galaxies,* ed. TX Thuan et al, pp. 257–264. Gif-sur-Yvette:Editions Frontieres

Leitherer C, Lamers HJGLM. 1991. *Ap. J.* 373:89–99

Leitherer C, Langer N. 1991. In *The Magellanic Clouds, IAU Symp. 148,* ed. RF Hanes, DK Milne, pp. 480–82. Dordrecht:Kluwer

Leitherer C, Robert C, Drissen L. 1992a. *Ap. J.* 401:596–617

Leitherer C, Wolborn NR, Heckman TM, Norman CA, eds. 1991. *Massive Stars in Starbursts.* Cambridge:Cambridge Univ. Press

Lennon DJ, Kudritzki RP, Becker ST, Butler K, Eber F, et al. 1991. *Astron. Astrophys.* 252:498–507

Lequeux J. 1986. In *Spectral Evolution of Galaxies,* ed. C Chiosi, A. Renzini, pp. 57–73. Dordrecht:Reidel

Luck RE, Lambert DL. 1985. *Ap. J.* 298:782–802

Lucy, L, Abbott DC. 1993. *Ap. J.* 405:738–46

Lundström I, Stenholm B. 1984. *Astron. Astrophys. Suppl.* 58:163–92

Maeder A. 1980. *Astron. Astrophys.* 92:101–10

Maeder A. 1981. *Astron. Astrophys.* 102:401–10

Maeder A. 1983. *Astron. Astrophys.* 120:113–29

Maeder A. 1985. *Astron. Astrophys.* 147:300–8

Maeder A. 1987a. *Astron. Astrophys.* 173:247–62

Maeder A. 1987b. *Astron. Astrophys.* 178:159–69

Maeder A. 1989. See Davidson et al 1989, pp. 15–26

Maeder A. 1990. *Astron. Astrophys. Suppl.* 84:139–77

Maeder A. 1991a. *Astron. Astrophys.* 242:93–111

Maeder A. 1991b. In *Evolution of Stars: The Photospheric Abundance Connection, IAU Symp. 145,* ed. G Michaud, A Tutukov, pp. 221–33. Dordrecht:Kluwer

Maeder A. 1991c. *Q. J. R. Astron. Soc.* 32:217–23

Maeder A. 1992a. *Astron. Astrophys.* 264:105–20

Maeder A. 1992b. See de Jager & Nieuwenhuijzen 1992, pp. 138–44

Maeder A, Lequeux J, Azzopardi M. 1980. *Astron. Astrophys.* 90:L17–20

Maeder A, Meynet G. 1989. *Astron. Astrophys.* 210:155–73

Maeder A, Meynet G. 1994. *Astron. Astrophys.* In press

Masegosa J, Moles M, del Olmo A. 1991. *Astron. Astrophys.* 224:273–79

Mas-Hesse JM. 1992. *Astron. Astrophys.* 253:49–56

Mas-Hesse JM, Kunth D. 1991a. *Astron. Astrophys. Suppl.* 88:399–450

Mas-Hesse JM, Kunth D. 1991b. See van der Hucht & Hidayat 1991, pp. 613–18

Massey P. 1981. *Ap. J.* 246:153–60

Massey P. 1985. *Publ. Astron. Soc. Pac.* 97:5–24

Massey P, Armandroff TE. 1991. See van der Hucht & Hidayat 1991, pp. 575–86

Massey P, Armandroff TE, Conti PS. 1986. *Astron. J.* 92:1303–33

Massey P, Armandroff TE, Conti PS. 1992. *Astron. J.* 103:1159–65

Massey P, Conti PS. 1983. *Ap. J.* 273:576–89

Massey P, Conti PS, Armandroff TE. 1987a. *Astron. J.* 94:1538–55

Massey P, Conti PS, Moffat AFJ, Shara MM. 1987b. *Publ. Astron. Soc. Pac.* 99:816–31

Massey P, Garmany CD, Silkey M, Degioia-Eastwood. 1989a. *Astron. J.* 97:107–30

Massey P, Johnson J. 1993. *Astron. J.* 105:980–1001

Massey P, Parker JW, Garmany CD. 1989b. *Astron. J.* 98:1305–34

Massey P, Thompson AB. 1991. *Astron. J.* 101:1408–28

Mateo M. 1988. *Astron. J.* 331:261–93

McLeod KK, Rieke GH, Rieke MJ, Kelley DM. 1993. *Ap. J.* 412:99–110

Meylan G, Maeder A. 1982. *Astron. Astrophys.* 108:148–56

Meylan G, Maeder A. 1983. *Astron. Astrophys.* 124:84–88

Meynet G. 1994a. *Ap. J. Suppl.* In press

Meynet G. 1994b. *Astron. Astrophys.* In press

Meynet G, Maeder A, Schaller G, Schaerer D, Charbonnel C. 1994. *Astron. Astrophys. Suppl.* 103:97–105

Meynet G, Mermilliod JC, Maeder A. 1993. *Astron. Astrophys. Suppl.* 98:477–504

Mihalas D. 1969. *Ap. J.* 156:L155–58

Moffat AFJ. 1988. *Ap. J.* 330:766–75

Moffat AFJ, Niemela VS, Marraco HG. 1990. *Ap. J.* 348:232–41

Moffat AFJ, Niemela VS, Phillips MM, Chu YH, Seggewiss W. 1987. *Ap. J.* 312:612–25

Moffat AFJ, Shara MM. 1983. *Ap. J.* 273:544–61

Moffat AFJ, Shara MM. 1987. *Ap. J.* 320:266–82

Moffat AFJ, Shara MM, Potter M. 1991. *Astron. J.* 102:642–53

Moffat AFJ, Vogt N, Paquin G, Lamontagne R, Barrera LH. 1986. *Astron. J.* 91:1386–91

Morgan DH, Good AR. 1985. *MNRAS* 216:459–465

Morgan DH, Vassiliadis E, Dopita MA. 1991.

MNRAS 251:51p–53p

Napiwotzki R, Rieschick A, Blöcker T, Schönberner D, Wenske V. 1993. In *Inside the Stars, IAU Colloq. 137,* ed. W Weiss, A Baglin. *ASP Conf. Ser.* 40:461–63

Nasi E, Forieri C. 1990. *Astrophys. Space Sci.* 166:229–58

Nieuwenhuijzen H, de Jager C. 1988. *Astron. Astrophys.* 203:355–60

Nieuwenhuijzen H, de Jager C. 1990. *Astron. Astrophys.* 231:134–36

Noels A, Magain E. 1984. *Astron. Astrophys.* 139:341–43

Noels A, Maserel C. 1982. *Astron. Astrophys.* 105:293–95

Nugis T. 1991. See van der Hucht & Hidayat 1991, pp. 75–80

Osmer PS. 1972. *Ap. J. Suppl.* 24:255–82

Owocki SP, Castor J, Rybicki GB. 1988. *Ap. J.* 335:914–30

Pagel BEJ, Simonson EA, Terlevich RJ, Edmunds MG. 1992. *MNRAS* 255:325–45

Palla F, Stahler SW, Parigi G. 1993. In *Inside the Stars, IAU Colloq. 137,* ed. W Weiss, A Baglin. *ASP Conf. Ser.* 40:437–39

Parker JW. 1991. *30 Doradus in the Large Magellanic Cloud: The Stellar Content and Initial Mass Function.* PhD thesis. Univ. Colo., Boulder

Parker JW. 1993. *Astron. J.* 106:560–77

Parker JW, Garmany CD. 1993. *Astron. J.* 106:1471–83

Parker JW, Garmany CD, Massey P, Walborn NR. 1992. *Astron. J.* 103:1205–23

Parker RAR. 1978. *Ap. J.* 224:873–84

Pauldrach A, Kudritzki RP, Puls J, Butler K, Hunsinger J. 1994. *Astron. Astrophys.* In press

Pauldrach A, Puls J, Kudritzki RP. 1986. *Astron. Astrophys.* 164:86–100

Pauldrach A, Puls J, Kudritzki RP. 1988. In *O-Stars and WR Stars, NASA SP-497,* ed. PS Conti, AB Underhill, pp. 173–99. Washington:NASA

Penny LR, Gies DR, Hartkopf WI, Mason BD, Turner NH. 1993. *Publ. Astron. Soc. Pac.* 105:588–94

Peimbert M. 1986. *Publ. Astron. Soc. Pac.* 98:1057–60

Phillips AC, Conti PS. 1992. *Ap. J. Lett.* 395:91–93

Podsiadlowski Ph, Joss PC, Hsu JJL. 1992. *Ap. J.* 391:246–64

Polcaro V, Viothi R, Rossi C, Norci L. 1992. *Astron. Astrophys.* 265:563–69

Prantzos N, Hashimoto M, Nomoto K. 1990. *Astron. Astrophys.* 234:211–29

Reitermann A, Baschek B, Stahl O, Wolf B. 1990. *Astron. Astrophys.* 234:109–18

Rieke G. 1991. See Leitherer et al 1991, pp. 205–16

Robert C, Leitherer C. Heckman TM. 1993. *Ap. J.* 418:749–59

Roberts MS. 1962. *Astron. J.* 67:79–85

Rubin VC, Hunter DA, Ford WK Jr. 1990. *Ap. J.* 365:86–92

Sargent WLW. 1970 *Ap. J.* 160:405–27

Sargent WLW, Searle L. 1970 *Ap. J. Lett.* 162:155–60

Scalo J. 1989. In *Windows on Galaxies,* ed. G Fabbiano et al, pp. 125–40. Dordrecht:Kluwer

Schaerer D, Charbonnel C, Meynet G, Maeder A, Schaller G. 1993b. *Astron. Astrophys. Suppl.* 102:339–42

Schaerer D, Maeder A. 1992. *Astron. Astrophys.* 263:129–36

Schaerer D, Meynet G, Maeder A, Schaller G. 1993a. *Astron. Astrophys. Suppl.* 98:523–27

Schaerer D, Schmutz W. 1994. *Astron. Astrophys.* In press

Schaller G, Schaerer D, Meynet G, Maeder A. 1992. *Astron. Astrophys. Suppl.* 96:269–331

Schild H, Lortet MC, Testor G. 1991. See van der Hucht & Hidayat, pp. 479–84

Schild H, Maeder A. 1984. *Astron. Astrophys.* 136:237–42

Schild H, Smith LJ, Willis AJ. 1990. *Astron. Astrophys.* 237:169–77

Schild H, Testor G. 1991. *Astron. Astrophys.* 243:115–17

Schild H, Testor G. 1992. *Astron. Astrophys.* 266:145–49

Schmutz W, Hamann WR, Wessolowski U. 1989. *Astron. Astrophys.* 210:236–48

Schmutz W, Leitherer C, Hubeny I, Vogel M, Hamann WR, Wessolowski U. 1991. *Ap. J.* 372:664–82

Schmutz W, Leitherer C, Gruenwald R. 1993. *Publ. Astron. Soc. Pac.* 104:1164–72

Schönberner D, Herrero A, Becker S, Eber F, Butler K, et al. 1988. *Astron. Astrophys.* 197:209–22

Schulte-Ladbeck RE. 1989. *Astron. J.* 97:1471–79

Schwarzschild M. 1958. *Structure and Evolution of Stars,* p. 296. Princeton:Princeton Univ. Press

Shara MM, Moffat AFJ, Smith LF, Potter M. 1991. *Astron. J.* 102:716–43

Shields GA. 1990. *Annu. Rev. Astron. Astrophys.* 28:525–60

Shields GA, Tinsley BM. 1976. *Ap. J.* 203:66–71

Signore M, Dupraz C. 1990. *Astron. Astrophys.* 234:L15–18

Smith LF. 1968a. *MNRAS* 138:109–21

Smith LF. 1968b. *MNRAS* 140:409–33

Smith LF. 1968c. *MNRAS* 141:317–27

Smith LF. 1973. In *WR and High-Temperature Stars, IAU Symp. 49,* ed. MKV Bappu, J Sahade, pp. 15–41. Reidel:Dordrecht

Smith LF. 1988. *Ap. J.* 327:128–38

Smith LF. 1991a. See van der Hucht & Hidayat, pp. 601–10

Smith LF. 1991b. In *The Magellanic Clouds, IAU Symp. 148,* ed. R Haynes, D Milne, pp.

267–72. Dordrecht:Kluwer
Smith LF, Hummer DG. 1988. *MNRAS* 230:511–34
Smith LF, Maeder A. 1989. *Astron. Astrophys.* 211:71–80
Smith LF, Maeder A. 1991. *Astron. Astrophys.* 241:77–86
Smith LF, Meynet G, Mermilliod J-C. 1994. *Astron. Astrophys.* In press
Smith LF, Shara MM, Moffat AFJ. 1990. *Ap. J.* 348:471–84
Sreenivasan SR, Wilson WJF. 1985. *Ap. J.* 290:653–59
St Louis N, Moffat AFJ, Drissen L, Bastien P, Robert C. 1988. *Ap. J.* 330:286–304
Stahl O. 1986. *Astron. Astrophys.* 164:321–27
Stahl O. 1987. *Astron. Astrophys.* 182:229–36
Stahl O, Wolf B, Klare G, Cassatella A, Krauther J, et al. 1983. *Astron. Astrophys.* 127:49–62
Stencel RE, Pesce JE, Bauer WH. 1989. *Astron. J.* 97:1120–38
Stothers RB. 1991a. *Ap. J.* 381:L67–70
Stothers RB. 1991b. *Ap. J.* 383:820–36
Stothers RB, Chin CW. 1977. *Ap. J.* 211:189–97
Stothers RB, Chin CW. 1979. *Ap. J.* 233:267–79
Stothers RB, Chin CW. 1983. *Ap. J.* 264:583–93
Stothers RB, Chin CW. 1990. *Ap. J. Lett.* 348:21–24
Stothers RB, Chin CW. 1991. *Ap. J.* 374:288–90
Stothers RB, Chin CW. 1992a. *Ap. J. Lett.* 390:L33–35
Stothers RB, Chin CW. 1992b. *Ap. J.* 390:136–43
Testor G, Schild H. 1990. *Astron. Astrophys.* 240:299–304
Thé PS, Arens M, van der Hucht KA. 1982. *Astrophys. Lett.* 22:109–18
Torres AV. 1988. *Ap. J.* 325:759–67
Truran JW, Weiss A. 1987. In *SN 1987A*, ed. IJ Danziger, pp. 271–282. Garching:ESO Workshop
Tuchman J, Wheeler JC. 1989. *Ap. J.* 344:835–43
Tuchman J, Wheeler JC. 1990. *Ap. J.* 363:255–64
Turolla R, Nobili L, Calvani M. 1988. *Ap. J.* 324:899–906
Tutukov AV, Yungelson LR. 1985. *Sov. Astron.* 29:352
Underhill AB. 1949. *MNRAS* 109:562–70
Vacca WD. 1991. *Wolf-Rayet Stars in the Milky Way, the Large Magellanic Cloud, and Emission-Line Galaxies.* PhD thesis. Univ. Colo., Boulder
Vacca WD. 1994. *Ap. J.,* in press
Vacca WD, Conti PS. 1992. *Ap. J.* 401:543–58
Vacca WD, Conti PS, Leitherer C, Robert C. 1994. In prep.
Vanbeveren D. 1988. *Astrophys. Space Sci.* 149:1–12

Vanbeveren D. 1991. *Astron. Astrophys.* 252:159–71
Vanbeveren D. 1994. *Astrophys. Space Sci.* In press
Vanbeveren D, Conti PS. 1980. *Astron. Astrophys.* 88:230–39
Vanbeveren D, de Loore C. 1993. In *Massive Stars : Their Lives in the Interstellar Medium,* ed. JP Cassinelli, ER Churchwell. *ASP Conf. Ser.* 35:257–59
van den Bergh S. 1968. *J. R. Astron. Soc. Can.* 62:69
van der Hucht KA. 1991. See van der Hucht & Hidayat 1991, pp. 19–36
van der Hucht KA. 1992. *Astron. Astrophys. Rev.* 4:123–59
van der Hucht KA, Hidayat B, eds. 1991. *Wolf-Rayet Stars and Interrelations with Other Massive Stars in Galaxies, IAU Symp. 143.* Dordrecht:Kluwer
van der Hucht KA, Hidayat B, Admiranto AG, Supelli KR, Doom C. 1988. *Astron. Astrophys.* 199:217–34
Venn KA. 1993. *Ap. J.* 414:316–32
Viotti R, Polcaro VF, Rossi C. 1993. *Astron. Astrophys.* 276:432–44
Voels SA, Bohannan B, Abbott DC, Hummer DG. 1989. *Ap. J.* 340:1073–190
Vrancken M, de Greve JP, Yungelson L, Tutukov A. 1991. *Astron. Astrophys.* 249:411–16
Walborn NR. 1976. *Ap. J.* 205:419–25
Walborn N. 1988. In *Atmospheric Diagnostics of Stellar Evolution, IAU Colloq. 108,* ed. K Nomoto, pp. 70–78. Berlin, Heidelberg:Springer-Verlag
Walsh JR, Roy J-R. 1987. *Ap. J. Lett.* 319:57–62
Whitmore BC, Schweizer F, Leitherer C, Borne K, Robert C. 1993. *Astron. J.* 106:1354–70
Willis AJ. 1987. *Q. J. R. Astron. Soc.* 28:217–24
Willis AJ. 1991. In *Evolution of Stars: The Photospheric Abundance Connection, IAU Symp. 145,* ed. G Michaud, A Tutukov, pp. 195–207. Dordrecht:Kluwer
Willis AJ. 1994. *Astrophys. Space Sci. Rev.* In press
Willis AJ, Schild H, Smith LJ. 1992. *Astron. Astrophys.* 261:419–32
Wolf B. 1989. *Astron. Astrophys. Suppl.* 217:87–91
Wolf B, Appenzeller I, Stahl O. 1981. *Astron. Astrophys.* 103:94–102
Wolf B, Stahl O, Smolinski J, Cassatella A. 1988. *Astron. Astrophys. Suppl.* 74:239–45
Wood DOS, Churchwell E. 1989. *Ap. J.* 340:265–72
Woosley SE, Langer N, Weaver TA. 1993. *Ap. J.* 411:823–39
Wray JD, Corso GJ. 1972. *Ap. J.* 172:577–82
Yorke HW. 1986. *Annu. Rev. Astron. Astrophys.* 24:49–87

Annu. Rev. Astron. Astrophys. 1994. 32: 277–318

COOLING FLOWS IN CLUSTERS OF GALAXIES

A. C. Fabian

Institute of Astronomy, Madingley Road, Cambridge CB3 0HA,
United Kingdom

KEY WORDS: clustering, galaxy formation, radio sources, X rays

INTRODUCTION

The gas clouds out of which galaxies, groups, and clusters form were heated by
the energy released during their initial gravitational collapse. Some of the gas
then cooled to form the objects readily observed today. In the case of ordinary
galaxies much of the gas cooled rapidly, but in massive galaxies and clusters
the gas cooled more slowly and a quasi-hydrostatic atmosphere formed. The
mass of uncooled hot gas exceeds that in visible stars in groups and clusters of
galaxies. The temperature of the atmosphere is close to the virial temperature
(typically several million K and greater) and is directly observable only in the
X-ray waveband. The hot atmosphere continues to lose energy by the emission
of radiation (principally X rays); in the central region, where the atmosphere is
naturally densest, a cooling flow forms. This region, which grows with time, is
where the cooling rate of the gas is sufficiently high that the gas particles lose
their energy to radiation. The weight of the overlying gas then causes a slow,
subsonic inflow; inhomogeneities in the gas cause cooled matter to drop out
and form cold clouds or stars throughout the flow. The net result is that cold
matter continues to be slowly deposited over a large volume in the core of the
massive galaxy, group, or cluster.

This idealized picture represents the main features of cooling flows, which
are found to be common in massive elliptical galaxies, groups, and clusters
of galaxies through X-ray observations. The surface brightness and spectrum
of the X rays show that the gas particles lose much of their thermal energy to
X-radiation in the central region of many of these objects, implying that the
cooling process necessary for the formation of galaxies has extended until the

0066–4146/94/0915–0277$05.00

present time and that the central massive galaxies are continuing to grow now. Most of this growth is not in terms of visible stars, however, since the radiative cooling rates in nearby rich clusters of 10s to 100s $M_\odot$ yr^{-1} would then make the central galaxies much bluer and brighter than even a superficial optical inspection allows. The cooled gas must somehow condense into optically dark objects, or some other process must come into play. Detailed observations at optical and other wavebands do reveal anomalies in the centers of cooling flows and confirm that the gas has a high density, a high pressure, and short cooling time, but so far do not reveal the cooled component. The only thing we know for certain is that gas is leaving the hot phase, where it is detectable in X rays, and is becoming some form of dark matter.

After a very brief overview of the intracluster medium, the X-ray properties of nearby cooling flows are discussed in some detail. We review the evidence, derived from X-ray images and spectra, that the gas really cools and that cooling flows are common. Therefore they must be long-lived and steady. Recent data clearly show the temperature of the gas dropping towards the center of many clusters. The spectra also require X-ray absorption, in excess of Galactic line-of-sight values, consistent with widespread cold gas which may be one signature of the cooled component of the flow. After reviewing the appearance of cooling flows in other wavebands, we discuss some theoretical issues on cooling flows, such as the lack of thermal conduction and the required multiphase nature of the gas. Much of the behavior of the cooling gas is very uncertain once it has dropped below X-ray emitting temperatures, with magnetic fields surely important, and in place of a secure theoretical discussion, we outline one possible picture of what is occurring. Finally we move onto distant cooling flows and their evolution before returning to galaxy formation.

The Hot Intracluster Medium

Enormous quantities of diffuse hot gas are observed in clusters and groups of galaxies. The gas can only support itself close to hydrostatic equilibrium in the gravitational field of a virialized cluster if its sound speed is similar to the typical velocity of a cluster galaxy (the velocity dispersion of the cluster). This is generally in the range 300 to 1200 km s^{-1}, implying that the gas temperature is 10^7–10^8 K. The main energy loss of gas at such high temperatures is through bremsstrahlung radiation, which produces the diffuse X-radiation from clusters of galaxies and is our principal source of information on their intracluster medium (ICM). Further indirect evidence for the gas is found in "head-tail" radio sources and from theories of the propagation of double-lobed radio sources. The general properties of the ICM have been reviewed by Sarazin (1986, 1988, 1992) and Fabian (1988b).

Most of the *observed* intracluster gas has an electron density, n_e, in the range of 10^{-4}–10^{-2} cm^{-3} and a temperature $T \sim 2 \times 10^7$–10^8 K, and is contained within a radius of 1 to 2 Mpc.[1] The total mass of gas in rich clusters ranges from 5×10^{13}–$5 \times 10^{14} M_\odot$ with an X-ray luminosity of $\sim 10^{43}$–3×10^{45} erg s^{-1}. Emission lines due to highly ionized iron are observed in all clusters that are bright enough to permit detection (e.g. see Rothenflug & Arnaud 1985, Edge 1989) showing that the gas has ~ 0.3 times solar abundance in iron (the abundance appears to be higher in low temperature clusters). The work of Canizares et al (1979, 1982, 1988), Mushotzky et al (1981), and Mushotzky (1992) on cooling regions in clusters shows that Si and S are also present at abundances close to solar, and that O may exceed solar abundance.

The origin of the gas is uncertain. Being metal-enriched, it cannot all be primordial. Some of the gas must have been processed through an early population of stars before being released back into intracluster space in supernova explosions (Larson & Dinerstein 1975, Mathews 1989, M. Arnaud et al 1992). It could also have been stripped from young galaxies during the formation of the cluster. In all cases, the gas receives a similar kinetic energy per unit mass as the galaxies, which makes the sound speed similar to the galaxy velocities. This energy is ultimately gravitational in origin, and there is no need for additional heating of the gas to account for its temperature and distribution. The presence of so much gas, comparable to, or exceeding, the total mass in observable stars and 10–30% of the virial mass of the cluster, suggests that galaxy formation is no more than 50% efficient (e.g. see David et al 1990).

OBSERVATIONAL EVIDENCE FOR COOLING FLOWS

The intracluster gas is, of course, densest in the core of a cluster and therefore the radiative cooling time, t_{cool}, due to the emission of the observed X rays, is shortest there. (For X-radiation from hot gas, $t_{cool} \propto T^\alpha / n$, where $-1/2 \lesssim \alpha \lesssim 1/2$ and n is the gas density.) A cooling flow is formed when t_{cool} is less than the age of the system, say $t_a \sim H_0^{-1}$. In the cases considered here, t_{cool} exceeds the gravitational free-fall time, t_{grav}, within the cluster, so

$$t_a > t_{cool} > t_{grav}, \tag{1}$$

and gas can be considered to be in quasi-hydrostatic equilibrium. The flow takes place because the gas density has to rise to support the weight of the overlying gas. It is essentially pressure-driven.

To see this clearly, consider the gaseous atmosphere trapped in the gravitational potential well of a cluster or galaxy to be divided into two parts at the radius r_{cool}, where $t_{cool} = t_a$. The gas pressure at r_{cool} is determined by the weight of the overlying gas, in which cooling is not important. Within r_{cool}, cooling reduces the gas temperature and the gas density must rise in order to maintain the pressure at r_{cool}. The only way for the density to rise (ignoring

---

[1] A Hubble constant of $H_0 = 50$ km s^{-1}Mpc^{-1} is assumed here and throughout this review.

matter sources within r_{cool}, which is a safe assumption in a cluster of galaxies) is for the gas to flow inward. This *is* the cooling flow. Although in principle the cooling could instead be balanced by sources of heat, we show below that this is neither physically nor astrophysically plausible, nor is it consistent with observations of cooler, X-ray line-emitting gas.

If the initial gas temperature exceeds the virial temperature of a central galaxy (which is generally the case for rich clusters but not for poor ones or individual galaxies) then the gas continues to cool as it flows inward. When the gas temperature has dropped to the virial temperature of the central galaxy, adiabatic compression of the inflowing gas under the gravitational field of the galaxy, i.e. the release of gravitational energy, counterbalances the radiative heat loss and can sustain, or even raise, the gas temperature as it flows further inward. Radiative heat loss causes a continual reduction in the entropy of the inflowing gas, but not necessarily in its temperature. The gas temperature can eventually drop catastrophically in the core of the galaxy if the gravitational potential flattens there. The net result is that the gas within r_{cool} radiates its thermal energy plus the $Pd V$ work done on it as it enters the region and the gravitational energy released within r_{cool}.

This is the behavior of an idealized, spherically symmetric, homogeneous cooling flow, in which the gas has a unique temperature and density at each radius. X-ray observations of real cooling flows indicate that they are inhomogeneous and must consist of a mixture of temperatures and densities at each radius. The homogeneous flow still gives fair approximations for many properties of the mean flow. The physics of the cooling flow mechanism is very simple, although the details of its operation are not.

The primary evidence for cooling flows comes from X-ray observations. There is less evidence at other wavelengths, and none of it supports the large mass deposition rates of 100s $M_\odot \, yr^{-1}$ inferred from the X-ray data for some clusters. We discuss this point more fully later, but it should be stressed that large amounts of distributed low-mass star formation need not be detectable at other wavelengths if the gas is initially at X-ray emitting temperatures. This is, perhaps, the crux of the controversial aspect of cooling flows: It is difficult to prove or disprove their existence with observations in wavebands other than the X-ray. The lack of clear evidence at other wavebands does not make the X-ray evidence any less compelling, though it does challenge our observational and theoretical ingenuity in those other wavebands.

It was *Uhuru* observations of clusters that first showed the mean cooling time of the gas in the cores of clusters to be close to a Hubble time (Lea et al 1973). These, and other early X-ray measurements described later, and theoretical considerations, led Cowie & Binney (1977), Fabian & Nulsen (1977), and Mathews & Bregman (1978) to independently consider the effects of significant cooling of the central gas, i.e. cooling flows. The process was noted by

Silk (1976) as a mechanism for the formation of central cluster galaxies from intracluster gas at early epochs and as a mechanism for general galaxy formation by Gold & Hoyle (1958). General reviews of cooling flows have been made by Fabian et al (1984b, 1991) and Sarazin (1986, 1988) and some other points of view may be found in the Proceedings of a NATO Workshop (Fabian 1988a).

X-Ray Imaging Evidence for Cooling Flows

A sharply peaked X-ray surface brightness distribution is indicative of a cooling flow, because it shows that the gas density is rising steeply towards the center of the cluster or group. Since the observed surface brightness depends upon the square of the gas density and only weakly on the temperature, this result is not model dependent. The high central density indicates a short cooling time.

Most of the images have been obtained with the *Einstein Observatory, EX-OSAT,* and *ROSAT,* although the peaked X-ray surface brightness was anticipated with data from the *Copernicus* satellite (Fabian et al 1974, Mitchell et al 1975), from rocket-borne telescopes (Gorenstein et al 1977), and from the modulation collimators on *SAS 3* (Helmken et al 1978).

The fraction of clusters with high central surface brightness is large, which means that cooling flows must be both common and long-lived. More than 30 to 50% of the clusters well-detected with the *Einstein Observatory* (Stewart et al 1984b, Arnaud 1988) have surface brightnesses that imply $t_{cool} < H_0^{-1}$ within the central 100 kpc or so. This fraction is certainly an underestimate, because the $\sim 1'$ angular resolution of the images dilutes the central surface brightness. Additional data from *EXOSAT* (Edge & Stewart 1991a,b) show that more than two thirds of the 50 X-ray brightest clusters in the sky (see list in Lahav et al 1989) have cooling flows (Edge et al 1992). Since this last sample (Table 1) is based on the total flux, to which the cooling flow makes only a minor contribution, the high fraction is not a selection effect. This fraction is also an underestimate, since many of the remaining clusters in the sample have not been imaged; thus their status is undefined. Whether H_0^{-1} should be used for t_a is debatable, but inspection of the results shows that reducing t_a by 2, say, does not much change the fraction of clusters containing cooling flows. The overall picture is that the prime criterion for a cooling flow, $t_{cool} < 10^{10}$ yr, is satisfied in a large fraction (~ 70–80%) of clusters. It is also satisfied in a number of poor clusters and groups (Schwartz et al 1980, Canizares et al 1983, Singh et al 1986, 1988, Schwartz et al 1991, Mulchaey et al 1993, Ponman & Bertram 1993). Cooling flow conditions also occur in large, isolated elliptical galaxies (Nulsen et al 1984, Canizares et al 1987). The largest catalogue of cooling flows to date is by White (1992) who has analyzed the images of most clusters observed with the *Einstein Observatory* (the selection criteria used for such observations means that this sample is not complete in the statistical sense).

Table 1   Cooling flow properties of the 55 X-ray brightest clusters[a]

Cluster	Redshift	O	R	Flux[b]	L_X[c]	Central Cooling Time (10^9 yr)	Bin Size (kpc)	Mass Flow Rate ($M_\odot$ yr^{-1})
A426	0.0183	√	√	75.0	110	0.48 ± 0.02	31 I	183
Ophiuchus	0.028	—	—	44.5	152	3.1 ± 7.4	29 L	75
Coma	0.0232	×	0	32.0	75	95 ± 239	47 I	0*
Virgo	18 Mpc	√	√	30.0	1	0.03 ± 0.01	3 L	10
A2319	0.0564	×	×	12.1	170	13.1 ± 2.2	73 I	66
A3571	0.0391	—	—	11.5	77	7.8 ± 12.9	35 L	79
Centaurus	0.0109	√	√	11.2	6	0.49 ± 0.10	10 L	18
Tri. Aust.	0.051	—	—	11.0	126	65 ± 254	160 L	0
3C129	0.022	—	√	9.59	20	5.1 ± 4.5	44 L	61
AWM7	0.0172	×	—	9.14	12	3.9 ± 1.2	65 I	42
A754	0.0542	×	×	8.53	105	9.5 ± 12.5	48 L	24
A2029	0.0767	×	√	7.52	197	3.8 ± 0.7	99 I	402
A2142	0.0899	×	—	7.50	272	3.0 ± 0.8	48 H	188
A2199	0.0300	√	√	7.12	30	2.4 ± 1.8	42 L	150
A3667	0.0530	×	—	6.68	83	165 ± 236	112 I	0*
A478	0.0882	√	√	6.63	241	2.3 ± 0.5	70 L	570
A85	0.0521	√	√	6.37	75	3.8 ± 0.6	69 I	236
A3266	0.0594	×	—	5.90	92	21.5 ± 17.3	70 H	10
A401	0.0748	×	×	5.88	147	20.1 ± 5.9	92 I	12
0745-191	0.1028	√	√	5.87	280	2.1 ± 0.5	32 L	702
A496	0.0330	√	√	5.67	25	2.1 ± 0.3	43 I	112
A1795	0.0627	√	√	5.30	89	2.5 ± 0.4	81 I	478
A2256	0.0581	×	0	5.20	83	175 ± 38	153 I	0*
Cygnus-A	0.057	√	√	4.78	69	4.2 ± 0.5	92 I	187
2A0335+096	0.0349	√	√	4.67	25	0.90 ± 0.11	35 L	142
A1060	0.0124	√	×	4.36	2	2.2 ± 3.3	16 L	9
A3558	0.0478	×	—	4.21	42	—	—	50
A644	0.0704	×	×	4.15	92	8.7 ± 1.1	146 I	326
A1651	0.0846	×	—	3.67	112	—	—	0
A3562	0.0499	×	—	3.52	39	4.5 ± 4.8	25 L	45
A1367	0.0215	×	0	3.45	7	21.5 ± 3.7	92 I	0
A399	0.0715	×	×	3.41	78	25.4 ± 5.0	139 I	0
A2147	0.0356	—	√	3.28	18	15.2 ± 24.5	83 L	54
A119	0.0440	×	—	3.03	26	14.6 ± 9.3	86 L	23
A3158	0.0590	×	—	2.99	46	23.2 ± 3.8	197 I	0
Hydra-A	0.0522	√	√	2.90	35	1.8 ± 0.7	50 L	315
A2052	0.0348	√	√	2.66	14	1.1 ± 0.2	40 L	90
A2063	0.0350	×	√	2.64	13	4.1 ± 1.0	68 I	45
A1644	0.0474	×	—	2.60	24	13.5 ± 4.6	59 I	19

Table 1 *continued*

Cluster	Redshift	O	R	Flux[b]	Lx[c]	Central Cooling Time (10⁹ yr)	Bin Size (kpc)	Mass Flow Rate ($M_\odot$ yr^{-1})
A2065	0.0721	—	—	2.79	65	33.3 ± 8.2	136 I	0
Klemola44	0.0283	—	—	2.55	9	—	—	—
A262	0.0164	√	√	2.35	3	0.87 ± 0.40	23 I	47
A2204	0.1523	—	—	2.20	235	—	—	—
A2597	0.0824	√	√	2.09	64	11.1 ± 2.0	185 I	480
A1650	0.0845	—	×	2.07	66	54 ± 103	160 I	0
A3112	0.0746	×	—	1.95	48	2.1 ± 1.0	82 L	430
A3532	0.0585	—	—	1.95	30	44.1 ± 14.9	141 I	0
A4059	0.0478	√	—	1.88	19	3.5 ± 8.5	35 L	124
A3391	0.0545	×	—	1.79	23	31.8 ± 5.7	140 I	0
MKW3s	0.0449	√	√	1.79	15	4.0 ± 0.9	77 I	151
A1689	0.181	×	—	1.75	268	13.8 ± 2.3	190 I	164
A576	0.0381	×	√	1.72	11	15.1 ± 3.3	77 I	6
A2244	0.1024	—	×	1.71	81	12.2 ± 2.0	120 I	82
A2255	0.0809	×	—	1.71	49	143 ± 268	97 I	0*
A1736	0.046	×	—	1.70	16	32.9 ± 7.5	115 I	0

[a]From Edge et al (1992). A dash in any column means that the information is not available. The statistical uncertainty in the estimate of $\dot{M}$ is typically about 20%; the systematic uncertainty, given errors on the gravitational potential in the core of a cluster, may be up to a factor of 2. Whether a (focused) cooling flow exists or not can be judged from the estimates of the central cooling time, which should be less than about 10^{10} yr for a flow. The cooling time depends on the size of the innermost bin used in analyzing the cluster image, which in turn depends on the signal-to-noise of the data, the distance to the cluster, and the resolution of the instrument (I for IPC, H for HRI, L for CMA). As discussed by Edge et al, an estimate of the cooling time of the gas within an image bin depends directly on the size of that bin (strongly if a cooling flow is present). Those clusters where the bin size is larger than about 100 kpc and $\dot{M}$ is listed as zero may still therefore contain a modest cooling flow. Further observations are needed there. A small $\dot{M}$ occurring in a gravitationally-unfocused way could occur in even the Coma-like clusters.

[b]2–10 keV flux in units of 10^{-11} erg cm^{-2} s^{-1}.

[c]2–10 keV luminosity in units of $10^{43}h_{50}^{-2}$ erg s^{-1}.

Asterisks in the final column indicate those clusters where there is no focused cooling flow taking place. A tick in the O column means optical line emission is detected, a tick in the R column means radio emission is detected from the central galaxy (Ball et al 1993 and references therein).

The mass deposition rate, $\dot{M}$, due to cooling (i.e. the accretion rate, although this is a poor term since most of the gas does not flow in far from r_{cool}) can be estimated from the X-ray images by using the luminosity associated with the cooling region (i.e. L_{cool} within r_{cool}) and assuming that it is all due to the radiation of the thermal energy of the gas, plus the PdV work done on the gas as it enters r_{cool}:

$$L_{cool} = \frac{5}{2} \frac{\dot{M}}{\mu m} kT, \tag{2}$$

where T is the temperature of the gas at r_{cool}. L_{cool} is similar (but not identical) to the central excess luminosity defined by Jones & Forman (1984); it ranges from $\sim 10^{42}$ to $> 10^{44}$ erg s^{-1} and generally represents $\sim 10\%$ of the total cluster luminosity. Values of $\dot{M} = 50$–$100 M_\odot$ yr^{-1} are fairly typical for cluster cooling flows. Some clusters show $\dot{M} \gtrsim 500 M_\odot$ yr^{-1} (e.g. A478, PKS0745, A1795, A2597, A2029, and Hydra A). The main uncertainties in the determination of $\dot{M}$ lie in the gravitational contribution to L_{cool} and the appropriate choice for t_a. Assuming $t_a \sim 10^{10}$ yr, the estimates of $\dot{M}$ are probably accurate to within a factor of 2 (Arnaud 1988). Empirically, we find that the deduced $\dot{M}$ is roughly proportional to $t_a^{1/3}$ (see Figure 2), which means that reducing t_a to 10^9 yr introduces only a factor ~ 2 reduction in $\dot{M}$.

Since we often measure an X-ray surface brightness profile for the cluster core (where the X-ray emission is well-resolved), we have $L_{cool}(r)$ which can be turned into $\dot{M}(r)$, the integral mass deposition rate within radius r. Generally, the surface brightness profiles are less peaked than they would be if all the gas were to flow to the center, giving roughly

$$\dot{M}(r) \propto r. \tag{3}$$

This means that the gas must be inhomogeneous, so that some of the gas cools out of the hot flow at large radii and some continues to flow inward.

The actual computation of $\dot{M}(r)$ is complicated, since we need to take into account how the gas cools and any gravitational work done on it. The thermal energy of the gas is generally more important than gravitational energy release in clusters (T_{gas} is several times greater than the virial temperature of the central galaxy), so a simple analysis gives a fair approximation to the profile (see Fabian et al 1986; Thomas et al 1987; White & Sarazin 1987a,b,c, 1988). Even if clusters have small core radii (say < 100 kpc), thereby significantly reducing L_{cool}, then the fraction with cooling flows would not be more than halved. In the case of A478, a comparison of the temperature decrease observed in the *ROSAT* Position Sensitive Proportional Counter (PSPC) spectrum with that inferred from the image shows that the core radius in that rich cluster is not small, and must exceed 200 kpc (Allen et al 1993).

The *ROSAT* High Resolution Imager (HRI) shows that the X-ray surface brightness profile in most clusters continues to rise inward (Figure 1) to within the inner 10 kpc. In A478, the cooling time is seen to drop below 4×10^8 yr at that radius and the total mass deposition rate $\dot{M} \approx 800 M_\odot$ yr^{-1} (Figure 2 taken from White et al 1994).

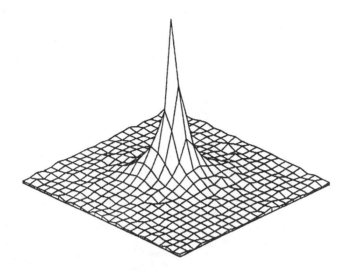

Figure 1   3-dimensional representation of the X-ray surface brightness of the A478 cluster as seen by the *ROSAT* HRI (White et al 1994). The pixel size is 24 arcsec which corresponds to about 60 kpc. The cooling flow extends to at least 200 kpc radius, incorporating most of the prominent peak in the figure. Much of the cluster emission lies beyond the 1.3 × 1.3 Mpc area shown here.

In the case of the Perseus cluster, the cooling flow peaks away from the nucleus of the central galaxy NGC 1275, in an arc about 15 kpc away (Figure 3 from Böhringer et al 1993). The radio source in the galaxy appears to be holding off the flow and creating two regions of low X-ray surface brightness either side of the nucleus, coincident with the radio lobes. A similar effect is seen in the tails of the radio lobes of Cygnus A (Harris et al 1994). Sarazin et al (1992a,b) have claimed the detection of X-ray filaments in HRI observations of the flows in A2029 and 2A0335 + 096. Although the central structure in 2A0335 + 096 is convincing, the filamentary structures in A2029 were not found to be significant in a re-analysis of the same image by White et al (1994). They may be artifacts of the small ellipticity of the emission since spurious features can appear significant if it is assumed that the underlying emission has circular symmetry. A linear structure identified as a cooling wake is seen in a PSPC image of the NGC5044 group (David et al 1994).

X-Ray Spectral Evidence for Cooling Flows

Key evidence that the gas actually does cool to lower temperatures than inferred from the X-ray images, is given by moderate to high resolution spectra of the cluster cores. Canizares et al (1979, 1982), Canizares (1981), Mushotzky et al (1981), and Lea et al (1982) used the Focal Plane Crystal Spectrometer (FPCS)

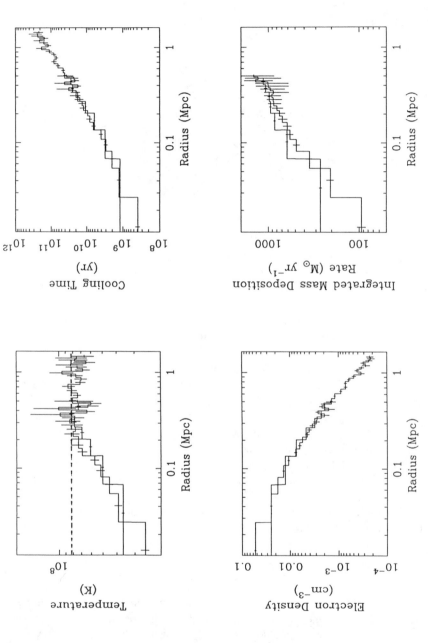

Figure 2 Properties of the ICM in A478 obtained by deprojecting the *ROSAT* HRI and PSPC images (small and larger bins, respectively) and solving for the density and temperature (White et al 1994, Allen et al 1994). The dotted line on the temperature plot indicates the mean temperature determined from the *Ginga*

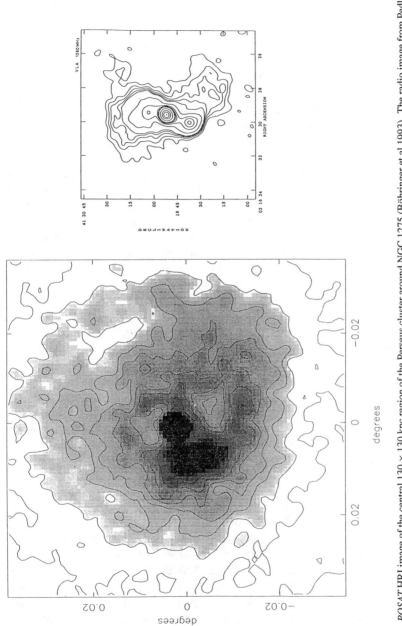

Figure 3 *ROSAT* HRI image of the central 130 × 130 kpc region of the Perseus cluster around NGC 1275 (Böhringer et al 1993). The radio image from Pedlar et al (1990) is shown at the right (to the same scale); the nuclei in both pictures superpose directly to show that the outer radio lobes fit in the X-ray dips N and S of the nucleus. The brightest X-ray patch lies about 15–20 kpc SE from the nucleus and does not correspond to features seen at other wavelengths, except that it may be the cause of the bend in the S radio lobe (note the radio contours) and lies just beyond the optical "blue loop" (see the figure in Sandage 1971).

and the Solid State Spectrometer (SSS) on the *Einstein Observatory* to show that there are low temperature components in the Perseus and Virgo clusters, consistent with the existence of cooling flows. Detailed examination of the line fluxes and of the emission measures of the cooler gas by Canizares et al (1988) and Mushotzky & Szymkowiak (1988) shows that, in the case of the Perseus cluster, the gas loses at least 90% of its thermal energy and that the mass deposition rates are in agreement with those obtained from the images. Good agreement is obtained also in several other clusters. The SSS and FPCS results for M87 show that the emission measure varies with temperature in the manner expected from a cooling gas. The spectral evidence generally contradicts alternative explanations in which the observed radiative cooling is balanced by some heat source.

The cooling time of the gas in the Perseus cluster which emits the FeXVII line ($T < 5 \times 10^6$ K) is less than 3×10^7 yr. Since the emission measure of this gas agrees with that inferred from the gas cooling at the higher temperatures that dominate the images and the SSS result, we conclude that the flow is steady and long-lived, i.e. $t_a \simeq H^{-1}$ (Nulsen 1988). The shape of the continuum and line spectrum observed with the SSS is consistent with the same mass deposition rate at all X-ray temperatures, further supporting this conclusion. Cooling flows cannot be some intermittent or transient phenomenon only a billion years old.

An important development has been the discovery of excess X-ray absorption in several cooling flows. This has arisen from spectral analysis of *Ginga* and the *Einstein Observatory* SSS data, particularly of A478 (White et al 1991, Johnstone et al 1992). The strong cooling flow in that cluster found from images should have led to an excess of soft emission detectable in the *Ginga* spectrum even above 2 keV. No such excess was apparent. Instead the SSS data showed X-ray absorption of the cooling flow, which considerably weakened the emergent soft excess. Excess absorption was required for good spectral fits to be obtained for many other cooling flows. The effect is one of line and continuum absorption and could not be explained by an increase in line emission (White et al 1991). To obtain fits to SSS data requires inclusion of a model for the ice that built up on the detector window, and was periodically removed by heating. Spectral fits to SSS data from active galaxies, for example, do not require excess absorption to be included for the objects themselves, so there is reasonable confidence that the values added to the cluster spectra represent absorbing material intrinsic to the clusters, although the precise values for the column density of absorbing matter have a large uncertainty.

Later data from the Broad Band X-ray Telescope (BBXRT) on the *ASTRO-1* mission confirm the need for excess X-ray absorption for several clusters (K. Arnaud et al 1992, Mushotzky 1992). The data for the Perseus cluster also show that the emission measure of the cooling component builds up towards the center of the flow in the manner expected from distributed mass deposition.

High-quality images from *ROSAT* pointed observations reveal the X-ray structure and spectra of cooling flows in new detail. Observations with the PSPC show that the gas temperature drops progressively inward within the cooling flow, with a gradient consistent with the results obtained from simple deprojection analyses (e.g. Schwarz et al 1992, Allen et al 1993, Allen & Fabian 1994). In the Centaurus cluster the temperature drops from the cluster mean of 3.5 keV to about 1 keV or less at the smallest radii measured. In the case of the massive flow in A478, the temperature drops from the cluster mean of $kT = 6.8\,keV$ at 300 kpc to below 3 keV within 50 kpc (Figure 4). At the same time the absorbing column density required for a good fit to the data increases to more than 10^{21} cm^{-2} above the value required at outer radii. Note that such a column density spread over a radius of 100 kpc implies a total mass exceeding 10^{11} $M_{\odot}$, assuming solar abundance ratios (the principal absorber in the PSPC band observed is oxygen). Simple considerations of the enormous emission expected if the absorber material were at intermediate temperatures

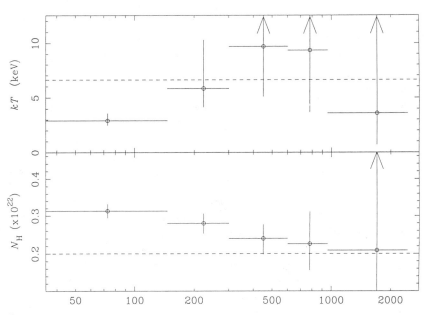

Figure 4 Gas temperature and column density measured as a function of radius from the *ROSAT* PSPC data of A478 (Allen et al 1993). The innermost gas in the cluster has clearly cooled well below the mean temperature obtained from the *Ginga* broad-beam spectrum (dashed line; Johnstone et al 1992). Excess absorption, corresponding to at least an extra 10^{21} cm^{-2}, is apparent in the central area.

(eg. 10^2–10^7 K) strongly indicates that it is very cold (White et al 1991).

Preliminary analysis of *ASCA* data on cooling flow clusters shows both the need for cooler spectral components in the cores and some excess absorption (Figure 5; Fabian et al, in preparation). The spectra are, however, sufficiently complex, as are any realistic models (with distributed absorption and possible nonsolar metal abundances), that uncertainties in the precise temperature distribution within the gas remain. Reasonably good agreement with the data is found if an absorbed cooling spectrum plus an isothermal spectrum (to represent the outer gas projected along the line of sight) are fitted to the data. The cooling flow spectrum is obtained by summing isothermal plasma models (such as those of Raymond & Smith 1977 or Mewe & Gronenschild 1981) inversely weighted by the cooling function at that temperature, $\Lambda(T)$. To see this, we recall (see e.g. Johnstone et al 1992) that as gas cools at constant pressure from T to $T - dT$ by emission of radiation, the bolometric luminosity is given by

$$dL_{\text{cool}} = n_e n_{\text{H}} \Lambda(T) dV = \frac{5}{2} \frac{\dot{M}}{\mu m} k dT, \qquad (4)$$

where $\Lambda(T)$ is the cooling function (see e.g. Raymond & Smith 1977). But since $dL_{\text{cool}}(\nu) = n_e n_{\text{H}} \epsilon_\nu(T) dV$, where $\epsilon_\nu(T)$ is the spectral emissivity of the gas, then the cooling spectrum

$$L_{\text{cool}}(\nu) = \frac{5}{2} \frac{\dot{M}}{\mu m} \int_0^{T_{\max}} \frac{\epsilon_\nu(T) dT}{\Lambda(T)}. \qquad (5)$$

The result is independent of the geometry. Such models are readily available in the spectral-fitting package XSPEC (Shafer et al 1991). See Wise & Sarazin (1993) for more complex cooling flow spectral models.

The presence of a cooling flow also appears to influence the overall spectrum of the cluster, as observed with broad-beam instruments such as were on *EXOSAT*, the *Einstein Observatory,* and *Ginga*. Such spectra enable the total cluster luminosity (L_X) and its (emission-weighted) mean temperature (T_X) and metallicity (Z) to be determined with good accuracy. Correlations of L_X and Z vs T_X show a significant nonstatistical scatter. Trends are seen in the correlations if the clusters are divided on central gas density (Edge & Stewart 1991a) or mass cooling rate, $\dot{M}$, (Yamashita 1992, Fabian et al 1994b) such that strong cooling flow clusters have a systematically higher luminosity and higher metallicity at a given temperature. The extent of the differences required to cause this effect is larger than can be explained by the cooling flow alone, and means that the whole cluster is in some way involved. Fabian et al (1994b) suggest that it could be due to widespread inhomogeneity of the intracluster medium, such as is necessary for producing the distributed mass deposition in the cooling flows. In other words, the intracluster medium can consist of a

NGC1275

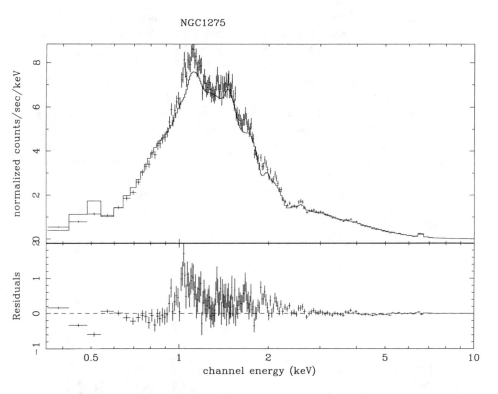

Figure 5 The spectrum of the inner core of the Perseus cluster (data points in the *upper panel*) from the CCD detectors on the Japanese–U. S. satellite *ASCA*. The residuals shown here (*lower panel*) are from the isothermal plasma model (shown as a solid line in the upper panel) which best fits the data above 2.5 keV. The temperature of the model gas is 3.65 keV, significantly less than the best-fitting *Ginga* value of 6.3 keV (Allen et al 1992b). Note that the residuals are mostly positive in the 0.9–2.5 keV band and negative from 0.4–0.9 keV. The positive residuals indicate more soft emission, the negative ones that excess absorption is required (absorption appropriate to the Galactic column density is included in the fit). A cooling flow spectrum (from > 4 keV) of $\sim$370 $M_\odot$ yr^{-1}, plus hot isothermal emission, gives a good overall fit to the data, provided that the cooling flow is absorbed by $\Delta N_H \sim 2.8 \times 10^{21}$ cm^{-2}.

range of densities and temperatures all in pressure equilibrium. This would not be detectable in even the best individual spectra from broad-beam instruments (Allen et al 1992b).

Summary of X-Ray Data

The overwhelming evidence of the images and spectra shows that cooling does occur at a steady rate over long times (at least several billion years). Since gas is then cooling out of the hot phase at rates of hundreds of solar masses per year, an inflow must occur. We do not expect direct evidence of any inward flow since the velocity is highly subsonic over most of its volume at $v_0 \approx 6\dot{M}_2^{1/3} t_{10}^{-1/3} T_8^{-1/6}$ km s^{-1}, where $\dot{M} = 100\,\dot{M}_2\,M_\odot$ yr^{-1}, t_{10} is the cooling time t_a at the edge of the flow in units of 10^{10} yr, and $T = 10^8 T_8$ K.

The consistency between estimates of $\dot{M}$ derived from spectral measurements and those derived from analysis of the surface brightness is shown in Figure 6. Agreement between these two methods gives further support for the existence of cooling flows. The spectral estimate uses emission lines and blends as a measure of the rate at which matter is cooling through a given temperature, whereas the estimate from the image is based on the rate with which matter must be cooling given its apparent density and temperature profiles.

The X-ray data also show that the cooling-flow emission is absorbed, providing evidence for widespread cooled gas. The inferred mass of the absorbing matter is comparable to that expected from a persistent cooling flow.

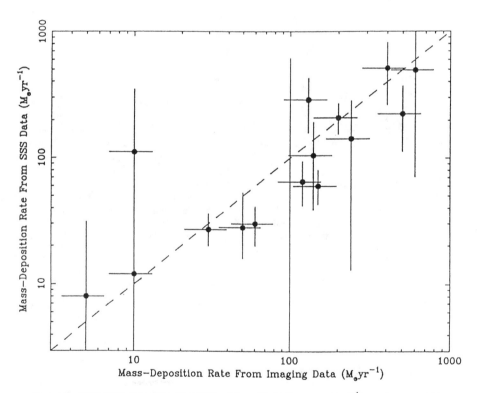

Figure 6   Total mass deposition rate obtained from fits to X-ray spectra ($\dot{M}_{\text{spectral}}$) compared to those from deprojections of the X-ray images ($\dot{M}_{\text{image}}$), from White et al (1991).

APPEARANCE OF COOLING FLOWS AT OTHER WAVEBANDS

Optical Data

In the optical waveband there is little direct evidence for cooling flows. If the cooling gas recombines only once then the expected line emission, which is spread out over several 100 kpc, would be undetectable. The *central* regions of flows, however, often do show optical-line nebulosity (Kent & Sargent 1979, Ford & Butcher 1979, Heckman 1981, Fabian et al 1981, Cowie et al 1983, Hu et al 1985, Johnstone et al 1987, Heckman et al 1989, Baum 1992, Crawford & Fabian 1992) which can be extensive (up to 10 kpc) and luminous (above 10^{43} erg s^{-1} in Hα) in a few cases. The emission-line spectra are characteristic of low-ionization plasmas with strong forbidden lines. The line widths indicate motions of up to a few $100\,\mathrm{km\,s^{-1}}$. Heckman et al (1989) classified the nebulosities into Types I and II according to the ionization state of the gas (e.g. whether the [NII]λ6584 line is stronger or weaker than Hα, respectively). The Type I objects tend to be associated with lower $\dot{M}$, whereas the type II objects tend to be in the more X-ray luminous and richer clusters. A more continuous progression of properties is now recognized (Crawford & Fabian 1992, Allen et al 1992a, Crawford et al 1994). The central galaxy in the cluster Sersic 159-03, for example, shows patches of nebulosity with both types of ionization.

The Hα emission in the nebulae, if due to recombination, would imply that each atom recombines $\sim$100–10,000 times. In the more luminous objects such as NGC 1275 in the Perseus cluster the total luminosity in UV and optical line emission (assuming it to be at least 30 times the Hα luminosity) is greater than 3×10^{44}erg s^{-1}. Such power is equivalent to the thermal energy content of the inner part of the flow around the nebulosity being released on a sound crossing time. Most of the observed nebulae are one to two orders of magnitude less luminous than this and some massive flows (e.g. A2029) show no detectable line emission. The emission-line nebulae are common in central galaxies with radio sources but are over-luminous when compared with radio galaxies of the same power that are not in clusters (Baum 1992). They are linked to the cooling flow but are not a necessary product of it.

The mass of ionized gas is small ($\ll 10^8\,M_\odot$) compared with the X-ray absorbing gas and the distribution of the emission is much more tightly peaked to the center of the flow (Heckman et al 1989) than that of the matter dropped by the flow itself. Studies of the source of ionization of the nebulae show that it must be distributed and is not, for example, due to a central active nucleus (Johnstone & Fabian 1988, Baum 1992). More plausibly, the nebulae are powered by the energy of the hot gas, or its turbulence (the emission lines often show a velocity spread of several $100\,\mathrm{km\,s^{-1}}$). The spectrum of the cooling gas is flat enough to create the Type I nebulae (Voit & Donahue 1990, Donahue & Voit

1991) but does not have the required power. Crawford & Fabian (1992) solve this problem using self-absorbed mixing layers (Begelman & Fabian 1990) produced as the surfaces of cold clouds embedded in the hot cooling gas are churned up by turbulence. The Type II nebulae are explained by the addition of the emission from shocks as the very dense clouds collide in flows where the central cloud density and turbulence are high. Shocks also explain the extended [FeII] emission observed around NGC 1275 (Rudy et al 1993).

The optical spectra do give one piece of direct evidence confirming the dense environment. The [SII] optical emission lines are density-sensitive. The pressures derived from the relative strength of the lines in the inner few kpc of several cooling flows are $P = nT \sim 10^6$–10^7 cm^{-3} K, which is in good accord with values expected from the cooling flow (Johnstone & Fabian 1988, Heckman et al 1989). Such pressures are at least 1000 times higher than those in our own interstellar medium and require an extensive cooling atmosphere around the central galaxy.

Many of the massive cooling flows also show a blue optical continuum, spatially extended over the central few kpc, in excess of that expected from the underlying old galaxy. In the case of NGC 1275 this emission is clearly from A (or late B) stars since strong Balmer absorption lines can be seen (Rubin et al 1977). *IUE* spectra of this region do however limit the total formation rate of massive stars to less than about $6 M_\odot$ yr^{-1} (Fabian et al 1984a, Nørgaard-Nielsen et al 1990). The amount of blue light correlates with the Balmer emission line luminosity (Johnstone et al 1987, Allen et al 1992a, Crawford & Fabian 1993b, Crawford et al 1994). Johnstone et al (1987) invoked massive young stars to produce the blue light and to ionize the nebulae, even when no stellar spectral features were evident in the blue light. McNamara & O'Connell (1989) found further examples of excess blue light from spectra and argued further for star formation. Studies of the visual and near-infrared colors of the central galaxies in cooling flows by Thuan & Puschell (1989) and Schombert et al (1993) show differences from normal elliptical galaxies. The first group find a possible correlation of color and visual magnitude with $\dot{M}$, but no correlation of K magnitude with $\dot{M}$. The blue colors are not explained by a starburst with a normal initial mass function (IMF).

Romanishin (1987) showed from imaging that the blue light was more concentrated than the old starlight. In an important step, McNamara & O'Connell (1993) have recently found that the excess blue light in A1795 and A2597 occurs in patches either side of the central galaxy along the radio axes (Figure 7). This is similar in appearance to many distant radio galaxies (to be discussed later) where optical polarization shows that much of the extended light is scattered. A study of the optical and *IUE* spectra of the excess blue light in several central galaxies (Crawford & Fabian 1993b) shows that they are fitted equally well by B5 stars or a featureless quasar-like power-law spectrum. The excess blue light

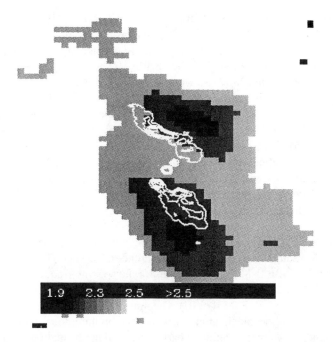

Figure 7 The excess blue light in A1795 mapped in U-I color (McNamara & O'Connell 1993) lies in 2 patches, each ~ 6 kpc, either side of the nucleus just beyond the radio lobes (*white contours*). The bar indicates U-I color.

may therefore be the continuum from an active nucleus in the central galaxy scattered into our line of sight by dust or electrons at larger radii. Sarazin & Wise (1993) argue that beamed emission from a BL Lac nucleus in the FR I host galaxy could be responsible. (The radio sources of most host galaxies are of the Fanaroff-Riley type I, FR I, thought to be associated with BL Lac objects.)

Whether the emission is due to scattered nucleus emission or star formation, the total rate of massive star formation is low, less than a few $M_\odot \, yr^{-1}$, and only detectable in the very centers of flows. The excess blue light occurs in objects with a luminous emission-line nebula and are thus linked in some way. They may both be indirect products of the dense cooling environment.

The outer envelopes of central cluster galaxies show no evidence for recent continued star formation. It is true that many such galaxies are cD and show a halo that can extend for many 100s of kpc, but there is no evidence that they are young (color gradients are small; see Mackie 1992 and references therein). Nor is there evidence, apart from the color and assumptions about the stellar IMF, that they are old. If it is assumed that the envelope around NGC 6166 in A2199 has been continuously formed at the present X-ray mass deposition rate, then (depending upon shape) the upper mass limit of the IMF must lie between 1 and $1.5\,M_\odot$ (Fabian et al 1991). The optical profile and mass-to-light ratios of central galaxies in cooling flows are generally consistent with continued low-mass star formation (Prestwich & Joy 1991). If, however, we require that all the dark matter around M87 (Stewart et al 1984a) is low-mass stars (whether due to an early massive cooling flow or not) then the upper mass limit must be about $0.2\,M_\odot$. One possible cooling flow that appeared to have an anomalous red envelope, GREG (Giant Red Envelope Galaxy; Maccagni et al 1988), has not yet been confirmed optically.

The optical continuum data show that most of the cooled gas from a flow does not form stars with a normal IMF. Most of the gas must remain as cold clouds, or form low-mass stars or objects (see e.g. Fabian et al 1982).

Globular clusters are another possible end-point of the cooled gas in clusters if the Jeans mass can be made appropriate to $> 10^4$ K (Fabian et al 1984b, Fall & Rees 1985). They do appear to be abundant around some central cluster galaxies (McLaughlin et al 1993 and references therein). The correlation is however more with B-M type than with $\dot{M}$; the specific frequency of globular clusters is highest where the magnitude difference between the first and second-ranked galaxy is small (B-M II or III). The most massive flows tend to be in B-M I or I-II clusters where the first ranked galaxy is more luminous. If cooling flows are involved, they must occur at an earlier stage in the formation of the central galaxies and/or clusters. Young globular clusters have been found around the center of NGC 1275 with the *HST* (Holtzmann et al 1992). These clusters have a range of colors and thus ages and can be explained in that case as due to the cooling flow (Richer et al 1993, see also Nørgaard-Nielsen et al 1994).

Clear evidence for the gas cooling below 10^6 K would be an important step in mapping the distribution of $\dot{M}(r)$, since the cooling time is then $\lesssim 10^7$ yr. Optical coronal lines such as FeX and FeXIV are good candidates to search for but the expected fluxes are very weak (Sarazin & Graney 1991). Anton et al (1991) have reported detection of weak coronal emission from [FeX]6374Å in the center of A1795, which is nevertheless much stronger than expected from the X-ray inferred, radius-dependent, $\dot{M}$. This detection has not been confirmed by Donahue & Stocke (1994), although they do claim a detection of [FeX] emission in PKS0745-191 with upper limits on three other clusters. Shields & Filippenko (1992) have conducted a deep search for FeX line emission in the Perseus cluster and obtained an upper limit that does not rule out cooling flows. Confirmation (or refutation) of cooling flows using optical coronal lines is a difficult task. Note that it is likely that the mechanisms that generate the optical nebulosities also create some gas at 10^5–10^6 K (e.g. mixing layers).

Optical absorption lines could give strong clues to the nature of the cooled gas. Little evidence has so far been obtained, although Carter & Jenkins (1992) have found narrow CaII and NaD lines in the spectrum of the nucleus of M87.

Some metal absorption lines in the spectra of quasars might be due to distributed cooled gas in clusters and groups along the line of sight (Crawford et al 1987). The discovery of a background quasar in the line of sight to a cooling flow would be important in this regard. Abraham et al (1993) argue that this is indeed the case for the strong foreground absorption in the spectrum of the BL Lac object $AO0235 + 164$.

The Thomson depth through a cooling flow, which measures the probability that photons are scattered by free electrons, is at least 1% in to a radius of 1 kpc in strong flows (Fabian 1989, White et al 1993). Wise & Sarazin (1990, 1992) have shown how detection of scattered (polarized) emission would be a helpful probe of the structure of the flow.

Infrared

IRAS sources at 60 and 100μm have been detected coincident with several cluster cooling flows (de Jong et al 1990, Bregman et al 1990, Grabelsky & Ulmer 1990). There is some possibility of confusion with Galactic cirrus at these wavelengths but if confirmed, the more luminous sources indicate a further major radiative energy loss from cluster cores (Bregman 1992).

Radio

Many galaxies with cooling flows have a strong central radio source (Valentijn & Bijleveld 1983, Jones & Forman 1984, Valentijn 1988, Zhao et al 1989, Burns 1990, Ball et al 1993; see Table 1). Well-known examples are Perseus A (NGC 1275), Cygnus A, PKS0745-191, Hydra A, and Virgo A (M87). Many are FR I sources (edge-darkened), although the classic FR II Cygnus A is also in a flow (Arnaud et al 1984). Of course such radio sources can only be produced from the relativistic outflow from an active nucleus if there is a dense surrounding medium to provide a working surface. *Cooling flows provide the most extensive and densest working surfaces possible.* The high density of the cooling flow can cause lower power jets to produce FR I sources (De Young 1993). In the case of NGC 1275 (Böhringer et al 1993), the radio lobes have been found to displace the X-ray emitting gas. NGC 1275 also has an outer halo of radio emission which may in part be due to magnetic fields and cosmic rays compressed in the flow (Fabian & Kembhavi 1982, Soker & Sarazin 1990, Becker 1992, Tribble 1993). It is interesting that the clusters that appear not to contain simple cooling flows (e.g. Coma, A2256) instead have distorted halo or wide angle tail sources which may indicate a recent merger (Fabian & Daines 1991, Böhringer et al 1992, Tribble 1993).

There is no simple correlation between radio power and $\dot{M}$. Burns (1990) finds that 71% of a sample of cD galaxies in cooling flows are radio-loud, compared with only 23% of cDs not in flows. Some of the most massive flows do not have powerful radio sources (e.g. A478). Amorphous radio sources such as A2052 (Burns 1990, Zhao et al 1993) and PKS0745-191 (Fabian et al 1985, Baum & O'Dea 1991) appear to be peculiar to cooling flows. Optical nebulosity is common in those cooling flows with a central radio source (see Table 1), but no correlation is found between optical and radio luminosities in objects with both (Crawford et al 1994).

The polarization of the radio emission provides a diagnostic with which to study the magnetic field in the ICM. Faraday rotation and depolarization of the emission have been mapped for the extended radio sources in Cygnus A (Dreher et al 1987), M87 (Owen et al 1990), Hydra A (Taylor et al 1990), 3C295 (Perley & Taylor 1990) and A1795 (Ge & Owen 1993), showing that the magnetic field and pressure in cooling flows increase inward (such that the field exceeds $30\mu G$) with a typical length scale for field reversal being 1–10 kpc. (The field could be much higher on smaller scales.) The implied gas pressures and radio equipartition pressures agree with those obtained from X-ray data. The lack of any strong hard X-ray emission due to inverse Compton radiation from the relativistic electrons in clusters that have halo radio sources (generally not cooling flows) shows that the magnetic field there is at least $0.1\mu G$ (Rephaeli & Gruber 1988).

Twenty-one cm absorption has been found in a few cooling flows. NGC 1275 has a strong feature at the rest wavelength of the cluster (Crane et al 1982, Jaffe 1990). The absorption indicates a column density of $N_H \sim 10^{21}T_1$ cm^{-2}, where the HI has a temperature of $10\,T_1$ K, spread over much of the inner 15 kpc. Strong limits on emission and absorption at 21 cm have been obtained on several other clusters (McNamara et al 1990; Jaffe 1991, 1992). The conversion of the limits on 21 cm optical depth (eg. $\tau_{obs} < 5 \times 10^{-4}$ for A2052; Jaffe 1991) to gas masses depends on assumptions about temperature, optical depth, covering fraction, and beam size (see Daines et al 1994). A simple interpretation rules out an X-ray absorbing column in which the hydrogen is wholly atomic.

CO has also been sought in cooling flows (Jaffe 1987, Bregman & Hogg 1988, Grabelsky & Ulmer 1990, O'Dea et al 1994, Braine & Dupraz 1994, Antonucci & Barvainis 1994) and apart from detections around NGC 1275 in the Perseus cluster (Lazareff et al 1989, Mirabel et al 1989), the limits are strong and rule out widespread warm CO.

Summary of Data from Non-X-Ray Wavebands

Many central galaxies in cooling flows show anomalies, such as strong, low-ionization, emission-line nebulae and diffuse blue light in the optical and high Faraday rotation and/or depolarization in the radio, which together are not seen in non-cooling-flow galaxies. These indicate that something unusual is occurring and are a link to the cooling flow but do not readily reveal the large mass cooling rates, or amounts of cooled gas, found in X rays. The optical anomalies are confined to the inner few kpc of the flow which corresponds to less than one-thousandth of its total volume. The bulk of the flow is undetected at non-X-ray wavelengths. Only a small fraction of the cooled gas can form stars with a normal IMF. Most must remain dark.

Pressures obtained from optical [SII] emission lines and from radio data are consistent with those found from X-ray data, confirming the first-order interpretation of the X-ray results and the short radiative cooling times of the hot gas. Velocity widths of the optical emission lines indicate that at least the inner region of a flow is turbulent.

THEORETICAL ISSUES ON COOLING FLOWS

The Global Structure

The original models for cooling flows (Cowie & Binney 1977, Fabian & Nulsen 1977, Mathews & Bregman 1978) assumed that the gas was homogeneous, i.e the gas has a single temperature and density at each radius. The gas density rises inward approximately $\propto r^{-1}$. The temperature drops inward in accord with pressure equilibrium, which means that its variation is weaker than $\propto r$ given the gravitational potential of the cluster and central galaxy within the flow. The thermal instability was invoked as a means of creating filaments near the center (Fabian & Nulsen 1977, Mathews & Bregman 1978). One problem with this instability, discussed later, is whether an overdense and therefore cooler (and more rapidly cooling) blob will fall and merge with regions of similar density or entropy (Cowie, Fabian & Nulsen 1980). This would suppress the instability. Instead, it was assumed that if the gas was unstable, then all phases moved together.

The *Einstein Observatory* X-ray images of the Perseus cluster (Fabian et al 1984b) and of several other clusters (Stewart et al 1984a,b) demonstrated that the mass deposition by the flow was distributed over 100 kpc or more. This requires that the gas is inhomogeneous. A clump must be overdense by about a factor of 2 at the cooling radius if it is to cool completely while flowing only halfway to the center. In a pioneering paper, Nulsen (1986) developed the idea of a multiphase, comoving cooling flow in which the major input parameter was the initial spectrum of density perturbations in the gas. Magnetic fields were invoked to bind the individual density perturbations or clouds both internally and to the mean flow. The clouds would otherwise break apart as they fell through the more tenuous phases.

A numerical multiphase model applied to the X-ray surface brightness profiles of several clusters (Thomas et al 1987) showed that the fractional volume distribution of the initial density perturbations had to have a characteristic shape, the origin of which was not clear but could itself be related to cooling. White & Sarazin (1987a,b,c, 1988) constructed models of cooling flows with distributed mass deposition which they compared to the data. They assumed various dependences for the rate at which matter cools out of the flow as functions of the local density and other properties and highlight the dependence of $\dot{M}$ on r as a function of the local gravitational profile. (There is general agreement between all approaches to the X-ray data that assume a cooling flow). These studies show that if one assumes the gas to be homogeneous when analyzing the data, the derived results are a good approximation to the (emission-weighted) mean values of temperature and density.

Whether or not the linear thermal instability grows in a gravitational field has been considered by Malagoli et al (1987), White & Sarazin (1987a), Balbus (1988), Tribble (1989a), Loewenstein (1989), and Balbus & Soker (1989) using both particle and wave approaches in Eulerian and Lagrangian systems. The work by Balbus & Soker (1989), which includes the effects of the background flow, indicates that any growth of the linear thermal instability is weak; the medium is only thermally unstable if it is convectively unstable (which a homogeneous cooling flow is not). Numerical work by Hattori & Habe (1990) and Yoshida et al (1991) demonstrate how some blob configurations might fall and mix.

Loewenstein (1990) and Balbus (1991) have shown that weak magnetic fields can destabilize cooling flows and allow the overdense gas to cool. The magnetic field helps to suppress the buoyancy which triggers convective motions in the flow. The magnetic field needs to be strong enough for the Alfvén velocity to exceed the phase velocity of buoyancy oscillations.

As already mentioned, the observations require that the gas initially has large density inhomogeneities, with fractional density $\delta\rho/\rho > r/R$ for a blob of size r at radius R in the flow (Nulsen 1986), which makes the amplification of infinitesimal perturbations a much less important issue for observed cooling flows. If denser blobs exist and survive dynamically in the flow, then they will cool well before reaching the center. The main issues (to be discussed later) are the origin and survival of such denser blobs.

Conduction is another effect that has been widely debated (see e.g. Takahara & Takahara 1979, Tucker & Rosner 1983, Friaca 1986, Gaetz 1989, Böhringer & Fabian 1989). The observations of cooling flows demand that cooler regions are immersed in hotter ones. The energy equation for a unit volume of gas in a constant-pressure cooling flow with conduction is

$$n\frac{d}{dt}\left(\frac{5kT}{2\mu}\right) = -n^2\Lambda + \nabla \cdot (\kappa \nabla T), \qquad (6)$$

where n, T, μ, Λ, and $\kappa \propto T^{5/2}$ are the gas density, temperature, mean molecular weight, cooling function, and Spitzer (1962) conductivity coefficient, respectively. It is implausible to make the two terms on the RHS balance, such that the LHS is zero (i.e no mass drop-out) over a wide temperature range; instabilities grow on a timescale shorter than the local cooling time. Either the gas then becomes isothermal, or conduction has a negligible effect relative to radiative cooling (Nulsen et al 1982, Stewart et al 1984b, Bregman & David 1988). For the conditions in most cooling flows, conduction must be suppressed by a factor of $< 10^{-2}$ (Binney & Cowie 1981, Fabian et al 1991) below the Spitzer value, and for an inhomogeneous flow to occur with small clouds the factor must be much larger.

The conduction models of Bertschinger & Meiksin (1986) fail to account for the cooler X-ray emitting gas that is observed, although they do predict a density rise in the core of the cluster. The conduction model of Sparks (1992) is not relevant to observed flows since the implied large-scale planar geometry would be readily observable in X-ray images (Fabian et al 1994a). His model is one variant of so-called warming flows that postulate a mass of cold matter heated, and possibly evaporated, by the conduction of heat from the surrounding hot gas in the core of a cluster (Bregman 1992, Sparks et al 1989, de Jong et al 1990). Such models do imply a wide range of temperature components but cannot explain the actual observed distribution of temperatures and, in particular, cannot account for the observed strength of the FeXVII line, without requiring impossibly high mass evaporation rates (Canizares et al 1993).

Magnetic fields are observed to occur in the ICM with a pressure of about 1% of the thermal pressure (Kim et al 1991). In a cooling flow the field should be amplified by compression. This is verified by observations of Faraday rotation in those flows surrounding extended luminous radio sources (Dreher et al 1987, Ge & Owen 1993). At small radii where the magnetic field is largest, it may reconnect and therefore inject energy into the hot gas (Soker & Sarazin 1990). Alternatively it may help stifle a homogeneous flow.

Tangled magnetic fields are usually invoked to suppress thermal conduction (as have plasma instabilities; Jafelice 1992). How this takes place is unclear but it seems plausible that mirroring can trap electrons in such a field and so reduce conduction to very low values (Borkowski et al 1990). Tribble (1989b) has studied models in which conduction takes place freely along highly tangled field lines. This leads to a different functional form for the conductivity than the usual one, involving a length-scale dependence. The plasma becomes inhomogeneous as cooling dominates some field lines but not others. It is possible that some such complex conduction model may account for some important aspects of the X-ray data, but it has not been developed so far.

The mass deposition rate can be reduced if a deep gravitational potential is assumed. The gravitational work done on the gas as it flows inward offsets some of the cooling luminosity. The largest factor that can be obtained in this way is about 2 (Arnaud 1988). Changing the potential on a slow timescale as the cluster evolves can have some effect, particularly in staving off the time at which a large flow develops (Meiksin 1990). The idea that the flows have not yet become steady and are only now about to happen is similar to the suggestion by Hu (1988) that strong cooling has only just begun (see also Murray & Balbus 1992). However, such models do not account for the good agreement of $\dot{M}_{\text{spectral}}$ with $\dot{M}_{\text{image}}$ or for the common occurrence of cooling flows.

Several authors have suggested that there is a heat source that counterbalances the effect of radiative cooling, so that the temperature of the gas does not actually decrease (in contradiction to the X-ray spectra showing a range of gas temperatures, cooling times—covering several orders of magnitude $< 10^{10}$ yr, and emission measures, all of which are consistent with simple cooling of the gas). No such process has yet been identified. Note that it requires 10^{61}–10^{62} erg to balance a strong flow over 10^{10} yr, much higher than the energy residing in the lobes of even the most powerful radio sources. Note too that most heating processes cause the gas to become thermally unstable.

It has been suggested that cosmic rays from a central engine or radio source, for example, could stop or heat the core of a flow (Tucker & Rosner 1983; see also Böhringer & Morfill 1988, Rephaeli 1987). This cannot be a general phenomenon that balances the radiative losses since then the cosmic-ray pressure would have to exceed the thermal pressure. No flow would then be inferred in the first place (Loewenstein et al 1991).

The motion of galaxies through the ICM may also act as a heat source (Miller 1986). Much of the drag energy may cause the core of a cluster to be a noisy place with many large-amplitude sound waves (Binney 1988) and/or internal gravity waves (Balbus & Soker 1990); the inward increasing density in a flow can focus the sound into its core (Pringle 1989). How that energy dissipates is not known, although phenomenological (Heckman et al 1989) and physical (Crawford & Fabian 1992) models for the origin of the optical line emission common in the centers of cooling flows require a source of chaotic or turbulent energy. The observed optical line widths require that the gas has motions of a few 100 km s^{-1}. A turbulent model for cooling flows, including star formation, has been proposed by Westbury & Henriksen (1992). Balbus & Soker (1990) find that the energy transported by gravity waves is unlikely to grossly affect a cooling flow. *No stable heating process yet devised is able to counteract the effects of radiative cooling and account for the observed X-ray images and spectra.*

The Local Structure of Cooling Flows

Spatially-distributed mass deposition implies inhomogeneous flows which in turn require that the gas is multiphase. The gas must consist of a multitude of cooling (and cooled) clouds at all radii. How such clouds are supported against gravitational infall and retain their integrity against breakup into the hotter phase is not clear (Loewenstein & Fabian 1990, Tribble 1991). Any dense gas cloud released into the ICM would fall toward the center of the cluster and fall apart under the ram pressure forces that develop unless something binds it together.

Magnetic fields may be responsible for this. The geometry of the field is uncertain and could range from field lines penetrating both clouds and hot ICM, to fields concentrated in the clouds, or some skin around the clouds (Daines et al 1994). The geometry of the clouds themselves is of course unknown and may be sheetlike rather than spherical. The sizes of clouds and in particular their column densities determine the terminal velocity. The smaller a cloud is, the slower it will fall. As mentioned above, conduction must also be suppressed, this time by very large factors.

The origin of the density inhomogeneities is unclear. They may be fossils of the past stripping of galaxies (Soker et al 1991) or of earlier mergers with denser cooler clusters. The spectrum of densities required to generate the observed $\dot{M} \propto r$ is close to that from cooling gas (Nulsen 1986, Thomas et al 1987, Thomas 1988, Tribble 1991) and may be generated by stirring of an old more homogeneous flow (Daines 1994). Chun & Rosner (1993) have studied the nonlocal behavior of thermal conduction in cluster halos and find that inhomogeneities may be expected.

We do not know how clouds evolve. They may coagulate if they collide at velocites below the internal sound speed of a cloud. They may break apart in falling relative to the hot phase and then mix into it, which has the effect of causing the cooling time of the newly mixed gas to be reduced and so promote formation of a new rapidly cooling cloud. In the center of the flow, cloud-cloud collisions may be common since the density of clouds will be highest there.

Despite these many uncertainties, we can identify some of the many stages in the cooling of a cloud. As the gas cools rapidly below $\sim 10^6$ K, it drops out of ionization equilibrium (Edgar & Chevalier 1986, Canizares et al 1988). Large clouds may drop out of pressure equilibrium with the surrounding gas at those temperatures also, since the sound crossing time of a cloud exceeds the cooling time (Cowie et al 1980). Smaller clouds lose pressure equilibrium at lower temperatures. The gas remains essentially optically thin in cooling; where lines are optically thick to resonance scattering (Gil'fanov et al 1987, Wise 1993) then they are also thick in the gas beyond the cooling flow. Numerical computations of collapsing clouds have been presented by David et al (1988) and David & Bregman (1989). Over the range $3 \times 10^4 < T < 3 \times 10^6$ K, the cooling time $t_{cool} \sim 2 \times 10^6 T_6^{5/2} P_5^{-1}$ yr, where $T = 10^6 T_6$ K and the thermal pressure $P = nT = 10^5 P_5$ cm^{-3} K ($P_5 \sim 1\text{--}100$ as the radius drops from ~ 100 kpc to 1 kpc). It therefore becomes very short as the gas cools. Observations of the same value $\dot{M}_{\text{spectral}}$ from different X-ray lines characteristic of different temperatures give good evidence that flows are steady. The sound crossing time of a cloud of size $r = r_0$ kpc is $t_{\text{cross}} \sim 10^7 r_0 T_6^{1/2}$ yr, so a cooling cloud can easily drop out of pressure equilibrium with the surrounding gas (whether it does so depends on its geometry and internal density distribution). If the cooling remained in ionization balance (which is unlikely) then only clouds smaller than 1 pc would be in pressure equilibrium below 10^5 K.

The mass of gas expected in a cooling flow as a function of temperature should roughly resemble the inverse of the cooling function, i.e. where the cooling is strong and the cooling times short, there is little gas. How the gas cools below 10^4 K as the gas recombines and forbidden and fine-structure lines dominate the cooling depends on the ionization state and opacity of the gas. Ferland et al (1994) show that a cloud exposed to the X-ray flux in a flow has a thin outer warm layer and a cold interior which is increasingly molecular with depth. The core quickly drops to the temperature of the microwave background. The longest timescale is the gravitational collapse time of the cloud (assuming it exceeds the Jeans mass) and is $t_J \sim 10^6 T_1^{1/2} P_5^{-1/2}$ yr. The minimum mass of cooled gas remaining in a steady flow is at least $\dot{M} t_J$ or $\sim 10^8 T_1^{-1/2} P_5^{-1/2} \dot{M}_2 M_\odot$, even if the efficiency of collapse was 100%.

The mass above which a cloud collapses under its own weight is the Jeans mass $M_J \sim 3 T_1^2 P_5^{-1/2} M_\odot$, only if thermal pressure opposes gravity. This is $\sim 0.1 M_\odot$ in cloud cores throughout much of a cooling flow, if $T \sim 3$ K. When a magnetic field dominates the pressure, the equivalent critical mass rises as $(n/n_0)^{-2}$ where n is the density in the cloud and n_0 is the density it would have attained at that temperature if the field had been absent. The mass can then be in the range of $M_{crit} \sim 10\text{–}100 M_\odot$, and perhaps more if the magnetic field came into equipartition with the thermal pressure at a much higher temperature. If $B \propto r^{-2}$ and the cloud collapses spherically then the magnetic pressure $P_B \propto T^{-4/3}$. If the magnetic field initially contributes about 1% of the total pressure at the mean cluster temperature, as observations indicate (Kim et al 1991), then it can easily dominate the total pressure by the time the gas has cooled to $\sim 10^4$ K (see also discussion by David & Bregman 1989). In this case, $M_{crit} \sim 10^6 M_\odot$, comparable to the mass range of globular clusters. The magnetic field is therefore important in cooled clouds, determining both their evolution and structural integrity.

The excess X-ray absorption suggests a low efficiency (or longer timescales for cloud collapse) by a factor of about 100 or more. What is unknown so far is whether this can be due to the rate at which clouds lose their magnetic field, or the time taken to accumulate enough mass to exceed the Jeans mass. There are also many other unknown or vague properties of cold clouds which we now briefly outline.

Even when the core of a cloud has cooled, recombined, and becomes mostly neutral, there is still residual ionization due to X-ray irradiation from the surrounding flow. This may provide the link for the magnetic field to bind the cloud; even when the field is unimportant in the core of the cloud it can still give the tension necessary to bind the warm layer at the surface. The low, but significant, ionization in the core of the cloud also enables H_2 to form through the presence of H^-, which in turn promotes the formation of other molecules. In this case the Jeans mass can be sub-stellar (Ferland et al 1994). In such conditions it may also be possible for grains to form (Fabian et al 1994b) perhaps through the higher acetylenes (B. Draine, private communication). Studies of emission-line nebulae often found at the centers of cooling flows, and continuum color maps do reveal the presence of dust (Hu 1992, Sparks et al 1993, Donahue & Voit 1993). The excess blue light is a further potential indicator of distributed grains, if the light is scattered (polarization measurements can test this). Molecules may then freeze onto cold grains, causing clouds to become very dust-rich and quite different from Galactic clouds.

Such possibilities are highly speculative until further observational data on the cooled gas can be obtained. The column density of gas derived from X-ray absorption, $N_H \sim 10^{20}-10^{21}$ cm^{-2}, corresponds to a thickness of only $10^{16}-10^{17} T_1 P_5$ cm, in gas at a temperature $10 T_1$ K and pressure $10^5 P_5$ cm^{-3} K, which is so much smaller than the size of the flow that it must consist of a mist of clouds around the central galaxy. At large radii a long-lived flow should be relatively quiet and undisturbed and the clouds may form grains and/or very low mass stars. The behavior of the core of a flow is different, since the hot gas within the inner 10 kpc or so can only support its own column density of cold clouds without being significantly affected by their weight. If matter drops out of the flow such that $\dot{M} \propto r$ then the mean density of cooled gas builds as r^{-2}, whereas the density in the hot gas is distributed as r^{-1}. This means that the hot gas in the core of any long-lived flow cannot support the full weight of the accumulated cold gas which dominates the dynamics of that region (Daines et al 1994). The clouds there may fall, collide, be shocked, and coagulate. This may be the source of the emission-line nebulae and of any normal star formation, since the bulk of the clouds may be warmer than those suspended in the hot gas at larger radii.

The tight HI and CO limits obtained for many cooling flows severely restrict the nature and form of accumulated cold clouds. Since HI can easily be optically thick (Loewenstein & Fabian 1990) and cold, its emission can therefore be negligible, but it ought to be detectable in absorption. The observational limits thus appear to rule out atomic hydrogen as a major constituent of the X-ray absorbing matter. The gas may therefore be almost completely molecular, which confronts the CO limits unless it either has not formed, or forms into more complex molecules. Alternatively, if the X-ray absorption is from very dust-rich clouds, the current observational limits may be overcome since the cooled matter may be mostly H_2 and dust, with most of the other elements adsorbed onto grain surfaces, embedded in a thin partially ionized and magnetic skin. Note that the dust cannot be similar to that in our Galaxy, since it would then have been detected if warmer than 10 K, and if its total mass exceeded $\sim 10^8 M_\odot$ (Annis & Jewitt 1993). In the above picture the grains would be larger, perhaps corresponding to a significant fraction of the X-ray absorption column density (a few μm).

No satisfactory theory or picture of cold, cooled clouds in cooling flows yet exists. The chemical and physical conditions are sufficiently different from those in our Galaxy that a simple extrapolation from the situation in our Galaxy may be inadequate.

Summary of the Conditions in a Cooling Flow

The picture that we have assembled of the inner ICM in a cluster with a cooling flow is tentative. The gas is inhomogeneous when hot, with densities ranging over at least a factor of 2 (and temperature ranging in the opposite sense to maintain pressure equilibrium). Conduction is highly suppressed and the cooler, denser clouds cool out of the flow at the largest radii, condensing into very dense blobs, and the hotter, more tenuous gas survives to the core of the flow. Weak magnetic fields bind the clouds together against the destructive ram-pressure forces and so determine the mass range of surviving clouds. As a cloud cools, its cooling time rapidly reduces to less than a million years and it drops out of pressure and ionization equilibrium. Depending on the geometry, the magnetic field is amplified and dominates the pressure and so supports the cloud against further compression. Cooling continues in the gas and slowly the magnetic field may be expelled from the cold core of the cloud by ambipolar diffusion. The gas temperature then drops to very low values as the gas becomes increasingly molecular and possibly dusty. If the cloud core is above the Jeans mass it may collapse and produce low-mass stars or brown dwarfs. This is the likely fate for most of the cooled gas, which left the flow at large radii. The efficiency of any star formation must be low in order that the X-ray absorbing column density remains. In the core of the flow, however, it is probable that a cloud collides with another one before forming this stage, leading to strong detectable optical line emission and more ionized, massive magnetic clouds. If above the critical mass, these may form some massive stars and possibly globular clusters. Small dusty shreds of cold gas and clouds may scatter the blue continuum from an active nucleus.

DISTANT COOLING FLOWS

The X-ray image of the cluster around the radio galaxy 3C295 at redshift $z = 0.46$ has the peaked shape of a cooling flow (Henry & Henriksen 1986) and many of the distant clusters found serendipitously by the *Einstein Observatory* and by *ROSAT* have a massive central galaxy with optical line emission (Donahue et al 1992, Allen et al 1992a, Crawford et al 1994) which often accompanies a cooling flow (all nearby central galaxies with line emission are in cooling flows but not all central galaxies in flows have optical line emission). The peaked X-ray emission also makes the clusters more detectable (Pesce et al 1990). A similar correlation of optical activity is found in spectra of part of the *ROSAT* Brightest Cluster Sample (BCS; Allen et al 1992a, Crawford et al 1994). The fraction of clusters with detected line emission does not appear to increase with z. The central galaxy with the highest Hα luminosity, Z3146, at a redshift of 0.3 is in the BCS. A *ROSAT* HRI image of this cluster (Figure 8) indicates a massive cooling flow of about $1000 M_\odot$ yr^{-1} (Edge et al 1994).

X-ray data (Edge et al 1990, Gioia et al 1990) and theoretical models (e.g. Evrard 1990, Katz & White 1993) show that clusters have evolved in a hierarchical manner and are continuing to do so such that the most luminous clusters are most numerous now. This has a profound effect on the evolution of cooling flows. The usual merger of a smaller cluster with a larger one introduces cooler gas from the smaller cluster and causes much stirring and turbulence in the gas in the core of the large one. Any cooling flow is strengthened (or initiated) and the turbulence dissipates by stirring mixing layers onto, and shocks between, cold clouds, thus enhancing the luminosity of the central optical nebulosity. The central galaxies also merge rapidly. The unusual merger of two large subclusters in which the central galaxies do not combine leads to the disruption of cooling flows, although cooling may continue in an unfocused manner (Fabian et al 1984b).

What is not immediately clear from this picture is whether cooling flows were stronger in the past or weaker. White (1988) has argued from the data that flows are increasing in strength with time, although this may in part be in a recovery sense since the last merger. Techniques developed by Bertschinger (1989) and Chevalier (1987) will be useful in estimating the theoretical evolution of flows. Henriksen (1993) has noted that cD clusters with a close and similar companion cluster do not have cooling flows and suggests that these pre-merger clusters are closer to the initial state and thus that the flows were generally weak in the past. McGlynn & Fabian (1984), on the contrary, argued from a similar consideration of the A399/401 pair that they may have already interacted and passed through each other before an eventual merger, so destroying any central flows. Much more observational and theoretical work on the behavior of the ICM, and of central cluster galaxies, during mergers is needed before any definitive statement can be made. Hierarchical cluster formation predicts that flows were much stronger in the past and more numerous, since there were then more subclusters.

Although the optical activity and the radio emission of the central galaxy generally involve only a small fraction of the cooling flow (in both radius and power sense), they can be used to identify and infer the properties of very distant cooling flows. Cygnus A, for example, has a well-detected cooling flow (Arnaud et al 1984) and indeed the high-pressure surrounding gas is probably necessary to cause such luminous radio emission (the gas is the working surface for the radio jet).

Most distant radio-loud quasars and galaxies have strong extended optical (and UV) line emission. The emission is so strong that it must be photoionized by the active nucleus of ionizing luminosity L_{ion}. The ionization state

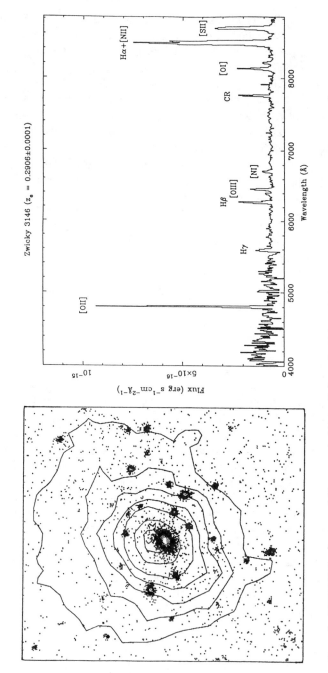

Figure 8   *ROSAT* HRI contours overlaid on an optical image (*left panel*) and the optical spectrum of the central galaxy in Z3146, from Edge et al (1994) and Allen et al (1992a), respectively. The cooling flow in this cluster, at $z = 0.29$, is about $1000 \, M_\odot \, yr^{-1}$. It has the most $H\alpha$-luminous nebula yet found in a cooling flow.

of the optical emitting gas can then be used to determine the pressure of the surroundings, e.g.

$$\frac{[\text{OIII}]}{[\text{OII}]} \propto f\left(\frac{L_{\text{ion}}}{Pr^2}\right), \tag{7}$$

where the gas pressure $P = nT$ exceeds $10^4 \, \text{cm}^{-3} \, \text{K}$ in a cooling flow and is 10^6–$10^7 \, \text{cm}^{-3} \, \text{K}$ at 10 kpc. The function f can be computed using a code such as Ferland's CLOUDY. Observations of radio-loud quasars, where optical/X-ray data constrain L_{ion} show that P decreases with radius around individual quasars and increases with redshift in a sample of quasars (Crawford et al 1988; Crawford & Fabian 1989; Forbes et al 1990; Heckman et al 1991a,b; Bremer et al 1992). The correlation could be more with L than with redshift, but it is clear that the environment of luminous radio-loud quasars has a higher pressure at $z \gtrsim 0.5$. For confining gas at a particular temperature, $t_{\text{cool}} \propto P^{-1}$, so higher pressures mean that gas in any reasonable potential well has a higher cooling rate. Indeed some of the objects must be at the limit where $t_{\text{cool}} \sim t_{\text{grav}}$, which we dub "maximal" cooling flows—the gas is cooling as fast as it is falling and cannot be any denser. (It is not a coincidence that the most radio luminous objects occur in such an environment.)

There may be some feedback between a cooling flow and a quasar, since the quasar radiation has a Compton temperature of about $10^6 \, \text{K}$. This locks the temperature of the gas in the innermost few 100 pc and increases fuel to the quasar since all gas phases flow together into the center. The accretion rate then rises to the Eddington limit. A cluster-cluster merger can however disrupt the feedback and so switch off the quasar (Fabian & Crawford 1990). The net effect is that the most luminous quasars are in the richest clusters that have not suffered a (roughly) equal collision.

There is considerable other evidence for a dense hot intracluster medium around powerful distant radio-loud objects. Direct galaxy counts (Yee & Green 1987, Yates et al 1989, Hill & Lilly 1991) show that these radio-loud objects are in clusters that appear richer at higher redshift and that their host galaxies are very luminous and hence massive (Romanishin & Hintzen 1989). Radio data show strong Faraday depolarization (Garrington et al 1988, Laing 1988), similar to that seen in Cygnus A (Dreher et al 1987) and other nearby cooling flows (e.g. Ge & Owen 1993), which requires a dense screen of electrons. This result has recently been extended to above a redshift of 3 by Carilli et al (1994) who have found that the most distant radio galaxies are surrounded by similar Faraday screens. The equipartition pressures obtained from the low surface brightness radio emission away from the heads of the radio lobes at $z \gtrsim 0.5$ is high. Also the radio sources are smaller for the same power (Barthel & Miley 1988). The optical-radio alignment effect, in which a blue optical continuum extends for tens of kpc along the radio axis and is also polarized (Tadhunter et al 1992 and references therein), is consistent with electron scattering in a dense intracluster medium (Fabian 1989) or widely distributed dust (such as could form in distributed cold clouds). The *ROSAT* PSPC detection of the $z = 1.079$ radio galaxy 3C356 (Crawford & Fabian 1993a) is consistent with a hot, dense gaseous halo, and *Einstein Observatory* HRI observations of the double quasar 0957 + 561 indicate that it has a neighboring, 8 kpc offset extended X-ray emission region magnified by the gravitational lens (Jones et al 1993). If due to hot gas, such a detectably-bright region must have a short cooling time and could be the peak of a surrounding flow. Its off-nucleus nature resembles that of the peak of emission in the Perseus cluster around NGC 1275 (Figure 3).

In summary, there is strong evidence that powerful radio-loud objects are surrounded by dense, cooling, intracluster (or intragroup) gas. Whether there are massive cooling flows at redshifts above 0.5 not associated with active objects is not known. Optically rich clusters at redshifts approaching 1 do not have X-ray luminosities much exceeding 2×10^{44} erg s^{-1} (Castander et al 1994), which limits any cooling flows to $\ll 1000 M_\odot$ yr^{-1} unless there is much absorption and/or the virial temperature is lower than 2 keV.

It is possible that X-ray absorption by the large quantities of cooled gas from a maximal flow smothers the X-ray emission from the flow. Only when powerful radio jets from the center clear a path from the nucleus is the emission from the inner regions easily detectable. In that case though, the luminosity of the (quasar) nucleus may outshine that from the cooling gas. Dusty cooling flows may occur in the infrared-luminous, optically polarized, radio galaxies IRAS 10214+4724 (Lawrence et al 1993) and IRAS 0914+4109 (Hines & Wills 1993).

COOLING FLOWS AND GALAXY FORMATION

We have seen that the most massive galaxies observed at $z > 0.5$ have many properties that can be interpreted as due to a surrounding cooling hot medium. Often $t_{cool} \sim t_{grav}$ and $\dot{M}$ is maximized (hence maximal cooling flow). Where $\dot{M}$ is a few thousand $M_\odot$ yr^{-1} a very large galaxy can then form in a few billion years:

$$M = 10^{12} \left(\frac{\dot{M}}{1000 M_\odot \, \mathrm{yr}^{-1}} \right) \left(\frac{t_a}{10^9 \, \mathrm{yr}} \right) M_\odot. \tag{8}$$

Cooling flows therefore must play some part in the formation of the most massive galaxies, i.e. the central cluster galaxies. Indeed, any theory of galaxy formation in which gas falls into potential wells, is heated to the virial temperature, and then cools with the possibility that $t_{cool} \gtrsim t_{grav}$ (e.g. Rees & Ostriker 1977, Silk 1977, White & Rees 1978, White & Frenk 1991) requires cooling flows.

In most hierarchical models for structure formation, mass overdensities begin with $t_{cool} < t_{grav}$ and form "normal" stars of which the more massive become supernovae. The energy feedback from these leaves most of the gas uncooled. The supernova ejecta also enriches the gas in metals. It is then incorporated into the next stage of the hierarchy. Many small perturbations do not proceed beyond this condition and appear as "normal" galaxies and loose groups. In larger perturbations, however, the total mass increases such that the object passes the cooling flow condition. This is where maximal cooling flows occur, which are thus expected in young clusters and massive groups before they merge to form richer clusters. The exact mass level at which this occurs depends on the role and fraction of any nonbaryonic dark matter (Figure 9).

What happens at this stage can be deduced from our studies of nearby cooling flows. The gas is multiphase at all radii and lays down cooled gas according to $M(< r) \propto r$, $(\rho \propto r^{-2})$. Much of the cooled gas is in the form of very cold clouds which may efficiently form low-mass stars. Star formation may therefore

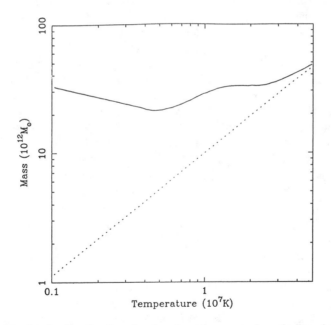

Figure 9   The domain of cooling flows in galaxy formation occurs above the line where $t_{cool} = t_{grav}$, shown as the dotted line (zero metal abundance) and the solid curve (0.4 solar abundance). The mass scale shown is for self-gravitating gas clouds, but reduces by a factor of at least 10 in a universe dominated by cold dark matter. In a simple hierarchical model, structures grow in successive stages proceeding from the lower left of the figure to the upper right; the metallicity of the gas increases as the structure grows, so the point at which an object passes into the cooling flow state depends upon the details of its evolution.

switch from the "normal" IMF which occurred during the earlier phases of the hierarchy (giving rise to the observed galaxy) to an almost exclusively low-mass mode (Thomas & Fabian 1990). A massive isothermal dark halo is thereby assembled (see Thomas 1988). While a cooling flow persists, the IMF of any star formation can only be "normal" for a small fraction of the cooled gas, or massive galaxies would appear even more luminous. The dark mass of the central galaxy is continuing to increase from the cooling flow.

Some such process is required in order to account for the upper luminosity of the largest galaxies. These are inferred to occur at the boundary where $t_{cool} = t_{grav}$ and some physical process is required to prevent more luminous galaxies from forming. Certainly more massive structures occur in the Universe, it is just that the most massive objects are mostly dark. Cooling flows are a mechanism for making the cooled gas dark. We do not know how or why, but the observations of nearby cooling flows reviewed here do show that this is so. Such a physical limit to the luminosity may help to explain the tight infrared K-band magnitude-redshift correlation noted for distant radio galaxies (Lilly & Longair 1984). If there is some alternative process that counteracts the cooling in some manner undetectable to X-ray observations, then it represents a major energy flow and determines the upper luminosity of galaxies.

The role of cooling flows in galaxy formation has also been studied by Ashman & Carr (1988, 1991). They consider both the cooling flows discussed above and ones that can occur at very low mass scales when the virial temperature is below 10^4 K. They investigate the total mass fraction that can be processed through cluster cooling flows and again argue that it is small unless significant heating occurs due to supernovae at the early stages of the hierarchy.

Note that it has not been argued here that most dark matter is baryonic and from cooling flows or that indeed any of it has been formed in this way except that around central cluster galaxies. If, however, it is found that much of the dark matter in lower mass systems (e.g. normal galaxies and loose groups) is baryonic, perhaps through the detection of gravitational microlensing, and similar to that in the cores of clusters, then a more general connection can be pursued.

The most massive structures in the present Universe, from giant galaxies and groups upward, should thereby have passed through a cooling-flow phase. Observationally this implies a soft X-ray background from the radiation of the cooling gas, although absorption in cooled gas may reduce the observed flux by large factors. Line absorption of background (or embedded) quasars by the dense cold clouds in the flow may explain some of the many absorption lines commonly seen in the spectra of distant quasars. The hypothesis predicts that massive protogalaxies and subclusters will be a turbulent, extended mess of rapidly cooling hot gas with a massive embedded population of dense cold clouds. The smallest clouds, or cloud fragments, may mix into the hot gas and produce absorption lines of high ionization, whereas the large clouds may be predominantly neutral and create damped Lyα lines. Lines from individual clouds may be narrow but the spread from all clouds may be up to 1000 km s^{-1}.

The formation of galaxies, and in particular massive galaxies, is therefore seen to be a complicated process with a stellar IMF that varies and depends in an indirect way on magnetic fields. Cooling-flow conditions must be common during the formation of massive galaxies and it is worth taking a careful look at nearby cooling flows to learn how they operate. These nearby examples show that much of the action is not directly detectable at visible wavelengths.

CONCLUSIONS

The X-ray observations of cooling flows form a consistent picture showing that much of the gas surrounding the central galaxy in a typical cluster is cooling and slumping inward. Inhomogeneities in the gas lead to spatially distributed mass deposition such that the cooled gas is widely spread throughout the flow. X-ray absorption is evidence for the cooled material, much of which must remain at radii of 100 kpc and more from the central galaxy. Observations at other wavebands support the high pressures derived from the X-ray data and clearly show anomalies in the central few kpc that are peculiar to these objects. There is no observational support from these wavebands for the absorbing or cooled matter, which must accumulate in the form of dense clouds, low-mass stars, or dust grains. The nature of embedded gas clouds is restricted by tight observational limits on HI and CO. There is, of course, much evidence for dark matter in these regions, so the total accumulated mass is no problem.

The behavior of a cooling flow requires that conduction is highly suppressed and that gas clouds retain some coherence. This requires tangled magnetic fields which complicate the geometry and evolution of cooled gas clouds. The gas is probably turbulent, especially in the center, and its overall state is complex. The ICM may be at least as complicated as our own interstellar medium.

There is growing evidence from observations of the most powerful distant radio objects such as the radio galaxies and quasars that cooling flows power, provoke, and shape luminous radio sources from the central galaxy. The distributed hot and cold gas in a surrounding cooling flow explains many of the phenomena associated with these objects. Finally, the hierarchical formation of structure in the Universe means that strong cooling flows were common in the past and that the processes involved in them have shaped the upper limit to the luminosity function of galaxies.

Acknowledgments

I am very grateful to Carolin Crawford and Claude Canizares for help and discussions. I also thank Steve Allen, Hans Böhringer, Alastair Edge, Keith

Gendreau, Roderick Johnstone, Paul Nulsen, Peter Tribble, and David White for comments and help with the preparation of figures, and Professor Y. Tanaka and his colleagues at ISAS for hospitality while much of this review was written. The Royal Society is thanked for support.

Literature Cited

Abraham RG, Crawford CS, Merrifield MR, Hutchings JB, McHardy IM. 1993. *Ap. J.* 415:101–12
Allen SW, Edge AC, Fabian AC, Böhringer H, Crawford CS, et al. 1992a. *MNRAS* 259:67–81
Allen SW, Fabian AC. 1994. *MNRAS*. In press
Allen SW, Fabian AC, Johnstone RM, Nulsen PEJ, Edge AC. 1992b. *MNRAS* 254:51–58
Allen SW, Fabian AC, Johnstone RM, White DA, Daines SJ, et al. 1993. *MNRAS* 262:901–14
Annis J, Jewitt D. 1993. *MNRAS* 264:593–96
Anton K, Wagner S, Appenzeller I. 1991. *Astron. Astrophys.* 246:L51–54
Antonucci R, Barvainis R. 1994. *Astron. J.* 107:448–50
Arnaud KA. 1988. See Fabian 1988a, pp. 31–40
Arnaud KA, Fabian AC, Eales SA, Jones C, Forman W. 1984. *MNRAS* 211:981–89
Arnaud KA, Serlemitsos PJ, Marshall FE, Petre R, Jahoda K, et al. 1992. In *Frontiers of X-Ray Astronomy*, pp. 481–84, ed. Y Tanaka, K Koyama. Tokyo: Universal Acad.
Arnaud M, Rothenflug R, Boulade O, Vigroux L, Vangioni-Flam E. 1992. *Astron. Astrophys.* 254:49–64
Ashman KM, Carr BJ. 1988. *MNRAS* 234:219–40
Ashman KM, Carr BJ. 1991. *MNRAS* 249:13–24
Balbus SA. 1988. *Ap. J.* 328:395–403
Balbus SA. 1991. *Ap. J.* 372:25–30
Balbus SA, Soker N. 1989. *Ap. J.* 341:611–30
Balbus SA, Soker N. 1990. *Ap. J.* 357:353–66
Ball R, Burns JO, Loken C. 1993. *Astron. J.* 105:53–66
Barthel PD, Miley GK. 1988. *Nature* 333:319–25
Baum SA. 1992. See Fabian 1992, pp. 171–98
Baum SA, O'Dea CP. 1991. *MNRAS* 250:737–49
Becker PA. 1992. *Ap. J.* 397:88–116
Begelman MC, Fabian AC. 1990. *MNRAS* 244:26P–29P
Bertschinger E. 1989. *Ap. J.* 340:666–78
Bertschinger E, Meiksin A. 1986. *Ap. J.*

Lett. 306:L1–5
Binney J. 1988. See Fabian 1988a, pp. 225–34
Binney J, Cowie LL. 1981. *Ap. J.* 247:464–72
Böhringer H, Fabian AC. 1989. *MNRAS* 237:1147–62
Böhringer H, Morfill GE. 1988. *Ap. J.* 330:609–19
Böhringer H, Schwarz RA, Briel UG, Voges W, Ebeling H, et al. 1992. See Fabian 1992, pp. 71–90
Böhringer H, Voges W, Fabian AC, Edge AC, Neumann DM. 1993. *MNRAS*. 264:L25–28
Borkowski KJ, Balbus SA, Fristrom CC. 1990. *Ap. J.* 355:501–17
Borkowski KJ, Shull JM. 1990. *Ap. J.* 348:169–85
Braine J, Dupraz C. 1994. *Astron. Astrophys.* In press
Bregman JN. 1992. See Fabian 1992, pp. 119–30
Bregman JN, David LP. 1988. *Ap. J.* 326:639–44
Bregman JN, David LP. 1989. *Ap. J.* 341:49–53
Bregman JN, Hogg DE. 1988. *Astron. J.* 96:455–57
Bregman JN, McNamara BR, O'Connell RW. 1990. *Ap. J.* 351:406–11
Bremer MN, Crawford CS, Fabian AC, Johnstone RM. 1992. *MNRAS* 254:614–26
Brinkmann W, Massaglia S, Müller E. 1990. *Astron. Astrophys.* 237:536–44
Burns JO. 1990. *Astron. J.* 99:14–30
Canizares CR. 1981. In *X-Ray Astronomy with the Einstein Satellite*, ed. R Giacconi, pp. 215–22. Dordrecht: Reidel
Canizares CR, Clark GW, Jernigan JG, Markert TH. 1982. *Ap. J.* 262:33–43
Canizares CR, Clark GW, Markert TH, Berg C, Smedira M, et al. 1979. *Ap. J. Lett.* 234:L33–38
Canizares CR, Fabbiano G, Trinchieri G. 1987. *Ap. J.* 312:503–13
Canizares CR, Markert TH, Donahue ME. 1988. See Fabian 1988a, pp. 63–72
Canizares CR, Markert RH, Markoff S, Hughes JP. 1993. *Ap. J. Lett.* 405:L17–20
Canizares CR, Stewart GC, Fabian AC. 1983.

Ap. J. 272:449–55
Carilli, CL, Owen FN, Harris DL. 1994. *Astron. J.* 107:480–93
Carter D, Jenkins CR. 1992. *MNRAS* 257:7P–12P
Castander F, et al. 1994. *Ap. J. Lett.* 424:L79–82
Chevalier RA. 1987. *Ap. J.* 318:66–77
Chun E, Rosner R. 1993. *Ap. J.* 408:678–88
Cowie LL, Binney J. 1977. *Ap. J.* 215:723–32
Cowie LL, Fabian AC, Nulsen PEJ. 1980. *MNRAS* 191:399–410
Cowie LL, Hu EM, Jenkins EB, York DG. 1983. *Ap. J.* 272:29–47
Crane P, van der Hulst J, Haschick A. 1982. In *Extra-Galactic Radio Sources, IAU Symp. 97*, pp. 307–8, ed. DS Heeschen, CM Wade. Dordrecht: Reidel
Crawford CS, Arnaud KA, Fabian AC, Johnstone RM. 1989. *MNRAS* 236:277–87
Crawford CS, Crehan DA, Fabian AC, Johnstone RM. 1987. *MNRAS* 224:1007–11
Crawford CS, Edge AC, Fabian AC, Allen SW, Böhringer H, et al. 1994. *MNRAS*. In press
Crawford CS, Fabian AC. 1989. *MNRAS* 239:219–45
Crawford CS, Fabian AC. 1992. *MNRAS* 259:265–80
Crawford CS, Fabian AC. 1993a. *MNRAS* 260:15P–19P
Crawford CS, Fabian AC. 1993b. *MNRAS* 265:431–48
Crawford CS, Fabian AC, Johnstone RM. 1988. *MNRAS* 235:183–92
Crusius-Wätzel A, Biermann PL, Lerche I, Schlickeiser R. 1990. *Ap. J.* 360:417–26
Daines SJ. 1994. *MNRAS*. In press
Daines SJ, Fabian AC, Thomas PA. 1994. *MNRAS*. In press
David LP, Arnaud KA, Forman W, Jones C. 1990. *Ap. J.* 356:32–40
David LP, Bregman JN. 1989. *Ap. J.* 337:97–107
David LP, Bregman JN, Seab CG. 1988. *Ap. J.* 329:66–81
David LP, Jones C, Forman W, Daines SJ. 1994. *Ap. J.* In press
de Jong T, Nørgaard-Nielsen HU, Jørgensen HE, Hansen L. 1990. *Astron. Astrophys.* 232:317–22
De Young DS. 1993. *Ap. J. Lett.* 405:L13–16
Donahue M, Stocke JT. 1994. *Ap. J.* 422:459–66
Donahue M, Stocke JT, Gioia IM. 1992. *Ap. J.* 385:49–60
Donahue M, Voit GM. 1991. *Ap. J.* 381:361–72
Donahue M, Voit GM. 1993. *Ap. J. Lett.* 414:L17–20
Dreher JW, Carilli CL, Perley RA. 1987. *Ap. J.* 316:611–25
Edgar RJ, Chevalier RA. 1986. *Ap. J. Lett.* 310:L27–30
Edge AC. 1989. *X-Ray Emission from Clusters of Galaxies*. PhD thesis. Univ. Leicester. 157 pp.
Edge AC. 1991. *MNRAS* 250:103–10
Edge AC, Fabian AC, Allen SW, Crawford CS, White DA, et al. 1994. *MNRAS*. Submitted
Edge AC, Stewart GC. 1991a. *MNRAS* 252:414–27
Edge AC, Stewart GC. 1991b. *MNRAS* 252:428–41
Edge AC, Stewart GC, Fabian AC. 1992. *MNRAS* 258:177–88
Edge AC, Stewart GC, Fabian AC, Arnaud KA. 1990. *MNRAS* 245:559–69
Evrard AE. 1990. *Ap. J.* 363:349–66
Fabian AC, ed. 1988a. *Cooling Flows in Clusters and Galaxies*. Dordrecht: Kluwer. 391 pp.
Fabian AC. 1988b. In *Hot Thin Plasmas in Astrophysics*, ed. R. Pallavicini. pp. 293–314. Dordrecht:
Fabian AC. 1988c. See Fabian 1988a, pp. 315–24
Fabian AC. 1989. *MNRAS* 238:41P–44P
Fabian AC, ed. 1992. *Clusters and Superclusters of Galaxies*. Dordrecht: Kluwer. 369 pp.
Fabian AC, Arnaud KA, Nulsen PEJ. 1984a. *MNRAS* 208:179–84
Fabian AC, Arnaud KA, Nulsen PEJ, Watson MG, Stewart GC, et al. 1985. *MNRAS* 216:923–32
Fabian AC, Arnaud KA, Thomas PA. 1986. In *Dark Matter in the Universe, IAU Symp. 117*, ed. TJ Kormendy, G Knapp. Dordrecht: Reidel
Fabian AC, Canizares CR, Böhringer H. 1994a. *Ap. J.* 425:40–42
Fabian AC, Crawford CS. 1990. *MNRAS* 247:439–43
Fabian AC, Crawford CS, Edge AC, Mushotzky RF. 1994b. *MNRAS*. 277:779–84
Fabian AC, Daines SJ. 1991. *MNRAS* 252:17P–19P
Fabian AC, Daines SJ, Johnstone RM. 1994. *MNRAS*. In press
Fabian AC, Kembhavi AK. 1982. In *Extragalactic Radio Sources*, ed. DS Heeschen, CM Wade, p. 453. Dordrecht: Reidel
Fabian AC, Ku WH-M, Malin DF, Mushotzky RF, Nulsen PEJ, Stewart GC. 1981. *MNRAS* 196:35P–37P
Fabian AC, Nulsen PEJ. 1977. *MNRAS* 180:479–84
Fabian AC, Nulsen PEJ, Canizares CR. 1982. *MNRAS* 201:933–38
Fabian AC, Nulsen PEJ, Canizares CR. 1984b. *Nature* 310:733–40
Fabian AC, Nulsen PEJ, Canizares CR. 1991. *Astron. Astrophys. Rev.* 2:191–226
Fabian AC, Zarnecki JC, Culhane JL, Hawkins FJ, Peacock A, et al. 1974. *Ap. J. Lett.* 189:L59–64
Fall SM, Rees MJ. 1985. *Ap. J.* 298:18–26
Ferland GJ, Fabian AC, Johnstone RM. 1994.

MNRAS 266:399–411

Forbes DA, Crawford CS, Fabian AC, Johnstone RM. 1990. *MNRAS* 244:680–90

Ford HC, Butcher H. 1979. *Ap. J. Suppl.* 41:147–72

Friaca ACS. 1986. *Astron. Astrophys.* 164:6–16

Gaetz TJ. 1989. *Ap. J.* 345:666–73

Garrington ST, Leahy JP, Conway RG, Laing RA. 1988. *Nature* 331:147–49

Ge JP, Owen FN. 1993. *Astron. J.* 105:778–87

Gil'fanov MR, Sunyaev RA, Churazov EM. 1987. *Sov. Astron. Lett.* 13:3–7

Gioia IM, Henry JP, Maccacaro T, Morris SL, Stocke JT, Wolter A. 1990. *Ap. J. Lett.* 356:L35–38

Gold T, Hoyle F. 1958. In *Paris Symp. on Radio Astronomy*, p. 574, ed. RN Bracewell. Stanford:Stanford Univ. Press

Goldman I, Rephaeli Y. 1991. *Ap. J.* 380:344–50

Gorenstein P, Fabricant D, Topka K, Tucker W, Harnden FR. 1977. *Ap. J. Lett.* 216:L95–100

Grabelsky DA, Ulmer MP. 1990. *Ap. J.* 355:401–09

Harris DL, Carilli CL, Perley RA. 1994. *Nature* 367:713–16

Hattori M, Habe A. 1990. *MNRAS* 242:399–418

Heckman TM. 1981. *Ap. J. Lett.* 250:L59–63

Heckman TM, Baum SA, van Breugel WJM, McCarthy P. 1989. *Ap. J.* 338:48–77

Heckman TM, Lehnert MD, van Breugel WJM, Miley GK. 1991a. *Ap. J.* 370:78–101

Heckman TM, Lehnert MD, Miley GK, van Breugel WJM. 1991b. *Ap. J.* 381:373–85

Helmken J, Delvaille JP, Epstein A, Geller MJ, Schnopper HW, Jernigan JG. 1978. *Ap. J. Lett.* 221:L43–48

Henriksen MJ. 1993. *Ap. J. Lett.* 407:L13–15

Henry JP, Henriksen MJ. 1986. *Ap. J.* 301:689–97

Hill GJ, Lilly SJ. 1991. *Ap. J.* 367:1–18

Hines DC, Wills BJ. 1993. *Ap. J.* 415:82–92

Holtzmann JA, Faber SM, Shaya TR, Lauer TJ, Groth EJ, et al. 1992. *Astron. J.* 103:691–702

Hu EM. 1988. See Fabian 1988a, pp. 73–86

Hu EM. 1992. *Ap. J.* 391:608–16

Hu EM, Cowie LL, Kaaret P, Jenkins EB, York DG, Roesler FL. 1983. *Ap. J. Lett.* 275:L27–32

Hu EM, Cowie LL, Wang Z. 1985. *Ap. J. Suppl.* 59:447–98

Jafelice LC. 1992. *Astron. J.* 104:1279–89

Jaffe W. 1987. *Astron. Astrophys.* 171:378–79

Jaffe W. 1990. *Astron. Astrophys.* 240:254–58

Jaffe W. 1991. *Astron. Astrophys.* 250:67–69

Jaffe W. 1992. See Fabian 1992, pp. 109–18

Johnstone RM, Fabian AC. 1988. *MNRAS* 233:581–99

Johnstone RM, Fabian AC, Edge AC, Thomas PA. 1992. *MNRAS* 255:431–40

Johnstone RM, Fabian AC, Nulsen PEJ. 1987. *MNRAS* 224:75–91

Jones C, Forman W. 1984. *Ap. J.* 276:38–55

Jones C, Stern C, Falco E, Forman W, David L, et al. 1993. *Ap. J.* 410:21–28

Katz N, White SDM. 1993. *Ap. J.* 412:455–78

Kent SM, Sargent WLW. 1979. *Ap. J.* 230:667–80

Kim K-T, Tribble P, Kronberg PP. 1991. *Ap. J.* 379:80–81

Lahav O, Edge AC, Fabian AC, Putney A. 1989. *MNRAS* 238:881–95

Laing RA. 1988. *Nature* 331:149–51

Larson RB, Dinerstein HL. 1975. *Publ. Astron. Soc. Pac.* 87:911–15

Lawrence A, Rowan-Robinson M, Oliver S, Taylor A, McMahon RG, et al. 1993. *MNRAS* 260:28–36

Lazareff B, Castets A, Kim D-W, Jura M. 1989. *Ap. J. Lett.* 336:L13–16

Lea SM, Silk J, Kellogg E, Murray S. 1973. *Ap. J. Lett.* 184:L105–12

Lea SM, Mushotzky RF, Holt SS. 1982. *Ap. J.* 262:24–32

Lilly SJ, Longair MS. 1984. *MNRAS* 211:833–55

Loewenstein M. 1989. *MNRAS* 238:15–41

Loewenstein M. 1990. *Ap. J.* 349:471–76

Loewenstein M, Fabian AC. 1990. *MNRAS* 242:120–34

Loewenstein M, Zweibel EG, Begelman MC. 1991. *Ap. J.* 377:392–402

Maccagni D, Garilli B, Gioia IM, Maccacaro T, Vettolani G, Wolter A. 1988. *Ap. J. Lett.* 334:L1–4

Mackie G. 1992. *Ap. J.* 400:65–73

Malagoli A, Rosner R, Bodo G. 1987. *Ap. J.* 319:632–36

Mathews WG. 1989. *Astron. J.* 97:42–56

Mathews WG, Bregman JN. 1978. *Ap. J.* 224:308–19

McGlynn TA. Fabian AC. 1984. *MNRAS* 208:709–18

McLaughlin DE, Harris WE, Hanes DA. 1993. *Ap. J. Lett.* 409:L45–48

McNamara BR, Bregman JN, O'Connell RW. 1990. *Ap. J.* 360:20–29

McNamara BR, O'Connell RW. 1989. *Astron. J.* 98:2018–43

McNamara BR, O'Connell RW. 1993. *Astron. J.* 105:417–26

Meiksin A. 1990. *Ap. J.* 352:466–94

Mewe R, Gronenschild HBM. 1981. *Astron. Astrophys. Suppl.* 45:11–52

Miller L. 1986. *MNRAS* 220:713–22

Mirabel IF, Sanders DB, Kazès I. 1989. *Ap. J. Lett.* 340:L9–12

Mitchell RJ, Charles PA, Culhane JL, Davison PJN, Fabian AC. 1975. *Ap. J. Lett.* 200:L5–8

Mulchaey JS, Davis DS, Mushotzky RF, Burstein D. 1993. *Ap. J. Lett.* 404:L9–12

Murray SD, Balbus SA. 1992. *Ap. J.* 395:99–112

Mushotzky RF. 1992. See Fabian 1992, pp. 91–108

Mushotzky RF, Holt SS, Smith BW, Boldt EA,

Serlemitsos PJ. 1981. *Ap. J. Lett.* 244:L47–52

Mushotzky RF, Szymkowiak AE. 1988. See Fabian 1988a, pp. 53–62

Nørgaard-Nielsen HU, Goudfrooij P, Jørgensen HE, Hansen L. 1994. *Astron. Astrophys.* In press

Nørgaard-Nielsen HU, Hansen L, Jørgensen HE. 1990. *Astron. Astrophys.* 240:70–77

Nulsen PEJ. 1986. *MNRAS* 221:377–92

Nulsen PEJ. 1988. See Fabian 1988a, p. 378

Nulsen PEJ, Stewart GC, Fabian AC. 1984. *MNRAS* 208:185–95

Nulsen PEJ, Stewart GC, Fabian AC, Mushotzky RF, Holt SS, et al. 1982. *MNRAS* 199:1089–100

O'Dea CP, Baum SA, Tacconi LJ, Maloney PR, Sparks WB. 1994. *Ap. J.* 422:467–79

Owen FN, Eilek JA, Keel WC. 1990. *Ap. J.* 362:449–54

Pedlar A, Ghataure HS, Davies RD, Harrison BA, Perley RA, et al. 1990. *MNRAS* 246:477–89

Perley RA, Taylor G. 1990. *Astron. J.* 101:1623–31

Pesce JE, Fabian AC, Edge AC, Johnstone RM. 1990. *MNRAS* 244:58–63

Ponman TJ, Bertram D. 1993. *Nature* 363:51–54

Prestwich AH, Joy M. 1991. *Ap. J. Lett.* 369:L1–4

Pringle JE. 1989. *MNRAS* 239:479–85

Raymond JC, Smith BW. 1977. *Ap. J. Suppl.* 35:419–39

Rees MJ, Ostriker JP. 1977. *MNRAS* 179:541–59

Rephaeli Y. 1987. *MNRAS* 225:851–58

Rephaeli Y, Gruber DE. 1988. *Ap. J.* 333:133–35

Richer HB, Crabtree DR, Fabian AC, Lin DNC. 1993. *Astron. J.* 105:877–85

Romanishin W. 1987. *Ap. J. Lett.* 323:L113–16

Romanishin W, Hintzen P. 1989. *Ap. J.* 341:41–48

Rothenflug R, Arnaud M. 1985. *Astron. Astrophys.* 144:431–42

Rubin VC, Ford, WK, Peterson CJ, Oort JH. 1977. *Ap. J.* 211:693–96

Rudy RJ, Cohen RD, Rossano GS, Erwin P, Puetter RC, et al. 1993. *Ap. J.* 414:527–34

Sandage A. 1971. In *Nuclei of Galaxies*, p. 271, ed. DJK O'Connell. Amsterdam:North Holland

Sarazin CL. 1986. *Rev. Mod. Phys.* 58:1–116

Sarazin CL. 1988. *X-Ray Emissions from Clusters of Galaxies.* Cambridge: Cambridge Univ. Press. 252 pp.

Sarazin CL. 1992. See Fabian 1992, pp. 131–50.

Sarazin CL, Graney CM. 1991. *Ap. J.* 375:532–43

Sarazin CL, O'Connell RW, McNamara BR. 1992a. *Ap. J. Lett.* 389:L59–62

Sarazin CL, O'Connell RW, McNamara BR. 1992b. *Ap. J. Lett.* 397:L31–34

Sarazin CL, Wise M. 1993. *Ap. J.* 411:55–66

Schombert JM, Barsony M, Hanlon PC. 1993. *Ap. J.* 416:L61–65

Schwartz DA, Bradt HV, Remillard RA, Tuohy IR. 1991. *Ap. J.* 376:424–29

Schwartz DA, Schwarz J, Tucker W. 1980. *Ap. J. Lett.* 238:L59–62

Schwarz RA, Edge AC, Voges W, Böhringer H, Ebeling H, Briel, UG. 1992. *Astron. Astrophys.* 256:L11–14

Shafer RA, Haberl F, Arnaud KA, Tennant AF. 1991. *XSPEC Users Guide.* Greenbelt, MD: NASA

Shields JC, Filippenko AV. 1992. *Astron. J.* 103:1443–50

Silk J. 1976. *Ap. J.* 208:646–49

Silk J. 1977. *Ap. J.* 211:638–48

Singh KP, Westergaard NJ, Schnopper HW. 1986. *Ap. J. Lett.* 308:L51–54

Singh KP, Westergaard NJ, Schnopper HW. 1988. *Ap. J.* 331:672–81

Soker N, Bregman JN, Sarazin CL. 1991. *Ap. J.* 368:341–47

Soker N, Sarazin CL. 1990. *Ap. J.* 348:73–84

Sparks WB. 1992. *Ap. J.* 399:66–75

Sparks WB, Ford HC, Kinney AL. 1993. *Ap. J.* 413:531–41

Sparks WB, Macchetto F, Golombek D. 1989. *Ap. J.* 345:153–62

Spitzer L. 1962. *Physics of Fully Ionized Gases.* New York:Wiley

Stewart GC, Canizares CR, Fabian AC, Nulsen PEJ. 1984a. *Ap. J.* 278:536–43

Stewart GC, Fabian AC, Jones C, Forman W. 1984b. *Ap. J.* 285:1–6

Tadhunter CN, Scarrott SM, Draper P, Rolph C. 1992. *MNRAS* 256:53P–58P

Takahara M, Takahara F. 1979. *Prog. Theor. Phys.* 62:1253–65

Taylor GB, Perley RA, Inoue M, Kato T, Tabara H, Aizu K. 1990. *Ap. J.* 360:41–54

Thomas PA. 1988. *MNRAS* 235:315–41

Thomas PA, Fabian AC. 1990. *MNRAS* 246:156–62

Thomas PA, Fabian AC, Nulsen PEJ. 1987. *MNRAS* 228:973–91

Thuan TX, Puschell JJ. 1989. *Ap. J.* 346:34–58

Tribble PC. 1989a. *MNRAS* 238:1–14

Tribble PC. 1989b. *MNRAS* 238:1247–60

Tribble PC. 1991. *MNRAS* 248:741–50

Tribble PC. 1993. *MNRAS* 263:31–36

Tucker WH, Rosner R. 1983. *Ap. J.* 267:547–50

Valentijn EA. 1988. *Astron. Astrophys.* 203:L17–20

Valentijn EA, Bijleveld W. 1983. *Astron. Astrophys.* 125:223–40

Voit GM, Donahue M. 1990. *Ap. J. Lett.* 360:L15–18

Westbury CF, Henriksen RN. 1992. *Ap. J.* 388:64–81

White DA. 1992. *The Multiphase Medium of Elliptical Galaxies and Clusters of Galaxies.* PhD thesis. Univ. Cambridge. 205 pp.

White DA, Fabian AC, Johnstone RM, Mushotzky RF, Arnaud KA. 1991. *MNRAS* 252:72–81

White DA, Fabian AC, Allen SW, Edge AC, Crawford CS, et al. 1994. *MNRAS*. In press

White RE. 1988. See Fabian 1988a, pp. 343–48

White RE, Sarazin CL. 1987a. *Ap. J.* 318:612–20

White RE, Sarazin CL. 1987b. *Ap. J.* 318:621–28

White RE, Sarazin CL. 1987c. *Ap. J.* 318:629–44

White RE, Sarazin CL. 1988. *Ap. J.* 335:688–702

White SDM, Frenk CS. 1991. *Ap. J.* 379:52–79

White SDM, Rees MJ. 1978. *MNRAS* 183:341–58

Wise MW. 1993. PhD thesis. Univ. Virginia

Wise MW, Sarazin CL. 1990. *Ap. J.* 363:344–48

Wise MW, Sarazin CL. 1992. *Ap. J.* 395:387–402

Wise MW, Sarazin CL. 1993. *Ap. J.* 415:58–74

Yamashita K. 1992. In *Frontiers of X-Ray Astronomy*, pp. 475–80, ed. Y Tanaka, K Koyama. Tokyo: Universal Acad.

Yates MG, Miller L, Peacock JA. 1989. *MNRAS* 240:129–66

Yee HKC, Green RF. 1987. *Ap. J.* 319:28–43

Yoshida T, Hattori M, Habe A. 1991. *MNRAS* 248:630–41

Zhao J-H, Burns JO, Owen FN. 1989. *Astron. J.* 98:64–107

Zhao J-H, Sumi DM, Burns JO, Duric N. 1993. *Ap. J.* 416:51–61

Annu. Rev. Astron. Astrophys. 1994. 32: 319–70

ANISOTROPIES IN THE COSMIC MICROWAVE BACKGROUND

Martin White, Douglas Scott, and Joseph Silk

Center for Particle Astrophysics and Departments of Astronomy and Physics, University of California, Berkeley, California 94720

KEY WORDS: background radiation, cosmology, theory, dark matter, early universe

1.   INTRODUCTION

In 1964, Penzias & Wilson (1965) serendipitously detected the microwave background as anomalous excess noise, coming from all directions and corresponding to a temperature of $\sim$3 K. This was immediately interpreted as being a relic of the Primeval Fireball by Dicke et al (1965), who had already been preparing an experiment in the hope of detecting it. Recently the remarkable success of the FIRAS instrument on the *COBE* satellite has confirmed that the cosmic microwave background radiation has a Planck spectrum with (Mather et al 1994)

$$T_0 = 2.726 \pm 0.010 \, \text{K} \, (95\% \, \text{CL}). \tag{1}$$

The blackbody nature of the cosmic microwave background (CMB) strongly suggests an origin in the early universe. In the standard Big Bang model thermalization occurred at an epoch $t \lesssim 1 \, \text{yr}$ or $T \gtrsim 10^7 \text{K}$. By 1975, the remote origin of the CMB was supported by the high degree of isotropy apart from the detection of the dipole anisotropy (Corey & Wilkinson 1976, Smoot et al 1977). The best-fitting dipole is $D_{\text{obs}} = 3.343 \pm 0.016 \, \text{mK} \, (95\% \, \text{CL})$ towards $(\ell, b) = (264°\!.4 \pm 0.3, 48°\!.4 \pm 0°\!.5)$ (Smoot et al 1991, 1992; Kogut et al 1993; Fixsen et al 1994). After correction for the motion of the Earth around the Sun, the Sun around the Galaxy, and the Galaxy relative to the center of mass of the Local Group, one infers (Smoot et al 1991, Kogut et al 1993) that our Local Group of galaxies is moving at a velocity of $627 \pm 22 \, \text{km s}^{-1}$ in a direction $(\ell, b) = (276° \pm 3°, 30° \pm 3°)$. Convergence of the local velocity vectors to the CMB dipole does not occur until a distance of $> 100 \, h^{-1} \text{Mpc}$

319

0066–4146/94/0915–0319$05.00

(herein the Hubble constant $H_0 = 100\,h\,\mathrm{km\,s^{-1}Mpc^{-1}}$), according to the dipole measured in the *IRAS* all-sky galaxy redshift survey (Strauss et al 1992), and possibly to $> 150\,h^{-1}\mathrm{Mpc}$ if the recent claim of a dipole in the nearby Abell cluster frame is confirmed (Lauer & Postman 1992, 1993; see also Plionis & Valdarnini 1991). Theoretical arguments actually suggest that convergence may only be logarithmic (Juszkiewicz et al 1990) if the large-scale density fluctuation spectrum has the Harrison-Zel'dovich form, $\delta\rho/\rho \propto \lambda^{-(n+3)/2}$ with $n = 1$ (Harrison 1970, Peebles & Yu 1970, Zel'dovich 1972).

Detection of anisotropy on smaller scales than that of the dipole has proved extremely difficult. The original detection paper set limits of about 10% on any anisotropy (Penzias & Wilson 1965). By 1968, the first simplistic theoretical predictions suggested that galaxy formation implied fluctuations in the CMB of the order of 1 part in 10^2 (Sachs & Wolfe 1967) or 10^3 (Silk 1967, 1968). As experimental sensitivity improved, the theoretical calculations grew more sophisticated (e.g. Peebles & Yu 1970, Doroshkevich et al 1978, Wilson & Silk 1981), predicting $\Delta T/T \sim 10^{-4}$ for universes containing predominantly baryonic matter. Claims of an electron neutrino mass of about 30 eV (Lyubimov et al 1980) stimulated interest in non-baryonic dark matter–dominated universes (e.g. Bond et al 1980, Doroshkevich et al 1980). Neutrinos as dark matter failed to account for structure formation (e.g. Kaiser 1983; White et al 1983, 1984) despite the fact that the neutrino window (now closed for ν_e but still open for ν_μ or ν_τ) allows neutrinos to be a plausible dark matter candidate (e.g. Steigman 1993).

After 1980, the inflationary cosmology (see Narlikar & Padmanabhan 1991 for a review) revived interest in non-baryonic dark matter, now considered more likely to be of the cold variety (e.g. Peebles 1982a, Blumenthal et al 1984, Frenk et al 1990). The hot/cold classification (Bond & Szalay 1983) amounts to the velocity dispersion of the candidate particle being much greater or much less than the canonical escape velocity of a typical galaxy: $\sim$300 km s^{-1} at the epoch of equal densities of matter and radiation, $1 + z_{eq} = 23{,}900\,(\Omega_0 h^2)$. Only at later times does substantial sub-horizon fluctuation growth occur.

As limits improved on small-scale fluctuations, to $\Delta T/T \sim 10^{-4}$ (Uson & Wilkinson 1984 a,b,c), refined theoretical estimates showed that, with the aid of dark matter, one could further reduce $\Delta T/T$ by an order of magnitude. (For a summary of the pre-*COBE* experimental situation see Partridge 1988, Readhead & Lawrence 1992). The experimental breakthrough came in 1992 (Smoot et al 1992) with the first detection of large angular scale anisotropies of cosmological origin in the CMB by the *COBE* DMR experiment (Smoot et al 1990). This has since been confirmed by at least one other experiment (Ganga et al 1993). Because of the sky coverage and frequency range spanned, one can now, with confidence, eliminate any Galactic explanation, as well as the possibility that nearby superclusters containing diffuse hot gas are imprinting

Sunyaev-Zel'dovich fluctuations on the CMB (Hogan 1992; Rephaeli 1993a,b; Bennett et al 1993). This conclussion is further strengthened by the lack of correlation with the X-ray background (Boughn & Jahoda 1993). Fluctuations are telling us about density perturbations at $z \sim 1000$. The first year DMR data represent a $> 7\sigma$ detection; the best determined measurement is the sky variance on scales of $10°$, $\sigma_{obs}(10°) = 30 \pm 5 \, \mu$K (Smoot et al 1992).

Several other experiments have subsequently reported detections on intermediate angular scales, $\sim 1°$ (with nine separate claims of detection of fluctuations by the end of 1993). These are all generally consistent with the *COBE* amplitude, given plausible extrapolation from the large angular scales, as described below, although there is cause for serious concern about foreground Galactic contamination as a consequence of limited sky and frequency coverage.

We are certainly now on the verge of a quantum leap in cosmological modeling. Large-scale "seed" power has been discovered at a level of $\Delta T / T \sim 10^{-5}$. These fluctuations are the fossil precursors of the largest structures we see today, which have scales of $\leq 50 \, h^{-1}$Mpc. On angular scales $\geq 10°$, they are also relics of the apparently noncausal initial conditions in the Big Bang, which can be accounted for by inflationary cosmology, and hence provide a possible verification of inflation. Gravity waves are another legacy from inflation, and can leave a distinguishable signature imprinted on the CMB [see Burke (1975) and Doroshkevich et al (1977) for a pre-inflation view]. Indeed, there have been recent proposals to utilize the CMB fluctuations on large scales to reconstruct the inflaton potential.

The connection between large-scale power in the matter distribution and that in the CMB is conceptually simple, if at early epochs one is in the linear regime. At large redshift, a comoving scale of 100 Mpc projects to an angular scale of approximately $\Omega_0 h$ degrees. Complications arise for several reasons. First, the statistical properties of the fluctuations are not known a priori. Inflation predicts that the fluctuations are Gaussian. However, in non-inflationary cosmologies, especially likely if $\Omega_0 < 1$ as favored by observations of the local universe, the intial conditions are non-Gaussian. Moreover, one may have non-linear topological defects as the source of seed density fluctuations. We cannot yet predict with much confidence the likely implications of such models for CMB anisotropies, largely because the connection with large-scale structure observations, to which the theory must ultimately be normalized, is tenuous.

Given an initial spectrum of density fluctuations, $\delta(k)$, one can calculate the transfer function to obtain the radiation power spectrum $P_{rad}(k)$. The scale $z_{eq} \propto (\Omega h^2)$ is imprinted, thereby inevitably guaranteeing a dependence proportional both to Ω_0 and, when normalized to an observed scale, to H_0. Curvature can complicate the matter further since, in a low Ω_0 universe, on scales larger than the curvature radius there is no unique definition of the matter fluctuation power spectrum.

Other cosmological model parameters that enter less directly are Ω_B, Ω_Λ, and Ω_ν, the contributions to Ω_0 in baryons, vacuum, and massive neutrinos, respectively. These all modify the detailed transfer function for a given $\delta(k)$. The ionization history of the Universe is yet another unknown. The intergalactic medium is highly ionized at $z = 5$. If it were even 90% ionized at $z \gtrsim 20$, the modification of the predicted $\Delta T/T$ can become significant, at the 10–20% level, on angular scales of a few degrees. If ionization occurred much earlier, there is strong smoothing of degree-scale fluctuations, but at the cost of regenerating them, together with subarcminute-scale fluctuations in second order, on the new last scattering surface.

This review is arranged as follows. Section 1 presents an overview of recombination, and introduces the various sources of temperature fluctuations. Section 2 summarizes structure formation theory, and the different fluctuation modes. The power spectrum formalism is described in Section 3. In Section 4, we review Gaussian autocorrelation function fitting, window functions, and alternative approaches to data analysis. Higher order effects (such as reionization) are described in Section 5. Problems arising from various types of uncertainties are summarized in Section 6. Section 7 discusses alternatives to the "standard" model and Section 8 describes some issues that a new generation of experiments will have to address.

1.1   *Recombination*

The photons we observe from the microwave background have traveled freely since the matter was highly ionized and they suffered their last Thomson scatterings. If there has been no significant early heat input from galaxy formation, then this happened when the Universe became cool enough for the protons to capture electrons (the recombination epoch). If the Universe was reionized early enough, then the photons will have been scattered more recently, the effects of which we discuss in Section 5.1. To understand the CMB fluctuations we observe, it is crucial to have a good picture of the recombination process.

The process of recombination would proceed via the Saha equation (see e.g. Lang 1980), except that recombinations to the ground state are inhibited by the recombination process itself (Novikov & Zel'dovich 1967). Thus recombination is controlled by the population of the first excited state, and the physical processes which either populate or depopulate it in the expanding Universe. This problem was first worked out in detail by Peebles (1968) and at about the same time by Zel'dovich et al (1968).

Any modification in our understanding of recombination would be crucial for microwave background anisotropies, but in fact little has changed since the seminal work of the late 1960s; the only significant improvement was made by Jones & Wyse (1985), who refined some of the earlier assumptions and included the possibility of non-baryonic matter. Matsuda, Sato & Takeda (1971)

Table 1   Parameters for the redshift of recombination for a range of cosmologies[a]

Ω_0	0.1					0.2			1				
Ω_B	0.1		0.05	0.01		0.05	0.01		0.1		0.05		0.01
h	0.5	1	0.5	1	1	0.5	1	1	0.5	1	0.5	1	1
z_{rec}	1060	1110	1070	1060	1100	1080	1060	1100	1080	1070	1100	1080	1170
Δz	81.8	85.1	85.8	82.5	98.4	89.5	84.4	106	91.5	85.5	104	92.1	135

[a]The location and width are obtained by fitting a Gaussian to $\exp(-\tau)d\tau/dz$.

considered the effects of collisional processes, which are negligible, and Krolik (1989, 1990) showed that two previously unconsidered scattering effects in the Ly α line almost completely cancel one another. Sasaki & Takahara (1993) showed that an accurate treatment of the stimulated rate lowers the ionization "freeze-out" but has no real effect at the recombination epoch.

Solving the coupled equations for the ionized fraction and matter temperature gives the evolution of the ionized fraction $x_e(z) \equiv n_e/n_B$ and the visibility function $g(z) \equiv e^{-\tau}d\tau/dz$, for Thomson scattering optical depth τ. This function measures the probability that the radiation was last scattered in a redshift interval dz. It is reasonably well approximated by a Gaussian with mean $z_{rec} \simeq 1100$ and width $\Delta z \simeq 80$, largely independent of Ω_0, Ω_B, and H_0, as shown in Table 1 (see also Scott 1991). Thus the epoch and thickness of the last scattering surface can be assumed to be independent of the cosmological model, although the amount of scattering will depend on $\Omega_B h^2$, the angular scales will depend on Ω_0, etc. Useful approximations to $x_e(z)$ and $g(z)$ are given by Sunyaev & Zeldovich (1970), Zabotin & Nasel'skiĭ (1982), Jones & Wyse (1985), Grachev & Dubrovich (1991), and Fink (1993). Note that $z_{rec} = 1100$ corresponds to $T_{rec} = 0.26\,\text{eV}$, and $t_{rec} = 5.6 \times 10^{12}(\Omega_0 h^2)^{-1/2}\,\text{sec}$. The thickness of the last scattering surface $\Delta z = 80$ at this epoch corresponds to a comoving scale of $6.6\,\Omega_0^{-1/2}\,h^{-1}\text{Mpc}$ and an angular scale of $3.\!'8\,\Omega_0^{1/2}$.

It is also worth pointing out that in the expanding Universe the fractional ionization approaches a constant, which is significantly different from zero: $x_e(\text{residual}) \propto (\Omega_0 h^2)^{1/2}(\Omega_B h^2)^{-1}$. For models with significant reionization, radiation drag may also be important (see e.g. Peebles 1965, Rees 1977, Hogan 1979, Peebles 1993). The fluctuations cannot grow until the photons release their hold on the matter which happens at $1 + z_{drag} \simeq 120\,(\Omega_0 h^2)^{1/5} x_e^{-2/5}$.

There is a prediction that there must be broad lines in the CMB spectrum due to the photons produced during H (and He) recombination at z_{rec}. These distortions may be large in the Wien region (Peebles 1968, Zel'dovich et al 1968, Lyubarski & Sunyaev 1983, Fahr & Loch 1991), but there are few photons out there, so this effect will be swamped by the background at $\sim 100\,\mu\text{m}$. In the Rayleigh-Jeans region, where there are more photons, the distortions are small (Dubrovich 1975, Bernstein et al 1977), although they may be enhanced if there is some extra energy injection during recombination (Lyubarski & Sunyaev 1983). If such lines could ever be detected they would be a direct probe of Δz and the physics of recombination. Recombination can also lead to trace amounts of the primordial molecules H_2, HD, LiH, etc (Lepp & Shull 1984, Puy et al 1993). Resonance scattering by intergalactic LiH molecules at $z \lesssim 400$ may possibly result in smoothing of CMB fluctuations up to degree scales (Dubrovich 1993, Melchiorri 1993, Maoli et al 1994) at long wavelengths, as the resonance line is redshifted.

1.2 Sources of $\Delta T / T$

In the standard recombination picture, the cosmic plasma becomes neutral, and the microwave background photons are last scattered at redshifts $z \sim 1100$. Hence any observed variation in the intensity of these photons gives us direct information about the Universe at that epoch, and potentially much earlier when the fluctuations were initially laid down. The theory behind microwave background anisotropies is reviewed by Kaiser & Silk (1986), Bond (1988), Efstathiou (1990), and others. Several effects contribute to fluctuations in the observed temperature of the radiation. Schematically, in rough order of importance with decreasing angular scale:

- $\Delta T / T = V_\odot / c$ dipole anisotropy, where $V_\odot$ is our motion relative to the radiation;
- $\Delta T / T = -\delta \phi$ gravitational potential or Sachs-Wolfe fluctuations;
- $\Delta T / T = \frac{1}{3} \delta \rho / \rho$ if the perturbations are adiabatic;
- $\Delta T / T = -\frac{1}{3} \delta S$ if the perturbations are isocurvature;
- $\Delta T / T = v / c$ Doppler shifts, when the photons were last scattered;
- $\Delta T / T = -2kT_e / (m_e c^2)$ Sunyaev-Zel'dovich fluctuations caused by scattering off hot electrons.

We are not concerned here with the dipole contribution (which up until *COBE* was the only measured temperature variation), since it is "extrinsic," being caused by our local motion (Peebles & Wilkinson 1968). [The "intrinsic" dipole is expected to be of similar amplitude to the quadrupole, i.e. $\sim$100 times smaller; similarly the "extrinsic" quadrupole has amplitude $\frac{1}{2}(v/c)$ of the dipole.] Sachs-Wolfe (Sachs & Wolfe 1967) fluctuations dominate at the largest scales, above that subtended by the horizon at last scattering. The sum of the gravitational redshift effect plus the intrinsic fluctuation leads to $\Delta T / T = -\frac{1}{3}\delta \varphi$ for adiabatic perturbations (see Appendix B; Section 2.4).

Large-scale anisotropies are higher in the case of isocurvature perturbations because the potential and intrinsic fluctuations add rather than partially canceling (Efstathiou & Bond 1986, Kodama & Sasaki 1986; see Section 2.4). These large-scale fluctuations are largely independent of any reionization, simply because no causal process during the scattering epoch can affect scales larger than the horizon size. At scales smaller than this, the radiation perturbations relax to a state of pressure equilibrium; therefore there are essentially no fluctuations caused by the peculiar gravity at small scales (Kaiser 1984).

Whether Sunyaev-Zel'dovich (Zel'dovich & Sunyaev 1969; Sunyaev & Zel'dovich 1970, 1972) fluctuations will be important depends on details of the distribution of hot electrons during a possible reionization. They will be largest in pancake (Szalay et al 1983, SubbaRao et al 1994), explosion (Hogan 1984), or other models with significant reheating. However $\Delta T / T \sim 10^{-5}$ is expected in nearby rich clusters on arc-minute scales (e.g. Sunyaev 1978, Rephaeli 1981, Bond 1988, Schaeffer & Silk 1988, Cole & Kaiser 1988, Trester & Canizares 1989, Cavaliere et al 1991, Bond & Myers 1991, Makino & Suto 1993), irrespective of reionization, with fluctuations in arbitrary directions an order of magnitude smaller (e.g. Markevitch et al 1991, Scaramella et al 1993). This effect causes a spectral distortion, which is easiest to observe as a decrement in the Rayleigh-Jeans part, and hence can be distinguished from the other effects, which are primeval in origin (Sunyaev & Zel'dovich 1972, Zel'dovich et al 1972). S-Z distortions have been seen in a few of the richest clusters (e.g. Gull & Northover 1976; Birkinshaw et al 1981, 1984; Uson 1986; Birkinshaw 1990; Klein et al 1991; Herbig et al 1992; Birkinshaw et al 1993; Wilbanks et al 1993; Jones et al 1993).

At small angular scales, the dominant mechanisms are the adiabatic fluctuations (Silk 1967) and scattering off moving electrons (Sunyaev & Zel'dovich 1970). In addition there is a complication at small scales, caused by gravitational lensing of the microwave background radiation by clustered matter (Chitre et al 1986; Blanchard & Schneider 1987; Kashlinsky 1988; Linder 1988a, 1990; Feng & Liu 1992), but Cole & Efstathiou (1989), Tomita & Watanabe (1989), and Sasaki (1989) have shown that the effects are negligible for most models of galaxy formation, except perhaps at sub-arcminute scales (Cayón et al 1993a,b, 1994). Lensing can enhance fluctuations on the smallest scales; the effect can be considered as observing the CMB with a beamwidth of order the dispersion of deflection angles. Furthermore, discrete sources at high redshift, with their radiation reprocessed by one of several emission mechanisms, could produce frequency-dependent fluctuations which would increase toward small angular scales (e.g. Dautcourt 1977; Sunyaev 1977, 1978; Hogan 1980, 1982; Korolëv et al 1986; Bond et al 1986, 1991a). Of course, faint radio sources, as well as diffuse emission from our own Galaxy, can give anisotropies, although these are normally regarded as a contaminant rather than a source (e.g. Danese et al

1983; Banday & Wolfendale 1990, 1991a,b; Banday et al 1991; Franceschini et al 1989; Masi et al 1991; Brandt et al 1994).

Schematically, calculations of the contribution from different sources are performed via an integral through the scattering surface (see, however, Section 5.1). For example, for the Doppler-induced variations, the total perturbation to the radiation temperature can be expressed as:

$$\frac{\Delta T}{T} = \int \left(\frac{v_\parallel}{c}\right) e^{-\tau} \frac{d\tau}{dt} dt, \tag{2}$$

where the integrand is evaluated at time t, with τ the optical depth measured from the present back to t, and $v_\parallel$ the component of peculiar velocity along the line of sight. The exponential factor is an approximation which allows for the effect of multiple scatterings—a very good approximation since the predicted fluctuations are so small.

In fact, to obtain accurate estimates of the fluctuations, and to determine the angular power spectrum, numerical calculations need to be done (see Section 3.2). However, there are semi-analytical methods for either standard recombination, using the tight-coupling limit [which gives both the matter and radiation spectra (Doroshkevich et al 1978; Bonometto et al 1983, 1984; Starobinskiĭ 1988; Doroshkevich 1988; Artio-Barandela et al 1991; Nasel'skiĭ & Novikov 1993; Jørgensen et al 1993, 1994; Dodelson & Jubas 1994, Atrio-Barandela & Doroshkevich 1994a)] or for reionization, using the free-streaming approximation [which assumes a given $T_m(k)$ (Vishniac 1987, Efstathiou 1988, Hu et al 1994, Atrio-Barandela & Doroshkevich 1994b)].

2. THEORY

2.1 *Inflation*

Perhaps the most well studied paradigm for producing density fluctuations in the early universe is the inflationary mechanism. A review of the inflationary predictions for primordial power spectra has recently been completed by Liddle & Lyth (1993). We will include here a brief overview for completeness (see also Narlikar & Padmanabhan 1991).

An inflationary phase drives the Universe towards flat spatial hypersurfaces, i.e. $\Omega + \Lambda/3H^2 = 1$, or $\Omega_0 = 1$ if there is no cosmological constant today (which is the standard assumption, also generally adopted in this review). In the inflationary paradigm, the fluctuations that cause temperature anisotropies in the CMB are generated by fluctuations in quantum fields during the inflationary phase. The wavelength of the fluctuations is stretched by the general expansion until they represent modes outside the horizon. For such modes the equation of motion is simple: The amplitude is a constant or, in the common jargon, they are "frozen in." When the period of inflation ends, the horizon grows faster than the scale factor, and eventually (today) these fluctuations "reenter" the horizon. For a massless field, since the amplitude of the fluctuation is frozen in while the energy of each quantum is redshifting with the general expansion, the number of quanta describing each state has to increase dramatically. Thus when the fluctuation reenters the horizon it can be considered to evolve classically.

During inflation, when the perturbations were generated, the scale is set by the Hubble constant, so we expect that the power spectrum of fluctuations is proportional to $(H_k/m_{Pl})^2$ where H_k is the Hubble constant at the time mode k left the horizon during inflation and m_{Pl} is the Planck mass. This is the entire story for isocurvature fluctuations. For fluctuations in the inflaton field, which are adiabatic and are the primary source of density perturbations from inflation, a solution of the perturbed Einstein equations shows that the energy density associated with the fluctuation grows while it is outside the horizon. The growth depends on the details of the inflaton potential V. The final result is that adiabatic fluctuations are enhanced over their isothermal counterparts by a factor $\propto (V/V')$ (see e.g. Kolb & Turner 1990, Mukhanov et al 1992).

If, during inflation, the Hubble constant changes only slowly, then we expect that the spectrum should be nearly scale invariant (in horizon crossing coordinates) or Harrison-Zel'dovich. If H_k is nearly constant then so is V (using the Friedmann equation) and we expect the enhancement of the adiabatic fluctuations to be large. As the potential becomes steeper the spectrum becomes more tilted, and the isocurvature fluctuations *can* become comparable to their adiabatic counterparts. Since in general V decreases with time, the slope of the power spectrum from inflation generically has more power on larger scales (see, however, Mollerach et al 1993).

The inflationary paradigm generally assumes that the inflaton field, or other fluctuating fields (e.g. axions, gravity), are weakly coupled during the epochs under consideration. In this case the fluctuations are predicted to be Gaussian. Models in which the inflaton field couples strongly to other fields or has nonnegligible self interactions can give rise to non-Gaussian fluctuations (Allen et al 1987). Also, if the evolution equations for the inflaton field are nonlinear, it is possible that fluctuations can be non-Gaussian (Kofman et al 1991, Moscardini et al 1991, Scherrer 1992).

2.2   *Structure Formation Theories*

The standard model of structure formation, which we shall use to explore the CMB fluctuations on degree scales, has been the Cold Dark Matter (CDM) model (e.g. Peebles 1982a, Blumenthal et al 1984, Davis et al 1985, Frenk 1991, Ostriker 1993). In this model, $\Omega_0 = 1$, with a variable fraction Ω_B residing in baryons and the rest in massive (nonrelativistic) dark matter. The initial fluctuations are assumed to be Gaussian distributed, adiabatic, scalar density fluctuations with a Harrison-Zel'dovich spectrum on large scales, i.e. $P_{mat}(k) \propto k^n$ with $n = 1$.

Under these assumptions the matter power spectrum can be calculated. It depends only on Ω_B and the Hubble constant. The main feature is a turnover from the $P_{mat}(k) \propto k$ form at large scales (small k) to a k^{-3} falloff at small scales (large k), which occurs at $k \simeq 0.03\,h\,\mathrm{Mpc}^{-1}$, the horizon size at matter-radiation equality. The reason for this turnover is that perturbations that enter the horizon before the universe becomes matter-dominated do not grow. This retards the growth of fluctuations on small scales, which spend longer periods inside the horizon during the radiation-dominated era. It is assumed the baryons fall into dark matter potentials that formed due to gravitational instability once the photon drag becomes small enough. The standard ($\Omega_0 = 1, n = 1, h = 0.5$) CDM matter spectrum (Holtzman 1989) has been plotted as the solid line in Figure 1 along with the power spectrum for the other models discussed in this section.

An alternative to CDM, now out of favor, is the Hot Dark Matter (HDM) model (e.g. Zel'dovich 1970, Bond et al 1980, Bond & Szalay 1983, Centrella et al 1988, Anninos et al 1991, Cen & Ostriker 1992). In this model, the difference between Ω_B and $\Omega_0 = 1$ is made up of relativistic particles. These relativistic particles have a minimum scale on which gravitational instability

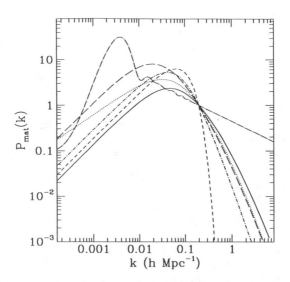

Figure 1 The matter power spectrum for a range of models of structure formation. The models are: CDM (*solid*), tilted CDM (*dotted*), HDM (*short dashed*), ΛCDM (*long dashed*), MDM (*dot-short-dashed*), and BDM (*dot-long-dashed*). All models have been arbitrarily normalized to 1 at $k = 0.2\,h\,\mathrm{Mpc}^{-1}$, which corresponds roughly to σ_8 normalization.

can cause overdensities due to their free streaming. This leads to a small-scale (large k) cutoff in the power spectrum as shown by the short-dashed line in Figure 1 (Holtzman 1989) for $\Omega_\nu = 0.94$ and $h = \frac{1}{2}$ ($m_\nu = 22\,\text{eV}$). In this model, galaxies form by the fragmentation of larger structures unless there is some extra small-scale power (e.g. Dekel 1983, Brandenberger et al 1990, Gratsias et al 1993).

A currently popular idea (e.g. Schaefer & Shafi 1992, M. Davis et al 1992, van Dalen & Schaefer 1992, Taylor & Rowan-Robinson 1992, Holtzman & Primack 1993, Klypin et al 1993) is the Mixed Dark Matter (MDM) model (Shafi & Stecker 1984, Achilli et al 1985, Umemura & Ikeuchi 1985, Valdarnini & Bonometto 1985, Bardeen et al 1987, Schaefer et al 1989, Holtzman 1989) which combines the above two scenarios, and may account for the extra power in large-scale structure measurements over the predictions of CDM (e.g. Maddox et al 1990, Kaiser et al 1991). The current favorite has an HDM component (of relic, massive neutrinos say, with $m_\nu \simeq 7\,\text{eV}$) of 30% of closure density and a CDM component (plus baryons) making up the rest. It is unclear whether this gives the required amplitude of fluctuations at smaller scales (e.g. Bartlett & Silk 1993, Kauffmann & Charlot 1994, Pogosyan & Starobinskiĭ 1993, Mo & Miralda-Escudé 1994). The *radiation* power spectrum for this model is very similar to that of CDM except on the very smallest scales where it is already strongly damped by the thickness of last scattering (see Section 3.2). An approximation to the matter power spectrum is shown by the dot-short-dashed line of Figure 1 (Holtzman 1989).

Alternatively, the difference between the CDM component and $\Omega_0 \simeq 1$ could be made up of a cosmological constant (e.g. Peebles 1984, Turner et al 1984, Efstathiou et al 1990, Kofman et al 1993). Including a nonzero Λ has little effect on large-scale structure, other than the enhancement of large-scale power due to lowering Ω, an effect that is used to account for the galaxy-galaxy correlations found in deep surveys (Maddox et al 1990, Loveday et al 1992) as well as to allow an older universe (e.g. Gunn & Tinsley 1975). The matter power spectrum for $\Omega_\Lambda = 0.8$ and $h = 1$ (Efstathiou et al 1992) is shown in Figure 1.

On all but the smallest scales, the CDM, HDM, ΛCDM, and MDM models give very similar radiation power spectra, showing that large- and intermediate-scale CMB fluctuations are not very sensitive to the form of the dark matter that makes $\Omega_0 = 1$ in these models. Since the shape of the matter spectra are changed, however, the relation between the large-scale power and the *bias* is changed, affecting comparison with large-scale structure studies.

Instead of modifying the matter content of the model, one variant sticks with $\Omega_0 = 1$ in CDM, but modifies the initial power spectrum away from the Harrison-Zel'dovich form (e.g. see Vittorio et al 1988, Salopek et al 1989, Cen et al 1992, Lucchin et al 1992, Fry & Wang 1992, Liddle et al 1992, Adams et al 1992, Cen & Ostriker 1993, Gottlober & Mucket 1993, Muciaccia et al 1993, Sasaki 1993, and Suto et al 1990 for CMB constraints). Such "tilted" spectra can be (but are not always) associated with a stochastic background of gravitational waves. The most commonly discussed model has a spectral slope of $n \simeq 0.7$ (to be compared with the standard $n = 1$ slope) and is shown by the dotted line in Figure 1.

The Baryonic Dark Matter model (BDM, also known as PBI or PIB, see Section 7.2) differs from its cold and hot dark matter rivals in that it does not assume $\Omega_0 = 1$ and postulates isocurvature rather than adiabatic fluctuations (Peebles 1987a,b; Cen et al 1993). The fluctuation spectrum is not taken to be Harrison-Zel'dovich, but the slope of the spectrum of entropy fluctuations ($P_{ent}(k) \propto k^m$) m is a free parameter that is adjusted to fit observational data. The transfer function $T(k) \rightarrow k^2$ at small k and $T(k) \rightarrow 1$ at large k. The principal feature of the (matter) power spectrum is a large "bump" at the matter-radiation Jeans length (Doroshkevich et al 1978, Hogan & Kaiser 1983, Jørgensen et al 1993), the height of which will depend on the ionization history and the values of Ω_B and h. The "bump" at large scales may help explain some of the large-scale velocity measurements (Groth et al 1989). Figure 1 shows a particular model (Cen et al 1993) which has $m = -0.5$, $\Omega_0 = \Omega_B = 0.1$, $h = 0.8$, and $x_e = 0.1$ (dot-dash line).

An orthogonal approach to structure formation is that of the defect models, e.g. global monopoles and textures or cosmic string models. In these models, field configurations known as "defects" which arise due to a phase transition in the early universe form the seeds for matter and radiation fluctuations. The prototypical defect model is the cosmic string model where the "defect" is one dimensional. In this model all the string properties are described by the mass per unit length of the string μ. Cosmologically interesting strings have $G\mu \sim 10^{-6}$. In this model most of the fluctuations are imprinted at high redshift. The fluctuations are non-Gaussian, having strong phase correlations which lead to sharp line discontinuities on the sky (Kaiser & Stebbins 1984, Brandenberger & Turok 1986, Stebbins 1988, Bouchet et al 1988). Most defect models being discussed now (e.g. D. Bennett et al 1992, Bouchet et al 1992, Veeraraghavan & Stebbins 1992, Hara et al 1993, Perivolaropoulos 1993a, Bennett & Rhie 1993, Pen et al 1994, Hindmarsh 1993, Coulson et al 1993, Vollick 1993, Durrer et al 1994, Stebbins & Veeraraghavan 1993, Brandenberger 1993) will produce a roughly scale-invariant CMB fluctuation spectrum as required. The major difference lies in their non-Gaussian nature. Since on large scales (e.g. *COBE*), many defects contribute to the observed fluctuations, the central limit theorem suggests that the fluctuations will look Gaussian. One has to go to smaller scales ($< 1°$) to observe significantly non-Gaussian structure. It is not clear whether there will be an easily detectable difference between the Gaussian and non-Gaussian models currently being considered.

2.3   *The Correlation Function*

It is conventional in CMB work to expand the temperature fluctuations in spherical harmonics

$$\frac{\Delta T}{T}(\theta, \varphi) = \sum_{\ell m} a_{\ell m} Y_{\ell m}(\theta, \varphi), \tag{3}$$

and work in terms of the multipole moments $a_{\ell m}$. One can define the sky correlation function

$$C_{\text{sky}}(\theta_{21}) = \left\langle \frac{\Delta T(\hat{n}_1)}{T} \frac{\Delta T(\hat{n}_2)}{T} \right\rangle_{21}, \tag{4}$$

where the average is taken over the (observed) sky with the separation angle θ_{21} held fixed. Using the properties of $Y_{\ell m}$ the correlation function $C_{\text{sky}}(\theta_{21})$ (hereafter we will drop the subscript "sky") is

$$C(\theta) = \frac{1}{4\pi} \sum_{\ell} a_{\ell}^2 P_{\ell}(\cos \theta), \tag{5}$$

where we have introduced the rotationally symmetric quantity $a_{\ell}^2 \equiv \sum_m |a_{\ell m}|^2$. The a_{ℓ}^2 are not to be confused with $(2\ell + 1)C_{\ell}$ (see later) which is often used to define the power spectrum of fluctuations in a Gaussian theory. At this point we have made no assumption about the underlying theory of fluctuations or the model of structure formation—the a_{ℓ}^2 are purely measured quantities on the sky. [Note that other authors' definitions of various quantities can differ from ours by 4π, $(2\ell + 1)$, or similar factors.] The *COBE* team (C. Bennett et al 1992, Kogut et al 1993, Fixsen et al 1994, Bennett et al 1994) quote results for the first two moments, i.e. $a_{\ell} T_0/\sqrt{(4\pi)}$ for $\ell = 1, 2$,

$$D_{\text{obs}} = 3.343 \pm 0.008 \,\text{mK} \qquad \text{and} \qquad Q_{\text{obs}} = 6 \pm 3 \mu\text{K}. \tag{6}$$

A common way of comparing theory and experiment is through the $a_{\ell m}$. Of course an actual measurement of a temperature difference on the sky involves finite resolution and specific measurement strategies modifying Equation 5. These are usually included in the theoretically predicted correlation function through a *window* or *filter function*, W_{ℓ}, described in Section 4.2.

2.4   *Fluctuations*

Both inflation and defect models predict that the fluctuation spectrum should be stochastic in nature. Thus we live in one sample of an ensemble of "possible universes" which was drawn from a distribution specified by the underlying theory. Due to the weak coupling nature of most inflationary theories, the distribution of fluctuations is predicted to be Gaussian. (The assumption of

Gaussian *fluctuations* is not to be confused with the further restrictive assumption of a Gaussian *shape* for the power spectrum.) In contrast, most defect models predict a non-Gaussian character for the fluctuations.

As discussed in Section 2.2, on large (e.g. *COBE*) scales fluctuations in almost all theories will look Gaussian. On smaller scales there is the possibility of detecting non-Gaussian phase correlations. There have been several tests for non-Gaussian fluctuations proposed, e.g. the total curvature or genus of isotemperature contours (Coles 1988b, Gott et al 1990, Brandenberger et al 1993, Smoot et al 1994), the distribution of peaks (Sazhin 1985, Zabotin & Nasel'skiĭ 1985, Bond & Efstathiou 1987, Vittorio & Juszkiewicz 1987, Coles & Barrow 1987, Coles 1988a, Gutiérrez de la Cruz et al 1993, Cayón et al 1993c, Kashlinsky 1993b), skewness and kurtosis of the observed temperature distribution (Scaramella & Vittorio 1991, 1993a; Luo & Schramm 1993a; Perivolaropoulos 1993b; Moessner et al 1993), or the 3-point function (Falk et al 1993, Luo & Schramm 1993b, Srednicki 1993, Graham et al 1993). Of these, the 3-point function currently seems the most promising, although more work is needed to see if it can detect departures from Gaussianness at the level predicted by the "defect" theories.

In general, the predictions of a theory are expressed in terms of predictions for the $a_{\ell m}$. If the fluctuations are Gaussian, the predictions are fully specified by giving the 2-point function for the $a_{\ell m}$ [just as the matter spectrum $P_{\text{mat}}(k)$ is predicted by giving $\xi(r)$]. Using rotational symmetry, it is conventional to write

$$\left\langle a_{\ell m}^* a_{\ell' m'} \right\rangle_{\text{ens}} \equiv C_\ell \, \delta_{\ell' \ell} \delta_{m' m}, \tag{7}$$

where the angle brackets here represent an average over the ensemble of possible universes. The prediction for CMB anisotropy measurements of a theory can thus be expressed as a series of C_ℓs. Alternatively one can work in k-space and define a (3-dimensional) power spectrum of fluctuations per wavelength interval. We describe this, and its relation to the C_ℓs, in Section 3.2. For large ℓ, the C_ℓs are approximately the same as the 2-dimensional power spectrum of fluctuations (Bond & Efstathiou 1987).

2.4.1 ADIABATIC   Any initial density perturbation may be decomposed into a sum of an adiabatic and an isocurvature perturbation. Since inflation naturally predicts adiabatic density perturbations, we will consider these first.

Adiabatic modes are fluctuations in the energy density, or the number of particles, such that the *specific entropy* is constant for any species i (assumed nonrelativistic here):

$$\delta \left(\frac{n_\gamma}{n_i} \right) \propto \frac{\delta \rho_i}{\rho_i} - \frac{3}{4} \frac{\delta \rho_\gamma}{\rho_\gamma} = 0. \tag{8}$$

In terms of the perfect fluid stress-energy tensor of general relativity the assumption of an adiabatic perturbation is equivalent to assuming that the pressure fluctuation is proportional to the energy density perturbation. For a discussion of the classification and behavior of adiabatic perturbations in relativity see Kolb & Turner (1990), Efstathiou (1990), Mukhanov et al (1992), and Liddle & Lyth (1993).

It has been the standard assumption that inflation predicts a "flat" or Harrison-Zel'dovich ($n = 1$) spectrum of adiabatic density fluctuations. Recently, with the advent of the *COBE* measurement, it has been emphasized that inflation generically predicts departures from the simple $n = 1$ form. In "new" inflation, the departure is logarithmic, e.g. $P_{\text{mat}}(k) \propto k \log^3(k/k_0)$, with $k_0 \sim e^{60} \text{Mpc}^{-1}$—this can lead to an effectively tilted spectrum between say, *COBE* scales and more traditional normalization scales (typically $n \approx 0.95$) even though the spectrum is "flat" on large scales. In "power-law," "chaotic," "natural," or "extended" inflation, one can have power law spectra with $n < 1$. Exotic models of inflation even allow $n > 1$ (Mollerach et al 1993).

On scales larger than the horizon size at last scattering (i.e. $\sim 2°$), the generation of temperature fluctuations from density inhomogeneities is straightforward to analyze. In addition to fluctuations in the temperature on the surface of last scattering (due to fluctuations in the radiation energy density), the matter perturbations give rise to potentials on the scattering surface and possibly time-dependent perturbations in the metric. Any time dependence (such as gravity waves, to be discussed later) leads to energy nonconservation along the photon line of sight. The potentials give rise to a "red"-shifting of photons as they leave the last scattering surface. To first order in the perturbing quantity, the total energy change of a photon (above and beyond the cosmological redshifting) from the time it leaves the last scattering surface (emission) is the integral along the unperturbed path of the (conformal) time derivative of the metric perturbation ($h_{\mu\nu}$), plus the change in the potential between last scattering and observation (see Appendix B):

$$\left(\frac{\Delta T}{T}\right)_{\text{SW}} = \left. \Phi \right|_e^o - \frac{1}{2} \int_e^o h_{\rho\sigma,0} \, n^\rho n^\sigma \, d\zeta, \tag{9}$$

where n^ρ is the direction vector of the photon and ζ is a parameter along the line of sight. If $h_{\mu\nu}$ is due solely to density perturbations, the integrand is basically $4\dot{\Phi}$, where the overdot represents a (conformal) time derivative. Either or both of these terms are known as the Sachs-Wolfe effect. The simplest part is the potential difference between the last scatterers and the observer. [The other (integral) term is usually associated with a background of gravitational waves, nonlinear effects, or $\Omega_0 < 1$ universes—see later.] To this energy shift must be added the temperature fluctuation on the last scattering surface itself. For fluctuations in the radiation field, we have

$$\frac{\Delta T}{T} = \frac{1}{4}\frac{\delta\rho_\gamma}{\rho_\gamma} = \frac{1}{3}\frac{\delta\rho}{\rho}, \tag{10}$$

where the second equality follows from the adiabatic condition. Because an overdensity gives a larger gravitational potential $\{\delta\rho/\rho = 2\delta\phi + \mathcal{O}[(k/H)^2]\}$ that a photon must climb out of, for adiabatic fluctuations, the two terms partially cancel. One finds that $\Delta T/T = -\frac{1}{3}\delta\phi$. The minus sign means that CMB hot spots are matter *under*-densities.

2.4.2 ISOCURVATURE   Isocurvature modes are fluctuations in the number density of particles which do not affect the total energy density. They perturb the specific entropy or the equation of state,

$$\delta s \equiv \delta\left(\frac{n_\gamma}{n_B}\right) \propto \frac{\delta\rho_B}{\rho_B} - \frac{3}{4}\frac{\delta\rho_\gamma}{\rho_\gamma} \neq 0. \tag{11}$$

While such perturbations are outside the horizon, causality precludes them from becoming an energy density perturbation. Inside the horizon, however, pressure gradients can convert an isocurvature perturbation into an energy density fluctuation.

The possibility of scalar isocurvature fluctuations is not well motivated by usual inflation models, although if more than one field contributes significantly to the energy density during inflation one can get isocurvature fluctuations (the energy density fluctuation is no longer proportional to the pressure fluctuation). For isocurvature fluctuations, a positive fluctuation in the matter density (and therefore the gravitational potential) is compensated by a negative fluctuation in the photon temperature. The Sachs-Wolfe effect and the initial temperature fluctuation therefore add (rather than cancel as in the adiabatic case), giving rise to six times more large-scale $\Delta T/T$ for a given "matter" perturbation. For this reason, isocurvature cold dark matter models that are normalized to give the observed peculiar velocities predict too large a temperature anisotropy in the CMB.

Specifically, CDM isocurvature models with roughly scale-invariant (i.e. $m = -3$) power spectra (e.g. in the axion model of Axenides et al 1993) are probably ruled out (Efstathiou & Bond 1986). The situation is similar for HDM (Sugiyama et al 1989). Scale-invariant baryon-dominated models are also in serious conflict with the microwave background anisotropies (Efstathiou 1988), and cannot be saved even by invoking a cosmological constant (Gouda & Sugiyama 1992). However, models with larger m are not as yet ruled out (see also Efstathiou & Bond 1987). Isocurvature fluctuations are these days only discussed in terms of the Baryonic Dark Matter model. This is an observationally-motivated model, with low Ω_0 in baryons only. The large fluctuations generated at small scales have to be erased by the reheating due to some early collapsed objects. The effects of such a reionization will be discussed later. Constraints from anisotropies on scales $\gtrsim 1°$ (Peebles 1987b, Sugiyama & Gouda 1992), from the Vishniac effect at small scales (Efstathiou 1988, Hu et al 1994), from spectral distortions (Daly 1991, Barrow & Coles 1991), and from the clustering properties of galaxies (Cen et al 1993) imply that only models with $-1 \lesssim m \lesssim 0$ are viable. High values of Ω_0 and high values of h which enhance the "bump" also tend to be ruled out. It has recently been shown (Sugiyama & Silk 1994) that the BDM picture generally leads to an effective slope $n_{\mathrm{eff}} \simeq 2$ for the radiation power spectrum on large scales. Fluctuations on smaller angular scales depend on a number of tunable parameters, making BDM complicated to constrain in practice (Hu & Sugiyama 1994b,c).

2.4.3 GRAVITATIONAL WAVES   Until now, we have focused on the anisotropies in the cosmic microwave background arising from density perturbations in the early universe. In many models, there is also the possibility that a stochastic background of long-wavelength gravitational waves (GW) can be produced (Starobinskiĭ 1979); for a discussion of inflationary models in this context see Rubakov et al (1982), Adams et al (1992), and Liddle & Lyth (1993). If such a background were to exist, it would leave an imprint on the CMB at large scales through the Sachs-Wolfe effect (Fabbri & Pollock 1983, Abbott & Wise 1984c, Starobinskiĭ 1985, Abbott & Schaefer 1986, Fabbri et al 1987, Linder 1988b, White 1992). With the advent of the *COBE* measurement of the power at large scales, many authors addressed the question of the interpretation in terms of scalar and tensor contributions (Krauss & White 1992, Liddle & Lyth 1992, Adams et al 1992, Salopek 1992, Lucchin et al 1992, Dolgov & Silk 1993).

If there is a sizable contribution from GW in the *COBE*-detected anisotropies, this would lower the predicted value of $(\Delta T / T)_{\mathrm{rms}}$ on smaller scales. This should be kept in mind when comparing degree-scale experiments or large-scale structure studies to power spectra normalized to *COBE* on large scales.

Unlike the anisotropies generated by scalar fluctuations (Section 3), those generated by (isocurvature) tensor perturbations, or GW, damp at scales comparable to the horizon (see e.g. Starobinskiĭ 1985, Turner et al 1993, Atrio-Barandela & Silk 1994), which means $\ell \sim \sqrt{1 + z_{\mathrm{rec}}} \simeq 30$ (see Appendix A). This can be understood as due to the redshifting of GW that entered the horizon before recombination. The maximal contribution to the anisotropy on some scale comes from gravitational waves with wavelengths comparable to that scale. GW begin to redshift after they enter the horizon; thus scales that are smaller than the horizon at last scattering are dominated by GW that have redshifted before the photon begins to travel to us. The different behavior at small scales leads one to hope that the two contributions could be disentangled. A detailed numerical analysis of the anisotropy generated by GW on both large and small scales has been carried out by Crittenden et al (1993a).

In general, GW provide a small contribution to $\Delta T / T$ on top of the scalar anisotropy. One requires a comparison of both large- and small-scale temperature anisotropies to isolate them. On large scales, one must deal with cosmic variance; on small scales one has sample variance and uncertainties due to cosmological parameters and history, which are far from orthogonal. The situation with regard to disentangling a gravitational wave signal is somewhat confused. White et al (1993) claim that cosmic variance and cosmological model uncertainty makes such a detection extremely difficult, while Crittenden et al (1993a) predict that a definitive detection is possible. [The analysis assumed a specific form for the relation between the spectral index and the ratio of scalar and tensor contributions to the quadrupole: $T/S = 7(1 - n)$ (R. Davis et al 1992). This form requires correction for most theories (Liddle & Lyth 1992, Kolb & Vadas 1993) and also biases the fit towards "detection" of a tensor component. In addition, recent work (Bond et al 1994) suggests that including uncertainties in cosmological history may alter Crittenden et al's conclusions regarding gravity waves.] This question is of some importance, since any possible GW signal will affect the power spectrum normalization inferred from *COBE*. Since the GW production predicted in most theories is very small [for $n \gtrsim 0.9$ as required by *COBE* and Tenerife (Hancock et al 1994) $T/S \lesssim 1$], perhaps their only observable effect for some time will be in generating large angular scale CMB anisotropies (Sahni 1990, Krauss & White 1992, Souradeep & Sahni 1992, Liddle 1994, Turner et al 1993). The possibility that GW lead to an observable polarization in the CMB (Polnarev 1985) has been shown to be very small (Crittenden et al 1993b; however, see Frewin et al 1993).

In theories of inflation, the normalization of the spectrum of scalar fluctuations depends on both the inflaton potential and its derivative at the epoch of fluctuation generation. In contrast, the tensor spectrum depends only on the value of the inflaton potential at the same epoch. This fact coupled with the *COBE* measurement can be used to limit the scale of inflation (Rubakov et al 1982, Lyth 1985, Krauss & White 1992, Liddle 1994). In principle, one can also derive information about the inflaton potential from both the tensor and scalar components of the CMB anisotropy (Liddle & Lyth 1992, R. Davis et al 1992, Salopek 1992). Recently, several authors have considered the possibility of reconstructing the "inflaton" potential from CMB observations (Hodges & Blumenthal 1990; Copeland et al 1993a,b, 1994; Lidsey & Tavakol 1993; Turner 1993; see also Carr & Lidsey 1993) or of using relations between observable parameters as "tests" of inflation (R. Davis et al 1992, Bond et al 1994, but see Liddle & Lyth 1992, Kolb & Vadas 1993).

3. POWER SPECTRUM

3.1 *Power Spectrum on Large Scales*

Let us take the power spectrum of primordial fluctuations to be a power law in comoving wavenumber k. In the "processed" radiation power spectrum, this simple power law is multiplied by a transfer function $T^2(k)$. On *COBE* scales

$T(k) \approx 1$, and we can write the temperature fluctuation power spectrum as

$$P_{rad}(k) = A (k\eta_0)^{n-1}, \tag{12}$$

where A is the amplitude for scalar perturbations and $\eta_0 \simeq 3t_0 = 2H_0^{-1}$ (for $\Omega_0 = 1$) is the conformal time today with scale factor normalized to unity. By using $k\eta_0$ as our fundamental variable, we have A as a dimensionless number multiplied by something of order 10^{n-1} on *COBE* scales. The connection between A and the normalization of the matter power spectrum is discussed in Section 3.3. [Another common convention is to define the matter power spectrum as $P_{mat}(k) = Bk^n$ on large scales, which means that the dimensions of B will depend on n (and will be length^4 for $n = 1$); see Equation (24).]

We can write the average over universes of the moments of the temperature anisotropy as

$$C_\ell \equiv \langle |a_{\ell m}|^2 \rangle \tag{13}$$

$$\simeq 16\pi \int_0^\infty \frac{dk}{k} A(k\eta_0)^{n-1} T^2(k)\, j_\ell^2(k\eta_0) \tag{14}$$

$$= 2^n \pi^2 A \frac{\Gamma(3-n)\Gamma\left(\ell + \frac{n-1}{2}\right)}{\Gamma^2\left(\frac{4-n}{2}\right)\Gamma\left(\ell + \frac{5-n}{2}\right)}, \qquad \text{if } T(k) = 1 \tag{15}$$

(see e.g. Peebles 1982c, Bond & Efstathiou 1987). For the special case of $n = 1$, we have $C_2/A = 4\pi/3$ and $C_\ell^{-1} \propto \ell(\ell + 1)$. This is often referred to as "flat" since potential fluctuations (and the amplitude of $\delta\rho/\rho$ at horizon crossing) are independent of scale, and it also makes $\ell(\ell + 1)C_\ell = $ constant.

In some older literature, the normalization of the power spectrum is given in terms of ϵ_H, the dimensionless amplitude of matter fluctuations at horizon crossing. For a flat spectrum this quantity is simply $\epsilon_H^2 = (4/\pi)A$.

The normalization convention used by the *COBE* group, Q_{rms-PS}, is obtained by a best fit to the correlation function assuming a flat spectrum of fluctuations and allowing the normalization to vary. In terms of C_2, this corresponds to

$$Q_{rms-PS} \simeq \langle Q_{rms}^2 \rangle^{0.5} = T_0 \left(\frac{5C_2}{4\pi}\right)^{1/2} \tag{16}$$

[For $n = 1$ the factor in parenthesis is $(5/3)A$, which allows a simple conversion from *quadrupole normalization* to our normalization in terms of A.] We would like to stress that Q_{rms-PS} is the *COBE* group's best estimate, measured from our sky, of the power spectrum normalization. It is *not* the quadrupole measured by the *COBE* team from their maps. The value quoted for Q_{rms-PS}, including the effects of systematic error, is (Smoot et al 1992, Wright et al 1994a, Bennett et al 1994)

$$Q_{rms-PS} = 17.6 \pm 1.5 \ \mu K,$$

which implies

$$A = (2.3 \pm 0.4) \times 10^{-11}, \quad B = (5.9 \pm 1.0) \times 10^5 (h^{-1}\text{Mpc})^4. \tag{17}$$

Since the analysis for $Q_{\text{rms-PS}}$ assumed a flat spectrum, one should not use (17) to normalize other spectra, although $\langle Q_{\text{rms}}^2 \rangle^{0.5}$ is still a valid way of quoting the normalization of the power spectrum.

For the first year data, a fit to the correlation function gives $n = 1.1 \pm 0.5$ (Smoot et al 1992). Including the second year data gives $n \approx 1.5 \pm 0.5$ (Bennett et al 1994, Wright et al 1994b) [by combining both *COBE* and Tenerife data, a stronger limit $n \gtrsim 0.9$ has been obtained (Hancock et al 1994)] and the inferred value of $Q_{\text{rms-PS}}$ is quite correlated with n (Seljak & Bertschinger 1993, Watson & Gutiérrez de la Cruz 1993). For $n \neq 1$ the best value for the normalization is (Smoot et al 1992)

$$\left. \frac{\Delta T}{T} \right|_{10°} = (1.1 \pm 0.1) \times 10^{-5} \tag{18}$$

which probes a range of ℓ centered around $\ell \approx 4$ (Wright et al 1994a). Note that for $n = 1$, these two normalizations differ by $\sim 10\%$, since the fit to $Q_{\text{rms-PS}}$ uses the full correlation function.

Another normalization sometimes used is the *bias*, defined through

$$b_\rho^{-2} = \sigma_8^2 = \int_0^\infty \frac{dk}{k} A(k\eta_0)^{n+3} T_m^2(k) \left. \left[\frac{3j_1(kr)}{kr} \right]^2 \right|_{r=8\,h^{-1}\text{Mpc}}, \tag{19}$$

where $T_m(k)$ is a matter transfer function (see later section) not to be confused with $T(k)$, and $\sigma^2(r)$ is the variance of the density field within spheres of radius r. The variance of galaxies, possibly biased relative to the matter ($\delta_{\text{gal}} = b\delta_\rho$), is roughly unity on a scale of $8\,h^{-1}$Mpc (Davis & Peebles 1983). Equation (19) is nontrivial to evaluate numerically because of the "ringing" of the j_1 and the final result is dependent on the transfer function assumed. For CDM, we will take (Efstathiou 1990)

$$T_m(k) = \left\{ 1 + \left[ak + (bk)^{3/2} + (ck)^2 \right]^\nu \right\}^{-1/\nu} \tag{20}$$

with $a = 6.4\Omega_0 h^{-2}$Mpc, $b = 3\Omega_0 h^{-2}$Mpc, $c = 1.7\Omega_0 h^{-2}$Mpc, and $\nu = 1.13$. We will set $\Omega_0 = 1$ and $h = 1/2$ unless otherwise noted. For $n = 1$ the *COBE* best fit gives $\sigma_8 \simeq 1.2$, i.e. an essentially unbiased model. However, this depends on the adopted values of Ω_0, h, etc. It is possible to have a nonstandard (e.g. $\Omega_\Lambda \simeq 0.8$) CDM model with the galaxies significantly biased on small scales as seems to be required (e.g. Davis et al 1985, Bardeen et al 1986, Frenk et al 1990, Carlberg 1991).

Large-scale flows also provide a measure of the power spectrum (Peebles 1993):

$$v_{rms}^2(r) \propto \int \frac{dk}{k} A(k\eta_0)^{n+1} T_m^2(k) e^{-k^2 r^2},$$ (21)

where v_{rms} is the 3-D velocity dispersion smoothed with a Gaussian filter of width r. This tends to probe scales similar to the degree-scale CMB experiments. Whether there is agreement between the two measures for a particular theory is still a matter of debate (see e.g. Vittorio & Silk 1985; Juszkiewicz et al 1987; Suto et al 1988; Atrio-Barandela et al 1991; Kashlinsky 1991, 1992, 1993a; Górski 1991, 1992).

3.2 Power Spectra on Smaller Scales

Although on large angular scales the transfer function is $T(k) \approx 1$, there is significant structure on small to intermediate scales. Generally the Sachs-Wolfe effect dominates the power spectrum on scales larger than the horizon size at last scattering and the power spectrum can be taken to be a power law. On smaller scales, however, causal interactions become important and the spectrum is modified.

For a given cosmological model, the shape of the fluctuation spectrum is fixed, and depends on the primordial spectrum (e.g. k^n) and its evolution as the waves enter the horizon. This makes the spectrum at smaller scales dependent on Ω_0, Ω_B, H_0, and the dark matter. Given a cosmological model, however, both the radiation and matter power spectra are well defined. For a fixed Ω_0, the radiation power spectrum depends only on the type of dark matter at the high k end.

The calculation proceeds by considering perturbations in the photon distribution function, $f(x, q, t)$, which can be written in terms of

$$\Delta = \int d^3q \, q \, \delta f \Big/ \int d^3q \, q \bar{f},$$ (22)

where $\bar{f}$ is the Planck function and q is the comoving photon momentum. Δ is the total energy, or brightness, perturbation, which is also $4 \, \Delta T / T$ for a uniform shift in temperature. Liouville's theorem tells us that the total phase space density is conserved for collisionless particles, i.e. the distribution function is constant along particle paths: $Df/Dt = 0$. Source terms ("collisions") add on the right hand side; the important sources are baryon velocities and photon density perturbations, coupled through Thomson scattering. The Boltzmann equation (or equation of radiative transfer, written here in the synchronous gauge) for Δ is

$$\dot{\Delta} + c\gamma_i \frac{\partial \Delta}{\partial x^i} - 2\gamma_i \gamma_j \dot{h}_{ij} = \sigma_T c n_e a (\Delta_0 - \Delta + 4\gamma_i v_B^i / c)$$ (23)

(see Peebles & Yu 1970, Hu et al 1994, Dodelson & Jubas 1994), where the dot denotes differentiation with respect to conformal time η ($= \int dt/a$), γ_i are the direction cosines defined by $\hat{q}$, n_e is the number density of free electrons, $a(t)$ is the cosmological scale-factor, v_B is the baryon peculiar velocity, Δ_0 is the isotropic part of Δ, h_{ij} is the metric perturbation, and isotropic scattering has been assumed. The Boltzmann equation is generally Fourier transformed, since in the linear approximation the different k-modes evolve independently, which makes the calculation tractable. It is also assumed that all fluctuations are still in the linear regime, which is a reasonable approximation for the relevant scales at the scattering epoch. Since the effects of the radiation on the matter cannot be ignored (except for late reionization), the equation of motion for the baryons needs to be solved simultaneously; this is the continuity equation for matter evolving freely, but also takes into account the Compton drag at early times.

The radiation and baryons evolve as a coupled fluid at early times, but need to be followed more accurately as the Universe recombines, and eventually the baryons decouple from the photons entirely. After this point, the photons can be assumed to free-stream to the observer, and the subsequent behavior of the anisotropies is often treated analytically. Dark matter evolves collisionlessly throughout, although its gravity feeds back into the baryon and photon evolution. Detailed calculations along these lines have been carried out for many different cosmological models, e.g. the work of Peebles & Yu (1970), Wilson & Silk (1981), Bond & Efstathiou (1984, 1987), Vittorio & Silk (1984, 1992), Holtzman (1989), Sugiyama & Gouda (1992), Dodelson & Jubas (1993a), Stompor (1993), and Crittenden et al (1993a,b) among others. The Boltzmann equation is usually solved by expanding the Δs in Legendre polynomials up to some high enough ℓ and numerically integrating the coupled equations.

We show in Figure 2 the power spectrum for CDM models as a function of wavenumber k, for a range of Ω_B consistent with Big Bang Nucleosynthesis (BBN) (Krauss & Romanelli 1990, Walker et al 1991, Smith et al 1993). To calculate the C_ℓ from these power spectra approximately, one integrates the power spectrum with measure $j_\ell^2(k\eta_0)dk/k$. Each C_ℓ is thus in effect an average of the power spectrum around $k \approx \ell H_0/2$. We have also included an $x_e = 1$ reionized BDM power spectrum (arbitrarily normalized) on the figure for comparison.

The plateau in Figure 2 at low k is the contribution from the Sachs-Wolfe effect (or gravitational redshift), which damps on the scale of the scattering surface thickness. The bumps and wiggles reflect the phase of the oscillating baryons and photons when recombination occurs, i.e. the number of oscillations a given mode has undergone in the time between entering the horizon and the switch-off of radiation pressure when the matter becomes neutral. These so-called "Doppler" peaks (physically they come from a combination of v and Δ sources which are difficult to separate) rise at $k \sim 0.01$ Mpc, the size of the

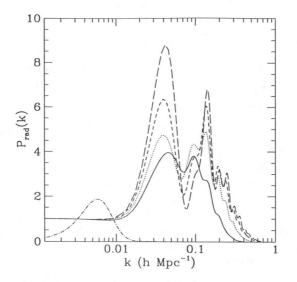

Figure 2   Power spectrum for "standard" CDM models ($h = 1/2$, $\Omega_0 = 1$ and $\Omega_\nu = \Omega_\Lambda = 0$) with $\Omega_B = 0.01$ (*solid*), 0.03 (*dotted*), 0.06 (*short-dashed*), and 0.10 (*long-dashed*) consistent with the range from BBN, from Sugiyama & Gouda (1992). The curves have been normalized to unity at small k. For comparison we also show a fully-ionized BDM model (*dot-dashed line*) with $\Omega_0 = 0.1$, $n = 0$, and $h = 0.5$, chosen (arbitrarily) to match at $k = 0.002\,h\,\mathrm{Mpc}^{-1}$.

horizon at the time of last scattering. In ℓ-space the rise to the first Doppler peak is quite gradual, which can lead an effective $n > 1$ even on relatively large scales, e.g. $n_{\mathrm{eff}} \simeq 1.15$ on *COBE* scales (Bond 1994). The height of the Doppler peaks is dependent on the number of scatterers or $\Omega_B h^2$ (scaling roughly as $\Omega_B^{1/3}$, which would be the dependence for an experiment like MAX; see also Fukugita et al 1990). The second and third peaks are harmonics of the first which is the fluctuation that underwent half an oscillation since it entered the horizon. The amplitude of these oscillations reflects the amount of growth before the perturbation enters the photon-baryon Jeans scale, which depends on $\Omega_B h^2$. There is a further dependence on h through z_{eq}. There is also an effect due to the baryons falling into the dark matter potential wells after recombination, although this happens largely after the photons have been scattered. CDM fluctuations that first entered the horizon during radiation domination suffer growth suppression, which partly accounts for the differing heights of the peaks. Note, also, the exponential cutoff in the power spectrum at large k, due to the effects of the thickness of the last scattering surface, as well as the familiar damping (Silk 1968) of baryon and photon fluctuations prior to decoupling. The damping scale is set by the thickness Δz (see Appendix A).

For a reionized model (see Section 5.1) the visibility function is centered around some z_* and has width $\Delta z \sim z_*$, so that the two important scales above are almost the same. The effect is that there is only one Doppler peak, which is at the scale of the horizon at z_* (i.e. smaller k or ℓ), with damping for higher wavenumbers. There are no extra bumps and wiggles, since at the time of last scattering the photons and baryons were no longer oscillating.

3.3   *The Observed Power Spectrum*

In this section, we consider the current status of measurements of the matter power spectrum (see Peacock 1991, Peacock & Dodds 1994), and the radiation power spectrum (see Bond 1993) in the context of inflation (see Liddle & Lyth 1993). In Figure 3 we compare the radiation power spectrum with the matter power spectrum measured by *IRAS*-selected galaxies (Fisher et al 1993, Feldman et al 1994), the CfA redshift survey (Vogeley et al 1992), and the APM galaxy survey (Baugh & Efstathiou 1993a,b).

Since one cannot make a theory-independent extrapolation from the radiation power spectrum (probed by CMB measurements) to the matter power spectrum (probed by large-scale structure work) in the following, we will assume CDM and write the matter power spectrum as

$$
\begin{aligned}
P_{\text{mat}}(k) &= 2\pi^2 \eta_0^4 A \, k T_m^2(k) \\
&= 2.55 \times 10^{16} A \, (k/h \, \text{Mpc}^{-1}) \, T_m^2(k) \, (h^{-1}\text{Mpc})^3,
\end{aligned} \tag{24}
$$

where A is as in (12) and T_m is a matter transfer function (see Equation 20), not to be confused with $T(k)$. The change in units and the index of the power law from k^0 to k^1 is a matter of convention. We follow the usual convention in large-scale structure (LSS) work that a "flat" spectrum has $P_{\text{mat}}(k) \propto k$ (see Liddle & Lyth 1993 for more discussion on the various definitions of power spectra) and work in units of h^{-1}Mpc. To convert from our normalization to that of the bias or σ_8 conventionally used in LSS work, see Equation (19); note that this conversion is senstive to any variation in the theory. The large-scale structure and CMB data are shown in Figure 3.

For the CMB anisotropy measurements, we have chosen some recent experiments for which we could estimate the best-fit normalization. Specifically, we show results from *COBE* (Smoot et al 1991, 1992; Wright et al 1994a), FIRS/MIT (Page et al 1990; Ganga et al 1993; Bond 1993, 1994), Tenerife (Davies et al 1992, Watson et al 1992, Hancock et al 1994), Python (Dragovan et al 1993), ARGO (de Bernardis et al 1993, 1994), SP91/ACME (13-point) (Schuster et al 1993), Saskatoon (Wollack et al 1993), MAX [MuP (Meinhold et al 1993) and GUM (Devlin et al 1993, Gundersen et al 1993)], and MSAM (Cheng et al 1994). We have concentrated here on those experiments that quote a detection, leaving out those that give only upper limits, e.g. Relikt (Klypin et al 1992), 19.2 GHz (Boughn et al 1992), SP89 (Meinhold & Lubin 1991),

SP91-9pt (Gaier et al 1992), ULISSE (de Bernardis et al 1992), White Dish (Tucker et al 1993), OVRO (Readhead et al 1989, Myers 1993), and VLA (Fomalont et al 1993). Note we have also avoided complicated issues involving redshift space corrections (e.g. Kaiser 1987) to the large-scale structure data.

The $P_{mat}(k)$ inferred from CMB anisotropies depends on the assumed theory (Ω_0 for example shifts the correspondence between θ and k, and shifts the amplitude for a given $\Delta T/T$). Consequently, the boxes in Figure 3 would have to be redrawn for each theory.

In addition to the survey data shown in Figure 3, there is information on large-scale flows (e.g. Kashlinsky & Jones 1991). Bertschinger et al (1990a) estimated the 3-D velocity dispersion of galaxies within spheres of radius $40\,h^{-1}$Mpc and $60\,h^{-1}$Mpc. After smoothing with a Gaussian filter on $12\,h^{-1}$ Mpc scales they found $\sigma_v(60) = 327 \pm 82\,{\rm km\,s^{-1}}$ and $\sigma_v(40) = 388 \pm$

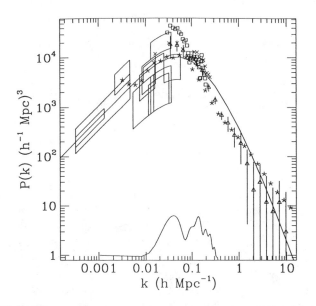

Figure 3 The CDM matter power spectrum on a range of scales as inferred from CMB and LSS data. The solid line is CDM normalized to *COBE*. Stars, crosses, squares, and triangles are the APM, CfA, *IRAS*-QDOT, and *IRAS*-1.2Jy surveys respectively, with *IRAS* surveys scaled to $\sigma_8^{DM} = 1$. Apart from scaling the *IRAS* surveys, we have avoided complicated issues related to normalization and redshift corrections. We show error bars on the *IRAS*-1.2Jy survey to indicate the approximate accuracy involved. The boxes are $\pm 1\sigma$ values of the matter power spectrum inferred from CMB measurements assuming CDM with $\Omega_B = 0.06$, with the horizontal extent taken to be between the half-peak points of the window functions for each experiment. From left to right the experiments are *COBE*, FIRS, Tenerife, SP91-13pt, Saskatoon, Python, ARGO, MSAM2, MAX-GUM & MAX-MuP, MSAM3. The radiation power spectrum for CDM assuming $\Omega_B = 0.06$ is shown at the bottom.

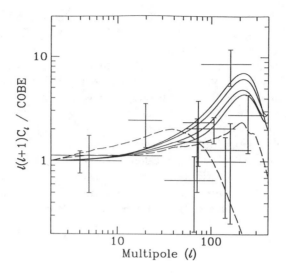

Figure 4 The normalization of a Harrison-Zel'dovich power spectrum of fluctuations required to reproduce the $\Delta T / T_{\text{rms}}$ quoted for each experiment. The horizontal error bar on each point gives the range of ℓ between the half-peak points of the window function. These points should be interpreted as only loose approximations to the results of a full analysis. From left to right the experiments are *COBE*, FIRS, Tenerife, SP91 (13 point), Saskatoon, Python, ARGO, MSAM (2-beam), MAX GUM & MuP, MSAM (3-beam). Also plotted are theoretical power spectra for CDM with $\Omega_B = 0.01, 0.03, 0.06, 0.10$ (*top*), and standard recombination (*solid*) and for $\Omega_B = 0.12$ and $z_{\text{rec}} = 150$ and 50 (*dashed*).

67 km s^{-1}. The unbiased CDM estimate for these quantities are 224 km s^{-1} and 287 km s^{-1} respectively. Translating this information into the power on scales of $40\,h^{-1}$Mpc and $60\,h^{-1}$Mpc gives roughly $\sigma_8 \simeq 1.15$ (Efstathiou et al 1992).

One can also consider the CMB data independently of theories of structure formation. In Figure 4, we show the current situation with regard to experiments that have quoted detections on degree scales or larger. We plot the normalization which, for an $n = 1$ power spectrum, would reproduce the quoted $\Delta T / T_{\text{rms}}$ for each experiment. If there is a Doppler peak in the power spectrum on degree scales, then this would show up as a higher required normalization for a flat spectrum to fit the data. All of these points should be interpreted as approximations to the results of a full analysis.

4.   DATA ANALYSIS

4.1   *The Gaussian Auto-Correlation Function*

It has become common in analyzing data from small-scale experiments to employ a Gaussian Auto-Correlation Function (GACF) as the assumed underlying "theory." The GACF is parameterized by two numbers, its amplitude C_0 and correlation angle θ_c:

$$C_{\text{GACF}}(\theta) \equiv C_0 \exp\left(-\frac{\theta^2}{2\theta_c^2}\right). \tag{25}$$

This "theory" is then convolved with the observing strategy and the predictions compared to the data. Usually limits or best-fit values are quoted on C_0 for a range of θ_c. In the language of the multipole moment expansion, the assumption of a GACF is equivalent to assuming

$$C_\ell = 2\pi C_0 \theta_c^2 \exp\left[-\tfrac{1}{2}\ell(\ell+1)\theta_c^2\right]. \tag{26}$$

Note that this is very different from CDM, where $\ell(\ell+1)C_\ell$ has a Sachs-Wolfe plateau followed by Doppler/adiabatic peaks.

Commonly a plot of C_0 vs θ_c is used to describe the sensitivity of a particular experiment to fluctuations on various scales, under the assumption that the sky correlation function is really a GACF. The GACF approximation is a simple way to understand the sensitivity of an experiment. By varying θ_c, one can match the "power spectrum" to the window function of the experiment (especially multi-beam experiments where the GACF power spectrum and the window function have similar shapes, which we will call "Gaussian"). The experiment is most sensitive to a GACF whose peak $\ell(\ell+1)C_\ell$ occurs at essentially the same place as the peak of its window function W_ℓ. The amplitude of fluctuations to which one is sensitive (or the "area" under the window function) is then parameterized by the minimum C_0.

For experiments in which the window function is similar to the GACF power spectrum, the approximation made in fitting with a GACF is numerically quite good. This is because, unless the underlying power spectrum varies rapidly on scales probed by the experiment, once the power spectrum is convolved with the window function, it has a "Gaussian" shape. The GACF retains its "Gaussian" shape when convolved with a Gaussian window function. Hence the two convolved spectra will look very similar (Bunn et al 1994b).

An analysis using a GACF will (roughly) take into account the peak and area of the window function of the experiment. The minimal values of $C_0^{1/2}$ will therefore be more comparable between experiments than the measured rms temperature fluctuations (which could be defined in an observer-dependent way). The correlations between nearby points will also be approximately correct for experiments with "Gaussian" window functions (and in which the window function approach is applicable; see Section 4.2). Although for some experiments, the GACF approximation may be used to give a fit to the data, the GACF assumption should be viewed with caution. One should bear in mind that the quoted $C_0^{1/2}$ is the best fit amplitude of fluctuations for a power spectrum that *is* a GACF with some fixed correlation angle θ_c. There is also little meaning in the values of $C_0^{1/2}$ at any point other than the minimum of the likelihood curve.

4.2   *Window Functions*

It has become conventional to describe the details of the instrument and the observing strategy in terms of a *window function* W_ℓ which describes the sensitivity of the experiment to the modes of the spherical harmonic decomposition of the CMB temperature fluctuations. The signal seen by any experiment can then be considered as the convolution of the sky power and the window function

$$\left(\frac{\Delta T}{T}\right)^2_{\text{rms}} \equiv C(0) = \frac{1}{4\pi} \sum_{\ell=2}^{\infty} a_\ell^2 W_\ell. \tag{27}$$

If one takes an ensemble average (over universes) of this expression, then $a_\ell^2 \to \langle a_\ell^2 \rangle = (2\ell + 1)C_\ell$. Often this ensemble average is assumed when the window function is computed.

The simplest and most common window function is that due to finite beam resolution. As expected, finite resolution introduces a high-ℓ cutoff. If the beam has a Gaussian response with a Gaussian width of σ, the window function is (see e.g. Silk & Wilson 1980, Bond & Efstathiou 1984, White 1992)

$$W_\ell = \exp\left[-\ell(\ell + 1)\sigma^2\right]. \tag{28}$$

For an experiment that measures temperatures by differencing 2- or 3-beam setups, the window functions, in addition to the beam smoothing factor, are (see e.g. Bond & Efstathiou 1987)

$$\exp\left[\ell(\ell+1)\sigma^2\right] W_\ell = \begin{cases} 2\left[1 - P_\ell(\cos\theta)\right] & \text{2-beam} \\ \frac{1}{2}\left[3 - 4P_\ell(\cos\theta) + P_\ell(\cos 2\theta)\right] & \text{3-beam} \end{cases} \tag{29}$$

where θ is the angle between the beams. Note that these types of experiments are not sensitive to the low-ℓ modes of the multipole expansion because of the differencing. Since the high-ℓ cutoff is controlled by the beam width while the separation (or *chop*) controls the low-ℓ behavior, one can increase both the width and height of the window function by separating these scales as much as possible.

Such a double- or triple-beam differencing strategy is often called a *square wave chop*. There are, however, other scan strategies that have been used. Several experiments (in particular South Pole, Saskatoon, and MAX) use a *sine wave chop*, moving the beam continuously back and forth across the sky, sinusoidally in time. Additionally, the temperature is weighted by ± 1 or by a harmonic of the chop frequency. The resulting time-integrated, weighted temperature is then the "difference" assigned to that point on the sky. Window functions for these experiments can be found in Bond et al (1991b), Dodelson & Jubas (1993), White et al (1993), and Bunn et al (1994b). [The window function for MAX, given in White et al (1993), should be multiplied by 1.13 to account for the finite size of the beam on the calibration: see Srednicki et al (1993).] There are also several interferometer experiments which make maps of the intensity of the radiation on small patches of the sky [e.g. ATCA (Subrahmanyan et al 1993), VLA (Fomalont et al 1993), and Timbie & Wilkinson (1990)]. The window function for these experiments can be measured as the Fourier transform of the beam pattern and for accuracy needs to be supplied by the experimenters.

We show the window functions vs ℓ for several experiments in Figure 5. Some numbers describing the functions shown here are given in Table 2. Note that the relative heights can have as much to do with the treatment of the data as with the sensitivity, i.e. the window function that is convolved with theory should be consistent with the observers' $\Delta T / T$. It is worth giving an example to illustrate this. Consider a triple-beam set-up, which consists of the difference of a difference of two temperatures. The experimenters could choose to assign a measurement of $T_1 - \frac{1}{2}(T_2 + T_3)$ to a point in direction "1," or they could have chosen to take $2T_1 - (T_2 + T_3)$.

In the latter case, the window function would be four times larger and the "measured" $(\Delta T / T)_{rms}$ would be two times bigger. The difference in height for the window function would be artificial. While in this case the difference is quite obvious, in some instances the effects can be more subtle. Experimentalists must therefore be explicit about their sampling, weighting, and calibrations before the correct window functions can be computed.

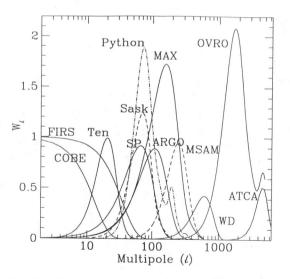

Figure 5 The window functions for large- and medium-scale experiments as a function of multipole. From left to right the experiments are *COBE* (with 10° smoothing), FIRS, Tenerife, SP91, Saskatoon (*dashed*), Python (*dot-dashed*), ARGO, MAX, MSAM (3-beam, *dashed*), White Dish (Method II, neglecting binning), OVRO, and ATCA. Some parameters of the window functions are displayed in Table 2.

Table 2   Parameters for the window functions[a]

Experiment	ℓ_0	ℓ_1	ℓ_2	Max
COBE	–	–	11	1.0
FIRS	–	–	30	1.0
Ten	20	13	30	1.0
SP91	66	32	109	0.9
Sask	71	44	102	1.2
Pyth	73	50	107	1.9
ARGO	107	53	180	0.9
MAX	158	78	263	1.7
MSAM2	143	69	234	2.1
MSAM3	249	152	362	0.9

[a] ℓ_0 represents the multipole at the maximum; ℓ_1 and ℓ_2 are the "half peak" points. The maximum value of the window function is also given. For MSAM we present results in 2-beam and 3-beam modes.

Common approximate formulae for the window functions or analysis procedures assume a square wave chop (e.g. Górski 1993, Gundersen et al 1993). This approximation usually does not reproduce the beam pattern on the sky all that well, although it works better for the window function. Even so, such approximations differ from the exact results, e.g. for MAX the difference between the exact result and (29) is $\sim$10% near the peak, and larger off-peak.

Given both a theory and the window function, it is straightforward to compute the expected rms temperature fluctuation. In Table 3, we show the predicted $\Delta T / T_{\text{rms}}$ for various experiments, normalized to $A = 1$. The predictions assume full sky coverage and an "average universe," though actual experiments may measure different values due to incomplete sky coverage or cosmic variance (to be discussed later).

It is sometimes possible to define window functions that correspond to off-diagonal elements of the correlation matrix or averages of the form

$$\langle T(\theta_1, \phi_1) T(\theta_2, \phi_2) \rangle_{\text{ens}} = \frac{1}{4\pi} \sum_{\ell=2}^{\infty} (2\ell + 1) C_\ell W_\ell(\theta_1, \phi_1; \theta_2, \phi_2) \qquad (30)$$

which are required when fitting data. (Note that this is different from the sky-averaged correlation function of the *COBE* group. It is not an average over our observed sky, but the covariance matrix required when computing likelihood functions, assuming Gaussian statistics for the temperature fluctuations.) In general, the window function approach works well for computing $\Delta T / T_{\text{rms}}$ or for experiments in which the data span only one dimension (such as the individual linear scans of the ACME South Pole experiment). In other cases, however, the data are two-dimensional on the sky and there can be strong anisotropies

Table 3 Predictions for CMB experiments in a CDM-dominated universe[a]

Experiment	Ω_B				
	0.00	0.01	0.03	0.06	0.10
COBE	2.57	2.62	2.63	2.63	2.64
FIRS	3.23	3.38	3.41	3.44	3.47
Ten	1.92	2.10	2.14	2.18	2.20
SP91	2.24	2.84	3.00	3.22	3.38
Sask	2.15	2.81	2.98	3.24	3.40
Pyth	2.69	3.84	4.10	4.49	4.76
ARGO	2.20	3.27	3.49	3.84	4.09
MAX	3.08	5.15	5.49	6.09	6.49
MSAM2	3.46	5.64	6.02	6.65	7.06
MSAM3	1.88	3.49	3.69	4.09	4.32

[a]We show the predicted $\Delta T / T_{\rm rms}$ for various experiments, normalized to $A = 1$. The predictions are for an all-sky average and an "average universe"; individual experiments may measure different values due to incomplete sky coverage or cosmic variance. For MSAM the predictions are shown for 2-beam and 3-beam modes. The column $\Omega_B = 0$ refers to an $n = 1$ power spectrum. All values assume CDM with $\Omega_0 = 1$ and $h = 0.5$.

in the theoretical covariance matrix which are difficult to include in this manner. Alternative approaches are then preferable (see e.g. Srednicki et al 1993). Also, if the scanning strategy or data analysis procedure is sufficiently tortuous, the window function approach is extremely complicated and simulations of the scanning, binning, and analysis become necessary. Coarse binning of data in an experiment which scans smoothly (rather than "stepping") across the sky is one example of this, where correlations introduced by the binning will be important.

4.3 *Fitting Data*

In comparing the theory of CMB fluctuations to the new measurements of the anisotropy on various scales, two primary techniques are used: the Bayesian *likelihood function* analysis and the frequentist *likelihood ratio* method (which is often calibrated using Monte-Carlo simulations). Both of these techniques have been discussed and compared in the review of Readhead & Lawrence (1992), and we will not discuss them in detail here.

It is perhaps important to emphasize, however, that the methods will lead to the same conclusions when the data set is "well-behaved," but can differ when the data are "unlikely" or atypical. Both methods have ways of testing for possible breakdowns in statistical assumptions (which are not always used). For example, conclusions of the Bayesian approach should be robust under changes of prior distribution, and conclusions of the frequentist approach should take into account the type-II error (the probability of accepting the hypothesis when the null hypothesis is true).

The computation of the likelihood function simplifies considerably if fluctuations are assumed to be Gaussian. This is because all the nonvanishing moments of the distribution can be related to the variance. For Gaussian-distributed fluctuations, the likelihood function is

$$\mathcal{L} \propto \frac{1}{\sqrt{\det C}} \exp\left[-\frac{1}{2} T_i C_{ij}^{-1} T_j\right], \tag{31}$$

where T_i are the measured temperatures and C_{ij} is the auto-correlation matrix which includes a theoretical and experimental piece. In the limit that the experimental errors σ_i are uncorrelated we have

$$C_{ij} = C_{ij}^{\text{th}} + \sigma_i \delta_{ij}, \tag{32}$$

where C_{ij}^{th} is computed using Equation (30) or its generalization (see e.g. Bond et al 1991b; Dodelson & Jubas 1993; Bunn et al 1994b; Srednicki et al 1993; Bond 1993; Vittorio et al 1989, 1991; Vittorio & Muciaccia 1991; Górski et al 1993). Notice that the term in the exponent is just the χ^2.

In the Bayesian approach, this function has to be multiplied by the "prior" to obtain the (relative) probability distribution for the parameters being fitted, while in the frequentist approach, the final distribution comes from ratios of likelihood functions or from Monte-Carlo simulation [see e.g. Berger (1985) for a discussion of these two philosophies].

It is also possible to fit more than one component to the data, in order to obtain the best constraints on the cosmic anisotropies. One example is in fitting an extra white noise component to the data (e.g. Bond 1993). Another example is the simultaneous fitting of a foreground signal of some given form (e.g. Dodelson & Stebbins 1993) to multichannel data. It is relatively straightforward to implement this idea (for either statistical approach), although there is some subjectivity in the choice of the number of parameters, which of them should be fixed (the frequency dependence of the foreground signal perhaps), and which to fit or integrate over. The fitting process becomes more computationally time consuming as more parameters are included.

5.   BEYOND LINEAR THEORY

5.1   *Reionization*

A knowledge of the variation of optical depth (or equivalently the ionized fraction x_e) and of the peculiar velocities associated with different scales are enough to calculate $\Delta T / T \, (\theta)$. Reionization will erase the fluctuations generated at $z \sim 1000$, by making the Universe optically thick at much lower redshifts, but in the process extra fluctuations will be generated due to the motions of the new last scatterers. Because reionization can be used to erase the primordial fluctuations in a model that would otherwise conflict with experimental upper limits, it is important to be able to calculate the secondary fluctuations so that the minimal level of anisotropy can be estimated.

The dominant contribution to anisotropies generated during reionization are Doppler shifts of the scatterers (Equation 23). The first (analytic) considerations of the velocity-induced perturbations in a reionized model were by Sunyaev (1977, 1978), Davis (1980), and Silk (1982). However Kaiser (1984) pointed out that an important term contributing to the fluctuations had been neglected in the previous work. The correct expression for $\Delta T / T$ could not be written quite as straightforwardly as in Equation (2). The evolution of the radiation fluctuations needs to be followed more accurately, and requires a numerical solution of the collisional Boltzmann equation for the photons (see Section 3.2).

Ostriker & Vishniac (1986) extended the discussion to second-order fluctuations (see Section 5.2); detailed calculations are presented by Vishniac (1987). Efstathiou & Bond (1987; also Efstathiou 1988) made further calculations of the effects of reionization, including details of the pattern of polarization and approximations valid for small angular scales. The important effects of reionization on anisotropies are at arc-minute scales, in particular, where primary anisotropies are expected to be erased, and approximate calculations can be used which are valid only for small angles and correspondingly for small wavelength perturbations. More recent calculations have been carried out for reionization in BDM models (e.g. Hu et al 1994), generic open universe models (Persi & Spergel 1993), CDM models (Hu et al 1994, Dodelson & Jubas 1994, Sugiyama et al 1993, Chiba et al 1993, Hu & Sugiyama 1994b,c), and for the case of decaying dark matter (Scott et al 1991).

An important epoch is the redshift at which the optical depth for photons becomes unity. Since Thomson scattering is independent of frequency, the evaluation of z_* (or equivalently η_*) is relatively simple:

$$\tau = -\int_0^z \bar{n}_e(z)\sigma_T c \left(\frac{dt}{dz}\right) dz, \tag{33}$$

which, for an $\Omega_0 = 1$ universe with a constant ionized fraction, becomes

$$\tau = \frac{\sigma_T c H_0}{4\pi G m_H} x_e \Omega_B (1 - Y_p) \left[(1+z)^{\frac{3}{2}} - 1\right] \simeq 0.035 \Omega_B h x_e z^{3/2}, \tag{34}$$

where Y_p is the primordial fraction of the baryonic mass in helium, which is assumed to be all neutral. Hence, for a particular model of the ionization history, the redshift for which $\tau = 1$ can be calculated. For example, for constant ionized fraction $z_* \simeq 69(h/0.5)^{-2/3} (\Omega_B/0.1)^{-2/3} x_e^{-2/3}$. For an open universe $z_* \propto \Omega_0^{1/3}$ very approximately, so that the last scattering surface can be at significantly lower redshift. If the universe is ionized as early as $z \gtrsim 100$, then the exact z of reionization is unimportant, since then $\tau \gg 1$.

Reionization need not be complete, i.e. the optical depth may not become very large back to the new scattering surface. Obviously if $\tau \lesssim 0.1$ say, then reionization has negligible effect. This would be the case for a standard CDM model fully ionized even up to $z \sim 20$. However if $0.1 \lesssim \tau \lesssim 1$, then a good approximation to the effects of erasure of the primary anisotropies (i.e. those generated at the standard recombination scattering surface) is that $\Delta T / T$ is reduced by the fraction of photons scattered at $z \sim 1000$ rather than at the new scattering surface. If $(1 - e^{-\tau_i})$ is the fraction of photons scattered since the Universe was reionized at z_i (i.e. this is the integral of the visibility function), then we simply have $(\Delta T / T)_{\mathrm{obs}} \simeq (\Delta T / T)_{\mathrm{prim}}(e^{-\tau_i})$. There will also be *secondary* anisotropies generated on the new last scattering surface, although these will generally be smaller (see below). However, second-order fluctuations at much smaller angular scales also need to be considered.

Early reionization is likely to have been inevitable for *COBE*-normalized CDM [Tegmark et al (1994), Sasaki et al (1993), and Fukugita & Kawasaki (1994), following the early studies by Couchman (1985) and Couchman & Rees (1985)]. On scales of order the horizon at z_* there will still be Sachs-Wolfe fluctuations and the polarization would be expected to be higher.

5.2 *Second-Order Anisotropies*

The idea that reionization would erase the primary anisotropies, while generating smaller secondary Doppler anisotropies (since the new scattering surface was so thick), was overturned by the realization that the Vishniac effect played an important role at small angular scales. The first-order Doppler effects mainly cancel when the redshifts and blueshifts are integrated through the thick last scattering surface. However there is a second-order term in the Boltzmann equation (23) coming from the $v\delta$ cross term in $\sigma_T \bar{n}_e (1 + \delta) v$. In solutions to the Boltzmann equation at small scales, this term is a convolution over velocity and density perturbations, which does not suffer from the cancellation effect. This leads to some models giving *larger* secondary anisotropies than the primary anisotropies which the reionization was invoked to erase.

We can obtain a scaling relation for the Doppler fluctuations from the new last scattering surface as follows. Notice that the last scattering shell is at η_* with width $\sim \eta_*$. The number of independent regions of scale λ lying across the thickness of the shell is $N \sim \eta_*/\lambda$. The optical depth through each region of comoving scale λ is $\tau_\lambda \sim 1/N \sim \lambda/\eta_*$. The fluctuations on scale θ ($\propto \lambda$) are about $N^{1/2}$ times the fluctuation due to a single lump. The fluctuation for each lump is second-order because of the approximate cancellation of redshifts and blueshifts through a lump, leading to an effect $\sim \tau_\lambda^2 v$ (see Kaiser 1984). From the continuity equation, the peculiar velocity is $v \sim \lambda \delta_k(\eta_*)/\eta_*$ and overdensities evolve as $\delta \propto \eta^2$. Therefore an order-of-magnitude estimate for the temperature fluctuations (i.e. $k^2\Delta$, see also Kaiser 1984, Vishniac 1987) is

$$(\Delta T/T)_\theta^2 \sim N(\Delta T/T)_{\text{lump}}^2 \sim k^{-3}P(k)/(\eta_* \eta_0^4) \propto (1+z_*)^{1/2}. \tag{35}$$

Hence earlier reionization generally leads to larger large-scale (first-order Doppler) fluctuations from the last scattering surface. Similarly for the second-order (e.g. Vishniac) fluctuations, the effect for each lump is $\sim \tau_\lambda v\delta$, giving $(\Delta T/T)_\theta^2 \sim k^2 P^2(k)(1+z_*)^{-5/2}$. Again, this argument is crude, but shows that second-order small-scale fluctuations from the secondary last scattering surface would be expected to be *smaller* for earlier reionization. There are several experimental limits at scales $\sim 1'$, which provide good constraints on these secondary fluctuations rather than the primary ones (e.g. Uson & Wilkinson 1984a,b,c; Readhead et al 1989; Fomalont et al 1993; Subrahmanyan et al 1993). The strong k dependence of the Vishniac effect means that the fluctuations will be very spiky, so that a double- or triple-beam experiment will be dominated by the zero-lag autocorrelation. Note, however, that nonlinear effects may also be important, e.g. if the universe is very clumpy, as may happen in BDM or other models with a large amount of small-scale power (Hu et al 1994), or if most of the baryons are trapped in galaxies at an early epoch (R. Juszkiewicz & P. J. E. Peebles, private communication).

In general, there are many second-order terms to consider once a fully second-order Boltzmann equation is derived (Hu et al 1994, Dodelson & Jubas 1994). For most of these terms, there will be a cancellation of redshifts and blueshifts (Kaiser 1984) through the last scattering surface, so that the major contribution to the fluctuations comes from modes with k-vector almost perpendicular to the lines of sight. The only significant term that remains is the Vishniac term, which is a convolution of v and δ, coupling large-scale velocity perturbations with small-scale density perturbations. In a model with significant reionization the primordial anisotropies are erased, a new Doppler peak is generated at smaller ℓ and of lower amplitude, and a Vishniac bump is generated at arc-minute scales. For CDM-type models, the Vishniac effect is negligible, but for models with a large amount of small-scale power and/or open models (e.g. BDM models, which also *require* reionization), the constraints can be restrictive. Generally the ionization history, $x_e(z)$ can be varied in such models, since there is no clear picture for the process of early reheating. The Vishniac anisotropies tend to become smaller for an earlier last scattering surface, while the first-order amplitude becomes bigger; so there are limits to how much the constraints can be avoided by invoking a different ionization history (see Hu & Sugiyama 1994b,c).

Another kind of second-order anisotropy is that coming from nonlinear effects in the growth of perturbations, which causes the potential to change with time even in an $\Omega_0 = 1$ universe. This change in potential in the light-crossing time across a fluctuation (Zel'dovich 1965, Rees & Sciama 1968, Dyer 1976) is known as the Rees-Sciama or integrated Sachs-Wolfe effect. It is small in almost all models (e.g. Argüeso & Martínez-González 1989; Martínez-González et al 1990, 1992, 1993, 1994; Anninos et al 1991), although large voids (Nottale 1984, Scaramella et al 1989, Thompson & Vishniac 1987, van Kampen & Martínez-González 1991, Arnau et al 1993), or structures such as the Great Attractor (Bertschinger et al 1990, Martínez-González & Sanz 1990, Hnatyk et al 1992, Goicoechea & Martin-Mirones 1992, Saez et al 1993) or the Great Wall (Atrio-Barandela & Kashlinsky 1992, Chodorowski 1992) etc might give observable signatures.

6. UNCERTAINTIES

If the underlying fluctuation spectrum is assumed to be stochastic in nature, then one is faced with the problem of trying to compare a theory which defines probability distributions as functions of the underlying parameters with just one sample drawn from these (our Universe). Only observations over an ensemble of universes would allow one to determine the parameters of the underlying theory unambiguously, even in principle. Since we can observe only our universe, there is an irremovable uncertainty in our ability to relate certain CMB measurements, no matter how precise, to parameters of the theory. The uncertainty introduced in determinations of theoretical parameters has been called "cosmic variance" or "theoretical uncertainty" (e.g. Abbott & Wise 1984a,b; Scaramella & Vittorio 1990, 1991, 1993b; Cayón et al 1991; White et al 1993).

In terms of the multipole moments the cosmic variance may be thought of as an uncertainty in relating

$$a_\ell^2 \leftrightarrow \left\langle a_\ell^2 \right\rangle_{\text{ens}} = (2\ell + 1)C_\ell. \tag{36}$$

In a theory with Gaussian fluctuations, the $a_{\ell m}$ are independent Gaussian random variables and the a_ℓ^2 are thus $\chi_{2\ell+1}^2$ distributed (Abbott & Wise 1984a,b; Abbott & Schaefer 1986; Vittorio et al 1988). The uncertainty in relating a_ℓ^2 and $\left\langle a_\ell^2 \right\rangle_{\text{ens}}$ is given by the width of the distribution, which scales as $(\ell + 1/2)^{-1/2}$. As one goes to smaller scales and probes larger ℓ, one thus becomes less sensitive to cosmic variance. On large scales, however, cosmic variance forms the limiting uncertainty in fixing (e.g.) the normalization of the power spectrum. For the quadrupole (with only 5 degrees of freedom), the actual amplitude in our Universe is likely to differ substantially from its expectation value (Gould 1993, Stark 1993).

While it is true that on smaller scales, which probe higher ℓ, the cosmic variance becomes negligible, this applies only to a full sky measurement. The variance from sampling only a fraction of the sky is larger than the cosmic variance one gets when sampling the whole sky. Since current small-scale experiments cover only small patches of the sky, this can be an important effect. In general for an experiment that samples a solid angle Ω, the cosmic variance is enhanced by a factor of $4\pi / \Omega$ (Scott et al 1994, see also Bunn et al 1994a). We should emphasize that this "sample variance" is completely independent of the experimental precision, and simply reflects the fact that the experiment has not covered enough of the sky to provide a good estimate of the rms value of $\Delta T / T$.

The spectrum of fluctuations depends not only on the primordial spectrum but on the evolution of perturbations as waves enter the horizon. This evolution introduces dependencies on Ω_0, H_0, and the dark matter. In addition, a knowledge of the ionization history of the universe is important since reionization would lead to both a reprocessing of primary fluctuations and generation of (small-scale) secondary fluctuations. (See Section 5.1.) A non-negligible ionized fraction during much of the history of the universe would also contribute a radiation drag which would affect the growth of baryon perturbations and alter the observed CMB anisotropy. In general, the uncertainties in the cosmological parameters and history produce very correlated shifts in the radiation power spectrum (Bond et al 1994, Hu & Sugiyama 1994b,c).

In addition, there are possible sources of "foreground" contamination such as synchrotron and free-free emission (at low frequency) and dust (at high frequency). These non-cosmological sources have been discussed in detail in the review of Readhead & Lawrence (1992) and we will not mention them further except to say that to discriminate between these sources it is necessary to have good frequency coverage and a wide range of angular resolutions. Note that, in particular, the galactic signal also makes it difficult to measure the quadrupole anisotropy (de Bernardis et al 1991, Bennett et al 1994).

7.   NON-STANDARD COSMOLOGIES

7.1   *Polarization*

Penzias & Wilson (1965) set an upper limit of 10% on the polarization of the CMB, and there have been several subsequent upper limits published (e.g. Nanos 1979, Caderni et al 1978, Lubin & Smoot 1981, Lubin et al 1983, Partridge et al 1988, Wollack et al 1993), but so far none that have been low enough to be cosmologically interesting.

Any quadrupole anisotropy in the radiation field will give rise to a linear polarization when it is Thomson scattered. This was first suggested in the case of an anisotropic universe (Rees 1968, Anile 1974, Dautcourt & Rose 1978), but can also arise, generally of smaller amplitude, from any inhomogeneity, and can moreover be enhanced by reionization (Negroponte & Silk 1980, Basko & Polnarev 1980, Stark 1981, Tolman & Matzner 1984, Nasel'skiĭ & Polnarev 1987, Harari & Zaldarriaga 1993, Ng & Ng 1993). A transfer equation similar to Equation (23) can be written for the polarization amplitude Δ_P (Bond & Efstathiou 1987). Solutions indicate that the level of anisotropy is generally small for standard recombination. Efstathiou (1988) has developed approximations for calculating the small angular scale autocorrelation function of Δ_P in reionization scenarios. Kaiser (1983) has shown that the rms polarization from density fluctuations is typically 10–20% of $\Delta T/T$ for adiabatic fluctuations on arc-minute scales. This means that it is still well below the detection levels of present-day experiments. The problem has been studied numerically, including both the contribution of density fluctuations and gravitational waves in Crittenden et al (1993a), where it was shown that the additional contribution due to gravity wave modes was too small to be resolved with current sensitivities (see also Frewin et al 1994). However, in future, as these experiments become more sensitive and it becomes more difficult to distinguish real fluctuations from those caused by confusion with faint radio sources and diffuse emission from the Galaxy, it may be that polarization information could be used as a discriminant.

The polarization pattern can also be a useful test of the ionization history of the universe, since it is expected to be coherent over scales corresponding to the thickness of the last scattering surface (Hogan et al 1982). Hence coherence on scales $\gg$ 10 arc minutes would imply that reionization had indeed taken place. Note that all of the statements above are for linear polarization—circular polarization is not expected to be generated unless there are primordial magnetic fields or strong anisotropies at the scattering epoch.

7.2 Open Universes

Until now, we have implicitly assumed that $\Omega_0 = 1$ as favored by inflationary models. However astronomical evidence generally favors $\Omega_0 \sim 0.2(\pm 0.1)$, based on large-scale structure ($<10\,\mathrm{Mpc}$) studies. The negatively-curved spatial hypersurfaces of a low density ($\Lambda = 0$) Friedmann model greatly complicate the analysis of large angular scale anisotropy in the CMB. One expects that the curvature radius will introduce a feature into the low-order multipoles, due to gravitational focusing of geodesics which shifts the power from lower to higher orders. The characteristic angular scale corresponds to the curvature radius of $\sim \frac{1}{2}\Omega_0$ radians. This effect is analogous to the "ring-of-fire" effect in weakly anisotropic cosmologies (Wilson & Silk 1981, Fabbri et al 1983). Early calculations include those of Kaiser (1982) and Peebles (1982b). This redistribution of the low-order multipole moments was realized in the elegant numerical computations of Wilson (Silk & Wilson 1981, Wilson & Silk 1981, Wilson 1983), who expanded the radiation power spectrum in terms of generalized wave-number $(k^2 + K)^{1/2}$, where K is the spatial curvature, defined as eigenvalues of the Laplace operator for density perturbations in a curved background.

Results similar to those of Wilson were subsequently obtained by Tomita & Tanabe (1983) and Abbott & Schaefer (1986). These latter authors used a gauge-invariant approach, expanded the power spectrum in wave-number k, and included cosmic variance in estimating the multipole moments. Górski & Silk (1989) computed the low-order multipoles for the case of primordial isocurvature (or entropy) fluctuations. Their results were generalized by Gouda et al (1991a,b), who included both primordial adiabatic and entropy fluctuations, and computed the contribution from the integrated Sachs-Wolfe term. This line-of-sight term vanishes in flat $\Omega_0 = 1$ models but gives an important contribution to all scales in an open model (Hu & Sugiyama 1994). Note that the location of the Doppler peak, $\ell \simeq 200/\Omega_0^{1/2}$, also depends on how open the Universe may be (Kamionkowski et al 1994).

Any simple interpretation of the results, however, is complicated by the fact that the concepts of wave number and power-law spectrum must be generalized to curved hyperspaces. They can no longer simply be interpreted in terms of constant curvature fluctuations associated with a spectrum of comoving scales if e.g. $n = 1$. Kamionkowski & Spergel (1993) have studied power spectra for primordial adiabatic fluctuations that are power laws in either volume (least large-scale power), distance, or eigenvalues of the Laplace operator (most large-scale power), and find that in all cases there is some suppression of the multipoles on scales larger than the curvature scale. In a flat universe, these three definitions of power spectrum would all coincide. There is large uncertainty in defining the spectrum because of the lack of any unique prescription for the initial conditions in a low-Ω universe. Indeed, much of the motivation for a Harrison-Zel'dovich spectrum is lost if $\Omega_0 < 1$, because one can no longer appeal to simple models of inflation. More complex models that predict fluctuation spectra in open models, whether for primeval curvature perturbations (Lyth & Stewart 1990) or for primordial entropy perturbations (Yokoyama & Suto 1991, Dolgov & Silk 1993), are not compelling.

Given the above, it appears that in low-Ω_0 universes, adiabatic CMB fluctuations are larger than their $\Omega_0 = 1$ counterparts for the same value of H_0 (of course when $\Omega_0 < 1$ the requirement of small H_0 is relaxed and the anisotropy can be somewhat reduced). The most detailed model of fluctuations in an open universe to date is the BDM model discussed in Section 2.2. Analysis of the spectrum produced by this model is complicated, as radiation drag (from the nonzero x_e) can alter the shape, and the integrated Sachs-Wolfe effect is important on large scales in addition to the secondary fluctuations induced on small scales (Gouda et al 1991a, Hu & Sugiyama 1994c). However, on very large scales ($\ell \lesssim 10$), there is an almost model-independent signature, $n_{\text{eff}} \simeq 2$, to the slope of the radiation power spectrum (Sugiyama & Silk 1994). This model appears to be tightly constrained by current observations (Chiba et al 1993, Hu & Sugiyama 1994b).

7.3   Effects of a Cosmological Constant

Due to the emerging, but still controversial, discrepancy between the lifetime of a matter-dominated $\Omega_0 = 1$ universe inferred from the Hubble constant, and the lifetime inferred by stellar evolution (see e.g. Demarque et al 1991) there has been interest in the possibility that the cosmological constant may be nonzero (see also Caroll et al 1992). This would allow $\Omega_0 < 1$ while retaining a "flat" cosmology and avoid the "Ω-problem" (see e.g. Kolb & Turner 1990). If a large vacuum energy drove an inflationary phase in the early universe, the idea that it may be nonzero today is not wildly implausible.

Note that a nonzero cosmological constant has little effect on the large-scale structure and dynamics of the Universe (e.g. Markevitch et al 1991, Lahav et al 1991), and will have negligible effect on the recombination process and the visibility function. However, microwave background fluctuations will be significantly different in $\Lambda \neq 0$ models. This is because of several effects. Firstly, there is a change in angular scales. The angle at recombination corresponding to a proper length λ can be approximated as $\theta \simeq 30'' \Omega_0^{1/3}(1 - \Omega_0)^{-1/4}\lambda(h^{-1}\text{Mpc})$ (Blanchard 1984, Stelmach et al 1990). Hence, for the same normalization, the $\Lambda \neq 0$ model measures anisotropies that correspond to smaller angular scales than in the $\Lambda = 0$ model. Secondly, since the growth of fluctuations is different in a $\Lambda \neq 0$ model, the potential fluctuations are no longer constant in time. Hence the Sachs-Wolfe temperature fluctuations are augmented by the integrated potential term (see Equation 9). This modifies the (scalar) Sachs-Wolfe formula (Górski et al 1992, Vittorio & Silk 1992, Sugiyama et al 1990, Hu & Sugiyama 1994, Stompor & Górski 1994). (The formula for the tensor mode contribution is essentially unchanged.) Generally the integrated Sachs-Wolfe effect for a flat $\Omega_0 < 1$ universe is less important than in the case of an open universe. Finally, the fluctuations (assuming they are linear) stop growing when the the cosmological constant becomes important, which happens at $1 + z \sim (\Omega_0^{-1} - 1)^{1/3}$. For a fixed power spectrum at $z = 0$, the potential in the early universe scales as Ω_0/D (Peebles 1984), where D is the growth factor and the extra factor of Ω_0 comes from the potential $\Phi \propto \Omega_0$. The effects of geometry tend to compensate the integrated Sachs-Wolfe contribution in an open model. Consequently, in a flat Λ-dominated universe, the effective value of n is less than unity on large scales. [Similar conclusions can also be drawn (Sugiyama & Sato 1992) for models in which the cosmological constant decays with time (Freese et al 1987, Ratra & Peebles 1988, Overduin et al 1993).]

7.4   Topology

The fluctuations at large angular scales can also be used to constrain the topology of the Universe, which in principle could be nontrivial (Zel'dovich 1973, Sokolov & Starobinskiĭ 1976, Fang & Houjun 1987, Fang 1991). The Sachs-

Wolfe spectrum of C_ℓs is an integral over the power spectrum (Equation 14) in an ordinary simply-connected universe, but becomes a sum over modes that are harmonics of the box-size in a universe that has the topology of a three-torus (i.e. has periodic boundary conditions). This sum fails to accurately approximate the integral on scales approaching that of the box. Then the fact that *COBE* measures a roughly flat power spectrum on large scales means that the box must be roughly the horizon size or bigger, unless the initial power spectrum has a pathological increase for small multipoles. More detailed comparisons indicate that the scale of any such topology is $\gtrsim 80\%$ of the horizon size (Stevens et al 1993, Sokolov 1993, Starobinskiĭ 1993).

For open universes, Gurzadyan and collaborators have argued that curvature effects may result in elongated shapes of anisotropies in the CMB (Gurzadyan & Kocharyan 1993, Gurzadyan & Torres 1993). The isotropy pattern of the CBR can also be used to place limits on the rotation of the Universe (Collins & Hawking 1973, Barrow et al 1985). Currently the limit on the dimensionless rotation is $\omega/H_0 \lesssim 10^{-6}$ (Smoot 1992), which is about $1''$ every 30 Gyr!

8.   CONCLUSIONS

The CMB is a unique laboratory for studying the initial conditions that gave rise to the observed Universe. In particular, the temperature fluctuations on scales from arc minutes to tens of degrees provide, at least in principle, a precise measure of the primordial density fluctuation power spectrum. The difficulties arise in large part because of the "dirty window" effect: foreground contamination, predominantly Galactic but possibly atmospheric and extragalactic, obscures our view of the last scattering surface. Improved sky coverage, angular resolution, and frequency coverage will eventually help surmount these obstacles, but we do not anticipate an early answer. Another problem arises with the proliferation of astrophysical parameters: These include Ω_0, baryon density, ionization history, primordial power spectrum shape, isocurvature or adiabatic fluctuation contribution, tensor mode strength, fraction of cold, warm and hot dark matter, Λ, the role of topological defects as seeds, non-Gaussian fluctuations, and decaying dark matter. In the future, we can hope that the sensitivity to all these parameters will be seen as a boon rather than as a "problem."

With post-*COBE* fluctuations being reported (in at least eight other experiments at the time of preparing this review), plots like Figures 3 and 4 are likely to become familiar. Several questions are immediately apparent: Are the experiments consistent with each other? Is there any evidence for a Doppler peak in the data? Problems with foreground contamination and other systematic effects may mean that some experiments should be assigned larger error bars. But given this extra leeway, and taking full account of the sample variance etc, will they prove to be consistent with Gaussian fluctuations on the sky? If the answer to this question is "no," then perhaps we will require non-Gaussian fluctuations, or patchy reionization, or perhaps even some new component of cold galactic dust. At the moment all of these avenues are worth exploring, but it is premature to say that any such ideas are necessary.

At present, the number of definitive conclusions one can draw is depressingly few. There is also a vigorous debate as to the optimal method of confronting experiment and theory, whether by Bayesian or frequentist techniques. It is clear that from the CMB measurements alone, one model, that of adiabatic fluctuations in a baryon-dominated universe, can be discarded. Dark matter has proved an invaluable foil for resurrecting flat models, and at present one cannot even eliminate the canonical cold dark matter model from consideration. However we are at an exciting moment in the history of this data-starved subject: Many results are being reported and we are on the verge of being able to eliminate, or to confirm, the $n = 1$ Harrison-Zel'dovich spectrum that was proposed in the earliest and simplest inflationary models. Almost all models predict Doppler/intrinsic fluctuation peaks over 10–60 arc min: Current experiments should be able to decide soon on the reality of these features as sky coverage is improved. If the Doppler peaks are not present then this would provide a strong argument in favor of reionization.

Theorists may rightly consider new directions. Late-time phase transitions (Hill et al 1989, Nambu et al 1991, Jaffe et al 1994, Luo & Schramm 1994) showed some brief promise of producing minimal fluctuations, but even this class of models is detectable at a level of $\Delta T/T \sim 10^{-5}$. Textures produce extreme hot spots, which can perhaps be alleviated with a high ($b \sim 4$) bias factor, and allow the possibility of living with hot or even baryonic dark matter, but remain a relatively soft target for theories of large-scale structure. Significant, large-scale, non-Gaussian behavior remains elusive in any model of structure formation, but might well be a useful weapon to bring to bear on the intermediate angular scale data. Perhaps a complex reionization history would generate non-Gaussian smoothing signatures. Galaxies might also contain unsuspectedly large amounts of cold dust in their halos, and provide an unavoidable contamination of the CMB fluctuation signal—a prospect that must seem less unlikely if halos indeed are baryonic. The fact that "point" sources are being reported in at least one CMB anisotropy experiment with spectra indistinguishable from that of the CMB must add to one's concern about foreground contamination that has not previously shown up in *IRAS* 100 μm maps. The large-scale bulk flows seen in the galaxy distribution, now observed to 15,000 km s^{-1}, imply observable signatures in the CMB on sub-degree scales: If these flows are confirmed and corresponding precursor $\Delta T/T$ fluctuations from the last scattering surface are not observed, a non-Gaussian fluctuation model would seem to be the only resolution.

The next few years promise to be a lively time in cosmology, both for theory and observation, as the $\Delta T/T$ measurements are refined. Observations on all angular scales, from tens of degrees to sub-arcminute scales, will undoubtedly play a role.

ACKNOWLEDGMENTS

We would like to thank the many people who communicated details of their experimental and other work, often ahead of publication. We had useful discussions with many colleagues, including Ted Bunn, Mark Devlin, Wayne Hu, Marc Kamionkowski, Lawrence Krauss, Andrew Liddle, Mark Srednicki, Albert Stebbins, Juan Uson, and particularly Naoshi Sugiyama.

APPENDIXES

A. *Some Useful Numbers*

Converting from multipole space ℓ to angles θ is accomplished via the approximate formula

$$\frac{\theta}{1^\circ} \simeq \frac{60}{\ell}. \tag{37}$$

It is this equality that is used to make statements that certain measurements are on a scale of θ. This correspondence, $\theta \simeq \ell^{-1}$ radians, seems most natural for large ℓ [cf the window function, Equation (28)]; for small ℓ, it may be better to think of $\theta \simeq \pi/\ell$ radians, but in any case the correspondence is not exact. At any $z \gg 1$, an angle θ degrees subtends a comoving distance of $105\,\theta\,(\Omega_0 h)^{-1}$Mpc.

To convert from ℓ to comoving wavenumber k we have $\ell \simeq (6{,}000\,h^{-1}\text{Mpc})k$. Additionally we have the physical scales (at horizon crossing)

$$\begin{aligned}
k_{\text{hor}}^{-1} &= 6000\ h^{-1}\text{Mpc} \\
k_{\text{curv}}^{-1} &= 3000\,(1 - \Omega_0)^{-1/2}\ h^{-1}\text{Mpc} \\
k_{\text{rec}}^{-1} &= 180\,\Omega_0^{-1/2}\ h^{-1}\text{Mpc} \\
k_{\text{eq}}^{-1} &= 39\,(\Omega_0 h)^{-1}\ h^{-1}\text{Mpc}.
\end{aligned} \tag{38}$$

The redshifts of equality, recombination, and decoupling (Compton cooling) for standard recombination are

$$\begin{aligned}
1 + z_{\text{eq}} &= 23{,}900\,(\Omega_0 h^2) \\
1 + z_{\text{rec}} &= 1100 \\
1 + z_{\text{dec}} &= 500\,(\Omega_B h^2)^{2/5},
\end{aligned} \tag{39}$$

and for reionization with constant ionized fraction x_e

$$1 + z_{dec} = 6.0 \, (\Omega_0 h^2)^{1/5} x_e^{-2/5}$$
$$1 + z_{drag} = 120 \, (\Omega_0 h^2)^{1/5} x_e^{-2/5}. \tag{40}$$

The following angles are also useful:

$$
\begin{aligned}
\text{Hubble radius} \quad \Delta\theta(H^{-1}, z \gg 1) &= \tfrac{1}{2}\Omega_0^{1/2} z^{-1/2} \\
&= 0\overset{\circ}{.}87 \Omega_0^{1/2} (z/1100)^{-1/2} \\
\text{Length scale} \quad \Delta\theta(\lambda, z \gg 1) &= \tfrac{1}{2}\Omega_0 H_0 a_0 \lambda \\
&= 34\overset{''}{.}4 \, (\Omega_0 h) \lambda_{Mpc} \\
&= 65\overset{''}{.}4 \, (\Omega_0^{2/3} h^{1/3})(M/10^{12} M_\odot)^{1/3} \\
\text{Curvature scale} \quad \Delta\theta(\mathcal{R} a_0/a, z \gg 1) &= \tfrac{1}{2}\Omega_0/(1 - \Omega_0)^{1/2} \\
&= 28\overset{\circ}{.}6 \, \Omega_0 (1 - \Omega_0)^{-1/2} \\
\text{Thickness scale} \quad \Delta\theta(\Delta z = 80) &= 3\overset{'}{.}8 \Omega_0^{1/2} \\
\text{Damping scale} \quad \Delta\theta(\lambda_{Damp}, z \gg 1) &= 1'.8 \Omega_B^{-1/2} \Omega_0^{3/4} h^{-1/2}.
\end{aligned} \tag{41}
$$

B.   *Sachs-Wolfe Effect*

In this appendix we give a brief derivation of the Sachs-Wolfe effect in a "flat" cosmology with no cosmological constant, using the metric perturbation approach. Our starting point is the metric (following Sachs & Wolfe 1967)

$$ds^2 = a^2(\eta) \left(g_{\mu\nu}^{(0)} + h_{\mu\nu} \right) dx^\mu dx^\nu = a^2(\eta) d\bar{s}^2, \tag{42}$$

where $a(\eta)$ is the scale factor, η is the conformal time and $g_{\mu\nu}^{(0)}$ is the unperturbed metric (Minkowski space). To understand the temperature fluctuations induced by the perturbations we need to study the photon trajectories in ds^2. For photon (null) geodesics $ds^2 = 0$, so by (42) there is a 1-to-1 correspondence between photon paths in ds^2 and $d\bar{s}^2$. This allows us to consider the problem first in $d\bar{s}^2$ and later translate our results in ds^2.

We can solve for the geodesics by extremizing the Lagrangian $g_{\mu\nu}\dot{x}^\mu\dot{x}^\nu$; the geodesic (Euler-Lagrange) equations for $d\bar{s}^2$ are

$$\frac{d}{d\zeta}\left((g_{\mu\nu}^{(0)} + h_{\mu\nu})\dot{x}^\nu \right) = \frac{1}{2} h_{\nu\rho,\mu}\dot{x}^\nu\dot{x}^\rho, \tag{43}$$

where ζ is a parameter along the photon trajectory and the overdot represents differentiation w.r.t. ζ. The term in parenthesis is the 4-momentum k_μ. Integrating, we find

$$k_0 = E + \frac{E}{2} \int_0^\zeta d\zeta' \, h_{\rho\sigma,0}\dot{x}^{(0)\rho}\dot{x}^{(0)\sigma} + \mathcal{O}(h^2) \quad \text{and} \quad \hat{k} = E\mathbf{e} + \mathcal{O}(h), \tag{44}$$

where E is the unperturbed energy and $x^{(0)} = (\text{const}+\zeta', \zeta'\mathbf{e})$ is the unperturbed photon path.

The photon energy seen by an observer with 4-velocity u ($|u^2| = 1$) is $k \cdot u$. Using $u = (1 - \frac{1}{2}h_{00}, \mathbf{v})$ with $|\mathbf{v}| \ll 1$ and (43)

$$\frac{(k \cdot u)_e}{(k \cdot u)_r} = \left(1 - \frac{1}{2}\int_e^r h_{\rho\sigma,0}\dot{x}^{(0)\rho}\dot{x}^{(0)\sigma}\,d\zeta + \frac{1}{2}h_{00}\bigg|_e^r - \mathbf{v}\cdot\mathbf{e}\bigg|_e^r\right) + \mathcal{O}\left(h^2\right), \quad (45)$$

where e and r refer to "emission" and "reception" respectively. The corresponding expression in ds^2 comes from multiplying the whole expression by $a(\eta_r)/a(\eta_e)$ to account for the cosmological redshift.

If we assume a uniform source and use the correspondence $h_{00} = 2\Phi$ between the metric perturbation and the Newtonian potential the temperature fluctuation induced is

$$\left(\frac{\Delta T}{T}\right)_{SW} = \Phi\bigg|_e^r - \mathbf{v}\cdot\mathbf{e}\bigg|_e^r - \frac{1}{2}\int_e^r h_{\rho\sigma,0}\dot{x}^{(0)\rho}\dot{x}^{(0)\sigma}\,d\zeta. \quad (46)$$

The three terms can be identified as the gravitational potential redshift, the Doppler effect due to motion of the emitter and receiver, and an extra effect due to the time dependence of the metric (see also Stebbins 1993). In a flat $\Lambda = 0$ universe Φ is constant in time in linear theory, so the last (integral) term vanishes and in the absence of Doppler shifts the potential change is known as the Sachs-Wolfe effect. In this limit the Sachs-Wolfe effect is simply the red-shifting of the photon as it climbs out of the potential on the surface of last scattering (assuming $\Phi = 0$ at the time of observation). In some cases, such as with gravitational waves, non-flat or Λ-dominated cosmologies, or nonlinear fluctuations, the integral term can also play a role.

Literature Cited

Abbott LF, Schaefer RK. 1986. *Ap. J.* 308:546

Abbott LF, Wise MB. 1984a. *Ap. J. Lett.* 282:L47

Abbott LF, Wise MB. 1984b. *Phys. Lett.* B135:279

Abbott LF, Wise MB. 1984c. *Nucl. Phys.* B244:541

Achilli S, Occhionero F, Scaramella R. 1985. *Ap. J.* 299:577

Adams FC, Bond JR, Freese K, Frieman JA, Olinto A. 1992. *Phys. Rev.* D47:426

Allen TJ, Grinstein B, Wise MB. 1987. *Phys. Lett.* B197:66

Anile AM. 1974 *Astrophys. Space Sci.* 29:415

Anninos P, Matzner RA, Tuluie R, Centrella J. 1991 *Ap. J.* 382:71

Argüeso F, Martínez-González E. 1989. *MNRAS* 238:1431

Arnau JV, Fullana MJ, Monreal L, Saez D. 1993. *Ap. J.* 402:359

Atrio-Barandela F, Doroshkevich AG. 1994a. *Ap. J.* 420:26

Atrio-Barandela F, Doroshkevich AG. 1994b. *Ap. J.* In press

Atrio-Barandela F, Doroshkevich AG, Klypin AA. 1991. *Ap. J.* 378:1

Atrio-Barandela F, Kashlinsky A. 1992. *Ap. J.* 390:322

Atrio-Barandela F, Silk J. 1994. *Phys. Rev. D.* 49:1126
Axenides N, Brandenberger R, Turner MS. 1983. *Phys. Lett.* 126B:178
Banday AJ, Wolfendale AW. 1990. *MNRAS* 245:182
Banday AJ, Wolfendale AW. 1991a. *MNRAS* 248:705
Banday AJ, Wolfendale AW. 1991b. *MNRAS* 252:462
Banday AJ, Giler M, Szabelska B, Szabelski J, Wolfendale AW. 1991. *Ap. J.* 375:432
Bardeen JM, Bond JR, Efstathiou G. 1987. *Ap. J.* 321:28
Bardeen JM, Bond JR, Kaiser N, Szalay AS. 1986. *Ap. J.* 304:15
Barrow JD, Juszkiewicz R, Sonoda DH. 1985. *MNRAS* 213:917
Barrow JD, Coles P. 1991. *MNRAS* 248:52
Bartlett J, Silk J. 1993. *Ap. J. Lett.* 407:L45
Basko MM, Polnarev AG. 1980. *MNRAS* 191:207
Baugh CM, Efstathiou G. 1993a. *MNRAS* 265:145
Baugh CM, Efstathiou G. 1993b. *MNRAS* In press
Bennett CL, Smoot GF, Hinshaw G, Wright EL, Kogut A, et al. 1992. *Ap. J. Lett.* 396:L7
Bennett CL, Hinshaw G, Banday A, Kogut A, Wright EL, et al 1993. *Ap. J. Lett.* 414:L77
Bennett CL, Kogut A, Hinshaw G, Banday AJ, Wright EL, et al. 1994. *Ap. J.* submitted
Bennett DP, Rhie SH. 1993. *Ap. J. Lett.* 406:L7
Bennett DP, Stebbins A, Bouchet FR. 1992. *Ap. J. Lett.* 399:L5
Berger JO. 1985. *Statistical Decision Theory and Bayesian Analysis.* New York: Springer-Verlag. 2nd ed.
Bernstein IN, Bernstein DN, Dubrovich VK. 1977. *Sov. Astron.* 54:727
Bertschinger E, Dekel A, Faber SM, Dressler A, Burstein D. 1990a. *Ap. J.* 364:370
Bertschinger E, Górski KM, Dekel A. 1990b. *Nature* 345:507
Birkinshaw M. 1990. In *The Cosmic Microwave Background: 25 Years Later,* ed. N Mandolesi, N Vittorio, p. 77. Dordrecht:Kluwer
Birkinshaw M, Gull SF, Hardebeck HE. 1984. *Nature* 309:34
Birkinshaw M, Gull SF, Moffet AT. 1981. *Ap. J. Lett.* 251:L69
Birkinshaw M, Gull SF, Hardebeck HE, Moffet AT. 1993. *Ap. J.* Submitted
Blanchard A. 1984. *Astron. Astrophys.* 132:359
Blanchard A, Schneider J. 1987. *Astron. Astrophys.* 184:1
Blumenthal GR, Faber SM, Primack JR, Rees MJ. 1984. *Nature* 311:517
Bond JR. 1988. In *The Early Universe,* ed. WG Unruh, GW Semenoff, p. 283. Dordrecht:Reidel
Bond JR. 1993. In *Proc. IUCAA Dedication Ceremonies,* ed. T Padmanabhan. New York:Wiley. In press
Bond JR. 1994. *Ap. Lett. Comm.* In press
Bond JR, Carr BJ, Hogan CJ. 1986. *Ap. J.* 306:428
Bond JR, Carr BJ, Hogan CJ. 1991a. *Ap. J.* 367:420
Bond JR, Crittenden R, Davis RL, Efstathiou G, Steinhardt PJ. 1994. *Phys. Rev. Lett.* 72:13
Bond JR, Efstathiou G. 1984. *Ap. J. Lett.* 285:L45
Bond JR, Efstathiou G. 1987. *MNRAS* 226:655
Bond JR, Efstathiou G, Lubin PM, Meinhold PR. 1991b. *Phys. Rev. Lett.* 66:2179
Bond JR, Efstathiou G, Silk J. 1980. *Phys. Rev. Lett.* 45:1980
Bond JR, Myers ST. 1991. In *Trends in Astroparticle Physics,* ed. D Cline, R Peccei, p. 262. Singapore:World Sci.
Bond JR, Szalay AS. 1983. *Ap. J.* 274:443
Bonometto SA, Caldara A, Lucchin F. 1983. *Astron. Astrophys.* 126:377
Bonometto SA, Lucchin F, Valdarnini R. 1984. *Astron. Astrophys.* 140:L27
Bouchet FR, Bennett DP, Stebbins A. 1988. *Nature* 335:410
Boughn SP, Cheng ES, Cottingham DA, Fixsen, DJ. 1992. *Ap. J. Lett.* 391:L49
Boughn SP, Jahoda K. 1993. *Ap. J. Lett.* 412:L1
Brandenberger RH. 1993. In *Current Topics in Astrofundamental Physics, Second International School of Physics 'D Chalonge,'* ed. N Sanchez, A Zichichi, p. 272. Singapore:World Sci.
Brandenberger RH, Kaplan DM, Ramsey SA. 1993. Preprint
Brandenberger RH, Perivolaropoulos L, Stebbins A. 1990. *Int. J. Mod. Phys.* 5:1633
Brandenberger RH, Turok N. 1986. *Phys. Rev.* D33:2182
Brandt WN, Lawrence CR, Readhead ACS, Pakianathan JN, Fiola TM. 1994. *Ap. J.* 424:1
Bunn EF, Hofman Y, Silk J. 1994. *Ap. J.* 425:359
Bunn EF, White M, Srednicki M, Scott D. 1994b. *Ap. J.* In press
Burke WL. 1975. *Ap. J.* 196:329
Caderni N, Fabbri R, Melchiorri F, Natale V. 1978. *Phys. Rev.* D17:1901
Carlberg RG. 1991. *Ap. J.* 367:385
Carr BJ, Lidsey JE. 1993. *Phys. Rev.* D48:543
Carroll SM, Press WH, Turner EL. 1992. *Annu. Rev. Astron. Astrophys.* 30:499
Cavaliere A, Menci N, Setti G. 1991. *Astron. Astrophys.* 245:L21
Cayón L, Martínez-González E, Sanz JL. 1991. *MNRAS* 253:599
Cayón L, Martínez-González E, Sanz JL. 1993a. *Ap. J.* 403:471
Cayón L, Martínez-González E, Sanz JL. 1993b. *Ap. J.* 413:10
Cayón L, Martínez-González E, Sanz JL. 1994. *Astron. Astrophys.* In press.
Cen R, Gnedin NY, Kofman LA, Ostriker JP. 1992. *Ap. J. Lett.* 399:L11

Cen R, Ostriker JP. 1992. *Ap. J.* 399:331
Cen R, Ostriker JP. 1993. *Ap. J.* 414:407
Cen R, Ostriker JP, Peebles PJE. 1993. *Ap. J.* 415:423
Centrella J, Gallagher JS, Melott AL, Bushouse HA. 1988. *Ap. J.* 333:24
Cheng ES, Cottingham DA, Fixsen DJ, Inman CA, Kowitt MS, et al. 1994. *Ap. J. Lett.* 422:L37
Chiba T, Sugiyama N, Suto Y. 1993. *Ap. J.* In press
Chitre SM, Narlikar JV, Padmanabhan T. 1986. *Phys. Lett.* A117:285
Chodorowski M. 1992. *MNRAS* 259:218
Cole S, Efstathiou G. 1989. *MNRAS* 239:195
Cole S, Kaiser N. 1988. *MNRAS* 233:637
Coles P. 1988a. *MNRAS* 231:125
Coles P. 1988b. *MNRAS* 234:509
Coles P, Barrow JD. 1987. *MNRAS* 228:407
Collins CB, Hawking SW. 1973. *MNRAS* 162:207
Copeland E, Kolb EW, Liddle AR, Lidsey JE. 1993a. *Phys. Rev. Lett.* 71:219
Copeland E, Kolb EW, Liddle AR, Lidsey JE. 1993b. *Phys. Rev.* D48:2529
Copeland E, Kolb EW, Liddle AR, Lidsey JE. 1994. *Phys. Rev.* D 49:1840
Corey BE, Wilkinson DT. 1976. *Bull. Am. Astron. Soc.* 8:351
Couchman HMP. 1985. *MNRAS* 214:137
Couchman HMP, Rees MJ. 1986. *MNRAS* 221:53
Coulson D, Ferreira P, Graham P, Turok N. 1994. *Nature* 368:27
Crittenden R, Bond JR, Davis RL, Efstathiou G, Steinhardt PJ. 1993a. *Phys. Rev. Lett.* 71:324
Crittenden R, Davis RL, Steinhardt PJ. 1993b. *Ap. J. Lett.* 417:L13
Daly RA. 1991. *Ap. J.* 371:14
Danese L, De Zotti G, Mandolesi N. 1983. *Astron. Astrophys.* 121:114
Dautcourt G. 1977. *Astron. Nachr.* 298:141
Dautcourt G, Rose K. 1978. *Astron. Nachr.* 299:13
Davies RD, Watson RA, Daintree EJ, Hopkins J, Lasenby AN, et al. 1992. *MNRAS* 258:605
Davis M. 1980. *Physica Scripta* 21:717
Davis M, Efstathiou G, Frenk CS, White SDM. 1985. *Ap. J.* 292:371
Davis M, Peebles PJE. 1983. *Ap. J.* 267:465
Davis M, Summers FJ, Schlegel D. 1992. *Nature* 359:393
Davis RL, Hodges HM, Smoot GF, Steinhardt PJ, Turner MS. 1992. *Phys. Rev. Lett.* 69:1856 (erratum:70:1733)
de Bernardis P, Masi S, Vittorio N. 1991. *Ap. J.* 382:515
de Bernardis P, Masi S, Melchiorri F, Melchiorri B, Vittorio N, 1992. *Ap. J. Lett.* 396:L57
de Bernardis P, Aquilini E, Boscaleri A, De Petris M, Gervasi M, et al. 1993. *Astron. Astrophys.* 271:683
de Bernardis P, Aquilini E, Boscaleri A, De Petris M, D'Andreta G, et al. 1994. *Ap. J.* 422:L33
Dekel A. 1983. *Ap. J.* 264:373
Dekel A, Bertschinger E, Yahil A, Strauss MA, Davis M, Huchra JP. 1993. *Ap. J.* 412:1
Demarque P, Deliyannis CP, Sarajedini A. 1991. In *Observational Tests of Cosmological Inflation,* ed. T Shanks, et al, p. 111. Dordrecht:Kluwer
Devlin M, Alsop D, Clapp A, Cottingham D, Fischer M, et al. 1993. In *Proc. NAS Colloquium on Physical Cosmology (Irvine).* In press
Dicke RH, Peebles PJE, Roll PG, Wilkinson, DT. 1965. *Ap. J.* 142:414
Dodelson S, Jubas JM. 1993. *Phys. Rev. Lett.* 70:2224
Dodelson S, Jubas JM. 1994. *Ap. J.* Submitted
Dodelson S, Stebbins A. 1993. *Ap. J.* Submitted
Dolgov A, Silk J. 1993. *Phys. Rev.* D47:2619
Doroshkevich AG. 1988. *Sov. Astron. Lett.* 14:125
Doroshkevich AG, Novikov ID, Polnarev AG. 1977. *Sov. Astron.* 21:523
Doroshkevich AG, Zel'dovich Ya B, Sunyaev RA. 1978. *Sov. Astron.* 22:523
Doroshkevich AG, Zel'dovich Ya B, Sunyaev RA, Khlopov MYu. 1980. *Sov. Astron. Lett.* 6:457
Dragovan M, Ruhl JE, Novak G, Platt SR, Crone B, et al. 1994. *Ap. J. Lett.* In press
Dubrovich VK. 1975. *Sov. Astron. Lett.* 1:3
Dubrovich VK. 1993. *Sov. Astron. Lett.* 19:132
Durrer R, Howard A, Zhou Z-H. 1994. *Phys. Rev. D* 49:681
Dyer CC. 1976. *MNRAS* 175:429
Efstathiou G. 1988. In *Large-Scale Motions in the Universe. A Vatican Study Week,* ed. VC Rubin, GV Coyne, p. 299. Princeton:Princeton Univ. Press
Efstathiou G. 1990. In *Physics of the Early Universe: Proc. 36th Scottish Universities Summer School in Physics,* ed. JA Peacock, AE Heavens, AT Davies, p. 361. New York:Adam Hilger
Efstathiou G, Bond JR. 1986. *MNRAS* 218:103
Efstathiou G, Bond JR. 1987. *MNRAS* 227:33 p
Efstathiou G, Bond JR, White SDM. 1992. *MNRAS* 258:1 p
Efstathiou G, Sutherland WJ, Maddox SJ. 1990. *Nature* 348:705
Fabbri R, Guidi I, Natale V. 1983. *Astron. Astrophys.* 122:151
Fabbri R, Lucchin F, Matarrese S. 1987. *Ap. J.* 315:1
Fabbri R, Pollock MD. 1983. *Ap. Phys. Lett.* B125:445
Fahr HJ, Loch R. 1991. *Astron. Astrophys.* 146:1
Falk T, Rangarajan R, Srednicki M. 1993. *Ap. J. Lett.* 403:L1
Fang L-Z. 1991. *Trends in Astroparticle Physics,* ed. D Cline, R Peccei, p. 34. Singapore:World Sci.

Fang L-Z, Houjun M. 1987. *Mod. Phys. Lett.* A2:229
Feldman HA, Kaiser N, Peacock JA. 1994. *Ap. J.* 426:23
Feng LL, Liu JM. 1992. *Astron. Astrophys.* 264:385
Fink E. 1993. PhD thesis. Univ. Florence
Fisher KB, Davis M, Strauss MA, Yahil A, Huchra JP. 1993. *Ap. J.* 402:42
Fixsen DJ, Cheng ES, Cottingham DA, Eplee RE, Isaacman RB, et al. 1994. *Ap. J.* 420:445
Fomalont EB, Partridge RB, Lowenthal JD, Windhorst RA. 1993. *Ap. J.* 404:8
Franceschini A, Toffolatti L, Danese L, De Zotti G. 1989. *Ap. J.* 344:35
Freese K, Adams FC, Frieman JA, Mottola E. 1987. *Nucl. Phys.* B287:797
Frenk CS. 1991. *Physica Scripta* T36:70
Frenk CS, White SDM, Efstathiou G, Davis M. 1990. *Ap. J.* 351:10
Frewin RA, Polnarev AG, Coles P. 1994. *MNRAS* 266:L21
Fry JN, Wang Y. 1992. *Phys. Rev.* D46:3318
Fukugita M, Kawasaki M. 1994. *MNRAS* In press
Fukugita M, Sugiyama N, Umemura M. 1990. *Ap. J.* 358:28
Gaier T, Schuster J, Gunderson J, Koch T, Seiffert M, et al. 1992. *Ap. J. Lett.* 398:L1
Ganga K, Cheng E, Meyer S, Page L. 1993. *Ap. J. Lett.* 410:L57
Goicoechea LJ, Martin-Mirones JM. 1992. *Astron. Astrophys.* 254:1
Górski KM. 1991. *Ap. J. Lett.* 370:L5
Górski KM. 1992. *Ap. J. Lett.* 398:L5
Górski KM. 1993. *Ap. J. Lett.* 410:L65
Górski KM, Silk J. 1989. *Ap. J. Lett.* 346:L1
Górski KM, Silk J, Vittorio N. 1992. *Phys. Rev. Lett.* 68:733
Górski KM, Stompor R, Juszkiewicz R. 1993. *Ap. J. Lett.* 410:L1
Gott JR, Park C, Juskiewicz R, Bies WE, Bennett DP, et al. 1990. *Ap. J.* 352:1
Gottlober S, Mucket JP. 1993. *Astron. Astrophys.* 272:1
Gouda N, Sugiyama N. 1992. *Ap. J. Lett.* 395:L59
Gouda N, Sugiyama N, Sasaki M. 1991a. *Prog. Theor. Phys.* 85:1023
Gouda N, Sugiyama N, Sasaki M. 1991b. *Ap. J. Lett.* 372:L49
Gould A. 1993. *Ap. J. Lett.* 403:L51
Grachev SJ, Dubrovich VK. 1991. *Astrophys.* 34:124
Graham P, Turok N, Lubin PM, Schuster JA. 1993. *Ap. J.* Submitted
Gratsias J, Scherrer RJ, Steigman G, Vilumsen JV. 1993. *Ap. J.* 405:30
Groth E, Juszkiewicz R, Ostriker JP. 1989. *Ap. J.* 346:558
Gull SF, Northover KJE. 1976. *Nature* 263:572
Gundersen JO, Clapp AC, Devlin M, Holmes W, Fischer ML, et al. 1993. *Ap. J. Lett.* 413:L1

Gunn JE, Tinsley BM. 1975. *Nature* 257:454
Gurzadyan VG, Kocharyan AA. 1993. *Europhys. Lett.* 22:231
Gurzadyan VG, Torres S. 1993. Preprint
Gutiérrez de la Cruz CM, Cayón L, Martínez-González E, Sanz JL. 1993. *MNRAS* In press
Hancock S, Davies RD, Lasenby AN, Gutiérrez de la Cruz CM, Watson RA, et al. 1994. *Nature* 367:333
Hara T, Mähönen P, Miyoshi S. 1993. *Ap. J.* 414:421
Harari DD, Zaldarriaga M. 1993. *Phys. Lett. B* 319:96
Harrison ER. 1970. *Phys. Rev.* D1:2726
Herbig T, Readhead ACS, Lawrence CR. 1992. *Bull. Am. Astron. Soc.* 24:1263
Hill CT, Schramm DN, Fry JN. 1989. *Comments Nucl. Part. Phys.* 19:25
Hindmarsh M. 1993. Preprint
Hnatyk BT, Lukash VN, Novosyadlyj BS. 1992. *Sov. Astron. Lett.* 18:563
Hodges HM, Blumenthal GR. 1990. *Phys. Rev.* D42:3329
Hogan CJ. 1979. *MNRAS* 188:781
Hogan CJ. 1980. *MNRAS* 192:891
Hogan CJ. 1982. *Ap. J. Lett.* 256:L33
Hogan CJ. 1984. *Ap. J. Lett.* 284:L1
Hogan CJ. 1992. *Ap. J. Lett.* 398:L77
Hogan CJ, Kaiser N. 1983. *Ap. J.* 274:7
Hogan CJ, Kaiser N, Rees MJ. 1982. In *The Big Bang and Element Creation,* ed. D Lynden-Bell,. *Philos. Trans. R. Soc. London Ser. A* 307:97
Holtzman JA. 1989. *Ap. J. Suppl.* 71:1
Holtzman JA, Primack JR. 1993. *Ap. J.* 405:428
Hu W, Scott D, Silk J. 1994. *Phys. Rev. D.* 49:648
Hu W, Sugiyama N. 1994a. *Phys. Rev. D* In press
Hu W, Sugiyama N. 1994b. *Ap. J.* Submitted
Hu W, Sugiyama N. 1994c. *Phys. Rev. D* Submitted
Jaffe A, Stebbins A, Frieman JA. 1994. *Ap. J.* 420:9
Jones BJT, Wyse RFG. 1985. *Astron. Astrophys.* 149:144
Jones M, Saunders R, Alexander P, Birkinshaw M, Dillon N, et al. 1993. *Nature* 365:320
Jørgensen HE, Kotok E, Nasel'skiǐ PD, Novikov ID. 1993. *MNRAS* 265:261
Jørgensen HE, Kotok E, Nasel'skiǐ PD, Novikov ID. 1994. *Astron. Astrophys.*, In press
Juszkiewicz R, Górski KM, Silk J. 1987. *Ap. J. Lett.* 323:L1
Juszkiewicz R, Vittorio N, Wyse RFG. 1990. *Ap. J.* 349:408
Kaiser N. 1982. *MNRAS* 198:1033
Kaiser N. 1983. *MNRAS* 202:1169
Kaiser N. 1984. *Ap. J.* 282:374
Kaiser N. 1987. *MNRAS* 227:1
Kaiser N, Efstathiou G, Ellis RS, Frenk CS, Lawrence A, et al. 1991. *MNRAS* 252:1
Kaiser N, Silk J. 1986. *Nature* 324:529
Kaiser N, Stebbins A. 1984. *Nature* 310:391
Kamionkowski M, Spergel DN. 1993. *Ap. J.*

submitted
Kamionkowski M, Spergel DN, Sugiyama N. 1994. *Ap. J. Lett.* In press
Kashlinsky A. 1988. *Ap. J. Lett.* 331:L1
Kashlinsky A. 1991. *Ap. J. Lett.* 383:L1
Kashlinsky A. 1992. *Ap. J. Lett.* 399:L1
Kashlinsky A. 1993a. *Ap. J.* 402:369
Kashlinsky A. 1993b. Preprint
Kashlinsky A, Jones BJT. 1991. *Nature* 349:753
Kauffmann G, Charlot S. 1994. *Ap. J. Lett.* Submitted
Klein U, Rephaeli Y, Schlickeiser R, Wielebinski R. 1991. *Astron. Astrophys.* 244:43
Klypin AA, Holtzman JA, Primack JR, Regös E. 1993. *Ap. J.* 416:1
Klypin AA, Strukov IA, Skulachev DP. 1992. *MNRAS* 258:71
Kodama H, Sasaki M. 1986. *Int. J. Mod. Phys.* A1:265
Kofman L, Blumenthal GR, Hodges HM, Primack JR. 1991. In *Large-Scale Structures and Peculiar Motions in the Universe*, ed. DW Latham, LA Nicolaci da Costa, ASP Conf. Ser., 15:339
Kofman L, Gnedin NY, Bahcall NA. 1993. *Ap. J.* 413:1
Kogut A, Lineweaver C, Smoot GF, Bennett CL, Banday A, et al. 1993. *Ap. J.* 419:1
Kolb EW, Turner MS. 1990. *The Early Universe*. Redwood City, CA:Addison-Wesley
Kolb EW, Vadas S. 1993. *Phys. Rev. D.*, Submitted
Korolëv VA, Sunyaev RA, Yakubtsev LA. 1986. *Sov. Astron. Lett.* 12:339
Krauss LM, Romanelli P. 1990. *Ap. J.* 358:47
Krauss LM, White M. 1992. *Phys. Rev. Lett.* 69:869
Krolik JH. 1989. *Ap. J.* 338:594
Krolik JH. 1990. *Ap. J.* 353:21
Lahav O, Lilje PB, Primack JR, Rees MJ. 1991. *MNRAS* 251:128
Lang KR. 1980. *Astrophysical Formulae*. New York:Springer-Verlag. 2nd ed.
Lauer TR, Postman M. 1992. *Ap. J. Lett.* 400:L47
Lauer TR, Postman M. 1993. Preprint
Lepp S, Shull JM. 1984. *Ap. J.* 280:465
Liddle AR. 1994. *Phys. Rev. D* 49:739
Liddle AR, Lyth DH. 1993. Phys. Rep. 231:1
Liddle AR, Lyth DH. 1992. *Phys. Lett.* B291:391
Liddle AR, Lyth DH, Sutherland WJ. 1992. *Phys. Lett.* B279:244
Lidsey JE, Tavakol RK. 1993. *Phys. Lett.* B309:23
Linder EV. 1988a. *Astron. Astrophys.* 206:199
Linder EV. 1988b. *Ap. J.* 326:517
Linder EV. 1990. *MNRAS* 243:353
Loveday J, Efstathiou G, Peterson BA, Maddox SJ. 1992. *Ap. J. Lett.* 400:L43
Lubin PM, Smoot GF. 1981. *Ap. J.* 245:1
Lubin PM, Melese P, Smoot GF. 1983. *Ap. J. Lett.* 273:L51

Lucchin F, Matarrese S, Mollerach S. 1992. *Ap. J. Lett.* 401:L49
Luo X, Schramm D. 1993a. *Ap. J.* 408:33
Luo X, Schramm D. 1993b. *Phys. Rev. Lett.* 71:1124
Luo X, Schramm DN. 1994. *Ap. J.* 421:393
Lyth DH. 1985. *Phys. Rev.* D31:1792
Lyth DH, Stewart ED. 1990. *Phys. Lett.* B252:336
Lyubarski YE, Sunyaev RA. 1983. *Astron. Astrophys.* 123:171
Lyubimov VA, Novikov EG, Nozik VZ, Tretyakov EF, Kosik VS. 1980. *Phys. Lett.* B94:266
Maddox SJ, Efstathiou G, Sutherland WJ, Loveday J. 1990. *MNRAS* 242:P43
Makino N, Suto Y. 1993. *Ap. J.* 405:1
Maoli R, Melchiorri F, Tosti D. 1993. *Ap. J.* 425:372
Markevitch M, Blumenthal GR, Forman W, Jones C, Sunyaev RA. 1991. *Ap. J. Lett.* 378:L33
Martel H. 1991. *Ap. J.* 366:353
Martínez-González E, Cayón L, Sanz JL. 1993. In *16th Texas Symp. on Relativistic Astrophysics*, ed. CW Akerlof, MA Srednicki,. *Ann. NY Acad. Sci.* 688:827
Martínez-González E, Sanz JL. 1990. *MNRAS* 247:473
Martínez-González E, Sanz JL, Silk J. 1990. *Ap. J. Lett.* 355:L5
Martínez-González E, Sanz JL, Silk J. 1992. *Phys. Rev.* D46:4193
Martínez-González E, Sanz JL, Silk J. 1994. *Ap. J. Lett.* In press
Masi S, de Bernardis P, De Petris M, Epitani M, Gervasi M, Guarni G. 1991. *Ap. J. Lett.* 366:L51
Mather JC, Cheng ES, Cottingham DA, Eplee RE, Fixsen DJ, et al. 1994. *Ap. J.* 420:439
Matsuda T, Sato H, Takeda H. 1971. *Prog. Theor. Phys.* 46:416
Meinhold PR, Clapp A, Cottingham DA, Devlin M, Fischer M, et al. 1993. *Ap. J. Lett.* 409:L1
Meinhold PR, Lubin PM. 1991. *Ap. J. Lett.* 370:L11
Melchiorri F. 1993. *Proc. Santander CMB Workshop* In press
Mo HJ, Miralda-Escudé J. 1994. *Ap. J. Lett.*, submitted
Moessner R, Perivolaropoulos L, Brandenberger, R. 1993. *Ap. J.* 425:365
Mollerach S, Matarrese S, Lucchin F. 1993. Preprint
Moscardini L, Matarrese S, Lucchin F, Messina, A. 1991. *MNRAS* 248:424
Muciaccia PF, Mei S, Degasperis G, Vittorio N. 1993. *Ap. J. Lett.* 410:L61
Mukhanov VF, Feldman HA, Brandenberger RH. 1992. *Phys. Rep.* 215:203
Myers ST, Readhead AC, Lawrence CR. 1993. *Ap. J.* 405:8
Nambu Y, Ishihara H, Gouda N, Sugiyama N.

1991. *Ap. J. Lett.* 373:L35
Nanos GP. 1979. *Ap. J.* 232:341
Narlikar JV, Padmanabhan T. 1991. *Annu. Rev. Astron. Astrophys.* 29:325
Nasel'skiĭ PD, Novikov ID. 1993. *Ap. J.* 413:14
Nasel'skiĭ PD, Polnarev AG. 1987. *Astrophys.* 16:543
Negroponte J, Silk J. 1980. *Phys. Rev. Lett.* 44:1433
Ng KL, Ng KW. 1993. Preprint:IP-ASTP-08-93
Nottale L. 1984. *MNRAS* 206:713
Novikov ID, Zel'dovich Ya B. 1967. *Annu. Rev. Astron. Astrophys.* 5:627
Ostriker JP. 1993. *Annu. Rev. Astron. Astrophys.* 31:689
Ostriker JP, Vishniac ET. 1986. *Ap. J. Lett.* 306:L51
Overduin JM, Wesson PS, Bowyer S. 1993. *Ap. J. Lett.* 404:L1
Page LA, Cheng ES, Meyer SS. 1990. *Ap. J. Lett.* 355:L1
Partridge RB. 1988. *Rep. Prog. Phys.* 51:647
Partridge RB, Nowakowski J, Martin HM. 1988. *Nature* 331:146
Peacock JA. 1991. *MNRAS* 253:1 p
Peacock JA, Dodds SJ. 1994. *MNRAS* 267:1020
Peebles PJE. 1965. *Ap. J.* 142:1317
Peebles PJE. 1968. *Ap. J.* 153:1
Peebles PJE. 1982a. *Ap. J.* 258:415
Peebles PJE. 1982b. *Ap. J.* 259:442
Peebles PJE. 1982c. *Ap. J. Lett.* 263:L1
Peebles PJE. 1984. *Ap. J.* 284:439
Peebles PJE. 1987a. *Nature* 327:210
Peebles PJE. 1987b. *Ap. J. Lett.* 315:L73
Peebles PJE. 1993. *Principles of Physical Cosmology.* Princeton:Princeton Univ. Press
Peebles PJE, Wilkinson DT. 1968. *Phys. Rev.* 174:2168
Peebles PJE, Yu JT. 1970. *Ap. J.* 162:815
Pen U-L, Spergel DN, Turok N. 1994. *Phys. Rev. D.* 49:692
Penzias AA, Wilson RW. 1965. *Ap. J.* 142:419
Perivolaropoulos L. 1993a. *Phys. Lett.* B298:305
Perivolaropoulos L. 1993b. *Phys. Rev.* D48:1530
Persi FM, Spergel DN. 1993. Preprint
Plionis M, Valdarnini R. 1991. *MNRAS* 249:46
Pogosyan DYu, Starobinskiĭ AA. 1993. *MNRAS* 265:507
Polnarev AG. 1985. *Sov. Astron.* 29:607
Puy D, et al. 1993. *Astron. Astrophys.* 267:337
Ratra B, Peebles PJE. 1988. *Phys. Rev.* D37:3406
Readhead ACS, Lawrence CR. 1992. *Annu. Rev. Astron. Astrophys.* 30:653
Readhead ACS, Lawrence CR, Myers ST, Sargent WLW, Hardebeck HE, et al. 1989. *Ap.J.* 346:566
Rees MJ. 1968. *Ap. J. Lett.* 153:L1
Rees MJ. 1977. In *The Evolution of Galaxies and Stellar Populations,* ed. BM Tinsley, RB Larson, p. 339. New Haven:Yale Univ. Obs.

Rees MJ, Sciama DW. 1968. *Nature* 217:511
Rephaeli Y. 1981. *Ap. J.* 245:351
Rephaeli Y. 1993a. In *16th Texas Symp. on Relativistic Astrophysics,* ed. CW Akerlof, MA Srednicki,. *Ann. NY Acad. Sci.* 688:818
Rephaeli Y. 1993b. *Ap. J.* 418:1
Rubakov VA, Sazhin MV, Veryaskin AV. 1982. *Phys. Lett.* B115:189
Sachs RK, Wolfe AM. 1967. *Ap. J.* 147:73
Saez D, Arnau JV, Fullana MJ. 1993. *MNRAS* 263:681
Sahni V. 1990. *Phys. Rev.* D42:453
Salopek DS. 1992. *Phys. Rev. Lett.* 69:3602
Salopek DS, Bond JR, Bardeen JM. 1989. *Phys. Rev.* D40:1753
Sasaki M. 1989. *MNRAS* 240:415
Sasaki M. 1993. *Prog. Theor. Phys.* 89:1183
Sasaki S, Takahara F. 1993. *Prog. Theor. Phys.* 45:655
Sasaki S, Takahara F, Suto Y. 1993. *Prog. Theor. Phys.* 90:85
Sazhin MV. 1985. *MNRAS* 216:25 p
Scaramella R, Baiesi-Pillastrini G, Chincarini G, Vettolani G, Zamorani G. 1989. *Nature* 338:562
Scaramella R, Cen R, Ostriker JP. 1993. *Ap. J.* 416:399
Scaramella R, Vittorio N. 1990. *Ap. J.* 353:372
Scaramella R, Vittorio N. 1991. *Ap. J.* 375:439
Scaramella R, Vittorio N. 1993a. *MNRAS* 263:L17
Scaramella R, Vittorio N. 1993b. *Ap. J.* 411:1
Schaefer RK, Shafi Q. 1992. *Nature* 359:199
Schaefer RK, Shafi Q, Stecker FW. 1989. *Ap. J.* 347:575
Schaeffer R, Silk J. 1988. *Ap. J.* 333:509
Scherrer RJ. 1992. *Ap. J.* 390:330
Schuster J, et al. 1993. *Ap. J. Lett.* 412:L47
Scott D. 1991. *Structure at high redshift.* PhD thesis. Univ. Cambridge
Scott D, Rees MJ, Sciama DW. 1991. *Astron. Astrophys.* 250:295
Scott D, Srednicki M, White M. 1994. *Ap. J. Lett.* 421:L5
Seljak U, Bertschinger E. 1993. *Ap. J. Lett.* 417:L9
Shafi Q, Stecker FW. 1984. *Phys. Rev. Lett.* 53:1292
Silk J. 1967. *Nature* 215:1155
Silk J. 1968. *Ap. J.* 151:459
Silk J. 1982. *Acta Cosmologica* 11:75
Silk J, Wilson ML. 1980. *Physica Scripta* 21:708
Silk J, Wilson ML. 1981. *Ap. J. Lett.* 244:L37
Smith MS, Kawano LH, Malaney RA. 1993. *Ap. J. Suppl.* 85:219
Smoot GF. 1992. In *The Infrared and Submillimetre Sky After COBE,* ed. M Signore, C Dupraz, p. 331. Dordrecht:Kluwer
Smoot GF, Gorenstein MV, Muller RA. 1977. *Phys. Rev. Lett.* 39:898
Smoot GF, Bennett CL, Kogut A, Aymon J, Backus C, et al. 1990. *Ap. J.* 360:685

Smoot GF, Bennett CL, Kogut A, Aymon J, Backus C, et al. 1991. *Ap. J. Lett.* 371:L1

Smoot GF, Bennett CL, Kogut A, Wright EL, Aymon J, et al. 1992. *Ap. J. Lett.* 396:L1

Smoot GF, Tenorio L, Banday AJ, Kogut A, Wright EL, et al. 1994. *Ap. J.* Submitted

Sokolov DD, Starobinskiĭ AA. 1976. *Sov. Astron.* 9:629

Sokolov IYu. 1993. *JETP Lett.* 57:617

Souradeep T, Sahni V. 1992. *Mod. Phys. Lett.* 7:3541

Srednicki M. 1993. *Ap. J. Lett.* 416:L1

Srednicki M, White M, Scott D, Bunn E. 1993. *Phys. Rev. Lett.* 71:3747

Stark PB. 1993. *Ap. J. Lett.* 408:L73

Stark RF. 1981. *MNRAS* 195:127

Starobinskiĭ AA. 1979. *JETP Lett.* 30:682

Starobinskiĭ AA. 1985. *Sov. Astron. Lett.* 11:113

Starobinskiĭ AA. 1988. *Sov. Astron. Lett.* 14:166

Starobinskiĭ AA. 1993. *JETP Lett.* 57:622

Stebbins A. 1988. *Ap. J.* 327:584

Stebbins A. 1993. In *16th Texas Symp. on Relativistic Astrophysics,* ed. CW Akerlof, MA Srednicki,. *Ann. NY Acad. Sci.* 688:824

Stebbins A, Veeraraghavan S. 1993. *Phys. Rev. D* 48:2421

Steigman G. 1993. *Nucl. Phys. B Proc. Suppl.* 31:343

Stelmach J, Byrka R, Dabrowski MP. 1990. *Phys. Rev.* D41:2434

Stevens D, Scott D, Silk J. 1993. *Phys. Rev. Lett.* 71:20

Stompor R. 1993. *Astron. Astrophys.* In press

Stompor R, Górski KM. 1994. *Ap. J. Lett.* 422:L41

Strauss MA, Yahil A, Davis M, Huchra JP, Fisher K. 1992. *Ap. J.* 397:395

SubbaRao MU, Szalay AS, Schaefer R, Gulkis S, von Gronefeld P. 1994. *Ap. J.* 420, 474

Subrahmanyan R, Ekers RD, Sinclair M, Silk J. 1993. *MNRAS* 263:416

Sugiyama N, Gouda N. 1992. *Prog. Theor. Phys.* 88:803

Sugiyama N, Gouda N, Sasaki M. 1990. *Ap. J.* 365:432

Sugiyama N, Sasaki M, Tomita K. 1989. *Ap. J. Lett.* 388:L45

Sugiyama N, Sato K. 1992. *Ap. J.* 387:439

Sugiyama N, Silk J. 1994. *Phys. Rev. Lett.*, Submitted

Sugiyama N, Silk J, Vittorio N. 1993. *Ap. J. Lett.* 419:L1

Sunyaev RA. 1977. *Sov. Astron. Lett.* 3:491

Sunyaev RA. 1978. In *IAU Symp. 79, Large-Scale Structure of the Universe,* ed. MS Longair, J Einasto, p. 393. Dordrecht:Reidel

Sunyaev RA, Zel'dovich YaB. 1970. *Astrophys. Space Sci.* 7:1

Sunyaev RA, Zel'dovich YaB. 1972. *Comments Astrophys. Space Phys.* 4:173

Sunyaev RA, Zel'dovich YaB. 1980. *Annu. Rev. Astron. Astrophys.* 18:537

Suto Y, Górski KM, Juszkiewicz R, Silk J. 1988. *Nature* 332:328

Suto Y, Gouda N, Sugiyama N. 1990. *Ap. J. Suppl.* 74:665

Szalay AS, Bond JR, Silk J. 1983. In *Formation and Evolution of Galaxies and Large Structures in the Universe,* ed. J Audouze, J Tran Thanh Van, p. 101. Dordrecht:Reidel

Taylor AN, Rowan-Robinson M. 1992. *Nature* 359:396

Tegmark M, Silk J, Blanchard A. 1994. *Ap. J.* 420:484

Thompson KL, Vishniac ET. 1987. *Ap. J.* 313:517

Timbie PT, Wilkinson DT. 1990. *Ap. J.* 353:140

Tolman BW, Matzner RA. 1984. *Philos. Trans. R. Soc. London Ser. A* 392:391

Tomita K, Tanabe T. 1983. *Prog. Theor. Phys.* 69:828

Tomita K, Watanabe K. 1989. *Prog. Theor. Phys.* 82:563

Trester JJ, Canizares CR. 1989. *Ap. J.* 347:605

Tucker GS, Griffen GS, Nguyen HT, Peterson JB. 1993. *Ap. J. Lett.* 419:L45

Turner MS. 1993. *Phys. Rev.* D48:5539

Turner MS, Steigman G, Krauss L. 1984. *Phys. Rev. Lett.* 52:2090

Turner MS, White M, Lidsey JE. 1993. *Phys. Rev.* D48:4613

Umemura M, Ikeuchi S. 1985. *Ap. J.* 299:583

Uson JM. 1986. In *Radio Continuum Processes in Clusters of Galaxies,* ed. C O'Dea, JM Uson, p. 255. Green Bank:NRAO

Uson JM, Wilkinson DT. 1984a. *Ap. J. Lett.* 277:L1

Uson JM, Wilkinson DT. 1984b. *Nature* 312:427

Uson JM, Wilkinson DT. 1984c. *Ap. J.* 283:471

Valdarnini R, Bonometto SA. 1985. *Astron. Astrophys.* 196: 235

van Dalen A, Schaefer RK. 1992. *Ap. J.* 398:33

van Kampen E, Martínez-González E. 1991. In *Second 'Recontres de Blois': Physical Cosmology,* ed. A Blanchard et al. Gif-sur-Yvette, France:Ed. Frontieres

Veeraraghavan S, Stebbins A. 1992. *Ap. J. Lett.* 395:L55

Vishniac ET. 1987. *Ap. J.* 322:597

Vittorio N, de Bernardis P, Masi S, Scaramella R. 1989. *Ap. J.* 341:163

Vittorio N, Juszkiewicz R. 1987. *Ap. J. Lett.* 314:L29

Vittorio N, Matarrese S, Lucchin F. 1988. *Ap. J.* 328:69

Vittorio N, Meinhold P, Muciaccia PF, Lubin PM, Silk J. 1991. *Ap. J. Lett.* 372:L1

Vittorio N, Muciaccia PF. 1991. In *After the First Three Minutes,* ed. SS Holt, CL Bennett, V Trimble, p. 141. New York:Am. Inst. Phys.

Vittorio N, Silk J. 1984. *Ap. J. Lett.* 285:L39

Vittorio N, Silk J. 1985. *Ap. J. Lett.* 293:L1

Vittorio N, Silk J. 1992. *Ap. J. Lett.* 385:L9

Vogeley MS, Park C, Geller MJ, Huchra JP. 1992. *Ap. J. Lett.* 391:L5

Vollick DN. 1993. *Phys. Rev. D* 48:3585

Walker PN, Steigman G, Schramm DN, Olive KA, Kang H-S. 1991. *Ap. J.* 376:51

Watson RA, et al. 1992. *Nature* 357:660

Watson RA, Gutiérrez de la Cruz CM. 1993. *Ap. J. Lett.* 419:L5

White M. 1992. *Phys. Rev.* D42:4198

White M, Krauss L, Silk J. 1993. *Ap. J.* 418:535

White SDM, Davis M, Frenk CS. 1984. *MNRAS* 209:27P

White SDM, Frenk CS, Davis M. 1983. *Ap. J. Lett.* 274:L1

Wilbanks TM, Ade PAR, Fischer ML, Holzapfel WL, Lange AE. 1993. In *16th Texas Symp. on Relativistic Astrophysics,* ed. CW Akerlof, MA Srednicki, *Ann. NY Acad. Sci.* 688:798

Wilson ML. 1983. *Ap. J.* 273:2

Wilson ML, Silk J. 1981. *Ap. J.* 243:14

Wollack EJ, Jarosik NC, Netterfield CB, Page LA, Wilkinson D. 1993. *Ap. J. Lett.* 419:L49

Wright EL, Smoot GF, Kogut A, Hinshaw G, Tenorio L, et al. 1994a. *Ap. J.* 420:1

Wright EL, Smoot GF, Bennett CL, Lubin PM. 1994b. *Ap. J.* Submitted

Yokoyama J, Suto Y. 1991. *Ap. J.* 379:427

Zabotin NA, Nasel'skiĭ PD. 1982. *Sov. Astron.* 59:447

Zabotin NA, Nasel'skiĭ PD. 1985. *Sov. Astron.* 29:614

Zel'dovich Ya B. 1965. *Adv. Astron. Astrophys.* 3:241

Zel'dovich Ya B. 1970. *Astron. Astrophys.* 5:84

Zel'dovich Ya B. 1972. *MNRAS* 160:1 p

Zel'dovich Ya B. 1973. *Comments Astrophys. Space Sci.* 5:169

Zel'dovich Ya B, Illarionov AF, Sunyaev RA. 1972. *Sov. Phys.–JETP* 33:644

Zel'dovich Ya B, Kurt VG, Sunyaev RA. 1968. *Zh. Eksp. Teor. Fiz.* 55:278, Engl. transl.: *Sov. Phys.–JETP*, 28:146 (1969)

Zel'dovich, Ya B, Sunyaev RA. 1969. *Astrophys. Space Sci.* 4:301

NOTE ADDED IN PROOF

Since we submitted this article the field of CMB anistropies has refused to stay still, and there have been several significant developments. Firstly, the situation with regard to the *COBE* value of n has been clarified. The best fit to the two-year data is 1.10 ± 0.32 including the quadrupole, and 0.87 ± 0.36 excluding the quadrupole (Górski 1994). Similarly the value of $Q_{\mathrm{rms-PS}}$ is 20 μK, rather than the earlier $17\mu K$. Secondly, the new MAX data (Devlin et al 1994, Clapp et al 1994) show consistently high detections at the half-degree scale in all three regions scanned. This has added support to the reality of the Doppler peaks. Lastly, there is a further claimed high detection on degree scales from an experiment based at the Italian Antarctic Base of which we have recently become aware (Piccirillo & Calisse 1993). There are also a couple of places where assumptions we made about experiments have turned out to have been wrong: the Python point in Figures 3 and 4 is too low by a factor of approximately 3; and the window function for ARGO in Figure 5 is too high by a factor of 4 (P. de Bernardis, private communication).

Clapp AC, Devlin MJ, Gundersen JO, Hagmann CA, Hristov VV, et al. 1994. *Ap. J. Lett.* Submitted

Devlin MJ, Clapp AC, Gundersen JO, Hagmann CA, Hristov VV, et al. 1994. *Ap. J. Lett.* Submitted

Gorski K, Hinshaw G, Banday AJ, Bennett, CL, Wright EL, et al. 1994. *Ap. J. Lett.* Submitted

Piccirillo L, Calisse P, 1993. *Ap. J.* 411:529

Annu. Rev. Astron. Astrophys. 1994. 32: 371–418

DYNAMICS OF COSMIC FLOWS

Avishai Dekel

Racah Institute of Physics, The Hebrew University of Jerusalem, Israel

KEY WORDS: cosmology, large-scale structure, peculiar velocities

1. INTRODUCTION

The editors suggested a review entitled "Are There Large-Scale Motions in the Universe?". The answer is "yes," in the sense that the interpretation of the data as motions is the simplest model, so far consistent with all other available data under the current "standard model" of physical cosmology. I review tests that could have ended up falsifying this model and failed, but the scope of this review is much extended as the field has developed far beyond the question of existence of motions. With the motions being accepted as a working hypothesis, the study of *large-scale dynamics* is becoming a mature scientific field where observation and theory are confronted in a quantitative way. It is this area of major activity in cosmology that is addressed here.

I make no attempt to provide a complete reference list, nor do I try to achieve a balanced discussion of all the issues of relevance and authors involved. My goal is to provide a critical account of some of the issues in this field that I find important, with emphasis on theoretical implications. In many cases I quote only a recent paper automatically implying "and references therein." The reader is referred to a comprehensive, observation-oriented review of large-scale motions in historical perspective by Burstein (1990b), a detailed review of distance indicators in a collection of essays by Jacoby et al (1992), and to *Principles of Physical Cosmology* by Peebles (1993).

The current phase of the field was seeded by two major developments. One was the confirmation of the dipole moment in the Cosmic Microwave Background (CMB) (Corey & Wilkinson 1976, Smoot et al 1977), indicating via Doppler shift that the Local Group of galaxies (LG) is *moving* at $\sim$600 km s^{-1} relative to the cosmological frame defined by the CMB. The other was the invention of methods for inferring distances *independent of redshifts* based on intrinsic relations between galaxy quantities (Section 3; Tully & Fisher 1977,

371

0066–4146/94/0915–0371$05.00

TF; Faber & Jackson 1976, FJ). The radial peculiar velocity of a galaxy (the "velocity" u) is the difference between its total radial velocity as read from the redshift (the "redshift" z) and the Hubble velocity at its true distance (the "distance" r). Improved versions of these methods reduced the distance errors to the level of 15–21% which, with several hundred measured galaxies across the sky, enabled modeling the large-scale velocity field in terms of few-parameter "toy" models (Section 4.1), starting with a Virgo-centric infall (Aaronson et al 1982b) and ending with spherical infall into a "Great Attractor" (GA) (Lynden-Bell et al 1988). The finding by the "seven samurai" (7S, Burstein et al 1986) that the LG participates in a large streaming motion launched the present high-intensity activity in this field. The toy modeling is gradually being replaced by nonparametric methods, where the full velocity *field* is reconstructed based on properties of gravitational flows (Section 4) and the associated mass-density fluctuation field is recovered from the spatial velocity derivatives (Section 2). With no simplified geometry imposed, the motions are not associated with single specific "sources"; the gravitational acceleration is an integral of a continuous density field consisting of swells and troughs simultaneously pulling and pushing.

A parallel major development has been of all-sky magnitude-limited redshift surveys with many thousands of galaxies, starting with the CfA and SSRS optical surveys and continuing with the very useful recent surveys based on the *IRAS* satellite (Section 5). The large-scale inhomogeneity in the galaxy distribution (e.g. de Lapparent et al 1986) provided a clear hint for associated motions. An all-sky redshift survey can be converted into a galaxy-density field and then integrated to derive a *predicted* velocity field under the assumption of gravity and a certain "biasing" relation between galaxies and mass. The comparison of the fields obtained from redshifts to those obtained from velocities is at the heart of the research of large-scale structure (LSS), and the results carry major implications (Sections 6.2 and 8.2).

Data of both types are rapidly accumulating and a major effort is directed at reducing the errors and carefully estimating those which remain, to enable quantitative testing of LSS formation theories. The standard theory consists of several *working hypotheses* which one tries to falsify by the observations or, if found consistent, uses to determine the characteristic model parameters. The hypotheses, which will be elaborated on later, can be listed as follows:

H1. The background *cosmology* is the standard homogeneous Friedman Robert-son Walker model, possibly with an Inflation phase, where the CMB defines a cosmological "rest frame." If so, then one wishes to determine the cosmological density parameter Ω (plus the cosmological constant Λ and the Hubble constant H).

H2. The structure originated from a random field of small-amplitude initial density *fluctuations*. If so, the goal is to find out whether they were Gaussian, whether the power spectrum (PS) was scale-invariant (power index $n = 1$), and whether the energy density was perturbed adiabatically or in an isocurvature manner.

H3. The spectrum of fluctuations was filtered during the radiation-plasma era in a way characteristic of the nature of the *dark matter* (DM) which dominates the mass density. The DM could be baryonic or non-baryonic. If non-baryonic it could be "hot" or "cold" depending on when it became nonrelativistic.

H4. The fluctuations grew by *gravitational instability* (GI) into the present LSS. This is a sufficient but not necessary condition for:

    (*a*) the quasilinear velocity field, smoothed over a sufficiently-large scale, is *irrotational*;

    (*b*) the galaxies trace a unique underlying velocity field, apart from possible "velocity bias" of $\sim 10\%$ on small scales.

H5. The density fluctuations of visible galaxies are correlated with the underlying mass fluctuations. If this relation is roughly linear, then the linearized continuity equation in GI implies a relation between velocity and galaxy density. If so, the characteristic parameter is the *density-biasing* factor b.

H6. The TF and $D_n - \sigma$ methods measure true *distances*, which allow the reconstruction of a large-scale velocity field with known and controlled systematic biases.

This review is geared toward the confrontation of observations with these hypotheses. The relevant observations may be classified into the following three major categories:

O1. Angular fluctuations in the CMB temperature at various angular scales.

O2. The distribution of luminous objects on the sky and in redshift space.

O3. Peculiar velocities of galaxies along the line of sight.

Note that O1 and O3 are related to the dynamical theoretical ingredients H1–H4 and H6, bypassing the uncertain nature of galaxy-density biasing H5. Also, O1 and O2 refer to the theory independently of H4a,b and H6, which address the velocities and their analysis.

2. GRAVITATIONAL INSTABILITY

This section provides a brief account of the standard theory of GI, and of the linear and quasi-linear approximations which serve the analysis of motions. Let $\mathbf{x}$, $\mathbf{v}$, and Φ_g be the position, peculiar velocity, and peculiar gravitational potential in comoving distance units, corresponding to $a\mathbf{x}$, $a\mathbf{v}$, and $a^2\Phi_g$ in physical units with $a(t)$ the universal expansion factor. Let the mass-density fluctuation be $\delta \equiv (\rho - \bar{\rho})/\rho$. The equations governing the evolution of fluctuations of a pressureless gravitating fluid in a standard cosmological background during the matter era are the *continuity* equation, the *Euler* equation of motion, and the *Poisson* field equation (e.g. Peebles 1980; 1993):

$$\dot{\delta} + \nabla \cdot \mathbf{v} + \nabla \cdot (\mathbf{v}\delta) = 0, \tag{1}$$

$$\dot{\mathbf{v}} + 2H\mathbf{v} + (\mathbf{v}\cdot\nabla)\mathbf{v} = -\nabla\Phi_g, \tag{2}$$

$$\nabla^2\Phi_g = (3/2)H^2\Omega\,\delta, \tag{3}$$

where H and Ω vary in time. The dynamics do not depend on the value of the Hubble constant H; it is set to unity by measuring distances in $\mathrm{km\,s^{-1}}$ ($1\,h^{-1}\mathrm{Mpc} = 100\,\mathrm{km\,s^{-1}}$).

In the *linear* approximation, the GI equations can be combined into a time evolution equation, $\ddot{\delta} + 2H\dot{\delta} = (3/2)H^2\Omega\delta$. The growing mode of the solution, $D(t)$, is irrotational and can be expressed in terms of $f(\Omega) \equiv H^{-1}\dot{D}/D \approx \Omega^{0.6}$ (see Peebles 1993, eq. 5.120). The linear relation between density and velocity is

$$\delta = \delta_0 \equiv -(Hf)^{-1}\nabla \cdot \mathbf{v}. \tag{4}$$

The use of δ_0 is limited to the small dynamical range between a *few* tens of megaparsecs and the $\sim 100\,h^{-1}\mathrm{Mpc}$ extent of the current samples. In contrast, the sampling of galaxies enables reliable dynamical analysis with smoothing scale as small as $\sim 10\,h^{-1}\mathrm{Mpc}$, where $|\nabla \cdot \mathbf{v}|$ takes on values ≥ 1 so that quasi-linear effects play a role. Unlike the strong nonlinear effects in virialized systems which erase any memory of the initial conditions, mild nonlinear effects carry crucial information about the formation of LSS, and should therefore be treated carefully. Figure 1 shows that δ_0 becomes a severe underestimate at large $|\delta|$. This explains why Equation (4) is invalid in the nonlinear epoch even where $\delta = 0$; the requirements that $\int \delta\, d^3x = 0$ by definition and $\int \nabla \cdot \mathbf{v}\, d^3x = 0$ by isotropy imply $-\nabla \cdot \mathbf{v} > \delta$ at $|\delta| \ll 1$. Fortunately, the small variance of $\nabla \cdot \mathbf{v}$ given δ promises that some function of the velocity derivatives may be a good local approximation to δ.

A basis for useful *quasi-linear* relations is provided by the *Zel'dovich* (1970) approximation. The displacements of particles from their initial, Lagrangian positions $\mathbf{q}$ to their Eulerian positions $\mathbf{x}$ at time t are assumed to have a universal time dependence,

$$\mathbf{x}(\mathbf{q}, t) - \mathbf{q} = D(t)\,\boldsymbol{\psi}(\mathbf{q}) = f^{-1}\mathbf{v}(\mathbf{q}, t). \tag{5}$$

For the purpose of approximating GI, the Lagrangian Zel'dovich approximation can be interpreted in Eulerian space, $\mathbf{q}(\mathbf{x}) = \mathbf{x} - f^{-1}\mathbf{v}(\mathbf{x})$, provided that the flow is laminar (i.e. that multi-streams are appropriately smoothed over). The solution of the continuity equation then yields (Nusser et al 1991)

$$\delta_c(\mathbf{x}) = \|I - f^{-1}\partial\mathbf{v}/\partial\mathbf{x}\| - 1, \tag{6}$$

where the bars denote the Jacobian determinant, and I is the unit matrix. The Zel'dovich displacement is first order in f^{-1} and $\mathbf{v}$, so δ_c involves second- and third-order terms (m_{v2}, m_{v3}) as well. The relation (6) is not easily invertible to provide $\nabla \cdot \mathbf{v}$ or $\mathbf{v}$ when δ is given, but a useful empirical approximation is $\nabla \cdot \mathbf{v} = -f\delta/(1 + 0.18\delta)$.

A modified approximation, which is derived by adding a second-order term to the Zel'dovich displacement (Moutarde et al 1991) and truncating all the

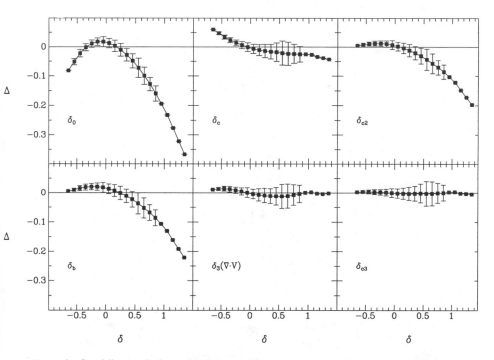

Figure 1 Quasi-linear velocity-to-density approximations. $\Delta \equiv \delta_{\text{approx}}(\mathbf{V}) - \delta_{\text{true}}$. The mean and standard deviation are from large standard-CDM N-body simulations normalized to $\sigma_8 = 1$, Gaussian-smoothed with radius $12\,h^{-1}\text{Mpc}$ (see Mancinelli et al 1994). Note the factor of 5 difference in scale between the axes.

expressions at second order while solving the continuity equation, is (Gramman 1993a)

$$\delta_{c2} = -f^{-1}\nabla \cdot \mathbf{v} + (4/7)f^{-2}m_{v2},$$
$$m_{v2} \equiv \sum_i \sum_{j>i}(\partial_i v_i \partial_j v_j - \partial_j v_i \partial_i v_j). \tag{7}$$

The factor $4/7$ replaces 1 in the second-order term of δ_c. Although terms are kept to second-order, it is still not an exact solution to the second-order equations of GI. This relation can be inverted in second-order to provide $\nabla \cdot \mathbf{v}$ given δ, with m_{v2} replaced by an analogous expression, m_{g2}, involving the gravitational acceleration $\mathbf{g}$.

Since the variance of δ given $\nabla \cdot \mathbf{v}$ is small, one expects that a nonlinear function of $\nabla \cdot \mathbf{v}$ that properly corrects for the systematic deviation can be a good quasi-linear approximation to δ (and vice versa). Assuming Gaussian initial fluctuations, Bernardeau (1993) found a solution in the limit of vanishing variance: $\delta_b = [1 - (2/3)f^{-1}\nabla \cdot \mathbf{v})]^{3/2} - 1$, which is easily invertible. A polynomial expansion with non-vanishing variance should have the form (Zehavi & Dekel, in preparation)

$$\delta_n(\nabla \cdot \mathbf{v}) = -f^{-1}\nabla \cdot \mathbf{v} + a_2 f^{-2}[(\nabla \cdot \mathbf{v})^2 - \mu_2]$$
$$+ a_3 f^{-3}[(\nabla \cdot \mathbf{v})^3 - \mu_3] + \cdots. \tag{8}$$

Because the first two terms vanish when integrated over a large volume, the moments $\mu_n \equiv \langle (\nabla \cdot \mathbf{v})^n \rangle$ must be subtracted off to make the nth-order term vanish as well. The coefficients can be crudely approximated analytically (e.g. Bernardeau 1993) or, using CDM simulations and Gaussian smoothing, the best coefficients are $a_2 \approx 0.3$ and $a_3 \approx -0.1$, tested for Ω values 0.1–1 and smoothing radii 5–12 h^{-1}Mpc at $\sigma_8 = 1$ (σ_8 is the rms of unsmoothed mass-δ in top-hat spheres of radius 8 h^{-1}Mpc). The structure of Eqation (8) makes it robust to uncertain features such as Ω, the shape of the fluctuation power spectrum, and the degree of nonlinearity as determined by the fluctuation amplitude and the smoothing. Such robustness is crucial when using a quasilinear approximation for determining Ω, for example (Section 8).

Figure 1 demonstrates the accuracy of the explicit quasi-linear approximations using CDM N-body simulations and 12 h^{-1}Mpc smoothing (Mancinelli et al 1994). The approximation δ_c, of scatter ~ 0.1, is an excellent approximation for $\delta \leq 1$ except that it is a slight overestimate at the negative tail. δ_{c2} and δ_b do better at the negative tail, but they are severe underestimates in the positive tail. $\delta_3(\nabla \cdot \mathbf{v})$ is an excellent robust fit over the whole quasi-linear regime. δ_{c3} is constructed from the three terms in the expansion of δ_c in powers of f^{-1} but with the numerical coefficients adjusted to achieve best fit in the simulation ($-1.05, 0.9, 1.5$ replacing unity, independent of Ω).

3.   MEASURING PECULIAR VELOCITIES

3.1   *Distance Indicators*

Measuring redshift-independent distances to many galaxies at large distances is the key to large-scale dynamics (for a review, see Jacoby et al 1992). The simplest method assumes that a certain class of objects is a "standard candle," in the sense that a distance-dependent observable is distributed intrinsically at random with small variance about a universal mean. The luminosity of an object ($\propto r^{-2}$) or its diameter ($\propto r^{-1}$), can serve as this quantity. In a pioneering study, Rubin et al (1976a,b) used the brightness of giant Sc spirals to discover a net motion for the shell at 35–60 h^{-1}Mpc that agrees within the errors with more modern results, but the large uncertainties in this simple distance indicator made this result controversial at the time.

So far, the most useful distance indicators for LSS have been of the TF-kind, based on intrinsic relations between two quantities: a distance-dependent quantity such as the flux $\propto L/r^2$, and a distance-independent quantity σ— the maximum rotation velocity of spirals (TF) or the velocity dispersion in ellipticals (FJ). The intrinsic relations are power laws, $L \propto \sigma^\beta$, i.e. $M(\eta) = a - b\eta$, where $M \equiv -2.5\log L + const$ is the absolute magnitude and $\eta \equiv \log\sigma$. The slope b can be determined empirically in clusters, where all the galaxies are assumed to be at the same distance, typically yielding $\beta \approx 3 - 4$, depending on the luminosity band (e.g. $\beta_I \approx 3$, $\beta_H \approx 4$). Then, for any other galaxy with observed η and apparent magnitude $m \equiv -2.5\log(L/r^2) + const$, one can determine a *relative* distance via $5\log r = m - M(\eta)$. There exists a fundamental freedom in determining the *zero point*, a, which fixes the distances at absolute values (in km s^{-1}, not to be confused with H which translates to Mpc). Changing a, i.e. multiplying the distances by a factor $(1 + \epsilon)$ while the redshifts are fixed, is equivalent to adding a monopole Hubble-like component $-\epsilon r$ to $\mathbf{v}$, and an offset 3ϵ to δ (Equation 4). It has been arbitrarily determined in several data sets, e.g. by assuming $u = 0$ for the Coma cluster, but a is better determined by minimizing the variance of the recovered peculiar velocity field in a large "fair" volume. The original TF technique has been improved by moving from blue to near-infrared photometry (H band, Aaronson et al 1979) and recently to CCD R and I bands, where spiral galaxies are more transparent and therefore the intrinsic scatter is reduced to $\sigma_m \sim 0.33$ mag, corresponding to a relative distance error of $\Delta = (\ln 10/5)\sigma_m \approx 0.15$.

A distance indicator of similar quality for ellipticals has proved harder to achieve. Minimum variance, corresponding to $\Delta = 0.21$, was found for a revised FJ relation involving three physical quantities: $DI^\alpha \propto \sigma^\beta$ with D the diameter and $I \propto L/D^2$ the surface brightness (Dressler et al 1987, Djorgovski & Davis 1987). The parameters were found to be $\alpha \approx 5/6$ and $\beta \approx 4/3$. By defining from the photometry a "diameter" at a fixed value of enclosed I, termed D_n, the relation returns to a simple form, $D_n \propto \sigma^\beta$, similar to FJ but with reduced variance.

The physical origin of the scaling relations is not fully understood, reflecting our limited understanding of galaxy formation. For the purpose of distance measurements, the mean empirical relation and its variance are what matters. However, one can point at an important physical difference between the two relations (Gunn 1989), which is relevant to the testing for environmental effects (Section 6.3). The $D_n - \sigma$ relation is naturally explained by virial equilibrium, $\sigma^2 \propto M/D$, and a smoothly varying $M/L \propto M^\gamma$, which together yield $DI^{1/(1+\gamma)} \propto \sigma^{2(1-\gamma)/(1+\gamma)}$, but the TF relation, involving only two of the three quantities entering the virial theorem, is more demanding—it requires an additional constraint which is probably imposed at galaxy formation.

There is some hope for reducing the error in the TF method to the $\sim 10\%$ range by certain modifications, e.g. by restricting attention to galaxies of normal morphology (Raychaudhury 1994). The most accurate technique to date uses the estimator based on surface-brightness fluctuations (SBF) in ellipticals (Tonry 1991), where the standard candle is the luminosity function of bright stars in the old population. These stars show up as distance-dependent fluctuations in sensitive surface-brightness measurements. The technique is being applied successfully out to $\sim 30\,h^{-1}$Mpc (e.g. Dressler 1994), with the improved accuracy of $\sim 8\%$ enabling high-resolution nonlinear analysis, and it can be of great value for LSS if applied at larger distances. The need to remove sources of unwanted fluctuations such as globular clusters requires high-resolution observations which could be achieved by *Hubble Space Telescope* (*HST*) or adaptive optics.

The prospects for the future can be evaluated by estimating the length scale over which LSS dynamics can be studied using a distance indicator of relative error Δ. The error in a velocity derived from N galaxies at a distance $\sim r$ is $\sigma_V \sim r\Delta/\sqrt{N}$. Let the mean sampling density be $\bar{n}$. Let the desired quantity be the mean velocity V in spheres of radius R, and assume that its true rms value is V_{20} at $R = 20\,h^{-1}$Mpc and $V_{20} \propto R^{-(n+1)}$ on larger scales, with n the effective power index of the fluctuation spectrum near R. Then the relative error in V is

$$\frac{\sigma_V}{V} \approx 0.033 \left(\frac{\bar{n}}{0.01}\right)^{-1/2} \left(\frac{\Delta}{0.15}\right) \left(\frac{V_{20}}{500}\right)^{-1} \left(\frac{R}{20}\right)^{n+1/2} \frac{r}{R}, \qquad (9)$$

where distances are measured in h^{-1}Mpc. The observations indicate that $V_{20} \sim 500\,\mathrm{km\,s^{-1}}$ and $n \sim -0.5$ for $R = 20\text{–}60\,h^{-1}$Mpc (Section 7.1). Thus, with ideal sampling of $\bar{n} \sim 0.01\,(h^{-1}\mathrm{Mpc})^{-3}$, the relative error is always only a few percent of r/R. This means that LSS motions can in principle be meaningfully studied at all distances r with smoothing $R \sim 0.1r$, as long as $n \sim -0.5$ at the desired R. Since n seems to be negative out to $\sim 100\,h^{-1}$Mpc (Section 7.1), dense deep TF samples promise to be useful out to several hundred megaparsecs. However, several technical difficulties pose a serious challenge at such distances. For example, the calibration requires faint cluster galaxies which are harder to identify, aperture effects become severe, and the spectroscopy capability is limited.

3.2 *Malmquist Biases*

The random scatter in the distance estimator is a source of severe systematic biases in the inferred distances and peculiar velocities, which are generally termed "Malmquist" biases but which should carefully be distinguished from each other (e.g. Lynden-Bell et al 1988; Willick 1994a,b).

The calibration of the TF relation is affected by the *selection bias* (or *calibration bias*). A magnitude limit in the selection of the sample used for calibration at a fixed *true* distance (e.g. in a cluster) tilts the "forward" TF regression line of M on η towards bright M at small η values. The bias extends to all values of η when objects at a large range of distances are used for the calibration. This bias is inevitable when the dependent quantity is explicitly involved in the selection process, and it occurs to a certain extent even in the "inverse" relation $\eta(M)$ due to existing dependences of the selection on η. Fortunately, the selection bias can be corrected once the selection function is known (e.g. Willick 1991, 1994a).

The TF inferred distance, d, and the mean peculiar velocity at a given d, suffer from an *inferred-distance bias*, which we term hereafter "M" bias. I comment later (Section 4.4) on a possible way to avoid the M bias by performing an inverse analysis in z-space, at the expense of a more complicated procedure and other biases. Here I focus on a statistical way for correcting the M bias within the simpler forward TF procedure in d-space. This bias can also be corrected in an inverse TF analysis in d-space, using the selection function $S(d)$ which is in principle derivable from the sample itself (Landy & Szalay 1992).

The current POTENT procedure uses the forward TF relation in d-space. If M is distributed normally for a given η, with standard deviation σ_m, then the TF-inferred distance d of a galaxy at a true distance r is distributed log-normally about r, with relative error $\Delta \approx 0.46\sigma_m$. Given d, the expectation value of r is (e.g. Willick 1991):

$$E(r|d) = \frac{\int_0^\infty r P(r|d)\, dr}{\int_0^\infty P(r|d)\, dr} = \frac{\int_0^\infty r^3 n(r) \exp\left\{-\frac{[\ln(r/d)]^2}{2\Delta^2}\right\} dr}{\int_0^\infty r^2 n(r) \exp\left\{-\frac{[\ln(r/d)]^2}{2\Delta^2}\right\} dr}, \tag{10}$$

where $n(r)$ is the number density in the underlying distribution from which galaxies were selected (by quantities that do not explicitly depend on r). The deviation of $E(r|d)$ from d reflects the bias. The homogeneous part (HM) arises from the geometry of space—the inferred distance d underestimates r

because it is more likely to have been scattered by errors from $r > d$ than from $r < d$, the volume being $\propto r^2$. If $n = const$, Equation (10) reduces to $E(r|d) = d \exp(3.5\Delta^2)$, in which the inferred distances are simply multiplied by a factor, 8% for $\Delta = 0.15$, equivalent to changing the zero-point of the TF relation. The HM bias has been regularly corrected this way since Burstein et al (1986).

Fluctuations in $n(r)$ are responsible for the inhomogeneous bias (IM), which is worse because it systematically enhances the inferred density perturbations and the value of Ω inferred from them. If $n(r)$ is varying slowly with r, and if $\Delta \ll 1$, then Equation (10) reduces to $E(r|d) = d[1 + 3.5\Delta^2 + \Delta^2 (d \ln n/d \ln r)_{r=d}]$, showing the dependence on Δ and the gradients of $n(r)$. To illustrate, consider a lump of galaxies at one point r with $u = 0$. Their inferred distances are randomly scattered to the foreground and background of r. For all galaxies with the same z, the inferred u on either side of r mimic a spurious infall towards r, which is interpreted dynamically as a spurious overdensity at r.

In the current data for POTENT analysis (Section 4.2) the IM bias is corrected in two steps. First, the galaxies are heavily grouped in z-space (Willick et al 1994), reducing the distance error of each group of N members to $\Delta/\sqrt{N}$ and thus significantly weakening the bias. Then, the noisy inferred distance of each object, d, is replaced by $E(r|d)$ (Equation 10), with an assumed $n(r)$ properly corrected for grouping. This procedure has been tested using realistic mock data from N-body simulations (Kolatt et al 1994), showing that IM bias can be reduced to a few percent. The practical uncertainty is in $n(r)$, which can be approximated by the high-resolution density field of *IRAS* or optical galaxies (Section 5), or by the recovered mass-density itself in an iterative procedure under some assumption about how galaxies trace mass. The second-step correction to δ recovered by POTENT is $< 20\%$ even at the highest peaks (Dekel et al 1994).

3.3 *Homogenized Catalogs*

Several samples of galaxies with TF or $D_n - \sigma$ measurements have accumulated in the past decade. Assuming that all galaxies trace the same underlying velocity field (Section 6.3), the analysis of large-scale motions greatly benefits from merging the different samples into one self-consistent catalog. The observers differ in their selection procedure, the quantities they measure, the method of measurement, and the TF calibration techniques, which cause systematic errors and make the merger nontrivial. The original merged set, compiled by D. Burstein (Mark II) and used in the first application of POTENT (Bertschinger et al 1990), consisted of 544 ellipticals and S0s (Lynden-Bell et al 1988, Faber et al 1989, Lucey & Carter 1988, Dressler & Faber 1991) and 429 spirals (Aaronson et al 1982a; Aaronson et al 1986, 1989; Bothun et al 1984). The

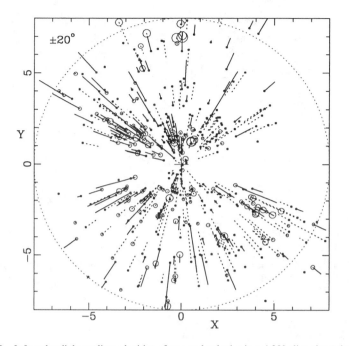

Figure 2   Inferred radial peculiar velocities of grouped galaxies in a ±20° slice about the Super-galactic plane from the homogonized Mark III catalog (Willick et al 1994). Distances and velocities are in 1000 km s^{-1}. The area of each circle marking the object position is proportional to the object richness. This slice contains 453 objects made of 1124 galaxies out of 1214 objects in the whole volume. Solid and dashed lines distinguish between outgoing and incoming objects. The positions and velocities are corrected for IM bias. Note the GA convergence (*left*) and the PP convergence (*right-bottom*).

current merged set (Willick et al 1994, Mark III of the Burstein series; Faber et al 1994) consists of ~2850 spirals (Mark II plus Han & Mould 1990, 1992; Mould et al 1991; Willick 1991; Courteau 1992; Mathewson et al 1992) and the ellipticals of Mark II. This sample enables a reasonable recovery of the dynamical fields with ~12 h^{-1}Mpc smoothing in a sphere of radius ~60 h^{-1}Mpc about the LG, extending to ~80 h^{-1}Mpc in certain regions (Section 4). Part of the data are shown in Figure 2.

As carried out by Willick et al (1994), merger of catalogs involves the following major steps: (*a*) Standardizing the selection criteria, e.g. rejecting galaxies of high inclination or low η which are suspected of large errors and sharpening any z cutoff. (*b*) Rederiving a provisional TF calibration for each data set using Willick's algorithm (1994a) which simultaneously groups, fits, and corrects for selection bias, and then verifying that inverse-TF distances to clusters are similar to the forward-TF distances. (*c*) Starting with one data set, adding each new set in succession using the galaxies in common to adjust the TF parameters of the new set if necessary. (*d*) Using only one measurement per galaxy even if it was observed by more than one observer to ensure well-defined errors, and using multiple observations for a "cluster" only if the overlap is small (e.g. < 50%). (*e*) Adding the ellipticals from Mark II, allowing for a slight zero-point shift (Section 6.3). Such a careful calibration and merger procedure is crucial for reliable results—in several cases it produced TF distances substantially different from those quoted by the original authors.

4.   ANALYSIS OF OBSERVED PECULIAR VELOCITIES

Given radial peculiar velocities u_i sparsely sampled at positions x_i over a large volume, with random errors σ_i, the first goal is to extract the underlying three-dimensional velocity field, $v(x)$. Under GI, this velocity field is subject to certain constraints, e.g. it is associated with a mass-density fluctuation field, $\delta(x)$—the other target for recovery. A field can be defined by a parametric model or by the field values at grid points. The number of independent parameters or grid points should be much smaller than the number of data points in view of the noisy data.

4.1   *Toy Models*

Given a model velocity field $v(\alpha_k, x)$, the free parameters α_k can be determined globally by minimizing a weighted sum of residuals, e.g.

$$-\log L \propto \sum_i W_i \left[u_i - \hat{x}_i \cdot v(\alpha_k, x_i) \right]^2. \tag{11}$$

If the errors are Gaussian and $W_i \propto \sigma_i^{-2}$, then L is the likelihood, and it is a useful approximation for log-normal errors as well. The model could first be a few-parameter "toy" model with simple geometry. Already the simplest bulk-flow model, $v(x) = B$ corresponding to $\delta = 0$, is of interest because the data clearly show a bulk flow component of several hundred $km\,s^{-1}$ in our neighborhood (Section 7.1). Another simple model is of spherical symmetry, expected to be a reasonable fit in voids and in regions dominated by one high density peak (Bardeen et al 1986). The velocity profile as a function of distance r from the infall center is not particularly constrained by GI, and a specific profile which proved successful is

$$v(r) = -v_{\lg} \left(\frac{r}{r_{\lg}} \right) \left[\frac{(r_{\lg}^2 + r_c^2)}{(r^2 + r_c^2)} \right]^{(n+1)/2} \tag{12}$$

The center is specified by its angular position and its distance $r_{\lg}$ from the Local Group, and the profile is characterized by its value $v_{\lg}$ at the LG, a core radius r_c, and a power index n. For $r \ll r_c$ the velocity rises $\propto r$ and for

$r \gg r_c$ it falls off $\propto r^{-n}$. The associated density profile is given in the linear approximation by the divergence $\delta = -f^{-1}r^{-1}\partial_r[r\,v(r)]$. The 7S ellipticals were modeled by a spherical infall model termed "The Great Attractor" (GA), with $r_{lg} = 42\,h^{-1}\text{Mpc}$ toward $(l, b) = (309°, +18°)$, $v_{lg} = 535\,\text{km s}^{-1}$, $r_c = 14.3\,h^{-1}\text{Mpc}$, and $n = 1.7$ (Faber & Burstein 1988). A similar model vaguely fits the local infall of spirals into Virgo (Aaronson et al 1982b). The merged Mark II data is well fitted by a multi-parameter hybrid consisting of a GA infall, a Virgo-centric infall, and a "local anomaly"—a bulk flow shared by the $\sim 10\,h^{-1}\text{Mpc}$-local neighborhood of $360\,\text{km s}^{-1}$ perpendicular to the supergalactic plane.

The toy models provide an intuitive picturing of the large-scale motions with clues about the associated mass sources, and they can be used as simple statistics for the comparison with theory (e.g. the bulk flow, Section 7.1; and the GA model, Bertschinger & Juszkiewicz 1988). However, toy modeling imposes an oversimplified geometry associated with assumed "sources" on a complex velocity field which actually arises from a continuous field of asymmetric density fluctuations (Section 4.5, Section 5). Moreover, the bulk velocity statistic computed globally suffers from a bias due to the large-scale sampling gradients. The monopole, involving the radial decrease of sampling density and rise of errors, has the effect of reducing the effective volume and thus reducing the apparent conflict between the high bulk flow and the theoretical expectations (Kaiser 1988). The sampling dipole, arising for example from oversampling in the GA direction, tends to enhance the component of the bulk velocity in that direction because it is dominated by the velocity in a smaller effective volume (Regos & Szalay 1989, Szalay 1988). The sampling quadrupole arising from the Galactic zone of avoidance (ZOA) introduces larger shot-noise into the component parallel to the Galactic plane, which could also result in an artificially high bulk velocity. These biases can be partially cured by equal-volume weighting (Section 4.2).

4.2 *Potential Analysis*

If the LSS evolved according to GI, then the large-scale velocity field is expected to be irrotational, i.e $\nabla \times \mathbf{v} = 0$. Any vorticity mode would have decayed during the linear regime as the universe expanded, and based on Kelvin's circulation theorem the flow remains vorticity-free in the quasi-linear regime as long as it is laminar. Irrotationality implies that the velocity field can be derived from a scalar potential, $\mathbf{v}(\mathbf{x}) = -\nabla\Phi(\mathbf{x})$, so the radial velocity field $u(\mathbf{x})$ should contain enough information for a full reconstruction. In the POTENT procedure (Bertschinger & Dekel 1989) the potential is computed by integration along radial rays from the observer, $\Phi(\mathbf{x}) = -\int_0^r u(r', \theta, \phi)\,dr'$, and the two missing transverse velocity components are then recovered by differentiation. Then $\delta(\mathbf{x})$ is approximated by δ_c (Equation 6). The nontrivial step is the *smoothing* of the data into $u(\mathbf{x})$. The aim in POTENT (Dekel et al 1990) is to reproduce the $u(\mathbf{x})$ that would have been obtained had the true $\mathbf{v}(\mathbf{x})$ been sampled densely and uniformly and smoothed with a spherical Gaussian window of radius R_s. With the data as available $u(\mathbf{x}_c)$ is taken to be the value at $\mathbf{x} = \mathbf{x}_c$ of an appropriate *local* velocity model $\mathbf{v}(\alpha_k, \mathbf{x} - \mathbf{x}_c)$ obtained by minimizing the sum (11) in terms of the parameters α_k within an appropriate local window $W_i = W(\mathbf{x}_i, \mathbf{x}_c)$ chosen as follows.

TENSOR WINDOW   Unless $R_s \ll r$, the u_is cannot be averaged as scalars because the directions $\hat{\mathbf{x}}_i$ differ from $\hat{\mathbf{x}}_c$, so $u(\mathbf{x}_c)$ requires a fit of a local 3-D model. The original POTENT used the simplest local model, $\mathbf{v}(\mathbf{x}) = \mathbf{B}$, for which the solution can be expressed explicitly in terms of a tensor window function.

WINDOW BIAS   The tensorial correction to the spherical window has conical symmetry, weighting more heavily objects of large $\hat{\mathbf{x}}_i \cdot \hat{\mathbf{x}}_c$. The resultant bias of a true infall transverse to the line of sight (LOS) is a flow towards the LG, e.g. ~ 300 km s^{-1} at the GA in the current reconstruction. A way to reduce this bias is by generalizing $\mathbf{B}$ into a *linear* velocity model, $\mathbf{v}(\mathbf{x}) = \mathbf{B} + \mathcal{L} \cdot (\mathbf{x} - \mathbf{x}_c)$, with $\mathcal{L}$ a symmetric tensor, which ensures local irrotationality. The linear terms tend to "absorb" most of the bias, leaving $\mathbf{v}(\mathbf{x}_c) = \mathbf{B}$ less biased. Unfortunately, a high-order model tends to peak undesired small-scale noise. The optimal procedure was found to be a first-order model fit out to $r = 40\,h^{-1}$Mpc, smoothly changing to a zeroth-order fit beyond $60\,h^{-1}$Mpc (Dekel et al 1994).

SAMPLING-GRADIENT BIAS (SG)   If the true velocity field is varying within the effective window, the nonuniform sampling introduces a bias because the smoothing is galaxy-weighted whereas the aim is equal-volume weighting. One should weight each object by the local volume it "occupies," e.g. $V_i \propto R_n^3$ where R_n is the distance to the nth neighboring object (e.g. $n = 4$). This procedure is found via simulations to reduce the SG bias in Mark III to negligible levels typically out to $60\,h^{-1}$Mpc but away from the ZOA. The $R_n(\mathbf{x})$ field can serve later as a flag for poorly-sampled regions, to be excluded from any quantitative analysis.

REDUCING RANDOM ERRORS   The ideal weighting for reducing the effect of Gaussian noise has weights $W_i \propto \sigma_i^{-2}$ but this spoils the carefully designed volume weighting, biasing u towards its values at smaller r_i and at nearby clusters where the errors are small. A successful compromise is to weight by both, i.e.

$$W(\mathbf{x}_i, \mathbf{x}_c) \propto V_i \, \sigma_i^{-2} \, \exp[-(\mathbf{x}_i - \mathbf{x}_c)^2/2R_s^2]. \tag{13}$$

Note that POTENT could alternatively vary R_s to keep the random errors at a constant level, but at the expense of producing fields not directly comparable to theoretical models of uniform smoothing. Another way to reduce noise is to eliminate badly-observed galaxies with large residuals $|u_i - u(\mathbf{x}_i)|$, where $u(\mathbf{x}_i)$ is obtained in a pre-POTENT smoothing stage. The whole smoothing procedure has been developed with the help of carefully designed mock catalogs "observed" from simulations.

ESTIMATING THE RANDOM ERRORS   The errors in the recovered fields are assessed by Monte-Carlo simulations, where the input distances are perturbed at random using a Gaussian distribution of standard deviation σ_i before being fed into POTENT. The error in δ at a grid point σ_δ (and similarly σ_v) is estimated by the standard deviation of the recovered δ (or v) over the Monte-Carlo simulations. In the well-sampled regions, which extend in Mark III out to 40–$60\,\mathrm{h}^{-1}\mathrm{Mpc}$, the errors are $\sigma_\delta \approx 0.1$–$0.3$, but they may blow up in certain regions at large distances. To exclude noisy regions, any quantitative analysis should be limited to points where σ_v and σ_δ are within certain limits.

Several variants of POTENT are worth mentioning. The potential integration is naturally done along radial paths using only $u(\mathbf{x})$ because the data are radial velocities, but recall that the smoothing procedure determines a 3-D velocity using the finite effective opening angle of the window $\sim R_s/r$. The transverse components are normally determined with larger uncertainty but the nonuniform sampling may cause the minimum-error path to be nonradial, especially in regions where $R_s/r \sim 1$. For example, it might be better to reach the far side of a void along its populated periphery rather than through its empty center. The optimal path can be determined by a *max-flow* algorithm (Simmons et al 1994). In practice, little can be gained by allowing nonradial paths because large empty regions usually occur at large distances where the transverse components are very noisy. Still, it is possible that the derived potential can be somewhat improved by averaging over many paths.

The use of the opening angle to determine the transverse velocities can be carried one step further by fitting the data to a power-series generalizing the linear model,

$$v_i(\mathbf{x}) = B_i + L_{ij}\tilde{x}_j + Q_{ijk}\tilde{x}_j\tilde{x}_k + C_{ijkl}\tilde{x}_j\tilde{x}_k\tilde{x}_l + \dots, \tag{14}$$

where $\tilde{\mathbf{x}} = \mathbf{x} - \mathbf{x}_c$. If the matrices are all symmetric, then the velocity model is automatically irrotational. It can therefore be used as the final result without appealing to the potential, with the density being automatically approximated by $\delta_c(\mathbf{x}) = \|I - L_{ij}\| - 1$ (Equation 6). The fit must be local because the model tends to blow up at large distances. The expansion can be truncated at any order, limited by the tendency of the high-order terms to pick up small-scale noise. The smoothing here is not a separate preceding step, the SG bias is reduced, and there is no need for numerical integration or differentiation, but the effective smoothing is again not straightforwardly related to theoretical models of uniform smoothing (Blumenthal & Dekel, in preparation).

Another method of potential interest without a preliminary smoothing step is based on *wavelet analysis*, which enables a natural isolation of the structure on different scales (Rauzy et al 1993). The effective smoothing involves no loss of information and no specific scale or shape for the wavelet, and the analysis is global and done in one step. How successful this method will be in dealing with noisy data and its comparison with theory and other data still remains to be seen.

4.3   *Regularized Multi-Parameter Models—Wiener Filter*

A natural generalization of the toy models of Section 4.1 is a global expansion of the fields in a discrete set of basis functions, such as a Fourier series of the sort

$$\delta = \sum_{\mathbf{k}} [a_{\mathbf{k}}\sin(\mathbf{k} \cdot \mathbf{x}) + b_{\mathbf{k}}\cos(\mathbf{k} \cdot \mathbf{x})],$$

$$\mathbf{v} = \sum_{\mathbf{k}} \frac{\mathbf{k}}{k^2} [a_{\mathbf{k}}\cos(\mathbf{k} \cdot \mathbf{x}) - b_{\mathbf{k}}\sin(\mathbf{k} \cdot \mathbf{x})], \tag{15}$$

with a very large number ($2m$) of free Fourier coefficients, and where δ and $\mathbf{v}$ are related via the linear approximation (4). The maximization of the likelihood involves a $2m \times 2m$ matrix inversion and, even if $2m$ is appropriately smaller than the number of data points, the solution will most likely follow Murphy's law and blow up at large distances, yielding large spurious fluctuations where the data are sparse, noisy, and weakly constraining. Thus the global fit requires some kind of regularization. Kaiser & Stebbins (1991) proposed to maximize the probability of the parameters subject to the data and an assumed *prior model* for the probability distribution of the Fourier coefficients; they assumed a Gaussian distribution with a power spectrum $\langle a_{\mathbf{k}}^2 \rangle = \langle b_{\mathbf{k}}^2 \rangle = P_0 k^n$.

This is in fact an application of the *Wiener filter* method (e.g. Rybicki & Press 1992) for finding the optimal estimator of a field $\delta(\mathbf{x})$ that is a linear functional of another field $u(\mathbf{x})$, given noisy data u_i of the latter and an assumed prior model:

$$\delta_{\mathrm{opt}}(\mathbf{x}) = \langle \delta(\mathbf{x})u_i \rangle \langle u_i u_j \rangle^{-1} u_j, \tag{16}$$

where the indices run over the data (Hoffman 1994, Stebbins 1994). If $u_i = u(\mathbf{x}_i) + \epsilon_i$ with ϵ_i independent random errors of zero mean, then the cross-correlation terms are $\langle \delta(\mathbf{x})u(\mathbf{x}_i) \rangle$, and the auto-correlation matrix is $\langle u_i u_j \rangle = \langle u(\mathbf{x}_i)u(\mathbf{x}_j) \rangle + \epsilon_i^2 \delta_{ij}$, both given by the prior model. δ_{opt} is thus determined by the model where the errors dominate, and by the data where the errors are small. If the assumed prior is Gaussian, the optimal estimator is also the most probable field given the data, which is a generalization of the conditional mean given one constraint (e.g. Dekel 1981). This conditional mean field can be the basis for a general algorithm to produce constrained random realizations (Hoffman & Ribak 1991). The same technique can also be applied to the inverse problem of recovering the velocity from observed density. Note (Lahav et al 1994) that for Gaussian fields the above procedure is closely related to maximum-entropy reconstruction of noisy pictures (Gull & Daniell 1978).

The maximum-probability method has so far been applied in a preliminary way to heterogeneous data in a box of side $200\,h^{-1}$Mpc (to eliminate periodic boundary effects) with $18\,h^{-1}$Mpc resolution. No unique density field came out as different priors ($-3 \le n \le 1$) led to different fits with similar χ^2. This means that the data used were not of sufficient quality to determine the fields this way; it would be interesting to apply this method to better future data. The method still lacks a complete error analysis, it needs to somehow deal with nonlinear effects, and it needs to correct for IM bias (perhaps as in Section 3.2).

A general undesired feature of maximum probability solutions is that they tend to be oversmoothed in regions of poor data, relaxing to $\delta = 0$ in the extreme case of no data. This is unfortunate because the signal of true density is modulated by the density and quality of sampling—a sampling bias which replaces the SG bias. The effectively varying smoothing length can affect any dynamical use of the reconstructed field (e.g. deriving a gravitational acceleration from a density field), as well as prevent a straightforward comparison with other data or with uniformly-smoothed theoretical fields. Yahil (1994) has recently proposed a modified filtering method which partly cures this problem by forcing the recovered field to have a constant variance.

4.4   *Malmquist-Free Analysis*

The selection bias (Section 3.2) can be practically eliminated from the calibration of an inverse TF relation, $\eta(M)$, as long as the internal velocity parameter η does not explicitly enter the selection process. An inverse analysis requires assuming a parametric model for the velocity field, $\mathbf{v}(\alpha_k, \mathbf{x})$ (Schechter 1980). Instead of (11), the sum minimized is then

$$\sum_i W_i \left[\eta_i(observed) - \eta_i(model)\right]^2, \tag{17}$$

with the model η given by the *inverse* TF relation,

$$\eta_i(model) = \tilde{a} + \tilde{b}\,M_i = \tilde{a} + \tilde{b}\,(m_i - 5\log r_i),$$
$$r_i = z_i - \hat{\mathbf{x}}_i \cdot \mathbf{v}(\alpha_k, \mathbf{x}_i). \tag{18}$$

The parameters are the inverse TF parameters $\tilde{a}$ and $\tilde{b}$ and the α_k characterizing the velocity model. This method is indeed free of inferred-distance M bias as long as r_i is uniquely derived from z_i, and not from the inferred distance.

An inverse method was first used by Aaronson et al (1982b) to fit a Virgocentric toy model to a local sample of spirals, and attempts to implement the inverse z-space method to an extended sample with a general velocity model are in progress (MFPOT by Yahil et al 1994). This is a nontrivial problem of nonlinear multi-parameter minimization with several possible routes. If the velocity model is expressed in z-space, then r_i is given explicitly by z_i but the results suffer from oversmoothing in collapsing regions. If in r-space, then r_i is implicit in the second equation of (18) requiring iterative minimization, e.g. carrying r_i from one iteration to the next one. The velocity model could be either global or local, with the former enabling a simultaneous minimization of the TF and velocity parameters and the latter requiring a sequential minimization of global TF parameters and local velocity parameters. While the results are supposed to be free of Malmquist bias, they suffer from other biases which have to be carefully diagnosed and corrected for. The inverse method is also being generalized to account for small-scale velocity noise (Willick, Burstein & Tornen 1994). The Malmquist-free results will have to be consistent with the M-corrected forward results before one can put to rest the crucial issue of M bias and its effect on our LSS results.

4.5   Fields of Velocity and Mass Density

Figure 3 shows supergalactic-plane maps of the velocity field in the CMB frame and the associated δ_c field (for $\Omega = 1$) as recovered by POTENT from the preliminary Mark III data. The data are reliable out to $\sim 60\,h^{-1}$Mpc in most directions outside the Galactic plane ($Y = 0$), and out to $\sim 70\,h^{-1}$Mpc in the direction of the GA (left-top) and Perseus-Pisces (PP, right-bottom). Both large-scale ($\sim 100\,h^{-1}$Mpc) and small-scale ($\sim 10\,h^{-1}$Mpc) features are important; e.g. the bulk velocity reflects properties of the initial fluctuations and of the DM (Section 7.1), while the small-scale variations indicate the value of Ω (Section 8).

The velocity map shows a clear tendency for motion from right to left, in the general direction of the LG-CMB motion ($L, B = 139°, -31°$ in supergalactic coordinates). The bulk velocity within $60\,h^{-1}$Mpc is 300–350 km s^{-1} towards ($L, B \approx 166°, -20°$) (Section 7.1) but the flow is *not coherent* over the whole volume sampled, e.g. there are regions in front of PP and at the back of the GA where the XY velocity components vanish, i.e. the streaming relative to the LG is opposite to the bulk flow direction. The velocity field shows local convergences and divergences which indicate strong density variations on scales about twice as large as the smoothing scale. The term "bulk velocity" refers to a useful statistic which measures large-scale aspects of the velocity field (Section 7.1) and should not necessarily be interpreted as a strongly coherent flow.

The GA at $12\,h^{-1}$Mpc smoothing and $\Omega = 1$ is a broad density peak of maximum height $\delta = 1.4 \pm 0.3$ located near the Galactic plane $Y = 0$ at $X \approx -40\,h^{-1}$Mpc. The GA extends towards Virgo near $Y \approx 10$ (the "Lo-

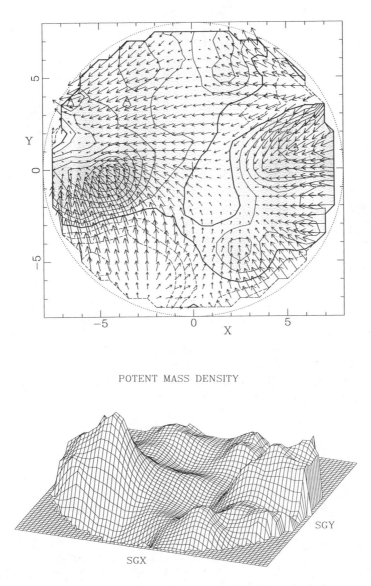

POTENT MASS DENSITY

Figure 3   The fluctuation fields of velocity and mass-density in the Supergalactic plane as recovered by POTENT from the Mark III velocities of ∼3000 galaxies with $12\,h^{-1}$Mpc smoothing. The vectors shown are projections of the 3-D velocity field in the CMB frame. Distances and velocities are in 1000 km s^{-1}. Contour spacing is 0.2 in δ, with the heavy contour marking $\delta = 0$ and dashed contours $\delta < 0$. The LG is at the center; the GA is on the left, PP on the right, and Coma is at the top. The gray-scale in the contour map and the height of the surface in the landscape map are proportional to δ (Dekel et al 1994).

cal Supercluster"), towards Pavo-Indus-Telescopium (PIT) across the Galactic plane to the south ($Y < 0$), and towards the Shapley concentration behind the GA. The structure at the top roughly coincides with the "Great Wall" of Coma, with $\delta \approx 0.5$. The PP peak which dominates the right-bottom is centered near Perseus with $\delta = 1.0 \pm 0.4$. PP extends towards Aquarius in the southern hemisphere, and connects to Cetus near the south Galactic pole, where the "Southern Wall" is seen in redshift surveys. Underdense regions separate the GA and PP, extending from bottom-left to top-right. The deepest region in the Supergalactic plane, with $\delta = -0.7 \pm 0.2$, roughly coincides with the galaxy-void of Sculptor (Kauffman & Fairall 1991).

One can still find in the literature statements questioning the very existence of the GA (e.g. Rowan-Robinson 1993), which simply reflect ambiguous definitions for this phenomenon. A GA clearly exists in the sense that the dominant feature in the local inferred velocity field is a coherent convergence, centered near $X \approx -40$. It is another question whether the associated density peak has a counterpart in the galaxy distribution or is a separate, unseen entity. The GA is ambiguous only in the sense that the good correlation observed between the mass density inferred from the velocities and the galaxy density in redshift surveys is perhaps not perfect (Section 6.2).

Other cosmographic issues of debate are whether there exists a back-flow behind the GA in the CMB frame, and whether PP and the LG are approaching each other. These effects are detected by the current POTENT analysis only at the 1.5σ level in terms of the random uncertainty. Furthermore, the freedom in the zero-point of the distance indicators permits adding a Hubble-like peculiar velocity which can balance the GA back-flow and make Perseus-Pisces move away from the Local Group. Thus, these issues remain debatable.

To what extent should one believe the recovery in the zone of avoidance which is empty of tracers? The velocities observed on the two sides of the ZOA are used as probes of the mass in the ZOA. The interpolation is based on the assumed irrotationality, where the recovered transverse components enable a reconstruction of the mass-density. However, while the SG bias can be corrected where the width of the ZOA is smaller than the smoothing length, the result could be severely biased where the unsampled region is larger. With $12\,h^{-1}$Mpc smoothing in Mark III the interpolation is suspected of being severely biased in $\sim$50% of the ZOA at $r = 40\,h^{-1}$Mpc (where $R_4 > R_s$), but the interpolation is pretty safe in the highly populated GA region, for example. Indeed, a deep survey of optical galaxies at small Galactic latitudes recently discovered that the ACO cluster A3627, centered at $(l, b, z) = (325°, -7°, 43\,h^{-1}Mpc)$, is an extremely rich cluster (Kraan-Korteweg & Woudt 1994), very near the central peak of the GA as predicted by POTENT, $\sim(320°, 0°, 40\,h^{-1}Mpc)$ (Kolatt et al 1994).

5. PREDICTED MOTIONS FROM THE GALAXY Z-DISTRIBUTION

All-sky complete redshift surveys provide extremely valuable data complementary to the peculiar velocity data, and the efficient techniques for measuring redshifts make them deeper, denser, and more uniform. Under the assumption of GI and an assumed biasing relation between galaxies and mass, a redshift survey enables an independent reconstruction of the density and velocity fields. The comparison of the smoothed fields recovered from redshifts and from velocities is a very important tool for testing the basic hypotheses and for determining the cosmological parameters.

The solution to the linearized GI equation $\nabla \cdot \mathbf{v} = -f\delta$ for an irrotational field is

$$\mathbf{v}(\mathbf{x}) = \frac{f}{4\pi} \int_{\text{all space}} d^3 x' \, \delta(\mathbf{x}') \frac{\mathbf{x}' - \mathbf{x}}{|\mathbf{x}' - \mathbf{x}|^3}. \tag{19}$$

The velocity is proportional to the gravitational acceleration, which ideally requires full knowledge of the distribution of mass in space. In practice (Yahil et al 1991) one is provided with a flux-limited, discrete redshift survey, obeying some radial selection function $\phi(r)$. The galaxy density is estimated by $1 + \delta_g(\mathbf{x}) = \sum n^{-1} \phi(r_i)^{-1} \delta^3_{\text{dirac}}(\mathbf{x} - \mathbf{x}_i)$, where $n \equiv V^{-1} \sum \phi(r_i)^{-1}$ is the mean galaxy density, and the inverse weighting by ϕ restores the equal-volume weighting. Equation (19) is then replaced by

$$\mathbf{v}(\mathbf{x}) = \frac{\beta}{4\pi} \int_{r < R_{\max}} d^3 x' \, \delta_g(\mathbf{x}') \, S(|\mathbf{x}' - \mathbf{x}|) \frac{\mathbf{x}' - \mathbf{x}}{|\mathbf{x}' - \mathbf{x}|^3}. \tag{20}$$

Under the assumption of linear biasing, $\delta_g = b\,\delta$, the cosmological dependence enters through $\beta \equiv f(\Omega)/b$. The integration is limited to $r < R_{\max}$ where the signal dominates over shot-noise. $S(\mathbf{x})$ is a small-scale smoothing window ($\geq 500 \text{ km s}^{-1}$) essential for reducing the effects of nonlinear gravity, shot-noise, distance uncertainty, and triple-value zones. The distances are estimated from the redshifts in the LG frame by

$$r_i = z_i - \hat{\mathbf{x}}_i \cdot [\mathbf{v}(\mathbf{x}_i) - \mathbf{v}(0)]. \tag{21}$$

Equations (20–21) can be solved iteratively: Make a first guess for the $\mathbf{x}_i$, compute the $\mathbf{v}_i$ by Equation (20), correct the $\mathbf{x}_i$ by Equation (21), and so on until convergence. The convergence can be improved by increasing β gradually during the iterations. Relevant issues follow.

SELECTION FUNCTION   An accurate knowledge of the probability that a galaxy at a given distance be included in the sample is essential, especially at large distances where ϕ^{-1} can introduce large errors. For a given flux limit, ϕ can be evaluated together with the luminosity function using a maximum-likelihood technique independent of density inhomogeneities.

ZONE OF AVOIDANCE   Regions in the sky not covered by the survey have to be filled with mock galaxies by some method of extrapolation from nearby regions. One way is to distribute these galaxies Poissonianly with the mean density of an adjacent volume, or to actually clone the adjacent region (Hudson 1993a). A more sophisticated extrapolation uses spherical harmonics and the Wiener filter method (e.g. Lahav et al 1994).

TRIPLE-VALUED ZONES   Galaxies in three different positions along a line of sight through a contracting region may have the same redshift. Given a redshift in a collapsing region where the problem is not resolved by the smoothing used, one can either take some average of the three solutions, or make an intelligent choice between them, e.g. by using the velocity field derived from observed velocities.

ESTIMATING SHOT-NOISE   This major source of error due to the finite number of galaxies can be crudely estimated using bootstrap simulations, where each galaxy is replaced with k galaxies, k being a Poisson deviate of $\langle k \rangle = 1$. For each realization one calculates ϕ and n, corrects for the ZOA, and solves for the linear velocity field. The mean and variance of the resulting density field are measures of the systematic and random errors. The bootstrap simulations demonstrate that the uncertainty in δ_g from the 1.2 Jy *IRAS* sample is typically less than 50% of the uncertainty in the density derived by POTENT from observed Mark III velocities.

NONLINEAR BIASING   Galaxies need not be faithful tracers of the mass (e.g. Dekel & Rees 1987), but there is growing evidence that they are strongly correlated (Section 6.2). This correlation can be crudely assumed to be a deterministic relation between the local smoothed density fields, e.g. linear biasing $\delta_g = b\delta$, which is one realization of the linear statistical relation between the variances of the fields predicted for linear density peaks in a Gaussian field (Kaiser 1984, Bardeen et al 1986). However, a more sophisticated analysis may require a more realistic biasing relation, e.g. an deviation from linear biasing which must be made for negative δ_g and $b < 1$ to prevent δ from falling unphysically below -1. The nonlinear generalization $1 + \delta_g = (1 + \delta)^b$ is useful (e.g. Dekel et al 1993); it fits quite well the biasing seen in simulations of galaxy formation in a CDM scenario involving cooling and gas dynamics (Cen & Ostriker 1993), exept that a small correction is needed to force the means of δ and δ_g to vanish simultaneously as required by definition.

QUASILINEAR CORRECTION   Even after $12\,h^{-1}$Mpc smoothing, δ_g is of order unity in places, necessitating a quasi-linear treatment. Local approximations from **v** to δ were discussed in Section 2, but the nonlocal nature of the inverse problem makes it less straightforward. A possible solution is to find an inverse relation of the sort $\nabla \cdot \mathbf{v} = F(\Omega, \delta_g)$, including nonlinear biasing and nonlinear gravity. This is a Poisson-like equation in which $-\beta\delta_g(\mathbf{x})$ is replaced by $F(\mathbf{x})$, and since the smoothed velocity field is irrotational for quasi-linear perturbations as well, it can be integrated analogously to Equation (20). With smoothing of $10\,h^{-1}$Mpc and $\beta = 1$, the approximation based on δ_c has an rms error $< 50\,\text{km s}^{-1}$ (Mancinelli et al 1994). Note that for very small b the δ associated with the observed δ_g could be nonlinear to the extent that the quasi-linear approximations break down. When δ is given, the Poisson equation can be integrated more efficiently by using grid-based FFT techniques than by straightforward summation. The r-space $\delta(\mathbf{x})$ deduced from the z-space galaxy distribution is not too sensitive to nonlinear effects or the value of Ω, so it is a reasonable shortcut to correct for nonlinear effects only in the final transformation from δ to $\mathbf{v}$ using FFT.

The *IRAS* Point Source Catalog served as the source for two very valuable redshift surveys, which have been carried out and analyzed in parallel. One contains the 5313 galaxies brighter than 1.2 Jy at 60μm with sky coverage (almost) complete for $|b| > 5°$ covering 88% of the sky (Strauss et al 1990, 1992a, to 1.9 Jy; extended by Fisher 1992). The other is a 1-in-6 sparsely-sampled survey of ≈ 2300 galaxies down to the *IRAS* flux limit of 0.6 Jy (Rowan-Robinson et al 1990, QDOT), which is now being extended to a fully-sampled survey (Saunders et al, in preparation). As for optical galaxies, Hudson (1993a) developed a clever way to reconstruct a statistically uniform density field out to $\sim 80\,h^{-1}$Mpc by combining the UGC/ESO diameter-limited angular catalogs and the ZCAT incomplete redshift survey. Figure 4 shows maps of the galaxy density fields and the associated predicted velocity fields, with the main features corresponding to those recovered from observed velocities (Figure 3), e.g. the GA, PP, and Coma superclusters and the voids in between (Section 6.2).

6.   TESTING BASIC HYPOTHESES

The data and analyses described above, combined with other astrophysical data, can help us evaluate some of the basic hypotheses laid out in the Introduction.

6.1   *CMB Fluctuations vs Motions*

If the CMB defines a standard cosmological frame then the established Heliocentric dipole pattern of $\delta T/T = (1.23 \pm 0.01) \times 10^{-3}$ is a direct measurement of peculiar velocity of the LG: $V(0) = 627 \pm 22\,\text{km s}^{-1}$ towards $(l, b) = (276° \pm 3°, +30° \pm 3°)$ (Kogut et al 1993). The Copernican hypothesis then implies that large peculiar velocities exist in general, and the question is only how coherent are they. A more esoteric interpretation tried to explain the CMB dipole by a horizon-scale gradient in entropy, relic of bubbly inflation, which can be made consistent with the smallness of the quadrupole and the achromaticity of the dipole (e.g. Gunn 1989; Paczynski & Piran 1990). In

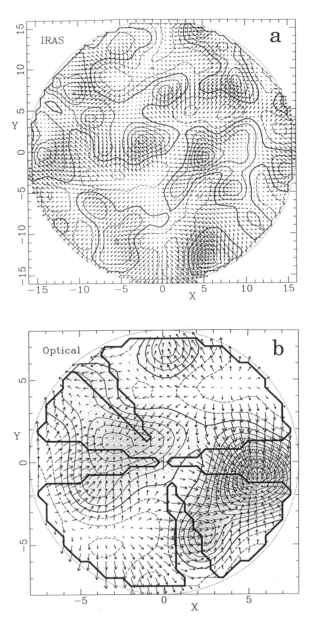

Figure 4 The fluctuation fields of galaxies in the Supergalactic plane as deduced from redshift surveys with $10\,h^{-1}$ Mpc smoothing. Distances and velocities are in $1000\,\mathrm{km\,s}^{-1}$. (*a*) Reconstructed by A. Yahil and M. Strauss using a power-preserving filter from the *IRAS* 1.2 Jy data in a sphere of radius $160\,h^{-1}$ Mpc. (*b*) Reconstructed by Hudson (1993a, 1994) from optical data within $80\,h^{-1}$ Mpc, extrapolated into the unsampled areas ouside the heavy contour.

addition to the general objection to global anisotropy based on the simplicity principle of Occam's razor, a non-velocity interpretation fails to explain the fact that the gravitational acceleration vector at the LG, $g(0)$, as inferred from the galaxy distribution in our cosmological neighborhood (Section 8.1), is within $20°$ of $V(0)$ and of a similar amplitude of several hundred km s^{-1} for any reasonable choice of Ω. This argument in favor of the reality of $V(0)$ and the standard GI picture is strengthened by the fact that a similar $g(0)$ is obtained from the POTENT mass field derived from velocity divergences in the neighborhood (Kolatt et al 1994).

The measurements of CMB fluctuations on scales $\leq 90°$ are independent of the local streaming motions, but GI predicts an intimate relation between them at the statistical level. The CMB fluctuations are associated with fluctuations in gravitational potential, velocity, and density in the surface of last scattering at $z \sim 10^3$, while similar fluctuations in our neighborhood have grown by gravity to produce the dynamical structure observed. The comparison between the two is therefore a crucial test for GI. Before *COBE,* the streaming velocities were used to predict the expected level of CMB fluctuations. The local surveyed region of $\sim 100\,h^{-1}$Mpc corresponds to a $\sim 1°$ patch on the last-scattering surface of an observer at a horizon distance. The major effect on scales $> 1°$ is the Sachs-Wolfe effect (1967), where potential fluctuations $\Delta\Phi_g$ induce temperature fluctuations via gravitational redshift, $\delta T/T = \Delta\Phi_g/(3c^2)$. Since the velocity potential is proportional to Φ_g in the linear and quasi-linear regimes, $\Delta\Phi_g$ is $\sim Vx$, where x is the scale over which the bulk velocity is V. Thus $\delta T/T \sim Vx/(3c^2)$. A typical bulk velocity of ~ 300 km s^{-1} across $\sim 100\,h^{-1}$Mpc corresponds to $\delta T/T \sim 10^{-5}$ at $\sim 1°$. If the fluctuations are scale-invariant ($n = 1$), then $\delta T/T \sim 10^{-5}$ is expected on all scales $> 1°$. Bertschinger et al (1990) produced a $\delta T/T$ map of the local region as seen by a distant observer, and predicted $\delta T/T \sim 10^{-5}$ from the local potential well associated with the GA. Now that CMB fluctuations of $\approx 10^{-5}$ have been detected in the range $1° - 90°$ (e.g. Smoot et al 1992, Schuster et al 1993, Cheng et al 1994), the argument can be reversed: If one assumes GI then the expected bulk velocity in the surveyed volume is ~ 300 km s^{-1}, i.e. the motions are likely to be real. If, alternatively, one accepts the velocities as real, then the CMB-POTENT agreement is a relatively sensitive test of gravitational instability (H4), truly addressing the specific time-evolution of structure predicted by GI.

6.2   Galaxies vs Dynamical Mass

The theory of gravitational instablilty plus the assumption of linear galaxy biasing predict a correlation between the dynamical density field and the galaxy density field, which can be addressed quantitatively based on the estimated errors in the two data sets. Figure 5 compares density maps in the Supergalactic plane for *IRAS* 1.2 Jy galaxies and POTENT Mark III mass, both Gaussian smoothed with radius $12\,h^{-1}$Mpc. The correlation is evident—the GA, PP, Coma and the voids all exist both as dynamical entities and as structures of galaxies. A quantitative comparison of these new data is in progress, but so far an elaborate statistical analysis (Dekel et al 1993) has been applied only to the earlier POTENT reconstruction based on Mark II data and *IRAS* 1.9 Jy survey. Noise considerations in POTENT limited that analysis to a volume $\sim(53\,h^{-1}$Mpc$)^3$ containing ~12 independent density samples. Monte-Carlo noise simulations showed that the POTENT density is consistent with being a noisy version of the *IRAS* density, i.e. the data are in agreement with the hypotheses of GI plus linear biasing (H4, H5).

What exactly can one learn from this observed correlation (Babul et al 1994)? First, it is hard to invoke any reasonable way to make the galaxy distribution and the TF measurements agree so well unless the velocities are real (H6), provided that IM bias has been properly corrected (Sections 3.2 and 4.4). Then, it is true that gravity is the only long-range force that could attract galaxies to stream toward density concentrations, but the fact that this correlation is predicted by GI plus linear biasing does not necessarily mean that it can serve as a sensitive test for either. Recall that converging (or diverging) flows tend to generate density hills (or valleys) simply as a result of mass conservation, independent of the source of the motions.

Let us assume for a moment that galaxies trace mass, i.e. the linearized continuity equation, $\dot{\delta} = -\nabla \cdot \mathbf{v}$, is valid for the galaxies as well. The observed correlation means $\delta \propto -\nabla \cdot \mathbf{v}$, and together they imply that $\dot{\delta} \propto \delta$, or equivalently

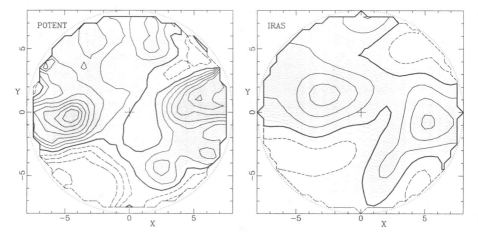

Figure 5 POTENT mass vs *IRAS* galaxy density fields in the Supergalactic plane, Gaussian smoothed with radius $12\,h^{-1}$Mpc. Contour spacing is 0.2 (a revised version from Dekel et al 1993).

that $\nabla \cdot \mathbf{v}$ is proportional to its time average. This property is not exclusive to GI; one can construct a counter example where the velocities are produced by a non-GI impulse. Even irrotationality does not follow from $\delta \propto -\nabla \cdot \mathbf{v}$; it has to be adopted based on theoretical arguments in order to enable reconstruction from radial velocities or from observed densities. Once continuity and irrotationality are assumed, the observed $\delta \propto -\nabla \cdot \mathbf{v}$ implies a system of equations that is identical in all its *spatial* properties to the equations of GI, but can differ in the constants of proportionality and their temporal behavior. It is therefore impossible to distinguish between GI and a non-GI model that obeys continuity plus irrotationality based only on snapshots of *present-day* linear fluctuation fields. This makes the relation between CMB fluctuations and velocities an especially important test for GI. On the other hand, the fact that the constant of proportionality in $\delta \propto -\nabla \cdot \mathbf{v}$ is indeed the same everywhere is a nontrivial requirement from a non-GI model. A version of the explosion scenario (Ostriker & Cowie 1981, Ikeuchi 1981), which tested successfully both for irrotationality and $\mathbf{v} - \delta$ correlation, requires certain synchronization among the explosions (Babul et al 1994).

While the sensitivity of the $\mathbf{v} - \delta$ relation to GI is only partial, this relation turns out to be quite sensitive to the validity of a continuity-like relation for the *galaxies*; when the latter is strongly violated all bets are off for the $\mathbf{v} - \delta_g$ relation. A nonlinear biasing scheme would make continuity invalid for the galaxies, which would ruin the $\mathbf{v} - \delta_g$ relation even if GI is valid. The observed correlation is thus a sensitive test for *density biasing*. It implies, subject to the errors, that the $\sim 10 \, h^{-1}$Mpc-smoothed densities of galaxies and mass are related via a simple, nearly linear biasing relation with b not far from unity.

6.3   *Environmental Effects: Ellipticals vs Spirals*

A priori, it is possible that the inferred motions are just a reflection of systematic variations in the distance indicators, e.g. due to environmental variations in intrinsic galaxy properties, or in the apparent quantities due to Galactic absorption (e.g. Silk 1989; Djorgovski et al 1989). Efforts to detect correlations between velocities and certain galaxy or environmental properties led so far to null or at most minor detections (Aaronson & Mould 1983; Lynden-Bell et al 1988; Burstein 1990a,b; Burstein et al 1990)—no correlation with absorption or with local galaxy density, and marginal correlations with absolute luminosity and with stellar population (Gregg 1994). Admittedly, these null results are only indicative as one cannot rule out a correlation with some other property not yet tested for.

Qualitative comparisons of the velocities of ellipticals (Es) and spirals (Ss) indicated general agreement (Burstein et al 1990), and now the Mark III data enable a quantitative comparison at the same positions (Kolatt & Dekel 1994a). Figure 6 compares the two fields as derived independently by POTENT, show-

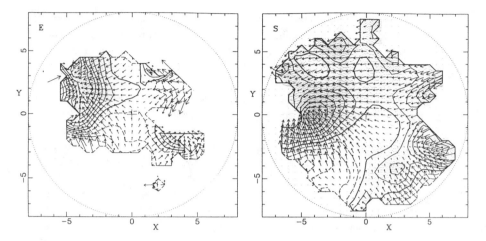

Figure 6 Ellipticals vs Spirals. Mass-density and velocity fields in the Supergalactic plane as recovered by POTENT independently for the two types. Smoothing is $12\,h^{-1}$Mpc. Contour spacing is 0.2. Distances and velocities are in $1000\,\mathrm{km\,s^{-1}}$ (Kolatt & Dekel 1994a).

ing a general resemblance. The radial velocities of each type were interpolated into a smoothed field $u(\mathbf{x})$ on a grid using $12\,h^{-1}$Mpc POTENT smoothing (Section 4.2), and then compared within a volume limited by the poor sampling of Es to $\simeq (50\,h^{-1}\mathrm{Mpc})^3$, containing ~ 10 independent sub-volumes. The two fields were found to be consistent with being noisy versions of the same underlying field, while the opposite hypothesis of complete independence is strongly ruled out, indicating that the consistency is not dominated by errors. With the improved data and bias-correction, a possible discrepancy indicated earlier by the Mark II data (Bertschinger 1991) is now gone. The strength of this test will improve further as the E sample grows in extent (e.g. project EFAR by Colless et al 1993) and in accuracy (e.g. the SBF method).

The E-S correlation is consistent with motions (H6), but it cannot rule out environmental effects beyond making this idea less plausible. If such effects were dominant, then the two distance indicators would have to vary coherently, which would require that the large-scale properties of the environment be different from the local properties determining the galaxy type (E in clusters, S in field). It would also require that the properties affecting the virial equilibrium which determines the $D_n - \sigma$ relation, and those affecting the additional constraint involved in the TF relation at galaxy formation (Section 2.1), vary together in space. The one quantity that could plausibly affect the two in a correlated way is M/L, but to test for this one will have to compare the TF/$D_n - \sigma$ velocities to those inferred from an independent indicator, e.g. SBF.

The E-S correlation is consistent with the hypothesis of GI (H4) because, following Galileo, the velocities of all test bodies in a given gravitational potential are predicted to be the same as long as they share the same initial conditions. However, the observed E-S correlation does not rule out any non-GI model where all objects obtain the same velocities independent of their type, e.g. cosmological explosions or radiation pressure instabilities (Hogan & White 1986).

A practical use of the independent derivation of $u(\mathbf{x})$ for Ss and Es is in matching the zero points of the distance indicators, otherwise determined in an arbitrary way (Section 2). A 5% Hubble-like outflow has to be added to the peculiar velocities of Es from Mark II to optimally match the S data of Mark III.

7. THE INITIAL FLUCTUATIONS

Having assumed GI, the large-scale structure can be traced backward in time to recover the initial fluctuations and constrain statistics which characterize them as a random field, e.g. the power spectrum (PS), and the probability distribution functions (PDFs). "Initial" here may refer either to the linear regime at $z \sim 10^3$ after the onset of the self-gravitating matter era, or to the origin of fluctuations in the early universe before being filtered on sub-horizon scales during the plasma-radiation era. The PS is filtered on scales $\leq 100 \, h^{-1} \mathrm{Mpc}$ by DM-dominated processes, but its shape on scales $\geq 10 \, h^{-1} \mathrm{Mpc}$ is little affected by recent nonlinear effects (because the rapid density evolution in superclusters roughly balances the slow evolution in voids at the same wavelength). The shape of the one-point PDF, on the other hand, is expected to survive the plasma era unchanged but it develops strong skewness even in the mildly-nonlinear regime. Thus, the present-day PS can be used as is to constrain the origin of fluctuations (large scale) and the nature of the DM (small scale), while the PDF needs to be traced back to the linear regime first. (For a review, see Efstathiou 1990.)

7.1 *Power Spectrum—Dark Matter*

The competing LSS formation scenarios are reviewed in Peebles (1993, Section 25). If the DM is all baryonic, then by nucleosynthesis constraints (see Kolb & Turner 1990, Section 4) the universe must be of low density, $\Omega < 0.2$, and a viable model for LSS is the Primordial Isocurvature Baryonic model (PIB) with several free parameters, typically of large relative power on large scales. With $\Omega \sim 1$ the non-baryonic DM constituents are either "hot" or "cold," and the main competing models are CDM, HDM, and MDM—a 7:3 mixture of the two (e.g. Blumenthal et al 1988, Davis et al 1992, Klypin et al 1993). The main difference in the DM effect on the PS arises from free-streaming damping of the "hot" component of fluctuations on galactic scales.

BULK VELOCITY   A simple and robust statistic related to the PS is the amplitude V of the vector average of the smoothed velocity field, $\mathbf{v}$, over a volume defined by a normalized window function $W_R(\mathbf{r})$ of a characteristic scale R (e.g. top-hat),

$$\mathbf{V} \equiv \int d^3x \, W_R(\mathbf{x}) \, \mathbf{v}(\mathbf{x}), \quad \langle V^2 \rangle = \frac{f^2}{2\pi^2} \int_0^\infty dk \, P(k) \, \tilde{W}_R^2(\mathbf{k}). \quad (22)$$

$\langle V^2 \rangle$ is predicted for a linear model with a density spectrum $P(k)$, where $\tilde{W}_R^2(\mathbf{k})$, the Fourier transform of $W_R(\mathbf{r})$, emphasizes waves $\geq R$. The bulk velocity is obtained from the observed radial velocities by minimizing Equation (11). The report by Dressler et al (1987) for the 7S Es sampled within $\sim 60\,h^{-1}$Mpc was $V = 599 \pm 104$ towards $(l, b) = (312°, +6°)$, which was interpreted prematurely as being in severe excess of the predictions of common theories. However, this measurement cannot be directly compared to the predictions for a top-hat sphere because the effective window is much smaller due to the nonuniform sampling (SG) and weighting (Kaiser 1988). The SG bias can be crudely corrected by volume weighting as in POTENT (Section 4.2), at the expense of large noise. Courteau et al (1993) find for the tentative Mark III data: $V_{40} = 335 \pm 38$ (295°, +35°) and $V_{60} = 360 \pm 40$ (279°, +11°), where V_R refers to a top-hat sphere of radius $R\,h^{-1}$Mpc. Alternatively, $\mathbf{V}$ can be computed from the POTENT $\mathbf{v}$ field by simple vector averaging from the grid. Figure 7 shows two results, one minimizing the SG bias by V_i weighting, and the other reducing the random errors by weighting $\propto \sigma_i^{-2}$ (Section 4.2). V_{60} is found to be in the range 270–360 km s^{-1} (296°, +11°)—smaller than previous estimates. The additional random error from Monte-Carlo noise simulations is typically 15%, not including cosmic scatter due to the fact that only one sphere has been sampled.

MACH NUMBER   The bulk velocity is robust but it relates to the normalization of the PS, not predicted from first principles by any of the competing theories but rather normalized by some other uncertain observation, e.g. the CMB fluctuations or the galaxy distribution with an unknown biasing factor. A statistic that measures the shape of the PS free of its normalization is the cosmic Mach number (Ostriker & Suto 1989), defined as $\mathcal{M} \equiv V/S$, where S is the rms deviation of the local velocity from the bulk velocity,

$$S^2 \equiv \int d^3x \, W_R(\mathbf{x}) \, [\mathbf{v}(\mathbf{x}) - \mathbf{V}]^2,$$

$$\langle S^2 \rangle = \frac{f^2}{2\pi^2} \int_0^\infty dk \, P(k) \, [1 - \tilde{W}_R^2(\mathbf{k})]. \quad (23)$$

$\mathcal{M}(R, R_s)$ measures the ratio of power on large scales $\gtrsim R$ to power on small scales $\gtrsim R_s$. Strauss et al (1993) derived $\mathcal{M} = 1.0$ for the local Ss (Aaronson et al 1982a) with $R \sim 20\,h^{-1}$Mpc and $R_s \to 0$, and found $\sim 5\%$ of their CDM simulations to have $\mathcal{M}$ as large—a marginal rejection or consistency depending on taste. On larger scales, using POTENT with $R = 60\,h^{-1}$Mpc (top-hat)

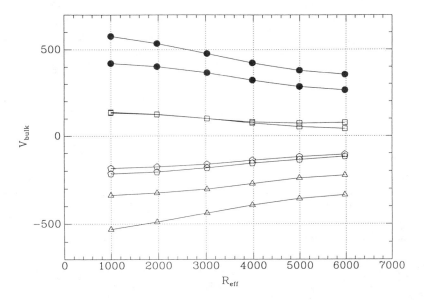

Figure 7   The bulk velocity in a top-hat sphere of radius R_{eff} about the LG as recovered by POTENT from the Mark III data. Shown are $|V|$ (●), V_x (△), V_y (*squares*), and V_z (*pentagons*). Distances and velocities are in km s^{-1}. The two results shown reflect the systematic uncertainty. The 1 σ uncertainty due to random distance errors is $\simeq 15\%$. (Dekel et al 1994)

and $R_s = 12\,h^{-1}\text{Mpc}$ (Gaussian), the tentative Mark III data yield $\mathcal{M} \sim 1$, which roughly coincides with the rms expected from CDM but has only ~5% probability to be that low for a PIB spectrum ($\Omega = 0.1$, $h = 1$, fully ionized, normalized to $\sigma_8 = 1$) (Kolatt et al, in preparation).

POWER SPECTRUM   The velocity field by POTENT enables a preliminary determination of the mass PS itself in the range $10–100\,h^{-1}\text{Mpc}$ (Kolatt & Dekel 1994b). It is best determined by the potential field, which is smoother and less sensitive to nonlinear effects than its derivatives **v** and δ. The PS was compared with the predictions of theoretical models which were N-body simulated, "observed," and fed into POTENT before the PS was computed using the same procedure applied to the observed data, thus eliminating the effect of systematic errors and estimating the random errors. The preliminary results indicate that the shape of the PS in the limited range sampled resembles a CDM-like PS with a shape parameter $\Gamma \sim 0.5$ (referring in CDM to Ωh), with the power index bending toward $n \simeq 0$ by $100\,h^{-1}\text{Mpc}$. The normalization for the mass PS, also obtained by Seljak & Bertschinger (1994), is $f\sigma_8 = 1.3 \pm 0.3$. This is not too sensitive to the PS shape, but still note that it is mostly determined by data at ~$30\,h^{-1}\text{Mpc}$. With $\sigma_{8\text{opt}} \approx 1$ for optical galaxies, this implies $\beta_{\text{opt}} \approx 1.3 \pm 0.3$

(compare Section 8). If $\Omega = 1$ the quadrupole in $\delta T/T$ by *COBE* corresponds to $\sigma_8 = 1.0, 0.6, 0.5$ with 45% error (including cosmic scatter) for CDM, MDM, and HDM spectra (yet to be calculated for an open universe), i.e. CDM is fine while HDM and MDM are $\sim 2\sigma$ low. In comparison, the PS of the different luminous objects are all well described by a CDM-like PS with $\Gamma \approx 0.25$, with the relative bias factors for Abell clusters, optical galaxies, and *IRAS* galaxies in the ratios 4.5:1.3:1 ($\pm 6\%$ rms, e.g. Peacock & Dodds 1994). If indeed $\sigma_{8opt} \sim 1$ for optical galaxies, then *IRAS* galaxies are slightly anti-biased and $\beta_i \approx 1.3\beta_{opt}$.

A DISCREPANCY ON A VERY LARGE SCALE?   The galaxy velocity field in the local $\sim 60\,h^{-1}$Mpc which has been studied quite accurately seems to be in general agreement with other observations and with the predictions of the common GI scenarios, but there is a disturbing hint for a possible discrepancy with our basic hypotheses on larger scales. Lauer & Postman (1993, LP) measured the bulk flow of the volume-limited system of 119 Abell/ACO clusters within $z < 150\,h^{-1}$Mpc, estimating distances from luminosity versus slope of the surface-brightness profile in brightest cluster galaxies (BCG) with claimed 16% error. The LG motion with respect to this system is found to be 561 ± 284 km s^{-1} towards $(l, b) = (220°, -28°)$, roughly 80° off the direction of the LG-CMB velocity $(276°, +30°)$. The inferred bulk velocity of the cluster system (of effective radius ~ 80–$110\,h^{-1}$Mpc) relative to the CMB is $V = 689 \pm 178$ km s^{-1} towards $(343°, +52°)$ ($\pm 23°$), in the general direction of the LG motion, the galaxy bulk flow, the GA, and the background Shapley concentration. Taken at face value this is a *high* velocity over a *large* scale; the rms bulk velocity within $R = 100\,h^{-1}$Mpc as predicted by the common theories normalized to *COBE* is below 200 km s^{-1}. However, the errors are large too.

The data have been subjected by LP to careful statistical tests. With a few tens of points contributing to the velocity in each direction, the shot-noise can clearly contribute a false signal of several hundred km s^{-1}. For assessing the theoretical implications, the actual observing scheme of LP was applied to N-body simulations of several competing scenarios (Strauss et al 1994). Clusters were placed at the 119 highest density peaks of appropriate mass, and Galactic extinction and observational errors were modeled. As noted by LP, the fact that the measured velocity vector lies away from the ZOA increases the statistical significance of this measurement. Taking the error ellipsoid into account, the probability by means of χ^2 of the LP result in the model universes simulated is 2.6–5.8%. Feldman & Watkins (1994), using a different technique, find the probabilities to be 6–10%. The LP bulk velocity is thus a $\sim 2\sigma$ deviation from several common theories. This is probably not enough for a serious falsification of these models, but it is certainly an intriguing result which motivates more accurate investigations. The BCG method is limited to one measurement per cluster, so an obvious strategy to reduce the errors would be to collect many TF and $D_n - \sigma$ distances per cluster.

A true $\sim$700 km s^{-1} velocity at $R \sim$ 100 h^{-1}Mpc would be in serious conflict with GI. First, if this velocity is typical then it predicts larger CMB fluctuations than observed on scales of $\sim$2°. Second, the gravitational acceleration on the LP sphere as estimated from the spatial distribution of clusters on even larger scales (Section 8.1, Scaramella et al 1991) predicts a flow of only $\sim$200 km s^{-1}. The way to interpret the LP result in the context of the conventional theories is either as a $\sim$2–2.5σ statistical fluke, or as a biased result due to a yet-unresolved systematic error in the BCG method or in the sample, e.g. a significant velocity biasing of the clusters (which is contrary to the naive expectations from GI, however).

7.2   Back in Time

The forward integration of the GI equations by analytic approximations or by N-body simulations cannot be simply reversed despite the time reversiblity of gravity. It is especially hopeless in collapsed systems where memory has been erased, but the case is problematic even for linear systems. When attempting backwards integration, the decaying modes (e.g. Peebles 1993, Section 5), having left no detectable trace at present, t_0, would amplify noise into dominant spurious fluctuations at early times. This procedure has a negligible probability of recovering the very special initial state of almost uniform density and tiny velocities which we assume for the real universe at t_i. This is a problem of mixed boundary conditions: Some of the six phase-space variables per particle are given at t_i and some at t_0. This problem can be solved either by eliminating the decaying modes or by applying the principle of least action.

ZEL'DOVICH TIME MACHINES   If the velocity field is irrotational, the Euler equation (2) can be replaced by the Bernoulli equation for the potentials, $\Phi_v - (\nabla\Phi_v)^2/2 = -2H\Phi_v + \Phi_g$. The Zel'dovich approximation, restricted to the growing mode, requires that each side vanish: One side relates Φ_v and Φ_g linearly and the other is the "Zel'dovich-Bernoulli" (ZB) equation (Nusser & Dekel 1992), $\dot\varphi_v - (\dot D/2)(\nabla\varphi_v)^2 = 0$, with the potentials in units of $a^2\dot D$. The ZB equation can be easily integrated backwards with a guaranteed uniform solution at t_i. φ_v at t_0 is extractable from observations of velocities (Section 4) or galaxy density (Section 5), and the initial $\mathbf{v}$ and δ can be derived from the initial φ_v using linear theory. While the ZB approximation conserves momentum (like δ_0) one can alternatively satisfy continuity under the Zel'dovich approximation (like δ_c), and obtain a second-order equation for φ_g which is somewhat more accurate (Gramman 1993b, Equations 2.24, 2.25). The recovered initial δ_i has deeper valleys and shallower hills compared to naive recovery using linear theory, e.g. the GA is less eccentric than assumed by Bertschinger & Juszkiewicz (1988).

RECOVERING THE IPDF   An important issue is whether or not the initial fluctuations were Gaussian. A Gaussian field is characterized by the joint PDFs of all order being generalized Gaussians (cf Bardeen et al 1986), and in particular the one-point probability of δ is $P(\delta) \propto \exp[-\delta^2/(2\sigma^2)]$. Common Inflation predicts Gaussian fluctuations but non-Gaussian fluctuations are allowed by certain versions of Inflation (cf Kofman et al 1990) and by models where the perturbations are seeded by cosmic strings, textures, or explosions (see Peebles 1993, Section 16). The present density PDF develops a log-normal shape due to nonlinear effects (Coles & Jones 1991, Kofman et al 1994): The tails become positively skewed because peaks collapse to large densities while the density in voids cannot become negative, and the middle develops negative skewness as density hills contract and valleys expand. On the other hand, the PDF of present-day velocity components is insensitive to quasi-linear effects (Kofman et al 1994).

The observed PDFs today agree with N-body simulations of Gaussian initial conditions (Bouchet et al 1993), but they have only limited discriminatory power against initial non-Gaussianities; the development of a density PDF with a general log-normal shape may occur even in certain cases of non-Gaussian initial fluctuations (e.g. Weinberg & Cole 1992), and the velocity PDF becomes Gaussian under general conditions due to the central limit theorem whenever the velocity is generated by several independent density structures. A more effective strategy seems to be to take advantage of the full dynamical fields at t_0, trace them back in time, and use the linear fields to discriminate between theories. The Eulerian Zel'dovich approximation can be used to directly recover the initial PDF (IPDF) as follows (Nusser & Dekel 1993). The tensor $\partial v_i/\partial x_j$ derived from $\mathbf{v}(\mathbf{x})$ is transformed to Lagrangian variables $\mathbf{q}(\mathbf{x})$. The corresponding eigenvalues $\mu_i \equiv \partial v_i/\partial x_i$ and $\lambda_i \equiv \partial v_i/\partial q_i$ are related via the key relation $\lambda_i = \mu_i/(1 - f^{-1}\mu_i)$. In the Zel'dovich approximation $\mathbf{v} \propto \dot{D}$ so the Lagrangian derivatives λ_i are traced back in time by simple scaling $\propto \dot{D}^{-1}$. The initial densities at $\mathbf{q}(\mathbf{x})$ can then be computed using linear theory, $\delta_{in} \propto -(\lambda_1 + \lambda_2 + \lambda_3)$, and the IPDF is computed by bin counting of δ_{in} values across the Eulerian grid, weighted by the present densities at the grid points.

A key feature of the recovered IPDF is that it is sensitive to the assumed value of Ω when the input data are velocities (Section 8.4), and is Ω-independent if the input is density. Thus it can be used to robustly recover the IPDF from the density field of the 1.2 Jy *IRAS* survey (Nusser et al 1994). The IPDF so determined is insensitive to galaxy biasing in the range $0.5 \leq b \leq 2$, at least for the power-law biasing relation assumed. Errors were evaluated using mock *IRAS*-like catalogs, and the IPDF was found to be consistent with Gaussian, e.g. the initial skewness S and kurtosis K are limited at the 3σ level to $-0.65 < S < 0.36$ and $-0.82 < K < 0.62$—useful for evaluating specific non-Gaussian models. The first-year *COBE* measurements are consistent with

Gaussian but noise limits their discriminatory power to strongly non-Gaussian models (Smoot et al 1994).

GAUSSIANIZATION   A simple method for recovering the initial fluctuations from the galaxy distribution under the assumption that they were Gaussian is based on the assertion that both gravitational evolution and biased galaxy formation tend to preserve the rank order of density in cells. The non-Gaussian distribution of galaxies in cells simply needs to be "Gaussianized" in a rank-preserving way (Weinberg 1991). The initial conditions can then be evolved forward using an N-body code and compared with the observed galaxy distribution in z-space until convergence. A self-consistent solution for a redshift survey in the PP region was found if $b_{\text{opt}} \sim 2$, while it was impossible to match the structure on small and large scales simultaneously with GI, Gaussian fluctuations, and no biasing. This result derived independently of Ω implies a high Ω once β_{opt} is determined by another method (Section 8), but note that the limited surveyed region may be an "unfair" sample.

LEAST ACTION   The general GI problem with mixed boundary conditions lends itself naturally to an application of Hamilton's action principle (Peebles 1989, 1990, 1993). The comoving orbit $x_i(t)$ of each mass point m_i is parametrized in a way that satisfies the boundary conditions $x_i(t_0) = x_{i0}$ and $\lim_{t \to 0} a^2 \dot{x}_i = 0$, and the action

$$S = \int_0^{t_0} L \, dt = \int_0^{t_0} dt \sum_i [\frac{1}{2} m_i a^2 \dot{x}_i^2 - m_i \Phi_g(x_i)] \qquad (24)$$

is minimized to determine the free parameters. The orbits can be $x_i(t) = x_i(t_0) + \sum_n f_n(t) C_{i,n}$, where $f_n(t)$ each satisfy the boundary conditions, and $C_{i,n}$ are the parameters. The problem can be solved to any desired accuracy by increasing n, and a proper choice of f_ns helps the series to converge rapidly to the desired solution. A generalization of the Zel'dovich approximation of the sort $f_n(t) = [D(t) - D(t_0)]^n$ is particularly efficient (Giavalisco et al 1993). In a preliminary application of this scheme to a redshift sample (Shaya et al 1994) the galaxies are assumed to trace mass and be self-gravitating. The complication caused by observing redshifts is solved by an iterative procedure: A tentative guess is made for the x_i at t_0 based on TF distances to a sub-sample of galaxies and a crude flow model, and the least-action solution provides peculiar velocities, i.e. redshifts, whose deviations from the observed redshifts are used to correct the x_i for the next iteration, until convergence. After the original study of the history of the Local Group (Peebles 1989), the method was applied to a sample based on 500 groups within $30 \, h^{-1} \text{Mpc}$ of the LG from the Nearby Galaxy Catalog. The results indicate that the unknown tidal forces from outside the sampled volume have a significant effect on the recovered fields and should be incorporated in future applications. The solution obtained depends on the assumed Ω, so a comparison with independent TF distances can in principle constrain Ω (Section 8.2).

8. THE VALUE OF Ω

Assuming that the inferred motions are real and generated by GI, they can be used to estimate Ω. Evidence from virialized systems on smaller scales suggests a low-density universe of $\Omega \sim 0.1$–0.2, but these values may be biased. The spatial *variations* of the large-scale velocity field allow measuring the mass density in a volume closer to a "fair" sample. One family of methods is based on comparing the dynamical fields derived from velocities to the fields derived from galaxy redshifts (Section 9.1, 9.2). Although these methods can be applied in the linear regime, they always rely on the assumed biasing relation between galaxies and mass often parametrized by b, so they provide an estimate of $\beta \equiv f(\Omega)/b$. Another family of methods measures β from redshift surveys alone, based on z-space deviations from isotropy (Section 9.3). Finally, there are methods that rely on nonlinear effects in the velocity data alone, and provide estimates of Ω independent of b (Section 9.4, 9.5). The various estimates of Ω and β are summarized in Table 1 at the end of the section. Note that the errors quoted by different authors reflect different degrees of sophistication in the error analysis, and are in many cases underestimates of the true uncertainty.

8.1 β from Galaxies vs the CMB Dipole

Equation (19) is best at estimating $\mathbf{v}(0)$, the linear velocity of the LG in the CMB frame due to the gravitational acceleration $\mathbf{g}(0)$ exerted by the mass fluctuations around it. A comparison with the LG velocity of 627 ± 22 km s^{-1} as given by the CMB dipole is a direct measure of β. One expects to obtain a lower bound on β because $v \sim 600$ km s^{-1} is very improbable in a low-density universe. One way to estimate $\mathbf{g}(0)$ is from a whole-sky galaxy survey where only the angular positions and the fluxes (or diameters) are observed, exploiting the coincidence of nature that both the apparent flux and the gravitational force vary as r^{-2}. If $L \propto M$, then the vector sum of the fluxes in a volume-limited sample is $\propto \mathbf{g}(0)$ due to the mass in that volume. This idea can be modified to deal with a flux-limited sample once the luminosity function is known, and applications to the combined UGC/ESO diameter-limited catalog of optical galaxies yield β_{opt} values in the range 0.3–0.5 (Lahav 1987, Lynden-Bell et al 1989). These estimates suffer from limited sky coverage, uncertain corrections for Galactic extinction, and different selection procedures defining the north and south samples. The *IRAS* catalog provides a superior sky coverage of 96% of the sky, with negligible Galactic extinction and with fluxes observed by one telescope, but with possible under-sampling of cluster cores (Kaiser & Lahav

1989). A typical estimate from the angular *IRAS* catalog is $\beta_1 = 0.9 \pm 0.2$ (Yahil et al 1986).

The redshift surveys provide the third dimension which could help in deriving $g(0)$ by Equation (19), subject to the difficulties associated with discrete, flux-limited sampling (Section 5). The question is whether $g(0)$ is indeed predominantly due to the mass within the volume sampled, i.e. whether $g(0)$ as computed from successive concentric spheres converges interior to R_{max}. This is an issue of fundamental uncertainty (e.g. Lahav et al 1990, Juszkiewicz et al 1990, Strauss et al 1992b). The $r - z$ mapping (21) could either compress or rarify the z-space volume elements depending on the sign of u in the sense that an outflow makes the z-space density δ_z smaller than the true density δ: $\delta_z(\mathbf{x}) \approx \delta(\mathbf{x}) - 2[\mathbf{v}(\mathbf{x}) - \mathbf{v}(0)] \cdot \hat{\mathbf{r}}/r$. The varying selection function adds to this geometrical effect [in analogy to $n(r)$ in the IM bias (Section 3.2)] and there is a contribution from $d\mathbf{v}/dr$ as well (Kaiser 1987). It is thus clear that the redshifts must be corrected to distances and that any uncertainty in $\mathbf{v}(\mathbf{x})$ at large $\mathbf{x}$ or at $\mathbf{x} = 0$ would confuse the derived $g(0)$. The latter is the Kaiser "rocket effect": If $\mathbf{v}(0)$ originates from a finite volume $r < r_o$ and the density outside r_o is uniform with $\mathbf{v} = 0$, then the measurements in z-space introduce a fake $g(0)$ in the direction of $\mathbf{v}(0)$ due to the matter outside r_o, and this $g(0)$ is logarithmically diverging with r. $\mathbf{v}(0)$ is uncertain because it is derived like the rest of $\mathbf{v}(\mathbf{x})$ from the density distribution—not from the CMB dipole. These difficulties in identifying convergence limit the effectiveness of this method in determining β and the PS on large scales. The hopes for improvement by increasing the depth are not high because the signal according to conventional PS models drops with distance faster than the shot-noise.

Attempting to measure β_1 from the *IRAS* data, Strauss et al (1992b) computed the probability distribution of $g(0)$ under several models for the statistics of fluctuations, via a self-consistent solution for the velocities and an ad hoc fix to the rocket effect, which enabled partial corrections for shot-noise, finite volume, and small-scale nonlinear effects. They confirmed that the direction of $g(0)$ converges to a direction only $\sim 20°$ away from the CMB dipole, but were unable to determine unambiguously whether $|g(0)|$ converges even within $100\,h^{-1}$Mpc. A maximum likelihood fit and careful error analysis constrained β_1 to the range 0.4–0.85 with little sensitivity to the PS assumed. Rowan-Robinson et al (1991, 1993) obtained from the QDOT dipole $\beta_1 = 0.8^{+0.2}_{-0.15}$. Hudson's (1993b) best estimate from the optical dipole is $\beta_{opt} = 0.72^{+0.37}_{-0.18}$.

The volume-limited Abell/ACO catalog of clusters with redshifts within $300\,h^{-1}$Mpc was used to compute $g(0)$ in a similar way under the assumption that clusters trace mass linearly (Scaramella et al 1991). An apparent convergence was found by $\sim 180\,h^{-1}$Mpc to the value $g(0) \approx 4860\beta_c$ km s^{-1}. A comparison with the LG-CMB motion of 627 km s^{-1} yields $\beta_c \approx 0.123$, which corresponds to $\beta_{\text{opt}} \approx 0.44$ and $\beta_{\text{I}} \approx 0.56$ if the ratios of biasing factors are 4.5:1.3:1 (Section 7.1). A similar analysis by Plionis & Valdarnini (1991) yielded convergence by $\sim 150\, h^{-1}$Mpc and β values larger by ~ 30–80%.

8.2 β from Galaxy Density vs Velocities

The linear correlation found between mass density and galaxy density (Section 6.2) can be used to estimate the ratio β. The density δ_v determined by POTENT from velocities assuming $\Omega = 1$ relates in linear theory to the true δ by $\delta_v \propto f(\Omega)\delta$, while linear biasing assumes $\delta = b^{-1}\delta_{\text{g}}$, so $\delta_v = \beta\delta_{\text{g}}$. Dekel et al (1993) carried out a careful likelihood analysis using the POTENT mass density from the Mark II velocity data and the density of *IRAS* 1.9 Jy galaxies, and found $\beta_{\text{I}} = 1.3^{+0.75}_{-0.6}$ at 95% confidence. A similar analysis based on the Mark III and *IRAS* 1.2 Jy is in progress. The degeneracy of Ω and b is broken in the quasi-linear regime, where $\delta(\mathbf{v})$ is no longer $\propto f^{-1}$. The compatible quasi-linear corrections in POTENT and in the *IRAS* analysis allow a preliminary attempt to separate these parameters, which yields for Mark II data $\Omega > 0.46$ (95% level) if $b_{\text{I}} > 0.5$. A correction for IM bias could reduce the 95% confidence limit to $\Omega > 0.3$ at most. These results are valid for linear biasing; possible nonlinear biasing may complicate the analysis because it is hard to distinguish from nonlinear gravitational effects.

The advantage of comparing densities is that they are *local,* independent of reference frame, and can be reasonably corrected for nonlinear effects. The comparison can alternatively be done between the observed velocities and those predicted from a redshift survey, subject to limited knowledge of the quadrupole and higher moments of the mass distribution outside the surveyed volume and other biases. Kaiser et al (1991) obtained from Mark II velocities versus QDOT predictions $\beta_{\text{I}} = 0.9^{+0.20}_{-0.15}$. An analysis by Roth (1994) using *IRAS* 1.9 Jy galaxies yielded $\beta_{\text{I}} = 0.6 \pm 0.3$ (2σ). Nusser & Davis (1994) implemented a novel method based on the Zel'dovich approximation in spherical harmonics to predict the velocity dipole of distant shells from the *IRAS* 1.2 Jy redshift survey and found in comparison to the dipoles derived from observed velocities $\beta_{\text{I}} = 0.6 \pm 0.2$.

Similar comparisons with the optical galaxy fields indicate a similar correlation between light and mass. A comparison at the velocity level gives (Hudson 1994) $\beta_{\text{opt}} = 0.5 \pm 0.1$, and a preliminary comparison at the density level with $12\, h^{-1}$Mpc smoothing indicates (Hudson et al 1994) $\beta_{\text{opt}} \approx 0.75 \pm 0.2$, in general agreement with the ratio of $b_{\text{opt}}/b_{\text{I}} \approx 1.3$–$1.4$ obtained by direct comparison. Shaya et al (1994) applied the least-action reconstruction method (Section 7.2) to a redshift survey of several hundred spirals within our local $30\, h^{-1}$Mpc neighborhood and by comparison to TF distances crudely obtained $\beta_{\text{opt}} \sim 0.4$.

8.3 *β from Distortions in Redshift Space*

Redshift samples, which contain hidden information about velocities, can be used on their own to measure β. The clustering, assumed isotropic in real space, $\mathbf{x}$, is anisotropic in z-space, $\mathbf{z}$, where $z = r + \hat{\mathbf{x}} \cdot \mathbf{v}$ displaces galaxies along the preferred direction $\hat{\mathbf{x}}$. While virial velocities on small scales stretch clusters into "fingers of god" along the line of sight, systematic infall motions enhance large-scale structures by artificially squashing them along the line of sight. The linear approximation $-\nabla \cdot \mathbf{v} = \beta \delta_g$ indicates that the effect is β-dependent because $-\nabla \cdot \mathbf{v}$ is related to the anisotropy in z-space while δ_g is isotropic, so the statistical deviations from isotropy can determine β (e.g. Sargent & Turner 1977).

Kaiser (1987) showed in linear theory that the anisotropic Fourier PS in z-space is related to the real-space PS of mass density, $P(k)$, via

$$P^z(k, \mu) = P(k)(1 + \beta\mu^2)^2, \tag{25}$$

where $\mu \equiv \hat{\mathbf{k}} \cdot \hat{\mathbf{x}}$. This relation is valid only for a fixed μ, i.e. in a distant volume of small solid angle (Zaroubi & Hoffman 1994), but there are ways to apply it more generally. The redshift PS can be decomposed into Legendre polynomials, $\mathcal{P}_l(\mu)$, with even multipole moments $P_l^z(k)$,

$$P^z(k, \mu) = \sum_{l=0}^{\infty} P_l^z(k)\mathcal{P}_l(\mu),$$

$$P_l^z(k) = \frac{2l + 1}{2} \int_{-1}^{+1} d\mu P^z(k, \mu)\mathcal{P}_l(\mu). \tag{26}$$

Based on Equation (25) the first two non-vanishing moments are

$$P_0^z(k) = \left(1 + \frac{2}{3}\beta + \frac{1}{5}\beta^2\right)P(k), \quad P_2^z(k) = \left(\frac{4}{3}\beta + \frac{4}{7}\beta^2\right)P(k), \tag{27}$$

so the observable ratio of quadrupole to monopole is a function of β and is independent of $P(k)$. A preliminary application to the 1.2 Jy *IRAS* survey yields $\beta_1 \sim 0.3$–0.4 at wavelength 30–40 h^{-1}Mpc, suspected of being an underestimate because of nonlinear effects out to $\sim 50\,h^{-1}$Mpc (Cole et al 1993). Peacock & Dodds (1994) developed a method for reconstructing the linear PS and they obtain $\beta_1 = 1.0 \pm 0.2$.

The distortions should be apparent in the z-space two-point correlation function, $\xi_z(r_p, \pi)$, which is the excess of pairs with separation π along the line of sight and r_p transversely (Davis & Peebles 1983). The contours of equal ξ, assumed round in r-space, appear in z-space elongated along the line of sight at small separations and squashed on large scales depending on β. Hamilton (1992, 1993) used the multiple moments of ξ_z, in analogy to Equations (26–27), and his various estimates from the 1.9 Jy *IRAS* survey span the range $\beta_1 = 0.25$–1. Fisher et al (1994a) computed $\xi^z(r_p, \pi)$ from the 1.2 Jy *IRAS* survey, and derived the first two pair-velocity moments. Their attempt to use the velocity dispersion via the Cosmic Virial Theorem led to the conclusion that this is a bad method for estimating Ω, but the mean, $\langle v_{12} \rangle = 109^{+64}_{-47}$ at $10\,h^{-1}$Mpc, yielded $\beta_1 = 0.45^{+0.27}_{-0.18}$. The drawbacks of using ξ versus PS are that (a) the uncertainty in the mean density affects all scales in ξ whereas it is limited to the $k = 0$ mode of the PS, (b) the errors on different scales in ξ are correlated whereas they are independent in a linear PS for a Gaussian field, and (c) ξ mixes different physical scales, complicating the transition between the linear and nonlinear regimes. Nonlinear effects tend to make all the above results underestimates.

A promising method that is tailored to deal with a realistic redshift survey of a selection function $\phi(r)$ and does not rely on the subtleties of Equation (25) is based on a weighted spherical harmonic decomposition of $\delta_z(\mathbf{z})$ (Fisher et al 1994b),

$$a^z_{lm} = \int d^3z\, \phi(r)\, f(z)\, [1 + \delta_z(\mathbf{z})]\, Y_{lm}(\hat{\mathbf{z}}),$$

$$\langle |a^z_{lm}|^2 \rangle = \frac{2}{\pi} \int_0^\infty dk\, k^2 P(k)\, |\psi^r_l(k) + \beta \psi^c_l(k)|^2. \tag{28}$$

The arbitrary weighting function $f(z)$ is vanishing at infinity to eliminate surface terms. The mean square of the harmonics is derived in linear theory assuming that the survey is a "fair" sample, and ψ^r and ψ^c are explicit integrals over r of certain expressions involving $\phi(r)$, $f(r)$, Bessel functions, and their derivatives. The first term represents real structure; the second is the correction embodying the z-space distortions. The harmonic PS in z-space, averaged over m, is thus determined by $P(k)$ and β, where the z-space distortions appear as a β-dependent excess at small l. The harmonic PS derived from the 1.2 Jy *IRAS* survey yields $\beta_1 = 1.0 \pm 0.3$ for an assumed $\sigma_8 = 0.7$ (motivated by the *IRAS* ξ, Fisher et al 1994a), with an additional systematic uncertainty of ± 0.2 arising from the unknown shape of the PS.

The methods for measuring β from redshift distortions are promising because they are relatively free of systematic errors and because very large redshift surveys will be achievable in the near future. With a sufficiently large redshift survey, one can even hope to be able to use the nonlinear effects to determine Ω and b separately.

8.4 Ω from PDFs Using Velocities

Assuming that the initial fluctuations are a random *Gaussian* field, the one-point PDF of smoothed density develops a characteristic skewness due to nonlinear effects early in the quasi-linear regime (Section 7.2). The skewness of δ is given in second-order perturbation theory by $\langle \delta^3 \rangle / \langle \delta^2 \rangle^2 \approx (34/7 - 3 - n)$, where n is the effective power index near the smoothing scale (Bouchet et al 1992).

Since this ratio for δ is practically independent of Ω, and since $\nabla \cdot \mathbf{v} \sim -f\delta$, the corresponding ratio for $\nabla \cdot \mathbf{v}$ strongly depends on Ω, and in second-order (Bernardeau et al 1994)

$$T_3 \equiv \frac{\langle (\nabla \cdot \mathbf{v})^3 \rangle}{\langle (\nabla \cdot \mathbf{v})^2 \rangle^2} \approx -f(\Omega)^{-1}(26/7 - 3 - n). \tag{29}$$

Using N-body simulations and $12\,h^{-1}$Mpc smoothing one indeed finds $T_3 = -1.8 \pm 0.7$ for $\Omega = 1$ and $T_3 = -4.1 \pm 1.3$ for $\Omega = 0.3$, where the quoted error is the cosmic scatter for a sphere of radius $40\,h^{-1}$Mpc in a CDM universe ($H_0 = 75, b = 1$). A preliminary estimate of T_3 in the current POTENT velocity field within $40\,h^{-1}$Mpc is -1.1 ± 0.8, where the error represents distance errors. With the two errors added in quadrature, $\Omega = 0.3$ is rejected at the $\sim 2\sigma$ level (somewhat sensitive to the assumed PS).

Since the PDF contains only part of the information stored in the data and is in some cases not that sensitive to the IPDF (Section 7.2), a more powerful bound can be obtained by using the detailed $\mathbf{v}(\mathbf{x})$ to recover the IPDF, and using the latter to constrain Ω. This is done by comparing the Ω-dependent IPDF recovered from observed velocities to an assumed IPDF (Nusser & Dekel 1993), most naturally a Gaussian as recovered from *IRAS* density (Section 7.2). The velocity out of POTENT Mark II within a conservatively selected volume was fed into the IPDF recovery procedure with Ω either 1 or 0.3, and the errors due to distance errors and cosmic scatter were estimated. The IPDF recovered with $\Omega = 1$ is found marginally consistent with Gaussian; the one recovered with $\Omega = 0.3$ shows significant deviations. The largest deviation bin by bin in the IPDF is $\sim 2\sigma$ for $\Omega = 1$ and $> 4\sigma$ for $\Omega = 0.3$, and a similar rejection is obtained with a χ^2-type statistic. The skewness and kurtosis are poorly determined because of noisy tails but the replacements $\langle x|x| \rangle$ and $\langle |x| \rangle$ allow a rejection of $\Omega = 0.3$ at the 5–6σ level.

8.5 Ω from Velocities in Voids

A diverging flow in an extended low-density region can provide a robust dynamical lower bound on Ω, based on the fact that large outflows are not expected in a low-Ω universe (Dekel & Rees 1994). The velocities are assumed to be induced by GI, but *no* assumptions need to be made regarding galaxy *biasing* or the exact statistical nature of the fluctuations. The derivatives of a diverging velocity field infer a nonlinear approximation to the mass density, $\delta_c(\Omega, \partial\mathbf{v}/\partial\mathbf{x})$ (Equation 6), which is an overestimate, $\delta_c > \delta$, when the true value of Ω is assumed. Analogously to $\delta_0 = -f(\Omega)^{-1}\nabla \cdot \mathbf{v}$, the δ_c inferred from a given diverging velocity field becomes more negative when a smaller Ω is assumed, and it may become smaller than -1. The value of Ω is bounded from below because mass is never negative, $\delta \geq -1$.

The inferred $\delta_c(\mathbf{x})$ smoothed at $12\,h^{-1}$Mpc and the associated error field σ_δ

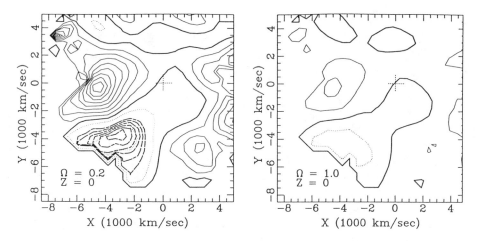

Figure 8 Maps of δ_c inferred from the observed velocities near the Sculptor void in the Super-galactic plane, for two values of Ω. The LG is marked by '+' and the void is confined by the Pavo part of the GA (*left*) and the Aquarius extension of PP (*right*). Contour spacing is 0.5, with $\delta_c = 0$ (*heavy*), $\delta_c > 0$ (*solid*), and $\delta_c < 0$ (*dotted*). The heavy-dashed contours mark the illegitimate downward deviation of δ_c below -1 in units of σ_δ, starting from zero (i.e. $\delta_c = -1$), and decreasing with spacing -0.5σ. The value $\Omega = 0.2$ is ruled out at the $2.9\,\sigma$ level (Dekel & Rees 1994).

are derived by POTENT from the observed radial velocities and, focusing on the deepest density wells, the assumed Ω is lowered until δ_c becomes significantly smaller than -1. The most promising test case provided by the Mark III data seems to be a broad diverging region centered near the supergalactic plane at the vicinity of $(X, Y) = (-25, -40)$ in h^{-1}Mpc—the "Sculptor void" of galaxies (Kauffman et al 1991) next to the "Southern Wall" (Figure 8). Values of $\Omega \approx 1$ are perfectly consistent with the data, but δ_c becomes smaller than -1 already for $\Omega = 0.6$. The values $\Omega = 0.3$ and 0.2 are ruled out at the 2.4-, and 2.9-σ levels in terms of the random error σ_δ. This is just a preliminary result. The systematic errors have been partially corrected for in POTENT, but a more specific investigation of the SG biases affecting the smoothed velocity field in density wells is required. For the method to be effective one needs to find a void that is (*a*) bigger than the correlation length for its vicinity to represent the universal Ω, (*b*) deep enough for the lower bound to be tight, (*c*) nearby enough for the distance errors to be small, and (*d*) properly sampled to trace the velocity field in its vicinity.

The estimates of Ω and β are sumarized in Table 1.

9.   DISCUSSION: ARE THE HYPOTHESES JUSTIFIED?

Are the motions real? The large-scale bulk flow rests upon the interpretation of the CMB dipole as motion of the Local Group in the CMB frame, based on the

Table 1 Ω and b^1

CMB dipole	vs galaxies angular	Yahil et al 86	$\beta_{\mathrm{I}} = 0.9 \pm 0.2^2$
	vs galaxies redshift	Strauss et al 92b	$\beta_{\mathrm{I}} = 0.4 - 0.85$
		Rowan-Rob. et al 91	$\beta_{\mathrm{I}} = 0.8^{+0.2}_{-0.15}$
	vs galaxies angular	Lynden-Bell et al 89	$\beta_{\mathrm{opt}} = 0.1 - 0.5$
	vs galaxies redshift	Hudson 93b	$\beta_{\mathrm{opt}} = 0.7^{+0.4}_{-0.2}$
	clusters	Scaramella et al 91	$\beta_{\mathrm{c}} \sim 0.13$
		Plionis et al 91	$\beta_{\mathrm{c}} \sim 0.17\text{–}0.22$
$\mathbf{v}$ vs δ_{g}	Potent-$IRAS$1.9 density	Dekel et al 93	$\beta_{\mathrm{I}} = 1.3^{+0.75}_{-0.6}$ (95%)
	Potent-$IRAS$1.2 v-dipole	Nusser & Davis 94	$\beta_{\mathrm{I}} = 0.6 \pm 0.2$
	TF-QDOT	Kaiser et al 91	$\beta_{\mathrm{I}} = 0.9^{+0.2}_{-0.15}$
	TF-QDOT clusters	Frenk et al 94	$\beta_{\mathrm{I}} = 1.0 \pm 0.3$
	TF inverse - $IRAS$1.9	Roth 94	$\beta_{\mathrm{I}} = 0.6 \pm 0.35$ (2σ)
	Potent-Optical density	Hudson et al 94	$\beta_{\mathrm{opt}} = 0.75 \pm 0.2$
	TF-Optical	Hudson 94	$\beta_{\mathrm{opt}} = 0.5 \pm 0.1$
	TF-Optical local	Shaya et al 94	$\beta_{\mathrm{opt}} \sim 0.4$
z-distortions	P_k $IRAS$1.2	Cole et al 93	$\beta_{\mathrm{I}} \gtrsim 0.3\text{–}0.4$
	P_k $IRAS$1.2	Peacock & Dodds 94	$\beta_{\mathrm{I}} = 1.0 \pm 0.2$
	ξ $IRAS$1.9	Hamilton 93	$\beta_{\mathrm{I}} \sim 0.25 - 1$
	ξ $IRAS$1.2	Fisher et al 94a	$\beta_{\mathrm{I}} = 0.45^{+0.3}_{-0.2}$
	Y_{lm} $IRAS$1.2	Fisher et al 94b	$\beta_{\mathrm{I}} = 1.0 \pm 0.3$
Velocities	Gaussian IPDF Potent	Nusser & Dekel 93	$\Omega > 0.3$ ($4 - 6\sigma$)
	Skew($\nabla \cdot \mathbf{v}$) Potent	Bernardeau et al 94	$\Omega > 0.3$ (2σ)
	Voids Potent	Dekel & Rees 94	$\Omega > 0.3$ (2.4σ)

$^1\beta = \Omega^{0.6}/b$, $b_C : b_{\mathrm{opt}} : b_I \approx 4.5 : 1.3 : 1.0$, see text.
^{2}All errors are 1σ unless stated otherwise.

assumption of cosmological isotropy (H1), and supported by the gravitational acceleration derived at the LG from the galaxy and mass distribution around it (Section 6.1). The detection of $\delta T/T \sim 10^{-5}$ is a clear indication, via GI, for ~ 300 km s^{-1} motions over $\sim 100\,h^{-1}$Mpc (Section 6.1). While this is reassuring, the LP hint for possible very-large coherence suggests caution (Section 7.1). Evidence that the TF-inferred motions about the LG are real (H6) are (*a*) the correlation $\delta_{\mathrm{g}} \propto -\nabla \cdot \mathbf{v}$, which is robustly predicted for true velocities based on continuity and would be hard to mimic by environmental effects (Section 6.2), (*b*) the failure to detect any significant correlation between velocities and the environment or other galaxy properties (Section 6.3), and (*c*) the similarity between the velocity fields traced by spirals (TF) and by ellipticals ($D_n - \sigma$) (Section 6.3).

Is *linear biasing* a good approximation (H5)? The galaxy-velocity correlation is most sensitive to it, and the observed correlation on scales $\gtrsim 10\,h^{-1}$Mpc is consistent with linear biasing properly modified in the tails (Section 6.2). However, it is difficult to distinguish nonlinear biasing from nonlinear gravitational effects, and the range of different estimates of β (Section 8) may indicate that the biasing parameter varies as a function of scale. The ratio of $\sim 10\,h^{-1}$Mpc-smoothed densities for optical and *IRAS* galaxies is $b_{\mathrm{opt}}/b_I \approx 1.3$–$1.5$.

Is *gravity* the dominant source of LSS (H4)? The observed velocity-density correlation (Section 6.2) is fully consistent with GI, but it is sensitive to continuity more than to the specific time dependence implied by gravity. Any non-GI process followed by a gravitating phase would end up consistent with this observation, and certain non-GI models may show a similar spatial behavior even if gravity never plays any role. The E-S correlation (Section 6.3) is also consistent with gravity as galaxies of all types trace the same velocity field (H4*b*), but any model where all galaxies are set into motion by the same mechanism could pass this test. The strongest evidence for gravitational origin comes from the statistical agreement between the fluctuations of today and those implied by the CMB at the time of recombination. A marginal warning signal for GI is provided by the ~ 700 km s^{-1} bulk velocity indicated by LP for rich clusters across $\sim 200\,h^{-1}$Mpc. Such a velocity at face value would be in conflict with the gravitational acceleration implied by the cluster distribution and with the $\delta T/T \sim 10^{-5}$ at $\sim 2°$, but the errors are large.

The property of *irrotationality* (H4*a*) used in the reconstruction from either velocities or densities is impossible to deduce solely from observations of velocities along the lines of sight from one origin. Irrotationality is assumed based on the theory of GI, or it can be tested against the assumption of isotropy by measuring the isotropy of the velocity field derived by potential analysis for a fair sample.

The observed CMB fluctuations provide evidence for *initial fluctuations* (H2), consistent with a scale-invariant $n \sim 1$ spectrum. The observed motions are also consistent with $n \sim 1$ (Section 7.1) (with the uncertain LP result as a possible exception). The indications for somewhat higher large-scale power in the clustering of galaxies (Maddox et al 1990) may reflect nontrivial biasing. The question of whether the fluctuations were Gaussian (H2) is not to be answered by observed velocities alone. The PDF of $\nabla \cdot \mathbf{v}$ is consistent with Gaussian initial fluctuations skewed by nonlinear gravity, but this is not a very discriminatory test. Nevertheless, the galaxy spatial distribution does indicate Gaussian initial fluctuations fairly convincingly (Section 7.2)

Can we determine the nature of the dark matter (H3)? In view of the tight nucleosynthesis constraints on baryonic density, the high Ω indicated by the motions requires non-baryonic DM. The mass-density PS on scales 10–$100\,h^{-1}$Mpc is calculable in principle but the current uncertainties do not allow a clear distinction between the possibilities of baryonic, cold, hot, or mixed DM. The mixed model seems to score best in view of the overall LSS data, as expected from a model with more free parameters, but CDM in fact does somewhat better in fitting the large-scale motions. I do not think that any of the front-runner models is significantly ruled out at this point, contrary to occasional premature statements in the literature about the "death" of certain models. I predict that were the DM constituent(s) to be securely detected in the laboratory, the corresponding scenario of LSS will find a way to overcome the $\sim 2\sigma$ obstacles it is facing now.

What can we conclude about the *background cosmology* (H1)? All the observations so far are consistent with large-scale homogeneity and isotropy (with the exception of the 2σ LP discrepancy). The motions say nothing about H or Λ (Lahav et al 1991), but they provide a unique opportunity to constrain Ω in several different ways. Some methods put a strong ($> 3\sigma$) lower bound of $\Omega > 0.2\text{--}0.3$. This is consistent with the theoretically-favored $\Omega = 1$ but "ugly" values near $\Omega \approx 0.5$ are not ruled out either. The range of β values obtained on different scales is partly due to errors, and any remaining difference may be explained by a scale-dependent nonlinear biasing relation between the different galaxy types and mass. The data is thus consistent with the predictions of Inflation: flat geometry and Gaussian, scale-invariant initial fluctuations. Recall however that $\Omega = 1$ predicts $t_0 = 6.3h^{-1}\text{Gyr}$, which will be in conflict with the age constraints from globular clusters, $t_0 = 15 \pm 3\text{Gyr}$, if the Hubble constant h is not close to 0.5.

The rapid progress in this field guarantees that many of the results and uncertainties discussed above will soon become obsolete, but I hope that the discussion of concepts will be of lasting value, and that the methods discussed can be useful as are and as a basis for improvements.

ACKNOWLEDGMENTS

I thank my students G. Ganon, T. Kolatt, S. Markoff and I. Zehavi for assistance, M. Hudson and A. Yahil for plots, and G. R. Blumenthal, S. M. Faber, Y. Hoffman, O. Lahav, M. Strauss, D. Weinberg, J. Willick, and A. Yahil for very helpful comments. This work has been supported by grants from the US-Israel Binational Science Foundation and the Israel Basic Research Foundation.

Literature Cited

Aaronson M, Bothun GD, Cornell ME, Dawe JA, Dickens RJ, et al. 1989. *Ap. J.* 338:654

Aaronson M, Bothun G, Mould J, Huchra J, Schommer RA, Cornell ME. 1986. *Ap. J.* 302:536

Aaronson M, Huchra J, Mould J. 1979. *Ap. J.* 229:1

Aaronson M, Huchra J, Mould J, Schechter PL, Tully RB. 1982b. *Ap. J.* 258:64

Aaronson M, Huchra J, Mould JR, Tully RB, Fisher JR, et al. 1982a. *Ap. J. Suppl.* 50:241

Aaronson M, Mould J. 1983. *Ap. J.* 265:1–17

Babul A, Weinberg D, Dekel A, Ostriker JP. 1994. *Ap.J.* In press

Bardeen J, Bond JR, Kaiser N, Szalay A. 1986. *Ap. J.* 304:15

Bernardeau F. 1992. *Ap. J. Lett.* 390:L61

Bernardeau F, Juszkiewicz R, Bouchet F, Dekel A. 1994. In preparation

Bertschinger E. 1991. In *Physical Cosmology*, ed. M Lachieze-Rey. Gif-sur-Yvette: Ed. Frontiere

Bertschinger E, Dekel A. 1989. *Ap. J. Lett.* 336:L5

Bertschinger E, Dekel A, Faber SM, Dressler A, Burstein D. 1990. *Ap. J.* 364:370

Bertschinger E, Gorski K, Dekel A. 1990. *Nature* 345:507

Bertschinger E, Juszkiewicz R. 1988. *Ap. J. Lett.* 334:L59

Blumenthal GR, Dekel A, Primack JR. 1988. *Ap. J.* 326:539

Bothun GD, Aaronson M, Schommer B, Huchra J, Mould J. 1984. *Ap. J.* 278:475

Bouchet F, Juszkiewicz R, Colombi S, Pellat R. 1992. *Ap. J. Lett.* 394:L5

Bouchet F, Strauss M, Davis M, Fisher KB, Yahil A, Huchra JP. 1993. *Ap. J.* 417:36

Burstein D. 1990a. In *Large Scale Structure and Peculiar Motions in the Universe*, ed. DW Latham, LN, Da Costa, Provo: ASP Conf. Ser.

Burstein D. 1990b. *Rep. Prog. Phys.* 53:421–81

Burstein D, Davies RL, Dressler A, Faber SM, Lynden-Bell D, Terlevich RJ, Wegner G. 1986. In *Galaxy Distances and Deviations from Universal Expansion*, ed. BF Madore, RB Tully, pp. 123–30. Dordrecht: Reidel

Burstein D, Faber SM, Dressler A. 1990. *Ap. J.* 354:18

Cen R, Ostriker JP. 1993. *Ap. J.* In press

Cheng ES, et al. 1994. *Ap. J. Lett.* In press

Cole S, Fisher KB, Weinberg D. 1993. *MNRAS* In press

Coles P, Jones B. 1991. *MNRAS* 248:1

Colless M, et al. 1993. *MNRAS* 262:475

Corey BE, Wilkinson DT. 1976. *Bull. Am. Astron. Soc.* 8:35

Courteau S. 1992. PhD thesis. Univ. Calif., Santa Cruz

Courteau S, Faber SM, Dressler A, Willick JA.

1993. *Ap. J. Lett.* 412:L51

Davis M, Peebles PJE. 1983. *Annu. Rev. Astron. Astrophys.* 21:109–30

Davis M, Summers FJ, Schlegel D. 1992. *Nature* 359:393

Dekel A. 1981. *Astron. Astrophys.* 101:79–87

Dekel A, Bertschinger E, Faber SM. 1990. *Ap. J.* 364:349

Dekel A, Bertschinger E, Yahil A, Strauss M, Davis M, Huchra J. 1993. *Ap. J.* 412:1

Dekel A, et al. 1994. *Ap. J.* In preparation

Dekel A, Rees MJ. 1987. *Nature* 326:455

Dekel A, Rees MJ. 1994. *Ap. J. Lett.* 422:L1–L4

de Lapparent V, Geller, MJ Huchra JP. 1986. *Ap. J. Lett.* 302:L1

Djorgovski S, Davis M. 1987. *Ap. J.* 313:59

Djorgovski S, de Carvalho R, Han MS. 1989. In *The Extragalactic Distance Scale,* ed. S van den Bergh, CJ Pritchet, p. 3. Provo: ASP

Dressler A, 1994. In *Cosmic Velocity Fields,* ed. F Bouchet, M Lachieze-Rey. Paris:IAP In press

Dressler A, Faber SM. 1991. *Ap. J.* 368:54

Dressler A, Lynden-Bell D, Burstein D, Davies RL, Faber SM, et al. 1987. *Ap. J.* 313:42

Efstathious G. 1990. In *Physics of the Early Universe,* ed. JA Peacock, AF Heavens, AT Davies. Edinburgh: SUSSP

Faber SM, Burstein D. 1988. In *The Vatican Study Week on Large Scale Motions in the Universe,* ed. GV Coyne, VC Rubin, pp. 116. Princeton: Princeton Univ. Press

Faber SM, et al. 1994. *Ap. J.* In preparation

Faber SM, Jackson RE. 1976. *Ap. J.* 204:668–83

Faber SM, Wegner G, Burstein D, Davies RL, Dressler A, Lynden-Bell D, Terlevich RJ. 1989. *Ap. J. Suppl.* 69:763

Feldman HA, Watkins R. 1994. In *Cosmic Velocity Fields,* ed. F Bouchet, M Lachieze-Rey. Paris:IAP In press

Fisher KB. 1992. PhD thesis. Univ. Calif., Berkeley

Fisher KB, Davis M, Strauss MA, Yahil A, Huchra JP. 1994a. *MNRAS* In press

Fisher KB, Scharf CA, Lahav O. 1994b. *MNRAS* In press

Frenk C, Kaiser N, Lucey J. 1994. In preparation

Giavalisco M, Mancinelli B, Mancinelli PJ, Yahil A. 1993. *Ap. J.* 411:9

Gramman M. 1993a. *Ap. J. Lett.* 405:L47

Gramman M. 1993b. *Ap. J.* 405:449

Gregg MJ. 1994. *Ap. J.*

Gull S, Daniell X. 1978. *Nature* 272:686

Gunn JE. 1989. In *The Extragalactic Distance Scale,* ed S. van den Berg, CJ Pritchet, p. 344. Provo: ASP

Hamilton AJS. 1992. *Ap. J. Lett.* 385:L5

Hamilton AJS. 1993. *Ap. J. Lett.* 406:L47

Han, MS, Mould JR. 1990. *Ap. J.* 360:448

Han, MS, Mould JR. 1992. *Ap. J.* 396:453

Hoffman Y. 1994. In *Cosmic Velocity Fields,*

ed. F Bouchet, M Lachieze-Rey. Paris:IAP In press

Hoffman Y, Ribak E. 1991. *Ap. J. Lett.* 380:L5

Hogan CJ, White SDM. 1986. *Nature* 321:575

Hudson M. 1993a. *MNRAS* 265:43

Hudson M. 1993b. *MNRAS* 265:72

Hudson M. 1994. *MNRAS* 266:475

Hudson M, et al. 1994. In preparation

Ikeuchi S. 1981. *Publ. Astron. Soc. Jpn.* 33:211

Jacoby GH, et al. 1992. *Publ. Astron. Soc. Pac.* 104:599–662

Juszkiewicz R, Vittorio N, Wyse RFG. 1990. *Ap. J.* 349:408

Kaiser N. 1987. *MNRAS* 227:1–21

Kaiser N. 1988. *MNRAS* 231:149–68

Kaiser N. 1984. *Ap. J. Lett.* 284:L9

Kaiser N, Efstathiou G, Ellis R, Frenk C, Lawrence A, et al. 1991. *MNRAS* 252:1

Kaiser N, Lahav O. 1989. *MNRAS* 237:129

Kaiser N, Stebbins A. 1991. In *Large Scale Structure and Peculiar Motions in the Universe*, ed. DW Latham, LN, Da Costa, pp. 111. Provo: ASP Conf. Ser.

Kauffman G, Fairall AP. 1991. *MNRAS* 248:313

Klypin A, Holtzman J, Primack JR, Regos E. 1993. *Ap. J.* 416:1

Kofman L, Bertschinger E, Gelb J, Nusser A, Dekel A. 1994. *Ap. J.* In press

Kofman L, Blumenthal GR, Hodges H, Primack JR. 1990. In *Large Scale Structure and Peculiar Motions in the Universe*, ed. DW Latham, LN, Da Costa, Provo: ASP Conf. Ser.

Kogut A, et al. 1993. *Ap. J.* 419:1–6

Kolatt T, Dekel A. 1994a. *Ap. J.* In press

Kolatt T, Dekel A. 1994b. *Ap. J.* Submitted

Kolatt T, Dekel A, Lahav O. 1994. *MNRAS* Submitted

Kolatt T, et al. 1994. In preparation

Kolb EW, Turner MS. 1990. *The Early Universe.* Menlo Park, CA: Addison-Wesley

Kraan-Korteweg RC, Woudt P. 1994. In *4th DAEC Meet. on Unveiling Large-Scale Structures behind the Milky Way*, ed. C Balkowski, RC Kraan-Korteweg. ASP Conf. Ser. In press

Lahav O. 1987. *MNRAS* 225:213–20

Lahav O, Fisher KB, Hoffman Y, Scharf CA, Zaroubi S. 1994. *Ap. J. Lett.* In press

Lahav O, Kaiser N, Hoffman Y. 1990. *Ap. J.* 352:448

Lahav O, Lilje PB, Primack JR, Rees MJ. 1991. *MNRAS* 251:128

Landy S, Szalay A. 1992. *Ap. J.* 391:494

Lauer TR, Postman M. 1993. *Ap. J.* In press

Lucey JR, Carter D. 1988. *MNRAS* 235:1177

Lynden Bell D, Faber SM, Burstein D, Davies RL, Dressler A, et al. 1988. *Ap. J.* 326:19

Lynden-Bell D, Lahav O, Burstein D. 1989. *MNRAS* 241:325–45

Maddox SJ, Efstathiou G, Sutherland WJ, Loveday J. 1990. *MNRAS* 242:43p

Mancinelli PJ, Yahil A, Ganon G, Dekel A. 1994. In *Cosmic Velocity Fields,* ed. F Bouchet, M Lachieze-Rey. Paris:IAP In press

Mathewson DS, Ford VL, Buchhorn M. 1992. *Ap. J. Suppl.* 81:413

Mould JR, et al. 1991. *Ap. J.* 383:467

Moutarde F, Alimi J-M, Bouchet FR, Pellat R, Ramani A. 1991. *Ap. J.* 382:377

Nusser A, Dekel A. 1992. *Ap. J.* 391:443

Nusser A, Dekel A. 1993. *Ap. J.* 405:437

Nusser A, Dekel A, Bertschinger E, Blumenthal GR. 1991. *Ap. J.* 379:6

Nusser A, Dekel A, Yahil A. 1994. *Ap. J.* In press

Nusser A, Davis M. 1994. *Ap. J. Lett.* In press

Ostriker JP, Cowie LL. 1981. *Ap. J. Lett.* 243:L127

Ostriker JP, Suto Y. 1990. *Ap. J.* 348:378

Paczynski B, Piran T. 1990. *Ap. J.* 364:341

Peacock JA, Dodds SJ. 1994. *MNRAS* In press

Peebles PJE. 1980. *The Large-Scale Structure of the Universe.* (Princeton: Princeton Univ. Press)

Peebles PJE. 1989. *Ap. J. Lett.* 344:L53

Peebles PJE. 1990. *Ap. J.* 362:1

Peebles PJE. 1993. *Principles of Physical Cosmology.* (Princeton: Princeton Univ. Press)

Plionis M, Valdarnini R. 1991. *MNRAS* 249:46

Rauzy S, Lachieze-Rey M, Henriksen RN. 1993. *Astron. Astrophys.* 273:357

Raychaudhury S. 1994. In *Cosmic Velocity Fields,* ed. F Bouchet, M Lachieze-Rey. Paris:IAP In press

Regos E, Szalay AS. 1989. *Ap. J.* 345:627

Roth JR. 1994. In *Cosmic Velocity Fields,* ed. F Bouchet, M Lachieze-Rey. Paris:IAP In press

Rowan-Robinson M. 1993. *Proc. Natl. Acad. Sci.* 90:4822

Rowan-Robinson M, Lawrence A, Saunders W, Crawford J, Ellis RS, et al. 1990. *MNRAS* 247:1

Rowan-Robinson M, Lawrence A, Saunders W, Leech K. 1991. *MNRAS* 253:485

Rubin VC, Ford WK Jr, Thonnard N, Roberts MS, Graham JA. 1976a. *Astron. J.* 81:687–718

Rubin VC, Thonnard N, Ford WK Jr, Roberts MS. 1976b. *Astron. J.* 81:719–37

Rybicki GB, Press WH. 1992. *Ap. J.* 398:169

Sachs RK, Wolfe AM. 1967. *Ap. J.* 147:73

Sargent WLW, Turner EL. 1977. *Ap. J. Lett.* 212:L3

Scaramella R, Vettolani G, Zamorani G. 1991. *Ap. J. Lett.* 376: L1

Schechter P. 1980. *Astron. J.* 85:801

Schuster J, et al. 1993. *Ap. J. Lett.* In press

Seljak U, Bertschinger E. 1993. *Ap. J.* In press

Shaya E, Peebles PJE, Tully B. 1994. In *Cosmic Velocity Fields,* ed. F Bouchet, M Lachieze-Rey. Paris:IAP

Silk J. 1989. *Ap. J. Lett.* 345:L11

Simmons JFL, Newsam A, Hendry MA. 1994. In *Cosmic Velocity Fields,* ed. F Bouchet, M Lachieze-Rey. Paris:IAP In press

Smoot GF, Gorenstein MV, Muller RA. 1977. *Phys. Rev. Lett.* 39:898

Smoot GF, et al. 1992. *Ap. J. Lett.* 396:L1

Smoot GF, Tenorio L, Banday AJ, Kogut A, Wright EL, et al. 1994. Preprint

Stebbins A. 1994. In *Cosmic Velocity Fields,* ed. F Bouchet, M Lachieze-Rey. Paris:IAP In press

Strauss MA, Cen R, Ostriker JP. 1993. *Ap. J.* 408:389

Strauss MA, Cen R, Ostriker JP. 1994. In *Cosmic Velocity Fields,* ed. F Bouchet, M Lachieze-Rey. Paris:IAP In press

Strauss MA, Davis M, Yahil A, Huchra JP. 1990. *Ap. J.* 361:49

Strauss MA, Huchra JP, Davis M, Yahil A, Fisher KB, Tonry J. 1992a. *Ap. J. Suppl.* 83:29

Strauss MA, Yahil A, Davis M, Huchra JP, Fisher KB. 1992b. *Ap. J.* 397:395

Szalay A. 1988. In *The Vatican Study Week on Large Scale Motions in the Universe,* ed. GV Coyne, VC Rubin, pp. 323–38. Princeton: Princeton Univ. Press

Tonry JL. 1991. *Ap. J. Lett.* 373:L1

Tully RB, Fisher JR. 1977. *Astron. Astrophys.* 54:661

Weinberg DH. 1991. *MNRAS* 254:315

Weinberg DH, Cole S. 1992. *MNRAS* 259:652

Willick J. 1991. PhD thesis. Univ. Calif., Berkeley

Willick J. 1994a. *Ap. J.* In press

Willick J. 1994b. *Ap. J.* In press

Willick J, et al. 1994. *Ap. J.* In preparation

Yahil A. 1994. *Ap. J. Lett.* Submitted.

Yahil A, Dekel A, Kolatt T, Blumenthal G. 1994. In preparation

Yahil A, Walker X, Rowan-Robinson M. 1986. *Ap. J. Lett.* 301:L1

Yahil A, Strauss MA, Davis M, Huchra JP. 1991. *Ap. J.* 372:380

Zaroubi S, Hoffman Y. 1994. Preprint

Zel'dovich, YaB. 1970. *Astron. Astrophys.* 5:20

Annu. Rev. Astron. Astrophys. 1994. 32: 419–63

COSMIC DUSTY PLASMA

D. A. Mendis and M. Rosenberg

Department of Electrical and Computer Engineering, University of California, San Diego, La Jolla, California 92093

KEY WORDS: dust-plasma interactions, charged dust, collective processes

1. INTRODUCTION

Different types of photometric observations in the 1930s (Trumpler 1930; Stebbins et al 1934; 1939) clearly showed that the dark "holes" in the Milky Way, observed by William Herschel almost 150 years earlier, were in fact regions of heavy obscuration by cosmic dust. Continuing observations since then have established that dust is an almost ubiquitous component of the cosmic environment. Remote sensing of dust in the interstellar, circumstellar, interplanetary, circumplanetary, and cometary environments has, more recently been complemented by in-situ detections of the last three. Furthermore, the inference of the existence of very small grains (so-called VSGs with dimensions of 10–100 Å) in the interstellar medium (Puget & Léger 1989) as well as their in-situ detection in the environment of comet P/Halley (Sagdeev et al 1989) reinforces the reasonable expectation that the transition from gas to large dust particles in the cosmic environment is a continuous one through macromolecules, clusters, and VSGs.

These dust grains are invariably immersed in ambient plasma and radiative environments. They must therefore be necessarily electrically charged and consequently coupled to the plasma through electric and magnetic fields, with the coupling becoming stronger as the grain size decreases.

While any plasma containing such charged dust grains is often loosely referred to as a dusty plasma, there are different regimes characterized by the relative magnitudes of three characteristic length scales, namely the dust grain size a, the plasma Debye length λ_D, and the average intergrain distance $d (\approx n_d^{-1/3}$, where n_d is the dust number density). In general cosmic plasma environments that are contaminated by dust can be characterized by either of two conditions 1. $a \ll \lambda_D < d$ or 2. $a \ll d < \lambda_D$. In the first case the dust may

419

Table 1 d/λ_D for a few cosmic environments

	$n_i(\text{cm}^{-3})$	$T(°K)$	$n_d(\text{cm}^{-3})$	d/λ_D	$a(\mu m)$
Interstellar clouds[a]	10^{-3}	10	10^{-7}	0.3	0.01–10
Noctilucent clouds[a]	10^3	150	10	0.2	~1
Saturn's E-ring[b]	10	10^5–10^6	10^{-7}–10^{-8}	$\lesssim 1$	~1
Halley's Comet					
1. inside ionopause[c]	10^3–10^4	$< 10^3$	10^{-3}	$\gtrsim 1$	.1~10
2. outside ionopause[d]	10^2–10^3	~10^4	5×10^{-9}–10^{-7}	$\gtrsim 10$	.01~10
Saturn's spokes[b]	0.1–10^2	2×10^4	1	$\lesssim 0.01$	$\lesssim 1$
Saturn's F-ring[b]	10–10^2	10^5–10^6	<30	$\lesssim 10^{-3}$	$\lesssim 1$

[a]Tsytovich et al (1990), [b]Goertz (1989), [c]Mendis et al (1985), [d]De Angelis et al (1988).

be considered as a collection of isolated screened grains ("dust-in-plasma"); in the second case the dust also participates in the screening process and therefore in the collective behavior of the ensemble (a true "dusty-plasma"). Estimates of the ratio d/λ_D in a few cosmic environments are shown in Table 1.

That cosmic dust would be electrically charged has been recognized for a long time (Spitzer 1941), yet theoretical studies of its many consequences was slow in coming. A strong impetus for the study of dust-plasma interactions in space came with the observations of curious features like the "radial spokes" across Saturn's B Ring by the *Voyager* spacecraft's cameras in the early 1980s (e.g. Hill & Mendis 1982a, Goertz & Morfill 1983). Presently, theoretical studies of cosmic dust-in-plasmas as well as dusty plasmas are proceeding at a rapidly increasing pace and dedicated laboratory studies, while lagging behind the theory, are beginning to make significant contributions in checking the predictions of various theories.

Several useful reviews of the present subject have appeared recently (Goertz 1989, Hartquist et al 1992, Northrop, 1992). The present review, in common with all of the above, does not pretend to be comprehensive. Rather than focusing on individual phenomena, we attempt to emphasize the basic physics that underlie both dust-in-plasma and dusty plasmas in space. We hope that this rather theoretical approach will prove useful to both the interested outsider as well as to a graduate research student who might contemplate entering this emerging new field.

We start (in Section 2) with a detailed discussion of the problem central to all the subsequent studies, which is the electrostatic charging of dust grains both when they may be regarded as being "isolated" and when they form an interacting ensemble. This is followed, in Section 3, with some of the physical consequences to the dust, such as electrostatic disruption, coagulation, and levitation. In Section 4, we consider the dynamics and orbital evolution of charged dust in cosmic environments where long-range intergrain electrical interactions are unimportant. In Section 5 we discuss the collective behaviors that are manifested: waves and instabilities. Here we distinguish between the cases in which the charged dust is considered to be massive and immobile and in which dust dynamics plays a role in wave behavior. We also discuss the role of charged dust in wave scattering.

2. ELECTROSTATIC CHARGING OF THE DUST

Central to all the studies that are discussed is the charge Q acquired by the dust grains. For a given dust grain at a given time, this is determined by the equation

$$\frac{dQ}{dt} = \frac{d}{dt} \cdot C(\phi - \overline{\phi}) = I, \tag{1}$$

where I is the total current to the grain, C is the grain capacitance, ϕ is the grain surface potential, and $\overline{\phi}$ is the average potential of the ambient plasma with a dust grain distribution. Contributions to I come from various electron and ion currents (see e.g. Whipple 1965) and include (a) electron and ion collection, (b) photo emission, (c) secondary electron emission due to energetic electron impact, (d) secondary electron emission due to energetic ion impact, (e) electric field emission, (f) thermoionic emission, (g) triboelectric emission, and (h) radioactive emission of electrons and α particles.

Of these, the most important in the cosmic environment are in general (a), (b), and (c), while (e) is important for very small grains. Friction-mediated (triboelectric) emission is known to be important in the charging of dust in volcanic plumes, and may also have played some role in the early protoplanetary nebula.

The currents to the grains depend on a number of properties of both the grains and the ambient medium. For instance, the electron and ion collection currents depend not only on the size and shape of the grains, but also on the electron and ion velocity distributions and densities, motion of the grain relative to the plasma, and the potential difference between the grain surface and the ambient dusty plasma, $(\phi_s - \overline{\phi})$. The photo-electron current depends on the electric properties of the grain, the grain surface potential, and of course, on the photo ionizing (uv) flux of radiation. The secondary electron emission current due to energetic electron impact depends on the secondary emission yield, which has recently been shown to be crucially dependent on the grain size (Chow et al 1993). The electric field emission current, which is caused by large surface electric fields, is negligible for the larger grains, but could become the dominant one when the grains become very small (Mendis & Axford 1974).

In a steady state, the grains may reach an equilibrium surface potential, ϕ_s, with respect to the plasma. Then $\frac{dQ}{dt} = 0$ and $Q = C(\phi_s - \overline{\phi})$. The values of both C and $\overline{\phi}$ depend on how closely the grains are packed together. For instance when $d \gg \lambda_D$, the grains may be regarded as being isolated. Then $C = C_{iso} = a(1 + a/\lambda_D)$ and $\overline{\phi}$ may be taken to be zero. On the other hand, when $d \ll \lambda_D$, C increases slightly above C_{iso} but $\bar{\phi}$ approaches ϕ_s and consequently the grain charge becomes very much smaller than Q_{iso} (Goertz & Ip 1984, Whipple et al 1985).

2.1   *Isolated Grain*

In order to illustrate some of these points, let us start with the simplest case of a small "isolated" ($a \ll \lambda_D \ll d$) grain at rest with respect to the plasma, and assume that the only currents of importance are electron and ion collection.

If $f_e(E)$ is the isotropic electron velocity distribution at infinity, where E is the kinetic energy $(= \frac{1}{2} m_e v_e^2)$, the electron current (density) to the grain is given by

$$J_e = -e \iiint f_e(E - e\phi_s) v_n d^3 \mathbf{v}. \tag{2}$$

Here v_n is the component of the velocity normal to the grain surface, and the integration is performed over all orbits that intersect the grain, i.e. for $E_{total} = E - e\phi_s \geq 0$. (e.g. see Laframboise & Parker 1973). Selecting spherical polar coordinates so that $d^3\mathbf{v} = v^2 \sin \theta \, d\theta \, d\phi \, dv$ and integrating over θ and ϕ one easily obtains

$$J_e = -\frac{2\pi e}{m_e^2} \int_{\max(0, e\phi_s)}^{\infty} E f_e(E - e\phi_s) \, dE. \tag{3}$$

Similarly, the ion current density to the grain is

$$J_i = +\frac{2\pi e}{m_i^2} \int_{\max(0, e\phi_s)}^{\infty} E f_i(E + e\phi_s) dE. \tag{4}$$

The forms of $J_{e,i}$ when f is Maxwellian are well known (e.g. see Goertz 1989). The equilibrium potential, obtained by equating the total current density $J_e + J_i$ to zero, in that case, is given by the well-known transcendental equation

$$\left(\frac{m_i}{m_e}\right)^{1/2} = R_T^{1/2} \left(1 - \frac{e\phi_s}{k_B T_i}\right) \exp\left(-R_T \frac{e\phi_s}{k_B T_i}\right), \tag{5}$$

where $R_T = T_i / T_e$.

If we also assume that the plasma consists of electrons and protons and further that $T_e = T_i = T$, the solution of Equation (5) gives $\phi_s \approx -2.51 \, k_B T / e$. The equilibrium potential is negative, in this case, because initially (i.e. when $\phi_s = 0$) more electrons (whose mobility is about 43 times that of the protons when $T_e = T_i$) than protons reach the grain surface and charge it negative. This causes the electron current to decrease and the ion current to increase, thereby increasing the grain potential, and the process will proceed until the grain acquires the equilibrium negative potential ϕ_s that makes these two currents equal in magnitude. For an electron-oxygen ion plasma the equilibrium potential ϕ_s

is more negative ($\approx -3.6 k_B T/e$) owing to the even smaller mobility of the heavier oxygen ions.

Astrophysical calculations generally assume that the plasma is indeed Maxwellian, a notable exception being that of Meyer-Vernet (1982), which includes suprathermal electrons. Most astrophysical plasmas, however, are observed to have non-Maxwellian high energy tails (e.g. Summers & Thorne 1991), and a generalized Lorenzian (κ) distribution has been often employed to fit the observed particle distributions (Gosling et al 1981, Leubner 1982, Armstrong et al 1983, Christon et al 1988). Consequently it is instructive to use such a distribution to recalculate the potential of an isolated small grain, as was done recently by Rosenberg & Mendis (1992). The normalized κ-distribution is given by

$$f_\kappa(E) = n \left(\frac{m}{2\pi \kappa E_0}\right)^{3/2} \frac{\Gamma(\kappa+1)}{\Gamma(\kappa-1/2)} \cdot \left(1 + \frac{E}{\kappa E_0}\right)^{-(\kappa+1)} \tag{6}$$

where κ is the spectral index, E is the particle energy, n is the particle number density, Γ is the gamma function, and E_0 is related to the temperature T by $E_0 = [(2\kappa-3)/2\kappa]k_B T$ for $\kappa > 3/2$. This distribution is also ideal for studying how the grain potential responds to increasing deviation from the Maxwellian plasma, because while it has a high energy tail with a power-law dependence $f_\kappa \propto E^{-(\kappa+1)}$ for $E \gg \kappa E_0$, it approaches the Maxwellian distribution as $\kappa \to \infty$. Following the same procedure as for a Maxwellian, described above, the equilibrium potential of the grain is now given by the equation

$$\left(\frac{m_i}{m_e}\right)^{1/2} = R_T^{1/2}(\kappa-1)\left(\frac{2}{2\kappa-3}\right)^\kappa \left[\frac{2\kappa-3}{2\kappa-2} - \frac{e\phi_s}{k_B T_i}\right]$$
$$\times \left[\frac{2\kappa-3}{2} - R_T \frac{e\phi_s}{k_B T_i}\right]^{\kappa-1}, \tag{7}$$

where it is assumed that $\kappa_e = \kappa_i = \kappa$. This result reduces to the Maxwellian result when $\kappa \to \infty$. Equation (7) also has a particularly simple solution when $\kappa = 2$ and $R_T = 1$, e.g.

$$\frac{e\phi_s}{k_B T_i} = \frac{1}{2}\left[1 - \left(\frac{m_i}{m_e}\right)^{1/4}\right]. \tag{8}$$

Although the grain potential ϕ_s is once again negative, its magnitude is found to be large as long as the electrons are described by a κ-distribution with this magnitude increasing with ion mass and decreasing with electron κ (see Figure 1). For example, for a pure oxygen Maxwellian plasma (with $T_e = T_i = T$), $\phi_s \simeq -3.6k_B T/e$, as stated earlier, whereas $\phi_s \approx -6.1k_B T/e$ when $\kappa_e = \kappa_i = \kappa = 2$ and $\phi_s \approx -4.5k_B T/e$ when $\kappa_e = \kappa_i = \kappa = 5$. This last case with $\kappa = 5$ is the one that appears to best describe space plasmas in several instances. The consequences of this increase in the magnitude of the grain potential will be discussed later.

So far we have assumed that the grain is at rest with respect to the plasma. However, grains (e.g. those in planetary rings) drift with respect to the plasma. The appropriate velocity distributions to be used in the calculation of the ion and electron collection currents are those in the frame of reference of the moving grain. Since drift speeds of these grains are very much smaller than the electron thermal speeds in the region of the planetary rings (e.g. Mendis et al 1984), they have negligible effect on the electron current. On the other hand, they are of the same order as the ion thermal speeds, so the ion current is strongly modified. The modified ion current for the case of a drifting Maxwellian in the grain frame is given in Whipple (1965, 1981). It was often assumed in the literature that this effect increases the ion current since one expects the moving grain to sweep up more ions. This in turn would make the grain potentials

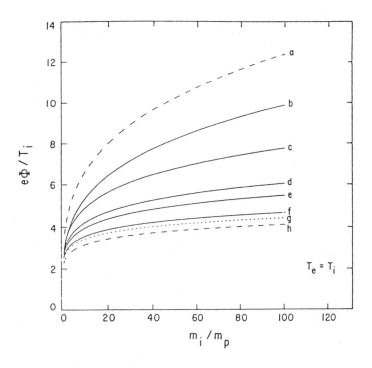

Figure 1   Plot of $e\Phi/T_i (\Phi = |\phi|, \phi < 0)$ as a function of ion mass (normalized to the proton mass) for $T_e = T_i$ in a Lorentzian plasma. The labels attached to the curves refer to the following spectral indices: (*a*) $\kappa_e = 2$, Maxwellian ions ($\kappa_i \to \infty$); (*b*) ($\kappa_e = \kappa_i = 2$; (*c*) $\kappa_e = \kappa_i = 3$; (*d*) $\kappa_e = \kappa_i = 5$; (*e*) $\kappa_e = \kappa_i = 7$; (*f*) $\kappa_e = \kappa_i = 25$; (*g*) Maxwellian electrons and ions ($\kappa_e = \kappa_i \to \infty$); and (*h*) $\kappa_i = 2$, Maxwellian electrons ($\kappa_e \to \infty$). Note that T_i is in energy units. (From Rosenberg & Mendis 1992.)

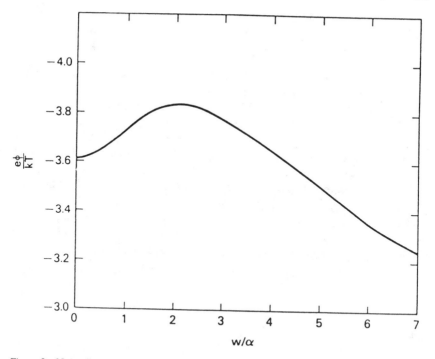

Figure 2   Normalized grain potential as a function of the ratio of the grain velocity w to the ion thermal velocity. (From Northrop et al 1989.)

less negative. A more recent study (Northrop et al 1989) has shown this to be so only when the Mach number M (grain speed/ion thermal speed) $\gtrsim 2$ (see Figure 2). When $0 < M \lesssim 2$ the grain potential turns out to be more negative. The reason for this is that the increase of ion flux on the leading face of the grain is more than offset by its decrease on the trailing face for this range of values of M. This is not the case when $M \gtrsim 2$.

A dominant process associated with grain charging at higher electron energies ($\gtrsim 50$ eV) is secondary electron emission, and such electron energies are common in certain regions of planetary magnetospheres. In this process, energetic electrons penetrate the grain and energize material electrons that may diffuse out of the grain. One defines the secondary electron yield $\delta(E)$, at a given energy E of the primary electrons, as the ratio of the secondary electron current to the primary electron current. Then the secondary electron current density J_s is given by

$$J_s = \frac{2\pi e}{m_e^2} \int_0^\infty E\delta(E)f_e(E - e\phi_s)\,dE, \quad \phi_s \leq 0 \tag{9}$$

$$J_s = \frac{2\pi e}{m_e^2} \exp\left(-\frac{e\phi_s}{k_B T_s}\right)\left(1 + \frac{e\phi_s}{k_B T_s}\right)\int_{e\phi}^{\infty} E\delta(E)f_e(E - e\phi_s)\,dE, \quad \phi_s \geq 0.$$

$$(10)$$

Here $k_B T_s$ is the thermal energy of the emitted secondary electrons (which are found to have a Maxwellian velocity distribution) with values in the range of 2–5 eV, regardless of the energy of the incoming primaries (e.g. Goertz 1989). The forms of J_s when $f_e(E)$ is either a Maxwellian or a generalized Lorentzian are given in Chow et al (1993), where the expressions for $\delta(E)$ for different grain sizes and grains materials are also calculated. Due to shape of $\delta(E)$, which first increases with energy to a maximum δ_m (which could be $\gg 1$) at some E_m, and then decreases monotonically (at least for grains typically $\gtrsim 0.1\mu$m), the current-energy curve of the grain subject only to electron and ion collection and secondary emission can have single or multiple real roots (Whipple 1965, Meyer-Vernet 1982): one negative root, and one positive root, or one negative and two positive roots, although the smaller positive root is unstable in the last case.

2.2   Grain Ensemble

We now consider the case of grain charging by plasma collection currents when the grain density in the plasma is high. In this case, the grain charge may be lower than its "isolated" value, as discussed earlier. There are two competing effects that lead to this result; one is that the capacitance of the grain increases, which tends to increase the charge, and the other is that the magnitude of the grain surface potential relative to the plasma potential decreases, which decreases the charge. The capacitance of a grain increases from its value in vacuum (where the capacitance is proportional to the grain radius) as the grain spacing becomes comparable to or less than the Debye length. In this case, the positive sheath (for a negatively charged grain) moves closer to the grain surface, thus essentially decreasing the capacitor gap, or the distance between the edge of the sheath and the grain surface, thereby increasing the capacitance (Whipple et al 1985, Northrop 1992). This effect, however, was found to be of the order of $< 0.1\%$ for parameters representative of the F Ring and the spokes of Saturn's ring system (Whipple et al 1985), where the dust density has been inferred to be comparable to the plasma density (e.g. Goertz 1989).

The more important effect at high dust density arises from electron depletion, when the dust grains carry a significant fraction of the (negative) charge density in the plasma, so that much of the electron charge resides on the grain surfaces. In this case, the surface potential of a grain does not have to be as negative with respect to the plasma potential as in the isolated grain case to balance the electron and ion currents to the grain. This leads to a decrease in the magnitude of the grain charge, since the charge is proportional to $\phi_s - \bar{\phi}$. Whipple et al

(1985) analyzed this effect by balancing the electron and ion current to the grain (considering only collection currents),

$$I_e = -e \left(\frac{8\pi k_B T_e}{m_e} \right)^{1/2} a^2 \bar{n}_e \exp[e(\phi_s - \bar{\phi})/k_B T_e] \tag{11}$$

$$I_i = e \left(\frac{8\pi k_B T_i}{m_i} \right)^{1/2} a^2 \bar{n}_i [1 - e(\phi_s - \bar{\phi})/k_B T_i], \tag{12}$$

taking into account the fact that the electron and ion densities $(\bar{n}_e, \bar{n}_i)$ in the plasma reservoir between the grain satisfy the condition of overall charge neutrality in the plasma

$$\bar{n}_e - \bar{n}_i = \eta_d Q/e. \tag{13}$$

Here ϕ_s is the grain surface potential, the reservoir of plasma between the grains is at an average potential $\bar{\phi}$, $\phi_s < 0$, Q is the grain charge, and $\eta_d = n_d/(1 - 4\pi a^3 n_d/3)$ is essentially the grain density. This leads to an equation for $y = e(\phi_s - \bar{\phi})/k_B T (T_e = T_i = T)$ as a function of a parameter $Z = 4\pi \lambda_D^2 \eta_d C(n_d)$, where $C(n_d)$ is the grain capacitance, and λ_D is the Debye length in the electron-ion plasma reservoir, $\lambda_D^2 = k_B T/[4\pi (\bar{n}_e + \bar{n}_i)e^2]$:

$$(1 - y)(1 - yZ) = \left(\frac{m_i}{m_e} \right)^{1/2} (1 + yZ) \exp(y). \tag{14}$$

Figure 3 shows the solution obtained from Equation (14) where it is seen that y, which is proportional to the grain charge $Q = C(\phi_s - \bar{\phi})$, decreases steeply with increasing Z, which is proportional to the dust density. Since the quantity Z is of the order of the ratio of the charge density on n_d "isolated" grains to the available charge density in the plasma, the depletion effect should affect grain charging when $Z \geq 1$, as shown. The condition $Z \geq 1$ implies that $4\pi n_d \lambda_D^3 \geq \lambda_D/a \gg 1$ for cosmic plasmas. The quantity yZ is essentially the ratio of the charge density carried by the grains to the charge density carried by the electrons and ions in the plasma. The decrease in grain charge arising from electron depletion can be quite large for certain plasmas; for example Whipple et al (1985) estimate that for a 1 micron grain in Saturn's F-ring, the magnitude of the negative grain charge would be $\sim 2.7 \times 10^{-4}$ times its "isolated" value. The maximum electron depletion, or minimum $\bar{n}_e/\bar{n}_i$, can be obtained from Equations (11) and (12) by solving $I_e + I_i = 0$ in the limit $y \ll 1$; this yields $\bar{n}_e/\bar{n}_i \to \sqrt{m_e/m_i}$ as $y \to 0$, and thus not all the electrons can be depleted in this approximation.

Goertz (1989) investigated how the grain potential $\phi(r)$ changes when dust grains are placed closer and closer to each other. A one-dimensional Poisson equation for the potential in the vicinity of a grain in the presence of finite electron, ion, and grain charge densities was solved numerically, coupled with an equation for electron and ion current balance to the grain (only plasma collection currents were considered). The electrons and ions were assumed to have Boltzmann distributions, while the grain spatial distribution was modeled by a number of equally spaced infinite sheets of grains. Figure 4 illustrates how the magnitude of the spatially varying potential decreases as the grains sheets are placed closer together.

Havnes et al (1987) studied the variation of the grain charge with dust density in a dust cloud embedded in a plasma, with the cloud at an average potential V_p, using coupled equations for current balance to the grain and overall charge neutrality in the cloud. The variation of the dust grain potential minus the cloud potential $U = (|\phi_s - V_p|)$ is shown in Figure 5 as a function of a parameter $P = n_d a T / n_p$ (Havnes et al 1987, Goertz 1989), where the grain radius a is in meters, T is the plasma temperature in eV and n_p is the plasma density outside the dust cloud. The grain charge is proportional to U. Again the decrease of $|U|$ with dust charge density n_d is evident from Figure 5. Values of the parameter P for various planetary rings are shown in Table 1 of the paper by Goertz (1989) to indicate where the electron depletion effect may play a significant role in determining the grain charge. Havnes et al (1990a) have also extended

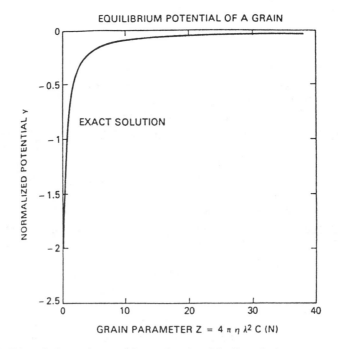

EQUILIBRIUM POTENTIAL OF A GRAIN

Figure 3   Dimensionless grain potential y as a function of the dimensionless parameter Z, which is quite closely proportional to the number density N of dust grains. (From Whipple et al 1985.)

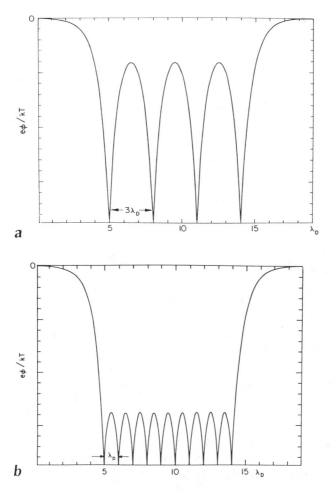

Figure 4 Steady state solution of the one-dimensional Poisson equation for grain sheets placed at regular intervals. The distance between grains is (*a*) $3\lambda_D$ and (*b*) λ_D. (From Goertz 1989.)

their analyses (Havnes et al 1984, 1987) to include a distribution of dust grain sizes and a photoemission charging current. A more detailed model developed by Wilson (1988) in connection with Saturn's rings includes many aspects of the rings' vertical structure as well as the proper solar illumination geometry. Besides photoelectron production, plasma absorption is also included. He finds that the electric field is more complicated in structure and stronger in magnitude within the cloud than at the edges.

The reduction of the magnitude of the grain charge when the intergrain spacing is reduced has recently been demonstrated experimentally. Xu et al

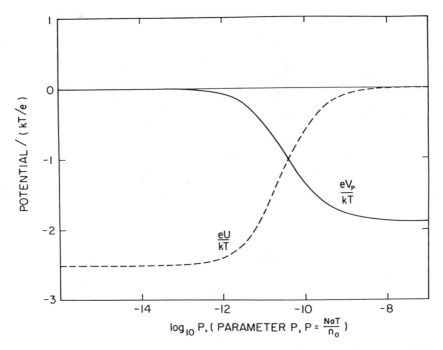

Figure 5   The variation of plasma potential V_p and grain potential U (proportional to the grain charge) as a function of P, i.e. dust density. T is the plasma temperature in electron volts, a is the grain radius in meters, and N and n_p are the dust and plasma densities, respectively. (From Goertz 1989.)

(1993) reported measurements showing how the average (negative) charge on a grain varies as the ratio of the intergrain spacing to the plasma Debye length is varied. For fixed dust density the dust charge density was found to decrease as the electron density decreases, i.e. as λ_D becomes $\gtrsim d$.

3.   COAGULATION, DISRUPTION, AND LEVITATION OF CHARGED DUST

We now consider several of the physical effects manifested by charged dust.

3.1   *Coagulation*

The fact that the current-energy curve can have multiple roots in the presence of secondary emission led Meyer-Vernet (1982) to propose that grains with identical electrical properties, immersed in the same plasma but having different charging histories, could achieve opposite potentials. This idea was further

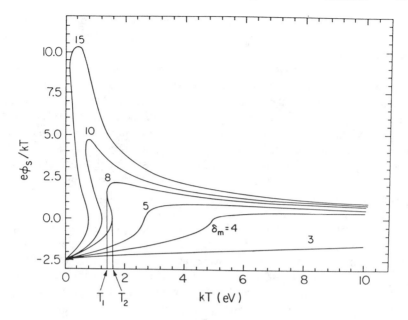

Figure 6 The equilibrium surface potential ϕ as a function of plasma temperature T for different values of the secondary yield parameters δ_m (the maximum yield). (From Goertz 1989.)

developed by Horanyi & Goertz (1990). The variation of the normalized grain surface potential $[\phi_s/(k_B T/e)]$ with plasma temperature (given in eV) for different values of δ_m is shown in Figure 6. Consider grains of different size immersed in the plasma and suppose that they all have the same δ_m (say 8). Now suppose that the temperature of the plasma is gradually increased. The potentials of all the grains will follow the $\delta_m = 8$ curve, but the large grains will respond faster. This follows from the fact that since $\frac{dQ}{dt} = I$, $\frac{a\phi_s}{\tau} \sim a^2$, and so the characteristic time τ required to reach the equilibrium potential corresponding to the instantaneous conditions $\sim 1/a$. Once the temperature increases beyond T_2 the grains clearly have to make a transition to the upper part of the curve, thereby changing the potential from a negative value to a positive value. Since the larger grains will collect the necessary positive charge to make this jump faster than the smaller grains, there will be oppositely charged grains in the plasma for some period of time. These authors studied the implication of this effect for grain coagulation; the attractive Coulomb force between the positively charged larger grains and the negatively charged smaller grains will naturally enhance the coagulation rate. Of course such coagulation can occur only during transient heating events, when the ambient plasma is temporarily heated, or when the grains move through a spatially confined hot plasma region.

Since the authors do not discuss the frequency and duration of such transient heating events in space, the efficiency of this process for grain coagulation remains uncertain.

Very recent work by Chow et al (1993) on secondary electron emission from small grains has shown that the dust grains of different sizes can acquire opposite charges in warm plasmas even in the absence of changes in the plasma environment. This is because the so-called Sternglass formula for the secondary emission yield $\delta(E)$, which is extensively used [including in the work of Meyer-Vernet (1982) and Horanyi & Goertz (1990)], applies only to the case of a semi-infinite planar slab, wherein the secondaries can escape only from one surface of the target, namely the side from which the primary electrons enter. For small grains (assumed spherical), secondary electrons are not limited to the point of entry of the primary electron; they can exit from all points on the grain surface. This increases the yield over that determined by using the Sternglass formula; the increase is very large for VSGs where the penetration depth of the primary electrons is comparable to its size, for a range of electron energies of interest (see Figure 7). There, the secondary electron yield obtained using the Sternglass formula would be very close to the curve marked "Jonker's" since the semi-empirical Sternglass formula (Sternglass 1954) approximates the theoretical derivation of Jonker (1952) for semi-infinite planar slabs. As expected the larger grains (diameter $D \gtrsim 1 \ \mu$m) follow the Jonker's yield curve. Chow et al (1993) have also calculated the equilibrium potential for conducting and insulating grains immersed in both Maxwellian and generalized Lorentzian plasmas. Due to this size effect on secondary emission they find that insulating grains with diameters 0.01 μm and 1 μm have opposite polarity (the smaller one being positive) when the plasma temperature is in the range 25–48 eV in a Maxwellian plasma. For Lorentzian plasmas this temperature range is shifted downward to 8–17 eV when $\kappa = 2$. These values may be in the range of the inferred values of $k_B T_e$ in many regions of the planetary ring systems (e.g. Goertz 1989) and comets (Mendis & Horanyi 1991) as well as in the interplanetary medium (Allen 1983), the local interstellar medium (Cox & Reynolds 1987), and supernova remnants (Raymond 1984). Furthermore, in the primordial protoplanetary nebula, if the kinetic energy of the gravitationally infalling gas was a source of ionization via the so-called critical ionization process (e.g. Alfven 1980) then electron thermal energies $\gtrsim 10$ eV may be possible at least in localized regions. Consequently, the existence of different sized grains of opposite polarity—negatively charged large grains and positively charged small grains—is possible in all of these environments. Enhanced coagulation à la Horanyi & Goertz (1990) would now take place, the only difference being that the larger grains will be always negative and the smaller ones always positive. Also, since transient heating effects are no longer necessary, the coagulation process should proceed continuously, and therefore more efficiently.

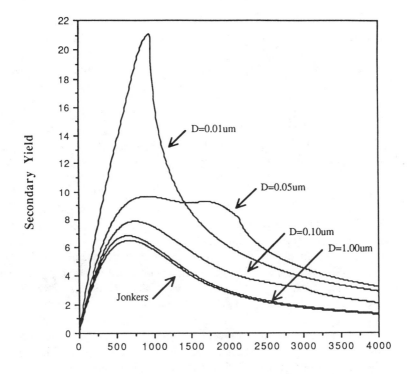

$$eVp \ (eV)$$

Figure 7 Secondary emission yield $\delta(E)$ as a function of the primary electron energy (eV_p) for different sized insulating grains (D). (From Chow et al 1993.)

In this connection it needs to be noted that photoemission too could cause grains to acquire opposite charges in the same plasma and radiative environment, even if they had the same size, provided they had widely different photoemission yields (Feuerbacher et al 1973). This also could lead to enhanced dust coagulation in certain regions of interstellar space as pointed out by the above authors.

3.2   Disruption

Coagulation is not the only physical effect of grain charging. Exactly the opposite effect, namely physical disruption of the grains could also occur if the grains acquire numerically very high potentials. This is a consequence of the electrostatic repulsion of like surface charges which produces an electrostatic tension in the body. If this electrostatic tension exceeds the tensile strength of the body across any section, the body will break up across that section. Considering a sphere of uniform tensile strength, F_t, integrating the component of the Maxwell stress $(E^2/8\pi)$ normal to the plane section of one hemisphere over its surface, and comparing it with the tensile force on that section, Opik (1956) showed that electrostatic distribution will occur across that section unless

$$F_t > F_e \left(= \frac{\phi^2}{8\pi a^2} \right), \tag{15}$$

where F_e is the electrostatic tension and a is the grain radius (all units are e.s.u. and c.g.s). For stability, this implies that

$$a > a_c (= 6.65|\phi| F_t^{-1/2}) \tag{16}$$

(where a_c is now measured in μm and ϕ is in volts, while F_t is in dynes/cm^2). It is clear from (15) that as a becomes smaller, the value of F_t required to prevent grain disruption increases rapidly. This also implies that as a grain begins to disrupt electrostatically it will continue to do so until reaching the smallest fragments for which the above macroscopic considerations apply (perhaps VSGs or clusters with $a \gtrsim 10$ Å). If this were so, it could also provide an insurmountable obstacle for grain growth in a plasma. What enables grains to circumvent this runaway disruption is the electric field emission of electrons from small grains, as was shown recently (Mendis 1991). This is because as the grain radius decreases, the surface electric field increases to such a value (typically $\gtrsim 10^7$ Vcm^{-1}) that rapid electron emission occurs from negatively charged grains and the grain potential decreases (numerically) to a value that is no longer given by the plasma environment but rather by the size alone, e.g. $|\phi| \approx 900a$ (Mendis & Axford 1974), where ϕ is in volts and a is in μm. Substituting the above value into Equation (15) we obtain the result that if $F_t > F_0 (= 3.6 \times 10^7$ dynes cm^{-2}), then the electric field emission limitation of the grain potential will prevent the electrostatic disruption of grains, regardless of their size. Consequently, materials such as iron ($F_t \sim 2 \times 10^{10}$ dyne cm^{-2}) and tektites ($F_t \sim 7 \times 10^{10}$ dyne cm^{-2}) are stabilized by this process against electrostatic disruption no matter how small the size. On the other hand, very fragile grains such as "cometary" grains $F_t (\approx 10^6$ dynes cm^{-2}) of very small radii (≈ 10 Å) will be stable only if $|\phi| \lesssim 0.15$ V, which corresponds to $T \lesssim 460$ K for a Maxwellian oxygen plasma and $T \lesssim 230$ K for a Lorentzian oxygen plasma with $\kappa = 2$.

It needs to be emphasized that the electric field emission effect will enable grain growth to take place in the above environments only if the grains are negatively charged. This means that if there is sufficient uv radiation to make the grain charge positive due to photoemission, grain growth will not proceed even in low temperature plasmas.

Fechtig et al (1979) suggested that the 2–3 orders of magnitude increase in the micrometeoroid flux (10^{-15}g $\lesssim m_d \lesssim 10^{-12}$g) observed within 10 $R_\oplus$ in the terrestrial auroral regions by the *HEOS-2* satellite was due to such electrostatic disruption of type III "fireballs" of mass $\gtrsim$ 10 g. This was shown to be unlikely since the tensile strength of these relatively large bodies would have to be improbably small for such a disruption to take place (Mendis 1981). On the other hand, since these bodies are expected to be highly irregular in shape, and since the electrostatic tension at any point in such a body is inversely proportional to the local radius of curvature, sharp edges and projections would get electrostatically chipped away as the body gets charged up (Hill & Mendis 1980a). Such a process of electrostatic "chipping" or erosion, rather than overall electrostatic disruption, has been proposed as the mechanism responsible for the aforementioned observation (Mendis 1981).

There are numerous examples of dust phenomena at comets that have been attributed to electrostatic disruption (see Mendis & Horanyi 1991 for a detailed review). These include the so-called pseudosynchronic bands or *striae* observed in cometary dust tails (Sekanina & Farrell 1980, Hill & Mendis 1980b); the discrete dust "packets" observed in the environment of comet P/Halley by the dust analyzers on the *VEGA* spacecrafts (Simpson et al 1987, 1989), and the peculiar spatial distribution of the VSGs (10^{-20}g $\lesssim m_d \lesssim 10^{-17}$g) also observed by the *VEGA 1* spacecraft at comet Halley (Sagdeev et al 1989, Fomenkova & Mendis 1992).

3.3 *Levitation*

Electrostatic charging can also lead to the levitation of fine dust lying on large surfaces. In this case, the charge Q acquired by the grain is proportional to its projected surface area (Singer & Walker 1962) and so $Q = \pi a^2 \sigma = \pi a^2 (E_n/4\pi) = a^2(\phi_s/4\lambda_D) = \left(\frac{a}{4\lambda_D}\right)(a\phi_s)$. Typically $a/\lambda_D \ll 1$, and so the charge on the grain is very much smaller than that it would acquire in free space. Mendis et al (1981) considered the charging of the bare cometary nucleus by the solar wind plasma and solar uv radiation at large heliocentric distances. They showed that while the subsolar point of the cometary surface acquires a positive potential of $\sim +5$ V due to the dominance of photoemission, the nightside could acquire a negative potential $\approx -m_p V_{sw}^2/2e$, (which ~ -1 kV when $V_{sw} \approx 600$ km/sec). Consequently, submicron-sized grains could overcome the gravitational attraction of the nucleus and levitate on the nightside of the comet, even when they had a deficit of just one electron charge. Since the nightside potential is highly modulated by the solar wind speed ($\sim V_{sw}^2$), grains are more likely to be electrostatically levitated and subsequently blown off the dark surface when the comet intercepts a high-speed solar wind stream. The sporadic brightness variations of comet Halley observed inbound at large heliocentric distances ($\sim$11–8 AU) were attributed to this effect by Flammer et al (1986). These authors showed that the comet encountered a corotating high-speed solar wind stream emanating from a southern coronal hole at the times of the observed brightness increases. More recently, a large brightness increase of comet Halley at a heliocentric distance ≈ 14.3 AU (outbound) has also been attributed to a solar flare–generated shock wave moving at a speed ~ 750 km/sec at that distance (Intrilligator & Dryer 1991). These authors suggest that the dust results from the pressure-induced rupture of the nucleus. This is very unlikely because the ram pressure at 14.3 AU, which is estimated $\sim 3 \times 10^{-9}$ dynes/cm^2, is far too small for the purpose. It is much more likely that the physical process responsible for the dust emission is electrostatic levitation from the nightside of the comet, as discussed earlier.

4. DYNAMICS

About 20 years ago Mendis & Axford (1974) discussed the basic nature of the trajectories of interplanetary dust grains entering planetary magnetospheres, getting charged to significant negative potentials within the plasmaspheres, and moving under the combined influence of planetary gravity and the Lorentz force. Since then there has been a growing body of literature on the dynamics of charged dust in planetary magnetospheres and the cometary environment.

The basic equation governing the dynamics of a charged dust grain of mass m_d, velocity $\mathbf{v}_d$, in the planetocentric inertial frame is

$$m_d \frac{d\mathbf{v}}{dt} = Q(t) \left[\mathbf{E} + \frac{\mathbf{v}_d \times \mathbf{B}}{c} \right] - \frac{GM_p m_d}{r^3} \mathbf{r} + \mathbf{F}_d + \mathbf{F}_r + \mathbf{F}_c. \tag{17}$$

In the region of the magnetosphere that is corotating with the planet of mass M_p with angular velocity Ω_p, $\mathbf{E} = -(\Omega_p \times \mathbf{r}) \times \mathbf{B}/c$, and $\mathbf{F}_c$, $\mathbf{F}_r$, and $\mathbf{F}_d$ are forces associated with collisions of the grain with plasma, radiation, and other grains respectively. The force $\mathbf{F}_c$, which arises from the relative motion between the grain and the plasma, has contributions both from the direct impacts with ions as well as from Coulomb interactions between the grain charge and the ions, and has been discussed at length by Northrop & Birmingham (1990). However, it was shown to be of negligible importance to the dynamics of charged dust in the magnetospheres of Jupiter and Saturn (e.g. Hill & Mendis 1980b, Mendis et al 1984), and plays only a marginal role in orbital evolution (e.g. Northrop et al 1989). The force $\mathbf{F}_r$ due to radiation pressure is also negligible in the Jovian and Saturnian magnetospheres, but is of crucial importance in the terrestrial magnetosphere, due to its greater proximity to the Sun (e.g. Horanyi & Mendis 1986a). Finally, $\mathbf{F}_d$ is negligible in all cases. The Lorentz force, due to the relative motion between the charged dust and the planetary magnetic field, however, can be comparable to the gravitational force for submicron-sized grains charged to reasonable potentials (e.g. see Mendis et al 1982).

Mendis et al (1982) discussed the epicyclic motion of charged dust in the equatorial plane of a planet whose magnetic and spin axes are parallel or antiparallel. In this case the corotational electric force is always radial. In the case in which Q/m_d is sufficiently large numerically that this motion may be described by the guiding-center approximation, and the net radial force is attractive, the guiding center will perform a circular orbit around the planet with an angular velocity Ω_G given by

$$\Omega_G^2 + \omega_0\Omega_G - \omega_0\Omega_p - \Omega_K^2 = 0, \tag{18}$$

where $\omega_0 = -QB/m_dc$ and $\Omega_K[= (GM_p/r^3)^{1/2}]$ is the Kepler angular velocity. The grains themselves will gyrate around the magnetic field lines with an angular velocity ω in an elliptical orbit whose minor axis is always radial, and whose axis ratio, b/a, is given by

$$\frac{b}{a} = \frac{-\omega}{\omega_0 + 2\Omega_G} \tag{19}$$

with

$$\omega^2 = \omega_0^2 + 4\omega_0\Omega_G + \Omega_G^2. \tag{20}$$

In the case of Saturn, where the magnetic and spin axes are known to be parallel (at least to within $0.5°$), if the grains are negatively charged the ellipse is performed in the retrograde sense while the guiding center motion is direct. It is easily seen from Equations (18), (19), and (20) that when $-Q/m_d \rightarrow \infty$, $\Omega_G \rightarrow \Omega_p$, $\omega \rightarrow \omega_0$, and $b/a \rightarrow 1$, and we recover the purely electrodynamic case. When $Q/m_d \rightarrow 0$, $\Omega_G \rightarrow \Omega_K$, $\omega \rightarrow +\Omega_K$, and $b/a \rightarrow 1/2$, and we recover Keplerian motion. In the general gravito-electrodynamic case: $1/2 < b/a < 1$ and $\Omega_K < \Omega_G < \Omega_p$ outside the synchronous radius (i.e. where $\Omega_K = \Omega_p$) and $\Omega_p < \Omega_G < \Omega_K$, inside.

As a consequence of the fact that $\Omega_G > \Omega_K$ outside the synchronous orbit Mendis et al (1982) pointed out that grains of a particular size in the F Ring could be in resonance with the inner "shepherding" satellite S27 of this ring. They consequently argued that the peculiar "wavy" nature of this ring, which is known to be composed largely of micron and submicron-sized dust, could be caused by such a novel "magneto-gravitational" resonance. However, it was subsequently shown (e.g. Grun et al 1984) that the charge on individual grains in the F Ring (where $\lambda_D \gtrsim$ the average integration distance) has a much smaller value than the "isolated" grain value assumed by Mendis et al (1982) and as a result could not lead to waves of the observed wavelength. In the analysis of Mendis et al (1982), the grain charge was assumed to be constant. However, more general treatment of the adiabatic motion of charged grains in planetary magnetospheres, which also takes into account the periodic variation of the grain charge, was given by Northrop & Hill (1983a). There are several causes for the periodic variation of the grain charge. One results from the variation of the total velocity of the grain at the gyrofrequency due to the gyration of the grain about the guiding center. This in turn causes a variation of the ion current (as discussed earlier) and thus of the grain potential with the same frequency. Others result from temperature gradients that may be present in the plasma as well as compositional variations, since both these result in periodic variations of the plasma collection currents at the gyrofrequency. Since all guiding center drifts may be regarded as being due to the variation of the gyroradius at the gyrofrequency, these could give rise to additional drifts of the guiding center. These drifts would also be azimuthal if the maximum and minimum grain potentials were achieved at the largest and smallest radial distances from the planet. This, however, is not the case, due to the small but finite electrical capacitance of the grain, which leads to a phase lag in the grain potential with respect to the radial oscillation. Thus the maximum and minimum potentials are reached at a small angular distance from the positions of maximum and minimum distances from the planet giving rise to a radial component in the drift. This drift has been dubbed the "gyrophase drift" by Northrop & Hill. Incidently, density gradients in the plasma periodically change the response time for charging and therefore the aforementioned phase lag. This in turn, changes the radial gyrophase drift rate. The direction of the drift depends not only of the polarity of the grain charge and on whether it is inside or outside the synchronous radius, but it also depends on the directions of the temperature and compositional gradients and on the ratio of the grain velocity with respect to the ion thermal velocity (see Figure 2). The variation of the grain charge at the gyrofrequency destroys the adiabatic invariance of the magnetic moment. However, the analysis of Northrop & Hill (1983a) led to the identification of another exact constant of the motion, which enabled them to find the so-called circulation radius where the magnetic moment became zero and stopped the radial drift. In the earlier numerical calculations of Hill & Mendis (1980b) of negatively charged grains in an assumed isothermal Jovian magnetosphere, the interplanetary grains that entered the magnetosphere and got charged did drift towards the synchronous radius, at the progressively slowing rate due to the decrease of the magnetic moment and consequent circularization of the orbits. However, what Northrop & Hill (1983a) showed was that circularization of the orbits and consequent cessation of the gyrophase drift depended on the geometry of the initial launch and could therefore take place before the synchronous radius was reached.

From a reexamination of one of the last *Voyager 2* photographs of Jupiter in forward-scattered light, as the spacecraft left the planet, Showalter et al (1985) discovered a very tenuous ring, composed largely of fine micron and submicron-sized dust, and extending outward from the brighter thin inner ring ($\sim 1.82 R_J$) to the vicinity of the satellite Thebe ($\sim 3.11 R_J$). The most interesting feature of this, so-called gossamer ring (see Figure 8) is the significant peak exactly at the synchronous radius (2.24 R_J). Showalter et al (1985) believed that the sources

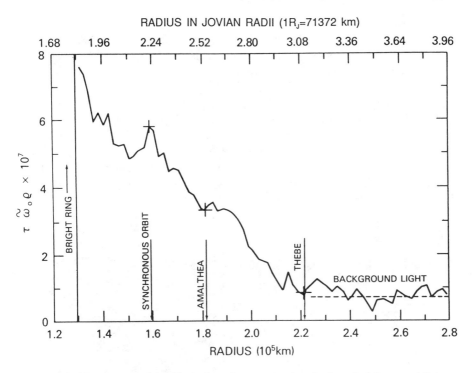

RADIUS IN JOVIAN RADII (1R$_J$=71372 km)

Figure 8 The gossamer ring. The ordinate is proportional to the intensity of scattered light. (Reproduced from Showalter et al 1985.)

of these rings were larger bodies ("mooms") straddling the synchronous orbit, and that the dust emitted from these mooms drifted away from the synchronous orbit in both directions due to plasma drag. A subsequent detailed analysis by Northrop et al (1989) showed that the gyrophase drift of small grains could greatly exceed the drift due to both the plasma drag and the Poynting-Robertson (radiation) "drag." While this gyrophase drift can cause the dust to drift both towards or away from the synchronous orbit, a maximum plasma temperature near the synchronous orbit could cause the grains to concentrate there since temperature-associated gyrophase drift is always towards the higher temperature. These authors suggest that the satellite Amalthea, rather than unobserved mooms, is the source of this dust. They also show that the satellite Io ($\sim 6R_J$), which is known to be a source of dust due to volcanic activity, is unlikely to be the source of at least the smaller dust particles in the gossamer ring, due to a very large plasma temperature gradient at the inner edge of the Io plasma torus. However, they concede that larger grains may penetrate this barrier due to a radial diffusion caused by the stochasticity of the number of charges residing on the grain at a given time, as proposed earlier by Morfill et al (1980).

One of the most intriguing features observed in the Saturnian ring system by both *Voyagers 1* and *2* was the near-radial spokes (Smith et al 1982), which more than anything else, provided the impetus for the study of dust-plasma interactions in planetary magnetospheres. These spokes are confined to the dense central B Ring with inner edge at 1.52 R_s and outer edge at 1.95 R_s. They have an inner boundary at ~1.72 R_s and an outer boundary at approximately the outer edge of the B Ring. Against the background of the B Ring they appear dark in backscattered light and bright in forward-scattered light, indicating that they are composed of micron and submicron-sized grains.

A typical spoke pattern is seen in Figure 9. The spokes exhibit a characteristic wedge shape (clearly apparent in the central spoke of the Figure); the vertex coincides with the position of the synchronous orbit at ~1.86 R_s. Movies of

Figure 9   Image of spokes in the B Ring. Note the distinct wedge shape of the spoke at the upper left portion of the picture. The vertex of the wedge is near the synchronous orbit. The rotation is counter clockwise with the leading edge being the slanted one. (*Voyager 2* photograph from Smith et al 1985.)

the observational data show the formation and dynamical evolution of these spokes. They are more easily seen and seem to be more sharply defined on the morning ansa as they are rotated out of Saturn's shadow. High-resolution *Voyager 2* images show that the leading and trailing edges of the spokes have distinctly different angular velocities. Inside the synchronous orbit where these spokes are seen most often, the leading tilted edge has essentially the Kepler value, whereas the trailing near-radial edge has approximately the corotational value. Based on the observing geometry, it has been argued that the material constituting the spokes is elevated above the ring plane (Smith et al 1982, Grun et al 1983). Several theories were quickly proposed to explain the formation and evolution of these spokes. While all of them involved electrical effects to produce the spokes, many of them specifically involved electrostatic levitation. [For a detailed review of these theories, see e.g. Mendis et al (1984).] The most detailed of these theories was the one due to Goertz & Morfill (1983). Since the intergrain distance in the B Ring is less than the Debye length, they treat the entire ring as a uniformly charged disc and assuming that the electron density near this disc is provided largely by photoelectrons they estimate that the ring potential is $\sim +5$ V. Under these circumstances only a small fraction of micron-sized grains will have even one electronic charge, as discussed in Section 3. Consequently, the grains are unlikely to be electrostatically levitated against the gravitational attraction of the ring. A much denser source of plasma is required to make the levitation process feasible.

Noting that spoke formation is a sporadic process, Goertz & Morfill (1983) proposed that such dense localized plasma columns are produced by meteor impacts with the ring particles. Initially, a very dense ($n_e \approx 10^{15} \mathrm{cm}^{-3}$) plasma cloud is formed, which expands rapidly while diamagnetically excluding the magnetic field until the diamagnetic factor $\beta \approx 1$. The neutral gas cloud produced by meteor impact will be photoionized. Losses through the loss cone to Saturn's atmosphere and via ring absorption will continue. The plasma column will initially move with the speed of the meteor target until $n_e \leq 10^2$ cm^{-3}, whereafter it will tend to corotate. During such conditions not only can small fine dust (typically $R_g \approx 0.1 \ \mu$m) be electrostatically levitated off the ring plane, but the plasma column itself will move radially. This is due to the fact that the relative motion between the charged dust (which is close to Keplerian) and the surrounding plasma (which is corotating) constitutes a current. Since this current is confined azimuthally to the plasma column, it must close through the ionosphere via a Pedersen current which is connected to the dust-ring current by a pair of field-aligned currents (Figure 10). The Pedersen current sets up an electric field in the ionosphere, which then maps back onto the ring plane and causes a radial $\mathbf{E} \times \mathbf{B}$ drift of the plasma cloud. Knowledge of the ionospheric Pedersen conductivity ($\Sigma_p \approx 0.1$ mho) leads to the calculation of the radial velocity V_R of the plasma column which ranges from ≈ 45 km s^{-1}

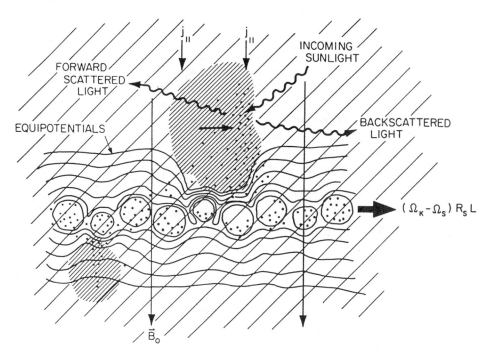

Figure 10 A schematic view of the model for formation of spokes. Underneath a dense plasma column (indicated by heavy shading) the equipotential contours are compressed and a large surface electric field exists. Dust particles are lifted off the rings and drift through the cloud which polarizes. A pair of field-aligned currents closes the current. (Figure from Goertz & Morfill 1983.)

at $L = 1.7$ (the inner edge of the spoke region) to zero at the synchronous orbit ($L = 1.86$). At $L = 1.9$, the outer edge of the B Ring, $V_R \approx 23$ km s^{-1}. As such a plasma column moves from its position of creation, it will leave behind a radial trail of dust much like a car racing across a dusty plain. Goertz & Morfill suggested that the observed spokes are formed in this way.

An alternative mechanism for spoke formation was proposed earlier by Hill & Mendis (1982a). While they too propose that small grains constituting the spokes are electrostatically levitated and blown off the surfaces of larger bodies when they are charged to high electrostatic potentials, they suggest that the sporadic field-aligned beaming of energetic electrons accelerated at electrostatic double layers high up in Saturn's ionosphere may produce the high surface potentials necessary for this process. They suppose that these double layers are highly confined in longitude but extended in latitude, so as to account for the radial extent of the thin spokes. While double layers are known to occur in the Earth's magnetosphere, they do so on field lines connected to the auroral region, where field-aligned currents flow from the magnetotail. The field lines threading the B Ring of Saturn are closed ones, and the question arises as to how such double layers could be generated there, because the flow of field-aligned currents constitutes a necessary condition for their formation on the field lines. The answer perhaps lies in the fact that the relative motion between the ring dust and the plasma constituted a novel type of dust ring current (Hill & Mendis 1982b). Ip & Mendis (1983) showed that this dust ring current is diurnally modulated by Saturn's shadow. Continuity of this current leads to field aligned currents that close through the ionosphere. This could lead to the formation of double layers high up in the ionosphere which will accelerate electrons down into the equatorial plane in the morning ansa, close to the terminator. This is consistent with observations of high spoke activity in this region.

While on the subject of the dust ring current, we note that Houpis & Mendis (1983) have attributed the observed fine ringlet structure of the ring to the resistive tearing of this dust ring current.

A very interesting and far-reaching consequence of the spoke formation due to meteorite bombardment was subsequently deduced by Shan & Goertz (1991), by asking the question as to what happens to the levitated dust. The component normal to the ring plane gives a charged dust particle a gravitoelectrodynamic trajectory which causes it to reintersect the ring plane about 180 degrees away in azimuth from its point of levitation and at different radial distance from the planet. If it collides with a ring particle at that point it will transfer angular momentum to it because its specific angular momentum is different from that of its target. This will cause the target body to move either inward or outward depending on whether its own specific angular momentum is larger or smaller than that of the projectile. Starting with an initial uniform distribution of mass in the region of the B Ring and assuming that the micrometeoroid bombardment rate remains constant at the present-day level, Shan & Goertz (1991) studied the redistribution of mass due to the above process. They found that the calculated profile closely matched the observed profile seen in Figure 11 after a period of $(4-8) \times 10^8$ yrs. If this mechanism is indeed responsible for the observed B Ring profile, and if one accepts the assumption of the constant rate of meteorite bombardment, one is led to the important conclusion that the observed ring system is significantly younger than the Solar System (age $\sim 4.5 \times 10^9$ yrs). Entirely different lines of argument (Borderies et al 1984, Connerney & Waite 1984, Northrop & Connerney 1986) lead to even smaller ages for the ring system ($\sim 4 \times 10^6$–7×10^7 yrs). These age estimates seem to contradict an earlier view that this ring system is a relic left from the formation of the Saturnian system $\sim 4.5 \times 10^9$ yrs ago (Alfven 1954). If indeed the rings were formed after the formation of Saturn the most likely scenario is the disruption of a moon ($\sim 3 \times 10^{22}$ g) by a large comet ($\sim 10^{18}$ g). However, Lissauer et al (1988) have concluded, from the cratering history of Saturn's present moons, that it is very unlikely that the rings were formed by such a disruption of a moon

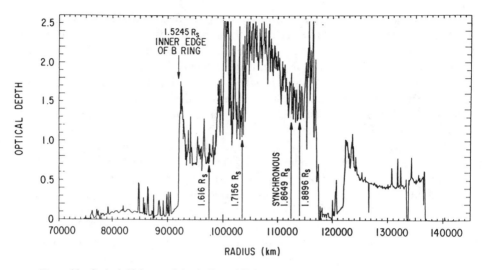

Figure 11 Optical thickness of the A, B, and C rings of Saturn. (A ring 122,000–137,000 km, B ring 92,000–118,000 km, and C ring 75,000–92,000 km.) (From Northrop & Hill 1983b.)

within the last 10^9 yrs. So the question of the age of the ring system remains unresolved.

Several studies based on dust-plasma interactions have also been used to explain the sharp discontinuities observed at $\sim 1.62 R_s$, and at the inner edge of the B-Ring at $\sim 1.524 R_s$ (Northrop & Hill 1982, 1983b; Ip 1983). These have been extensively reviewed elsewhere (e.g. Mendis et al 1984) and are not further discussed here, except to mention that all these involve the upward siphoning of dust impact-produced plasma or very small charged grains, with large numerical values of Q/m_d inside some critical radius within a corotating magnetosphere.

Horanyi et al (1990; see also Horanyi et al 1991) studied the dynamics of submicron-sized dielectric grains ejected from the small Martian satellite Phobos due to micrometeoroid bombardment. They found that these grains were influenced not only by gravity but also by solar radiation pressure and Lorentz forces. They concluded that any grains that remain in the vicinity of the orbit of Phobos are the ones that are governed mainly by gravity and are thus much larger than 1 μm. Very small ($a < 0.2 \mu$m) grains are removed from the Martian-environment very rapidly, whereas larger submicron-sized grains remain in orbit around Mars for several months forming a nonuniform, time-dependent dust halo.

Horanyi & Burns (1991) studied the effect of planetary shadows on the dynamics of charged dust grains in motion around the planets. Since the photoelectron current is periodically shut off in the shadow, the charge in the grain potential and thus the electromagnetic perturbation resonates with the orbital period, and can rapidly change the size and eccentricity of the grain orbits. In an earlier study, Hill & Mendis (1982c) came to a similar conclusion. They attributed the unusually large eccentricities of isolated ringlets, located in relatively transparent regions of Saturn's ring system, where presumably the plasma density is relatively high, to such a shadow resonance. They also argued that very small grains of certain sizes could be rapidly removed from these regions, due to so-called gyro-orbital resonances, i.e. when $\Omega_G/\omega = 2$, 3, etc (see Equations 18 and 20).

When the magnetic field of a planet is not an aligned and centered dipole, a charged dust particle orbiting the planet will experience a time-variable Lorentz force even if the orbit is circular and in the equatorial plane and the grain charge is constant. If the orbital frequency of the dust particle is commensurate with the frequency of this variable Lorentz force it can undergo large out-of-plane and radial excursions as shown by Burns et al (1985) for charged Jovian ring particles. In particular, they show that the outer edge of the lenticular Jovian dust halo (see also Consolmagno 1983) coincides with a low-order resonance of the above type.

Recently Horanyi et al (1992) have studied the motion of charged dust launched from the Saturnian satellite Enceladus ($\sim 3.95 R_s$). In this study they take into account the oblateness of Saturn's gravity field as well as the obliquity of Saturn's orbit around the Sun. While giving rise to novel effects, the calculated dust distribution has many of the observed characteristics of Saturn's E-Ring ($3R_s \lesssim r \lesssim 8R_s$).

One of the most unexpected observations of the recent *Ulysses* mission to Jupiter was the detection of quasi-periodic (~ 28 days) high-speed (20–56 km/sec) streams of submicron-sized grains (1.6×10^{-16} g $< m_d < 1.1 \times 10^{-14}$ g), during its distant Jovian encounter. Horanyi et al (1993) conclude that the initial source of these grains, which appear to come from the Jovian magnetosphere, are the observed volcanic eruptions on the Jovian satellite Io. They find that the Lorentz force on these charged grains of radius $\lesssim 0.1$ μm is sufficient to overcome Io's gravity and inject them into the Jovian magnetosphere. There the grains acquire a potential $\sim +3$ V (outside the Io plasma torus) due to the dominance of secondary emission, and while the smallest grains ($a \lesssim 0.02$ μm) remain tied to the magnetic field and tend to corotate with it, those in the size range $0.02 \lesssim a \lesssim 0.1$ μm get slung out from the magnetosphere (see Figure 12). These particles are in the correct mass range of the detected particles; their calculated velocities (30 km s^{-1} $\lesssim v \lesssim 100$ km s^{-1}) also roughly correspond to the observed velocities. Horanyi et al also provide a plausible explanation to the observed periodicity of these streams on the basis of the dependence of the exit direction of the escaping grains on both the geographic and the geomagnetic coordinates of Io.

The dusty cometary environment is an ideal cosmic laboratory for the study

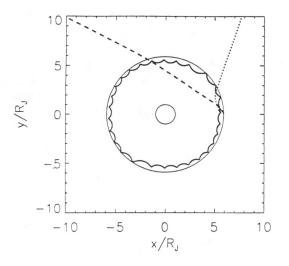

Figure 12 The trajectories of dust particles with size $a = 0.02$ (*solid curve*), 0.03 (*dotted line*) and 0.1 μm (*dashed line*) started from Io. At $t = 0$ the longitude of Io is zero in both the inertial frame and in magnetic coordinates. The two circles represent the surface of Jupiter and Io's orbit. (From Horanyi et al 1993.)

of the physical and dynamical consequences of dust-plasma interactions. Indeed, there exists a growing number of both remote and in-situ observations of cometary phenomena that attest to such interaction. Some of the physical consequences have already been discussed in Section 3. Among the dynamical consequences are (*a*) the asymmetry of sunward dust envelopes of certain comets (Wallis & Hassan 1983), (*b*) the overall spatial distribution of dust at comets P/Giacobini-Zinner and P/Halley (Horanyi & Mendis 1986a, b), and (*c*) the "wavy" appearance of the dust tail of comet Ikeya-Seki (1965f) (Horanyi & Mendis 1987).

The cometary dust particles, which are entrained by the gas sublimating from the nucleus, are immersed in a plasma and radiative environment, and are thus electrically charged. Outside the cometary ionopause (which separates the outflowing cometary ions from the inflowing, contaminated solar wind plasma), the plasma is magnetized. The equation governing the motion of a charged dust particle in the cometocentric inertial frame is given by an equation analogous to (17). The main difference from the planetary case is that the gravitational force of the small cometary nucleus is negligible, whereas the radiation pressure force $\mathbf{F}_r$, is not. $\mathbf{E}$ represents the convectional electric field $-\mathbf{v} \times \mathbf{B}/c$, where $\mathbf{v}$ is the velocity of the contaminated solar wind plasma relative to the cometary nucleus. As in the planetary case, $\mathbf{F}_c$ and $\mathbf{F}_d$ are negligible. In the dense cometary ionopause where $T_e \lesssim 10^3$ K and $n_e \gtrsim 10^3$ cm^{-3}, the grain potential $\phi \sim -0.2$ V. In the region of largely undisturbed solar wind, where photoemission is dominant $\phi \sim 2$–5 V, whereas in the region between the outer cometary bow shock and the ionopause the grain potentials could have both positive and negative values. While conducting grains (e.g. C) may remain negative throughout ($\phi \sim -10$ to -20 V), dielectric grains (e.g. silicates) could switch from being negative closer to the ionopause to being positive closer to the bow shock due to the importance of secondary electron emission in the later region (e.g. Wallis & Hassan 1983). The electrostatic acceleration of the grain is given by $\mathbf{g_e} = -Q(\mathbf{v} \times \mathbf{B})/cm_d$, since $|\mathbf{v_d}| \ll |\mathbf{v}|$. The acceleration due to solar radiation pressure is given by $\mathbf{g_r} = (Q_{pr}/c) \cdot (L_0/4\pi r^2) \cdot (\pi a^2/m_d)$, where Q_{pr} is the scattering efficiency for radiation pressure and L_0 is the mean solar luminosity ($\approx 3.90 \times 10^{33}$ erg s^{-1}). To a good approximation $\mathbf{E}$ is constant outside the ionopause along the sun-comet axis. Assuming that $|\mathbf{B}| = 5\gamma$ and that $\mathbf{B}$ is inclined at $45°$ to V_{SW} at 1 AU, one obtains

$$X = \frac{|\mathbf{g_e}|}{|\mathbf{g_r}|} = \frac{10^{-2}|\phi|}{Q_{pr}a(\mu)}, \qquad (24)$$

where ϕ is in volts.

The values of X for different sized grains are given in Table 2, where it is assumed that $|\phi(V)| = 5$. It is seen that $X \ll 1$ when $a = 1\mu$, but $X \approx 0.28$ for Magnetite and $X \sim 1$ for Olivine when $a \approx 0.1\mu$. When $a = 0.03\mu$, $X \gg 1$ for Olivine. As Wallis & Hassan (1983) pointed out this is the reason why the sunward dust envelopes (which are the envelopes of the trajectories of the grains emitted from the nucleus and which are moving under the acceleration $\mathbf{g_e} + \mathbf{g_r}$) are skewed away from the sun-comet axis. If $\mathbf{g_g} = 0$, clearly the apex of these envelopes will lie on the sun-comet axis.

Horanyi & Mendis (1986a,b) subsequently performed numerical simulations to calculate the distribution of charged dust grains in the environments of comments P/Giacobini-Zinner and P/Halley, using plasma and magnetic models of calculate $\phi(t)$. In these calculations, the orbital motion of the comet was also taken into account. In the case of comet P/Giacobini-Zinner, a scatter

Table 2 Ratio of X for different sized grains in the cometary environment[a]

$a(\mu m)$	X	
	Olivine	Magnetite
1.0	0.05	0.04
0.5	0.85	0.06
0.1	1.00	0.28
0.03	16.5	3.35

[a]From Mendis & Horanyi (1991).

plot of the distribution of the grains of various sizes in a plane normal to the sun-comet axis, at a distance of 10^4 km behind the nucleus, is shown in Figure 13. In this case, the interplanetary magnetic field is assumed to be in the orbital plane of the comet, so that the convectional electric field is normal to this plane.

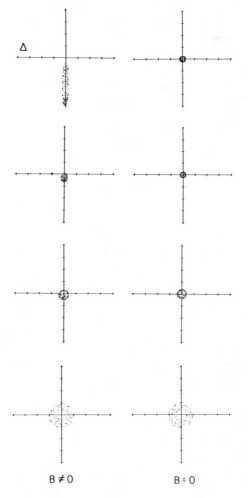

Figure 13 The distribution of dust grains of various sizes (top to bottom, 0.03 μm, 0.1 μm, 0.3 μm, and 1.0 μm) in a plane normal to the Sun-comet axis at a distance 10^4 km down the tail of comet P/Giacobini-Zinner. The left column shows the scatter plots of the dust distributions when the interplanetary magnetic field is included. The right column shows the corresponding scatter plots when the interplanetary magnetic field is excluded. (From Horanyi & Mendis 1986b.)

The grain sizes increase from 0.03 μ at the top of the Figure to 1.0 μm at the bottom. The first column shows the distributions when the electromagnetic effects are taken into account ($\mathbf{B} \neq \mathbf{0}$); the second column shows the corresponding distributions when the electromagnetic effects are neglected (e.g. $\mathbf{B} = \mathbf{0}$). The nucleus in each plot is located at the origin of coordinates. The elongation of the distribution normal to the orbital plane is obvious, particularly for the smallest grains (0.03 μm). In that case, it is also seen that the grains are concentrated well below the orbital plane. The larger grains have a more symmetrical distribution about the axis. In the absence of electromagnetic effects (e.g. $\mathbf{B} = \mathbf{0}$), it is seen, once more, that the larger ("older") grains concentrate away from the axis, while the smaller ("younger") grains concentrate closer to the axis. These distributions are, of course, axially symmetric in this case.

The NASA *ICE* spacecraft intercepted the tail of comet Giacobini-Zinner on September 11, 1985, at a distance of about 8×10^3 km from the nucleus, moving in a generally south to north direction in the comet's reference frame. Although the spacecraft did not carry a dust detector, the plasma wave instrument detected impulsive signals that were attributed to dust impacts on the spacecraft (Gurnett et al 1986). Although an asymmetry in the impact rate between the inbound and outbound legs consistent with the prediction for charged grains was indeed observed (see Figure 14), the asymmetry was rather small and may also be consistent with nonisotropic emission of grains from the nucleus, as suggested by Gurnett et al (1986). However, this observation coupled with the observation that the smaller particles were encountered farther away than the larger ones and also had a greater asymmetry between the two legs (Gurnett et al 1986) strongly supports the charged dust model. If the grains were not charged, the larger ones would have been encountered before the smaller ones.

Besides naturally occurring dust, there is now a significant component of anthropogenic dust in the terrestrial magnetosphere. Small (0.1–10 μm)Al_2O_3 spherules are dumped into the Earth's lower magnetosphere during solid rocket motor burns used for transfer of satellites from low earth to geosynchronous orbit. The flux from one such burn could exceed the natural micrometeoroid flux in that size range (Mueller & Kessler 1985). Although solar radiation pressure plays the dominant role in the orbital evolution of much of this dust, electromagnetic effects, as described for the more distant magnetospheres of Jupiter and Saturn, play a crucial role at the lower end of the dust mass spectrum ($a \lesssim 0.1$ μm). In that case, these electromagnetic forces conspire with solar radiation pressure to eliminate these grains from the magnetosphere in a comparatively short time, while significantly changing the residence time of large ($0.1\mu \lesssim a \lesssim 1$ μm) grains (Horanyi & Mendis 1986c, Horanyi et al 1988). The authors have also discussed the hazards that these grains may pose to artificial satellites in or near geosynchronous orbit.

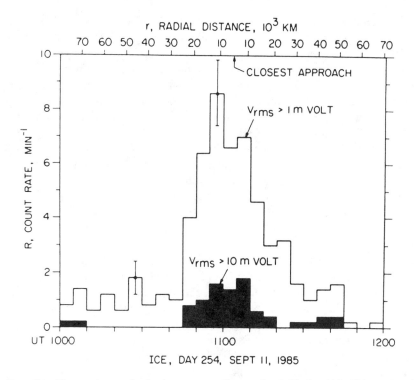

Figure 14   The counting rate for dust impacts exceeding two thresholds, 1 and 10 mVolt, measured by the *ICE* spacecraft. Most of the impacts occur within 30,000 km of closest approach to comet P/Giacobini-Zinner. Note that the amplitude of the voltage pulse (V_{rms}) is proportional to the grain mass. (From Gurnett et al 1986.)

5.   WAVES, INSTABILITIES, AND WAVE SCATTERING

This section discusses collective effects in a dusty plasma as manifested by the behavior of waves and instabilities, and wave scattering on dust grains in a plasma.   An early paper on the collective oscillations of a "microparticle plasma" used a multifluid model to study waves arising from a distribution of fluid velocities in a cold plasma comprised of a number of negatively and positively charged microparticles (James & Vermeulen 1968). As reviewed by Goertz (1989), several papers appeared on wave behavior in dusty astrophysical plasma in the 1980s, including studies of electrostatic dust beam-driven waves in Saturn's rings (Bliokh & Yarashenko 1985) and the effect of charged dust on Alfven wave propagation in interstellar clouds (Pilipp et al 1987). During the past few years there has been a substantial increase in theoretical studies of the collective behavior of dusty plasmas that address how the presence of charged dust grains in a plasma can affect wave dispersion properties, instabilities, and wave scattering. We review some recent theoretical work in this rapidly developing, relatively new field of research.

5.1   Waves

When $a \ll d \ll \lambda_D$, the charged dust grains may be considered as point particles, similar to multiply-charged positive or negative ions (Fortov & Iakubov 1990). When the dust particles can be treated as just another component of the plasma, there are similarities with plasmas comprising electrons, ions, and additional negative (or positive) ions (D'Angelo 1990); however, there are differences from negative ion or two-ion component plasmas in several respects. First, typical dust grains have charge-to-mass ratios that are generally orders of magnitude smaller than those of ions. For example, a dust grain of mass density 0.5 gm/cm^2 and radius $a \sim 1$ micron has a mass of about 10^{12} proton masses; the charge $q_d = -Z_d e$ on such a grain in a thermal 10 eV oxygen plasma, when only plasma collection is considered, is of the order of $\sim 2.5 \times 10^4$ elementary electron charges. Since the plasma frequency of species α is $\omega_{p\alpha} = (4\pi n_\alpha q_\alpha^2/m_a)^{1/2}$ and the gyrofrequency of species α is $\Omega_\alpha = |q_\alpha B/m_\alpha c|$, typical frequencies associated with the dynamics of dust grains are very low compared with typical ion wave frequencies in standard electron-ion plasmas. Second, the grain charge depends on both the properties of the dust particles and the ambient plasma and radiative properties, as discussed in Section 2; thus the charge to mass ratio can differ in different environments even for a grain of fixed size and electrical properties. Third, it has been inferred or observed in situ that dust grains generally in space plasmas are polydisperse, that is, they have a size distribution, with the number density of grains generally given by a power law $n_d da \propto a^{-p} da$, with p ranging from ~ 0.9 to 4.5 in various space and astrophysical environments (e.g. Havnes 1990a and references therein). Since the dust mass $m_d \propto a^3$ and the dust charge q_d is generally $\propto a$ (see Section 2), the frequencies associated with the dust grains, such as the dust plasma frequency and the dust gyrofrequency, may be continuous variables (Goertz 1989).

Some basic properties of various waves in dusty plasmas have been obtained from multifluid analyses (e.g. Shukla 1992, D'Angelo 1990), which treat the dust grains as a component of a three-component plasma comprising electrons, ions, and (negatively or positively) charged dust of uniform mass and charge (and therefore of uniform size). The fluid equations for each species α are the equations of continuity and momentum,

$$\frac{\partial n_\alpha}{\partial t} = -\nabla \cdot (n_\alpha \mathbf{v}_\alpha) \tag{25}$$

$$n_\alpha m_\alpha \left(\frac{\partial}{\partial t} + \mathbf{v}_\alpha \cdot \nabla \right) \mathbf{v}_\alpha = -\nabla P_\alpha + q_\alpha n_\alpha \left(\mathbf{E} + \frac{\mathbf{v}_\alpha}{c} \times \mathbf{B} \right) \tag{26}$$

coupled with Maxwell's equations. In addition there is an equation expressing overall charge neutrality in the plasma,

$$Z_i n_i + \varepsilon_d Z_d n_d = n_e \qquad (27)$$

(where $\varepsilon_d = 1, -1$ for positively, negatively, charged grains respectively). Here q_α, Z_α are the charge, charge state of each species α, $\mathbf{E}$ and $\mathbf{B}$ are the electric and magnetic fields, and n_α, $\mathbf{v}_\alpha$, m_α, and P_α are the density, fluid velocity, mass, and (isotropic) pressure of each species α (α = e, i, d for electrons, ions, and dust grains, respectively) (e.g. Shukla 1992). To obtain dispersion relations for linear waves, the fluid equations are solved for small perturbations about an equilibrium steady state (denoted by subscript o), assuming that first-order quantities [denoted by superscript (1)] vary as $\exp[i(\mathbf{k} \cdot \mathbf{x} - \omega t)]$ (see e.g. Krall & Trivelpiece 1973). For electrostatic waves, $\mathbf{E}^{(1)} = -\nabla \phi^{(1)}$, where $\phi^{(1)}$ is the perturbed scalar potential, and $\mathbf{B}^{(1)} = \mathbf{0}$.

The presence of charged dust has been shown to both modify the usual linear modes known in an electron-ion plasma and lead to the presence of new modes associated with the dust grains in low frequency and low phase velocity regimes (e.g. Shukla 1992). For example, ion modes can be modified due to charge imbalance between the electrons and ions in the equilibrium state, that is, due to $n_{eo} \neq Z_i n_{io}$: Referring to the fluid equations, when the dust is assumed to be immobile (with the dust mass $m_d \rightarrow \infty$), the perturbed dust velocity $\mathbf{v}_d^{(1)} = 0$ from Equation (26) and dust contributes to the dispersion relation through the charge neutrality condition (27). New low-frequency modes associated with the response of the dust grains can arise when the dust grain dynamics are included via the momentum and continuity Equations (25 and 26) for the dust species.

In an unmagnetized dusty plasma, dust can modify the linear dispersion relation for ion-acoustic waves (Shukla & Silin 1992). In the phase velocity regime where electron inertia is negligible $v_{td}, v_{ti} \ll v_{ph} \ll v_{te}$ (where $v_{t\alpha}$ is the thermal speed of species α and $v_{ph} = \omega/k$), and in the limit of very large dust mass (e.g. considering the dust grains to be immobile) the dispersion relation for dust ion-acoustic waves in a plasma with electrons, singly-charged ions, and charged dust is (Shukla & Silin 1992)

$$\omega^2 = \delta \frac{k^2 c_s^2}{1 + k^2 \lambda_{De}^2}, \qquad (28)$$

where $\delta = n_{io}/n_{eo} > 1$ for negatively charged dust, and $c_s = (k_B T_e/m_i)^{1/2}$ is the usual ion sound speed. Since the phase velocity of this mode for long wavelengths (i.e. $k\lambda_{De} \ll 1$) is $v_{ph} = (\delta k_B T_e/m_i)^{1/2}$, the dust ion-acoustic mode can exist as a normal mode of the system even for $T_e = T_i$ as long as dust grains carry most of the negative charge in the plasma, i.e. $\delta \gg 1$ [with $\delta < (m_i/m_e)$ in order to have $v_{ph} < v_{te}$], because in that case ion Landau damping is small. This is in contrast to the electron-ion plasma, where $T_e \gg T_i$ is required for propagation of ion-sound waves (Krall & Trivelpiece 1973).

This mode may be relevant in astrophysical situations where the plasma is isothermal but where dust can carry much of the negative charge, such as in the F-Ring of Saturn (Shukla & Silin 1992). This mode may also be relevant to the possible existence of an electrostatic shock just inside the ionopause of comet Halley. While observations of the ion density profile just inside the cometry ionopause are consistent with the existence of a shock (Damas & Mendis 1992), the exact nature of the shock is unclear. Körösmezey et al (1987) speculated on the existence of such an electrostatic shock due to the rapid increase in the calculated electron temperature just inside the ionopause. However, the values of T_e/T_i, calculated using different assumptions, are in the range ~2–5, which may not be sufficient to prevent Landau damping of the ion-acoustic wave which mediates the electrostatic shock. But the presence of a large quantity of dust, particularly in the form of VSGs, within the dusty cometary ionosphere may prevent this damping and allow such a shock to form.

In the lower phase velocity regime $v_{td} \ll v_{ph} \ll v_{ti}, v_{te}$ the presence of dust can lead to a new dust-acoustic wave associated with the grain dynamics (Rao et al 1990). Assuming that electron and ion inertia are negligible in this phase velocity regime (so that the left-hand side of Equation (26) = 0 for electrons and ions), and that the electrons and singly charged ions are in Boltzmann equilibrium, and also assuming that the dust fluid is cold and is described by Equations (25) and (26), the dispersion relation for the dust-acoustic wave was obtained as

$$\omega^2 = Z_d(\delta - 1)\frac{m_i}{m_d}\frac{k^2 c_s^2}{(1 + k^2\lambda_{De}^2 + \delta T_e/T_i)} \tag{29}$$

for negatively charged grains with $\delta > 1$. In this case, the electrons and ions provide the pressure while the dust mass provides the inertia (Rao et al 1990). Excitation of the dust-acoustic mode by streaming electrons and ions may have relevance to planetary rings, and is discussed in Section 5.2.

In a magnetized, homogeneous dusty plasma the presence of charged dust can similarly affect electrostatic waves such as acoustic waves, electrostatic ion cyclotron (EIC) waves, and lower hybrid waves. A multi-fluid analysis of a three-component dusty magnetized plasma (D'Angelo 1990) showed that there are two ion-acoustic waves and two EIC waves associated with the positive ions and the negative (or positive) dust grains. For negatively charged dust, frequencies of both acoustic waves increase with dust density [see also expressions (28) and (29)], as does the frequency of the positive ion EIC wave. For positively charged grains, the frequency of the ion acoustic mode decreases with increasing dust density, while the frequency of the ion EIC mode approaches the ion gyrofrequency as n_d increases. These results are shown in Figure 15: They are analogous to those obtained previously for either two-ion component (e.g. Suszcynsky et al 1989) or negative ion plasmas (e.g. D'Angelo et al 1966, Song et al 1989) apart from the different charge-to-mass ratio of the dust

(D'Angelo 1990). The dispersion relation of the lower hybrid mode in the frequency regime $\Omega_d, \Omega_i \ll \omega \ll \Omega_e$, is also modified by the presence of charged dust: In the limit $m_d \to \infty$ in a dense plasma with $\omega_{pe}^2/\Omega_e^2 \gg 1$ (Shukla 1992), the lower hybrid wave frequency is

$$\omega^2 \approx \delta Z_i \Omega_e \Omega_i,$$

which increases as the negative dust density increases. We note that a cross-field current-driven lower hybrid instability has been invoked to provide anomalous resistivity in an electron-ion plasma in a protostellar cloud (Norman & Heyvaerts 1985). Since dust grains may carry most of the charge in the weakly ionized plasma of protostellar clouds during certain stages in their evolution (e.g. Nishi et al 1991, Umebayashi & Nakano 1990), the effect of dust on the lower hybrid instability may need to be examined.

The effect of negatively charged dust grains on electrostatic drift waves in an

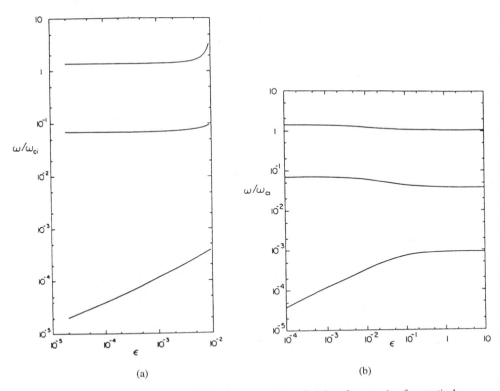

(a)   (b)

Figure 15   (a) Wave frequency normalized to $\omega_{ci}(=\Omega_i)$ as a function of $\varepsilon = n_{do}/n_{io}$ for negatively charged grains. Here $m_d/m_i = 10^8$, $Z_d = 100$, $T_d/T_i = 0.1$, $T_i/T_e = 0.2$, $k_x\rho_i = 0.4$, and $k_z\rho_i = 0.04$, where $\mathbf{B} = (0, 0, B)$, $\mathbf{k} = (k_x, 0, k_z)$. (b) Same as (a), but for positively charged grains. (From D'Angelo 1990.)

inhomogeneous magnetized low $\beta (= 8\pi n_{io} k_B T / B^2)$ plasma was considered by Shukla et al (1991) using a multifluid analysis. The dynamics of negatively charged dust grains was shown to both modify the dispersion properties of the usual electrostatic drift waves, and lead to the appearance of new low frequency dust drift waves in the regime $\omega \ll \Omega_d$. In the latter case both the ions and the electrons are assumed to be in Boltzmann equilibrium, and a new drift wave appears which couples the dust drift and dust-acoustic waves (Shukla et al 1991).

Dust can also modify the dispersion properties of low frequency electromagnetic waves, including Alfven waves (Shukla 1992, Rao 1993) and magnetosonic waves (Rao 1993). For example, the Alfven wave spectrum in a cold plasma in the very low frequency regime $\omega \ll \Omega_d$ becomes (Shukla 1992)

$$\omega^2 = \frac{k^2 v_A^2}{[1 + (v_A/c)^2 + n_{do}m_d/n_{io}m_i]},$$

where the usual Alfven speed is $v_A = (B^2/4\pi n_{io}m_i)^{1/2}$. We note that for many astrophysical and space dusty plasmas, the last term in the denominator $n_{do}m_d/n_{io}m_i$ may be $\gg 1$ owing to the large ratio of the dust to ion mass [for example the ionized gas in cool, dense interstellar clouds e.g. Pilipp et al (1987)], so that the dust dynamics decreases the phase speed in this regime. Dust also leads to new branches of the dispersion curve in a cold plasma in analogy with negative ion plasmas (e.g. Teichmann 1966, Shawhan 1966). For example, for parallel (to **B**) propagation in a cold magnetized plasma with negatively charged grains, the right circularly polarized mode has a resonance at Ω_d and a cutoff at a higher frequency which depends on the fraction of charge density and mass density carried by the grains (Mendis & Rosenberg 1992, Shukla 1992). In addition, under certain conditions there can be whistler mode propagation for the left-handed circularly polarized mode in the regime $(\delta - 1)\Omega_i/\delta \gg \omega \gg \Omega_d$ (Mendis & Rosenberg 1992).

Nonlinear acoustic waves in dusty plasmas have also been investigated using multifluid analyses. Bharuthram & Shukla (1992) investigated the formation of large amplitude ion-acoustic solitons in a dusty unmagnetized plasma with negatively charged grains, cold ions, and Boltzmann distributed electrons. The presence of dust grains was found to lead to the appearance of rarefactive (negative potential) solitons, which do not exist in the absence of dust, as well as compressional (positive potential) solitons. Nonlinear ion-acoustic waves may be relevant to the formation of an electrostatic shock inside the ionopause at comet Halley, as discussed above in relation to the dust ion-acoustic wave. Rao et al (1990) showed that dust-acoustic waves could also propagate nonlinearly as solitons of either negative or positive potential in a three-component plasma with electrons, ion, and negatively charged, cold dust grains. Verheest (1992) extended the latter analysis to a multispecies dusty plasma, allowing for both hot and cold electrons and for a number of cold dust grain species, and found also that both rarefactive and compressive solitons could propagate.

Other approaches have been used to study how charged dust grains affect waves in dusty plasmas in the regime $d \gtrsim \lambda_D$ (e.g. de Angelis et al 1988, 1989). Kinetic analyses have shown that dust can significantly affect wave propagation, even when the charge density carried by the grains is small compared with the electron or ion charge density, due to inhomogeneities arising from the dust grains and their screening clouds which modify the plasma equilibrium. De Angelis et al (1989) considered the parameter regime $|q_d n_{do}/e n_{eo}| \ll 1$ and showed that the presence of a random spatial distribution of charged, massive, fixed dust grains can modify the stationary equilibrium electron and ion distributions due to the dust-generated background electrostatic potential. Using a Vlasov analysis, it was shown that electron plasma waves could be damped in the "cold plasma" regime ($\omega \gg k v_{te}$), where Landau damping is negligible. This was interpreted as being due to the interaction of electrons with a low phase velocity wave ($\omega, \mathbf{k} - \mathbf{q}$), which resulted from the beating of an electron plasma wave characterized by ($\omega, \mathbf{k}$) with a zero-frequency wave ($0, \mathbf{q}$) arising from plasma density inhomogeneities around the charged grains. The wave damping rate was found to be of the order of $(q_d n_{do}/e n_{eo})^2 \omega_{pe}$ (de Angelis et al 1989). Following the method of de Angelis et al (1989), Salimullah & Sen (1992) investigated the low frequency response of a dusty plasma, finding both changes to the usual ion-acoustic wave and the appearance of a new mode with phase velocity $\ll v_{ti}$.

5.2   *Instabilities*

The presence of charged dust in a plasma can both modify the behavior of usual plasma instabilities and lead to the appearance of new instabilities. Since dust grains are subject to non-electromagnetic forces such as gravity, friction, or radiation pressure, there can be new sources of free energy to drive instabilities, including relative drifts between the charged dust and the lighter plasma particles (electrons and ions) in cosmic dusty plasma environments (e.g. Tsytovich et al 1990). In planetary rings for example, the dust grains, whose orbits are to first order Keplerian, move azimuthally around the planet with a speed between the Kepler and corotation speeds (see Section 4), while the plasma ions and electrons tend to corotate with the planet; this leads to a relative azimuthal drift between dust and plasma. Another example is in a dusty cometary environment, where there can be relative streaming between the dust and the solar wind plasma flow (Havnes 1988).

Several hydrodynamic and kinetic instabilities in dusty plasmas have been investigated. Havnes (1988; see also Hartquist et al 1992) used a kinetic analysis to study streaming instabilities in a cometary environment where there is a relative drift between the cometary dust and the solar wind plasma flow, with flow speed u in the range $v_{te} > u > v_{ti}$. It was found that an instability driven by streaming dust in a three-component unmagnetized dusty plasma required small dust velocity dispersion and that the presence of a sufficiently high neutral density could quench the instability due to ion-neutral or electron-neutral collisions. The conditions for growth were found to be more probable for comets at large distances from the Sun. Two-stream instabilities driven by either ion or dust beams in an unmagnetized dusty plasma have also been studied by Bharuthram et al (1992), who found that the dust affects both growth rates and ranges of drift speeds for which instability occurs.

The Kelvin-Helmholtz instability can occur in a plasma when there is a gradient in the fluid flow speed between adjacent fluid layers, as happens for example in the cometary environment (e.g. Ershkovich & Mendis 1986). D'Angelo & Song (1990) considered the effect of either negatively or positively charged dust on the Kelvin-Helmholtz instability in a magnetized, low β plasma with shear in the ion field-aligned flow, using a multifluid analysis. The dust was assumed to be immobile and of uniform mass and charge. The charged dust alters the critical shear for the onset of instability from that in an electron-ion plasma, where a relative speed between adjacent flows of the order of the ion sound speed is required for instability (D'Angelo 1965). The critical shear increases with dust charge density in a plasma with negatively charged grains and decreases with dust charge density in a plasma with positively charged grains (D'Angelo & Song 1990).

Ion-acoustic and dust-acoustic instabilities were investigated using a standard Vlasov analysis for a dusty unmagnetized plasma with electrons, ions, and dust of uniform mass and charge (Rosenberg 1993). When the electrons have a weak drift u in a range $v_{ti} < u < v_{te}$, dust ion-acoustic waves can be excited if $\delta T_e / T_i \gg 1$ (in which case ion Landau damping is small), when u is greater than the phase velocity of the mode $\delta^{1/2} c_s$. This instability may be particularly relevant to cosmic plasmas, where generally it is assumed that $T_e \sim T_i$, in environments where dust carries much of the negative charge. For example, it has been shown that dust grains in weakly ionized plasmas in protostellar clouds may carry much of the negative charge by electron sticking during certain stages of cloud contraction (Umebayashi & Nakano 1990, Nishi et al 1991). Ion-acoustic instability in such a dusty plasma might lead to anomalous resistivity and anomalous magnetic field diffusion as contrasted with the case of isothermal electron-ion plasmas in protostellar clouds for which the instability may not be operative (Norman & Heyvaerts 1985).

The dust-acoustic mode may be excited under certain conditions when the plasma electrons and ions drift together with speed u relative to the charged dust component. For example, in planetary rings beyond the corotation radius, the dust grain azimuthal speed is controlled primarily by gravity, while the lighter plasma particles corotate with the planet [at Saturn's G Ring, which

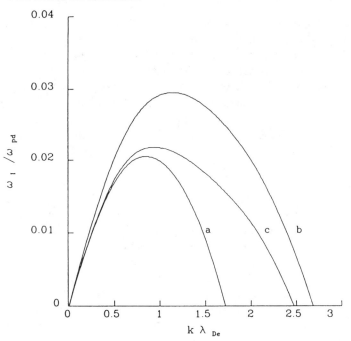

Figure 16 Growth rate of dust-acoustic instability ω_{I} normalized to dust plasma frequency ω_{pd} as a function of $k\lambda_{\mathrm{De}}$ with $T_{\mathrm{e}} = T_{\mathrm{i}} = T_{\mathrm{d}}$, $m_{\mathrm{i}}/m_{\mathrm{d}} = 1.6 \times 10^{-11}$, and $u/c_{\mathrm{s}} = 0.2$. (*a*) $Z_{\mathrm{d}} = 4 \times 10^4$, $n_{\mathrm{do}}/n_{\mathrm{io}} = 3 \times 10^{-8}$, $\varepsilon_{\mathrm{d}} = -1$, (*b*) $Z_{\mathrm{d}} = 10^2$, $n_{\mathrm{do}}/n_{\mathrm{io}} = 5 \times 10^{-3}$, $\varepsilon_{\mathrm{d}} = -1$, (*c*) $Z_{\mathrm{d}} = 10^3$, $n_{\mathrm{do}}/n_{\mathrm{io}} = 10^{-4}$, $\varepsilon_{\mathrm{d}} = +1$. (From Mendis et al 1993.)

is located at about 2.8 radii from the planet, the Kepler speed is $\sim$15 km/s, while the corotation speed is $\sim$29 km/s (e.g. Mendis & Axford 1974); the ion thermal speed for oxygen in a 50 eV plasma is $v_{\mathrm{i}} \sim 17$ km/s]. When the drift is weak ($v_{\mathrm{td}} < u < v_{\mathrm{ti}}$), dust-acoustic waves can be excited in the long wavelength limit ($k^2\lambda_{\mathrm{De}}^2 < 1$) in an isothermal plasma if $R = Z_{\mathrm{d}}|1-\delta|/(1+\delta) \gg 1$ (in which case dust Landau damping is small), when u is greater than the phase velocity of the mode $(Rm_{\mathrm{i}}/m_{\mathrm{d}})^{1/2}c_{\mathrm{s}}$. When $\delta \gtrsim 1$ (corresponding for example to dust parameters in the outer Saturnian rings) and the plasma is isothermal, the instability is driven primarily by the ions with maximum growth rate $\gamma \sim \sqrt{\pi/8}(u/c_{\mathrm{s}})\omega_{\mathrm{pd}}$ (Rosenberg 1993). Growth rates for different values of Z_{d} and $n_{\mathrm{do}}/n_{\mathrm{io}}$ are illustrated in Figure 16; the behavior with k is analogous to that of the usual ion-acoustic instability apart from the shift in k for maximum growth arising from the ion contribution to the effective Debye length. Typical dust plasma frequencies in the outer Saturnian rings, where $a \sim 1~\mu$m, $n_{\mathrm{d}} \sim 3 \times 10^{-7}$ cm^{-3} (e.g. Goertz 1989), and the "isolated" grain charge value is $q_{\mathrm{d}} \sim 5 \times 10^4$ elementary electron charges (assuming only plasma collection currents) yields an estimated dust plasma frequency, $\omega_{\mathrm{pd}} \sim 0.04$ s^{-1}, which is large compared with an ion-dust collision frequency $\tau_{\text{id}}^{-1} \sim \pi a^2 n_{\text{d}} v_{\text{ti}} \approx 10^{-7} \text{ s}^{-1}$, using values from Table 1. Melandsø et al (1993) have considered further the excitation of dust-acoustic waves including a model for wave damping by grain charge perturbations, and also found a regime of instability for parameters typical of Saturn's outer rings. Calculations that include the effect of the planetary magnetic field on these instabilities have yet to be done.

5.3 *Wave Scattering*

It has been shown that there can be enhanced wave scattering from dust grains in a plasma under certain conditions (e.g. Tsytovich et al 1989). Thomson scattering by dust grains, with a scattered power proportional to $q_{\text{d}}^4/m_{\text{d}}^2$, is negligible owing to the very large dust mass (Tsytovich 1992). However, an "isolated" charged dust grain in a plasma is surrounded by a cloud of plasma particles which screens out the grain charge over a distance of the order of a Debye length: For a negatively (positively) charged grain the screening cloud is comprised of both attracted ions (electrons) and repelled electrons (ions). A wave propagating through the plasma can be scattered coherently by the electrons in the screening cloud surrounding the charged grain when the wavelength is larger than the screening length. [The power scattered from the ions in the cloud would be smaller by a factor $\sim(m_{\text{e}}/m_{\text{i}})^2$ (e.g. Tsytovich 1992, La Hoz 1992)]. This can lead to an increase in the scattering cross section as compared with scattering by free electrons or by fluctuations in the plasma (Tsytovich et al 1989; Bingham et al 1991, 1992; de Angelis et al 1992; La Hoz 1992; Hagfors 1992). When the scattering is coherent, all the electrons in the screening cloud move in phase, yielding a scattered power proportional to the square of the electron screening charge (Tsytovich 1992).

However, when n_{d} is large enough so that the grain charge (considering the contribution from plasma collection currents) is reduced due to the electron depletion effect, as discussed in Section 2.2, the enhancement decreases since there are fewer electrons in the screening cloud (de Angelis et al 1992). Calculations of electromagnetic wave scattering from charged dust particles in a plasma using "dressed test particle" approaches (La Hoz 1992, Hagfors 1992) found that for enhanced scattering d should be larger than a Debye length. In this case there is negligible electron depletion in the screening cloud surrounding the charged grain.

In the regime $d \gg \lambda_{\text{D}} \gg a$, each grain is screened in a distance of the order of the Debye length and the dust can be considered as a collection of individual scatterers (Bingham et al 1991). In this regime, for long incident wavelengths $\lambda_0 > d$, the scattering cross section for electromagnetic waves by dust grains is given by $\sigma = n_{\text{d}} \sigma_{\text{s}}$, where σ_{s} is the cross section for scattering from a single grain, which for $\lambda_0 \geq \lambda_{\text{D}}$ is given by $\sigma_{\text{s}} \simeq \sigma_0 Z_{\text{d}}^2$, where σ_0 is the Thomson cross section (Bingham et al 1991, 1992). Comparing this σ with the scattering cross section from plasma fluctuations when no dust is present, $\sigma_p \simeq n_e\sigma_0/2$, implies that $n_d Z_d^2/n_e \gg 1$ is required for the scattering by "isolated" dust grains to dominate the scattering by plasma fluctuations (de Angelis et al 1992).

Havnes et al (1990b; see also Havnes et al 1992) suggested that the scattering from positively charged dust in the Earth's mesopause might explain the strong radar backscatter from the high latitude, summer mesopause observed at certain frequencies referred to as PSME (Polar Mesopheric Summer Echoes). The backscatter region occurs at an ionospheric height of about 80–90 km, which is also the altitude range of noctilucent clouds (e.g. Gadsden & Schröder 1989), which are thought to be comprised of solid particulates, probably mainly of ice (e.g. La Hoz 1992, Havnes et al 1990b). Havnes et al (1990b) suggest that if the photoelectric work function of mesospheric dust particles are lowered due to contaminants or impurities, the grains could attain surface potentials of a few volts positive even in the mesosphere where most of the solar photons have energy less than the photoelectric work function of amorphous ice. However, further work indicated that radar backscatter from dust particles may not fully explain PSME (La Hoz 1992, Hagfors 1992, Havnes et al 1992).

6. CONCLUSIONS

The study of cosmic dusty plasma is a relatively new area of research both with respect to theoretical development as well as specific applications. The study of the physics and dynamics of charged dust grains in certain solar system environments (discussed in Sections 3 and 4) arose in response to the observation (both in-situ and remote) of intriguing phenomena associated with fine dust. Here theory and observation continue to remain closely connected. While laboratory experimentation is lagging behind, recent experiments (e.g. Xu et al 1993) are beginning to clarify such basic questions as the relation of grain charge to the ratio d/λ_D, which had led to some early theoretical disagreements (e.g. see Goertz & Ip 1984, Whipple et al 1985).

The study of waves, instabilities, and scattering processes in dusty cosmic plasmas (discussed in Section 5) arose, however, not in response to specific observations, but from the realization that they are necessary consequences of dust-plasma interactions. As such this is an area where theory is leading observation. This is an area where theory is also leading experiment; at present we are unaware of any dedicated laboratory experiments on wave behavior in dusty plasmas [although observations of low frequency noise in a dust device were reported by Sheehan et al (1990)].

Waves associated with the presence of dust would not be directly observable in cosmic plasmas except perhaps via satellite observations in the solar system. Nonetheless they could have important ramifications that can be inferred. Wave turbulence generated by instabilities (discussed in Section 5) can lead to anomalous resistivity and viscosity and associated anomalous transport processes. This would have implications for the macroscopic dynamics of certain cosmic environments. For example, if dust ion-acoustic waves can be excited in protostellar plasma clouds (see Section 5.2), then anomalous resistivity may play a role in the dissipation of magnetic fields in these clouds. This process of magnetic field decoupling from contracting cloud material is an outstanding issue in star formation scenerios (e.g. see Shu et al 1987). As another example, if a dust-acoustic instability is important in certain planetary rings (see Section 5.2), then the associated anomalous viscosity could have implication for the dynamics of dust grains in ring environments where the classical collisional drag between ions and grains appears to be unimportant (e.g. Mendis et al 1982).

We expect that as research in the area of dusty plasmas continue, it would stimulate further applications to phenomena in astrophysics and space science. The recent discovery of so-called VSGs in certain cosmic plasmas is an added impetus to these studies. Since these dust grains have large specific charges they would be most susceptible to electrodynamic forces. Furthermore, while they add negligibly to the dust/gas mass ratio, they significantly enhance the cummulative surface area of the dust in the plasma, which in turn would enhance the occurence of collective dusty-plasma processes.

ACKNOWLEDGMENTS

Support from the following grants are gratefully acknowledged: NASA NAGW-2252 and NSF AST-9213836 (M. R.); NASA NAGW-1502, NSF AST-9200981 and LANL/IGPP 93-132 (D. A. M.).

Literature Cited

Alfven H. 1954. *On the Origin of the Solar System*, p. 79. Oxford: Clavenden
Alfven H. 1980. *Cosmic Plasma.* Dordrecht:Reidel
Allen CW. 1983. *Astrophysical Quantities*, p. 160. London: Athlone. 3rd ed.
Armstrong TP, Paonessa MT, Bell, EV II, Krimigis SM. 1983. *J. Geophys. Res.* 88:8893
Bharuthram R, Saleem H, Shukla PK. 1992. *Phys. Scr.* 45:512
Bharuthram R, Shukla PK. 1992. *Planet. Space Sci.* 40:973
Bingham R, de Angelis U, Tsytovich VN, Havnes O. 1991. *Phys. Fluids* B3:811
Bingham R, de Angelis U, Tsytovich VN, Havnes O. 1992. *Phys. Fluids* B4:282

Bliokh PV, Yaroshenko VV. 1985. *Sov. Astron.* 29:330
Borderies N, Goldreich R, Tremaine S. 1984. In *Planetary Rings*, ed. R Greenberg, A Brahie, p. 713. Tucson: Univ. Ariz. Press
Burns JA, Schaffer LE, Greenberg RJ, Showalter MR. 1985. *Nature* 316:115
Chow VW, Mendis DA, Rosenberg M. 1993. *J. Geophys. Res.* 98:19065
Christon SP, Mitchell DG, Williams DJ, Frank L, Huang CY, Eastman TE. 1988. *J. Geophys. Res.* 97(E9):14,773
Connerney JEP, Waite JH. 1984. *Nature* 312:136
Consolmagno GJ. 1983. *J. Geophys. Res.* 88:5607
Cox DP, Reynolds RJ. 1987. *Annu. Rev. Astron.*

Astrophys. 25:303
Damas MC, Mendis DA. 1992 *Astrophys. J.* 396:704
D'Angelo N. 1965. *Phys. Fluids* 8:1748
D'Angelo N. 1990. *Planet. Space Sci.* 38:1143
D'Angelo N, Song B. 1990. *Planet. Space Sci.* 38:1577
D'Angelo N, von Goeler S, Ohe T. 1966. *Phys. Fluids* 9:1605
De Angelis U, Bingham R, Tsytovich VN. 1989. *J. Plasma Phys.* 42:445
De Angelis U, Forlani A, Tsytovich VN, Bingham R. 1992. *J. Geophys. Res.* 97:6261
De Angelis U, Formisano V, Giardano M. 1988. *J. Plasma Phys.* 40:399
Ershkovich AI, Mendis DA. 1986. *Astrophys. J.* 302:849
Fechtig H, Grun E, Morfill GE. 1979. *Planet. Space Sci.* 27:511
Feuerbacher B, Willis RF, Fitton B. 1973. *Astrophys. J.* 181:102
Flammer KR, Jackson B, Mendis DA. 1986. *Earth, Moon and Planets* 35:203
Fomenkova MN, Mendis DA. 1992. *Astrophys. Space Sci.* 189:327
Fortov VE, Iakubov IT. 1990. *Physics of Nonideal Plasma,* chap. 8. New York: Hemisphere
Gadsden M, Schröder W. 1989. *Noctilucent Clouds.* New York:Springer-Verlag
Goertz CK. 1989. *Res. Geophys.* 27:271
Goertz CK, Ip W-H. 1984. *Geophys. Res. Lett.* 11:349
Goertz CK, Morfill GE. 1983. *Icarus* 53:219
Gosling JT, Asbridge JR, Bame SJ, Feldman WC, Zwickl RD, et al. 1981. *J. Geophys. Res.* 86:547
Grun E, Morfill GE, Mendis DA. 1984. In *Planetary Rings,* ed. R Greenberg, A Brahick, p. 275. Tucson: Univ. Ariz. Press
Grun E, Morfill GE, Terrile RJ, Johnson TV, Schweln GT. 1983. *Icarus* 54:227
Gurnett DA, Averkamp TF, Scarf FL, Green E. 1986. *Geophys. Res. Letts.* 13:291
Hagfors T. 1992. *J. Atmos. Terr. Phys.* 54:333
Hartquist TW, Havnes O, Morfill GE. 1992. *Fundam. Cosmic. Phys.* 15:107
Havnes O. 1988. *Astron. Astrophys.* 193:309
Havnes O, Aanesen TK, Melandsø, F. 1990a. *J. Geophys. Res.* 95:6581
Havnes O, de Angelis U, Bingham R, Goertz CK, Morfill GE, Tsytovich V. 1990b. *J. Atmos. Terr. Phys.* 52:637
Havnes O, Goertz CK, Morfill GE, Grun E, Ip W-H. 1987. *J. Geophys. Res.* 92:2281
Havnes O, Melandsø F, La Hoz C, Aslaksen TK, Hartquist T. 1992. *Phys. Scr.* 45:535
Havnes O, Morfill GE, Goertz CK. 1984. *J. Geophys. Res.* 89:10999
Hill JR, Mendis DA. 1980a. *Can. J. Phys.* 59:897
Hill JR, Mendis DA. 1980b. *Moon and Planets* 23:53

Hill JR, Mendis DA. 1980c. *Astrophys. J.* 242:395
Hill JR, Mendis DA. 1982a. *J. Geophys. Res.* 87:7413
Hill JR, Mendis DA. 1982b. *Geophys. Res. Lett.* 9:1069
Hill JA, Mendis DA. 1982c. *Moon and Planets* 26:217
Horanyi M, Burns JA. 1991. *J. Geophys. Res.* 96:19,283
Horanyi M, Burns JA, Hamilton D. 1992. *Icarus* 97:248
Horanyi M, Burns JA, Tatrallyay M, Luhman JG. 1990. *Geophys. Res. Letts.* 17(6):853
Horanyi H, Goertz CK. 1990. *Astrophys. J.* 361:105
Horanyi M, Houpis HLF, Mendis DA. 1988. *Astrophys. Space Sci.* 144:215
Horanyi M, Mendis DA. 1986a. *Astrophys. J.* 307:800
Horanyi M, Mendis DA. 1986b. *J. Geophys. Res.* 91:335
Horanyi M, Mendis DA. 1986c. *Adv. Space Res.* 6(7):127
Horanyi M, Mendis DA. 1987. *Earth Moon and Planets* 35:203
Horanyi M, Morfill G, Grun E. 1993. *Nature* 363:1993
Horanyi M, Tatrallyay M, Juhaśz A, Luhmann JG. 1991. *J. Geophys. Res.* 96:11,283
Houpis HLF, Mendis DA. 1983. *The Moon and Planets* 29:39
Intrilligator DS, Dryer M. 1991. *Nature* 353:407
Ip W-H. 1983. *J. Geophys. Res.* 88:819
Ip W-H, Mendis DA. 1983. *Geophys. Res. Letts.* 10:207
James CR, Vermeulen F. 1968. *Can. J. Phys.* 46:855
Jonker JH. 1952. *Phillips Res. Rep.* 7:1
Körösmezey A, Cravens TE, Gombosi TI, Nagy AF, Mendis DA, et al. 1987. *J. Geophys. Res.* 92:7331
Krall NA, Trivelpiece AW. 1973. *Principles of Plasma Physics.* New York:McGraw Hill
La Hoz C. 1992. *Phys. Scr.* 45:529
Laframboise JG, Parker LW. 1973. *Phys. Fluids* 16:629
Leubner MP. 1982. *J. Geophys. Res.* 87:6335
Lissauer JJ, Squyers SW, Hartman WK. 1988. *J. Geophys. Res.* 93:13776
Melandsø F, Aslaksen T, Havnes O. 1993. *J. Geophys. Res.* 98:13315
Mendis DA. 1981. In *Investigating the Universe,* ed. FD Kahn, p. 353. Dordrecht: Reidel
Mendis DA. 1991. *Astrophys. Space Sci.* 176:163
Mendis DA, Axford WI. 1974. *Rev. Earth Planet. Sci.* 2:419
Mendis DA, Hill JR, Houpis HLF, Whipple ECJ. 1981. *Astrophys. J.* 249:787
Mendis DA, Hill JR, Ip W-H, Goertz CK, Grun E, 1984. In *Saturn,* ed. T. Gehrels, MS Matthews, p. 54. Tucson: Univ. Ariz. Press

Mendis DA, Horanyi M. 1991. In *Cometary Plasma Processes,* ed. AJ Johnston, p. 17. Geophys. Monogr. No. 6. Washington, DC: AGU

Mendis DA, Houpis HLF, Hill JR. 1982. *J. Geophys. Res.* 87:3449

Mendis DA, Houpis HLF, Marconi ML. 1985. *Fundam. Cosmic. Phys.* 10:1

Mendis DA, Rosenberg M. 1992. *IEEE Trans. Plasma Sci.* 20:929

Mendis DA, Rosenberg M, Chow VV. 1993. In *Dusty Plasma, Noise and Chaos in Space and in the Laboratory,* ed. H. Kikuchi. New York: Plenum. In press

Meyer-Vernet N. 1982. *Astron. Astrophys.* 105:98

Morfill GE, Grun E, Johnson TV. 1980. *Planet. Space Sci.* 28:1087

Mueller AC, Kessler DJ. 1985. *Adv. Space Res.* 5(2):77

Nishi R, Nakano T, Umebayashi T. 1991. *Astrophys. J.* 368:181

Norman C, Heyvaerts J. 1985. *Astron. Astrophys.* 147:247

Northrop TG. 1992. *Physica Scr.* 45:475

Northrop TG, Birmingham. 1990. *Planet. Space Sci.* 38:319

Northrop TG, Connerney JEP. 1986. *Icarus* 70:124

Northrop TG, Hill JR. 1982. *J. Geophys. Res.* 87:6045

Northrop TG, Hill JR. 1983a. *J. Geophys. Res.* 83:1

Northrop TG, Hill JR. 1983b. *J. Geophys. Res.* 88:6102

Northrop TG, Mendis DA, Schaffer L. 1989. *Icarus* 79:101

Opik EJ. 1956. *Irish Astron. J.* 4:84

Pilipp W, Hartquist TW, Havnes O, Morfill GE, 1987. *Ap. J.* 314:341

Puget JL, Leger A. 1989. *Annu. Rev. Astron. Astrophys.* 27:161

Rao NN, Shukla PK, Yu MY. 1990. *Planet. Space Sci.* 38:543

Rao NN. 1993. *Planet. Space Sci.* 41:21

Raymond JC. 1984. *Annu. Rev. Astron. Astrophys.* 22:75

Rosenberg M. 1993. *Planet. Space Sci.* 41:229

Rosenberg M, Mendis DA. 1992. *J. Geophys. Res.* 97:14,773

Sagdeev RZ, Evlanov EN, Fomenkova MN, Prilutskii OF, Zubov BV. 1989. *Adv. Space. Sci.* 9:263

Salimullah M, Sen A. 1992. *Phys. Lett.* A 163:82

Sekanina Z, Farrell JA. 1980. In *Solid Particles in the Solar System,* ed. J. Halliday, BA McIntosh, p. 267. Dordrecht: Reidel

Shan, L-H, Goertz CK. 1991. *Astrophys. J.* 367:350

Shawhan SD. 1966. *J. Geophys. Res.* 71:5585

Sheehan DP, Carillo M, Heidbrink W. 1990. *Rev. Sci. Instrum.* 61:3871

Showalter MR, Burns JA, Cuzzi JN, Pollack JB, 1985. *Nature* 316:526

Shu FH, Adams FC, Lizano S. 1987. *Annu. Rev. Astron. Astrophys.* 25:23

Shukla PK. 1992. *Phys. Scr.* 45:504

Shukla PK, Silin VP. 1992. *Phys. Scr.* 45:508

Shukla PK, Yu MY, Bharuthram R. 1991. *J. Geophys. Res.* 96:21,343

Simpson JA, Rabinowitz D, Tuzzolino AJ, Ksanfomality LV, Sagdeev RZ. 1987. *Astron. Astrophys.* 187:742

Simpson JA, Tuzzolino AJ, Ksafomality LV, Sagdeev RZ, Vaisberg OL. 1989. *Adv. Space. Sci.* 9:259

Singer SF, Walker EH. 1962. *Icarus* 1:112

Smith BA, Soderblom LA, Beebe RF, Boyce J, Briggs GA, et al. 1982. *Science* 212:163

Song B, Suszcynsky D, D'Angelo N, Merlino RL. 1989. *Phys. Fluids* B1:2316

Spitzer L. 1941. *Astrophys. J.* 93:396

Stebbins J, Huffer CH, Whitford AE. 1934. *Publ. Washburn Obs.* 15:(V)

Stebbins J Huffer CH, Whitford AE. 1939. *Astrophys. J.* 90:209

Sternglass EJ. 1954. *Sci. Pap. 1772,* Westinghouse Res. Lab., Pittsburgh

Summers D, Thorne RM. 1991. *Phys. Fluids* 83:1835

Suszcynsky D, D'Angelo N, Merlino RL. 1989. *J. Geophys. Res.* 94:8966

Teichmann J. 1966. *Can. J. Phys.* 44:2973

Trumpler RJ. 1930. *Lick Obs. Bull.* 14:154

Tsytovich VN. 1992. *Phys. Scr.* 45:521

Tsytovich VN, de Angelis U, Bingham R. 1989. *J. Plasma Phys.* 42:429

Tsytovich VN, Morfill GE, Bingham R, De Angelis U. 1990. *Comments Plasma Phys. Controlled Fusion* 13:153

Umebayashi T, Nakano T. 1990. *MNRAS* 243:103

Verheest F. 1992. *Planet. Space Sci.* 40:1

Wallis MK, Hassan MHA. 1983. In *Cometary Exploration,* Vol. II, ed. TI Gombosi, p. 57. Budapest: Hungarian Acad. Sci.

Whipple EC Jr. 1965. *The equilibrium electric potential of a body in the upper atmosphere and in interplanetary space. NSASA-GSFC Publ. X-615-65-296*

Whipple EC Jr. 1981. *Rep. Prog. Phys.* 44:1197

Whipple EC Jr, Northrop TG, Mendis DA. 1985. *J. Geophys. Res.* 90:7405

Wilson GR. 1988. *J. Geophys. Res.* 93:12,771

Xu W, D'Angelo N, Merlino RL. 1993. *J. Geophys. Res.* 98:7843

Annu. Rev. Astron. Astrophys. 1994. 32: 465–530

PRE-MAIN-SEQUENCE BINARY STARS

Robert D. Mathieu

Department of Astronomy, University of Wisconsin-Madison, Madison, Wisconsin 53706

KEY WORDS: infrared companions, orbital eccentricity, protostellar disks, star formation, stellar ages, stellar masses

INTRODUCTION

The observational study of pre-main-sequence (PMS) binary stars is in many ways a very young field; most PMS binaries known today were discovered in the past decade. Nonetheless, T Tauri stars have been under study for more than a half century, and the serendipitous discovery of visual pairs has always been a by-product of their observation (e.g. Joy & Van Biesbroeck 1944). The acceleration of discovery in recent years has at least two stimuli, one technical and one sociological. First, the frequency of binaries among main-sequence solar-type stars peaks at semimajor axes of order 50 AU, projecting to less than 0.5″ at the distance of the nearest star-forming regions. The requisite high-angular-resolution techniques (and near-infrared detectors) have only recently permitted the major surveys for PMS binaries now coming to fruition. Second, and equally important, the formation and early evolution of binaries has attracted increasing attention from those studying star formation. This can perhaps be attributed to both a growing confidence in our general picture for single-star formation (e.g. Shu et al 1987) and recognition that however correct our theories of single-star formation may be, the usual product of a star-formation event is a multiple-star system.

Until recently, the primary observational constraint on the mechanisms of binary formation and early evolution have been provided by main-sequence (MS) binaries in the solar vicinity, acting as a surrogate for the zero-age-main-sequence (ZAMS) binary population. As the ultimate product of the binary formation process, ZAMS binaries do supply essential constraints. However,

465

the ZAMS binary population represents only a boundary constraint; its properties need not be matched by the binary population at any earlier time. Hence, it is essential to investigate younger binary populations in order to develop an understanding of early binary evolution and, ultimately, to study the formation of binaries in situ. In fact, binary formation occurs prior to the PMS phase, so that while study of PMS binaries pushes the observational boundary condition to earlier times it does not reveal formation in progress. Nonetheless, the PMS phase is likely a time of important evolutionary processes which are only now being revealed by observation. It is these rapidly advancing observations that are the subject of this review.

For complete generality this review might have undertaken discussion of all binaries containing stars not yet contracted to the ZAMS. However, such a definition encompasses a dauntingly large range of stellar masses and evolutionary states. In practice, the sample of binaries considered here includes only low-mass, PMS stars. Except for a brief discussion of candidate protobinaries, all of the binaries discussed will have at least one optically detected star. This operational definition of "pre-main-sequence" limits the sample to binaries in which at least one star has evolved past the stellar birthline (Stahler 1983), or more generally, which has largely shed its natal envelope. In addition, the sample will include only binaries having spectral types later than A. This restriction intentionally excludes the Herbig Ae/Be stars—a population ripe for significant study but not yet for review. Together these constraints limit the sample to binaries with stars having theoretical stellar masses less than $\approx 3\ M_\odot$ (e.g. Palla & Stahler 1990), and in practice masses between a few tenths and two solar masses predominate. Similarly, the theoretical ages of the PMS binaries to be discussed range from 10^5–10^7 yr. (A representative sample is shown in Figure 8.)

The word "binary" also requires some discussion, since star formation produces higher levels of multiplicity as well. A binary star is seemingly well defined as two gravitationally bound stars in orbit. However, this strict definition can be limiting in practice. For example, in a hierarchical triple an issue may only be of interest on the size scale of the binary-tertiary separation. As such the triplicity is irrelevant, and for the purposes of discussion the wider system is most easily considered (and referred to) as a binary star. It is fortunate that most PMS triples, quadruples, etc are found to be already hierarchical, and typically the assorted pairings can be considered independently. Hence, in this review the word "binary" is used in reference to specific pairings, usually pairs of single stars but sometimes higher order pairings within hierarchical systems. When higher order multiplicity is important to the discussion, the choice of words reflects it (e.g. "triples," "tertiary," etc). Most generally, each independent grouping referred to here—single or otherwise—is a "system," and a non-single system is referred to as a "multiple system." (See also the discussion of these issues by Reipurth & Zinnecker 1993.)

The goals of this review are threefold. First and foremost is to provide a review of our present observational knowledge of PMS binaries. References to the early work upon which we build today and to the flourishing theoretical work are therefore incomplete. Second, the review is written with someone new to the field in mind, so some subtler technical issues have been sacrificed for comprehensiveness. Also in this spirit, an Appendix provides samples of PMS binaries from which the exploration of ideas might begin. Third, the review is structured so as to clearly distinguish observational results from their interpretation and discussion. It would be naive to suggest that the acquisition and recognition (e.g. in reviews) of observations are not guided by theoretical perceptions, and progress is arguably much enhanced as a result. Nonetheless, high-quality observations often outlive their immediate interpretations and bear truths not at first perceived. Hence, a comprehensive review of the observations is presented first, followed by discussions of some of their implications. Finally, the reader will profit by also reading the recent reviews of Reipurth (1988), Zinnecker (1989), and Bodenheimer et al (1993).

PRE-MAIN-SEQUENCE BINARY FREQUENCY

Investigation of binary frequency among MS field stars has a long history (cf Heintz 1969, Batten 1973, Abt 1983, Duquennoy & Mayor 1991). In contrast, the first extensive surveys for multiple systems among PMS stars have been only recently completed. The MS binary population extends over more than seven decades in semimajor axis, so that comprehensive study of the PMS binary population demands complementary investigations using widely varying techniques. Spectroscopic, lunar occultation, speckle, and solid-state array studies of PMS stars have been underway over the past decade.

Although PMS binaries have been detected in many nearby star-forming regions, the Taurus-Auriga (Tau-Aur) association has been the most extensively surveyed. Joy noted the visual pairs UX Tau, UZ Tau, RW Aur, and UY Aur in his discovery studies of T Tauri stars; at the time these four binaries comprised more than a third of all known T Tauri stars (Joy & van Biesbroeck 1944). Two decades later, Herbig (1962) listed 14 pairs in the association. Cohen & Kuhi (1979) listed an additional 9 binaries (as well as several small groups) found serendipitously with the Lick acquisition TV during the course of their spectroscopic survey. Spurred by the speckle detection of T Tau as a subarcsecond binary (Dyck et al 1982), infrared high-angular-resolution techniques have greatly accelerated the discovery of PMS binaries in Tau-Aur. At present, the number of resolved binaries with angular separations less than 10" stands at 66 (in 59 multiple systems), as tabulated in Table A1. To this list can be added three spectroscopic binaries, V826 Tau, 045251+3016 and LkCa 3 (Tables A2 and A3). Clearly, even in just this one association, investigators of young binaries have a rich population to explore. Indeed the excess of riches is perhaps the most unexpected discovery so far.

A measure of the binary frequency in the Tau-Aur association has recently been gained from several infrared high-angular-resolution surveys (Simon et al 1992, Simon 1992, Ghez et al 1993, Leinert et al 1993, Richichi et al 1994). Leinert et al (1993) have compiled 2.2 μm speckle and InSb-array observations for a sample of 104 stars in Tau-Aur. In the projected separation range of 0.13" to 13", or 18 AU to 1800 AU,[1] they list 44 multiple systems (39 binaries, 3 triples, and 2 quadruples) for a frequency of multiple systems of 42% ± 6%. Leinert et al (1993) argue that very few companions were missed in the range of separation to which they were sensitive, but this count is certainly incomplete outside that range. An additional 9 objects in this sample are known to be multiple from the higher resolution speckle survey of Ghez et al (1993), the lunar occultation observations of Simon et al (1992 and personal communication) and Richichi et al (1994), and spectroscopic studies (Mundt et al 1983, Mathieu et al 1989). Hence the union of these studies places a lower limit of 51% on the frequency of multiple systems in this Tau-Aur sample. Richichi et al (1994) estimate the frequency of multiple systems in this sample to be 60% ± 9% between separations of 0.013" to 13", or 1.8 AU to 1800 AU.

Clearly multiple systems comprise a large fraction of the PMS population in the Tau-Aur association. Indeed, these surveys yield PMS binary frequencies roughly twice that found for field MS solar-type binaries in the same range of separations. The Tau-Aur PMS binary frequency distribution[2] as a function of period is shown in Figure 1. Also shown is the binary frequency distribution among MS solar-type stars, after correction for undetected binaries (Duquennoy & Mayor 1991). The observed excess of PMS binaries at intermediate periods is evident. Leinert et al (1993) find that their observed PMS binary frequency in the Tau-Aur region, integrated between projected separations of 18 AU and 1800 AU, exceeds the MS solar-type binary frequency by a factor 1.9 ± 0.3.

Similar results are found by Ghez et al (1993), who used 2.2 μm speckle observations at the Hale 5-m telescope to search for binaries among 69 young stars in both the Tau-Aur (45 stars) and the Scorpius-Ophiuchus (24 stars; Sco-Oph) star-forming regions. In the combined sample they found 31 binaries and 1 triple system. They argue that detection is complete for binaries with projected separations between 16 AU and 252 AU and with flux ratios of $\Delta K < 2$ mag. Restricting analysis to the 22 binaries satisfying these criteria, they find a binary frequency of 34% ± 7%, a factor of two larger than the MS solar-type binary frequency in the same separation range. Ghez et al (1993) then derive frequencies

---

[1]Distances used in the papers under discussion will be adopted here.

[2]The distributions show the frequency of all pairings of a given separation, whether in binaries or hierarchical multiple systems. Thus, a triple system will contribute twice to the count, and a quadruple will contribute three times. The ordinate represents the fraction of all systems, whether single or multiple. In the text, "binary frequency" refers to integration under this distribution.

of binaries with larger flux ratios from subsamples of stars for which they have more sensitive observations. In total they find a frequency of 60% ± 17% for binaries with projected separations between 16 AU and 252 AU and with flux ratios of $\Delta K < 4$ mag. (The sample included only 1 binary with $\Delta K > 3$ mag, which contributed 14% to the derived frequency.) Ghez et al (1993) conclude that the PMS binary frequency is four times that among MS solar-type stars, and that the frequencies are different at the 2.5 σ level of confidence. In addition, they find no significant difference in the binary frequencies of the Tau-Aur and Sco-Oph star-forming regions. Even so, the binary frequency found in Sco-Oph alone cannot be distinguished from the MS binary frequency at the 2 σ confidence level. Unfortunately, in all PMS binary surveys to date meaningful comparisons of interesting subgroups are hampered by small sample sizes.

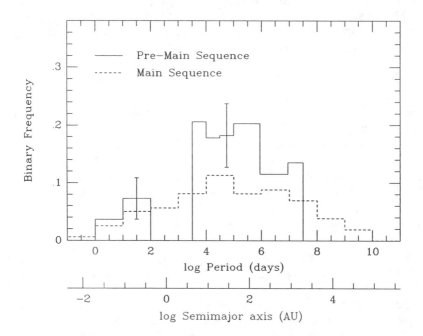

Figure 1 Frequency distributions as a function of period for pre-main-sequence binaries and main-sequence solar-type binaries. The intermediate-period pre-main-sequence data are from the Tau-Aur region (compiled by Richichi et al 1994); the shortest period data are from the Tau-Aur, Sco-Oph, and Corona Australis star-forming regions (Mathieu 1992b). The period regimes where no pre-main-sequence data are shown are incompletely surveyed at present. The ordinate is defined as the number of binaries per logarithmic period interval divided by the number of systems (single plus multiple). Triples contribute two binaries to the figure, and so on for higher-order systems. The relation of the period and semimajor axis abscissa scales is appropriate for a total mass of $1 M_\odot$. The sampling errors are large, particularly for the pre-main sequence data; two representative 1 σ error bars are shown.

A possible excess of binaries among PMS stars also has been suggested by Reipurth & Zinnecker (1993) from their CCD ($\lambda \approx 0.9\,\mu$m, resolution limit $\approx 1''$) imaging survey of stars in 11 southern nearby star-forming regions. In total, they list 87 PMS binaries. Among a systematically surveyed sample of 238 PMS stars, they identify 37 binaries and 1 triple between projected separations of 150 AU to 1800 AU. Thus, in this range they find a frequency of multiple systems of 16% ± 3% (which they argue is very nearly complete and uncontaminated with false detections). They also note that this frequency is the same as in the Tau-Aur study of Leinert et al (1993) over the same range of projected separation. Extrapolating to separations outside this range with the functional form of the MS separation distribution (Duquennoy & Mayor 1991), they find a total binary frequency of 80%. This frequency is to be compared with values of 60% for MS solar-type stars (Duquennoy & Mayor 1991) and 45% for field K-dwarfs (Mayor et al 1992). More directly however, the frequency of MS solar-type binaries between 150 AU and 1800 AU is 12% ± 3%, so the observed PMS binary frequency is not distinct from that of solar-type stars at a statistically significant level.

In contrast to these PMS binary frequencies found in associations and small groups, Prosser et al (1994) find no enhancement in the observed binary frequency in the dense Trapezium cluster. Based on V- and I-band *HST* Planetary Camera images of the cluster, they identify 35 binaries with projected separations between 0.06″ and 1.2″, or 26 AU and 528 AU. They argue that this sample includes only a few chance alignments, and find a binary frequency in this separation range of 12%. Taking an average system mass for these binaries of 1 $M_\odot$, Prosser et al (1994) find that this binary frequency is equal to that expected from the Duquennoy & Mayor (1991) study of solar-type stars. They conclude that there is no evidence that the binary frequency in the Trapezium cluster is unusual. Prosser et al (1994) note that their sensitivity to faint companions varied with separation, from maximum magnitude differences of 4 mag at wide separations to roughly 1.5 mag at separations of 0.1″ – 0.2″. The implications of these limitations with respect to the comparison with the MS binary frequency have not yet been explored.

Radial-velocity surveys for spectroscopic binaries explore the shortest period PMS binaries. Until recently, definitive spectroscopic binaries (i.e. with orbit determinations) had been notably absent among PMS stars. As discussed later, arguably this could be linked to the disruption of inner disks and consequent reduced Hα emission in close binaries. Thus, early Hα surveys for young stars may have selected against spectroscopic binaries. In addition, spectroscopic surveys are less sensitive to velocity variation in stars with strong emission lines, extreme veiling, or rapid rotation. Observations of weak-lined PMS stars minimize the technical difficulties and provide at least an upper limit on the total PMS short-period binary frequency, depending on whether classical T

Tauri stars are in fact deficient in short-period binaries. [Hereinafter PMS stars with strong emission lines, specifically with Hα equivalent widths in excess of 10 Å, will be referred to as "classical T Tauri" stars. Strom et al (1989a) suggested an Hα equivalent width of 10 Å as a reasonable division of PMS stars with and without substantial accretion activity.]

A high-precision radial-velocity survey of 55 X-ray selected stars in the Tau-Aur, Sco-Oph, and Corona Australis associations revealed 8 binaries with orbital solutions (Mathieu et al 1989, updated in Mathieu 1992b and Table A3 here). For periods less than 100^d (which should be nearly complete) the consequent binary frequency is 11% ± 4%. As can be seen in Figure 1, this is indistinguishable from the frequency found among solar-type stars of 7% ± 2% [Duquennoy & Mayor (1991); Abt & Levy (1976) found a frequency of 12%, see Mathieu (1992b)]. However, the uncertainties are such that this result cannot exclude an enhancement in short-period binary frequency similar to that being found at intermediate periods.

Alternatively, Mathieu (1992b; see also Herbig 1977) attempted to statistically constrain the PMS short-period binary frequency by examining the distribution of radial-velocity measurements obtained in a survey of 36 PMS stars in the Tau-Aur association. Most stars in the sample were classical T Tauri stars that nonetheless provided precise radial-velocity measurements. Mathieu (1992b) found that the number of high-amplitude velocity variables was somewhat lower than, but not significantly distinct from, that predicted by a MS solar-type binary population. However, such analyses are severely compromised by the small expected frequency of short-period binaries (e.g. Figure 1). If it is fair to maximize the sample size by combining classical T Tauri stars and weak-lined PMS stars from several star-forming regions, the union of these two studies gives 6 binaries with periods less than 100 days among 91 PMS stars for a frequency of 7% ± 3%, the same as found for MS solar-type stars. Thus, in this large sample there is no evidence for an excess of short-period PMS binaries over that found in the field, but studies of larger, carefully selected samples are much needed.

There is very limited information regarding binary frequencies at the longest orbital periods. MS solar-type binaries have estimated periods as long as 10^7 yr, corresponding to semimajor axes of more than 10^4 AU and separations of arcminutes at Tau-Aur, for example. At such separations spatial association is difficult to assess, and whether the pairs are gravitationally bound or stable is impossible to prove with present techniques. At separations an order of magnitude smaller, Reipurth & Zinnecker (1993) note a deficiency of binaries with projected separations between 1200 AU and 2400 AU compared to MS solar-type binaries, but the expected number in their sample is only a few so that the statistical significance is marginal.

A thorough synthesis of the findings of different techniques remains to be done. One of the more interesting areas of investigation will be the hierarchical structuring within PMS multiple systems. Ghez et al (1993) find that triples and quadruples comprise 14% of their sample. Correcting for the presently incomplete coverage in period, they estimate that the true frequency is ≈35%. In contrast, Duquennoy & Mayor (1991) found triple and higher order systems comprised only 5% of all MS solar-type systems, similar to the findings of Abt & Levy (1976). Interestingly, significantly higher estimates have been made for the frequency of triples among main sequence stars. For example, Herczeg (1988) argues that at least a third of MS spectroscopic binaries have tertiaries at visual binary separations—a substantially higher frequency than found by the survey studies.

Only one of the triples and quadruples in the Ghez et al (1993) speckle sample would have been detected as such without additional direct imaging or spectroscopic studies that explored very different domains of binary separation. Similarly, the direct imaging study of Reipurth & Zinnecker (1993), sensitive to just one decade in period, revealed only one triple among 238 targets. Evidently hierarchical organization of multiple systems is established prior to the PMS phase (but see Abt 1986).

Finally, we consider the shape of the PMS binary period distribution. For longer period binaries both Reipurth & Zinnecker (1993) and Leinert et al (1993; Figure 1) find the frequency of PMS binaries to decrease with increasing separation. At short periods the PMS spectroscopic binary frequency is also less than that found for intermediate-period binaries (Mathieu 1992b; Figure 1). Hence, the PMS binary frequency distribution as a function of semimajor axis is at least qualitatively consistent with the log-normal-like distribution found for MS binaries (Duquennoy & Mayor 1991). Leinert et al (1993) and Reipurth & Zinnecker (1993) further argue that their observed distributions are formally consistent with the functional form found by Duquennoy & Mayor (1991), but their period ranges are not adequate to discriminate against other distributions.

ORBITAL ECCENTRICITY DISTRIBUTION

At present, our knowledge of the orbital eccentricity distribution of PMS binaries rests entirely upon spectroscopic orbital elements. Orbital elements have been published for 11 PMS spectroscopic binaries and 2 additional candidate PMS binaries. Unpublished orbital elements for an additional 12 PMS binaries are known to the author. These 25 PMS binaries are listed in Tables A2 and A3, along with a selection of orbital elements. Most of the binaries in this list were discovered serendipitously or were gleaned from radial-velocity surveys not yet completed; the only completed and analyzed survey is that of Mathieu et al (1989). Nonetheless, the number of measured orbital eccentricities is now sufficient to allow discussion of the eccentricity distribution of PMS binaries.

The distribution of orbital eccentricity with period for the PMS spectroscopic binaries is shown in Figure 2a. Several significant conclusions can be drawn. First, orbital eccentricities from $e = 0$ to $e = 0.8$ have been established by the PMS phase of stellar evolution. The binaries shown in Figure 2a typically have theoretical ages between 10^6 yr and 10^7 yr. However, several of the binaries, with both large and small eccentricities, lie near the stellar birthline. Evidently the range of orbital eccentricities is established in young binaries prior to their unveiling at the stellar birthline.

Second, the morphology of the PMS eccentricity distribution has several features:

1. Six of the eight shortest period ($P < 6.4^d$) binaries have circular orbits.

2. Circular or near-circular ($e < 0.1$) orbits are not found at longer periods, with the one notable exception of the classical T Tauri binary GW Ori.

3. The maximum orbital eccentricity increases with increasing period to at least periods $\approx 1000^d$.

None of these features is due to selection effects; the most likely bias in the sample is against discovery of long-period, high-eccentricity orbits. However, given the small numbers, sampling uncertainties are a serious concern.

Third, the PMS and MS eccentricity distributions are very similar over those periods where both have been observed. Figure 2b shows the eccentricity distribution of nearby field MS solar-mass stars (Duquennoy & Mayor 1991 and personal communication; see also Abt & Willmarth 1992 and Mayor et al 1992). As with the PMS binaries, the shortest period MS orbits are typically circular, there are no near-circular orbits at longer periods, and the maximum orbital eccentricity rises with increasing period.

The similarity of the PMS and MS eccentricity distributions suggests that the orbital eccentricity distribution for periods $\lesssim 1000$ days is largely established by $\approx 10^6$ yr. Nonetheless, several of these PMS binaries show evidence for massive circumbinary disks, so that orbital evolution may be ongoing in these cases. In this regard, the binary GW Ori is notable in that it has an orbital eccentricity of $e = 0.04 \pm 0.06$ and a very high submillimeter luminosity, which suggests an unusually massive disk (Mathieu et al 1994). Such a low eccentricity is unusual at a period as long as 242 days, although the uncertainty on the eccentricity measurement is sufficiently large that the orbit is not securely anomalous. Clearly, an outstanding observational goal is to compare the eccentricity distributions of binaries with and without massive disks, which will require a substantial increase in the number of orbit determinations for classical T Tauri stars.

At periods longer than ≈ 1000 days the presently known PMS binary population can add little to the discussion. Spectroscopic and high-angular-resolution

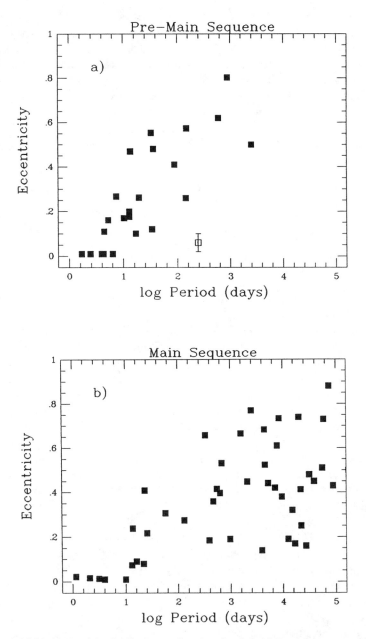

Figure 2   Orbital eccentricity distribution as a function of period. (*a*) Pre-main-sequence binaries (see Tables A2 and A3). The open square is the binary GW Ori. (*b*) Field main-sequence solar-type binaries (Duquennoy & Mayor 1991).

observations, only begun recently, will have to be continued for many years before orbital solutions for such PMS binaries can be obtained (e.g. Ghez et al 1994). Until then, consideration of eccentricity distributions at longer periods must continue to rely on MS binaries. Discussion of the MS long-period eccentricity distribution has a rich history among visual binary observers, which we will not review here (cf Heintz 1969). We note only that in their study of the solar-mass binary population Duquennoy & Mayor (1991) found that (after incompleteness corrections) the MS long-period eccentricity distribution was weighted toward higher eccentricities.

The predominance of circular orbits at the shortest periods merits some discussion. This feature is characteristic of most binary populations, and has typically been attributed to tidal circularization processes. The longest-period circular orbit has been defined as the "circularization cutoff period" (Duquennoy et al 1992), and recently much effort has focused on the evolution of the cutoff period as a function of binary age. Observations have delineated cutoff periods for solar-mass binaries in four MS populations: 7.05^d in the Pleiades (0.1 Gyr), 8.5^d in the Hyades (0.8 Gyr), 12.4^d in M67 (4 Gyr), and 18.7^d in the halo field (16 Gyr). These cutoff periods increase with age, consistent with active main-sequence tidal circularization. However, this trend is not as evident between the Pleiades and Hyades clusters, which perhaps suggests that circular orbits among binaries with periods less than $7–8^d$ are established at formation. Clearly, the PMS orbital eccentricity distribution is essential for establishing to what extent circularization of shorter period binaries occurs prior to the main sequence. (References, and a more detailed review contrasting binary observations and tidal circularization theory, can be found in Mathieu et al 1992.)

Presently, the PMS binary W134, with a period of 6.4^d, has the longest-period circular orbit (Padgett & Stapelfeldt 1993). Since, by definition, observed cutoff periods are lower limits, this binary would indicate that the cutoff periods of PMS binaries and binaries in young clusters are very similar. However, the components of W134 have theoretical masses of $2.2\,M_\odot$ and $2.3\,M_\odot$, and are destined to become A stars on the main sequence (Padgett & Stapelfeldt 1993); hence, comparison of W134 with the solar-mass MS binary populations is not well founded. If W134 is excluded, the PMS cutoff period becomes 4.3^d, set by the circular orbit of the solar-mass binary ORI569. Between 4.3^d and 10.4^d inclusive three solar-mass PMS binaries are known, all having eccentric orbits. A fourth binary, EK Cep, also has an eccentric orbit at a period of 4.4 days, but its interpretation is complicated by the pairing of a ZAMS $2\,M_\odot$ star with a solar-mass PMS secondary. While the membership of the known short-period PMS binaries is not as homogeneous as desirable, given the present sample the circularization cutoff period of solar-mass PMS binaries appears to be significantly shorter than that found in any MS solar-mass binary population.

SECONDARY MASS DISTRIBUTION

The determination of secondary mass distributions of PMS binaries is in its infancy. To date only angularly resolved systems have been considered (Simon et al 1992, Leinert et al 1993, Reipurth & Zinnecker 1993). A general finding is that the distribution of secondary-to-primary flux ratio ranges from unity to the limits of detection ($\ll 0.1$). As one example, we show in Figure 3 the 2.2 μm flux-ratio distribution of the Tau-Aur binaries with projected separations of 18–1800 AU (Leinert et al 1993). The large range of flux ratios is an indication that the PMS secondary mass distribution may be similar to that of wide binaries in the field. (The distribution of 2.2 μm flux-ratios among all Tau-Aur binaries compiled in Table A1 is little different, and indeed the range of flux ratios is large even among those binaries with separations less than 18AU.)

Detailed comparison of PMS flux-ratio distributions with initial mass functions are made difficult by the time dependence of the PMS mass-luminosity relation, theoretical mass-luminosity calibrations which are less secure than for MS stars, the availability of a relative flux at only one wavelength for many binaries, possible contamination of the fluxes from one or both stars by disk emission (especially at 2 μm), and variability. In addition, the primaries included in the PMS surveys are not all the same mass, a concern if the mass-luminosity relation or the secondary mass distribution is mass dependent. In practice, Simon et al (1992) and Reipurth & Zinnecker (1993) have argued that

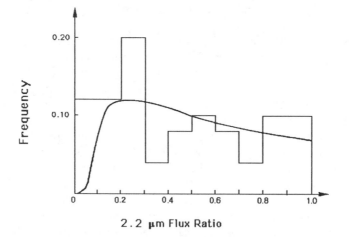

2.2 μm Flux Ratio

Figure 3   Distribution of 2.2 μm flux ratio for Tau-Aur binaries with projected separations between 18 AU and 1800 AU. The solid curve is the distribution obtained by uncorrelated association of components with masses less than 1 $M_\odot$ from a Miller-Scalo (1979) initial mass function. (Adapted from Leinert et al 1993.)

a linear relation of K brightness with mass is a reasonable approximation to the mass-luminosity law, in the absence of contamination from a disk.

With this approximation, both Leinert et al (1993) and Reipurth & Zinnecker (1993) find that their observed flux-ratio distributions are consistent with random pairings of stars from a single parent population similar to the initial mass function of the field. The predicted distribution of one such model is shown in Figure 3. This finding is consistent with the secondary mass distribution of field MS solar-type binaries of similar separation (Abt & Levy 1976, Duquennoy & Mayor 1991), which suggests that the membership in binaries is established prior to an age of order 10^6 yr.

DISKS AND ACCRETION IN YOUNG BINARY ENVIRONMENTS

Introductory Comments

The observational evidence for circumstellar and circumbinary material in PMS binary environments is strong. Excess emission at near-infrared through millimeter wavelengths indicates extended material having a wide range of temperatures and distances from the stars. For a few binaries, observations at high angular resolution have resolved extended emission on size scales from a few hundred AU to several thousand AU. For PMS stars in general, such observations are commonly associated with disks (i.e. rotationally supported, highly flattened structures). Comprehensive reviews of the evidence for disks around PMS stars can be found in Basri & Bertout (1993), Beckwith & Sargent (1993), and Strom et al (1993). Observational estimates for disk radii cover a wide range; unresolved observations of dust continuum emission give radii of order 100 AU (e.g. Adams et al 1990), while interferometric line observations have found disk-like structures extending to radii of order 1000 AU (e.g. Koerner et al 1993b).

Although developed in the context of single stars, the arguments for disks are typically applicable to binary systems as well. Indeed, some of the stronger cases put forth as stars with disks were later discovered to be binaries (e.g. DF Tau; Bertout et al 1988, Simon et al 1992). Nonetheless, the conceptual framework that has been developed for disks around single stars must be transferred to binary systems with care. Perhaps most importantly, a binary introduces its semimajor axis as a scale-length in the problem. We cannot expect disks in binaries with separations of thousands of AU to mimic disks in binaries with separations of tens of AU. Models developed for disks around single stars *may* transfer reasonably well to disks associated with the very closest or widest binaries, but in general we are unlikely to be so fortunate.

Furthermore, the concept of a single disk must be discarded. Obviously, two stars may have two associated disks. More fundamentally, a binary embedded within a coplanar disk can be expected to dynamically clear a region on the scale length of its semimajor axis. Consequently, in such an environment three distinct disks may be identified. Each star can have an associated disk within its Roche lobe; these are referred to generally as *circumstellar* disks and specifically as *circumprimary* and *circumsecondary* disks. In addition, a *circumbinary* disk may be present external to the binary orbit. One or all of these disk components may be absent in any given system, particularly at the extremes of binary separation. (This nomenclature can be loosely generalized to non-coplanar and non-disk-like distributions of extended material as well.)

Finally, coplanarity of stellar and disk orbits cannot be assured, and indeed may be unlikely for wider systems. Non-coplanar binary and tertiary orbits have been found in main-sequence triple systems (Fekel 1981, Hale 1994). In the HH 111 and HH 1-2 regions, pairs of jets have been found with common origins but very different alignments, which may arise from non-coplanar circumstellar disks in young binaries (Gredel & Reipurth 1993, Reipurth et al 1993b, Zinnecker 1989). In addition, both fragmentation and capture formation scenarios easily produce orbital planes inclined with respect to associated disks. Consequences of such noncoplanarity include warping, twisting, and dispersal of disks (Terquem & Bertout 1993, Clarke & Pringle 1993, Heller 1993), so that disks around PMS binaries may not be azimuthally or equatorially symmetric. Thus, while shorter period binaries may resemble the Sun-Jupiter system embedded in the protosolar nebula, pairs of spiral galaxies may be a better analogy for longer period binary-disk systems.

Disk Frequency

Clearly a basic datum is the frequency of disk material in binary systems. Excess near-infrared (e.g. 2.2 μm) emission is used as a tracer of material within roughly 0.1 AU from a star; for single stars such material is often presumed to be associated with more extensive disks. Numerous PMS binaries studied with resolved, multi-wavelength observations show excess near-infrared emission from one or both components. These include Glass I and DoAr24E (Chelli et al 1988), Haro 6-10 (Leinert & Haas 1989, Menard et al 1993), XZ Tau (Haas et al 1990), GG Tau (Leinert et al 1991), T Tau (Dyck et al 1982, Ghez et al 1991), Z CMa (Koresko et al 1991), UX Tau, HK Tau, UZ Tau EW, and Haro 6-37 (Moneti & Zinnecker 1991), and DD Tau (Tessier et al 1994). In many cases these near-infrared excesses have been taken to be indicative of at least one circumstellar disk, although in several the excess is associated only with an "infrared companion" whose physical interpretation is not clear (see below).

A large sample of the binaries in Table A1 have been surveyed for 2.2 μm excesses by Strom et al (1989a). In that study the photometric data were typically for the combined light of the systems and the analyses treated the (at the time unknown) binaries as single stars. Fischer & Mathieu (in preparation) have reanalyzed the near-infrared fluxes of the recently discovered binaries in the Strom et al sample, incorporating the measured flux ratios at K into the analyses. In at least 80% of the cases for which Strom et al inferred optically thick disks from the magnitudes of the infrared excesses, the conclusion is unchanged by the binarity of the system. Consequently, 40–50% of the binaries in the Strom et al study have near-infrared excesses indicative of at least one optically thick circumstellar disk. The stellar associations or extents of these disks are unknown, though. Interestingly, even very short period binaries also show near-infrared excesses. Examples include AK Sco (Andersen et al 1989), W134 (Padgett & Stapelfeldt 1993), and 162814-2427 (Jensen & Mathieu, in preparation). The periastron separations in these binaries are less than 0.14 AU, a few tens of stellar radii at most. In these cases the interpretation of near-infrared excesses as due to circumstellar disks may be problematic.

Mid-infrared (e.g. 10 μm) emission is a more robust diagnostic for extended material since it is less sensitive to photospheric contributions than 2.2 μm emission (Skrutskie et al 1990). Emission at 10 μm typically derives from material within roughly 1 AU of a star, so that sources of 10 μm excess emission are unambiguously circumstellar for all but the shortest period binaries. A sample of 22 binaries in the Tau-Aur association have been surveyed for 10 μm excesses by Skrutskie et al (1990). The derived excesses ΔN, measured as the logarithmic difference of observed and photospheric fluxes in the N band, are given in Table A1. For single stars Skrutskie et al (1990) argue that values of $\Delta N > 1.2$ dex indicate optically thick inner disks; this would be a conservative criterion for the combined light of binaries, in which only one of the two stars contributing to the optical light may have an associated disk. Based on this criterion, 16 of the 22 multiple systems in the Tau-Aur association with measured ΔN have detected optically thick circumstellar disks. Many of these binaries also have strong Hα emission, indicative of accretion from circumstellar disks (Skrutskie et al 1990). These results place a lower limit of 28% on the frequency of optically thick, circumstellar disks among *all* known multiple systems in Tau-Aur. A comparable number of Tau-Aur binaries without computed values of ΔN were detected by *IRAS* at 12 μm (Table A1), also typically indicative of optically thick circumstellar material (Skrutskie et al 1990). Thus, this incomplete survey of mid-infrared emission from the Tau-Aur binaries suggests that circumstellar disks are present in at least 50% of the binaries. The presence of these mid-infrared diagnostics for circumstellar disks does not evidently depend on binary separation. (Again, the attribution of these excesses to disks must be tempered by the existence of "infrared companions.")

Determining the frequency of circumbinary disks is more problematic. Two likely circumbinary disks have been resolved (GG Tau, Dutrey et al 1994; T Tau, Weintraub et al 1989, 1992). In addition, many PMS binaries show unresolved continuum emission from far-infrared to millimeter wavelengths. Unfortunately, the relation of each emission to circumbinary disks can be unclear.

First, given simple single-star disk models substantial millimeter emission is found to originate within a few tens of AU of a PMS stellar surface (Beckwith et al 1990), so that for binaries with periastron separations $\approx$ 50 AU or larger, millimeter emission may be purely circumstellar. Second, the spatial distribution of emission within binary environments is likely to be more complex, a case in point being the detection of a circumbinary ring of continuum emission around GG Tau (Dutrey et al 1994). Third, the radial extent of disk material inferred from millimeter line observations can be substantially larger than that inferred from continuum data. Thus based on CO line maps Weintraub et al (1989) argue for a circumbinary disk around T Tau with an outer radius of at least 800 AU, substantially larger than the typical disk radii of $\approx$ 100 AU derived from continuum spectral energy distribution arguments (e.g. Adams et al 1990). Finally, extended envelopes provide further ambiguity in attributing long-wavelength emission to circumbinary disks [cf recent observations of T Tau by Weintraub et al (1992)].

Thus, in the absence of high-angular-resolution images the confidence with which long-wavelength emission can be associated with circumstellar or circumbinary material depends on binary separation. At present detection limits, far-infrared and submillimeter continuum emission from binaries with separations of a few AU or less can be confidently interpreted as having circumbinary origin (e.g. Mathieu et al 1994). Thus, several PMS spectroscopic binaries with mid-infrared to submillimeter flux detections clearly have circumbinary material: V4046 Sgr (de la Reza et al 1986, Jensen et al, in preparation), AK Sco (Andersen et al 1989, Jensen et al, in preparation), GW Ori (Mathieu et al 1991, 1994), and 162814-2427 and 162819-2423S (Mathieu 1992c, Jensen & Mathieu, in preparation). For wider binaries, modeling of disk temperature and surface-density distributions is required to locate the source of unresolved long-wavelength emission. In this way Marsh & Mahoney (1993) argued for a circumbinary disk around DF Tau (projected separation of 13 AU), as did Koresko et al (1991) for the much wider binary Z CMa (115 AU) (but see Haas et al 1993). A different approach to the problem has been taken by Menard et al (1993) who on the basis of polarization measurements found that Haro 6-10 (170 AU) is surrounded by a highly flattened circumbinary structure. Finally, at the widest separations (e.g. thousands of AU) it is likely that long-wavelength continuum emission can be safely attributed to circumstellar disks, although the possibility of yet more extended disks has been suggested theoretically (e.g. Lin & Pringle 1990).

Disk Masses

Because of the large opacity of dust in the infrared, the optically thick infrared emission excesses do not necessarily imply large disk masses (Strom et al 1993). Disk masses are better measured from fluxes at submillimeter and millimeter wavelengths, where dust opacities are sufficiently low that the outer regions of disks may be optically thin yet thermal emission remains detectable. Unfortunately, the derived masses have several large uncertainties. First, the emissivity of dust at submillimeter wavelengths is poorly known, which introduces uncertainties as large as an order of magnitude (e.g. Beckwith & Sargent 1993). Second, dust is expected to be a minor constituent of disk material (e.g. $\approx 1\%$ in the interstellar medium; Draine & Lee 1984). Hence, the inferred total masses require a large correction for the gas-to-dust ratio, and are sensitive to abundance variations such as found by Dutrey et al (1994) for GG Tau. Both of these uncertainties depend sensitively on the processing of interstellar dust and gas in the protostellar environment (e.g. grain growth). Third, derived disk masses depend upon assumed radial surface-density and temperature distributions (e.g. Adams et al 1990, Beckwith et al 1990). This uncertainty is particularly problematic for binary environments where disk structure can be complex. Finally, submillimeter emission from sources other than disks (e.g. envelopes) can be a significant source of confusion.

A survey of 1.3 mm emission from PMS stars in the Tau-Aur association was undertaken by Beckwith et al (1990), included in which were 38 of the binaries in Table A1 (although most were not recognized as binaries at the time). Eleven of the binaries were detected. Modeled as single stars with geometrically thin disks having power-law temperature and surface-density distributions, Beckwith et al (1990) derived disk masses in the range of 0.004 $M_{\odot}$ to 0.3 $M_{\odot}$. Such disk masses are similar to the disk masses that they derived for apparently single stars, and are in most cases small fractions of the stellar masses in these binaries. For the remaining binaries, observations place typical upper limits on the disk masses of 0.01 $M_{\odot}$ or less. Almost all of the binaries for which disk masses were measurable had projected separations of greater than 50 AU, a point discussed in more detail below.

Single-dish millimeter and submillimeter observations of PMS binaries are rapidly accumulating. A partial list of additional observations of PMS binaries includes Skinner et al (1991), Zinnecker & Wilking (1992), Reipurth & Zinnecker (1993), Andre & Montmerle (1993), and Jensen et al (1994). Observations of two binaries have suggested intriguingly large masses of circumbinary material. Reipurth et al (1993a) suggest that at least 1.8 $M_{\odot}$ of material surrounds Z CMa, and Mathieu et al (1994) derive a mass estimate of 1.5 $M_{\odot}$ for a circumbinary disk around GW Ori. Both results are uncertain by at least a factor of a few and are model dependent, but nonetheless suggest that very young binaries may reside within massive circumbinary environments.

Disk Structure

For coplanar binary-disk systems, dynamical theory predicts that binaries will tidally truncate both circumstellar and circumbinary disks (e.g. Lin & Papaloizou 1993, Artymowicz & Lubow 1994). In the presence of both circumstellar and circumbinary disks this truncation leads to gaps between the disks with size scales similar to the binary orbit. An important corollary prediction is that material will not flow from a circumbinary disk across a binary orbit to circumstellar disks, or ultimately to stellar surfaces (Artymowicz et al 1991, Pringle 1991a). Thus, continued accretion of a circumstellar disk without replenishment may lead to a binary residing in a central hole within a circumbinary disk. Companions may accelerate this evolution by enhancing accretion rates of circumstellar disks (Ostriker et al 1992). On the other hand, without depletion by accretion the lifetimes of circumbinary disks may be lengthened.

This scenario is supported by recent interferometric observations of GG Tau. Dutrey et al (1994; see also Kawabe et al 1993) mapped GG Tau in both the ^{13}CO (1→0) line and millimeter continuum, achieving ≈ 2 arcsec resolution in both. Figure 4 shows maps at three velocities superimposed on their high-resolution continuum map. Also shown in Figure 4 is a contour presentation of the continuum map. The continuum emission is resolved and has a radial extent of at least 800 AU. The line emission is also resolved, and the velocity field is consistent with a disk in Keplerian rotation about a binary system of $\approx 1.2\,M_\odot$ and inclined $\approx 43°$ to the line of sight. Perhaps most exciting, the continuum map shows a central depression in the millimeter emission. Dutrey et al interpret this as a central cavity with a radius of ≈ 180 AU in a disk. The size of this cavity is substantially larger than the projected binary separation of 41 AU. They argue that the larger hole is the consequence of tidal clearing by an eccentric binary orbit with semimajor axis ≈ 70 AU. They find the emission distribution to be best fit by a ring of material at the inner edge of the circumbinary disk and containing $\approx 90\%$ of the circumbinary disk mass.

An indirect approach to the study of disk structure is analysis of infrared spectral energy distributions (SEDs). Spatial structure in unresolved binary disks should be reflected in wavelength structure of binary SEDs. In principle the spatial structure can be recovered by comparison of observed SEDs with predictions of physical models for disk material in binary environments; in practice physical models do not yet exist, so only simple parameterized models have been considered. Typically, a binary system has been modeled by a single stellar photosphere and an axisymmetric, geometrically thin disk from which an annular, (nearly) evacuated gap is removed. The consequent circumstellar and circumbinary disks are both described by the same power-law temperature and spatial radial distributions (as well as heating by the photosphere). In such a model, emission is reduced at those wavelengths characteristic of the temperatures in the gap. The central wavelength of the resulting dip in the SED depends on the radius of the gap; the width in wavelength (or the depth) of the dip gives a measure of the gap width. Central holes in disks can be similarly identified, except that the short-wavelength side of the dip is bounded by the

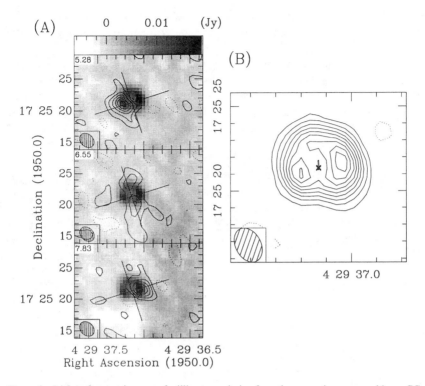

Figure 4   (*a*) Interferometric maps of millimeter emission from the pre-main-sequence binary GG Tau. The gray scale shows 3-mm continuum emission. The contour maps show ^{13}CO (1→0) line emission at three LSR velocities. The velocity gradient is consistent with Keplerian rotation about the binary for a system mass of 1.2 $M_\odot$. The cross indicates the apparent disk plane and axis. (*b*) A contour presentation of the continuum emission. A central depression in the image, very near the binary position is evident. Dutrey et al (1994) infer that the binary resides in a central cavity of radius ≈ 180 AU. (Adapted from Dutrey et al 1994.)

SED of the stellar photosphere. Such models are clearly oversimplified, particularly in cases of active accretion disks or binaries with mass and luminosity ratios near unity. Nonetheless, they provide crude estimates of the scale lengths of structure in binary environments.

This approach is particularly effective for shorter period binaries, where the small binary separations predict structure in the optically thick near- and mid-infrared domains of SEDs. It was first applied to the classical T Tauri binary GW Ori, a single-lined spectroscopic binary having a circular orbit with separation ≈ 1 AU (Mathieu et al 1991). The SED of GW Ori shows a broad dip around 10 μm, as well as a very strong 10 μm silicate emission feature. Mathieu et al (1991) were able to model the SED with a disk having an annular gap from

0.17 AU to 3.3 AU. This gap encompassed the independently determined orbit of the secondary, consistent with the secondary having cleared a gap in the associated disk.

Several other spectroscopic binaries also have complex SEDs, including 162814-2427, 162819-2423S, and V4046 Sgr (Figure 5). Jensen & Mathieu (in preparation; Figure 5) have modeled the SED of the binary 162814-2427 (projected semimajor axis of 0.27 AU, $e = 0.5$), which shows very little excess emission in the near-infrared but large excesses at 10 μm and 20 μm. They find the binary to reside within a nearly evacuated central hole in a circumbinary disk. They find the inner radius of the circumbinary disk to be ≈ 0.3 AU, while the projected maximum distance of the stars from the system center-of-mass is 0.2 AU. Thus, the derived scale length is again reasonably consistent with the binary dynamically setting the inner edge of the circumbinary disk.

Similarly, Marsh & Mahoney (1993) argue that the SED of DF Tau indicates an evacuated region from 0.1 AU to 17 AU (with uncertainties on both values

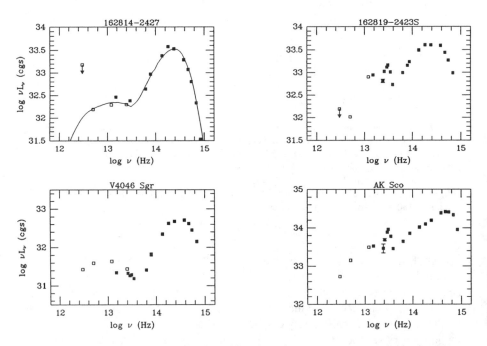

Figure 5 The spectral energy distributions of four pre-main-sequence spectroscopic binaries. The filled squares are ground-based data from the literature and unpublished observations; open squares are *IRAS* data. Silicate emission features are seen at 10 μm in 162819-2423S and AK Sco. The solid curve fit to the observations of 162814-2427 is a theoretical model derived from a spatially thin, optically thick disk with a nearly evacuated central hole of radius 0.3 AU (Jensen & Mathieu, in preparation).

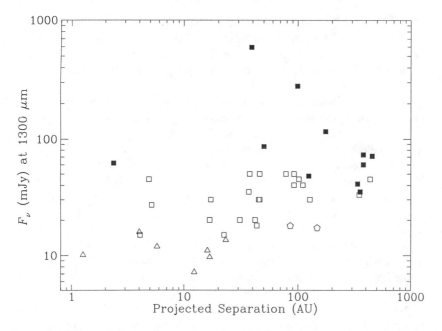

Figure 6 1300 μm continuum fluxes F_ν as a function of projected separation for pre-main-sequence binaries in the Tau-Aur region. Filled symbols are detections, open symbols are 3 σ upper limits. Squares are measurements from Beckwith et al (1990); triangles and pentagons are fluxes inferred from 800 μm measurements of Jensen et al (1994) and 1100 μm measurements of Skinner et al (1991), respectively, both scaled assuming optically thick emission. Note the weaker emission from the binaries with smaller projected separations. (From Jensen et al 1994.)

of a factor $\approx$ 2). The projected separation of the binary is 12 AU. Given the uncertainty, the inner radius of the circumbinary disk might plausibly be set by tidal truncation (Artymowicz & Lubow 1994). However, the small circumstellar disk radius is curious; unless the binary orbit is very eccentric it is difficult to attribute such a small outer radius to tidal truncation. On the other hand, attributing it to depletion of the circumstellar disk by accretion poses the question of why any circumstellar disk remains at all.

Mathieu et al (1991) also suggested that the strong silicate emission feature of GW Ori might be due to very small amounts (less than one Earth mass) of warm optically thin dust in the gap. Thus, silicate emission may be another diagnostic for cleared regions in disks, particularly within a few AU of the stars where dust is warm. DF Tau also has a silicate emission feature (Cohen & Witteborn 1985, Marsh & Mahoney 1993), as do the binaries AK Sco, 162819-2423S, V853 Oph, and SR24N (Jensen & Mathieu, unpublished data; Figure 5). Based on the presence of a silicate emission feature, Ghez et al (1991) argue that the disk around T Tau N is optically thin at 10 μm (but see Weintraub et al

1992). However, strong silicate emission has not been detected from 162814-2427 (Jensen & Mathieu, unpublished data) or V4046 Sgr (Hutchinson et al 1990). Furthermore, strong silicate emission is not unique to binary systems (Cohen & Witteborn 1985) so that attributing its origin to dynamically cleared regions is not secure.

The impact of binaries on disk structure may also be evident in submillimeter emission. Beckwith et al (1990; see also Beckwith & Sargent 1993) noted that binaries with projected separations less than 140 AU showed very little disk mass, while wider systems had typical disk masses. Hence, they suggested that the presence of close companions inhibits large disk masses. Since that study, numerous "single" stars for which they had measured disk masses have been discovered to be binaries. In addition, more stringent upper limits have been set on the submillimeter fluxes of some Tau-Aur binaries at the smallest separations (Jensen et al 1994). The distribution of 1.3 mm flux with projected separation for binaries in Tau-Aur is shown in Figure 6. A marked change in detection frequency is evident at a projected separation of $\approx$ 50 AU, with most of the upper limits for binaries with separations between roughly 5 AU and 50 AU falling below the detections among wider binaries.

Adopting the simple disk model described above and temperatures and surface densities typical for 0.01 $M_\odot$ disks around single stars, Jensen et al (1994) found that the lower fluxes of the closer binaries could be explained by the presence of wide gaps, for example, as might be cleared by binaries with substantial orbital eccentricities (e.g. $e \approx 0.4$; Artymowicz & Lubow 1994). The present upper limits press hard on this simple model because of predicted circumbinary disk emission in the closer binaries; still lower flux limits would rule it out. Nonetheless, the more general point is that the weak submillimeter emission from binaries with separations of tens of AU does not necessarily imply the absence of large surface-density disks everywhere in the binary environments. Rather, it may be evidence of reduced disk surface area resulting from dynamical redistribution of disk material by the binaries. Indeed, the submillimeter data comfortably permit most binaries with infrared excesses to have tidally truncated circumstellar disks with surface densities similar to disks around single stars.

Interestingly, several submillimeter detections have been made of binaries with separations of a couple of AU or less. Beckwith et al (1990) derived a disk mass of 0.005 $M_\odot$ for the occultation binary HP Tau with a projected separation of 2.4 AU. GW Ori, with a separation of 1 AU, is among the most luminous PMS stars at submillimeter wavelengths. Mathieu et al (1994) place a lower limit on circumbinary material of 0.3 $M_\odot$; if the emission is taken to originate in a disk they find a disk mass of 1.5 $M_\odot$ with an uncertainty of a factor of 3 due to dust opacity. Jensen et al (in preparation) have also detected emission at 800 μm and 1100 μm from AK Sco (projected semimajor axis of 0.14 AU) and V4046 Sgr (0.024 AU). DQ Tau, CW Tau, and UX Tau A are candidate spectroscopic binaries (Mathieu 1992b) with strong submillimeter detections (Beckwith et al 1990). All other definite spectroscopic binaries in Tau-Aur and Sco-Oph have also been observed, without detection (Skinner et al 1991, Jensen et al, in preparation).

Disk Accretion

As noted in the previous section, if binaries prevent a replenishing flow of material from circumbinary disks, then continued accretion will exhaust circumstellar disks and end accretion at the stellar surfaces. The exhaustion timescale will depend, among other things, on the masses of circumstellar disks and on accretion rates. A reasonable presumption is that the reservoir of circumstellar disk material tends to decrease with decreasing binary separation. In addition, dynamical enhancement of accretion rate increases with shorter period (Ostriker et al 1992). Consequently, without replenishment the lifetimes of both massive circumstellar disks and active accretion at stellar surfaces would be expected to decrease with decreasing binary period. Hence, a critical observation is to assess whether substantial accretion continues in shorter period binaries.

Because it is a widely available datum, $H\alpha$ emission strength is often used as a diagnostic for active accretion (e.g. Hartigan et al 1990). Simon et al (1992) do not find any difference in the distribution of $H\alpha$ equivalent width between single and multiple systems, considering each sample as a whole. Leinert et al (1993) reach the same conclusion by comparing the frequencies of classical T Tauri stars and weak-lined PMS stars, both in their total sample and as a function of projected separation between 18 and 1800 AU. On the other hand, Ghez et al (1993) find that the relative frequency of classical T Tauri stars does decrease significantly with decreasing binary separation over the range of 16 to 252 AU. They conclude that nearby companion stars shorten the time of active accretion. Resolution of the differing conclusions of Leinert et al (1993) and Ghez et al (1993) is critical to understanding disk evolution. Although a careful comparison of the analyses is merited, it is likely that resolution will be best served with larger samples of PMS stars observed at comparable or higher angular resolution. A preliminary look can be obtained by considering the 11 resolved binaries in Tau-Aur and Sco-Oph with projected separations between 1 AU and 16 AU [see Table A1 plus SR 20 (Ghez et al 1993)]. Based on the classifications of Herbig & Bell (1988), except for GN Tau which was only recently discovered to have an $H\alpha$ equivalent width of 62 Å (Briceno et al 1993), this binary sample includes five classical T Tauri stars, four weak-lined PMS stars, HQ Tau with unknown status, and V807 Tau, a classical T Tauri hierarchical triple in which the emission cannot yet be associated with any particular component. Thus, in this small sample of shorter period PMS binaries there is no evident paucity of classical T Tauri stars.

Spectroscopic techniques provide a binary sample of yet smaller separations. Somewhat surprisingly given the intensive study of classical T Tauri stars over the years, very few spectroscopic binaries have been definitively found among them.[3] Spectroscopic orbits have been determined for at most three—V4046 Sgr (de la Reza et al 1986), AK Sco (Andersen et al 1989), and GW Ori (Mathieu et al 1991). [The ambiguity in the count is due to some uncertainty in the classical T Tauri status of V4046 Sgr and AK Sco (see Mathieu 1992b)]. However, systematic high-precision radial-velocity searches for classical T Tauri spectroscopic binaries have also been few (Herbig 1977, Hartmann et al 1986, Mathieu 1992b). In addition, sample sizes have been small, and possibly biased due to the difficulty in measuring accurate radial velocities for the more rapidly rotating or heavily veiled classical T Tauri stars. Mathieu (1992b) analyzed a well-observed sample of 25 classical T Tauri stars in the Tau-Aur association. A few candidate velocity variables have been identified, but neither periods nor orbits have yet been determined. A statistical analysis of the sample, using Monte Carlo techniques to compare the observed velocity distributions with those expected from the field main-sequence solar-type binary population, showed the sample size to be too small to meaningfully test whether the frequency of short-period classical T Tauri binaries differed from the MS frequency. Still, the difficulty in uncovering short-period classical T Tauri binaries stands in intriguing contrast to studies of X-ray-selected weak-lined PMS stars. The first PMS spectroscopic binary, V826 Tau, was discovered serendipitously among only five X-ray detected PMS stars, and in a subsequent study of 55 such stars 7 more spectroscopic binaries were easily found (Mathieu et al 1989; Mathieu 1992b).

Case studies can be as enlightening as statistical results. For example, the binary DF Tau, with a projected separation of 12 AU (Ghez et al 1993), is one of the more active T Tauri stars, showing a large ultraviolet excess, heavy spectral veiling, and an Hα equivalent width of 54 Å. Bertout et al (1988) derived an accretion rate of a few times $10^{-7} M_\odot$ yr^{-1} from the SED. Evidently, accretion is occurring in this binary system at a rate comparable to the rates seen in single T Tauri stars. DF Tau is also very young ($\approx 10^5$ yr; Simon et al 1993). Consequently, the presently accreting material might arguably have its origin in circumstellar disks even if unreplenished, requiring these disks to have initial masses of a few hundredths of a solar mass given a constant accretion rate. Such masses are not a priori implausible at separations of order 12 AU, although they are in marked contrast to the upper limit of 0.01 $M_\odot$ on the present disk mass of DF Tau (Beckwith et al 1990), as well as the model of Marsh & Mahoney (1993) in which the DF Tau circumstellar disk is only tenths of an AU in size.

---

[3] As is so often the case in the field of pre-main-sequence stars, this point was anticipated in a sentence written 30 years previously by Herbig (1962): "If one considers the fraction of normal dwarfs that occur in binaries, it is rather curious that no rapid spectroscopic binary has been detected among the T Tauri stars, unless contractive evolution of this type is limited to single stars, or very wide pairs."

GN Tau, with a minimum projected separation of 6 AU, is likely similar to DF Tau as it has a continuum spectrum at low dispersion, Hα equivalent width of 62 Å, and is detected by *IRAS* at 12 μm, 25 μm, and 60 μm (Briceno et al 1993, Strom et al 1989a).

GW Ori, another very young binary, also has strong Hα emission (28–57 Å in the literature). In addition, Mathieu et al (1991) found the near-infrared luminosity to be greatly in excess of that which could be provided by a reprocessing disk, from which they inferred a large accretion rate. However, with a stellar separation of only 1 AU, the large near-infrared luminosity cannot plausibly be explained in terms of steady accretion from an unreplenished circumstellar disk. Either replenishment or another energy source for the luminosity than viscous accretion is needed. Interestingly, Basri & Batalha (1990) found no evidence for spectral veiling or boundary layer emission from GW Ori.

A third case is the binary 162814-2427. As discussed above, this binary appears to reside within a central hole in a circumbinary disk. 162814-2427 has an Hα equivalent width of only a few Å, consistent with little accretion from a circumstellar disk. Nonetheless, the Hα line profile is at times strongly inverse-P-Cygni, with a velocity width in the redshifted absorption of several hundred km s^{-1} (Walter et al 1994). Typically, such line profiles have been interpreted as diagnostics for infalling material (e.g. Calvet & Hartmann 1992).

These cases indicate that accretion can occur in even the short-period binary environment. In addition, near- and mid-infrared emission excesses are found in these and other short-period PMS binaries (Table A1), which shows that binary circumstellar environments are not entirely exhausted. Evidence for substantial accretion rates in close binaries has thus far been found primarily in rather young systems, and infrared excesses require relatively small amounts of circumstellar material. The challenge to observers is to better understand the distribution of accretion diagnostics with *both* binary separation and age, and to clarify the nature of circumstellar disks in binaries.

INFRARED COMPANIONS

One of the most intriguing findings in the past decade has been the discovery of PMS binaries with "infrared companions." The first infrared companion was found associated with T Tau, and this system remains the best studied; we consider it here in detail as a case study. Using one-dimensional speckle interferometric techniques, Dyck et al (1982) found T Tau to be an infrared pair, oriented very nearly north-south with a separation of 0.6″. An ambiguity as to which of the pair was associated with the optical star was resolved by subsequent optical astrometric and infrared speckle observations which showed the northern source to be the optical star (Hanson et al 1983, Schwartz et al 1984). Recent speckle observations measure an angular separation of 0.73″ ± 0.01″, or a projected separation of 102 AU (Ghez et al 1991). Both components are thermal radio sources; the southern star is more luminous at radio wavelengths. Schwartz et al (1986) concluded that it is the infrared companion that drives the associated molecular flows and Herbig-Haro objects. They also detected an emission bridge between the two sources at 2 cm wavelength, the origin of which remains unclear but which provides further support for association of the two sources. A third optical component was reported by Nisenson et al (1985), but the companion has not been detected in later observations (Gorham et al 1992, Ghez et al 1991).

Ghez et al (1991) and Gorham et al (1992) have obtained resolved photometry of the T Tau system between wavelengths of 0.6 μm and 20 μm; the SEDs of both components are shown in Figure 7. The SED of T Tau N is similar to other classical T Tauri stars. More notably, essentially none of the optical luminosity derives from T Tau S, the infrared companion. However, at wavelengths longward of 3 μm the energy density of T Tau S exceeds that of T Tau N by as much as a factor of four. Integrated from 0.8 μm to 20 μm the luminosity of the infrared companion is twice that of the optical component (Ghez et al 1991). Interestingly, Ghez et al (1991) find T Tau S to show a two-magnitude increase in infrared fluxes over a five-year interval, which they suggest is color-independent.

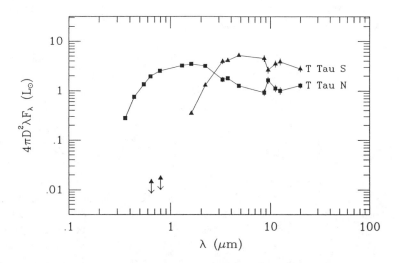

Figure 7 The spectral energy distributions of T Tau N (*squares*) and T Tau S (*triangles*). The infrared companion T Tau S is undetected in the optical but contributes twice as much luminosity as T Tau N in the wavelength range 0.8 μm to 20 μm. The 10 μm silicate feature is in emission for T Tau N and in absorption for T Tau S, suggesting a large extinction toward the latter. (From Ghez et al 1991.)

The exinction toward the infrared companion is difficult to assess given the absence of light from an optical photosphere. Bertout (1983) argues for a visual extinction between 8 mag and 19 mag. Presuming the companion to be an embedded stellar source, Whitney & Hartmann (1993) suggest an extinction of $\approx$ 5 mag in the J band (1.25 μm), or $A_V \approx$ 18 mag (Savage & Mathis 1979). Ghez et al (1991) detect a 10 μm silicate absorption feature associated with T Tau S, from which they derive a smaller extinction of $A_V \approx$ 4.6 mag. While they differ, all of these estimates are substantially larger than the visual extinction of $A_V = 1.4$ toward T Tau N (Cohen & Kuhi 1979). Hence Ghez et al (1991) argue that the material dimming T Tau S must be located around T Tau S alone. It is also noteworthy that while the SED of T Tau S is much redder than a typical T Tauri star and suggests a source having high extinction, it nonetheless peaks in the near infrared and thus is not characteristic of a protostellar SED.

The optical/near-infrared properties of T Tau are characteristic of the sample of infrared companions. At present, a list of infrared companions must include Glass I and DoAr 24E (Chelli et al 1988), Haro 6-10 (Leinert & Haas 1989), XZ Tau (Haas et al 1990), Z CMa (Koresko et al 1991, Haas et al 1993), and UY Aur (Tessier et al 1994). An additional five companions for consideration are listed by Zinnecker & Wilking (1992): UZ Tau (Moneti & Zinnecker 1991), SSV 63, WSB 4 (Meyer et al 1993), Ser/G1, and VV CrA. Zinnecker & Wilking (1992) estimate the frequency of infrared companions to be approximately 10% of all T Tauri binary systems.

Large luminosities of infrared companions relative to their optical companions are not unusual. Chelli et al (1988) find the infrared companion of Glass I to be nearly a factor of six more luminous than its optical companion (attributing all of the *IRAS* flux to the infrared component). For XZ Tau Haas et al (1990) bracket the luminosity ratio of the infrared companion to the optical companion between 3 and 10, and for Haro 6-10 Leinert & Haas (1989) similarly limit the luminosity ratio to between near 1 and 7. The infrared companion of Z CMa is more luminous than the optical FU Ori star, which led Koresko et al (1991) to conjecture that the infrared companion is also an FU Ori star. However, based on optical spectropolarimetric data showing highly polarized emission lines, Whitney et al (1993) propose that the infrared companion may be an Ae/Be star. Only DoAr24E is notably different; the luminosity of its infrared companion is only a fifth that of its optical companion. These numbers can only be taken as representative, since large near-infrared variability is a characteristic trait of systems with infrared companions. The variability may be associated with either the optical or infrared companions, or both (Chelli et al 1988, Ghez et al 1991, Haas et al 1993, Menard et al 1993, personal communications to Zinnecker & Wilking 1992).

That infrared companions are in fact stellar and physically associated with their optical companions is argued in several of the discovery papers (e.g. Leinert & Haas 1989). The presence of ice or silicate absorption features in several infrared companions (T Tau, XZ Tau, Haro 6-10, and DoAr24E; see references above) also proves that infrared companions are being seen through large column densities of dust grains. In the case of T Tau, silicate emission is seen from the optical component, which emphasizes the very different conditions along the lines of sight to the two stars. Interestingly, Ch. Leinert (personal communication) reports that the ice feature of Haro 6-10 is variable.

While the lack of optical emission may be ascribed to extinction, the spectral energy distributions (SEDs) of infrared companions cannot be solely attributed to reddened photospheres. The presence of an infrared excess is clearly evident in the SED of T Tau S (Figure 7), as well as the SEDs of other infrared companions. If the *IRAS* fluxes are attributed primarily to the infrared companions, then the infrared excesses from the near-infrared through 100 μm indicate circumstellar material from near the stellar surfaces to distances of at least several AU. Many of the optical companions also have near-infrared excesses from associated circumstellar material [e.g. T Tau N (Figure 7), XZ Tau (Haas et al 1990), Haro 6-10 (Leinert & Haas 1989)].

The ages derived from theoretical tracks for the optical components do not necessarily place them among the youngest PMS binaries. Chelli et al (1988) derive an age of 1–3 million years for DoAr24E, Leinert & Haas (1989) make a similar age estimate for Haro 6-10, and Simon et al (1993) find an age of 0.8 million years for T Tau, typical for their sample of Taurus PMS stars. Chelli et al (1988) and Simon et al (1993) do find Glass I and XZ Tau, respectively, to be very young. Based on their frequency estimate of $\approx$ 10%, Zinnecker & Wilking (1992) find typical ages $\approx 10^5$ yr *if* systems with infrared companions are precursors to binary T Tauri stars (typical age $\approx 10^6$ yr). They also note a somewhat higher frequency (75%) of 1.3 mm continuum emission detections among binaries with infrared companions than among all stars surveyed by Beckwith et al (1990), also suggestive of youth. However, this frequency is not measurably different from the detection rates of Beckwith et al for wide binaries (64%, projected separations greater than 100 AU) or for single stars (55%).

The reader may have noticed that an "infrared companion" was not defined above, but rather was introduced by the case of T Tau. With the advent of extensive near-infrared high-resolution surveys for binaries the definition of an infrared companion has become somewhat ambiguous. On the purely semantic side, many optical PMS stars have been newly discovered in the infrared to be binaries. Occasionally such stars will be said to have "infrared companions." However, the fact that they were discovered to be binaries in the infrared need not reflect anything further than the wavelength used for the observations;

in most cases it is not yet known whether either of the components have unusually small optical fluxes. Furthermore, the term "infrared companion" has recently also been applied to optical secondaries having redder infrared colors than their primaries. For example, Zinnecker & Wilking (1992) include the long-known optical binaries UZ Tau and Ser/G1 in their list of PMS binaries with infrared companions, despite the fact that in both cases the primary and secondary optical fluxes differ by only a factor of a few. Such a breadth of definition likely masks important physical distinctions. For example, unusual near-infrared colors of optical PMS stars may be a reflection of the amount or radial distribution of disk material (Moneti & Zinnecker 1991), while very red SEDs more typical of the original infrared companions may be indicative of more embedded sources, perhaps at earlier stages of evolution. When angularly resolved SEDs are available for a larger sample of PMS binaries, a classification of PMS binaries relying on the SEDs of each component, akin to the Classes I–III of Adams et al (1987), will provide an organization with more physical significance.

AGES

The distribution in the H-R diagram of PMS binary stars is not evidently different from apparently single stars (Figure 8). Perhaps most importantly, binary stars of all periods have been found among the very youngest PMS stars, i.e. at the upper envelope of PMS stars in the theoretical H-R diagram. If this upper envelope is taken to define a "stellar birthline" (Stahler 1983), the simple but nonetheless important conclusion is that binary formation occurs prior to the pre-main-sequence phase of evolution. This conclusion is further supported by the discovery of binaries still embedded in molecular cores, such as SVS 20 (Eiroa et al 1987) and several cases discussed by Zinnecker (1989), and perhaps even protobinaries (discussed below).

In fact, Simon et al (1993) noticed that multiple systems preferentially populate the very upper envelope of PMS stars in the H-R diagram, and consequently have been assigned systematically younger ages than single stars. They attribute these trends to overestimated luminosities for the multiples due to their multiple contributions of light. After correction by the measured K flux ratios, Simon et al (1993) find that the ages of single and multiple systems are indistinguishable. Taking the ages derived from single stars to be more secure, they find that the average age of stars in the Taurus star-forming region is a factor 2–3 times older than previously thought. This age increase translates directly into numerous important characterizations of PMS stars, including disk lifetimes, total mass accretion, angular momentum deposition, etc. The empirical location of the stellar birthline, previously fit well by theory, is also lowered in luminosity.

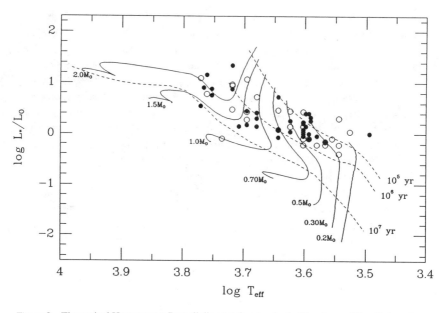

Figure 8 Theoretical Hertzsprung-Russell diagram for stars in the Tau-Aur and Sco-Oph regions. The open circles represent stars detected as multiple at speckle resolutions (projected separations 16–252 AU); filled circles represent stars without speckle companions. For the multiple systems the estimated contributions from each companion star have been removed to obtain the luminosities, so the positions are those of the primary stars. Also shown are evolutionary tracks from D'Antona & Mazzitelli (1994). (Adapted from Ghez et al 1993. Note that the sample is magnitude limited at K.)

In terms of binary formation theory, the relative ages of binary components are as critical as the absolute ages. Components of binaries with separations greater than a few arcseconds can be independently compared to theoretical tracks from which ages can be derived. It should be noted that such ages are measured from the stellar birthline and thus contain no measure of the duration of earlier phases. Walter et al (1988) found good agreement between the estimated ages of the components of five naked T Tauri pairs. Hartigan et al (1994) have recently completed a major study of 39 PMS binaries. For 26 of these they were able to determine effective temperatures and luminosities for both components, which they compared to three sets of PMS evolutionary tracks. They find no measurable age differences between components of two thirds of the binaries. For the remaining third they do find significant age differences, usually in the sense that the less massive star is the younger. There is no evident correlation of this non-coevality with any other stellar or binary property. The lack of more pronounced age differences in the presence of active accretion is particularly interesting in light of the suggestion that accretion will delay evolution down the Hayashi tracks (Hartmann & Kenyon 1990).

MASSES OF PRE-MAIN-SEQUENCE STARS

Historically, one of the foremost roles of binaries in astrophysics has been in the measurement of stellar masses. Such measurements are very much needed for PMS stars; at present the absolute mass calibrations of theoretical evolutionary tracks are unconstrained by dynamical mass determinations. Until very recently, astrometric mass determinations have not been technically feasible given distances of at least 100 pc to the nearest star-forming regions. In addition, only one eclipsing binary having an approximately solar-mass PMS star has been discovered: EK Cep (Tomkin 1983, Hill & Ebbighausen 1984, Popper 1987). This system consists of a 2.03 $M_\odot$ ZAMS primary and a 1.12 $M_\odot$ secondary. Popper, following in detail a suggestion of Tomkin, has argued that the secondary is oversized for a ZAMS star. Given that the more massive primary is on the ZAMS, he concludes that the secondary is a PMS star. His preliminary analysis finds the position of the secondary in the theoretical H-R diagram to be between the 1.0 $M_\odot$ and 1.25 $M_\odot$ evolutionary tracks of Iben at an age of 2×10^7 yr, consistent with the dynamical mass determination. Martin & Rebolo (1993) have repeated the comparison with newer theoretical models and find a similar age. They find the secondary to be slightly more luminous than predicted from the 1.1 $M_\odot$ theoretical tracks, but nonetheless in agreement given the measurement errors. Unfortunately, the star is very near the ZAMS so that it places only weak constraints on PMS evolutionary tracks.

An alternative approach is to test the *relative* mass calibrations of PMS evolutionary tracks with double-lined PMS spectroscopic binaries. Such binaries provide accurate dynamical mass ratios. If theoretical masses can be determined for each component of such a binary, then the theoretical and dynamical mass ratios can be immediately compared. Marschall & Mathieu (1988) first used this approach for the binary P1540, using the PMS evolutionary tracks of Cohen & Kuhi (1979). They were only able to bring the theoretical masses into agreement with the dynamical mass ratio by assuming non-coeval formation of the binary components. A more sophisticated study has recently been done by Lee (1992) for five double-lined binaries (including P1540). He found that the relative mass and age calibrations of the pre-main-sequence tracks of both Mazzitelli and VandenBerg are consistent with the measured dynamical mass ratios and similar ages of the binary components (within a couple Myr in most cases). These evolutionary tracks are for single stars, and so this conclusion presumes that on timescales of Myrs the internal evolution of the stars in close binaries is similar to that of isolated stars.

DISCUSSION

The PMS Binary Population

The recent discoveries of PMS binaries have yielded several basic insights. Foremost among them is that a rich population of binaries exists by the pre-main sequence stage of evolution. Indeed, binaries are found among the very youngest PMS stars. Evidently at least some, and quite likely most, binaries form prior to the stellar birthline. Second, PMS binaries have been found to have semimajor axes ranging over five decades. Only the precursors of the very widest binaries in the field have not yet been securely identified in the PMS population. In fact, PMS pairs with appropriately large projected separations are known (e.g. Hartigan et al 1994), but whether they are bound is not certain. These widest binaries represent a particularly interesting regime of separation given their fragility, and the suggested paucity of such binaries in the survey of Reipurth & Zinnecker (1993) is intriguing but not at all definitive. Third, the log-normal shape of the period distribution found among MS binaries is already at least approximately present among PMS binaries. Finally, hierarchical organizations of triple and higher order systems are also established early in the PMS phase.

The discovery of PMS binary frequencies substantially in excess (factors of 2–4) of MS solar-type binary frequencies was unexpected, and critical inquiry has only just begun. As to statistical significance, Leinert et al (1993) argue that the observed excess found in the Tau-Aur association is significant at the $3\,\sigma$ confidence level. Ghez et al (1993) find the discrepancy between the PMS and MS binary frequencies to be a $2.5\,\sigma$ effect in their combined Tau-Aur and Sco-Oph sample. The enhancement in the Sco-Oph sample alone is at less than the $2\,\sigma$ confidence level. Reipurth & Zinnecker (1993) conclude only that there is "possibly an excess" among stars drawn from a large sample of star-forming regions. Hence, while a high intermediate-period binary frequency in the Tau-Aur association seems likely, the extension of that result to low-mass binary formation in general is not yet established. Still, the measurement of enhanced binary frequencies in several associations and over several decades of semimajor axis should caution against overly quick attribution to the vagaries of sampling statistics.

Along with evolutionary state, the MS and PMS binary surveys are quite different in sample selection and observational techniques. This raises concerns about the direct comparison of their results. An issue that has not yet been analyzed carefully is whether the high PMS binary frequencies might be artifacts of 1. the MS studies being done in the visual as compared to 2.2 μm for several PMS surveys, or 2. the flatter mass-luminosity relationship of PMS stars on convective tracks compared to MS stars. Both effects would be expected to enhance the detectability of lower-mass companions to PMS stars. Thus, it may be significant that the *HST* Trapezium survey (Prosser et al 1994) and the CCD survey of southern regions (Reipurth & Zinnecker 1993), both done at wavelengths shortward of 1 μm, show the least enhancement in PMS binary frequency. On the other hand, the evolution of the mass-luminosity relation of PMS stars is complex, and especially when the primary is on the radiative track detection of secondaries may not be enhanced for PMS stars (Reipurth & Zinnecker 1993). Similarly, disks may preferentially enhance the brightness of the primaries at 2.2 μm (Moneti & Zinnecker 1991), which, arguably, makes detection of a companion more difficult. Finally, the PMS surveys to date have been magnitude limited at 2 μm, possibly introducing biases caused by the interplay of such limits with the PMS luminosity function. However, both Ghez et al (1993) and Leinert et al (1993) argue that the magnitude limit does not significantly influence their result. [In contrast, the MS survey of Duquennoy & Mayor (1991) was parallax limited.] Clearly, a critical analysis convolving the PMS luminosity evolution of binary components with actual detection limits is needed in order to more confidently compare PMS and MS binary frequencies.

Implicit in these comparisons with field MS binary populations is the assumption that a "ZAMS binary frequency" can be well defined for comparison to the PMS population. In fact, most existing investigations of MS binary populations have sampled field stars in the solar vicinity, within which are included blends of ages and sites of origin that cannot be easily disentangled. Furthermore, the PMS samples contain a range of masses from a few tenths to a couple solar masses, so that a single-mass MS population is unlikely to be an appropriate comparison. Indeed, Hartigan et al (1994) find typical masses of PMS binary components to be substantially less than one solar mass. [See Figure 8. Comparison of this figure with Figure 11 in Ghez et al (1993) shows vividly the dependence of inferred stellar masses on choice of evolutionary models.] Nonetheless, lower masses will not account for the observed excess of PMS binaries, since surveys of MS stars of less than a solar mass yield somewhat lower binary frequencies than the usual comparison population of MS solar-type stars (e.g. Mayor et al 1992, Fischer & Marcy 1992). Also, among the PMS binaries Leinert et al (1993) find no dependence of binary frequency on K magnitude. Ghez et al (1993) find a slightly higher binary frequency among lower-mass stars but the enhancement is not statistically significant.

If the observed enhancements of PMS binary frequencies at intermediate separations are taken at face value, the implications depend critically on the shape of the PMS period distribution. Duquennoy & Mayor (1991) derive a total binary frequency among MS solar-mass stars of 61%, and find relatively few (5%) triple or higher order systems. Consider first the premise that PMS and MS binaries have the same period distributions. Then an enhancement of a factor two in PMS binary frequency requires that compared to MS binaries either more PMS systems are multiple, more PMS multiple systems are higher order than binary, or a combination of the two. In the Tau-Aur association, the last appears to be the case. Even larger enhancements, such as the factor four argued by Ghez et al (1993), demand that essentially all PMS systems be multiple. It remains to be shown empirically whether this is the case; to date the frequency of actually detected multiple systems in any studied sample of PMS stars has been no larger than $\approx 50\%$.

Alternatively, Reipurth & Zinnecker (1993) hypothesize that the total frequencies of MS and PMS binaries may be the same, but that the PMS semimajor axis distribution is narrower than that of MS binaries and broadens with time. Ghez et al (1993) and Simon (1992) both note that the numbers of known PMS binaries in their samples (detected by all techniques) are already comparable to the numbers expected from the MS binary population integrated over all periods. Yet their surveys are not yet complete in period coverage, implying that the number of PMS companions must exceed that on the MS. Richichi et al (1994) come to a similiar conclusion, and also point out the breadth of the PMS period distribution with the addition of lunar occultation binaries (Figure 1). Finally, there is no evidence that the PMS short-period binary frequency is reduced relative to the MS (Figure 1), as might be expected for a narrower period distribution.

If binary formation processes produce binaries at a higher frequency than found on the MS, then evolutionary processes must reduce the number of binaries. Ghez et al (1993) have suggested that young triple and higher order systems are disrupted by encounters with other stars, noting the observed higher frequency of triple systems among the PMS stars. They also comment that in such encounters it is the wide pairs that are most fragile, so that the PMS frequency of the shortest period binaries would be most likely to survive into the MS binary population, consistent with the PMS spectroscopic binary results. A critical issue facing any such disruption mechanism is how to obtain sufficiently high encounter rates to break apart binaries with semimajor axes of only 10–1000 AU. The local stellar densities around PMS stars in Tau-Aur and Sco-Oph are very low and the available time short. Equally importantly, the binaries observed to have enhanced frequencies are not soft with respect to the typical velocity dispersions of associations (≈ 2 km s^{-1}; e.g. Hartmann et al 1986, Mathieu 1986). Disruption in dense clusterings might be considered, although most of the PMS binaries detected to date are not in such groups. Very wide binaries ($\approx 10,000$ AU) may be disrupted in the Galactic field over 10 Gyr, but such disruption is not expected for those at issue here (Weinberg et al 1987).

High binary frequencies in regions like Tau-Aur only represent a problem to be resolved if a large fraction of field stars derive from such associations. It has been suggested that most stars form in OB associations (e.g. Miller & Scalo 1978). In the same spirit, Lada et al (1991) have argued that a large fraction of stars form in high-stellar-density cluster environments. Such environments might produce a substantially different binary population, due to both formation processes and subsequent dynamical influences. In this context the finding of Prosser et al (1994) that the binary frequency in the Trapezium cluster is similar to that in the field is intriguing. Reipurth & Zinnecker (1993) also note a marginal trend toward higher binary frequency in dark clouds with fewer stars, although Leinert et al (1993) see no similar trend among the sub-associations of the Tau-Aur region. In total, the results to date would be consistent with the hypothesis that the formation of binaries is enhanced in the lower density environments of T Tauri associations, but that the binary population of the field derives largely from higher density cluster environments or OB associations. The binary frequencies of the low-mass populations of OB associations and young clusters are critically in need of investigation.

Whatever may be the final resolution of the observed high PMS binary frequencies, there is no doubt that many more than half of the usually studied PMS stars in the Tau-Aur population are binaries. Ultimately one of the most significant consequences of this result will be to introduce substantial ambiguities into the interpretation of the many observations, past and future, that do not resolve the binary components. Analyses of such data in the context of single stars may be misleading.

Dynamical Evolution of Disks in the Young Binary Environment

A binary embedded within a disk will establish a disk structure characterized by dynamically cleared regions on size scales of the binary semimajor axis (for a review see Lin & Papaloizou 1993). In addition, the transfer of energy and angular momentum between binary and disk can dictate the evolution of both the disk and the binary orbit (e.g. Artymowicz et al 1991, Pringle 1991a, Ostriker et al 1992, Bonnell & Bastien 1992). Unfortunately, the details of both disk structure and evolution in young binary environments are intimately linked to the conditions set up by formation processes (coplanarity, circumstellar and circumbinary disk masses, etc). As such, predictions for the nature of disks by the PMS phase are both varied and uncertain (e.g. Bonnell & Bastien 1993, Clarke & Pringle 1993). Of course, that variety may reflect reality.

Nonetheless, some framework is needed within which to begin consideration of observations of PMS binaries and associated disks. In this spirit we develop here a simple cartoon for the evolution of an isolated, coplanar, binary-disk system. We anticipate that the assumption of coplanarity will be valid for the shortest period systems and less secure with increasing orbital period (e.g. Hale 1994). We consider initially the case of a binary with a circular orbit and a small secondary/primary mass ratio. We ignore the details of binary formation, and ad hoc begin with the binary embedded in a continuous coplanar disk having an outer radius much greater than the binary separation and a mass less than that of the secondary star. Given the posited binary mass ratio, the primary star lies near the center of this disk. On a dynamical timescale the companion will tidally clear an annular gap containing its orbital path. Thus, circumstellar disks are established, with the circumprimary disk being the larger of the two (Artymowicz & Lubow 1994, Lin & Papaloizou 1993). The inner radius of the circumbinary disk is set by a balance between the angular momentum transport in the disk and the torque provided by the binary. Artymowicz et al (1991; see also Clarke 1992b) argue that this torque is achieved at narrow resonances in the circumbinary disk whose strengths decrease rapidly with increasing radius. Consequently, they assert that the inner radius of the circumbinary disk is insensitive to disk viscosity. Their SPH calculations find the inner radius to be approximately twice the binary semimajor axis. However, Lin & Papaloizou (1993) have argued that the influence of the resonances is not so localized within the disk, and that given smaller viscosities and higher sound speeds, larger inner radii of circumbinary disks can exist.

Once this annular gap is established, disk evolution continues on slower (e.g. viscous) timescales. Accretion may proceed in the circumstellar disks, and indeed for sufficiently massive disks it may be dynamically enhanced by the companions (Ostriker et al 1992, Bonnell & Bastien 1992). Because of the torques provided by the binary, material is not expected to flow from the circumbinary disk to the circumstellar disks (Artymowicz et al 1991, Pringle 1991a). Thus, unless replenished from otherwise infalling material, accreting circumstellar disks will exhaust themselves. Depending on the timescale of this exhaustion with respect to the circumbinary disk lifetime, the binary may then reside in a hole within the circumbinary disk. Outside of the binary orbit, Pringle (1991a; see also Artymowicz & Lubow 1994) finds that the circumbinary disk will adjust its surface density distribution so as to match the angular momentum drawn from the binary with the viscous angular momentum flow of the disk. Once the appropriate surface density distribution has been established, a prolonged phase ensues during which the circumbinary disk is expelled.

Such disk evolution requires a response from the binary. From energy arguments the semimajor axis of the binary must decrease. The direction in which the orbital eccentricity will evolve depends on the relative rates of energy and angular momentum extraction. Recent simulations have found that circumbinary disks will increase the eccentricity of low-eccentricity orbits (Artymowicz et al 1991, Lin & Papaloizou 1993). Indeed, a particularly interesting theoretical result is the high efficiency with which a circumbinary disk of relatively small mass (e.g. $10^{-2}\ M_\odot$, the minimum mass solar nebula) will drive orbital evolution (Pringle 1991a, Artymowicz et al 1991, Artymowicz 1992, Clarke 1992b). Binary orbital elements can be expected to change significantly in only thousands of orbital periods—a timescale less than PMS disk lifetimes for binaries with semimajor axes less than 100 AU and indeed short enough to be observable in the shortest period binaries. We further consider this point in the discussion of orbital evolution below.

This binary-disk system is the simplest case. Among other things, the detailed distribution of disk material will depend on both the binary mass ratio and the orbital eccentricity. For example, stars in a circular binary with unit mass ratio are equidistant from the center of mass of the system. Consequently, the morphology of the dynamically cleared region changes qualitatively, but the essential nature of the disk evolutionary history will not change. The interaction of a disk with an eccentric binary is a more difficult problem. In terms of resonant clearing the primary modification is that for given semimajor axis the inner radius of the circumbinary disk increases with eccentricity and the outer radii of the circumstellar disks decrease (Artymowicz & Lubow 1994). Finally, the evolution of disks in non-coplanar systems, an important issue, is not addressed further here.

Observational tests of this cartoon picture have only recently begun to appear. Most broadly, the reduced submillimeter continuum emission from binaries with separations between $\approx$ 5 AU and $\approx$ 50 AU (Figure 6) clearly indicates that disks in which binary stars are embedded are different from disks in wider binaries or disks around single stars. The most straightforward interpretation is simply that disk surface densities are everywhere lower in the presence of an embedded binary, due to binary formation processes or later dynamical dispersal of disk material from the system. As such, the cartoon picture described above would have little application. However, the lower submillimeter emission also can be explained as the result of reduced emitting surface area resulting from dynamical clearing. Given present observations, disk material outside the dynamically cleared regions need not differ greatly from disks around single stars. More sensitive submillimeter observations will enable us to distinguish between these two interpretations.

A specific prediction of this picture is that circumbinary disk material will not extend inward to within a few times the semimajor axis from the center of mass. This configuration should be established rapidly and maintained for the lifetime of a disk. Modeling of the recent interferometric maps of GG Tau jibe with this prediction (Figure 4 and Dutrey et al 1994); the binary has a projected separation of 41 AU and resides within a cavity of outer radius $\approx$ 180 AU. The models also suggest that most of the circumbinary material is distributed in a narrow ring, which if confirmed would place interesting constraints on the dynamical processes by which the cavity was cleared (e.g. Pringle 1991a). The presence of GG Tau/c at a projected separation of 1400 AU (10.3″ ; Leinert et al 1993) is likely relevant. Koerner et al (1993a) find the ^{13}CO (2→1) emission to be elongated toward GG Tau/c, which suggests tidal interaction, and have since detected ^{13}CO (2→1) emission at the GG Tau/c position (DW Koerner, personal communication).

Analyses of the SEDs of several PMS binaries with separations of $\lesssim$ 10 AU also suggest depletion of disk material on size scales comparable to the binary separations (cf discussions above of GW Ori, 162814-2427, and DF Tau). Quantitatively, the inner radii for circumbinary disks derived from observations do not agree with predictions from resonant dynamics to better than factors of a few (e.g. Artymowicz & Lubow 1994). This in itself is not a critical concern, for the disk emission models used are very simple, particularly in the characterizations of the temperature and surface-density distributions in the vicinity of any gaps. The expected thermal and density distributions of disk material in the young binary environment is a critical gap in our present theoretical understanding, and leaves observers without a secure framework within which to interpret results.

A more serious criticism is that structure in infrared SEDs of PMS stars are not uniquely associated with known binaries. Marsh & Mahoney (1992, 1993) have pointed out several PMS stars with SEDs akin to that of GW Ori, of which only DF Tau is a known binary. While they show that these SEDs can be fit with disks having annular gaps, which they attribute to undetected low-mass companions, the successful fits may simply reflect the flexibility of the disk-emission model of Mathieu et al (1991). As an alternative, Boss & Yorke (1993) have argued that the structure in the SED of T Tau can be produced by a complex temperature distribution in a spatially continuous disk. Also, long-wavelength features in the SEDs of a wide variety of objects have been attributed to reprocessed photospheric light in non-disk-like distributions of material (e.g. Mathieu et al 1991; but see Mathieu et al 1994 and Marsh & Mahoney 1993). At the least, the observation that some short-period binaries show complex infrared SEDs reflects similarly complex environments. That this complexity is closely linked to the dynamical modification of disk structure is quite plausible, but the details of both the surface-density and temperature distributions remain a challenge to observers.

A second prediction of our cartoon is that binaries restrict accretion of material from circumbinary disks to circumstellar disks and ultimately to the surfaces of the stars. At present there is no secure evidence that accretion diagnostics are less frequent among binaries, although tantalizing suggestions are provided by the paucity of classical T Tauri stars among the closest binaries in the Taurus survey of Ghez et al (1993) and by the lack of known spectroscopic binaries among the classical T Tauri stars. Certainly, some PMS binaries of all separations show accretion diagnostics, so accretion can occur in binary environments. The issue remains whether the reservoirs for the accreting material—the circumstellar disks—are replenished.

In a similar vein, the high frequency of infrared excesses among PMS binaries of all separations shows that the circumstellar environments of many binaries are not entirely evacuated. Indeed, the available upper limits on submillimeter and millimeter emission from binaries can not rule out the presence of substantial circumstellar disks. As an illustrative example, GG Tau shows large excesses at all infrared wavelengths (and evidence for a large accretion rate; Basri & Batalha 1990), indicative of at least one circumstellar disk. If we adopt for specificity the binary parameters suggested by Dutrey et al (1994), the maximum radius of a circumprimary disk is less than $\approx$ 25 AU. Following the analysis of GG Tau by Beckwith et al (1990), the 1.3 mm flux from a 0.001 $M_{\odot}$ disk of that radius would be only 15 mJy. Thus, despite the central decrease in its millimeter emission, the GG Tau system may nonetheless include substantial circumstellar disks. Defining the nature of such circumstellar disks in all binaries with infrared excesses will be a key clue to circumstellar accretion and replenishment rates in binary environments. The problem is made yet more intriguing by those binaries whose infrared luminosities exceed that which can be provided by reprocessing of photospheric light in a passive thin disk (e.g. GW Ori and DF Tau).

Lastly, it would be of interest to determine observationally at what binary separations the evidence for binary-disk interaction fades. Wide binaries with periastron separations greater than a few times the radii of any associated disks are unlikely to dynamically influence disk evolution (E. Ostriker, personal communication). Moneti & Zinnecker (1991) found that both components of wide pairs (typical separation $\approx$ 400 AU) showed the same near-infrared emission characteristics as (apparently) single T Tauri stars, from which they concluded that at such separations there was little influence of one star on the disk of the other. In addition, among wide binaries (projected separations $\gtrsim$ 300 AU) there is no tendency for the (visual) primaries or secondaries to have larger Hα equivalent widths (Table A1), which suggests that accretion rates also are not influenced by companion stars at those separations. (Indeed, the Hα equivalent width of the 50-AU binary UZ Tau W is approximately twice that of the distant tertiary UZ Tau E.) However, both of these diagnostics reflect conditions in disks well inside of the binary separations. The extension to 800 AU of millimeter continuum emission from GG Tau suggests that PMS binaries with separations of several hundred AU may still influence the evolution of the outermost disk regions. Perhaps it is only in the very widest binaries that disk evolution might be expected to be little different from that of disks around isolated stars. Certainly, this must be true for binaries whose orbital timescales are comparable to disk lifetimes (e.g. semimajor axes greater than 25,000 AU for lifetimes of 3 Myr).

Infrared Companions

The nature of infrared companions is perhaps the least understood phenomenon among PMS binaries. At first glance the pairing of optical and infrared companions challenges our intuition of coeval formation and evolution of binary components. SEDs peaking in the mid- to far-infrared are characteristic of very young, embedded stellar objects, the result of large extinctions through infalling envelopes (e.g. Adams et al 1987). With continued infall a protostellar SED becomes bluer as the young star and associated disk are revealed. Thus, in this scenario, infrared companions would be at substantially earlier evolutionary states than their optical companions.

Attribution of infrared companions to protostellar infalling envelopes can be problematic, however, since the SED of an infrared companion must be produced without at the same time producing a large extinction toward the optical component. Yet infrared companions cannot be extincted by infalling envelopes with radii less than the binary separations, since the lifetimes of such envelopes would be very short. For binary separations of hundreds of AU or less, the infall timescales would be less than 10^4 yr—more than an order of magnitude smaller than either typical isochronal ages of the optical companions or the statistical age estimate for infrared companions by Zinnecker & Wilking (1992).

Thus more sophisticated evolutionary scenarios are likely needed if infrared companions are to be taken to be in early evolutionary states. Rotational support can extend the lifetime of circumstellar material. Thus Hanson et al (1983) suggest that T Tau has a low-mass companion enveloped in a rapidly accreting massive disk. The large ratio of disk-to-companion mass is required so that the inflowing circumbinary disk would not be truncated by the companion. Similarly, Ghez et al (1991) conjecture that the two magnitude increase in near-infrared flux seen in T Tau is the result of a variable accretion rate, as also suggested for FU Ori stars. Continuing the FU Ori analogy, they propose that the SED of T Tau S derives from a rapidly accreting disk whose luminosity dominates the stellar photosphere. A similar scenario has been suggested for Z CMa (Koresko et al 1991, Haas et al 1993; but see Whitney et al 1993).

Alternatively, the SEDs of infrared companions might result from observation of PMS stars through highly inclined circumstellar disks. In terms of infrared colors, the (apparently single) classical T Tauri stars HL Tau and R Mon are analogues to infrared companions (e.g. Leinert & Haas 1989). HL Tau has been noted as a likely case for a PMS star viewed through a highly inclined circumstellar disk (Cohen 1983, Sargent & Beckwith 1991). A similar explanation for infrared companions requires either that the optical companions no longer have circumstellar disks, or that the disks of each component are not coplanar. Since the SEDs of some optical companions indicate the presence of circumstellar disks (e.g. T Tau N), non-coplanarity would be required in at least some cases. Note that explanations of infrared companions based on disk extinction place little constraint on the evolutionary state of the extincted stars. However, if the large luminosities of infrared companions were due to rapid disk accretion, arguably the associated disks would be geometrically thicker and have a greater likelihood of intercepting the line of sight. An interesting variant on a non-coplanar disk picture is that an infrared companion lies behind the disk of the optical companion, although Barth et al (1994) argue that this cannot be the case for Z CMa since optical light from the infrared companion is detected in scattered light.

A nonspherical infalling circumbinary envelope with proper orientation may also preferentially extinguish one binary component relative to the other. Based on polarization maps, Whitney & Hartmann (1993) argue that T Tau has a residual circumbinary envelope with polar holes centered on the optical companion. Seen nearly face-on [for which there is independent evidence in the case of T Tau (Herbst et al 1986)], they argue that such an envelope might provide a clear line of sight to the optical star but a highly extincted line of sight to the infrared companion 100 AU distant. Again, this hypothesis does not constrain the nature of the secondary, and as a general explanation does not predict large luminosities.

In contrast, Bonnell & Bastien (1993) envision the infrared companion as the more massive star of a very young binary. They argue that since the more massive star will lie at the center of mass of an infalling envelope, the line of sight to a secondary 100 AU distant will be much less extincted by the envelope. They also argue that circumsecondary disks will be smaller and shorter lived than circumprimary disks. The asserted low extinctions to the optical secondaries are in contradiction with the models of Whitney & Hartmann (1993) for the T Tau case and need to be established. Nonetheless, the large luminosities of many infrared companions may be naturally explained by the greater photospheric luminosities and accretion rates associated with more massive primary stars.

From an observational point of view, the primary difference between these hypotheses is geometrical and the most critical observations are those which can determine distribution of material around infrared companions. Two promising directions are millimeter interferometry and near-infrared polarization, both of which may be able to address the presence and orientation of disks or envelopes in binaries with infrared companions. More broadly, the place of infrared companions in the context of the entire PMS binary population will be much better understood when systematic study of the SEDs of a large unbiased sample of binary components is completed. High-resolution multi-wavelength (optical and infrared) observations of the rich, recently discovered samples of sub-arcsecond-separation binaries should be enlightening. Of particular interest will be exploration of how the SEDs of companions ("infrared" and otherwise) vary with separation. While most of the infrared companions listed in Zinnecker & Wilking (1992) have projected separations of 100 AU or greater, XZ Tau is notable in having a projected separation of only 42 AU.

Finally, the possible presence of an infrared companion—and indeed "normal" companions—must be a consideration in any study involving SEDs of

PMS stars. Unknown cool companions can influence luminosity derivations through both the extinction estimates and the measures of total fluxes. More directly, an unknown infrared companion can mislead an analysis of SED structure. Thus, in some cases the inferred near-infrared excesses for composite systems are entirely attributable to the infrared companions, with the light from the optical companions being consistent with reddened photospheres (e.g. Glass I and DoAr 24E). At longer wavelengths the flat infrared SEDs found for some classical T Tauri stars—including T Tau—have been a puzzle, since there is more far-infrared luminosity than can be accounted for by simple reprocessing or accretion disks (e.g. Adams et al 1988). Confusion in the photometry from close infrared companions has been suggested as a solution, with T Tau being offered as an example (Kenyon & Hartmann 1987, Bonnell & Bastien 1993). While possible, this is not the entire explanation for the flat spectrum of T Tau. Ghez et al (1991) find the SED of the optical component (T Tau N) to still be flat (Figure 7), and they fit the infrared excess with a power-law temperature distribution having an exponent of $q = 0.42$, which is smaller than any found by Beckwith et al (1990) for other T Tauri stars. So in the case of T Tau, the essential problem of the infrared energy production is not resolved by the presence of the companion. (Interestingly, the mid-infrared SED of T Tau S is also flat, and appears remarkably similar to that of T Tau N despite their greatly differing optical/near-infrared SEDs.) A similar conclusion is drawn by Tessier et al (1994) who have recently completed a 3–5 μm speckle survey of eight flat-spectrum T Tauri stars and found no new infrared companions.

Orbital Evolution

Numerous processes capable of modifying orbital elements are at play during early binary evolution. Among these are:

1. *Tidal circularization.* The rate of tidal circularization is very sensitive to the ratio of periastron distance to stellar radius, and is expected to be strongly enhanced early in the PMS phase.

2. *Circumstellar/circumbinary material.* Observed disk masses of $\gtrsim 10^{-2} M_\odot$ can drive rapid orbital evolution, given appropriate radial surface-density distributions.

3. *Accretion and mass loss.* If accretion occurs in PMS binaries at rates of order 10^{-6}–$10^{-7} M_\odot$ yr^{-1}, then accreted mass over intervals as short as 10^5 yr can be dynamically significant. Mass losses through outflows and winds are also present.

4. *Encounters with external stars.* While infrequent at the densities of associations, encounters may play a key role for binaries in young dense clusters

and for multiple loosely bound stars in dense molecular cores (Larson 1990, Clarke & Pringle 1991b, Clarke 1992b).

In light of even this partial list, it is plausible that the orbital elements of every PMS binary have evolved. With the possible exception of stellar encounters, these evolutionary processes would be most significant at the beginning of the PMS stage or earlier. Since known PMS binaries typically have ages $\gtrsim$ 10^6 yr, the observed similarity of MS and PMS orbital distributions is not in contradiction with such orbital evolution.

To date, variability of orbital elements has not been detected, so observational evidence of orbital evolution must be indirect. The strongest evidence is perhaps the existence of very short period PMS binaries. Mathieu et al (1989) note that the separation of the components in the binary 155913-2233 is only 10 $R_\odot$. The separations within V4046 Sgr, V826 Tau, ORI569, and EK Cep are not much larger. The components of these binaries fit comfortably within their Roche radii. Yet solar-mass stars at the stellar birthline have radii of order 5 $R_\odot$ (Stahler 1983). Clearly such stars would not even remain detached in binaries of such short period. Either the internal evolution of the stars in these binaries was very much different than for isolated stars (which cannot be ruled out), or the orbits of these binaries evolved to their present short periods.

Similar arguments for orbital evolution have been made for binaries of somewhat longer periods. Mathieu (1992a) noted that the upper envelope of the orbital eccentricity distribution (Figure 2) also corresponds to small periastron separations ($\approx$ 15 $R_\odot$ for eccentricities smaller than 0.5). Stars near the stellar birthline in binaries with eccentricities greater than the observed upper envelope would have frequent encounters at periastron separations of less than a few stellar radii. The orbital elements of such binaries would evolve rapidly, presumaby to shorter period and lower eccentricity, and indeed the independent survival of the two stars may be in doubt. Thus, the existence today of both PMS and MS eccentric binaries with such small periastron separations suggests that they evolved to their present orbits from orbits with larger periastron separations. Alternatively, Bonnell & Bastien (1993) propose that increasing orbital eccentricity can be linked to increasing separation of the binary fragments in an elongated cloud. That the trend is seen at separations of only a few AU in MS binaries may pose difficulties for this approach.

As an aside, these arguments suggest that we might expect to find PMS interacting binaries. The relative rates of stellar contraction and periastron reduction will dictate how often such binaries are produced; their detection frequency will further depend on the duration of their interaction. No evident examples of such objects are known.

The action of tidal circularization is strongly indicated by the predominance of circular orbits at short periods. Given the large radii and deep surface convection zones of PMS stars, effective tidal circularization would be expected, even though the available time is brief (Plavec 1970, Mayor & Mermilliod 1984). In order to evaluate this idea quantitatively, Zahn & Bouchet (1989) convolved tidal circularization theory with PMS stellar evolution models to calculate the expected ZAMS circularization cutoff period for a variety of stellar mass combinations ($0.5\ M_\odot$–$1.25\ M_\odot$). Beginning the calculations at the stellar birthline, they find that most PMS tidal circularization occurs very soon thereafter (within 10^5 yr for solar-mass stars) because of the strong dependence on stellar radius. They conclude that tidal circularization during the PMS phase should produce circular orbits in ZAMS binaries with periods less than 7.3–8.5 days (and that no further circularization should occur on the main sequence). The presently known PMS binaries with periods of less than 20 days all lie well below the stellar birthline, and so in this picture the circularization process should be largely completed for them. The binaries EK Cep and P2486, with eccentric orbits at periods of 4.4 days and 5.2 days respectively, stand in contrast to this prediction, although the case of EK Cep is not a strong one (Duquennoy et al 1992, Mathieu 1992a).

Although the circular orbit of W134 at a period of 6.4 days demonstrates that PMS tidal circularization at periods approaching those predicted by Zahn & Bouchet can occur (Padgett & Stapelfeldt 1993), given stellar masses of $2.3\ M_\odot$ and $2.2\ M_\odot$ the applicability of the Zahn & Bouchet calculations is not clear. Still, W134 can be directly compared with MS binaries having A-type stars. Such binaries have circular orbits with periods as long as 10 days (e.g. Matthews & Mathieu 1992). Such long-period circular orbits in a comparatively young field MS population are an interesting problem, particularly since radiative envelopes may be less effective for tidal circularization than the convective envelopes of solar-mass stars. The case of W134 strongly argues that the answer lies in the circularization of A-type binaries *prior* to arrival at the ZAMS.

Tidal processes modify orbital period and stellar rotation as well as orbital eccentricity. The sign of period change is sensitive to details of angular momentum redistribution as the stars come to synchronization. Generally, in cases of initially subsynchronous stellar rotation, the orbital period will decrease.

Rotational synchronization in PMS binaries has been little explored by either observation or theory. The evolution of stellar rotation in close binaries can be complex; stars may come in and out of synchronization depending on the interplay of tidal effects and their rotational evolution (e.g. in response to contraction; Zahn & Bouchet 1989). In addition, Edwards et al (1993) have suggested that disks also act to regulate stellar angular momentum. Stellar rotation in close PMS binaries is ripe for observational study. The results will place constraints on the timescale and nature of angular momentum redistribution during tidal circularization, and more generally on the evolution of angular momentum within PMS stars.

At periods much longer than tens of days, tidal circularization will be ineffective in modifying orbital elements. Recent attention has focused on orbital
evolution through resonant interactions of binaries with associated disks. In
particular, circumbinary disks are predicted to both decrease orbital period and
excite eccentricity in nearly circular orbits (Artymowicz et al 1991, Lubow
& Artymowicz 1992). The binary GW Ori presents an interesting contrast
to these theoretical predictions. Artymowicz (1992) finds the characteristic
growth time $e/\dot{e}$ for the eccentricity of GW Ori to be $\approx 10^4$ yr in the presence
of a circumbinary disk with a mass comparable to the minimum mass solar
nebula. Mathieu et al (1994) find GW Ori to have a much more massive circumbinary disk ($> 0.3 \, M_{\odot}$). Yet the orbital eccentricity of GW Ori is small;
Mathieu et al (1991) found $e = 0.04 \pm 0.06$. Similarly, using the formalism of
Artymowicz et al (1991), Mathieu et al (1994) find the characteristic reduction
time $a/\dot{a}$ for the semimajor axis to also be of order ten thousand years. Given
a stellar age $\gtrsim 10^5$ yr, this would imply an initial binary separation larger than
typical disk radii. Looked at in a different way, the rapid rate of semimajor
axis evolution indicated by the dynamical theory places in question the very
survival of a binary embedded within a disk (Clarke 1992a).

Evidently, in the case of GW Ori a large increase in eccentricity has not been
driven by the suggested circumbinary disk. The rate of resonant excitation of
eccentricity is very sensitive to the detailed distribution of mass at the strongest
resonances (located at distances of a factor of a few times the binary separation).
Lubow & Artymowicz (1992) note that the lack of eccentricity in the GW Ori
orbit can be consistent with theory if there is a gap in the disk with an outer
radius of 3 AU, as suggested by Mathieu et al (1991). In this case, no disk mass
resides at the strongest resonances and there is little coupling between the binary
and the circumbinary disk. As Lubow & Artymowicz (1992) note, a physical
mechanism for producing such a large gap is not known; Artymowicz & Lubow
(1994) argue that such a large gap cannot be created by resonant interactions.
In addition, the observational constraints on the specific gap dimensions are
not strong. Nonetheless, GW Ori displays all of the signatures of a massive
circumbinary disk. Thus, the small eccentricity may be a clue to the distribution
of disk matter in the vicinity of the binary orbit, about which there is very little
empirical information, and consequently may provide insight into the viscous
evolution of circumbinary disks (Pringle 1991a).

Mathieu (1992a) has suggested that the lack of circular and near-circular
orbits ($e < 0.1$) among both PMS and MS binaries ($P > 10^d$) can be naturally explained if massive disks are typically associated with young binaries,
even if only very early in their evolution. Lubow & Artymowicz (1992) argue
that eccentricity growth will be particularly strong at the smallest eccentricities
($e < 0.2$). (Indeed, the presence of numerous binaries with orbital eccentricities
as low as ≈ 0.2 may be an interesting constraint on either this theory or the fre-

quency of massive circumbinary disks.) Thus, given a variety of disk structures and evolutionary histories, a distribution of eccentricities greater than $e \approx 0.2$ would be expected, if not specified. This picture presumes that disks associated with binaries do indeed drive orbital eccentricity—a not entirely secure premise given the case of GW Ori. In addition, resonant interactions with disks must diminish for binaries with separations greater than typical protostellar disk radii, perhaps of order 100 AU (or $P \approx 10^5$ days).

Alternatively, most binary formation mechanisms currently under consideration tend to produce highly eccentric orbits, at least initially (Adams & Benz 1992, Burkert & Bodenheimer 1993, Bonnell & Bastien 1993, Boss 1993a, Pringle 1989), so orbital evolution may not be necessary to explain the relative absence of low-eccentricity binaries. From the point of view of binary formation theories the challenge may actually be somewhat the reverse, that is to produce binaries with eccentricities of a few tenths. In the case of fragmentation, calculations to date have followed a protobinary for at most a few orbits, so the final orbital eccentricity produced by the formation process remains uncertain. Dissipation through disk-disk encounters, interactions with envelopes, and accretion have all been invoked to provide a distribution of eccentricities (Burkert & Bodenheimer 1993, Bonnell & Bastien 1993, Boss 1993a, Clarke & Pringle 1991a), perhaps even circularizing orbits (Boss 1993a).

Encounters of a binary with other single or binary stars will also produce orbital evolution (as well as possible exchanges of members). The equilibrium distribution $f(e) \approx 2e$ established by stellar encounters has few low-eccentricity orbits, and Duquennoy & Mayor (1991) have noted that disruption of small stellar groups tends to produce similar eccentricity distributions. Even single encounters typically produce eccentric orbits in the binary products (Heggie 1975). Similarly, Clarke & Pringle (1991a) have found that a wide range of eccentricities can be produced by encounters within loosely bound triple systems. The typical local stellar density of PMS binaries in associations is too low for stellar encounters to be significant for orbital evolution in 10^7 yr. Whether most PMS binaries could have formed in small groups which have since disintegrated, or whether most MS binaries derive from more densely populated nurseries (e.g. Lada et al 1991) remains to be considered.

Given the ambiguities in theoretical predictions for "initial" eccentricity and period distributions, it will be difficult to draw secure conclusions regarding PMS orbital evolution from comparisons with observed PMS distributions. Progress is more likely to be achieved by observational definition of the periods and eccentricities of binaries at the stellar birthline (or earlier). At present it is difficult to achieve significant time resolution among the PMS binary sample, simply because of the small numbers at the youngest ages. With larger surveys, and the determination of longer period orbits with very high-angular-resolution techniques, statistical studies of orbital evolution during the PMS phase will be possible. Equally important will be detailed case studies of binaries embedded within disks. Disk-driven evolution of orbital elements (line of apsides, eccentricity, period) may be detectable in the shortest period binaries.

Implications for Binary Formation

It was only shortly after double stars were proven to be physically associated by Herschel and his contemporaries in the late 1700s that Laplace put forth the first theory of their origin. Numerous ideas have been pursued in the meantime which can be broadly categorized as capture, fission of a protostar, independent bound condensations, cloud fragmentation during collapse, and disk fragmentation. Recently, several review papers have described and evaluated the relative merits of these theories, in particular with respect to observed properties of the MS binary population (Bodenheimer 1992, Bodenheimer et al 1993, Bonnell & Bastien 1993, Boss 1993a, Clarke 1992a, Pringle 1991b). At present, capture and fission are out of favor on theoretical grounds, except perhaps for disk-enhanced capture in very-high-density environments such as the Trapezium cluster (Larson 1990, Clarke & Pringle 1991b). Dynamically induced fragmentation, for example through cloud collisions, has been suggested to produce several independent condensations in molecular cores (Pringle 1989, Chapman et al 1992). Given dissipative encounters these condensations may reconfigure into more tightly bound multiple systems (Clarke & Pringle 1991a). Fragmentation during cloud collapse has been studied extensively in the past decades (Bodenheimer et al 1993). A promising recent development has been the investigation of fragmentation of elongated clouds (Zinnecker 1991, Bonnell & Bastien 1993 and papers therein, Boss 1993b). Finally, the investigation of disk instabilities by Adams et al (1989) has revitalized the idea of companion formation in disks. Adams & Benz (1992) speculate that disk fragmentation may occur at an early stage of infall, and thereby produce two protostellar cores with substellar masses. Calculations have not yet actually produced a structure of stellar mass (Adams & Benz 1992). In summary, these assorted fragmentation pictures differ primarily in the time of fragmentation with respect to cloud collapse: 1. prompt fragmentation of a cloud, after which the fragments collapse to form stars, 2. fragmentation during cloud collapse, and 3. fragmentation of disks composed of infallen material.

Unfortunately, attempts to test specific theories with observations run quickly into a "theory gap," a term coined by Clarke (1992a) to describe the fact that for none of the hypotheses do existing calculations proceed sufficiently far in time to produce a binary isolated from its natal material. Indeed, fragmentation calculations are seldom carried beyond a few initial dynamical timescales. Consequently, significant processes such as infall, disk interactions, and the like have not yet come to completion, and prediction of the state of the system by the MS, or even PMS, evolutionary phase can only be rather speculative.

Characterization of the situation as a "theory gap" perhaps places too much of the burden on theorists; observers must continue to push into the protobinary domain in order to create the interface.

In addition, increasingly sophisticated numerical techniques have now permitted exploration of more free parameters and initial conditions. For example, the recent studies of fragmentation of elongated clouds with arbitrary angular momentum vectors have found a great variety of evolutionary paths to a diverse assortment of proto-binary systems (Bonnell & Bastien 1993, Boss 1993b). A similarly large variety of products have been found from study of binary formation through cloud collision (Chapman et al 1992). Although these studies can produce many of the observed properties of the MS binary population—and from initial conditions similar to those seen in molecular clouds—it remains true that the actual MS population is only a subset of the theoretical possibilities. Observational tests critically distinguishing between the presently popular hypotheses are difficult to find. Being wary of lengthy but premature and inconclusive discourse, we leave evaluation of formation theories to another time and restrict the discussion here to *new* constraints on binary formation mechanisms provided by recent study of PMS binaries.

The most secure result is simply that PMS binaries are present in large numbers, perhaps surprisingly large. Furthermore, the youngest PMS binaries lie very near the stellar birthline, and candidates for yet younger, embedded protobinary systems have been found. The evidence strongly suggests that binary formation occurs on timescales shorter than or comparable to the formation of the stars themselves, and that the star-formation process is in large part a binary-formation process.

In addition, when viewed broadly many properties of the MS binary population are established by the PMS phase. The presently known PMS binary population includes seven decades of orbital period, a bell-shaped period distribution, essentially all orbital eccentricities, a wide range of luminosity ratios, and hierarchical multiple systems. That these properties are established as early as the stellar birthline is not definite, but certainly a range of periods, luminosity ratios, and orbital eccentricities, as well as hierarchical multiples, exist among the few such very young binaries known. The implication is that these general properties of the MS binary population are also established during or shortly after binary formation.

Nonetheless, the recent studies of PMS binaries should imply caution in using the MS binary population in detail as the target for binary formation theories. For example, these studies suggest that the efficiency of formation of binary and multiple systems can be higher than indicated by studies of MS stars. The implication would be that the ensemble of binary formation mechanisms must produce multiple systems at a higher efficiency than previously thought—perhaps near 100%. Furthermore, if single stars are rare but triple and higher order systems are frequent, the implication is that few multiple systems have decayed by the PMS stage. This may be an indication that multiple systems typically form in stable hierarchical configurations, although dissipative encounters (e.g. with disks) may also tend to stabilize initially unstable configurations (Clarke & Pringle 1991a). Multiple fragmentation in either colliding or collapsing clouds can provide both high efficiency and hierarchical systems, given appropriate initial conditions (e.g. Boss 1991, Chapman et al 1992, Bonnell & Bastien 1993). However, why hierarchical systems should decay during or after the PMS phase so as to produce single stars in the field MS population is a critical problem for such highly efficient formation of multiple systems.

The relative ages of binary components have the potential to strongly constrain binary formation mechanisms. A critical ambiguity in any such discussion is the meaning of "age" for young stars (Hartigan et al 1994). Recent ages derived from PMS theoretical tracks have zero points at the stellar birthline. Hence, this age does not include prior collapse or accretion phases, which for single stars may last of order a million years if typically observed accretion rates of order 10^{-6}–$10^{-7} M_\odot$ yr^{-1} are steady. This uncertainty remains even for relative ages since time prior to the stellar birthline may depend on stellar mass. Thus, the finding of Hartigan et al (1994) that about one third of their wide-binary sample show significant age differences between components, and typically in the sense that the less massive star is younger, is notable in terms of binary formation. They argue that the systematic sense of the age differences does not support capture formation. Rather, they suggest that components in wide binaries form as independent but bound fragments; the age difference is then the result of slower accretion onto the less massive secondary.

The recent discovery of several unaligned outflows originating from the same site is very intriguing evidence for non-coplanarity of disks and binary orbits. Even among MS triples, investigations of non-coplanarity are few, so here PMS binaries can contribute more generally to our understanding of the lower-mass binary population. The HH1-2 source is a known VLA double source with a projected separation of 1380 AU (Reipurth et al 1993b); otherwise there is little evidence regarding at what separations non-coplanarity can occur. Assessment of non-coplanarity is of particular interest for binaries with separations of a few hundred AU or less. Although fragmentation and capture scenarios can easily—and perhaps preferentially—produce non-coplanar systems, binaries formed by disk fragmentation would very likely be coplanar.

The PMS binary environment may still bear imprints of the time of formation. Observations show that circumstellar disks, optically thick at 2 μm and 10 μm, are a frequent product of binary formation. The production of such disks is implicit in any formation theory that creates stars through disk accretion; even disk-enhanced capture does not disrupt circumstellar disks (Clarke & Pringle 1993). On the other hand, the survival of circumstellar disks, especially in shorter period binaries, may be a substantial constraint on binary-disk evolution shortly after formation.

Circumbinary material is likely to be a more significant clue to the binary formation process. For example, in the case of GW Ori (separation $\approx$ 1 AU) we can be reasonably confident that binary formation produced a massive circumbinary disk (Mathieu et al 1994). The disk must have a much greater radius than the binary separation, which suggests that it formed from infall after formation of the binary. This would be consistent with formation from fragmentation of a smaller disk early in the protostellar phase, although how material further accreted onto the protostars is an open question. Bonnell & Bastien (1993) also predict circumbinary disks after bar or disk fragmentation, although their simulations cannot produce such a close binary. Indeed, even given the early state of binary formation theory, the formation of short-period binaries stands out as a problem for which we have only the thinnest conjectures.

Finally, secondary mass distributions are a key constraint on binary formation mechanisms. The initial studies do not indicate any difference between the secondary mass distributions of wide PMS and MS binaries, although in the case of the PMS binaries the distribution is not well defined. The issue of secondary mass distributions among MS binaries has been controversial and merits an update here. Abt & Levy (1976) suggested that the secondary mass functions of solar-type binaries depended on period. In particular, binaries with periods greater than $\approx$ 100 yr had a secondary mass function similar to the Van Rhijn function (a reasonable approximation to the initial mass function at these masses); binaries of shorter periods had a flatter secondary mass distribution, in the sense of tending toward more binaries with comparable mass components. The result stimulated analyses of literature data with conclusions pro and con regarding the reality of the dependence on period (e.g. Halbwachs 1987, Trimble 1990); typically these discussions have focused on the impact of selection effects. The diversity of opinion has grown of late. In their observational study of solar-type stars, Duquennoy & Mayor (1991) found no dependence of the secondary mass function on period. However, in the most recent report Mazeh et al (1992) improved on their analysis and found that for binaries with periods of less than 3000^d (i.e. the Duquennoy & Mayor sample of spectroscopic binaries), the secondary mass distribution was very nearly flat—in marked contrast to the distribution for visual binaries and to the initial mass function. This result is critical for binary formation theory since, as noted by Abt & Levy (1976), it strongly suggests different formation histories for short- and long-period binaries. One speculation might be that longer period binaries form through a cloud fragmentation process while shorter period binaries form through disk fragmentation.

With the exception of N-body capture (Clarke 1992b), at present binary formation theories cannot make precise predictions for secondary mass distributions; specific distributions of mass ratios depend upon initial conditions. Nonetheless, Bonnell & Bastien (1993) make the important general point that in the fragmentation of elongated clouds even slight asymmetries in density distributions provide an appropriate range of mass ratios, and production of unit mass ratio requires unexpectedly good symmetry. Thus, a tendency toward non-unit mass ratios is to be expected. Leinert et al (1993) argue that the observed luminosity ratios of PMS binaries cannot be attributed to capture, but can be achieved simply by associating stars independently chosen from the initial mass function. (As an aside, if most stars do form in binaries the initial mass function is not independent of the binary formation process.)

PROTOBINARIES

Evidently, binary star formation can occur prior to the PMS phase of evolution. Furthermore, many of the properties of MS binaries are already at least approximately present in the PMS phase, and likely were established earlier. With the PMS binaries now opened to observational study, the forefront of discovery will soon move to yet younger, embedded protobinaries. While these objects lie outside the purview of this review, their brief mention will provide the reader with a glimpse of what is to come. In particular, we introduce the reader to IRAS 16293-2422, at present the most extensively studied candidate protobinary.

IRAS 16293-2422 (hereafter 16293), in the ρ Oph star-forming region, was recognized as an important source even before its binary nature was realized. The SED is very cold, and is undetected at wavelengths shorter than 25 μm. On the basis of CS line profiles, Walker et al (1986) suggested that it is a deeply embedded, actively accreting protostar. Mundy et al (1986) obtained interferometric 3 mm continuum maps and found an elongated structure. With the addition of ^{13}CO maps showing velocity gradients, they argued instead for a disk-like structure surrounding a protostar. Further CS and molecular line observations led Menten et al (1987) to conclude that rotation was present, and that a more complex model than spherical accretion was necessary.

The perception of 16293 changed significantly when Wootten (1989) resolved two components separated in projection by 5″ (800 AU at 160 pc) at 2 cm and 6 cm wavelengths. Both components lay within the elongated millimeter continuum emission. Recently, the two components have also been resolved at millimeter wavelengths; this assures the presence of two sites of warm dust (Mundy et al 1992, Walker et al 1993) and strengthens the picture of two protostars in a protobinary system. The southeast source is the origin of a molecular outflow, a strong wind, and a cluster of water masers (Walker et al 1993, Estralla et al 1991, Wootten 1989), all indicative of extreme youth. The northwest source has a higher peak luminosity at millimeter wavelengths.

Since the source of the millimeter emission from the northwest component is unresolved at a resolution of 400 AU, Mundy et al (1992) conclude that the thermal emission from both components arises from circumstellar disks. On length scales of several thousand AU, Walker et al (1993) used several CS transitions to map a velocity gradient along the axis of the two components. They concluded that a slowly rotating circumbinary disk is also present, although the rotation may not be sufficient to provide support against gravity. Ammonia observations also reveal a similarly aligned, elongated structure with a radius ≈ 4000 AU (Mundy et al 1990).

Estimates of the circumstellar mass associated with each source are highly uncertain (studies differ even in the identification of the more massive source), but are found to be between $\approx 0.4\,M_\odot$ and $\approx 0.9\,M_\odot$ (Mundy et al 1992, Walker et al 1993). Dynamical mass estimates of both stellar and circumstellar mass, also highly uncertain, are comparable to the total circumstellar masses derived from millimeter emission. Thus, even given the large uncertainties, it is likely that the protostellar sources presently have masses $\lesssim 1\,M_\odot$, and that, if bound, ultimately this system will be a PMS binary similar to those reviewed here. Interestingly, Walker et al (1993) argue that the active southeast source is the more massive on the basis of the kinematic center being shifted toward it.

Given these mass estimates, Mundy et al (1992) argue that the large total luminosity of 30–$40\,L_\odot$ cannot be attributed to stellar photospheres but may plausibly be attributed to spherical accretion at rates $\approx 10^{-5}\,M_\odot\ \mathrm{yr}^{-1}$. Walker et al (1993) find possible evidence for such infall in their CS line profiles. Even infalling envelopes with such large accretion rates cannot account for the extreme extinction in the near-infrared, so Mundy et al associate the extinction with accretion disks.

The 16293 system lies at the origin of several outflow lobes with different position angles (e.g. Walker et al 1990, Mizuno et al 1990). Walker et al (1993) suggest that one well-collimated outflow is associated with the southeast component and a second less collimated outflow might be associated with an earlier outflow event from the northwest component. In this picture the lack of alignment in the bipolar outflows again might indicate that the circumstellar disks of each component are not coplanar, although Mizuno et al (1990) explore several other mechanisms for modifying the orientation of an outflow.

Although IRAS 16293-2422 is at present the most extensively studied protobinary, the discovery rate of additional candidates is increasing. Protobinary candidates include NGC1333 IRAS 4 (Sandell et al 1991), IRAS 19156+1906 in L723 (Anglada et al 1991), and perhaps ρ Oph B1 (Sasselov & Rucinski 1990). At present none of these are entirely convincing as being both protostellar and bound, but they merit further study. At a somewhat later evolutionary state, but still embedded, is SVS 20 (Eiroa et al 1987, Eiroa & Leinert 1987). The SEDs of both components of SVS 20 resemble those of infrared companions.

Finally, several recently discovered pairs of unaligned Herbig-Haro sources and outflows may originate in very young binary stars (Gredel & Reipurth 1993, Reipurth et al 1993b).

SUMMARY AND THE FUTURE

Given the present rapid and exciting progress in observations of PMS binaries, it is perhaps optimistic to put forth a summary description of the PMS binary population. Still, the state of the observations has improved remarkably in the past decade with numerous important results in which we can have some confidence.

1. The PMS stellar population is rich in multiple systems, with a frequency of at least 50%. Binary formation is the *primary* branch of the star-formation process. The frequency of PMS binaries and higher order systems may be a factor of two or more in excess of that found among field MS solar-type stars, but this recent finding requires further investigation.

2. Binaries are found throughout the domain of PMS stars in the H-R diagram, indicating a range of ages extending to as young as the stellar birthline. Embedded binaries and protobinary candidates are also being discovered. Every indication is that binaries form prior to the PMS phase of stellar evolution.

3. Many properties of field MS solar-type binaries are already established in PMS binaries. These include (*a*) a range in periods from days to at least 10^5 yr, (*b*) a bell-shaped period distribution with a median of roughly a few hundred years, (*c*) circular orbits for the shortest period binaries, (*d*) a broad range of orbital eccentricities at longer periods, with few circular orbits and increasing maximum eccentricity with period, (*e*) a large range in secondary masses in the longer period binaries, with a distribution that arguably mimics the initial mass function, and (*f*) hierarchical organization of triples and quadruples. This general similarity of the PMS and MS binary populations does not imply that the two are identical. Several differences in detail, discussed in the main text, are critical clues to the formation and early evolution of binaries.

4. Circumstellar and circumbinary material is common in young binary environments. The frequency of $2\,\mu$m and $10\,\mu$m emission excesses among PMS binaries is $\approx$ 50%, comparable to the frequency among PMS single stars (Strom et al 1993). Such excesses are indicative of circumstellar material, which based on indirect arguments is typically taken to have a disk distribution. Far-infrared and submillimeter emission is also seen from many systems. For the closest binaries and in the cases of GG Tau and T Tau, where the millimeter emission is resolved, this light clearly origi-

nates in circumbinary material. In several other cases the data also strongly point to circumbinary disks, but as with single stars distinguishing between circumbinary disks and envelopes (or dispersed material) is difficult.

5. Accretion onto stellar surfaces occurs in PMS binaries. Strong Hα emission, spectral veiling, ultraviolet excesses, and large infrared luminosities are seen in some binaries at all periods. There are hints that accretion rates decrease in binaries with separations of less than tens of AU, but these results are not secure. Clarification of the nature of accretion in binary environments will be essential to understanding the evolution of circumstellar disks.

6. The discovery of infrared companions poses one of the more interesting puzzles of PMS binaries. These binaries include one ordinary PMS star and a companion with spectral energies peaking in the near- to mid-infrared, typically without detectable optical emission. Although their explanation is unclear, they may possibly be critical cases through which we can study the emersion of a binary from an embedded or rapid accretion phase.

The past decade has been primarily a time of *discovery* of PMS binaries. The surveys—both spectroscopic and imaging—must continue, for the confidence in many of the resulting conclusions is hampered by large sampling errors. Significantly larger samples will be required before we can begin to meaningfully search for dependences of the binary population on age, stellar mass, disk mass, formation site, etc. At the same time, the search for *evolution* in PMS binary-disk systems will become a central theme of observations in the next decade. Much of this review has presented the first steps taken in this search, with intriguing but not yet definitive results.

The speckle and imaging surveys have provided a wealth of targets; now, each component of these binaries and their environments must be thoroughly studied. A key contributor to future progress will be the increased angular resolution soon available from X-ray to millimeter wavelengths. A basic goal must be the characterization of the stellar SEDs throughout this wavelength range. This community-wide endeavor will drive important progress in many critical issues touched on in this review, including secondary mass distributions, coevality of formation, circumstellar disk frequency and accretion rates (especially boundary layer emission) as a function of binary separation, the role of infrared companions, etc.

Of equal importance will be the resolution of extended continuum and line emission at wavelengths from the optical to the radio. These observations are essential in providing direct images of the binary environment on scales as small as a few AU, and should substantially clarify the distribution of circumbinary material in young binary environments. The study of polarization is also rapidly rising in importance for understanding of environment geometry.

In a different but equally significant direction, the wide range of theoretical estimates in the literature for the mass of any given PMS star cries out for dynamical mass determinations. The lack of dynamical mass determinations for PMS stars remains one of the fundamental holes in star-formation study; increasing the number of measured masses by all available techniques, particularly for binary components with ages of 10^6 yr or less, must be a primary goal. With the present rapid advances in ground-based high-angular-resolution techniques, as well as astrometry with the *Hubble Space Telescope,* astrometric orbits should soon come to the fore in measuring PMS stellar masses (e.g. Ghez et al 1993). However, the sensitivity to distance in the derivation of astrometric masses will require similar advances in parallax measurements. These efforts will complement on-going spectroscopic studies, and ultimately overlap of the two will remove large uncertainties introduced by poorly determined distances. In addition, the number of PMS eclipsing binaries should be substantially increased by dedicated high-frequency imaging of young clusters and groups.

Finally, observations need to push further into the era of binary formation itself in order that they may more directly confront theories of binary formation. As the past decade was one of discovery of PMS binaries, the next may provide discoveries of binaries in the process of formation. We can only hope that such protobinaries do not prove as elusive as have protostars.

ACKNOWLEDGMENTS

I thank E. Jensen for much generous assistance in the preparation of this review. I am very grateful to the numerous colleagues who provided manuscripts, data, and ideas in advance of publication, with particular thanks to A. Dutrey, D. Fischer, A. Ghez, P. Hartigan, J. King, D. Koerner, D. Latham, Ch. Leinert, M. Mayor, B. Reipurth, A. Richichi, M. Simon, and F. Walter. I also benefited greatly from conversations with P. Artymowicz, P. Bodenheimer, C. Clarke, D. Koerner, D. Lin, E. Ostriker, F. Shu, C. Walker, A. Wootten, and H. Zinnecker. P. Bodenheimer, E. Jensen, and H. Zinnecker provided valuable reviews of the entire manuscript. I am indebted to D. Wuolu, the UW astronomy librarian, for bibliographic assistance. Most of the manuscript was written while I was on sabbatical leave at Lick Observatory, where the warm hospitality and scientific stimulation were a great pleasure. Funding support was provided by a Guggenheim Fellowship, the NASA Center for Star Formation Studies, the Wisconsin Alumni Research Foundation, Lick Observatory, and a Presidential Young Investigator Award.

APPENDIX

Two compilations of PMS binaries are presented here. Table A1 lists all known angularly resolved PMS binaries in the Taurus-Auriga star-forming region with separations of less than $10''$. A separation of $10''$ is a rough limit at $2.2\,\mu$m, above which false pairings with background stars become significantly frequent (e.g. Simon et al 1992, Leinert et al 1993). With detailed observation of each component yet wider binaries can be confidently identified, and the reader is directed to Moneti & Zinnecker (1991) and Hartigan et al (1994) for lists of likely pairs in Tau-Aur with projected separations greater than $10''$. Note, however, that at very wide separations ($\gtrsim 20''$) even spatially associated pairs of young stars may not be bound.

Tau-Aur is at present the most comprehensively studied star-forming region in terms of both numbers of stars and observational techniques. However, the reader is cautioned that the thoroughness of its study (in particular the historical development of the stellar sample) is very much a result of its proximity and declination; there is no guarantee that results derived from this region are typical for the formation of binaries in general. Similarly, Table A1 is by no means a comprehensive list of all resolved PMS binaries; the reader is recommended to Simon et al (1987), Zinnecker (1989), Ghez et al (1993), and Reipurth & Zinnecker (1993) for additional resolved binaries in southern star-forming regions.

Tables A2 and A3 are compilations of all PMS spectroscopic binaries known to the author. (Note that three are in Tau-Aur.) The identifications as PMS are not secure for two binaries listed. Both V4046 Sgr and HD155555 show strong lithium absorption, characteristic of PMS stars (de la Reza et al 1986, Pasquini et al 1991). V4046 Sgr also shows a classical T Tauri emission spectrum and excess emission from the near-infrared to the submillimeter (Byrne 1986, de la Reza et al 1986, Jensen et al, in preparation). Nonetheless, neither is associated with a dark cloud and both have unknown distances, so their PMS status is uncertain (e.g. Hutchinson et al 1990). EK Cep is also an unusual entry in that it contains a 2 $M_\odot$ ZAMS primary (spectral type A1) with an oversized 1.1 $M_\odot$ secondary, likely to be PMS (see discussion of masses of PMS stars).

In all tables the binary parameters are taken from that study citing the highest precision. (Particularly for the resolved binaries, these may not be the discovery references.) There is substantial overlap in the studies of Ghez et al (1993), Leinert et al (1993), and Simon et al (1992). These studies agree well, except in the flux ratios derived by the two speckle studies and, occasionally, in $180°$ differences in position angle. These differences might be attributable to photometric variability, but this has not yet been established.

Along with binary parameters, Tables A1–A3 include an assortment of observable quantities, chosen to convey photospheric properties, evidence for accretion and extended material, and information regarding observability. The specific values presented were selected on the basis of achieving as much uniformity in their origin as possible. For the Tau-Aur sources (Table A1), the primary references for all but ΔN and the 1.3 mm fluxes are Hartigan et al

(1994) and Strom et al (1989a). When a star is not included there, the observed quantities are taken from Herbig & Bell (1988), from the referenced binary paper, or as otherwise noted. The values of ΔN are taken from Skrutskie et al (1990) and the 1.3 mm fluxes are taken from Beckwith et al (1990). Except for some of the widest binaries, all values given are for the combined light of each system. Note that flux variability (both at broadband and spectroscopic resolutions) is a common phenomenon among PMS stars at optical/near-infrared wavelengths, and possibly at longer wavelengths as well. In addition, for the more active stars spectral type can depend on wavelength. Finally, the quantities L_* and ΔN are derivative and hence depend on the assumptions in their derivation. The reader is referred to the original references, and for ΔN to the discussion in the main text on disks.

Masses and ages are not given in the tables. These quantities are highly dependent on theoretical models, from which very different values can be obtained, and in many cases numbers have not yet been derived for individual components. With this caution, the reader is referred to Beckwith et al (1990), Hartigan et al (1994), and Simon et al (1993) for recent listings of masses and ages.

The spectroscopic binaries have been less uniformly studied. Unless noted otherwise, the values given in Tables A2 and A3 were taken either from the referenced binary paper or from Herbig & Bell (1988), in which the original references can be traced. For the double-lined systems the entries are for the combined light, except for the eclipsing binary EK Cep.

Table A1 Resolved pre-main-sequence binaries in the Taurus-Auriga star-forming region

HBC	m	Name	d (")	p (°)	Flux Ratio (at K)	Ref.	Spec. Class	L_* ($L_\odot$)	Hα (Å)	ΔN (dex)	12 μm (Jy)	1.3 mm (mJy)	V (mag)	K (mag)
66		HQ Tau [a]	0.009*	—	0.14	1	K2	2.4	35	1.95	1.70	<45	12.6 [a]	7.4
414	m	HP Tau [a]	0.017*	—	0.26	2	K5	1.3	2		2.41	62	13.5	7.3
404	m	HP Tau G3	0.022*	—	0.40	2	M0	1.8	5		<0.04		14.6	8.7
46		V807 Tau S	0.023*	—	1.0	3	M0	0.7	16		<0.04		11.3	8.3
418		ZZ Tau	0.029*	—	0.44	3	M4		4.5		0.16	<15	14.3	8.5
409	m	HV Tau Aa	0.035*	—	0.58	1	M1		1.4	1.03	<0.04	<45	14.0:	7.8
36		FF Tau	0.037*	—	0.40	1	K7				<0.07	<27	13.7	8.7
367		GN Tau	0.041*	—	0.63	1	cont. [b]		62 [b]		0.59	<50	15.7 [a]	8.4
39		DF Tau	0.088	329	0.47	4	M0.5	1.7	54	1.55	1.16	<25	12.1	6.8
81		V773 Tau	0.112	295	0.47	4	K2	7.8	3	1.48	2.80	42 [c]	10.6	6.5
29		DI Tau	0.12	294	0.13	4	M0.5	1.0	1	0.73	0.10	<35	12.8	8.4
398	m	RW Aur BC	0.120	111	–	4					2.60	<20 [d]	12.7:	8.7
369		V410 Tau	0.123	218	0.16	4	K7	1.9	2.1	0.37	0.9:	<30	10.9	7.5
423		FW Tau	0.16	160	1.0	1	M3-5 [b]		17 [b]		0.11:	<15	17.1 (B) [a]	9.2
59		V928 Tau	0.165	125	0.60	4	M0.5		1.2		<0.09		13.7:	8.0
383		FO Tau	0.166	182	0.65	4	M2	1.9	117		0.49	<50	15.0:	8.1
420		LkHα 332/G1	0.215	77	0.56	4	M1	3.1 [e]	4		<0.02		15.4	8.1
54		IS Tau	0.221	92	0.16	4	K2	1.0 [e]	12		0.32	<20	14.3	8.0
47		FS Tau	0.265	60	0.12	1	M1	1.37	57		1.61	<35	15.7	7.3
422		IW Tau	0.27	177	0.91	5	K7	1.8	4.0		<0.09	<50	12.5	8.3
50		GG Tau Aa	0.288	3	0.25	4	M0	0.5	56	1.74	1.37	593 [f]	12.2	7.5
55		V927 Tau	0.30	290	0.73	5	M5.5	1.4	5		<0.04	<20	14.6	8.7
69		LkHα 332/G2	0.30	243	0.60	5	M0	0.7	3		<0.02		14.9	8.2
31		XZ Tau	0.311	153	0.51	4	M3	0.8	274	2.33	3.36	<18	14.9	7.1
53		GH Tau	0.314	299	0.56	4	M1.5	1.0	10	1.55	0.69	<30	13.0	7.8
404		V955 Tau	0.33	204	0.22	5	M0	0.4	13	1.50	0.73:	<50	14.2	8.0
368		CZ Tau	0.33	84	0.46	5	M3	0.6	6		0.96:	<30	15.4	9.3
30	m	UZ Tau W	0.360	359	0.46	4	M3	1.8	79		1.51	172 [g]	13.7	7.5
351	m	V807 Tau NS	0.375	330	0.37	4	M0	2.56	5		<0.04		11.3	7.0
416		LkCa3	0.491	77	0.96	4	M1	0.8	2.5	0.02	<0.05		12.1	7.5
68		DD Tau	0.56	186	0.52	4	M4	0.49	120	1.31	0.23	<50	14.1	7.9
		034903+2431	0.61	317	0.22	5	K5		1.6		<0.09	<25 [h]	12.2	9.2
		Haro 6-28	0.66	246	0.63	5	M5		92		2.10	<40	17.3:	9.3
		VY Tau	0.66	317	0.26	5	M0	0.5	4.9		0.11:	<50	13.8	8.8

Table A1 (continued)

HBC	Name	d (″)	p (°)	Flux Ratio (at K)	Ref.	Spec. Class	L_* ($L_\odot$)	$H\alpha$ (Å)	ΔN (dex)	12 μm (Jy)	1.3 mm (mJy)	V (mag)	K (mag)
412	043230+1746	0.70	68	1.00	5	M2	0.65	9.0		<0.07		14.1	9.1
35	T Tau	0.71	176	0.09	4	K1	12.5	38	2.12	16.48	280	9.9	5.4
386 m	FV Tau	0.73	92	0.64	4	K5	1.8	19		1.93	< 45	15.4:	7.5
387 m	FV Tau/c	0.74	293	0.17	1	M3	0.7	29		1.93		17.2:	8.5
377	FQ Tau	0.79	69	0.90	5	M2		114		<0.11	< 40	15.6:	9.3
76	UY Aur	0.89	225	0.28	5	K7	1.4	73	2.21	3.85	48	12.4	7.0
44	FX Tau	0.91	292	0.55	5	M1	0.9	9.6	1.45	0.49	< 30	13.9	8.1
379	LkCa7	1.05	25	0.56	5	K7	0.93	4.0	0.25	<0.06	< 24[h]	12.6	8.3
389	Haro 6-10	1.21	355	0.13	6	K3				16.65	116[i]	17.2:	7.0
	GG Tau Bb	1.4	135	0.19	5,7	M5	0.2	83				17.3:	10.0
80/81 m	RW Aur AB	1.50	258	0.23	5	cont.	0.4	84		2.60	< 20[d]	11.1/12.7:	6.9
358	040047+2603W	1.58	226	0.60	5	M3		13		<0.04		15.2	9.4
356/357	040012+2545	2			8	K2	0.33	0.9/1.8		<0.04		12.9	10.2
411	CoKu Tau/3	2.04	177	0.29	5	M1		4.6		0.45		17.3:	8.2
48	HK Tau	2.4	175	0.059	9	M0.5	0.7	29	1.47	0.31	41	15.8	8.7/11.8[j]
57	GK Tau	2.4	66	0.03[k]	11,12	K7	1.1	15		<0.15		13.5/17.0	7.6
	IT Tau	2.48	225	0.23	1	K0[1]	5.4[1]	4.3[1]		0.27	< 33	14.3[1]	8.2
45	DK Tau	2.53	115	0.30	1,3,5	K7/K7	1.6	19/34	1.93	2.24	35	12.4	7.0
43 m	UX Tau AC	2.7	181	0.069	9	K5	1.0	20	1.31	0.30	73	11.4/15.1(R)	7.7/10.5[j]
73/424	Haro 6-37	2.7	37	0.45	9	K7/M1	0.9/0.4	21/166	1.75	1.20	60:[d]	13.5/14.7	7.5/8.9
	J4872	3	225	0.49[k]	10,11	K7	1.7	2.0		<0.15[1]		13.0/13.9	8.2
60/406	HN Tau	3.1	215	0.042	9	K5/M4	0.6/0.03	138/108		1.59	< 45	14.2/18.1:	8.2
51/395	V710 Tau	3.24	357	0.83	5	M0.5/M2	0.7/0.8	89/11	1.15	0.35	71	14.2/14.4	8.6/8.6
52/53 m	UZ Tau EW	3.78	273	0.49	1	M2/M3	1.0/0.6	41/79	1.83 (E)	1.51	172[g]	13.1/13.7	8.1/7.5
418 m	HV Tau AB	4.00	45	0.029	1	M1	1.0/0.6	4.5		<0.04	< 45	14.0:	7.8[n]
43/42 m	UX Tau AB	5.9	269	0.29	1	K5/M2	1.0/0.6	20/8	1.31 (A)	0.30	73	11.4/13.7	7.7/8.9
355/354	035135+2528	6.3	298	0.46	12	K0/K2	0.3/0.2	0/0		<0.07		12.7/13.9	10.3/11.1
64	HO Tau[o]	6.9	109		12	M0.5	0.2	115		<0.11		14.4/16:	9.6
75	DS Tau[o]	7.1	294	0.045	9	K2/F8	1.0/0.2	70/0	1.37	0.34	< 30	12.4/13.6	8.0/11.3
360/361	040142+2150	7.2	65	0.89	12	M3/M3	0.2/0.2	10/8		0.07	23	14.9/15.1	10.0/10.1
352/353	035120+3154	8.7	71	0.77	12	G8/K0	0.7/0.6	3/2		<0.05		11.8/12.4	9.6/9.9
415/414	HP Tau G2/G3[o]	10.0	243	0.24	5,9	G2/K5	9.3/1.3	2/2		<0.04		11.0/14.6	7.2/8.7

HBC: Herbig & Bell (1988); an 'm' indicates multiple entries in table of same data (i.e. for unresolved observations of multiple system).

d: Projected angular separation; * means projection against lunar limb from occultation measurement.

p: Position angle, measured east of north.

$L*$:   Photospheric luminosity, taken from Hartigan et al (1994), Strom et al (1989a), or binary reference in the order; represents combined light from all stellar components, except in widely separated cases.

$H\alpha$:   Equivalent width (positive = emission); taken from Hartigan et al (1994), Strom et al (1989a), or binary reference in that order.

ΔN: Logarithmic excess flux ratio over photosphere at 10 μm, taken from Skrutskie et al (1990).

12 μm: Continuum $IRAS$ measurement from Strom et al (1989a) or Weaver & Jones (1992), in that order; ':' means uncertainty >50%.

1.3 mm: Continuum flux from Beckwith et al (1990) unless otherwise noted.

V: Combined light of all components, except in widely separated cases; taken from Hartigan et al (1994), Strom et al (1989a), or Herbig & Bell (1988), in that order.

K: Combined light of both components; taken from Hartigan et al (1994), Strom et al (1989a), or binary reference in that order, or reference noted; in triples, K of close pair is derived from flux ratio of wide pair, K of wide pair is light of entire system.

aa   HQ Tau was not confirmed to be binary by Richichi et al (1994); HP Tau was not found to be binary by Simon et al (1987).

a   Jones & Herbig (1979).

b   Briceno et al (1993).

c   Jensen et al (1994) did not detect at 800 μm.

d   Reipurth & Zinnecker (1993).

e   Beckwith et al (1990).

f   See also Dutrey et al (1994).

g   See also Simon & Guilloteau (1992).

h   At 1100 μm; Skinner et al (1991).

i   Reipurth et al (1993).

j   Moneti & Zinnecker (1991).

k   At 0.9 μm; Reipurth & Zinnecker (1993).

l   Hartmann et al (1991).

n   Simon et al (1992).

o   Moneti & Zinnecker (1991) identify the companion to DS Tau as a background field star; Hartigan et al suspect that the very faint companion to HO Tau is a background star; Moneti & Zinnecker take HP Tau G3 to be a companion to HP Tau (separation 17.5''), with the nature of HP Tau G2 uncertain.

1: Simon et al (1992)
2: Richichi et al (1994)
3: M. Simon, personal communication
4: Ghez et al (1993)
7: Leinert et al (1991)
8: Walter et al (1988)
9: Moneti & Zinnecker (1991)
10: Hartmann et al (1991)

Table A2 Pre-main-sequence double-lined spectroscopic binaries

HBC	Name	P (days)	γ (km s⁻¹)	e	$a \sin i$ (AU)	$m \sin^3 i$ ($M_\odot$)	Q	Ref.	Spec. Class	L_* ($L_\odot$)	$H\alpha$ (Å)	$12\,\mu$m (Jy)	V mag	K mag	Location
	HD155555	1.681652	2.7	0	0.013	0.964	1.07	1	G5	3.3:	abs		6.73		isolated
662	*V4046 Sgr*	2.42131	-6.8	0	0.024	0.30	1.07	2	K5		>100	0.45	10.40	7.34 [a]	isolated
400	V826 Tau	3.88776	17.5	0	0.013	0.0203	1.02	3	K7	1.1	1.6	<0.07	12.11	8.32	Tau-Aur
	OriNTT 569 [b]	4.25	29.	0	0.04	0.61	1.0	4	K4 [c]	1.5 [c]	0.5 [c]		13.61 [c]	10.63 [c]	Orion Belt
	EK Cep	4.42782	-10.9	0.109	0.077	3.15 [d]	1.81	5	A1,G5	15.1,1.5	3		7.85		isolated
	P2486	5.1882	20.0	0.161	0.066	1.37	1.04	4	G5 [e]	5.25 [f]	-3.2 [e]		11.38 [f]	9.92 [f]	Trapezium
536	W134	6.353	26.	0	0.099	3.2	1.04	6	G5	16.1	1	0.34	12.44	9.81	NGC2264
271	OriNTT 429	7.46	25.	0.27	0.10	2.2	1.0	4	K3 [c]	3.8 [c]	0.7 [c]		12.82 [c]	9.85 [c]	Orion Belt
487	AK Sco	13.6093	-1.1	0.469	0.143	2.12	1.01	7	F5	16	wk em	2.60	8.82	6.59	Sco-Cen?
447	P2494	19.4815	24.0	0.262	0.146	1.089	1.41	8	K0 [e]	>11.5 [f]	-0.3 [e]	<0.19	10.74 [f]	8.54 [f]	Trapezium
	P1540	33.73	20.2	0.12	0.188	0.79	1.32	9	K3	16:		<2.8	11.33	8.02	Trapezium
	162814-2427 [b]	35.95	-6.1	0.48	0.267	1.96	1.1	10	K7	1	wk em	0.49	12.23	7.19	ρ Oph

HBC: Herbig and Bell (1988).
Name: The pre-main-sequence status of italicized stars is ambiguous.
γ: Center-of-mass velocity.
$a \sin i$: Projected semimajor axis of relative orbit.
$m \sin^3 i$: m = total mass.
Q: Mass ratio.
L_*:   Total luminosity from binary reference unless otherwise noted.
$H\alpha$: Equivalent width (positive = emission) from binary reference unless otherwise noted.
$12\,\mu$m: Continuum *IRAS* measurement from *IRAS* Point Source Catalog or Weaver & Jones (1992).

[a] Hutchinson et al (1990).
[b] Constant-velocity tertiary spectrum also seen.
[c] Lee (1992) and FM Walter, personal communication.
[d] Total mass, derived from eclipse light-curve solution for inclination angle.
[e] King (1993 and personal communication).
[f] Rydgren & Vrba (1984).

1 : Pasquini et al (1991).
2 : de la Reza et al (1986 and personal communication), Byrne (1986).
3 : Mundt et al (1983), Reipurth et al (1990).
4 : Preliminary orbit of author and collaborators
5 : Tomkin (1983), Hill & Ebbighausen (1984).
6 : Padgett & Stapelfeldt (1993).
7 : Andersen et al (1989).
8 : B Reipurth, H Lindgren, J-C Mermilliod, M Mayor, personal communication
9 : Marschall & Mathieu (1988).
10 : Mathieu et al (1989).

Table A3 Pre-main-sequence single-lined spectroscopic binaries

HBC[a]	Name	P (days)	γ (km s^{-1})	e	$a_1 \sin i$ (AU)	$f(m)$ ($M_\odot$)	Ref.	Spec. Class	L_* ($L_\odot$)	Hα (Å)	12μm (Jy)	V mag	K mag	Location
	155913-2233 [b]	2.42378	-2.3	0	0.014	0.064	1	K5	0.9	0.7 [c]		11.21	8.08	Sco-Cen
	160905-1859	10.400	-6.4	0.17	0.015	0.0042	1	K1	1.2	-0.27 [c]		11.67	8.08	Sco-Cen
	VSB126	12.924	17.0	0.18	0.017	0.0040	2	K0		-1.9		13.43	11.4 [d]	NGC2264
	Lk Ca 3	12.941	14.9	0.20	0.032	0.027	3	M1 [e]	2.6 [e]	2.5 [e]	<0.05 [e]	12.10 [e]	7.53 [e]	Tau-Aur
	155808-2219 [b]	16.925	-5.0	0.10	0.048	0.051	3	M3 [c]	0.34 [c]	3.32 [c]		13.74 [c]	8.80 [c]	Sco-Cen
459	P1925	32.94	25.9	0.55	0.045	0.011	2	K3				13.38	8.5 [f]	Trapezium
	162819-2423S	89.1	-4.9	0.41	0.10	0.017	1	G8	2.4	wk em [c]	2.54	10.57	6.70	ρ Oph
	160814-1857	144.7	-6.6	0.26	0.19	0.048	1	K2	1.3	0.71 [c]		11.94	7.68	Sco-Cen
	P1771	149.5	25.6	0.57	0.16	0.024	2	K4				13.47	9.5 [f]	Trapezium
85	GW Ori	241.9	28.1	0.04±0.06	0.10	0.0026	4	G5	66	27.6	7.87	9.80	6.19	B30
644	Haro 1-14c	591	-8.7	0.62	0.36	0.018	5	K3			< 0.04	12.29		ρ Oph
	VSB111	879	26.3	0.80	0.6	0.04	2	G8		-0.3		12.37	10.2 [d]	NGC2264
	045251+3016	2530.	14.9	0.48	1.6	0.09	1,3	K7	1.4	0.7 [c]	0.11: [e]	11.60	8.27	Tau-Aur

$a_1 \sin i$: Projected semimajor axis of primary orbit.

$f(m)$: Mass function; $f(m) = (m_2 \sin i)^3 / (m_1 + m_2)^2$ where m_2 is mass of secondary.

[a] Other table headings as in Table A2.

[b] Constant-velocity tertiary spectrum also seen.

[c] Walter et al (1988, 1994).

[d] F Piche, personal communication.

[e] Strom et al (1989b).

[f] M McCaughrean, personal communication.

1: Mathieu et al (1989).

2: RD Mathieu, LM Marschall, DW Latham, in preparation.

3: Preliminary orbit by author and collaborators.

4: Mathieu et al (1991).

5: B Reipurth, H Lindgren, M Mayor, personal communication.

Literature Cited

Abt HA. 1983. *Annu. Rev. Astron. Astrophys.* 21:343–72

Abt HA. 1986. *Astrophys. J.* 304:688–94

Abt HA, Levy S. 1976. *Astrophys. J. Suppl.* 30:273–306

Abt HA, Willmarth DW. 1992. See McAlister & Hartkopf 1992, p. 82

Adams FC, Benz W. 1992. See McAlister & Hartkopf, p. 185

Adams FC, Emerson JP, Fuller GA. 1990. *Astrophys. J.* 357:606–20

Adams FC, Lada CJ, Shu FH. 1987. *Astrophys. J.* 312:788–806

Adams FC, Lada CJ, Shu FH. 1988. *Astrophys. J.* 326:865–83

Adams FC, Ruden SP, Shu FH. 1989. *Astrophys. J.* 347:959–75

Andersen J, Lindgren H, Hazen ML, Mayor M. 1989. *Astron. Astrophys.* 219:142–50

Andre P, Montmerle T. 1993. *Astrophys. J.* 420:837–62

Anglada G, Estralla R, Rodriguez LF, Torrelles JM, Lopez R, Canto J. 1991. *Astrophys. J.* 376:615–17

Artymowicz P. 1992. *Publ. Astron. Soc. Pac.* 104:769–74

Artymowicz P, Clarke CJ, Lubow SH, Pringle JE. 1991. *Astrophys. J. Lett.* 370:L35–38

Artymowicz P, Lubow SH. 1994. *Astrophys. J.* 421:651–67

Barth W, Weigelt G, Zinnecker H. 1994. *Astron. Astrophys.* In press

Basri G, Batalha C. 1990. *Astrophys. J.* 363:654–69

Basri G, Bertout C. 1993. See Levy & Lunine 1993, pp. 543–66

Batten AH. 1973. *Binary and Multiple Systems of Stars.* Oxford:Pergamon. 278 pp.

Beckwith SVW, Sargent AI. 1993. See Levy & Lunine 1993, pp. 521–42

Beckwith S, Sargent A, Chini R, Gusten R. 1990. *Astron. J.* 99:924–45

Bertout C. 1983. *Astron. Astrophys.* 126:L1–4

Bertout C, Basri G, Bouvier J. 1988. *Astrophys. J.* 330:350–73

Bodenheimer P. 1992. In *Evolutionary Processes in Interacting Binary Stars, IAU Symp. 151,* ed. Y Kondo, RF Sistero, RS Polidan, p. 9. Dordrecht:Kluwer

Bodenheimer P, Ruzmaikina T, Mathieu RD. 1993. See Levy & Lunine, pp. 367–404

Bonnell I, Bastien P. 1992. *Astrophys. J. Lett.* 401:L31–34

Bonnell I, Bastien P. 1993. *Astrophys. J.* 406:614–28

Boss AP. 1991. *Nature* 351:298–300

Boss AP. 1993a. In *The Realm of Interacting Binary Stars,* ed. J Sahade, GE McCluskey, Y Kondo, p. 355. Dordrecht: Kluwer

Boss AP. 1993b. *Astrophys. J.* 410:157–67

Boss AP, Yorke HW. 1993. *Astrophys. J. Lett.* 411:L99–102

Briceno C, Calvet N, Gomez M, Hartmann LW, Kenyon SJ, Whitney BA. 1993. *Publ. Astron. Soc. Pac.* 105:686–92

Burkert A, Bodenheimer P. 1993. *MNRAS* 264:798–806

Byrne PB. 1986. *Irish. Astron. J.* 17:294–300

Calvet N, Hartmann L. 1992. *Astrophys. J.* 386:239–47

Chapman S, Pongracic H, Disney M, Nelson A, Turner J, Whitworth A. 1992. *Nature* 359:207–10

Chelli A, Zinnecker H, Carrasco L, Cruz-Gonzalez I, Perrier C. 1988. *Astron. Astrophys.* 207:46–54

Clarke CJ. 1992a. See Duquennoy & Mayor 1992, p. 38

Clarke CJ. 1992b. See McAlister & Hartkopf 1992, p. 176

Clarke CJ, Pringle JE. 1991a. *MNRAS* 249:584–87

Clarke CJ, Pringle JE. 1991b. *MNRAS* 249:588–95

Clarke CJ, Pringle JE. 1993. *MNRAS* 261:190–202

Cohen M. 1983. *Astrophys. J. Lett.* 270:L69–71

Cohen M, Kuhi LV. 1979. *Astrophys. J. Suppl.* 41:743–843

Cohen M, Witteborn FC. 1985. *Astrophys. J.* 294:345–56

D'Antona F, Mazzitelli I. 1994. *Astrophys. J. Suppl.* 90:467–500

de la Reza R, Quast G, Torres CAO, Mayor M, Meylan G, Llorente de Andres F. 1986. In *New Insights in Astrophysics,* pp. 107–11. ESA SP-263

Draine BT, Lee HM. 1984. *Astrophys. J.* 285:89–108

Duquennoy A, Mayor M. 1991. *Astron. Astrophys.* 248:485–524

Duquennoy A, Mayor M, eds. 1992. *Binaries as Tracers of Stellar Formation.* Cambridge:Cambridge Univ. Press

Duquennoy A, Mayor M, Mermilliod J-C. 1992. See Duquennoy & Mayor 1992, p. 52

Dutrey A, Guilloteau S, Simon M. 1994. *Astron. Astrophys.* In press

Dyck HM, Simon T, Zuckerman B. 1982. *Astrophys. J. Lett.* 255:L103–6

Edwards S, Strom SE, Hartigan P, Strom KM, Hillenbrand LA, et al. 1993. *Astron. J.* 106:372–82

Eiroa C, Leinert Ch. 1987. *Astron. Astrophys.* 188:46–48

Eiroa C, Lenzen R, Leinert Ch, Hodapp K-W. 1987. *Astron. Astrophys.* 179:171–75

Estralla R, Anglada G, Rodriguez LF, Garay G. 1991. *Astrophys. J.* 371:626–30

Fekel FC Jr. 1981. *Astrophys. J.* 246:879–98

Fischer DA, Marcy GW. 1992. *Astrophys. J.* 396:178–94

Ghez AM, McCarthy DW Jr, Weinberger AJ, Neugebauer G, Matthews K. 1994. In *Infrared Astronomy with Arrays: The Next Generation,* ed. IS McLean. In press

Ghez AM, Neugebauer G, Gorham PW, Haniff CA, Kulkarni S, et al. 1991. *Astron. J.* 102:2066–72

Ghez AM, Neugebauer G, Matthews K. 1993. *Astron. J.* 106:2005–23

Gorham PW, Ghez AM, Haniff CA, Kulkarni SR, Matthews K, Neugebauer G. 1992. *Astron. J.* 103:953–59

Gredel R, Reipurth B. 1993. *Astrophys. J. Lett.* 407:L29–32

Haas M, Christou JC, Zinnecker H, Ridgeway ST, Leinert Ch. 1993. *Astron. Astrophys.* 269:282–90

Haas M, Leinert Ch, Zinnecker H. 1990. *Astron. Astrophys.* 230:L1–4

Halbwachs JL. 1987. *Astron. Astrophys.* 183:234–40

Hale A. 1994. *Astron. J.* 107:306–32

Hanson R, Jones BJ, Lin DNC. 1983. *Astrophys. J. Lett.* 270:L27–30

Hartigan P, Hartmann L, Kenyon SJ, Strom SE, Skrutskie MF. 1990. *Astrophys. J. Lett.* 354:L25–28

Hartigan P, Strom KM, Strom SE. 1994. *Astrophys. J.* In press

Hartmann LW, Hewett R, Stahler S, Mathieu RD. 1986. *Astrophys. J.* 309:275–93

Hartmann LW, Jones BF, Stauffer JR, Kenyon SJ. 1991. *Astron. J.* 101:1050–62

Hartmann LW, Kenyon SJ. 1990. *Astrophys. J.* 349:190–96

Heggie DC. 1975. *MNRAS.* 173:729–87

Heintz WD. 1969. *J. R. Astron. Soc. Can.* 63:275–98

Heller CH. 1993. *Astrophys. J.* 408:337–46

Herbig GH. 1962. *Adv. Astron. Astrophys.* 1:47–103

Herbig GH. 1977. *Astrophys. J.* 214:747–58

Herbig GH, Bell KR. 1988. *Lick Obs. Bull. 1111*

Herbst W, Booth JF, Chugainov PF, Zajtseva GV, Barksdale W, et al. 1986. *Astrophys. J. Lett.* 310:L71–75

Herczeg TJ. 1988. *Astrophys. Space Sci.* 142:89–95

Hill G, Ebbighausen EG. 1984. *Astron. J.* 89:1256–60

Hutchinson MG, Evans A, Winkler H, Jones JS.

1990. *Astron. Astrophys.* 234:230–32

Jensen ELN, Mathieu RD, Fuller GF. 1994. *Astrophys. J. Lett.* In press

Jones BF, Herbig GH. 1979. *Astron. J.* 84:1872–89

Joy AH, van Biesbroeck G. 1944. *Publ. Astron. Soc. Pac.* 56:123–24

Kawabe R, Ishiguro M, Omodaka T, Kitamura Y, Miyama, SM. 1993. *Astrophys. J. Lett.* 404:L63–66

Kenyon SJ, Hartmann L. 1987. *Astrophys. J.* 323:714–33

King JR. 1993. *Astron. J.* 105:1087–95

Koerner DW, Sargent AI, Beckwith SVW. 1993a. *Astrophys. J. Lett.* 408:L93–96

Koerner DW, Sargent AI, Beckwith SVW. 1993b. *Icarus* 106:2–10

Koresko CD, Beckwith SVW, Ghez AM, Matthews K, Neugebauer G. 1991. *Astron. J.* 102:2073–78

Lada EA, DePoy DL, Evans NJ, Gatley I. 1991. *Astrophys. J.* 371:171–82

Larson RB. 1990. In *Physical Processes in Fragmentation and Star Formation,* ed. R Capuzzo-Dolcetta, C Chiosi, A di Fazio, p. 389. Dordrecht:Kluwer

Lee C-W. 1992. *Double-lined pre-main-sequence binaries: a test of pre-main-sequence evolutionary theory.* PhD thesis. Univ. Wis., Madison

Leinert Ch, Haas M. 1989. *Astrophys. J. Lett.* 342:L39–42

Leinert Ch, Haas M, Richichi A, Zinnecker H, Mundt R. 1991. *Astron. Astrophys.* 250:407–19

Leinert Ch, Weitzel N, Zinnecker H, Christou J, Ridgeway S, et al. 1993. *Astron. Astrophys.* 278:129–49

Levy EH, Lunine JI, eds. 1993. *Protostars and Planets III.* Tucson:Univ. Ariz. Press

Lin DNC, Papaloizou JCB. 1993. See Levy & Lunine 1993, pp. 749–836

Lin DNC, Pringle JE. 1990. *Astrophys. J.* 358:515–24

Lubow SH, Artymowicz P. 1992. See Duquennoy & Mayor 1992, p. 145

Marschall LM, Mathieu RD. 1988. *Astron. J.* 96:1956–64

Marsh K, Mahoney MJ. 1992. *Astrophys. J. Lett.* 395:L115–18

Marsh K, Mahoney MJ. 1993. *Astrophys. J. Lett.* 405:L71–74

Martin EL, Rebolo R. 1993. *Astron. Astrophys.* 274:274–78

Mathieu RD. 1986. In *Highlights of Astronomy,* ed. J-P Swings, p. 481. Dordrecht:Reidel

Mathieu RD. 1992a. See Duquennoy & Mayor 1992, p. 155

Mathieu RD. 1992b. See McAlister & Hartkopf 1992, p. 30

Mathieu RD. 1992c. In *Evolutionary Processes in Interacting Binary Stars, IAU Symp. 151,* ed. Y Kondo, RF Sistero, RS Polidan, p. 21.

Dordrecht:Kluwer

Mathieu RD, Adams FC, Latham DW. 1991. *Astron. J.* 101:2184–98

Mathieu RD, Adams FC, Fuller GF, Jensen ELN, Koerner DW, Sargent AI. 1994. *Astron. J.* In press

Mathieu RD, Duquennoy A, Latham DW, Mayor M, Mazeh T, Mermilliod J-C. 1992. See Duquennoy & Mayor 1992, p. 278

Mathieu RD, Walter FM, Myers PC. 1989. *Astron. J.* 98:987–1001

Matthews L, Mathieu RD. 1992. See McAlister & Hartkopf 1992, p. 244

Mayor M, Duquennoy A, Halbwachs J-L, Mermilliod J-C. 1992. See McAlister & Hartkopf 1992, p. 73

Mayor M, Mermilliod J-C. 1984. In *IAU Symp. 105, Observational Tests of Stellar Evolution Theory,* ed. A Maeder, A Renzini, p. 145. Dordrecht:Reidel

Mazeh T, Goldberg D, Duquennoy A, Mayor M. 1992. *Astrophys. J.* 401:265–68

McAlister HA, Hartkopf WI, eds. 1992. *Complementary Approaches to Double and Multiple Star Research, IAU Colloq. 135.* San Francisco:Astron. Soc. Pac.

Menard F, Monin J-L, Angelucci F, Rouan D. 1993. *Astrophys. J. Lett.* 414:L117–20

Menten KM, Serabyn E, Gusten R, Wilson TL. 1987. *Astron. Astrophys.* 177:L57–60

Meyer MR, Wilking BA, Zinnecker H. 1993. *Astron. J.* 105:619–29

Miller GE, Scalo JM. 1978. *Publ. Astron. Soc. Pac.* 90:506–13

Mizuno A, Fukui Y, Iwata T, Nosawa S, Tokano T. 1990. *Astrophys. J.* 356:184–94

Moneti A, Zinnecker H. 1991. *Astron. Astrophys.* 242:428–32

Mundt R, Walter FM, Feigelson ED, Finkenzeller U, Herbig GH, Odell AP. 1983. *Astrophys. J.* 269:229–38

Mundy LG, Wilking BA, Myers ST. 1986. *Astrophys. J. Lett.* 311:L75–79

Mundy LG, Wootten HA, Wilking BA. 1990. *Astrophys. J.* 352:159–66

Mundy LG, Wootten A, Wilking BA, Blake GA, Sargent AI. 1992. *Astrophys. J.* 385:306–13

Nisenson P, Stachnik RV, Karovska M, Noyes R. 1985. *Astrophys. J. Lett.* 297:L17–20

Ostriker E, Shu FH, Adams FC. 1992. *Astrophys. J.* 399:192–212

Padgett D, Stapelfeldt K. 1993. *Astron. J.* 107:720–28

Palla F, Stahler SW. 1990. *Astrophys. J. Lett.* 360:L47–50

Pasquini L, Cutispoto G, Gratton R, Mayor M. 1991. *Astron. Astrophys.* 248:72–80

Plavec M. 1970. In *Stellar Rotation,* ed. A Slettebak, p. 133. Dordrecht:Reidel

Popper DM. 1987. *Astrophys. J. Lett.* 313:L81–83

Pringle JE. 1989. *MNRAS* 239:361–70

Pringle JE. 1991a. *MNRAS* 248:754–59

Pringle JE. 1991b. In *The Physics of Star Formation and Early Evolution,* ed. CJ Lada, ND Kylafis, p. 437. Dordrecht:Kluwer

Prosser CF, Stauffer JR, Hartmann L, Soderblom DR, Jones BF, Werner MW, McCaughrean MJ. 1994. *Astrophys. J.* 421:517–41

Reipurth B. 1988. In *Formation and Evolution of Low Mass Stars,* ed. AK Dupree, MTVT Lago, pp. 305–18. Dordrecht:Reidel

Reipurth B, Chini R, Krugel E, Kreysa E, Sievers A. 1993a. *Astron. Astrophys.* 273:221–38

Reipurth B, Heathcote S, Roth M, Noriega-Crespo A, Raga AC. 1993b. *Astrophys. J. Lett.* 408:L49–52

Reipurth B, Lindgren H, Nordstrom B, Mayor M. 1990. *Astron. Astrophys.* 235:197–204

Reipurth B, Zinnecker H. 1993. *Astron. Astrophys.* 278:81–108

Richichi A, Leinert Ch, Jameson R, Zinnecker H. 1994. *Astron. Astrophys.* In press

Rydgren AE, Vrba FJ. 1984. *Astron. J.* 89:399–405

Sandell G, Aspin C, Duncan WD, Russell APG, Robson EI. 1991. *Astrophys. J. Lett.* 376:L17–20

Sargent AI, Beckwith SVW. 1991. *Astrophys. J. Lett.* 382:L31–35

Sasselov DD, Rucinski SM. 1990. *Astrophys. J.* 351:578–82

Savage BD, Mathis JS. 1979. *Annu. Rev. Astron. Astrophys.* 17:73–113

Schwartz PR, Simon T, Campbell R. 1986. *Astrophys. J.* 303:233–38

Schwartz PR, Simon T, Zuckerman B, Howell, RR. 1984. *Astrophys. J. Lett.* 280:L23–26

Shu FH, Adams FC, Lizano S. 1987. *Annu. Rev. Astron. Astrophys.* 25:23–81

Simon M. 1992. See McAlister & Hartkopf 1992, p. 41

Simon M, Chen WP, Howell RR, Slovik D. 1992. *Astrophys. J.* 384:212–19

Simon M, Ghez AM, Leinert Ch. 1993. *Astrophys. J. Lett.* 408:L33–36

Simon M, Guilloteau S. 1992. *Astrophys. J. Lett.* 397:L47–49

Simon M, Howell RR, Longmore AJ, Wilking BA, Peterson DM, Chen W-P. 1987. *Astrophys. J.* 320:344–55

Skinner SL, Brown A, Walter FM. 1991. *Astron. J.* 102:1742–48

Skrutskie MF, Dutkevich D, Strom SE, Edwards S, Strom KM, Shure MA. 1990. *Astron. J.* 99:1187–95

Stahler S. 1983. *Astrophys. J.* 274:822–29

Strom KM, Strom SE, Edwards S, Cabrit S, Skrutskie MF. 1989a. *Astron. J.* 97:1451–70

Strom KM, Wilkin FP, Strom SE, Seaman RL. 1989b. *Astron. J.* 1444–50

Strom SE, Edwards S, Skrutskie M. 1993. See Levy & Lunine 1993, pp. 837–66

Terquem C, Bertout C. 1993. *Astron. Astrophys.* 274:291–303

Tessier E, Bouvier J, Lacombe F. 1994. *Astron. Astrophys.* In press

Tomkin J. 1983. *Astrophys. J.* 271:717–24

Trimble V. 1990. *MNRAS* 242:79–87

Walker CK, Carlstrom JE, Bieging JH. 1993. *Astrophys. J.* 402:655–66

Walker CK, Carlstrom JE, Bieging JH, Lada CJ, Young ET. 1990. *Astrophys. J.* 364:173–77

Walker CK, Lada CJ, Young ET, Maloney PR, Wilking BA. 1986. *Astrophys. J. Lett.* 309:L47–51

Walter FM, Brown A, Mathieu, RD, Myers PC, Vrba FJ. 1988. *Astron. J.* 96:297–325

Walter FM, Vrba FJ, Mathieu RD, Brown A, Myers PC. 1994. *Astron. J.* 107:692–719

Weaver WB, Jones G. 1992. *Astrophys. J. Suppl.* 78:239–66

Weinberg MD, Shapiro SL, Wasserman I. 1987. *Astrophys. J.* 312:367–89

Weintraub DA, Kastner JH, Zuckerman B, Gatley I. 1992. *Astrophys. J.* 391:784–804

Weintraub DA, Masson CR, Zuckerman B. 1989. *Astrophys. J.* 344:915–24

Whitney BA, Clayton GC, Schulte-Ladbeck RE, Calvet N, Hartmann L, Kenyon SJ. 1993. *Astrophys. J.* 417:687–96

Whitney B, Hartmann LW. 1993. *Astrophys. J.* 402:605–22

Wootten A. 1989. *Astrophys. J.* 337:858–64

Zahn J-P, Bouchet L. 1989. *Astron. Astrophys* 223:112–18

Zinnecker H. 1989. In *Low Mass Star Formation and Pre-Main Sequence Objects,* ed. B Reipurth, p. 447. Garching:ESO

Zinnecker H. 1991. In *Fragmentation of Molecular Clouds and Star Formation, IAU Symp. 147,* ed. E Falgarone, F Boulanger, G Duvert, p. 526. Dordrecht:Kluwer

Zinnecker H, Wilking BA. 1992. See Duquennoy & Mayor 1992, p. 526

Annu. Rev. Astron. Astrophys. 1994. 32: 531–590

BARYONIC DARK MATTER

Bernard Carr

Schools of Mathematical Sciences, Queen Mary and Westfield College, Mile End Road, London E1 4NS, England

KEY WORDS: black holes, Population III stars, background radiation, gravitational lensing, brown dwarfs

1. INTRODUCTION

The evidence for dark matter, on all scales from star clusters ($10^6 M_\odot$) to the Universe itself ($10^{22} M_\odot$), has built up steadily over the past 50 years (Faber & Gallagher 1979, Trimble 1987, Turner 1991, Ashman 1992). Although the strength of the evidence on different scales varies considerably, there is now little doubt that only a small fraction of the mass of the Universe is in visible form. However, we remain uncertain as to the identity of the dark material. Proposed candidates span the entire mass range from 10^{-5}eV to $10^{12} M_\odot$, with a dichotomy between those—primary particle physicists—who would like the dark matter to be some sort of elementary particle and those—primarily astrophysicists—who would prefer it to be some sort of astrophysical object. In the first case, the dark matter would have to be nonbaryonic, with the particles being relics from the hot Big Bang; in the second case, it would have to be baryonic, with the dark objects being made out of gas which has been processed into the remnants of what are sometimes termed "Population III" stars.

During the 1970s the dark matter was usually assumed, at least implicitly, to be baryonic (e.g. Ostriker et al 1974, White & Rees 1978), but in the 1980s attention veered towards the nonbaryonic candidates. This was partly because of developments in particle physics, but also because it was realized that there are good cosmological reasons for believing that not everything can be baryonic (Hegyi & Olive 1983, 1986). For a while "hot" dark matter was popular, but soon "cold" dark matter took center stage and many people still regard this as the "standard" model. In the past few years, however, attention has returned to the baryonic candidates—partly because of perceived problems in the cold model and also because there may now be direct evidence for baryonic dark

0066–4146/94/0915–0531$05.00

matter. Entire conferences are now devoted to the topic (e.g. Lynden-Bell & Gilmore 1990), and there seems to be a growing realization that there are so many dark matter problems that one probably needs *both* baryonic and nonbaryonic solutions.

This review focuses almost exclusively on baryonic dark matter. Section 2 presents the observational evidence for dark matter in various contexts; the discussion here is rather brief because it is only necessary to highlight those issues that relate to baryonic dark matter in particular. Section 3 reviews the general arguments for baryonic and nonbaryonic dark matter, concluding—as indicated above—that one probably needs both. Section 4 discusses *why* one expects baryonic dark matter to form; this involves a brief review of the "Population III" scenario. Section 5 summarizes the constraints on the Population III scenario which come from background light and nucleosynthetic considerations. These limits are essentially as described by Carr et al (1984) but Sections 6 and 7 focus on topics—dynamical effects and lensing effects—that have seen important recent developments. The most plausible candidates seem to be the black hole remnants of high mass stars or low mass objects, so we focus on these candidates in more detail in Sections 8 and 9, respectively. We conclude in Section 10 with a reappraisal of these and other candidates and we assess the prospects of finding or excluding them.

2.   EVIDENCE FOR DARK MATTER

The observational evidence for dark matter arises in many different contexts and baryonic dark matter is not necessarily inplicated in all of these. We therefore begin by identifying the observational issues most relevant to the baryonic versus nonbaryonic dilemma.

2.1   *Local Dark Matter*

Measurements of the stellar velocity and density distribution perpendicular to the Galactic disk provide an estimate of the total disk density. This turns out to be about 0.1 $M_\odot pc^{-3}$ and it has long been suspected (Oort 1932) that this exceeds the density in visible stars. *The possibility of disk dark matter is very important in the present context because—of all the dark matter problems—this is the one most likely to have a baryonic solution.* Unfortunately, the evidence is very controversial. Bahcall (1984a,b,c) used counts of F dwarfs and K giants to conclude that the density of unseen material must be at least 50% that of the visible material. He also concluded that the disk dark matter must have an exponential scale height of less than 700 pc, so that it must itself be confined to a disk. However, Bahcall assumed a particular model and Bienayme et al (1987), using a different model, found a best-fit dark matter density of only $0.01 M_\odot pc^{-3}$, and even this could be removed if the halo was slightly flattened.

Knapp (1988) came to the same conclusion by studying the velocity dispersion and scale height of molecular hydrogen. Further doubt was cast in a series of papers by Kuijken & Gilmore (1989), who used the full distribution function for the velocities and distances of K dwarfs rather than assuming a particular model. Although Gould (1990) used a maximum likelihood analysis to conclude that Kuijken & Gilmore's data were not inconsistent with the Bahcall et al claim, Kuijken & Gilmore (1991) disagreed with this. More recently, Bahcall et al (1992a) have concluded from another analysis of K giants that the no-disk-dark-matter hypothesis is only consistent with the data at the 14% level and their best-fit model has a dark density of 0.15 $M_\odot pc^{-3}$, which corresponds to 53% more dark matter than visible matter. For present purposes the existence of disk dark matter will be regarded as an open question.

2.2   *Spiral Galaxies*

The best evidence for dark matter in galaxies comes from the rotation curves of spirals, since the dependence of the rotation speed V upon galactocentric distance R is a measure of the density profile $\rho(R)$. An important feature of our own and many other spiral galaxies is that the rotation speed, after an initial rise, remains approximately constant with increasing R (Rubin et al 1980). This implies that the mass within radius R increases like R, which is faster than the increase of visible mass. [Valentjin (1990) has claimed that spiral galaxies have sufficient dust to be opaque, thereby increasing the stellar mass content (cf Disney et al 1989), but Burstein et al (1991) disagree with this and the possibility will be neglected here.] Although the dark matter does not dominate within the optical galaxy (at least for bright galaxies), neutral hydrogen observations suggest that V continues to remain constant well beyond the visible stars (Sancisi & van Albada 1987). *In considering the baryonic contribution to galactic halos, the crucial issue is how far the halos extend.* For our galaxy the minimum halo radius consistent with rotation curve measurements, the local escape speed, and the kinematics of globular clusters and satellite galaxies is 35 kpc; the dynamics of the Magellanic Stream and the Local Group of galaxies may require a halo radius of 70 kpc (Fich & Tremaine 1991). We will see later that these values are marginally consistent with a baryonic halo. However, Zaritsky et al (1993) argue from observations of satellite systems that spiral galaxies typically have 200 kpc halos and this would be inconsistent with their being composed of baryons.

One indication that halos are dominated by nonbaryonic material may come from the fact that V has the same value in the optical region (where the bulge and disk dominate) as it does well beyond (where the dark matter dominates). This "conspiracy" may require that the ratio of baryonic to nonbaryonic dark mass be comparable to the dimensionless rotation parameter expected for protogalaxies as a result of tidal spinup (Fall & Efstathiou 1981, Blumenthal et al 1986);

both are of order 0.1. A recent calculation of this effect, allowing for the response of the dark halo to the dissipative infall of the luminous material, implies a baryonic to nonbaryonic ratio of 0.05 (Flores et al 1993). However, this would not apply if the baryons went dark before galaxy formation. Also, the conspiracy is only required for bright galaxies because only for these is the disk dynamically dominant in the central regions.

Another relevant issue concerns the *roundness* of galactic halos. If galactic halos are baryonic, one would expect their formation to involve dissipation, in which case they should be flatter than in the nonbaryonic case: N-body experiments show that dissipationless collapse does give some flattening but the resultant triaxial halos are rarely flatter than E6 (Frenk et al 1988, Dubinski & Carlberg 1991). Thus, evidence for halos flatter than this would be evidence for baryonic dark matter. Polar ring galaxies probably provide the best probe of halo shape, and these do seem to indicate triaxiality (Whitmore et al 1987), sometimes (e.g. for NGC 4650A) as high as E6 (Sackett & Sparke 1990). The existence of warped disks may also require triaxial surrounding halos (Teuben 1991), and such disks seem to be ubiquitous (Bosma 1991). Triaxiality in our own halo could also explain the asymmetries of the HI distribution (Blitz & Spergel 1991). Nevertheless, it is not clear whether there is enough triaxiality in these cases to imply baryonic halos.

2.3   *Elliptical Galaxies*

The mass distribution in ellipticals can be probed by measuring the velocity dispersion of the stars and globular clusters. Unfortunately, the velocities do not determine the density profile uniquely and this method gives no evidence for dark matter within the *central* regions of ellipticals (de Zeeuw 1992), although the dynamics of globular clusters does provide evidence for dark matter around M87 (Huchra & Brodie 1987, Mould et al 1990). The best information therefore comes from X-ray observations of hot gas. These do, in fact, provide evidence for dark matter and, in many cases, one finds the same $M \sim R$ law that characterizes spirals (Forman et al 1985, Sarazin 1986). Although these analyses assume that the gas is isothermal, usually one only has poor information on the temperature profile. However, Fabian et al (1986) obtain even larger minimum masses on the assumption that the halo is confined by a hydrostatic outer atmosphere. Giant ellipticals are sometimes the focus of cooling flows and this suggests that at least some of the dark matter in ellipticals may be baryonic (Fabian 1994).

2.4   *Dwarf Galaxies*

Some of the dwarf irregulars are extremely gas-rich, which means that their HI rotation curves can be traced to many optical scale lengths. Many of them seem to have much higher dark mass fractions than bright spirals, with their dark halos dominating even within the optical regions. Particularly striking examples are DDO 154 (Carignan & Freeman 1988), for which the dark-to-luminous mass ratio exceeds 10 at the last measured point of the rotation curve, GR 8 (Carignan et al 1990), and DDO 170 (Lake et al 1990). Dwarf spheroidals also seem to have dark halos (Lin & Faber 1983, Aaronson 1983, Aaronson & Olszewski 1987). This claim is based on measurements of velocity dispersions and tidal radii for the six dwarf spheroidals within the Local Group. Originally the dispersions had to be inferred from the individual velocities of only a dozen or so objects per galaxy, but higher resolution velocity measurements now provide much better data (Mateo et al 1991) and seem to confirm the results of the earlier work. *The presence of dark matter in dwarf galaxies is crucial in the present context because it requires that halos consist either of baryonic or cold nonbaryonic dark matter.* Lake (1990) has argued that the observations are more consistent with the first possibility: If the formation of the halos were dissipationless, their central densities imply that the galaxies need to form at a redshift exceeding 30, whereas they should form at a redshift of 10 in the Cold Dark Matter (CDM) scenario.

2.5   *Groups and Clusters of Galaxies*

Galaxies are clumped on various scales (as members of binaries, small groups, and rich clusters) and velocity dispersion measurements indicate that the dynamical mass exceeds the visible mass on all these scales. Binaries can only be studied statistically (because one does not know the orbital inclination in any individual case), so the data are less clear-cut here; however, there is compelling evidence for dark mass in clusters of galaxies. This is confirmed by X-ray data on the gas temperature (which provide an independent measure of the gravitational potential). In rich clusters, the dark mass dominates by at least a factor of 10 and the recent discovery of hot gas in two *small* groups of galaxies (HG92 and HCG62) by *ROSAT* shows that there are comparable amounts of dark matter there (Mulchaey et al 1993, Ponman & Bertram 1993).

In assessing whether the dark mass in groups and clusters can be baryonic, it is important to determine whether it is the same as the halo dark matter. Although the cluster dark mass cannot all be associated with individual galaxies *now*— for then dynamical friction would result in the most massive galaxies being dragged into the cluster center (White 1976)—it may still have derived from the galaxies originally. Indeed, in the hierarchial clustering picture one would expect the galaxies inside a cluster to be stripped of their individual halos to form a collective halo (White & Rees 1978). However, this would only suffice to explain all the cluster dark matter if the original galactic halos were larger than about 200 kpc and, in this case, we will see that they could not be purely baryonic unless one invokes inhomogeneous Big Bang nucleosynthesis.

2.6   *Background Dark Matter*

None of the forms of matter discussed above can have the critical density required for the Universe to recollapse: $\rho_{crit} = 3H_o^2/8\pi G = 2 \times 10^{-29}h^{-2}$ g cm^{-3}, where $h = H_o/(100 \text{ km s}^{-1} \text{ Mpc}^{-1})$. As discussed later, disk dark matter could only have $\Omega_d \sim 0.001$, while the halo and cluster dark matter could only have $\Omega_h \sim 0.01-0.1$ and $\Omega_c \sim 0.1-0.2$, respectively. However, according to the currently popular inflation theory (Guth 1981), in which the Universe undergoes an exponential expansion phase at some early time, the total density should have almost exactly the critical value ($\Omega = 1$). [See, however, Ellis (1988) and Ellis et al (1991) for a different point of view.] This would have two possible implications: 1. There is another dark component, or 2. galaxy formation is biased (Kaiser 1984, Dekel & Rees 1987) in the sense that galaxies form preferentially in just a small fraction of the volume of the Universe. Although the second possibility avoids a proliferation of dark matter species, some people now invoke a mixture of hot and cold dark matter anyway (e.g. Taylor & Rowan-Robinson 1992).

In either case, one would expect the mass-to-light ratio to increase as one goes to larger scales, and there is some indication of this from dynamical studies. One can probe the density on scales above 10 Mpc, for example, by analyzing large-scale streaming motions (Dressler et al 1987, Bertschinger & Dekel 1989) or by determining the dipole moment of the *IRAS* sources (Rowan-Robinson et al 1990). In all these analyses, the inferred density depends on the bias parameter b (dynamical effects depending on the product $\Omega^{0.6}b^{-1}$). More sophisticated analyses are needed to determine Ω and b separately (Peacock & Dodds 1994, Nusser & Dekel 1993). The *IRAS* dipole suggests a critical density if the *IRAS* sources are unbiased ($b = 1$); however, this conclusion would be erroneous if there was a significant contribution to the dipole from distances beyond 100 Mpc (Scaramella et al 1991). For a recent review of the evidence for and against $\Omega = 1$, see Coles & Ellis (1994).

3.   BARYONIC VS NONBARYONIC DARK MATTER

Candidates for the dark matter may be grouped into nonbaryonic and baryonic types. These will be referred to these as "Inos" and "Population III", respectively, and the candidates are listed explicitly in Table 1 in order of increasing mass. Some of the ino candidates are elementary particles and—depending on their mass—these are usefully classified as "hot" or "cold" since this affects their clustering properties. The term *Weakly Interacting Massive Particle* or *WIMP* is often used to describe these particles, though some people restrict this term to particles that are massive enough to be cold. The other inos are more exotic relics from the Big Bang and, for present purposes, primordial black holes are included in this category. [For a comprehensive review of the ino

Table 1   Baryonic and nonbaryonic matter candidates

	INOS		POPULATION III
Axions	(10^{-5} eV)	Snowballs	?
Neutrinos	(10 eV)	Brown dwarfs	$(<0.08\ M_\odot)$
Photinos	(1 GeV)	M-dwarfs	$(0.1\ M_\odot)$
Monopoles	(10^{16} GeV)	White dwarfs	$(1\ M_\odot)$
Planck relics	(10^{19} GeV)	Neutron stars	$(2\ M_\odot)$
Primordial holes	$(>10^{15}\text{ g})$	Stellar holes	$(\sim 10\ M_\odot)$
Quark nuggets	$(< 10^{20}\text{ g})$	VMO holes	$(10^2\text{--}10^5\ M_\odot)$
Shadow matter	?	SMO holes	$(>10^5\ M_\odot)$

candidates, see Turner (1991).] Table 1 illustrates that there are many forms of nonluminous matter, so it is naive to assume that all the dark matter problems will have a single explanation. Even though some of the candidates in Table 1 can probably be rejected, many viable ones remain.

3.1   Cosmological Nucleosynthesis

The main argument for nonbaryonic dark matter is associated with Big Bang nucleosynthesis. This is because the success of the standard picture in explaining the primordial light element abundances $[X(^4\text{He}) \approx 0.24, X(^2\text{D}) \sim X(^3\text{He}) \sim 10^{-5}, X(^7\text{Li}) \sim 10^{-10}]$ only applies if the baryon density parameter Ω_b is strongly constrained. Walker et al (1991) find that it must lie in the range

$$0.010\, h^{-2} < \Omega_b < 0.015\, h^{-2}, \tag{3.1}$$

where the upper and lower limits come from the upper bounds on ^4He and $^2\text{D} + ^3\text{He}$, respectively. The upper limit implies that Ω_b is well below 1, which suggests that no baryonic candidate could provide the critical density required in the inflationay scenario. The standard scenario therefore assumes that the total density parameter is 1, with only the fraction given by (3.1) being baryonic. Until recently, cold inos seemed to be most compatible with large-scale structure observations; this led to the popularity of the CDM scenario.

Recently, X-ray data on the mass of gas in groups and clusters of galaxies suggest that the standard CDM picture may not be satisfactory. Although the gas does not suffice to explain all the dark matter, the ratio of the visible baryon mass (i.e. the mass in the form of stars and hot gas) to total mass is still anomalously high compared to the mean cosmic ratio implied by Equation (3.1). For example, the baryon fraction is 13% for the small group HCG62 (Ponman & Bertram 1993) and it tends to be in the range 20–30% for rich clusters. In particular, ROSAT observations of Coma suggest that the baryon fraction within the central 3 Mpc is about 25%, which is five times as large as the standard cosmological ratio (White et al 1993). It is hard to understand how the extra baryon concentration would come about since dissipation should be unimportant on these scales and most other astrophysical processes (such as winds and supernovae) should *decrease* the local baryon fraction. [See, however, Babul & Katz (1993) for a contrary view.] Unless one invokes a cosmological constant, this suggests that either the cosmological density is well below the critical value or the baryon density is much higher than implied by the standard cosmological nucleosynthesis scenario.

In the past few years considerable work has focused on the question of whether one can circumvent condition (3.1) by invoking a first-order phase transition at the quark-hadron era. The idea is that the transition would generate fluctuations in the baryon density. Neutrons would then diffuse from the overdense regions (because their cross-section is less than that of the protons), which would lead to variations in the neutron-to-proton ratio. One can then produce deuterium in the regions where the density is low, without appreciably modifying the average helium production (Applegate et al 1987, Alcock et al 1987). However, there is still a problem getting the observed lithium abundance. This arises because, as one varies Ω_b, $X(^7Li)$ has a minimum at around $\Omega_b \sim 0.01$, and the observed abundance almost exactly matches this minimum. Any fluctuations in the baryon density will therefore tend to lead to an overabundance of lithium.

Interest in the effects of the quark-hadron transition was revived by the suggestion of Malaney & Fowler (1988) that neutrons could diffuse back into the overdense regions and destroy lithium, provided that the separation between the nucleation sites was finely-tuned ($d \sim 10$ m). However, in this case, helium may be overproduced. A detailed numerical investigation of the effects of simultaneously varying Ω_b, d, the amplitude of the baryon density fluctuations (R), and the volume fraction at high density (f_v) by Kurki-Suonio et al (1990) suggested that, although values of R as large as 100 are compatible with observation if $d < 300$ m, one can never have $\Omega_b = 1$. More recently, Mathews et al (1993) have argued that the largest possible value for the baryon density is $\Omega_b = 0.09h^{-2}$, so a critical density of baryons still seems to be excluded unless $H_o < 35$. For an up-to-date review of inhomogeneous nucleosynthesis, see Malaney & Mathews (1993).

3.2   *Microwave Anisotropies*

A second argument for nonbaryonic dark matter is associated with the upper limits on and detections of anisotropies in the cosmic microwave background (CMB). To form the observed large-scale structure through purely gravitational processes, the amplitude of the fluctuations in the matter density at decoupling must have exceeded a minimum value; this implies a minimum amplitude for the CMB anisotropies which may contravene observations for a purely baryonic model. The anisotropies are reduced in a model dominated by nonbaryonic dark matter ($\Omega \gg \Omega_b$), partly because the density fluctuations start growing earlier (from when the dark matter dominates the density) and partly because they continue growing for a longer period (fluctuations freezing out at a redshift $z \approx \Omega^{-1}$). Despite this argument, it is not clear that the anisotropy constraints require Ω to be as large as 1—especially if one relinquishes scale-invariant fluctuations—because both the amplitude and angular scale of the anisotropies are reduced in a low density Universe owing to the effects of radiation pressure at decoupling (Coles & Ellis 1994). In the past few years, therefore, much attention has focused on baryon-dominated models with "primeval isocurvature" fluctuations (Peebles 1987a,b). The fluctuations are assumed to have a power-law form and the problem is to determine whether one can choose a spectral index n which simultaneously matches the *COBE* anisotropies at $10°$–$90°$ (Smoot et al 1992) and the large-scale structure data (Cen et al 1993). One can already place strong constraints on the combination of Ω_b and n (Efstathiou et al 1992, Gouda & Sugiyama 1992), and some researchers claim that baryon-dominated models are already excluded (Chiba et al 1993).

3.3 *Arguments for Baryonic Dark Matter*

The cosmological nucleosynthesis argument is a two-edged sword: It requires both baryonic and nonbaryonic dark matter (Pagel 1990). This is because the value of Ω_b allowed by Equation (3.1) almost certainly exceeds the density of visible baryons Ω_v. A careful inventory by Persic & Salucci (1992) shows that the contributions to Ω_v are 0.0007 in spirals, 0.0015 in ellipticals and spheroidals, $0.00035\,h^{-1.5}$ in hot gas within an Abell radius for rich clusters, and $0.00026\,h^{-1.5}$ in hot gas out to a virialization radius in groups and poor clusters. This gives a total of $(2.2+0.6\,h^{-1.5}) \times 10^{-3}$, so Equation (3.1) implies that the fraction of baryons in dark form must be in the range 70%–95% for $0.5 < h < 1$. Note, however, that the Persic-Salucci estimate does not include any contribution from low surface brighteners galaxies (McGaugh 1994) or dwarf galaxies (Bristow & Phillipps 1994).

The discrepancy between Ω_b and Ω_v could be resolved if there were an appreciable density of intergalactic gas. We know there must be some neutral gas in the form of Lyman-α clouds, but the density parameter associated with the "damped" clouds is probably no more than $0.003\,h^{-2}$ (Lanzetta et al 1991)—comparable to the density in galaxies, and consistent with the idea that these are protogalactic disks. Although the missing baryons could conceivably be in the form of a hot intergalactic medium (either never incorporated into galaxies or expelled by supernovae and galactic winds), the temperature would need to be finely tuned (Barcons et al 1991). The Gunn-Peterson test requires $\Omega(HI) < 10^{-8}h^{-1}$ (Sargent & Steidel 1990), while the *COBE* limit on the Compton distortion of the microwave background ($y < 3 \times 10^{-5}$) requires that, for a temperature T at redshift z,

$$\Omega(\text{HII}) < 0.03 \left(\frac{T}{10^8 \text{K}}\right)^{-1} \left(\frac{y}{10^{-5}}\right) [(1+z)^{3/2} - 1]^{-1} h^{-1} \text{K} \qquad (3.2)$$

(Mather et al 1994). The latter limit implies that a smooth intergalactic medium (IGM) cannot generate the observed X-ray background, although there is still a temperature range beween 10^4 K and 10^8 K in which one could have $\Omega_{\text{IGM}} \sim \Omega_b$. Whether one could *expect* so much gas to remain outside galaxies depends on its thermal history (Blanchard et al 1992).

The other possibility is that the missing baryons are inside galactic halos. The halo dark matter cannot be in the form of hot gas for it would generate too many X rays. Recently, however, Pfenniger et al (1993) have argued that it could be in the form of cold molecular gas. In their model, the gas is initially in the form of dense cloudlets with mass $10^{-3} M_\odot$ and size 30 AU in a rotationally supported disk. The cloudlets then build up fractally to larger scales. Their model is motivated by the claim that spirals evolve along the Hubble sequence from Sd to Sa and that their mass-to-light ratio decreases in the process, which requires that the dark matter be progressively turned into stars. It also explains why the surface density ratio of dark matter and HI gas is constant outside the optical disk (Carignan et al 1990).

The final possibility—and the one that is the focus of the rest of this review— is that the dark baryons have been processed into stellar remnants. Even if stellar remnants have enough density to explain the alleged dark matter in the Galactic disk, this would be well below the value required by Equation (3.1), for if all disks have the 60% dark component envisaged for our Galaxy by Bahcall et al (1992a), this only corresponds to $\Omega_v \approx 0.001$. The more interesting question is whether the baryonic density could suffice to explain the dark matter in galactic halos; the term "Massive Compact Halo Object" or "MACHO" has been coined in this context. If our Galaxy is typical, the density associated with galactic halos would be $\Omega_h \approx 0.01 h^{-1} (R_h/35 \text{ kpc})$ where R_h is the halo radius. [The mass-to-light ratio for our Galaxy is $(14$–$24)$ $(R_h/35 \text{ kpc})$ (Fich & Tremaine 1991) corresponding to $\Omega_h = (0.008$–$0.014) h^{-1} (R_h/35 \text{ kpc})$; a more precise calculation would involve integrating over galaxies of all masses but then one would need to know the mass-dependence of R_h (Ashman et al 1993).] Thus Equation (3.1) implies that *all* the dark matter in our halo could be baryonic only for $R_h < 50 h^{-1}$kpc. We saw in Section 2.2 that the minimum size of our halo is 70 kpc, which would just be compatible with this. If it is larger, the baryonic fraction could only be $(R_h/50 h^{-1} \text{kpc})^{-1}$. The cluster dark matter has a density $\Omega_c \approx 0.1$ and Equation (3.1) implies that this matter cannot be purely baryonic unless one invokes inhomogeneous nucleosynthesis.

We note that there is no necessity for the Population III stars to form before galaxies just as long as some change in the conditions of star formation makes their mass different from what it is today. However, the epoch of formation will be very important for the relative distribution of baryonic and nonbaryonic dark matter, especially if the nonbaryonic dark matter is "cold" so that it can cluster in galactic halos. In this case, if the Population III stars form before galaxies, one might expect their remnants to be distributed throughout the Universe (White & Rees 1978), with the ratio of the baryonic and nonbaryonic densities being the same everywhere and of order 10. If they form at the same time as galaxy formation, perhaps in the first phase of protogalactic collapse, one would expect the remnants to be confined to halos and clusters. In this case, their contribution to the halo density could be larger since the baryons would probably dissipate and become more concentrated. Angular momentum considerations suggest that the local baryon fraction must be increased by at least a factor of 10 (Fall & Efstathiou 1981). If the WIMPs are hot and cannot cluster in halos, then halos would consist exclusively of MACHOs. These possibilities are illustrated in Figure 1.

3.4   Variants of the Baryonic Dark Matter Scenario

One may consider three variants of Baryonic Dark Matter (BDM) scenario, depending on how strongly one wishes to retain homogeneous primordial nu-

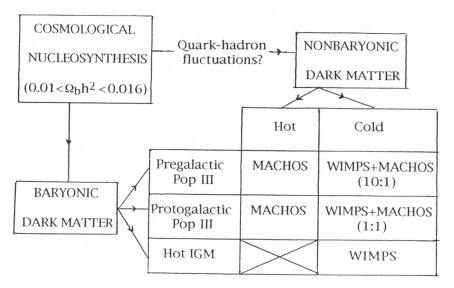

Figure 1   The relative contributions of WIMPs and MACHOs to the halo density in various scenarios. Halos can consist exclusively of WIMPs only if the dark baryons are in a hot intergalactic medium and they can consist exclusively of MACHOs only if the WIMPs are hot. The most natural hypothesis is that they contain both.

cleosynthesis, inflation, and nonbaryonic dark matter:

- In the *standard* BDM scenario, one retains all three assumptions, which requires $\Omega_b \approx 0.01\,h^{-2}$ and $\Omega = 1$. In this case, we conclude that 70% to 95% of baryons are dark but MACHOs alone can provide galactic halos only if $R_h < 50\,h^{-1}$kpc. The cluster and critical density dark matter must be WIMPs and, if they are cold, the halo dark matter is expected to be a mixture of MACHOS and WIMPS. This conclusion pertains even for $\Omega < 1$, as may be required by the large baryon fraction in clusters.

- In the *maximal* BDM scenario, one assumes that the Universe has a critical density of baryons ($\Omega_B = \Omega = 1$), thereby relinquishing the need for nonbaryonic dark matter without giving up inflation (Fowler 1990). The scenario is inconsistent with homogeneous nucleosynthesis unless one invokes unrealistically low values of H_o. When the upper limit on Ω_b was $0.06\,h^{-2}$ s (Yang et al 1984), it was possible to have $\Omega_b = 1$ by invoking the only moderately extreme value $H_o = 25$ (Shanks 1985), but the new upper limit would require $H_o = 10$, which is probably absurd [see, however, Harrison (1993) for a contrary view].

- In the *baryon-dominated* scenario, one only assumes the existence of the dark matter for which there is direct dynamical evidence and attributes this solely to baryons ($\Omega_b = \Omega \approx 0.1$). In this case, one has to give up both inflation and homogeneous nucleosynthesis. In order to explain the observed light element abundances, one then has to invoke some exotic astrophysical process, such as the spallation of primordial helium by high energy photons from accreting black holes (Gnedin & Ostriker 1992). The viability of this scenario also depends on whether the isocurvature baryon-dominated model is compatible with the CMB anisotropy constraints.

Most of the emphasis in this review is on the standard BDM scenario but, in assessing which baryonic candidates are viable, it is important to bear in mind the more radical proposals.

4.   POPULATION III STARS

There is some confusion in the literature because the term "Population III" has been used in two distinct contexts. It has been applied to describe: 1. the stars that generate the first metals; and 2. the stars hypothesized to provide the dark matter in galactic halos. In either case, the stars only warrant a special name if they are definitely distinct from Population II stars, i.e. if they form at a distinct epoch or if the initial mass function (IMF) of the first stars is bimodal (with distinct populations of stars forming in different locations). We will see that this may not be the case for stars of type 1, but it probably is for stars of type

2. If one requires *both* kinds of "Population III" stars, it is not obvious which ones come first. One could envisage situations in which the dark objects form before, after, or contemporaneously with the stars that make the first metals.

4.1   *Population III as the First Metal Producers*

Stars of type 1 must exist because heavy elements can only be generated through stellar nucleosynthesis. However, the most natural assumption is that these are merely the ones at the high mass end of the Population II mass spectrum, since in this case they would generate the first metals because they evolved fastest. This is already sufficient to explain most of the abundance characteristics of Population I and II stars (Truran 1984, Wheeler et al 1989, Rana 1991, Pagel 1992). At one point, there appeared to be a metallicity cut-off of order 10^{-5} below which no stars were found (Bond 1981); this suggested that the first stars were more massive than those forming today. However, the evidence for the cut-off has now gone away: Beers et al (1992) find that the Z distribution for Population II stars extends well below 10^{-6}, and there exists one object with $Z = 6 \times 10^{-7}$ (Bessel & Norris 1984). In any case, the number of low-Z objects is not necessarily incompatible with the assumption that the IMF has always been the same (Pagel 1987), so there is no obvious reason for supposing that the first stars were qualitatively different from Population II. However, one cannot be sure that there are not abundance anomalies at some level (cf Kajino et al 1990, Suntzeff 1992). This is important because, if the dark baryons are in the remnants of massive stars, one might expect some nucleosynthetic consequences.

4.2   *Population III as Dark Matter Producers*

We have seen that it is possible that most of the baryons were processed through a first generation of pregalactic or protogalactic stars and henceforth the term "Population III" is used specifically in this sense. However, it should be stressed that the cosmological interest in Population III stars is not confined to the dark matter issue. They would also be expected to produce radiation, explosions, and nucleosynthesis products, and each of these could have important cosmological consequences (Carr et al 1984). Although there are no observations which unambiguously *demand* that most of the baryons were processed through Population III stars, there are theoretical reasons for anticipating their formation. This is because the existence of galaxies and clusters of galaxies implies that there must have been density fluctuations in the early Universe and, in many scenarios, these fluctuations would also give rise to a population of pregalactic stars. The precise way in which this occurs depends on the nature of the fluctuations and the nature of the dominant dark matter, as we now discuss.

In a baryon-dominated universe with isothermal or isocurvature density fluctuations, the first bound objects usually have a mass corresponding to the baryonic Jeans mass at decoupling. This is $M_{Jb} \approx 10^6 \Omega_b^{-1/2} M_\odot$, where Ω_b is the baryon density parameter, and clouds of this mass would bind at a redshift ~ 100, depending on the form of the spectrum of fluctuations at decoupling. Larger bound objects—like galaxies and clusters of galaxies—would then build up through a process of hierarchical clustering (Peebles & Dicke 1968). Regions smaller than M_{Jb}, even though their initial overdensity might be higher, would not begin to collapse until they were larger than the Jeans length and by then they would generally have been erased either by viscous damping prior to decoupling or by nonlinear processes during the oscillatory period after decoupling (Carr & Rees 1984). However, more exotic possibilities arise if the fluctuation spectrum is sufficiently steep for the fluctuations to be *highly* nonlinear on smaller scales because, in this case, very small regions could collapse well before recombination (Hogan 1978). Indeed, this is expected in the primordial isocurvature baryon-dominated model (Hogan 1993).

In the Cold Dark Matter scenario, in which the density of the Universe is dominated by cold particle relics, structure also builds up hierarchically (Blumenthal et al 1984). In this case, one expects bound clumps of the particles to form down to very small scales (Hogan & Rees 1988), but baryons would only fall into the potential wells, forming bound clouds, on baryon scales above $M_{Ja} \approx 10^6 \Omega_b \Omega_a^{-3/2} M_\odot$, where Ω_a is the cold particle density (Carr & Rees 1984, de Araujo & Opher 1990). In fact, the formation of the pregalactic clouds is even easier in this case because the cold particle fluctuations grow by an extra factor of $10 \Omega_a$ between the time when the cold particles dominate the density and decoupling.

In a baryon-dominated universe with adiabatic density fluctuations, the first objects to form are pancakes of cluster size (Zeldovich 1970) because adiabatic fluctuations are erased by photon diffusion for $M < 10^{13} \Omega_b^{-5/4} M_\odot$ (Silk 1968). Galaxies and smaller scale structures therefore have to form as a result of fragmentation. This scenario appears to be excluded by CMB anistropy constraints but a similar picture applies if one has adiabatic fluctuations in a Hot Dark Matter scenario, in which the Universe's mass is dominated by a particle like the neutrino. In this case, the fluctuations are erased by neutrino free-streaming for $M < 10^{15} \Omega_\nu^{-2} M_\odot$ (Bond et al 1980), so the first objects to form are pancakes of supercluster scale. In both scenarios one expects the pancakes to initially fragment into clumps of mass $10^8 M_\odot$; these clumps must then cluster in order to form galaxies. Even in this case, therefore, one might expect pregalactic clouds to form, albeit at a relatively low redshift ($z < 10$).

All of these scenarios would be modified if the Universe contained topological relics such as strings or textures (Cen et al 1991). Such relics could induce the formation of smaller scale bound regions than usual. For example, Silk & Stebbins (1993) find that in the CDM picture with strings, up to 10^{-3} of the mass of the Universe could go into cold dark matter clumps at the time of matter-radiation equilibrium. These clumps would then accrete baryonic halos, forming globular-cluster type objects.

In the explosion scenario (Ostriker & Cowie 1981, Ikeuchi 1981), the first objects to form are explosive seeds (stars or clusters of stars). These generate shocks which sweep up vast shells of gas; when the shells overlap, most of the gas gets compressed into thin sheets (Carr & Ikeuchi 1985). The sheets then fragment either directly into galaxies or into lower-mass systems, depending on the cooling mechanism (Bertschinger 1983, Wandel 1985). Although the explosion scenario was originally invoked to explain large-scale structure, this now seems to be incompatible with the upper limit on the y-parameter permitted by *FIRAS*. However, one can still envisage this as a mechanism for amplifying the fraction of the gas going into stars—an idea applicable in models with or without nonbaryonic dark matter (Scherrer 1993).

4.3   *Expected Mass of Population III Stars*

In all these scenarios, an appreciable fraction of the Universe may go into subgalactic clouds before galaxies themselves form. What happens to these clouds? In some circumstances, one expects them to be disrupted by collisions with other clouds because the cooling time is too long for them to collapse before coalescing. However, there is usually some mass range in which the clouds survive. For example, the range is 10^6–$10^{11} M_\odot$ in the hierarchical clustering scenario. In this case, they could face various possible fates. They might just turn into ordinary stars and form objects like globular clusters. On the other hand, the conditions of star formation could have been very different at early times and several alternatives have been suggested.

Some people argue that the first stars could have been much smaller than at present. Fairly general arguments suggest that the minimum fragment mass could be as low as 0.007 $M_\odot$ (Low & Lynden-Bell 1976, Rees 1976) and it is possible that conditions at early epochs—such as the enhanced formation of molecular hydrogen (Palla et al 1983, Yoshii & Saio 1986, Silk 1992)—could allow the formation of even smaller objects. One might also invoke the prevalence of high-pressure pregalactic cooling flows (Ashman & Carr 1988, Thomas & Fabian 1990), analogous to the cluster flows observed at the present epoch (Fabian et al 1984) but on a smaller scale. This possibility is discussed in detail in Section 9.2.

Other people argue that the first stars could have been much larger than at present. For example, the fragment mass could be increased before metals formed because cooling would be less efficient (Silk 1977). There is also observational evidence that the IMF may become shallower as metallicity decreases (Terlevich 1985), thereby increasing the fraction of high mass stars. Another possibility is that the characteristic fragment mass could be increased by the effects of the microwave background (Kashlinsky & Rees 1983) or by the absence of substructure in the first bound clouds (Tohline 1980).

One could also get a mixture of small and large stars. For example, Cayrel (1987) has proposed that one gets the formation of massive exploding stars in the core of the cloud, followed by the formation of low mass stars where the gas swept up by the explosions encounters infalling gas. Kashlinsky & Rees (1983) have proposed a scheme in which angular momentum effects lead to a disk of small stars around a central very massive star. Salpeter & Wasserman (1993) have a scenario in which one gets clusters of neutron stars and asteroids.

In the baryon-dominated isocurvature scenario, with highly nonlinear fluctuations on small scales, the collapse of the first overdense clouds depends on the effects of radiation diffusion and trapping. Hogan (1993) finds that sufficiently dense clouds collapse very early into black holes with a mass of at least $1M_\odot$, while clouds below this critical density delay their collapse until after recombination and may produce neutron star or brown dwarf remnants. One of the attractions of this idea is that it allows a baryon density parameter higher than that indicated by Equation (3.1) because the nucleosynthetic products in the high density regions are locked up in the remnants, leaving the products from the low density regions outside (cf Gnedin et al 1994).

It is possible that the first clouds collapse directly to form supermassive black holes (Gnedin & Ostriker 1992). Usually clouds will be tidally spun up by their neighbors as they become gravitationally bound and the associated centrifugal effects then prevent direct collapse. However, just after recombination, Compton drag could prevent this tidal spin-up, especially if the gas becomes ionized or contains dust (Loeb 1993). More detailed numerical hydrodynamical studies of this situation have been presented by Umemura et al (1993), who allow for different ionization histories and for different ratios of baryonic to nonbaryonic density. For a fully ionized gas, the baryonic disk loses angular momentum very effectively and shrinks adiabatically. Even if rotation is important, one could still get a supermassive disk which slowly shrinks to form a black hole due to angular momentum transport by viscous effects (Loeb & Rasio 1993). One might even end up with a supermassive binary system.

While there is clearly considerable uncertainty as to the fate of the first bound clouds, our discussion indicates that they are likely to fragment into stars that are either larger or smaller than the ones forming today. Theorists merely disagree about the direction! One certainly needs the stars to be very different if they are to produce a lot of dark matter. One also requires the clouds to fragment very efficiently. Although this might seem rather unlikely, there are circumstances even in the present epoch where this occurs; for example, in starburst galaxies or cooling flows. This is also a natural outcome of the hierarchical explosion scenario (Carr & Ikeuchi 1985).

We note that there is no necessity for the Population III stars to form before galaxies. It is possible that the Population III clouds just remain in purely gaseous form and become Lyman-α clouds (Rees 1986), in which case the formation of the dark-matter-producing stars would need to be postponed until the epoch of galaxy formation. Nevertheless, there is at least the possibility that the Population III stars were pregalactic, and this would have various attractions. For example, it would permit the Universe to be reionized at high redshifts (Hartquist & Cameron 1977), thereby hiding small-scale anisotropies in the microwave background (Gouda & Sugiyama 1992), and it might help to explain why the intergalactic medium appears to be ionized back to redshifts of at least 5 (Schneider et al 1991). Pregalactic stars might also be invoked to explain pregalactic enrichment (Truran & Cameron 1971) and the existence of substantial heavy element abundances in intergalactic clouds at redshifts above 3 (Steidel & Sargent 1988) and in intracluster gas at low redshifts (Hatsukade 1989).

4.4 Population III Remnants

Even if a large fraction of the baryons are processed through Population III stars, this does not necessarily guarantee dark matter production. However, most stars ultimately produce dark remnants and we now list the various possibilities.

LOW MASS OBJECTS   We will see in Section 9.3 that stars in the range 0.08–0.8 $M_\odot$ (which are still on the main-sequence) are probably excluded from explaining any of the dark matter problems. However, objects in the range 0.001–0.08 $M_\odot$ would never burn hydrogen and would certainly be dim enough to escape detection. [Note that Salpeter (1993) argues that the critical mass for hydrogen burning could be higher for Population III stars because slow protostellar accretion could lead to degenerate cores with lower central temperatures than usual.] Such brown dwarfs (BDs) represent a balance between gravity and degeneracy pressure. Those above 0.01 $M_\odot$ could still burn deuterium; Shu et al (1987) have argued that this may represent a lower limit to a BD's mass but this conclusion is not definite. The evidence for stars in the brown dwarf mass range (e.g. Simon & Becklin 1992, Steele et al 1993) is controversial, but this merely reflects the fact that they are hard to find (Stevenson 1991) and it would be very surprising if the IMF happened to cut off just above $0.08 M_\odot$. Most searches have focused on BDs in binary systems with M-dwarfs; however, we already know that the BDs making up the dark matter could not be in such binaries else the M-dwarfs would have more than the dark density (cf McDonald & Clarke 1993). Objects below $0.001 M_\odot$ are held together by intermolecular rather than gravitational forces (i.e. they have atomic density) and may be described as snowballs. We will see in Section 10.1 that such objects are unlikely to constitute the dark matter.

INTERMEDIATE MASS OBJECTS   Stars in the range 0.8–4 $M_\odot$ would leave white dwarf remnants, while those between $8M_\odot$ and some mass M_{BH} would leave neutron stars remnants. In either case, the remnants would eventually cool and become dark. (Stars in the mass range 4–8 $M_\odot$ could be disrupted entirely during their carbon-burning stage.) Stars more massive than M_{BH} could evolve to black holes; the value of M_{BH} is uncertain but it may be as high as $50M_\odot$ (Schild & Maeder 1985) or as low as $25M_\odot$ (Maeder 1992). Only intermediate mass remnants definitely form at the present epoch; this is why some theorists favor them as dark matter candidates (Silk 1991, 1992, 1993). However, we will see in Section 5.2 that their nucleosynthetic consequences may make them poor dark matter candidates.

VERY MASSIVE OBJECTS   Stars in the mass range above 100 $M_\odot$, which are termed "Very Massive Objects" or VMOs, would experience the pair-instability during their oxygen-burning phase (Fowler & Hoyle 1964). This would lead to disruption below some mass M_c but complete collapse above it (Woosley & Weaver 1982, Ober et al 1983, Bond et al 1984). VMO black holes may therefore be more plausible dark matter candidates than ordinary stellar black holes. In the absence of rotation, $M_c \approx 200M_\odot$; however, M_c could be as high as $2 \times 10^4 M_\odot$ if rotation were maximal (Glatzel et al 1985). Note that stars with an initial mass above 100 $M_\odot$ are radiation-dominated and therefore unstable to pulsations during hydrogen burning. These pulsations would lead to considerable mass loss but are unlikely to be completely disruptive. Nevertheless, there is no evidence that VMOs form at the present epoch, so they are invoked specifically to explain dark matter.

SUPERMASSIVE OBJECTS   Stars larger than $10^5 M_\odot$ are termed "Supermassive Objects" or SMOs. If they are metal-free, they would collapse directly to black holes on a timescale $10^4 (M/10^5 M_\odot)^{-1} y$ before any nuclear burning (Fowler 1966). They would therefore have no nucleosynthetic consequences, although they could explode in some mass range above $10^5 M_\odot$ if they had nonzero metallicity (Fricke 1973, Fuller et al 1986). SMOs would also generate very little radiation, emitting only 10^{-11} of their rest-mass energy in photons. The existence of SMOs is rather less speculative than that of VMOs since supermassive black holes are thought to reside in some galactic nuclei and to power quasars (Blandford & Rees 1991). However, these would only have a tiny cosmological density.

Note that Population III stars are likely to span a range of masses, so the remnants need not be confined to one of the candidates listed above. From the point of view of the dark matter problem, one is mainly interested in where *most* of the mass resides. However, the other components could also have important observational consequences, as in the Salpeter & Wasserman (1993) scenario,

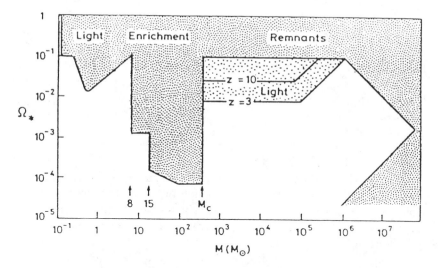

Figure 2 Summary of the various constraints on the density of Population III stars of mass M (an updated version of Figure 6 in Carr et al 1984). All the limits except the "light" one assume that the objects are inside galactic halos. The light limit is only interesting for stars that burn at $z < 30$ unless the light is reprocessed by dust. The nucleosynthesis limits for $M > 10^2 M_\odot$ assume that the fraction of mass lost during the hydrogen and helium burning phase is not too large.

where the small number of neutron stars is invoked to explain gamma-ray bursts.

5.   CONSTRAINTS ON POPULATION III

In this section we review the constraints that can be placed on Population III stars and their remnants by considering their generation of background light and nucleosynthetic products. The discussion is only brief because these limits are essentially the same as they were a decade ago. The constraints on $\Omega_*(M)$, the density of stars of mass M in units of the critical density, are summarized in Figure 2, which is an updated version of Figure 6 from Carr et al (1984). The remnant constraints, which are also shown in Figure 2, are discussed in detail in Section 6.

5.1   *Background Light Constraints*

An effect that constrains the number of Population III objects over every mass range between 0.1 and $10^5 M_\odot$ is the generation of background light during the stellar main-sequence phase. The fact that the observed background radiation density over all wavebands cannot exceed $\Omega_{RO} \approx 10^{-4}$ in units of the critical density (and it is smaller in most bands) permits a general constraint on $\Omega_*(M)$. One can obtain more precise constraints by using information about the waveband in which the radiation is expected to reside (Peebles & Partridge 1967, Thorstensen & Partridge 1975, Carr et al 1984, McDowell 1986, Negroponte 1986) but an integrated background light limit has the virtue of generality.

For stars larger than $0.8 M_\odot$, which have already burnt their nuclear fuel, one just compares their total light production to Ω_{RO} to obtain a constraint on $\Omega_*(M)$. Since 7 MeV per baryon is released in burning hydrogen to helium, the background light density generated should be $\Omega_R = 0.007 \, \Omega_* f_b (1 + z_*)^{-1}$ in units of the critical density, where z_* is the redshift at which the stars burn their fuel (the minimum of the formation redshift z_f and the redshift z_{MS} at which the age of the Universe equals their main-sequence time) and f_b is the fraction of the star's mass burnt into helium. By using the known dependence of f_b and z_{MS} on M, one can predict the value of Ω_R as a function of Ω_*, M, and z_f. Since the observed background density over all wavebands does not exceed 10^{-4}, this implies a constraint on Ω_* as function of M and z_f. For stars with $M < 0.8 M_\odot$, which are still burning, one compares the product of the luminosity $L(M)$ and the age of the Universe to Ω_{RO}. The resulting constraints on $\Omega_*(M, z_*)$ are shown in Figure 2; these are somewhat stronger than indicated by Carr et al (1984) because the limits on Ω_{RO} have improved (See Section 6.1). Peebles & Partridge (1967) used this argument to preclude stars in the mass range $0.3–2.5 M_\odot$ from having the critical density.

5.2   Enrichment Constraints

One of the strongest constraints on the spectrum of Population III stars comes from the fact that stars in the mass range $4 M_\odot$ to $M_c \approx 200 M_\odot$ should produce an appreciable heavy element yield (Arnett 1978), either via winds in their main-sequence phase or during their final supernova phase. We take the yield Z_{ej} to be 0.01 for $8 M_\odot < M < 15 M_\odot$, $0.5 - (M/6 M_\odot)^{-1}$ for $15 M_\odot < M < 100 M_\odot$, and 0.5 for $100 M_\odot < M < M_c$ (Wheeler et al 1989). Carr et al (1984) took $z_{ej} = 0.2$ for $4 M_\odot < M < 8 M_\odot$ on the assumption that such stars explode as a result of degenerate carbon burning. However, it seems possible that they evolve to white dwarfs without exploding, so Figure 2 assumes there is no enrichment for $M < 8 M_\odot$. Whether Population III stars are pregalactic or protogalactic, the enrichment they produce cannot exceed the lowest metallicity observed in Population I stars ($Z = 10^{-3}$). One infers that the density of the stars must satisfy

$$\Omega_* < 10^{-3} Z_{ej}^{-1} \Omega_g = 10^{-4} \left(\frac{Z_{ej}}{0.1} \right)^{-1} \left(\frac{\Omega_g}{0.01} \right), \tag{5.1}$$

where Ω_g is the gas density before the stars form, assumed to be around 0.01

in view of the Equation (3.1). This enrichment constraint is shown in Figure 2 [note that it is stronger than indicated by Carr et al (1984) because they assumed $\Omega_g = 0.1$.] It immediately excludes neutron stars or ordinary stellar black holes from explaining any of the dark matter problems unless the Population III precursors are clumped into clusters whose gravitational potential is so high that ejected heavy elements cannot escape (Salpeter & Wasserman 1993). If one wants to produce the dark matter without contravening the enrichment constraint, the most straightforward solution is to assume that the spectrum either starts above M_c (as in the black hole scenario) or ends below $8M_\odot$ (as in the brown dwarf or white dwarf scenario), so that there is no pregalactic enrichment at all.

5.3   *Helium Constraints*

Although stars return helium to the background Universe in most mass ranges, the associated constraints on the fraction of the Universe going into Population III stars are only weak because of the uncertainties in the primordial helium abundance. However, the helium limit is important in the $M > M_c$ range because there may be no heavy element yield here. Because the pulsational instability leads to mass-shedding of material convected from its core, a VMO is expected to return helium to the background medium during core-hydrogen burning (Bond et al 1983). The net yield depends sensitively on the mass loss fraction ϕ_L. If this is very high, the yield will be low because most of the mass will be lost before significant core burning occurs. However, for ϕ_L below the critical value $(1-Y_i)/(2-Y_i)$, the mass loss is always slower than the shrinkage of the convective core and one can show that the fraction of mass returned as new helium is

$$\Delta Y = \left(1 - \frac{Y_i}{2}\right) \phi_L^2 \leq 0.25(1 - Y_i)^2 \left(1 - \frac{Y_i}{2}\right)^{-1} \tag{5.2}$$

Here Y_i is the initial (primordial) helium abundance and the equality sign on the right applies only if ϕ_L has the critical value. This does not impose a useful constraint on the number of VMOs if ϕ_L is well below the critical value since ΔY is then very small. However, there is some indication from numerical calculations that hydrogen-shell burning may produce a super-Eddington luminosity which completely ejects the stellar envelope (Woosley & Weaver 1982, Bond et al 1984). This would guarantee the maximal helium production permitted by Equation (5.2) and have profound cosmological implications. If $Y_i = 0.23$, corresponding to the conventional primordial value, $\Delta Y = 0.17$, so one would substantially overproduce helium if much of the Universe went into VMOs. In this case, only black holes in the mass range above $10^5 M_\odot$ could be viable candidates for the dark matter. On the other hand, if $Y_i = 0$, then $\Delta Y = 0.25$, which is tantalizingly close to the standard primordial value. This raises the question of whether the Population III VMOs invoked to produce the dark matter might also generate the helium usually attributed to cosmological nucleosynthesis. Of course, the added attraction of the hot Big Bang model is that it predicts the observed abundances of other light elements. One might conceivably generate these elements by invoking high energy photons from accreting black holes to spallate helium—either within the surrounding accretion tori (Rees 1984, Ramadurai & Rees 1985; Jin 1989, 1990) or in the background Universe (Gnedin & Ostriker 1992, Gnedin et al 1994); however, these models seem somewhat contrived.

5.4 Black Hole Accretion Constraints

Any black hole remnants of Population III stars would tend to generate radiation through accretion; this could be important at both the present and pregalactic epochs. In particular, if we assume that halo or disk black holes accrete ambient gas at the Bondi rate and that the accreted material is converted into radiation with efficiency η, then one may impose interesting constraints on the density of the black holes $\Omega_B(M)$ merely by requiring that the radiation density generated since the epoch of galaxy formation does not exceed the observed density in the appropriate waveband. For example, if we assume that the radiation emerges at 10 keV and that $\eta = 0.1$, we infer $\Omega_B(M) < (M/10^5 M_\odot)^{-1}$ for halo holes and $\Omega_B(M) < (M)/10 M_\odot)^{-1}$ for disk holes (Carr 1979). These limits have also been studied by Hegyi et al (1986). Stronger limits may come from constraints on the number of *individual* sources in our own Galaxy. Thus Ipser & Price (1977), using a particular accretion model, preclude $10^5 M_\odot$ holes from comprising the halo because of the non-observation of suitable infrared and optical sources.

One might expect the background light constraints to be even stronger for pregalactic black holes since the background gas density would have been higher at early times. If we assume Bondi accretion, then the luminosity will exceed the Eddington value for some period after decoupling if $M > 10^3 \eta^{-1} M_\odot$. However, the pregalactic limit is actually weaker: It takes the form $\Omega_B(M) < (M/10^6 M_\odot)^{-1}$ for $\eta = 0.1$ with only a weak dependence on the photon energy (Carr 1979). This is a consequence of two factors: 1. a large fraction of the emitted radiation goes into heating the matter content of the Universe rather than into background light; and 2. the heating of the Universe will boost the matter temperature well above the usual Friedmann value and this will reduce the accretion rate (Meszaros 1975, Carr 1981a, Gnedin & Ostriker 1992). Nevertheless, the effect on the thermal history of the Universe could be of great interest in its own right. For example, accreting black holes could easily keep the Universe ionized throughout the period after decoupling. The sort of background generated by the pregalactic accretion phase of a population of $10^6 M_\odot$ black holes is indicated in Figure 6.

6.   DYNAMICAL CONSTRAINTS

A variety of constraints can be placed on the mass of any dark compact objects in the disk and halo of our own Galaxy by considering their dynamical effects. The constraints are usually calculated on the assumption that the objects are black holes but, as emphasized in Section 6.4, most of them also apply for dark clusters of smaller objects. There are also constraints for dark objects in clusters of galaxies or in the intergalactic medium, though these are weaker. The limits are summarized as upper limits on the density parameter $\Omega_B(M)$ for black holes of mass M in Figure 3, where the disk, halo, and cluster dark matter are assumed to have densities of 0.001, 0.1, and 0.2, respectively.   Figure 3 updates and—in some respects corrects—Figure 1 of Carr (1978).

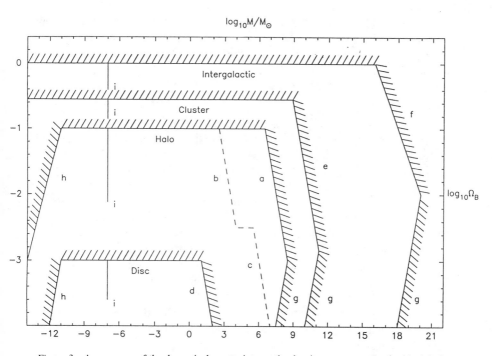

Figure 3   A summary of the dynamical constraints on the density parameter Ω_B for black holes of mass M located in the Galactic disk, the Galactic halo, clusters of galaxies, and the intergalactic medium. The total dark matter density in these cases is taken to be 0.001, 0.1, 0.2, and 1, respectively. The limits come from : (*a*) disk heating; (*b*) globular cluster disruption; (*c*) dynamical friction; (*d*) binary disruption; (*e*) galaxy distortions; (*f*) galaxy peculiar velocities; (*g*) one black hole per disk/halo/cluster/Universe; and (*h*) comet observations. Limits (*b*) and (*c*) are shown by broken lines because they are not so secure. The evaporation limit (*i*) is nondynamical.

6.1   *Disk Heating by Halo Holes*

As halo objects traverse the Galactic disk, they will impart energy to the stars there. This will lead to a gradual puffing up of the disk, with older stars being heated more than younger ones. Lacey & Ostriker (1985) have argued that black holes of around $10^6 M_\odot$ could provide the best mechanism for generating the observed amount of puffing. In particular, this explains: 1. why the velocity dispersion of the disk stars, σ, scales with age as $t^{1/2}$; 2. the relative velocity dispersions in the radial, azimuthal, and vertical directions; and 3. the existence of a high energy tail of stars with large velocity (Ipser & Semenzato 1985). In order to normalize the $\sigma(t)$ relationship correctly, the number density of the holes n must satisfy $nM^2 \approx 3 \times 10^4 M_\odot^2 pc^{-3}$. Combining this with the local halo density $\rho_h = nM \approx 0.01 M_\odot pc^{-3}$ gives $M = 2 \times 10^6 M_\odot$.

This argument is no longer compelling because more recent measurements give smaller velocity dispersions for older stars, so that σ may no longer rise as fast as $t^{1/2}$ (Carlberg et al 1985, Stromgren 1987, Gomez et al 1990). Heating by a combination of spiral density waves and giant molecular clouds may now give a better fit to the data (Lacey 1991). Nevertheless, one can still use the Lacey-Ostriker argument to place an upper limit on the density in halo objects of mass M (Carr et al 1984):

$$\Omega_B < \Omega_h \min\left[1, \left(\frac{M}{M_{heat}}\right)^{-1}\right], \qquad M_{heat} = 3 \times 10^6 \left(\frac{t_g}{10^{10}y}\right)^{-1}, \quad (6.1)$$

where t_g is the age of the Galaxy. Otherwise the disk would be more puffed up than observed. This limit is shown in Figure 3, along with the line that corresponds to having at least one black hole of mass M within the Galaxy.

Although the dependence is not shown explicitly in Equation (6.1), M_{heat} also scales as σ^2 and ρ_h^{-1}. Thus, by applying the disk-heating argument to galaxies with higher dark matter density, lower stellar velocity dispersion, or smaller age, one can obtain stronger constraints. For the gas-rich dwarf galaxy DD0154 (which has $\sigma = 17$ km s^{-1}, an age of at least 1.5 Gyr, and a central dark matter density of $0.009 M_\odot pc^{-3}$), Rix & Lake (1993) find $M < 7 \times 10^5 M_\odot$. For the dwarf galaxy GR8 (which has $\sigma = 4$ km s^{-1}, an age of at least 1 Gyr, and a central dark matter density of $0.07 M_\odot pc^{-3}$), they find $M < 6 \times 10^3 M_\odot$. Of course, unless the black holes form pregalactically, there is no reason for expecting the halo objects to have the same mass in different galaxies, so these limits are not shown in Figure 3.

6.2   *Disruption of Stellar Clusters by Halo Objects*

Another type of dynamical effect associated with halo objects would be their influence on bound groups of stars (in particular, globular clusters and loose clusters). Every time a halo object passes near a star cluster, the object's tidal field heats up the cluster and thereby reduces its binding energy. Over a sufficiently large number of fly-bys this could evaporate the cluster entirely. This process was first discussed by Spitzer (1958) for the case in which the disrupting objects are giant molecular clouds. Carr (1978) used a similar analysis to argue that the halo objects must be smaller than $10^5 M_\odot$ or else loose clusters would not survive as long as observed—but this argument neglected the fact that sufficiently massive holes will disrupt clusters by single rather than multiple fly-bys. The correct analysis was given by Wielen (1985) for halo objects with the mass of $2 \times 10^6 M_\odot$ required in the Lacey-Ostriker scenario and by Sakellariadou (1984) and Carr & Sakellariadou (1994) for halo objects of general mass.

By comparing the expected disruption time for clusters of mass m_c and radius r_c with the typical cluster lifetime t_L, one finds that the local density of halo holes of mass M must satisfy (cf Ostriker et al 1989)

$$\rho_B < \begin{cases} \dfrac{m_c V}{G M t_L r_c} & \text{for} \qquad M < m_c \left(\dfrac{V}{V_c}\right) \\[2ex] \left(\dfrac{m_c}{G t_L^2 r_c^3}\right)^{1/2} & \text{for } m_c \left(\dfrac{V}{V_c}\right) < M < m_c \left(\dfrac{V}{V_c}\right)^3 \\[2ex] \dfrac{m_c^{2/3} M^{1/3}}{(V t_L r_c^2)} & \text{for} \qquad M > m_c \left(\dfrac{V}{V_c}\right)^3 . \end{cases} \qquad (6.2)$$

Here $V_c \sim (G m_c / r_c)^{1/2}$ is the velocity dispersion within the cluster, V is the speed of the halo objects (~ 300 km s^{-1}) and we have neglected numerical factors of order unity. The increasing mass regimes correspond to disruption by multiple encounters, single encounters, and nonimpulsive encounters, respectively. Any lower limit on t_L therefore places an upper limit on ρ_B. The crucial point is that the limit is independent of M in the single-encounter regime, so that the limit bottoms out at a density of order $(\rho_c / G t_L^2)^{1/2}$. The constraint is therefore uninteresting if this exceeds the observed halo density ρ_h. In particular, if the clusters survive for the lifetime of the Galaxy, which is essentially the age of the Universe t_0, the limiting density is just $(\rho_c \rho_0)^{1/2}$, where ρ_0 is the mean cosmological density. If t_L is much larger, than t_0, the fraction of clusters disrupted within t_0 is $f_c \sim t_0 / t_L$ and so the limiting density is reduced by the factor f_c.

The strongest limit is associated with globular clusters, for which we take $m_c = 10^5 M_\odot$, $r_c = 10$ pc, $V_c = 10$ km s^{-1}, and $t_L > 10^{10}$y. We also assume that the holes have a speed $V = 300$ km s^{-1}. Rather remarkably, due to the "coincidence" that the halo density is the geometric mean of the cosmological density and the globular cluster density, the upper limit on ρ_B is comparable to the actual halo density; this suggests that halo objects might actually *determine* the characteristics of surviving globular clusters (cf Fall & Rees 1977). Numerical calculations for the disruption of globular clusters by Moore (1993) confirm the general qualitative features indicated above: gradual mass loss for small halo objects and sudden disruption for larger ones. However, using data for nine particular globular clusters, Moore infers an upper limit of $10^3 M_\odot$. This is in the multiple-encounter regime and considerably stronger than the limit implied by Equation (6.2) with $t_L = t_o$, presumably because his clusters are very diffuse. Because of the uncertainties, the line corresponding to Moore's result is only shown dotted in Figure 3.

6.3   Effect of Dynamical Friction on Halo Objects

Another important dynamical effect is that halo objects will tend to lose energy to lighter objects and consequently drift toward the Galactic nucleus (Chandrasekhar 1964). In particular, one can show that halo objects will be dragged into the nucleus by the dynamical friction of the Spheroid stars from within a Galactocentric radius

$$R_{df} = \left(\frac{M}{10^6 M_\odot}\right)^{2/3} \left(\frac{t_g}{10^{10}y}\right)^{2/3} \text{kpc}, \tag{6.3}$$

and the total mass dragged into the Galactic nucleus is therefore

$$M_N = 9 \times 10^8 \left(\frac{M}{10^6 M_\odot}\right)^2 \left(\frac{t_g}{10^{10}y}\right)^2 \left(\frac{a}{2 \text{ kpc}}\right)^{-2} M_\odot, \tag{6.4}$$

where a is the halo core radius (Carr & Lacey 1987). This exceeds the upper observational limit of $3 \times 10^6 M_\odot$ (Sellgren et al 1990, Spaenhauer et al 1992) unless

$$\Omega_B < \Omega_h \left(\frac{M}{3 \times 10^3 M_\odot}\right)^{-1} \left(\frac{a}{2 \text{ kpc}}\right) \left(\frac{t_g}{10^{10}y}\right)^{-1} \tag{6.5}$$

This is certainly stronger than the disk-heating limit; it may also be stronger than the cluster disruption limit.

Although this argument would seem to preclude the Lacey-Ostriker proposal, there is an important caveat in this conclusion (Hut & Rees 1992). Equation (6.4) implies that about 10^3 holes of $10^6 M_\odot$ would have drifted into the Galactic nucleus by now, corresponding to one arrival every $10^7 y$. Once two black holes have reached the nucleus, they will form a binary, which will eventually coalesce due to loss of energy through gravitational radiation. If a third hole arrives before coalescence occurs, then the "slingshot" mechanism could eject one of the holes and the remaining pair might also escape due to the recoil (Saslaw et al 1974). Hut & Rees estimate that the time for binary coalescence is shorter than the interval between infalls, which suggests that the slingshot is ineffective. However, there is another problem with Equation (6.5): Dynamical friction will also deplete the number of stars in the nucleus and this will eventually suppress dynamical friction unless there is an efficient mechanism to replenish the loss-cone (Begelman et al 1980). Limit (6.5) is clearly not completely firm, so it is only shown dotted in Figure 3.

6.4   Is the Halo made of Dark Clusters?

We have seen that both the cluster disruption and dynamical friction constraints may be incompatible with the Lacey & Ostriker proposal that $2 \times 10^6 M_\odot$ halo black holes generate the observed disk-heating. There is also the problem that supermassive halo black holes might generate too much radiation through accretion as they traverse the disk (Ipser & Price 1977). To circumvent these objections, Carr & Lacey (1987) have proposed that the disk heaters are $2 \times 10^6 M_\odot$ *clusters* of smaller objects rather than single black holes. The accretion luminosity is then reduced by a factor of order the number of objects per cluster and the dynamical friction problem is avoided, provided the clusters are disrupted by collisions before they are dragged into the Galactic nucleus by dynamical friction.

One can extend this idea to a more general cluster scenario (Wasserman & Salpeter 1993, Kerins & Carr 1994, Moore & Silk 1994). If we assume that the clusters all have the same mass M_c and radius R_c, then they will be disrupted by collisions within the Galactocentric radius (6.3) at which dynamical friction operates, providing

$$R_c > 1.4 \left(\frac{a}{2 \text{ kpc}} \right)^2 \left(\frac{t_g}{10^{10} y} \right)^{-1} \text{ pc.} \tag{6.6}$$

If this condition is not satisfied, then M_c must be less than the value indicated by Equation (6.5). In order to avoid the evaporation of clusters as a result of 2-body relaxation, one also requires

$$R_c > 0.04 \left(\frac{m}{0.01 M_\odot} \right)^{2/3} \left(\frac{t_g}{10^{10} y} \right)^{2/3} \left(\frac{M_c}{10^6 M_\odot} \right)^{-1/3} \text{ pc,} \tag{6.7}$$

where m is the mass of the components. An *upper* limit on R_c comes from requiring that the clusters do not disrupt at our own Galactocentric radius $R_0 \sim 10$ kpc which implies

$$R_c < 35 \left(\frac{R_0}{10 \text{ kpc}} \right)^2 \left(\frac{t_g}{10^{10} y} \right)^{-1} \text{ pc.} \tag{6.8}$$

These dynamical limits, together with the disk-heating limit (6.1), are indicated by the bold lines in Figure 4, which show that the values of M_c and R_c are constrained to a rather narrow range. The cluster-disruption upper limit on M_c is not shown because it is rather model-dependent but it could further reduce the range.

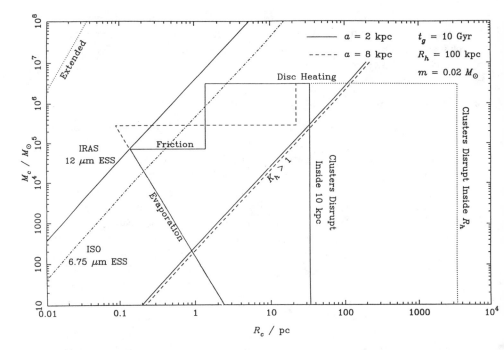

Figure 4 Dynamical constraints on the mass M_c and radius R_c of clusters that provide the dark mass in the Galactic halo. Outside the bold lines the clusters are either disrupted by collisions or produce excessive disk heating or evaporate or produce an excessive buildup of mass in the Galactic nucleus as a result of dynamical friction. Also shown are the extended source sensitivity of *ISO* (for an integration time of 100 s) and *IRAS*, assuming that the brown dwarfs have the optimal mass of $0.02M_\odot$, and the region where the clusters cover the sky.

There is some uncertainty in the positions of the boundaries in Figure 4. If one merely requires that the clusters do not disrupt at the edge of the halo, the upper limit (6.8) is increased by a factor of $(R_h/R_o)^2$, as indicated by the dotted line in Figure 4. The dynamical friction limits are sensitive to the value of a: the limits are shown for $a = 2$ kpc and $a = 8$ kpc since this spans the range of likely values. The evaporation limit given by Equation (6.7) depends on the value of m: Figure 4 assumes $m = 0.02M_\odot$. Note that together Equations (6.1), (6.7), and (6.8) require the cluster components to be smaller than $10M_\odot^2$, which probably excludes their being VMO black holes.

6.5   Constraints on Dark Objects Outside Halos

Dynamical constraints on dark objects in the Galactic disk are generally stronger than the halo limits. In particular, Bahcall et al (1985) have argued that the disk dark matter could not comprise objects larger than $2M_\odot$ or else they would disrupt the wide binaries observed by Latham et al (1984). [This limit can be deduced from Equation (6.2) by identifying m_c and r_c with the total mass and separation of the binary.] This is an important constraint because, if correct, it rules out disk dark matter comprising stellar black holes. However, the Bahcall et al conclusion has been disputed by Wasserman & Weinberg (1987) on the grounds that there is no sharp cut-off in the distribution of binary separations above 0.1 pc. The limit is therefore weakened (somewhat arbitrarily) to $10M_\odot$ in Figure 3.

Dynamical constraints on dark objects in clusters of galaxies are weaker than the halo limits. For example, one does not get an interesting constraint by applying Equation (6.2) to the disruption of cluster galaxies by cluster black holes because the upper limit on ρ_B exceeds the cluster density. However, one does get an interesting constraint from upper limits on the fraction of galaxies f_g with unexplained tidal distortions. Equation (6.2) can also be applied in this case, except that the limits are weakened by a factor λf_g^{-1}, where the parameter $\lambda (\sim 2)$ represents the difference between distortion and disruption. Van den Bergh (1969) applied this argument to the Virgo cluster and inferred that black holes binding the cluster could not be bigger than $10^9 M_\odot$. If we assume that Virgo is typical, we obtain the limit indicated in Figure 3. We also show the limit corresponding to the requirement that there be at least one black hole of mass M within the cluster.

The dynamical constraints on intergalactic black holes are even weaker. The most interesting one comes from the fact that, if there were a population of huge intergalactic black holes, each galaxy would have a peculiar velocity due to its gravitational interaction with the nearest one (Carr 1978). If the holes were smoothly distributed and had a number density n, one would expect every galaxy to have a peculiar velocity of order $GMn^{2/3}t_g$. Since the CMB dipole anisotropy shows that the peculiar velocity of our own Galaxy is only 600 km s^{-1}, one infers a limit $\Omega_B < (M/10^{16}M_\odot)^{-1/2}$ and this is also shown in Figure 3. The limit on the bottom right corresponds to the requirement that there be at least one object of mass M within the current particle horizon.

7. GRAVITATIONAL LENSING EFFECTS

One of the most useful signatures of baryonic dark matter candidates is undoubtedly their gravitational lensing effects. Indeed, it is remarkable that lensing could permit their detection over the entire mass range of $10^{-7}M_\odot$ to $10^{12}M_\odot$. All sorts of astronomical objects can serve as lenses (Blandford & Narayan 1992) but the crucial advantage of Population III objects is that they are compact and spherically symmetric, which makes their effects very clean. To search for them, one requires sources that are numerous, small, bright, and have predictable intrinsic variations (Nemiroff 1991a). The most useful sources to date have been quasars, galaxies, radio jets, gamma-ray bursts, and stars; all of these are discussed below. Other possibilities include radio sources (Blandford & Jarosynski 1981), supernovae (Schneider & Wagoner 1987, Linder et al 1988, Rauch 1991), and pulsars (Krauss & Small 1991). There are two distinct lensing effects and these probe different but nearly overlapping mass ranges: *macrolensing* (the multiple-imaging of a source) can be used to search for objects larger than $10^4 M_\odot$, while *microlensing* (modifications to the intensity of a source) can be used for objects smaller than this. The current constraints on the density Ω_c of compact objects in various mass ranges are brought together in Figure 5.

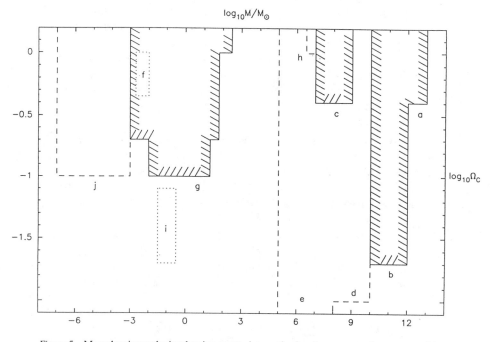

Figure 5   Macrolensing and microlensing constraints on the density parameter for compact objects of mass M. Current limits are shown by shaded lines and potential ones by broken lines. (*a*) VLA limit of Hewitt (1986); (*b*) optical and *HST* limit of Surdej et al (1993); (*c*) VLBI limit of Kassiola et al (1991); (*d*) and (*e*) potential speckle interferometry and VLBA limits; (*f*) region required to explain the quasar variations claimed by Hawkins (1993); (*g*) Dalcanton et al (1994) quasar line-continuum limit; (*h*) gamma-ray burst limit of Nemiroff et al (1993), assuming these are at a cosmolgical redshift; (*i*) corresponds roughly to the range of values required to explain the MACHO and EROS microlensing results; (*j*) potential limit associated with the null results from the EROS CCD study.

7.1   *Macrolensing Constraints on Compact Objects*

If one has a population of compact objects with mass M and density parameter Ω_c, then the probability P of one of them image-doubling a source at redshift $z \approx 1$ and the separation between the images θ are given by

$$P \approx (0.1-0.2)\Omega_c, \qquad \theta \approx 6 \times 10^{-6} \left(\frac{M}{M_\odot} \right)^{1/2} h^{1/2} \text{ arcsec} \qquad (7.1)$$

(Press & Gunn 1973). One can therefore use upper limits on the frequency of macrolensing for different image separations to constrain Ω_c as a function of M. Although optical searches, VLA, and the *Hubble Space Telescope* (*HST*) can only constrain objects down to $10^{10} M_\odot$ (corresponding to a resolution of 0.1 arcsec), speckle cameras (with a resolution of 10^{-2} arcsec) can get down to $10^8 M_\odot$, while VLBI and VLBA (with resolutions of 1 and 0.1 milliarcsec) can search for objects as small as $10^6 M_\odot$ and $10^4 M_\odot$. The best strategy is to look for dim images near bright objects (Nemiroff & Bistolas 1990), which requires a large dynamic range, but one can also look for circular distortions and gravity rings (Saslaw et al 1985, Turner et al 1990). The usual approach is to derive the "detection volume," defined as the volume between the source and observer within which the lens would need to lie in order to produce an observable effect (Nemiroff 1989, Kassiola et al 1991). Limits are then obtained by adding the detection volume for each source and comparing this to the volume per source expected for a given Ω_C.

There have been several optical and radio surveys to search for multiply-imaged quasars (Hewitt et al 1989, Bahcall et al 1992b). In particular, Hewitt (1986) used VLA observations to infer $\Omega_C(10^{11}-10^{13} M_\odot) < 0.4$, Nemiroff (1991b) used optical QSO data from Crampton et al (1989) to infer $\Omega_C(M > 10^{9.9} M_\odot) < 1$ and $\Omega_C(M > 10^{10.3} M_\odot) < 0.25$, and Surdej et al (1993) used data on 469 highly luminous quasars (including *HST* observations) to infer $\Omega_C(10^{10}-10^{12} M_\odot) < 0.02$. To probe smaller scales, one must use high resolution radio sources: Kassiola et al (1991) have used lack of lensing in 40 VLBI objects to infer $\Omega_C(10^7-10^9 M_\odot) < 0.4$, while a study by Patnaik et al (1992) of 200 flat spectrum radio sources may lead to a limit $\Omega_C(10^6- 10^9 M_\odot) < 0.01$ (Henstock et al 1993). (Flat spectrum sources are dominated by a single core and are therefore more likely to be lensed; this limit assumes that no sources are identified and is not included in Figure 5.)

Future observations could strengthen these constraints considerably: Speckle interferometry could push $\Omega_C(10^8-10^{10} M_\odot)$ down to 0.01, while VLBA could push $\Omega_C(10^5-10^8 M_\odot)$ down to 0.001 (Surdej et al 1993). These two limits are shown as broken lines in Figure 5. Another interesting possibility is to search for lensing distortions in radio jets (Kronberg et al 1991); this would permit the detection of objects with mass around $10^6 M_\odot$ since the Einstein radius for such objects is of order milliarcsecs and therefore comparable to the characteristic jet scale. Of course, jets may be intrinsically kinky but Wambsganss & Paczynski (1992) have pointed out that this poses no problem if one uses VLBI and VLBA maps of the jets in image-doubled quasars because only one of the images would then be kinked. Their numerical simulations show that the effects of supermassive black holes would be numerous and obvious. Lenses between 0.3 and $3 \times 10^6 M_\odot$ would certainly be noticeable for a dynamic range of 100:1 and may have already been excluded (Heflin et al 1991, Garrett et al 1994).

7.2   Microlensing in Macrolensed Quasars

Even if a lens is too small to produce resolvable multiple images of a source, it may still induce detectable intensity variations. In particular, one can look for microlensing in quasars that are already macrolensed. This possibility arises because, if a galaxy is suitably positioned to image-double a quasar, then there is also a high probability that an individual halo object will traverse the line of sight of one of the images (Gott 1981); this will give intensity fluctuations in one but not both images. Although the effect would be observable for objects bigger than $10^{-4} M_\odot$, the timescale of the fluctuations is around $40(M/M_\odot)^{1/2}$y, and this would exceed a decade for $M < 0.1 M_\odot$.

There is already evidence of this effect for the quasar 2237+0305 (Irwin et al 1989). This has four images at a redshift of 1.7 and the lens is a galaxy at redshift 0.04. The brightest image brightened by 0.5 magnitudes from September, 1987 to August, 1988 and then dimmed by 0.15 magnitudes by September, 1988. There was no variation in the other images, even though the difference in light-travel time is only hours. The observed timescale for the variation indicates a mass in the range $0.001 M_\odot$ to $0.1 M_\odot$, although Wambsganss et al (1990) argue that it might be as high as $0.5 M_\odot$, the mass where a standard IMF gives the dominant contribution. (The variable image is almost exactly aligned with the center of the lensing galaxy, where the density should be dominated by ordinary stars). Analysis of more extensive data (Corrigan et al 1991) has strengthened the evidence for microlensing with a mass below $0.1 M_\odot$ (Webster et al 1991).

7.3   The Effect of Microlensing on Quasar Luminosity

Evidence for the microlensing of unmacrolensed quasars may come from study-ing their luminosity variations (Peacock 1986, Kayser et al 1986, Schneider & Weiss 1987, Refsdal & Starbell 1991, Lewis et al 1993), and there may already be cases of this. In particular, Nottale (1986) claims that lensing by low mass objects may explain some optically violently variable quasars. For example, the quasar 0846+51 brightened by 4 magnitudes in a month and then dimmed by 1 magnitude in a few days. The fact that its line of sight is only 12 arcsec from a galaxy suggests that the variation may result from microlensing by one of the halo objects, in which case the mass of the halo object must be in the range 10^{-4} to $10^{-2} M_\odot$.

More dramatic, but no less controversial, evidence for the effect of microlensing on quasar luminosity comes from Hawkins (1993), who has been monitoring 300 quasars in the redshift range 1–3 over the past 17 years using a wide-field Schmidt camera. He finds quasi-sinusoidal variations of amplitude 0.5 m on a 5y timescale and he attributes this to lenses with mass $\sim 10^{-3} M_\odot$. The crucial point is that the timescale decreases with increasing z, which is the opposite to what one would expect for intrinsic variations (and these would be on a shorter timescale anyway). The timescale also increases with the luminosity of the quasar. He tries to explain this by noting that the luminosity should increase with the size of the accretion disk, but this only works if the disk is larger than the Einstein radius of the lens (about 0.01 pc), which is questionable. Another worrisome feature of Hawkins' claim (cf Schneider 1993) is that he requires the density of the lenses to be close to critical (so that the sources are being transited continuously). In this case, Big Bang nucleosynthesis constraints require the lenses to be nonbaryonic, so he is forced to invoke primordial black holes.

7.4   The Effect of Microlensing on Quasar Density

Quasars that are not bright enough to be included in a flux-limited sample may be amplified by quasar microlensing, thereby bringing them above the detection threshold (Turner 1980, Canizares 1981, Peacock 1982, Schneider et al 1992) and modifying the apparent number density. This effect depends strongly on the quasar luminosity function, which may itself be influenced by lensing (Vietri & Ostriker 1983). There are several indications that this happens. For example, Webster et al (1988) found that faint galaxies were 4.4 times as numerous as usual within 6 arcsec of high redshift quasars and attributed this to the galaxies enhancing the quasar density. However, to explain such a high enhancement, they had to attribute to the galaxies unrealistically massive halos (Hogan et al 1989), so the origin of this effect is not well understood.

A similar result was found by Hammer & Le Fevre (1990), who found the quasar density within 5 arcsec of $z > 1$ radio galaxies to be nine times greater than expected; Bartleman & Schneider (1993) claim that this can be explained if the quasar luminosity function is sufficiently steep. Rix & Hogan (1988) have claimed a *lower* limit of $\Omega_C(0.001\text{--}10^{10} M_\odot) > 0.25$ from an excess of quasar-galaxy pairs in the Einstein Medium Source Survey, but Dalcanton et al (1994) argue that they underestimate the amplification (and hence overestimate Ω_C) by underestimating the steepness of the quasar luminosity function. Kovner (1991) obtains constraints on $\Omega_C(0.001\text{--}10^{10} M_\odot)$ by studying the slope of the bright quasar counts.

Rodrigues-Williams & Hogan (1994) have found an excess of quasars in the direction of clusters. Their sample comprises 129 quasars with $1.4 < z < 2.2$

and 70 clusters at $z \sim 0.2$; they find an overdensity of 1.7. Unfortunately, this does not seem to be consistent with the most plausible mass distribution. Note that there is also a lensing effect that reduces the number of quasars near clusters because the background area is expanded. Which effect wins depends on the steepness of the quasar luminosity function. The amplification effect wins when the luminosity function is steep, but the spread effect wins when it is shallow. There may also be evidence for the second effect: Boyle et al (1988) have found a 30% deficit of high redshift quasars within 4 arcsec of clusters, although they attribute this to the effect of dust.

7.5   Line-to-Continuum Effects of Quasars

In some circumstances, only part of the quasar may be microlensed. In particular, the line and continuum fluxes may be affected differently because they may come from regions that act as extended and pointlike sources, respectively. [For a lens at a cosmological distance, the Einstein radius is $0.05(M/M_\odot)^{1/2}h$ pc, whereas the size of the optical continuum and line regions are of order 10^{-4} pc and 0.1–1 pc, respectively.] This effect can be used to probe individual sources. For example, the variations in the line-to-continuum ratio for different images of the same macrolensed quasar can be used to constrain the mass of the objects in the lensing galaxy. Evidence for such an effect may already exist in the case of the double quasar 2016 + 112, where variations in the intensity ratios for the different images suggest that the lensing objects have a mass in the range $3 \times 10^4 M_\odot$ to $3 \times 10^7 M_\odot$ (Subramanian & Chitre 1987).

The line-continuum effect can also show up in statistical studies of many quasars and there is one particularly important effect in this context. One would expect the characteristic equivalent width of quasar emission lines to decrease as one goes to higher redshift because there would be an increasing probability of having an intervening lens. Indeed, a third of quasars should have equivalent widths smaller by 2–3 at only a moderate redshift if $\Omega_C = 1$. This idea was first studied by Canizares (1982). More recently, Dalcanton et al (1994) have compared the equivalent widths for a high and low redshift sample comprising 835 Einstein Medium Source Survey quasars and 92 Steidel-Sargent absorption systems and find no difference. They infer the following limits:

$$\Omega_C(0.001\text{–}60M_\odot) < 0.2, \quad \Omega_C(60\text{–}300M_\odot) < 1,$$
$$\Omega_C(0.01\text{–}20M_\odot) < 0.1. \tag{7.2}$$

The mass limits come from the fact that the amplification of even the continuum region would be unimportant for $M < 0.001M_\odot$, while the amplification of the broad-line regions would be important (cancelling the effect) for $M > 20M_\odot$ if $\Omega_c = 0.1$, for $M > 60M_\odot$ if $\Omega_c = 0.2$ or for $M > 300M_\odot$ if $\Omega_c = 1$. (These limits are indicated in Figure 5). This compares with the earlier Canizares

(1982) constraint of $\Omega_C(0.01-10^5 M_\odot) < 1$; his upper mass limit was larger because the size of the broad-line region was thought to be larger then. Note that Equation (7.2) is incompatible with Hawkins' claim that $\Omega_C(10^{-3} M_\odot) \sim 1$, although one would only need to reduce Ω_C or M slightly.

7.6 Microlensing of Gamma-Ray Bursts

Another method of seeking evidence for compact objects in the mass range $10^6-10^8 M_\odot$ is to look for echoes from gamma-ray bursts (on the assumption that these are at cosmological distances). The images can be resolved temporally but not spatially (Paczynski 1987). This effect has been considered by many people (Webster & Fitchett 1986, Krauss & Small 1991, Blaes & Webster 1991, Mao 1992, Gould 1992, Narayan & Wallington 1992). The most recent analysis is that of Nemiroff et al (1993), who find no evidence for echoes in data for 44 bursts discovered by the *Gamma Ray Observatory*. Using the detection volume technique and theoretical redshifts for the bursts, they infer a limit $\Omega_C(10^{6.5}-10^{8.1} M_\odot) < 1$, which is shown in Figure 5. However, it must be stressed that the redshifts of the bursts are quite uncertain (they may not even be cosmological) and, for any particular burst, all one can strictly infer is a constraint on M as a function of redshift.

7.7 Microlensing of Stars by Halo Objects in our own Galaxy

Attempts to detect microlensing by objects in our own halo by looking for intensity variations in stars in the Magellanic Clouds and the Galactic Bulge have now been underway for several years and may already have met with success. In this case, the timescale for the variation is $P = 0.2(M/M_\odot)^{1/2}$y, so one can seek lenses over the mass range $10^{-8}-10^2 M_\odot$, but the probability of an individual star being lensed is only $\tau \sim 10^{-6}$, so one has to look at many stars for a long time (Paczynski 1986). The likely event rate is $\Gamma \sim N\tau P^{-1} \sim (M/M_\odot)^{-1/2}$y^{-1}, where $N \sim 10^6$ is the number of stars. Thus, small masses give frequent short-duration events (e.g. $0.01 M_\odot$ events would last a week and occur a few times a year) and are best sought with CCDs, while large masses give rare long-duration events (e.g. $10 M_\odot$ events would last a year and occur every few years) and are best sought with photographic plates. The key feature of these microlensing events is that the light-curves are time-symmetric and achromatic; this may allow them to be distinguished from intrinsic stellar variations (Griest 1991).

Three groups are involved; each now claims to have detected lensing events. The American group (MACHO) has used a dedicated telescope at Mount Stromlo to study 10^7 stars in red and blue light in the LMC, the SMC, and the Galactic Bulge. After analyzing 4 fields near the center of the LMC (2×10^6 stars with 250 observations per star), they have obtained one event (Alcock et al 1993): The duration is 34 days (corresponding to a mass of $0.1 M_\odot$) and the amplification is $A = 6.8$. The French group (EROS) has been studying stars in the LMC and their approach is two-pronged: They are seeking 1–100 day events (corresponding to 10^{-4}–$1 M_\odot$ lenses) with digitized red and blue Schmidt plates obtained with the ESO telescope in Chile and 1 hour to 3 day events (corresponding to 10^{-7}–$10^{-3} M_\odot$ with CCDs taken at the Observatoire de Haute Provence. The CCD searches have given no results, which presumably implies a limit $\Omega_C(10^{-7}$–$10^{-3} M_\odot) < 0.1$, but analysis of 3×10^6 stars on the Schmidt plates yields two events (Auborg et al 1993): One is associated with a main-sequence star and has $A = 2.5$ and $P = 54$ d (corresponding to a mass of $0.2 M_\odot$); the other is associated with a star between the main-sequence and the giant branch and has $A = 3.3$ and $P = 60$ d (corresponding to a mass of $0.3 M_\odot$). They have also confirmed the MACHO event in red light. The Polish collaborative (OGLE) are using the Las Companas telescope in Chile to look at 7×10^5 stars in the Galactic bulge (Udalski et al 1993). They have claimed one event with $A = 2.4$ and $P = 42$ d (corresponding to a mass of $0.3 M_\odot$) which they attribute to a disk M-dwarf, but they only have data in one color. The rough values of M and Ω_c for these events are indicated in Figure 5, but there is considerable uncertainty in both these values.

8.   THE BLACK HOLE SCENARIO

One of the most important signatures of the black hole scenario would be the infrared/submillimeter background generated by the stellar precursors. In Section 5.1 we discussed a general constraint on $\Omega_*(M)$, which depended only on the fact that the background light must appear in *some* waveband. Here we discuss more precise constraints for VMOs, exploiting the fact that we can predict the waveband in this case very exactly. The calculation can be extended to cover the mass range below $100 M_\odot$, but that range may be excluded by nucleosynthetic constraints anyway. We also consider the generation of gravitational radiation by VMO or SMO black holes.

8.1   *Background Light Observations*

The detection of cosmological background radiation in the IR and submillimeter bands is difficult because of foregrounds from scattered zodiacal light (ZL), interplanetary dust (IPD), and interstellar dust (ISD). Estimates of these competing backgrounds are shown in Figure 6, where the background light intensity $I(\lambda)$ has been expressed in critical density units by defining a quantity $\Omega_R(\lambda) \equiv 4\pi \lambda I(\lambda)/c^3 \rho_{crit}$. One sees that there are minima at around 4μ, 100μ, and 400μ, so these are the best "windows" in which to search for an extragalactic background. Although positive detections have been claimed in all of these windows, none has been subsequently confirmed, so only upper limits on $\Omega_R(\lambda)$

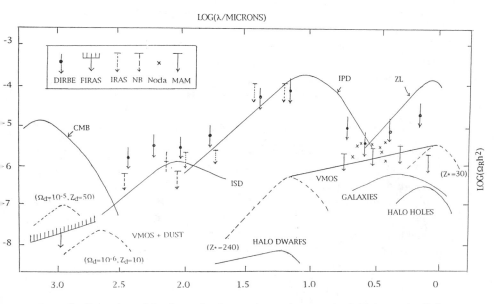

Figure 6 Comparison of the observational constraints on the extragalactic background radiation density from *DIRBE, FIRAS, IRAS*, Nagoya-Berkeley (NB), Matsumoto et al (MAM), and Noda et al with the background expected in the VMO scenario for different dust abundances. Also shown are the CMB, the local foregrounds, and the background from galaxies, accreting $10^6 M_\odot$ halo black holes, and $0.08 M_\odot$ halo BDs.

are currently available; we begin by summarizing these. For comparison, the CMB peaks at $\lambda_{peak} = 1400\mu$ with a density $\Omega_R = 2 \times 10^{-5} h^{-2}$.

The *FIRAS* results (Mather et al 1990, 1994) imply that the CMB is so well fit by a black-body spectrum that any extra background must have an intensity less than 0.03% of the CMB density over the range $500-5000\mu$. This implies $\Omega_R(\lambda) < 6 \times 10^{-9} h^{-2} (\lambda/\lambda_{peak})^{-1}$. The *DIRBE* results at the south ecliptic pole (Hauser et al 1991) give upper limits in the J, K, L, M, 12μ, 25μ, 60μ, 100μ, $120-200\mu$ and $200-300\mu$ bands. However, the limits indicated in Figure 6 are very conservative since they do not include any subtraction for the foreground backgrounds from interstellar and interplanetry dust. Careful modeling of these foreground contributions may improve the limits. Figure 6 includes the limits derived by Oliver et al (1992) by using *IRAS* and *DIRBE* data in conjunction with detailed dust models. It also shows the limits obtained from an analysis (Lange et al 1991) of the Nagoya-Berkeley rocket data (Matsumoto et al 1988a). At one stage *IRAS* data seemed to indicate a 100μ background with $\Omega_R(100\mu) = 3 \times 10^{-6} h^{-2}$ (Rowan–Robinson 1986) but this is inconsistent with the *DIRBE* results.

The *DIRBE* and *IRAS* limits are very weak around 12μ and 25μ because the interplanetary dust emission is so large. In the near-IR, a Japanese rocket experiment (Matsumoto et al 1988b) gave a limit $\Omega_R(1-5\mu) < 3 \times 10^{-5}h^{-2}$ with the possible detection of a "line" at 2.2μ with $\Omega_R(2.2\mu) = 3 \times 10^{-6}h^{-2}$. However, this claim was always controversial because of the problem of subtracting starlight and rocket exhaust. Recent observations by Noda et al (1992) give $\Omega_R(1.6-4.7\mu) < 3 \times 10^{-6}h^{-2}$, which seems to exclude such a line.

8.2   Infrared Background from VMOs

We now compare these limits with the background expected from a population of VMOs. This can be predicted very precisely since all VMOs have a surface temperature T_s of about 10^5K and generate radiation with efficiency $\varepsilon \approx 0.004$. We normalize the VMO density parameter to the value $\Omega_* \approx 0.1$ required to explain galactic halos and assume that they produce black-body radiation with temperature T_s. If the radiation is affected only by cosmological redshift, its density and peak wavelength at the present epoch should be

$$\Omega_R(\lambda_{peak}) = 4 \times 10^{-6} \left(\frac{\Omega_*}{0.1}\right) \left(\frac{1+z_*}{100}\right)^{-1},$$

$$\lambda_{peak} = 4 \left(\frac{1+z_*}{100}\right) \mu, \tag{8.1}$$

where z_* is the redshift at which the VMOs burn. We can place an *upper* limit on z_* by noting that the main-sequence time of a VMO is $t_{MS} \approx 2 \times 10^6$y (independent of mass), so that z_* cannot exceed the redshift when the age of the Universe is t_{MS}. This implies $z_* < 240h^{-2/3}$ and so $\lambda_{peak} < 15\mu$ and $\Omega_R > 10^{-6}$ for $h > 0.5$. We can place a lower limit on z_* from UV/optical background light limits. As discussed by McDowell (1986) and Negroponte (1986), these imply a constraint on the density of VMOs burning at any redshift z_*. If one requires $\Omega_* \approx 0.1$, this places a lower limit on z_*, mainly because one needs the radiation to be redshifted into the near-IR band, where the background light limits are weaker. In the absence of neutral hydrogen absorption, one requires $z_* > 30$, which implies $\lambda_{peak} > 1\mu$ and $\Omega_R < 10^{-5}$. The peak of the VMO background must then lie somewhere on the heavy line in Figure 6 and the spectrum must lie within the region bounded by the broken line. Note that the observational limits are only just beginning to constrain the VMO scenario and they may never be able to exclude it if z_* is so large (>200) that most of the VMO light is pushed beyond 10μ, where it would be hidden by interstellar dust.

The constraints on the VMO scenario would be much stronger if the light was reprocessed by dust as discussed by many workers (McDowell 1986, Negroponte 1986, Bond et al 1986 Wright & Malkan 1987, Lacey & Field 1988, Adams et al 1989, Draine & Shapiro 1989). Such dust could either be pregalactic in origin or confined to galaxies themselves if galaxies cover the sky. If the dust cross-section for photons of wavelength λ is assumed to be geometric (πr_d^2 for a grain radius r_d) for $\lambda \gg r_d$, but to fall off as λ^{-1} for $\lambda \gg r_d$ then the spectrum should peak at a present wavelength (Bond et al 1986)

$$\lambda_{\text{peak}} = 400 \left(\frac{1+z_*}{100}\right)^{1/5} \left(\frac{r_d}{0.01\mu}\right)^{1/5} \left(\frac{1+z_d}{10}\right)^{1/5} \mu, \qquad (8.2)$$

where z_d is the epoch of dust production and we have used Equation (8.1) with $\Omega_* \approx 0.1$ to express Ω_R in terms of z_*. The crucial point is that the wavelength is very insensitive to the various parameters appearing in Equation (8.2) because the exponents are so small. At one time the Nagoya-Berkeley experiment (Matsumoto et al 1988a) appeared to indicate a submillimeter excess peaking at almost exactly the wavelength predicted. However, the Nagoya-Berkeley excess has now been disproved by *FIRAS* and the question arises of whether the VMO-plus-dust scenario is still compatible with *COBE* results.

It should be stressed that one does not necessarily expect dust reprocessing anyway. Pregalactic dust with density Ω_d would only absorb UV photons for

$$z_d > 10 \left(\frac{\Omega_d}{10^{-5}}\right)^{-2/3} \left(\frac{r_d}{0.1\mu}\right)^{2/3}, \qquad (8.3)$$

where Ω_d is normalized to the sort of value appropriate for galaxies. It is not clear whether this condition can be satisfied. One has no direct evidence for pregalactic dust but in any hierarchical clustering picture one would expect at least some pregalactic dust production (Najita et al 1990). For example, one could envisage the dust produced by the first dwarf galaxies being blown into intergalactic space because the gravitational potential of the dwarfs would be so small. The dust in galaxies themselves would suffice to reprocess the VMO background only if galaxies cover the sky which—for galaxies like our own—requires the redshift of galaxy formation to exceed about 10 (Ostriker & Heisler 1984, Heisler & Ostriker 1988, Ostriker et al 1990). Even if galaxies do cover the sky, the analysis of Fall et al (1989) indicates that the dust-to-gas ratio in primordial galaxies may only be 5–20% that of the Milky Way for $2 < z < 3$, which makes the opaqueness condition difficult to satisfy.

In general, one would expect there to be both a far-IR dust background and a near-IR attenuated starlight background, with the relative intensity reflecting the efficiency of dust reprocessing. By changing the amount of dust, one can redistribute the light between the near-IR and far-IR in an attempt to obviate the constraints. In order to examine the issue in more detail, Bond et al (1991) have carried out a more sophisticated analysis, in which the dust cross section is assumed to scale as $\lambda^{-\alpha}$ at infrared wavelengths. They also introduce a more realistic model for the source luminosity history, allowing for both "burst" and "continuous" models. Comparison with the far-IR and *COBE* constraints is shown in Figure 6 for two of their models with $\alpha = 1.5$ and $z_* = 100$. One has $\Omega_d = 10^{-5}$ and $z_d = 50$ (which is above the *FIRAS* constraint); the other has $\Omega_d = 10^{-6}$ and $z_d = 10$ (which is below it). This shows that the VMO-plus-dust scenario is only viable for models with a high redshift of energy release ($z_* = 100$) and small amounts of dust ($\Omega_d = 10^{-6}$). Of the models considered by Bond et al (1991), Wright et al (1994) claim that only their model 12 still survives.

8.3   Generation of 3K Background

If the dust-reprocessed radiation is itself absorbed by the dust, then the radiation could be completely thermalized, leaving no residual distortions at all. Some people have therefore proposed that the *entire* CMB is grain-thermalized starlight (Layzer & Hively 1973, Rees 1978). This is possible in principle—and Equation (8.1) shows that this idea is certainly not precluded energetically—but the grains would have to form at a high redshift and be very elongated in order to thermalize at long wavelengths (Wright 1982, Hoyle & Wickramsinghe 1989 Hawkins & Wright 1988, Arp et al 1990). The *FIRAS* results now make this model rather hard to sustain. An alternative proposal is that black hole accretion generates the CMB at a somewhat higher redshift ($z \sim 10^3$), when thermalization by free-free processes is possible (Carr 1981b). Of course, any scheme that envisages the CMB deriving from Population III stars or black holes also requires that the early Universe be cold or tepid (with the primordial photon-to-baryon ratio being much less than its present value of 10^9). In this case, one must also invoke VMOs or their remnants to generate the observed light element abundance, as discussed in Section 5.2.

8.4   Gravitational Radiation from Black Holes

The formation of a population of black holes of mass M at redshift z_B would be expected to generate bursts of gravitational radiation with a characteristic period and duration:

$$P_o \approx 10GM\frac{(1+z_B)}{c^3} \approx 10^{-2}\left(\frac{M}{10^2 M_\odot}\right)(1+z_B) \text{ s}. \tag{8.4}$$

One can show that the expected time between bursts (as seen today) is less than their characteristic duration provided that $\Omega_B > 10^{-2}\Omega^{-2}$, where Ω is the total density parameter. (Bertotti & Carr 1980). If the holes make up galactic halos, one would therefore expect the burst to form a background of waves with present density $\Omega_g = \varepsilon_g\Omega_B(1+z_B)^{-1}$, where ε_g is the efficiency with which the collapsing matter generates gravity waves. If ε_g were as high as 0.1, the background could be detectable by ground-based laser interfreometers (e.g. LIGO) for M below $10^3 M_\odot$, by Doppler tracking of interplanetary spacecraft (e.g. *Cassini*) for M in the range 10^5–$10^{10} M_\odot$, and by pulsar timing for M

above $10^9 M_\odot$. The observable domains are indicated in Figure 7 and the dotted lines indicate how the predicated backgrounds depend on M and z_B. Note that the value of ε_g is very uncertain and it is probably well below 0.1 for isolated collapse.

The prospects of detecting the gravitational radiation would be much better if the holes formed in binaries (Bond & Carr 1984). This is because two sorts of radiation would then be generated: (*a*) continuous waves as the binaries spiral inward due to quadrupole emission; and (*b*) a final burst of waves when the components finally merge. The burst would have the same characteristics as that associated with isolated holes but it would be postponed to a lower redshift and ε_g would be larger (~ 0.08) because of the larger asymmetry; both factors would increase Ω_g. The continuous waves would also be interesting since they would extend the spectrum to longer periods, thus making the waves detectable by a wider variety of techniques. Over most wavebands, the spectrum of the waves would be dominated by binaries whose initial separation is such that they are coalescing at the present epoch. This corresponds to a separation $a_{\text{crit}} = 10^2 (M/10^2 M_\odot)^{3/4} R_\odot$. The total background generated by the binaries

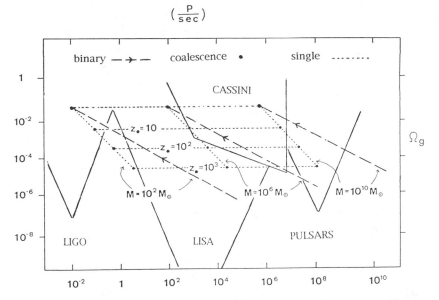

Figure 7   The spectrum of background gravitational waves generated by isolated black holes and coalescing binary black holes. In the first case, we assume that the holes have $\Omega_B = 1$ and that they form at a redshift z_*. In the second case, we assume that the binaries have the separation a_{crit} such that they coalesce at the present epoch. Also shown are the (Ω_g, P) domain accessible to ground-based interometry, Doppler tracking of interplanetary spacecraft, pulsar timing, and space-based interferometry.

is also shown in Figure 7: For each value of M, $\Omega_g(P)$ goes as $P^{-2/3}$, as indicated by the broken lines. Providing the fraction of binaries $f_{\rm crit}$ with around the critical separation is not too small, the background should be detectable by LIGO for $M < 10^3 M_\odot$, by *Cassini* for $10^5 M_\odot < M < 10^{10} M_\odot$, by LISA for $M_\odot < M < 10^{10} M_\odot$, and by pulsar timing for $M > 10^6 M_\odot$.

One could also hope to observe coalescences occurring at the present epoch. For our own halo, the average time $t_{\rm burst}$ between bursts and their expected amplitude $h_{\rm burst}$ would be

$$t_{\rm burst} = 10 \left(\frac{M}{10^2 M_\odot} \right) f_{\rm crit}^{-1} h^{-1} {\rm y}, \quad h_{\rm burst} = 7 \times 10^{-17} \left(\frac{M}{10^2 M_\odot} \right). \qquad (8.5)$$

Although the time would be uncomfortably long, one could also detect bursts from the Virgo cluster every $4(M/10^2 M_\odot)$ days with somewhat improved sensitivity. Haehnelt (1994) has argued that LISA could detect coalescence bursts throughout the Universe for M in the range 10^3–$10^6 M_\odot$.

9.   LOW MASS OBJECTS

In this section, we focus specifically on the Low Mass Object (LMO) scenario. There are several reasons why LMOs currently seem to be the most plausible option. Firstly, there may be direct evidence from cluster cooling flows that baryons can turn into low mass stars with high efficiency even at the present epoch. [This topic is reviewed by Fabian (1994) in this volume, so I merely summarize the key points below and omit references.] Secondly, recent data on the stellar IMF in our own Galaxy suggests there may be a higher fraction of LMOs when the metallicity is low. Thirdly, as we saw in Section 7, microlensing data may already indicate that there is dark matter in the form of LMOs.

9.1   *Cooling Flows in Clusters*

X-ray observations suggest that the cores of many clusters contain hot gas which is flowing inwards because the cooling time is less than the Hubble time. This condition is satisfied in 70–80% of *EXOSAT* clusters and in some poor clusters and groups as well. Direct evidence for cooling comes from Fe XVII line emission, since this shows that the temperature decreases as one goes inwards. The mass flow rates are typically in the range 50–100 $M_\odot {\rm y}^{-1}$, extending up to $10^3 M_\odot {\rm y}^{-1}$ in some cases, and they seem to have persisted for at least several billion years. There is consistency between the flow rates derived from spectral measurements and those derived from surface brightness analysis.

The mass appears to be deposited over a wide range of radii with a roughly $M \propto R$ distribution (which requires that the gas be very inhomogeneous), but it cannot be going into stars with the same mass spectrum as in the solar neighborhood, or else the central regions would be bluer and brighter than observed. Some cooling flows do exhibit a blue optical continuum over the central few kpc, but the associated massive star formation rate must be less than a few $M_\odot y^{-1}$, which is only a fraction of the total inflow rate. This suggests that the cooling flows produce very low mass stars, possibly because the high pressure (of order $10^6 \text{cm}^{-3}\text{K}$) reduces the Jeans mass. An important feature of a cooling flow is that it is quasi-static, in the sense that the cooling time exceeds the local dynamical time, and it is this condition which is supposed to preserve the high pressure. The Jeans mass could be as low as $0.1 M_\odot$ if the cloud gets as cool as the microwave background radiation ($T = 3$ K); this is not inconceivable because observations suggest that the gas is mainly molecular, which could allow grains to form abundantly.

A recent twist in this scenario has been the detection of large amounts of cold X-ray absorbing material in many clusters (White et al 1991). The cold gas extends out to 100 kpc and the mass involved is usually around $10^{12} M_\odot$ (comparable to that expected from a cooling flow that has persisted for a cosmological time). This raises the question of whether we still need low mass stars, especially in view of the Pfenniger et al (1994) proposal that the dark matter in galactic halos could be cold gas. Of course, if cooling flows do make low mass stars, one might expect some cold gas as an intermediate state. This issue has yet to be resolved.

9.2 Pregalactic and Protogalactic Cooling Flows

Although cooling flows provide a natural way of turning gas into low mass stars with high efficiency, those observed in the centers of clusters could not themselves be responsible for either the cluster dark matter (since this is distributed throughout the cluster) or the halo dark matter in galaxies outside clusters. In order to account for the usual dark matter problems, one therefore needs cooling flows on the scale of galaxies or below. Only the most massive cluster galaxies exhibit cooling flows at the present epoch—but it would not be surprising if smaller scale cooling flows occurred at earlier cosmological epochs since X-ray data already suggest that cooling flows evolve hierarchically to larger scales (Evrard 1990, Katz & White 1993).

These considerations prompted Ashman & Carr (1988) and Thomas & Fabian (1990) to consider the circumstances in which one could expect high-pressure quasi-static flows to occur at pregalactic and protogalactic epochs. The situation is best illustrated for the hierarchical clustering scenario, in which, as time proceeds, increasingly large gas clouds bind and virialize. The mass fraction of a cloud cooling quasi-statically is maximized when the cooling time t_c is comparable to the free-fall time t_f: Collapse does not proceed at all for $t_c \gg t_f$, whereas it is not quasi-static for $t_c \ll t_f$. In any particular variant of the hierarchical clustering scenario, one can specify the mass binding as

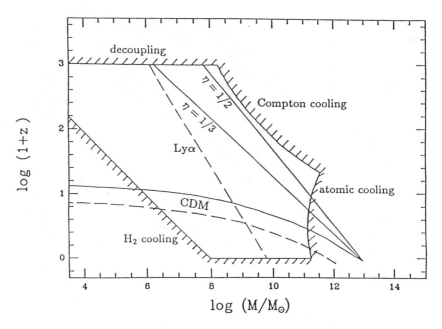

Figure 8 The (*M, Z*) region within which a cloud of mass *M* binding at a redshift *z* will cool within a Hubble time. The other lines show how the binding mass evolves with redshift for the CDM scenario with or without bias (*solid* and *broken lines*) and for the isocurvature model. A lot of gas may be processed through a cooling flow where the binding curve hits the cooling curve. This happens at both a pregalactic and protogalactic era in the CDM case, but the pressure is too low to make LMOs in the former case. In the isocurvature case, only protogalactic cooling flows occur.

a function of redshift. For a cloud of mass *M*, the dynamical time will just be of order the Hubble time at that redshift, whereas the cooling time will depend upon the density and virial temperature of the cloud (which are themselves determined by *M* and *z*). Thus, one can specify a region in the (*M, z*) plane of Figure 8 in which bound clouds will cool within a dynamical time. This applies above a lower mass limit associated with molecular hydrogen or Lyman-α cooling and below an upper mass limit associated with atomic hydrogen cooling (Rees & Ostriker 1977) or the Compton cooling of the microwave background.

The condition $t_c \sim t_f$ will be satisfied at the boundary of the region (shown shaded) and the intersection of this boundary with the binding curve $M(z)$ singles out two characteristic mass-scales and redshifts. These correspond to what Ashman & Carr term "Pervasive Pregalactic Cooling Flows" (PPCFs) and what Thomas & Fabian term "Maximal Cooling Flows" since the amount of gas cooling quasi-statically is maximized. The associated mass-scales are always of order 10^4–$10^8 M_\odot$ and $10^{11} M_\odot$, but the redshifts depend on the particular scenario. Figure 8 shows the binding curves corresponding to the Cold Dark Matter scenario, the broken curve corresponding to the biased version, and the baryon-dominated isocurvature scenario (with η specifying the exponent in the mass dependence of the density fluctuations at decoupling).

One might anticipate most of the dark matter being made on the smaller scale because much of the gas will have been consumed by the time atomic cooling becomes important. However, this does not happen in the Cold Dark Matter picture because the spectrum of fluctuations is very flat on subgalactic scales and, in the isocurvature models, $M(z)$ may never be small enough for low mass PPCFs to occur after decoupling. Both these features are indicated in Figure 8. Ashman & Carr (1992) therefore argue that most of the dark matter would need to be made by high mass protogalactic PPCFs. Another argument in favor of the protogalactic PPCFs is that the pressure is probably too low to make LMOs on the smaller scale.

One problem with invoking protogalactic cooling flows to make the dark matter is that one might expect most of the gas to have gone into clouds with $t_c < t_f$ and such clouds should make ordinary stars. This "cooling catastrophe" raises the question of whether there could be enough gas left over to make ordinary galaxies (White & Frenk 1991, Blanchard et al 1992). One way around this is to invoke supernovae to reheat the gas so that most of it can avoid cooling until the protogalactic epoch (Thomas & Fabian 1990). Another way is to argue that even clouds with $t_c < t_f$ can make a lot of dark matter (Ashman 1990). The idea here is that gas always drops out at such a rate as to preserve the PPCF condition $t_c \sim t_f$ for the surviving gas. One thus gets a two-phase medium, with cool dense clouds embedded in hot high-pressure gas. This was originally proposed as a mechanism to make globular clusters at a protogalactic epoch (Fall & Rees 1985), but Ashman (1990) argues that sufficiently small clouds would fragment into dark clusters rather than visible clusters in the presence of molecular hydrogen. By applying the same idea to other galaxies, he predicts that the fraction of dark mass in spirals should increase with decreasing disk mass and this may be observed (Persic & Salucci 1990).

9.3   M-Dwarfs vs Brown Dwarfs

In determining how small LMOs would need to be to provide the disk or halo dark matter, important information comes from red and infrared observations. From searches for sources in our own halo, Richstone et al (1992) find that the halo mass-to-light ratio from stars between $0.5M_\odot$ and $0.8M_\odot$ exceeds 400, while Bahcall & Soneira (1984) find that the ratio from stars down to $0.15M_\odot$ must exceed 650. This implies that stars in these mass ranges can only contribute a small fraction to the halo density. Even stronger constraints come from Gilmore & Hewett (1983), who find that the local number density of stars in the mass range 0.08–$0.1M_\odot$ can be at most 0.01 pc^{-3}. This is a hundred times too small to explain the local dark matter problem and ten times too small to explain the halo problem.

A similar conclusion is indicated by infrared observations of other spiral galaxies. For example, the K-band mass-to-light ratio exceeds 50 for NGC 4565 (Boughn et al 1981), 100 for M87 (Boughn & Saulson 1983), 64 for NGC 5907 (Skrutskie et al 1985), and 140 for NGC 100 (Casali & James 1994). Since the mass-to-light ratio is less than 60 for stars bigger than $0.08M_\odot$, the lower limit for hydrogen-burning (D' Antona & Mazzitelli 1985), this suggests that any hydrogen-burning stars are excluded. Lake (1992) has criticized some of these limits on the grounds that they involve attributing all the dynamical mass to the halo objects but the correction to the mass-to-light ratio for M87 and NGC 100 could hardly get it below 60. These observations therefore suggest that the halo dark matter must be in the form of brown dwarfs.

9.4   Evidence from Population I and Population II

Although it is difficult to observe brown dwarfs (BDs) themselves, one can study the IMF of stars in the mass range above the hydrogen-burning limit and infer whether its extrapolation would permit a lot of BDs. If one assumes that the IMF has the power-law form

$$\frac{dN}{dm} \sim m^{-x} \quad \text{for} \quad m_{min} < m < m_{max} \tag{9.1}$$

(at least over some mass range), then most of the mass is in the smallest stars for $x > 2$ and in the largest ones for $x < 2$. Determining the value of x in the LMO range is difficult, partly because obtaining the luminosity function is hard and partly because there are large uncertainties in the mass-luminosity relation as one approaches the hydrogen-burning limit. Nevertheless, there does now seem to be a convergence of opinion (Bessell & Stringfellow 1993).

Let us first consider the possibility that the disk dark matter (if it exists) is in the form of BDs. Early studies of the luminosity function for nearby stars (Reid & Gilmore 1982, Gilmore & Reid 1983, Gilmore et al 1985) suggested that the IMF is too shallow for BDs to have an interesting density. These results were initially contradicted by the results of Hawkins (1985) and Hawkins & Bessell (1988), who went to somewhat fainter magnitudes and claimed that the observations were consistent with an IMF which steepened enough to put all the dark mass in BDs. However, the data of Tinney et al (1992, 1993) make it quite clear that the IMF flattens off below 0.2 $M_\odot$ and, unless it rises again below $0.08M_\odot$, the contribution to the local dark matter must be small (Tinney 1993). This is also consistent with the results of Kroupa et al (1993), Comeron et al (1993), and Hu et al (1994). In particular Kroupa et al (1993), find $x = 2.7$ for $m > 1M_\odot$, $x = 2.2$ for $0.5 < m < 1M_\odot$, and $0.7 < x < 1.8$ for $0.08M_\odot < m < 0.5M_\odot$. This suggests that stars of $0.5M_\odot$ should dominate the disk density. BDs may dominate the number density but, unless the value of x changes below $0.08M_\odot$, they cannot contain more than 1% of the disk mass.

The situation is less clear-cut when one considers Population II stars. Richer et al (1991) claim that metal-poor globular clusters have $x = 3.6$ below $0.5M_\odot$ down to at least $0.14M_\odot$, while Richer & Falman (1992) claim that stars in the Galactic Spheroid have $x = 4.5 \pm 1.2$ in the same mass range. This does allow the possibility that most of the mass is in the smallest objects; indeed, BDs could explain all the halo dark matter if the IMF extended down to $M_{min} \sim 0.01M_\odot$. However, Richer & Falman also point out that the rotation curve of the Galaxy requires that the total spheroid mass cannot exceed $7 \times 10^{10}M_\odot$, which implies that the IMF cannot extend below $0.05M_\odot$. It is therefore unlikely that Population II stars themselves could explain the halo dark matter. As stressed by Lake (1992), the main point of these results is that they lend support to the suggestion that low metallicity enhances the fraction of mass in low mass objects.

It should be stressed that there is an important difference between attributing the disk and the halo dark matter to BDs. If the disk dark matter comprises BDs, one would expect them to represent the low mass tail of the Population I IMF since all disk stars presumably form at the same time. However, there may be no connection between the dark halo stars and Population II stars because they probably form at a different time and place. One should therefore be wary of attempts to exclude the halo from comprising BDs on the grounds that Population II stars have a particular IMF, as do Hegyi & Olive (1983, 1986, 1989).

9.5   Infrared Searches for Brown Dwarfs

Even though brown dwarfs do not burn hydrogen, they still generate some luminosity in the infrared. They radiate first by gravitational contraction (for about $10^7 y$) and then by degenerate cooling. If the disk or halo dark matter is in the form of brown dwarfs, it is therefore important to consider whether they can be detected via this infrared emission. Current constraints on BDs are rather weak (Low 1986, van der Kruit 1987, Beichmann et al 1990, Nelson et al 1993) but the prospects of detection will be much better with impending space satellites such as *ISO* and *SIRTF*.

The problem has been addressed in various contexts by several authors. Karimabadi & Blitz (1984) have calculated the expected intensity from BDs with a discrete IMF comprising an $\Omega = 1$ cosmological background. Adams & Walker (1990) have discussed the possibility of detecting the collective emission of the brown dwarfs in our own Galactic halo for both a discrete and power-law IMF. Daly & McLaughlin (1992) have considered the prospects of detecting the emission of individual halo brown dwarfs of a given mass and age in the Solar vicinity, as well as the collective emission of brown dwarfs in other galaxy halos.

Kerins & Carr (1994) have considered the possibility that the BDs are assembled into dark clusters and also discuss how infrared observations at different wavelengths could be used to probe the mass spectrum of the brown dwarfs.

As an illustration of the feasibility of detecting radiation from BDs, let us consider the prospects of detecting the nearest one in our halo. If the BDs all have the same mass m, then the local halo density ($\rho_o = 0.01 M_\odot \text{pc}^{-3}$) implies that the expected distance to the nearest one is $0.55(m/0.01 M_\odot)^{1/3}$pc. The expected spectra are shown in Figure 9 and compared to the sensitivities of *IRAS* and *ISO*. This assumes the temperature and luminosity of Stevenson (1986) where the BD age and opacity are taken to be 10^{10}y and 0.01 cm^{2}g^{-1} (corresponding to electron-scattering). Although *IRAS* gives no useful constraints (it is too weak by a factor of 2 even for the optimal mass of $0.07 M_\odot$), the ISOCAM instrument on *ISO* could detect $0.08 M_\odot$ BDs in a few hours, $0.04 M_\odot$ BDs in a few days, and $0.02 M_\odot$ BDs in a few months. Note that disk BDs, would be younger, locally more numerous, and more opaque than halo BDs, increasing the peak flux by 6 and decreasing the peak wavelength by 0.6. *IRAS* results

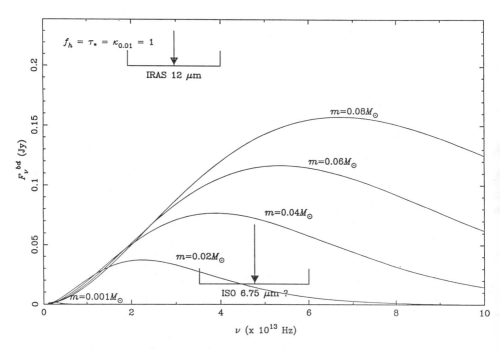

Figure 9 Expected flux from the nearest halo BD for various values of BD mass. The *IRAS* point source sensitivity at 12μ is shown; this is a factor of two above the predicted flux even in the optimal case. The expected 3σ *ISO* 6.75μ sensitivity is also shown, assuming an observation time of 10 days and a 100 s integration time.

already imply that BDs with a discrete IMF could provide the disk dark matter only if their mass is below $0.01 M_\odot$.

One might expect the BDs to be easier to detect if they are in clusters. This is because, although the distance to the nearest source is increased by a factor $(M_c/m)^{1/3}$, the luminosity is increased by (M_c/m), giving an increase in flux of $(M_c/m)^{1/3}$. Rix & Lake (1993) have already used this to exclude the cluster scenario. However, they assume that the clusters are point sources and, as illustrated in Figure 4, the dynamical constraints discussed in Section 6.4 imply that the clusters will always be extended sources. In fact, the *IRAS* extended source sensitivity (EES) at 12μ is too low to permit the detection of clusters. The *ISO* extended source sensitivity at 6.75μ will suffice, but the time required to find these clusters is very sensitive to their mass and radius. Note that the halo clusters will cover the sky if they are large enough, corresponding to the line $K_h > 1$ in Figure 4, in which case detecting the clusters is equivalent to detecting the halo background. *ISO* would take several months to detect the Galactic background, even in the optimal case with $m = 0.08 M_\odot$ (Kerins & Carr 1994). The background spectrum in this case is also indicated in Figure 6.

10.   CONCLUSIONS AND FUTURE PROSPECTS

10.1   *Reappraisal of Baryonic Dark Matter Candidates*

By way of summarizing the key points of this review, and also because it provides an opportunity to mention some candidates that have not yet been covered, we conclude with a reappraisal of the various baryonic dark matter candidates (cf Carr 1990, Dalcanton et al 1994).

SNOWBALLS   Condensations of cold hydrogen can be excluded in most mass ranges. In order to avoid being disrupted by collisions within the age of the Universe, they must have a mass of at least 1g (Hegyi & Olive 1983, Wollman 1992). Constraints in the mass range above this have been discussed by Hills (1986): Snowballs are excluded by the upper limit on the frequency of encounters with interstellar meteors between 10^{-3}g and 10^7g, by the number of impact craters on the Moon between 10^7g and 10^{16}g, and by the fact that no interstellar comet has crossed the Earth's orbit in the last 400 years between 10^{15}g and 10^{22}g. The limits are marginally stronger for halo objects, because of their larger velocities, and are shown in Figure 3. Hegyi & Olive (1983) have argued that snowballs would be evaporated by the microwave background, but Phinney (1985) has pointed out that this only happens below a mass of 10^{22}g. De Rujula et al (1992) have claimed an even stronger limit on the grounds that snowballs smaller than $10^{-7} M_\odot \sim 10^{26}$g would be evaporated within the age of the Universe by their own heat; this is also indicated in Figure 3. Another argument against snowballs is that, since one would expect only hydrogen to condense, the cosmic helium abundance would be increased to an unacceptably high value if the fraction of the Universe going into them were more than $(1 - Y_{min}/Y_{max})$, where Y_{min} is the minimum primordial abundance (≈ 0.2) and Y_{max} is the maximum presolar helium abundance (≈ 0.3). This suggests that the fraction must be less than 30%.

BROWN DWARFS   Fragmentation could in principle lead to objects smaller than $0.08 M_\odot$ and there may be evidence that such brown dwarfs form prolifically in cooling flows (Section 9.1). Such objects might be detectable as infrared sources; it is not surprising that *IRAS* has not found them but *ISO* or *SIRTF* could be expected to detect brown dwarfs with masses down to $0.01 M_\odot$ (Section 9.5). Another important signature of brown dwarfs, in either our own or other galactic halos, is the intensity fluctuations in stars or quasars induced by their microlensing effects. This effect would be observable for objects over the entire brown dwarf mass range and may have already been found (Sections 7.2, 7.3, and 7.7). Observations of microlensing on different timescales could also give information about the mass spectrum of the brown dwarfs (De Rujula et al 1991). The brown dwarf scenario currently appears to be the most plausible. In any case, the combination of infrared and microlensing searches should soon either confirm or eliminate it.

M-DWARFS   Stars in the range 0.3–0.8 $M_\odot$ are excluded from solving any of the dark matter problems by background light limits (Section 5.1). Lower mass hydrogen-burning stars would also seem to be excluded by source count constraints and infrared measurements of other galaxy halos (Section 9.3).

WHITE DWARFS   These would be the natural end-state of stars with initial mass in the range of $0.8–8 M_\odot$ and they could certainly fade below detectability if they formed sufficiently early in the history of the Galaxy. The fraction of the original star that is left in the white dwarf remnant is low but one could still produce a lot of dark matter if there were many generations of stars (Larson 1986). In some sense white dwarfs are the most conservative candidates, since we know that they form prolifically today. The problem is that one needs a very contrived mass spectrum if they are presumed to make up galactic halos: The IMF must be restricted to between 2 and 8 $M_\odot$ to avoid producing too much light or too many metals (Ryu et al 1990) and even then one must worry about excessive helium production (Section 5.3). However, this scenario would have many interesting observational consequences, such as an abundance of cool white dwarfs (Tamanaha et al 1990) and a large number of X-ray sources formed from white dwarf binaries which have coalesced into neutron stars (Silk 1993). A potential problem is that the fraction of white dwarfs in binaries might produce too many type 1a supernovae (Smecker & Wyse 1991), although this might actually be required to explain the high-velocity pulsars moving towards the disk in our own Galaxy (Eichler & Silk 1992). Even if white dwarfs do not have a high enough density to explain the halo dark matter, they could still explain the dark matter in the Galactic disk (if this exists).

NEUTRON STARS   Although neutron stars would be the natural end-state of stars in some mass range above 8 $M_\odot$, the fact that the poorest Population I stars have metallicity of order 10^{-3} places an upper limit on the fraction of the Universe's mass that can have been processed through the stellar precursors—this probably precludes their explaining any of the dark matter problems (Section 5.2). The only way out is to adopt the proposal of Wasserman & Salpeter (1993) in which the neutron stars are in clusters, so that their nucleosynthetic products are trapped within the cluster potentials. Even in this scenario, the neutron stars contain only 1% of the halo dark matter; most of the mass is in asteroids. Nevertheless, the small admixture of neutron stars has an intriguing consequence since collisions between the neutron stars and asteroids are supposed to explain gamma-ray bursts.

STELLAR BLACK HOLES   Stars larger than some critical mass $M_{\mathrm{BH}} \approx 25\text{--}50 M_\odot$ may leave black hole rather than neutron star remnants, with most of their nucleosynthetic products being swallowed. However, they will still return a substantial amount of heavy elements through winds prior to collapsing (Maeder 1992), so normal stellar black holes are probably excluded. In any case, stellar black holes could not provide the disk dark matter because the survival of binaries in the disk requires that the local dark objects are smaller than $2M_\odot$ (Section 6.5). Stellar black holes could also be detected by their lensing effects on the line-to-continuum ratio of quasars; this already excludes black holes from having a critical density below $300M_\odot$ or a tenth critical density (required for halos) below $20M_\odot$ (Section 7.5).

VMO BLACK HOLES   Since stars larger than some critical mass $M_c \approx 200M_\odot$ undergo complete collapse, they may be better candidates for the dark matter than ordinary stars. However, VMOs are radiation-dominated and therefore unstable to pulsations; these pulsations are unlikely to be completely disruptive, but they could lead to considerable mass loss and possible overproduction of helium (Section 5.3). Another important constraint on the number of stellar black hole remnants is provided by background light limits. Although these can be obviated if the stars burn at a sufficiently high redshift, the scenario is becoming increasingly squeezed by the *FIRAS* data (Section 8.2). However, VMO black holes are relatively unconstrained by lensing effects, since the line-to-continuum constraint only applies below $300M_\odot$ (Section 7.5). Laser interferometry might just detect the gravitational wave background generated by a large population of VMO black holes, especially if they form in binary systems (Section 8.4).

SUPERMASSIVE BLACK HOLES   We have seen that SMOs larger than $10^5 M_\odot$ would collapse directly to black holes without any nuclear burning due to relativistic instabilities. However, halo black holes would heat up the disk stars more than is observed unless they were smaller than about $10^6 M_\odot$ (Section 6.1), so they would have to lie in the narrow mass range $10^5 – 10^6 M_\odot$, and the survival of globular clusters (Section 6.2) and dynamical friction effects (Section 6.3) probably exclude even this range. If the dark matter in clusters comprises black holes, then the absence of unexplained tidal distortions in the visible galaxies implies that they must be smaller than $10^9 M_\odot$ (Section 6.5). The number of SMO black holes is also constrained by macrolensing searches: Their density parameter must be less than 0.4 between 10^7 and $10^9 M_\odot$ and less than 0.02 between 10^{11} and $10^{13} M_\odot$ (Section 7.2). The background gravitational waves generated by the formation of SMO black holes could in principle be detected by space interferometers or the Doppler tracking of interplanetary spacecraft (Section 8.4).

10.2   Best Bet Candidates

The various constraints on the form of baryonic dark matter discussed in this review are brought together in Table 2; the shaded regions are excluded by either dynamical, nucleosynthetic, lensing, or light constraints. This assumes that the objects all have the same mass, so that one does not have extra constraints associated with assumptions about the IMF. The dotted regions may also be excluded but this is less certain. Table 2 does not include the dynamical constraints on the nonbaryonic candidates, but it should be noted that only cold inos could explain the presence of dark matter in galaxies. However, hot inos could explain the cluster and background dark matter; large-scale structure and microwave anisotropy observations may even require a mixture of hot and cold inos (Taylor & Rowan Robinson 1992).

Whether the dotted region in Table 2 is excluded depends on whether one believes that the primordial nucleosynthesis constraint permits the cluster dark matter to be baryonic. This is possible only if one invokes inhomogeneous cosmological nucleosynthesis, but this scenario should still be taken seriously. It would be remarkable if the Universe came through the quark hadron phase transition with no fluctuations at all; the surprise is that the resulting light element abundance are relatively insensitive to these. On the other hand, it seems clear that even inhomogeneous nucleosynthesis will not permit baryons to have the critical density. A critical baryon density is excluded in most mass ranges anyway. The prospects for the "maximal BDM" scenario therefore seem bleak.

The prime message of Table 2 is that one could not expect any single candidate to explain all four dark matter problems. On the other hand, the table does constrain the possible solutions:

Table 2 Constraints on baryonic dark matter candidates[a]

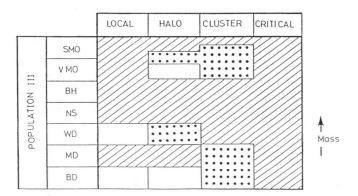

[a]The shaded regions are excluded by at least one of the limits discussed in the text and the dotted regions are improbable. SMO, VMO, and BH refer to the black hole remnants of Supermassive Objects, Very Massive Objects, and ordinary stars respectively; WD = white dwarf; MD = M-dwarf; BD = brown dwarf.

1. The local dark matter (if it exists) could be white dwarfs or brown dwarfs but observations of the Population I IMF gives no reason for expecting this; presumably it could not be inos since these are nondissipative and so would not settle into a disk.

2. The halo dark matter could be brown dwarfs, white dwarfs, or VMO black holes; all of these possibilities require a departure from the standard Population II IMF, but the first probably requires the least radical departure since observations may already indicate a preponderance of low mass stars at small metallicity (white dwarfs—although in a sense the most conservative candidate—require cutting the IMF off at both ends).

3. The cluster dark matter may be partly baryonic, especially if galactic halos are baryonic, but we have seen that it could only be dominated by baryonic dark matter if one invokes inhomogeneous cosmological nucleosynthesis.

4. The background dark matter (if it exists) would have to be inos; if the inos are cold, one would expect both the halo and cluster dark matter to be a mixture of WIMPs and MACHOs.

Finally, we should comment on the possibility of Primordial Black Holes (PBHs). Although these are not baryonic (since they form mainly from radiation rather than from gas), they share many of the features of their baryonic counterparts. These could certainly contribute to the dark matter in principle, and those smaller than $10^5 M_\odot$ (which form before the time of cosmological nucleosynthesis) could have the critical density. However, whether they form from initial inhomogeneities (Hawking 1971, Carr 1975), from phase transitions (Hawking et al 1982, Crawford & Schramm 1982, Dolgov & Silk 1993), or from the collapse of cosmic loops (Polnarev & Zemboricz 1988, Hawking 1989), fine-tuning is required to get an interesting cosmological density because the fraction of the Universe going into PBHs at time t must be $\sim 10^{-6} t^{1/2}$, where t is in seconds. Therefore, despite the invocation of PBHs to explain gravitational lensing effects (Section 7.3), this does not seem very likely.

10.3 Future Prospects

Some of the most exciting developments in this field have come from gravitational lensing studies and we can expect a proliferation of these efforts in the next few years. Macrolensing searches are placing ever more stringent limits of the number of high mass baryonic objects; microlensing searches may have already provided evidence for low mass ones. It is surely significant that microlensing evidence from both quasars and stars all point towards lens masses in the range 0.001–$0.1 M_\odot$. Admittedly, the masses indicated by the MACHO and EROS results are marginally too high; the most likely values are all in the M-dwarf range, whereas light constraints require the halo objects to be smaller than $0.08 M_\odot$. Nevertheless, there is a fairly broad probability distribution for the lensing masses, depending on the assumed velocity and spatial distribution of the halo objects (Kerins 1994), so this puzzle may yet be resolved.

Although we have not reviewed here the plethora of theoretical papers that have appeared since the microlensing results were announced, it should be stressed that, even if the lensing results are genuine, they do not preclude WIMPs from providing some or even most of the dark halo. This is because the microlensing searches only probe the part of the halo at Galactocentric radii from 10–20 kpc, whereas the halo itself could extend much further than this. There could therefore be plenty of WIMPs further out, especially if the dark baryons are preferentially concentrated as a result of dissipation. Even within the 10–20 kpc region, the number of MACHO and EROS events observed merely suggests that the fraction of halo mass in MACHOs must exceed 10% (Gates & Turner 1994). It is therefore important that WIMP searchers should not be too discouraged by the success of the MACHO searchers. It still seems a fair bet that the world needs both MACHOs and WIMPs.

Another source of exciting developments in this field has been *COBE*. We have seen that the *FIRAS* constraints on the microwave spectral distortions and the *DIRBE* measurements of the infrared background density already severely restrict any scenario in which the dark matter is in the relics of massive stars. This is especially true if the radiation has been reprocessed into the far-IR by dust. Indeed, the only hope for these scenarios may be that the radiation remains in the near-IR where it may be hidden by interplanetary dust emission. The *FIRAS* constraints on the Compton y-parameter may also exclude the supermasive accreting black hole scenario. The *COBE* DMR constraints on the microwave anisotropies (though not treated in detail here) are also highly pertinent to the baryonic dark matter scenarios. Some people claim that these already rule out baryon-dominated models, but this conclusion is sensitive to assumptions about the form of the initial density fluctuations, so this has not been stressed here.

Looking further to the future, two more developments will have an important impact. If the halo dark matter is in brown dwarfs, then the next generation of infrared space satellites will either detect these or push their mass down to below $0.001 M_\odot$. In this case, we have stressed the importance of knowing whether the brown dwarfs are clustered because this determines whether one is seeking discrete or extended sources. If the halo dark matter is in black holes, then the next generation of gravitational wave detectors (either ground-based or space-based interferometers) will have an excellent chance of detecting the associated gravitational radiation—the period and amplitude of the waves will indicate the mass and formation redshift of the black holes. It seems likely that MACHOS will have been identified or excluded by the end of the millenium!

ACKNOWLEDGMENTS

Much of my own work over the past decade has focused on the topic of this review, so I hope my own prejudices have not shown too glaringly. I would like to thank all my baryonic-dark-matter collaborators over this period for many enjoyable and stimulating interactions: Dave Arnett, Keith Ashman, Bruno Bertotti, Dick Bond, Wolfgang Glatzel, Craig Hogan, Satoru Ikeuchi, Eamonn Kerins, Cedric Lacey, Jonathan McDowell, Joel Primack, Martin Rees, Michael Rowan-Robinson, Mary Sakellariadou, Humitako Sato, Joe Silk, and Mike Turner. I would also like to thank Cathy Clarke, Andy Fabian, Gerry Gilmore, Mike Hawkins, George Lake, Ben Moore, Adi Nusser, Bernard Pagel, and Sasha Polnarev for helpful discussions. Finally, I would like to thank Eamonn Kerins for helping me in my literature search and Jonathan Gilbert for his assistance in producing the figures.

Literature Cited

Aaronson M. 1983. *Ap. J. Lett.* 266:L11
Aaronson M, Olszewski E. 1987. In *Dark Matter in the Universe,* ed. J Kormendy, GR Knapp, p. 153. Dordrecht:Reidel

Adams FC, Freese K, Levin J, McDowell J. 1989. *Ap. J.* 360:24
Adams FC, Walker TP. 1990. *Ap. J.* 359:57
Alcock C, Fuller GM, Mathews GJ. 1987. *Ap.*

J. 320:439

Alcock C, Akerlof CW, Allsman RA, Axelrod TS, Bennett DP, et al. 1993. *Nature* 365:621

Applegate JH, Hogan CJ, Scherrer RJ. 1987. *Phys. Rev. D* 35:1160

Arnett WD. 1978. *Ap. J.* 219:1008

Arp HC, Burbidge G, Hoyle F, Narlikar JV, Wickramasinghe NC. 1990. *Nature* 346:807

Ashman KA. 1990. *MNRAS* 247:662

Ashman KA. 1992. *Publ. Astron. Soc. Pac.* 104:1109

Ashman KM, Carr BJ. 1988. *MNRAS* 234:219

Ashman KM, Carr BJ. 1991. *MNRAS* 249:13

Ashman KM, Salucci P, Persic M. 1993. *MNRAS* 260:610

Auborg E, Bareyre P, Bréhin S, Gros M, Lachièze-Ray M, et al. 1993. *Nature* 365:623

Babul A, Katz N. 1993. *Ap. J. Lett.* 406:L51

Bahcall JN. 1984a. *Ap. J.* 276:156

Bahcall JN. 1984b. *Ap. J.* 276:169

Bahcall JN. 1984c. *Ap. J.* 287:926

Bahcall JN, Flynn C, Gould A. 1992a. *Ap. J.* 389:234

Bahcall JN, Hut P, Tremaine S. 1985. *Ap. J.* 290:15

Bahcall JN, Soneira RM. 1984. *Ap. J. Suppl.* 55:67

Bahcall JN, Maoz D, Doxsey R, Schneider DP, Bahcall NA, et al. 1992b. *Ap. J.* 387:56

Barcons X, Fabian AC, Rees MJ. 1991. *Nature* 350:685

Bartleman M, Schneider P. 1993. *Astron. Astrophys.* 268:1

Beers TC, Preston GW, Shectman SA. 1992. *Astron. J.* 103:1987

Begelman MC, Blandford RD, Rees MJ. 1980. *Nature* 287:307

Beichman CA, Chester T, Gillett FC, Low FJ, Matthews K, Neugebauer G. 1990. *Astron. J.* 99:1569

Bertotti B, Carr BJ. 1980. *Ap. J.* 236:1000

Bertschinger E. 1983. *Ap. J.* 268:17

Bertschinger E, Dekel A. 1989. *Ap. J. Lett.* 336:L5

Bessel MS, Norris J. 1984. *Ap. J.* 285:622

Bessel MS, Stringfellow GS. 1993. *Annu. Rev. Astron. Astrophys.* 31:433

Bienayme O, Robin AC, Creze M. 1987. *Astron. Astrophys.* 180:94

Blaes OM, Webster RL. 1992. *Ap. J. Lett.* 391:L63

Blanchard A, Valls-Gabaud D, Mamon GA. 1992. *Astron. Astrophys.* 264:365

Blandford RD, Jaroszynski M. 1981. *Ap. J.* 246,1

Blandford RD, Narayan R. 1992. *Annu. Rev. Astron. Astrophys.* 30:311

Blandford RD, Rees MJ. 1991. In *Testing the AGN Paradigm,* ed. S Holt, S Neff, C Urry, p. 3. Woodbury, NY:Am. Inst. Phys.

Blitz L, Spergel DN. 1991. *Ap. J.* 370:205

Blumenthal GR, Faber SM, Flores R, Primack JR. 1986. *Ap. J.* 301:27

Blumenthal GR, Faber SM, Primack JR, Rees MJ. 1984. *Nature* 311:517

Bond HE. 1981. *Ap. J.* 248:606

Bond JR, Arnett WD, Carr BJ. 1984. *Ap. J.* 280:825

Bond JR, Carr BJ. 1984. *MNRAS* 207:585

Bond JR, Carr BJ, Arnett WD. 1983. *Nature* 304:514

Bond JR, Carr BJ, Hogan CJ. 1986. *Ap. J.* 306:428

Bond JR, Carr BJ, Hogan CJ. 1991. *Ap. J.* 367:420

Bond JR, Efstathiou G, Silk J. 1980. *Phys. Rev. Lett.* 45:1980

Bosma A. 1991. In *Warped Disks and Inclined Rings Around Galaxies,,* ed. S Casertano, PD Sackett, F Briggs, p. 40. Provo:Brigham Young Univ. Press

Boughn SP, Saulson PR. 1983. *Ap. J. Lett.* 265:L55

Boughn SP, Saulson PR, Seldner M. 1981. *Ap. J. Lett.* 280:L15

Bristow P, Phillipps S. 1994. *MNRAS* 267:13

Boyle BL, Fong R, Shanks T. 1988. *MNRAS* 231:897

Burstein D, Haynes MP, Faber SM. 1991. *Nature* 353:515

Canizares C. 1981. *Nature* 291:620

Canizares CR. 1982. *Ap. J.* 263:508

Carignan C, Beaulieu S, Freeman KC. 1990. *Astron. J.* 99:178

Carignan C, Freeman KC. 1988. *Ap. J. Lett.* 332:L33

Carlberg RG, Dawson PC, Hsu T, VandenBerg DD. 1985. *Ap. J.* 294:674

Carr BJ. 1975. *Ap. J.* 201:1

Carr BJ. 1978. *Comm. Astrophys.* 7:161

Carr BJ. 1979. *MNRAS* 189:123

Carr BJ. 1981a. *MNRAS* 195:669

Carr BJ. 1981b. *MNRAS* 194:639

Carr BJ. 1990. *Comm. Astrophys.* 14:257

Carr BJ, Bond JR, Arnett WD. 1984. *Ap. J.* 277:445

Carr BJ, Ikeuchi S. 1985. *MNRAS* 213:497

Carr BJ, Lacey CG. 1987. *Ap. J.* 316:23

Carr BJ, Rees MJ. 1984. *MNRAS* 206:315

Carr BJ, Sakellariadou M. 1994. Preprint

Casali M, James P. 1994. *MNRAS* In press

Cayrel R. 1987. *Astron. Astrophys.* 168:81

Cen RY, Ostriker JP, Peebles PJE. 1993. *Ap. J.* 415:423

Cen RY, Ostriker JP, Spergel DN, Turok N. 1991. *Ap. J.* 383:1

Chandrasekhar S. 1964. *Ap. J.* 140:417

Chiba T, Sugiyama N, Suto Y. 1993. Preprint

Coles P, Ellis. GFR. 1994. *Nature* In press

Comeron F, Rieke GH, Burrows A, Rieke M. 1993. *Ap. J.* 416:185

Corrigan RT, Irwin MJ, Arnaud J, Fahlman GG, Fletcher JM, et al. 1991. *Astron. J.* 102:34

Crampton D, McClure RD, Fletcher JM, Hutchings JB. 1989. *Astron. J.* 98:1188

Crawford M, Schramm DN. 1982. *Nature*

298:538
Dalcanton J, Canizares C, Granados A, Steidel CC, Stocke JT. 1994. *Ap. J.* In Press
Daly RA, McLaughlin GC. 1992. *Ap. J.* 390:423
D'Antona F, Mazzitelli I. 1985. *Ap. J.* 296:502
de Araujo JCN, Opher R. 1990. *Ap. J.* 350:502
Dekel A, Rees MJ. 1987. *Nature* 326:455
De Rujula A, Jetzer Ph, Masso E. 1991. *MNRAS* 250:348
De Rujula A, Jetzer Ph, Masso E. 1992. *Astron. Astrophys.* 254:99
de Zeeuw PT. 1992. In *Morphological and Physical Classification of Galaxies*, ed. G Busarello, M Capacciolli, G Longo. Dordrecht:Kluwer
Disney M, Davies JI, Phillipps S. 1989. *MNRAS* 239:939
Dolgov A, Silk J. 1993. *Phys. Rev. D* 47:4244
Draine BT, Shapiro PR. 1989. *Ap. J. Lett.* 344:L45
Dressler A, Faber SM, Burstein D, Davies RL, Lynden-Bell D, Terlevich RL, Wegner G. 1987. *Ap. J. Lett.* 313:L37
Dubinski J, Carlberg RG. 1991. *Ap. J.* 378:496
Efstathiou G. Bond JR, White SDM. 1992. *MNRAS* 258:1
Eichler D, Silk J. 1992. *Science* 257:937
Ellis GFR. 1988. *Class. Quant. Grav.* 5:891
Ellis GFR, Lythe DH, Mijic MB. 1991. *Phys. Lett. B* 271:52
Evrard AE. 1990. *Ap. J.* 363:349
Faber SM, Gallagher JS. 1979. *Annu. Rev. Astron. Astrophys.* 17:135
Fabian AC. 1994. *Annu. Rev. Astron. Astrophys.* 32:277
Fabian AC, Nulsen PEJ, Canizares CR. 1984. *Nature* 310:733
Fabian AC, Thomas PA, Fall SM, White RE. 1986. *MNRAS* 221:1049
Fall SM, Efstathiou G. 1981. *MNRAS* 193:189
Fall SM, Pei YC, McMahon RG. 1989. *Ap. J. Lett.* 341:L5
Fall SM, Rees MJ. 1977. *MNRAS* 81:37P
Fall SM, Rees MJ. 1985. *Ap. J.* 298:18
Fich M, Tremaine S. 1991. *Annu. Rev. Astron. Astrophys.* 29:409
Flores R, Primack JR, Blumenthal GR, Faber SM. 1993. *Ap. J.* 412:443
Forman W, Jones C, Tucker W. 1985. *Ap. J.* 293:102
Fowler W. 1966. *Ap. J.* 144:180
Fowler W. 1990. See Lynden-Bell & Gilmore 1990, p. 257
Fowler W, Hoyle F. 1964. *Ap. J. Suppl.* 9:201
Frenk CS, White SDM, Davis M, Efstathiou G. 1988. *Ap. J.* 327:507
Fricke KJ. 1973. *Ap. J.* 183:941
Fuller GM, Woosley SE, Weaver TA. 1986. 307:675
Garrett MA, Calder RJ, Porcas RW, King LJ, Walsh D, Wilkinson PN. 1994. *MNRAS* In press
Gates E, Turner MS. 1994. *Phys. Rev. Lett.*

72:2520
Gilmore G, Hewett P. 1983. *Nature* 306:669
Gilmore G, Reid N. 1983. *MNRAS* 202:1025
Gilmore G, Reid N, Hewett P. 1985. *MNRAS* 213:257
Glatzel W, El Eid MF, Fricke KJ. 1985. *Astron. Astrophys.* 149:419
Gnedin NYu, Ostriker JP. 1992. *Ap. J.* 400:1
Gnedin NYu, Ostriker JP, Rees MJ. 1994. *Ap. J.* In press
Gomez AE, Delhaye J, Grenier S, Jaschek C, Arenou F, Jaschek M. 1990. *Astron. Astrophys.* 236:95
Gott JR. 1981. *Ap. J.* 243:140
Gouda N, Sugiyama N. 1992. *Ap. J. Lett.* 395:L59
Gould A. 1990. *MNRAS* 244:25
Gould A. 1992. *Ap. J. Lett.* 386:L5
Griest K. 1991. *Ap. J.* 366:412
Guth AH. 1981. *Phys. Rev. D* 23:347
Haehnelt MG. 1994. *MNRAS* In press
Hammer F, LeFevreo. 1990. *Ap. J.* 357:38
Harrison ER. 1993. *Ap. J. Lett.* 405:L1
Hartquist T, Cameron AGW. 1977. *Astrophys. Space Sci.* 48:145
Hatsukade I. 1989. PhD thesis. Miyazaki Univ.
Hauser MG, Kelsall T, Moseley SH, Silverberg RF, Murdock T, et al. 1991. In *After the First Three Minutes*, ed. S Holt, C Bennet, V Trimble, p. 161. Woodbury, NY: Am. Inst. Phys.
Hawking SW. 1971. *MNRAS* 152:75
Hawking SW. 1989. *Phys. Lett. B* 231:237
Hawking SW, Moss IG, Stewart JM. 1982. *Phys. Rev. D* 26:2681
Hawkins I, Wright EL. 1988. *Ap. J.* 324:46
Hawkins MRS. 1985. *Phil. Trans. R. Astron. Soc.* 320:553
Hawkins MRS. 1993. *Nature* 366:242
Hawkins MRS, Bessell MS. 1988. *MNRAS* 234:177
Heflin MB, Gorenstein MV, Lawrence CR, Burke BF. 1991. *Ap. J.* 378:519
Hegyi DJ, Kolb EW, Olive KA. 1986. *Ap. J.* 300:492
Hegyi DJ, Olive KA. 1983. *Phys. Lett. B.* 126:28
Hegyi DJ, Olive KA. 1986. *Astrophys. J.* 303:56
Hegyi DJ, Olive KA. 1989. *Astrophys. J.* 346:648
Heisler J, Ostriker JP. 1988. *Ap. J.* 332:543
Henstock DR, Wilkinson PN, Browne IWA, Patnaik AR, Taylor GB, et al. 1993. In *Gravitational Lenses in the Universe*, ed. J Surdej, D Fraipent-Caro, E Gosset, S Refsdal, M Remy, p. 325. Univ. Liege.
Hewett PC, Webster RL, Harding ME, Jedrezewski RI, Foltz CB. 1989. *Ap. J. Lett.* 346:L61
Hewitt JN. 1986. PhD thesis. Mass. Inst. Technol. Cambridge
Hewitt JN, Burke BF, Turner EL, Schneider DP, Lawrence CR, et al. 1989. In *Gravitational Lenses*, ed. JM Moran, JN Hewitt, KY Loh, p. 147. Berlin:Springer-Verlag

Hills JG. 1986. *Astron. J.* 92:595
Hogan CJ. 1978. *MNRAS* 185:889
Hogan CJ. 1993. *Ap. J. Lett.* 415:L63
Hogan CJ, Narayan R, White SDM. 1989. *Nature* 339:106
Hogan CJ, Rees MJ. 1988. *Phys. Lett. B* 205:228
Hoyle F, Wickramasinghe NC. 1989. *Astrophys. Space Sci.* 154:143
Hu EM, Huang JS, Gilmore G, Cowie LL. 1994. *Nature* In press
Huchra J, Brodie J. 1987. *Astron. J.* 93:779
Hut P, Rees MJ. 1992. *MNRAS* 259:27P
Ikeuchi S. 1981. *Publ. Astron. Soc. Jpn.* 33:211
Ipser JR, Price RH. 1977. *Ap. J.* 216:578
Ipser JR, Semenzato R. 1985. *Astron. Astrophys.* 149:408
Irwin MJ, Webster RL, Hewett PC, Corrigan RT, Jedrzewski RI. 1989. *Astron. J.* 98:1989
Jin L. 1989. *Ap. J.* 337:603
Jin L. 1990. *Ap. J.* 356:501
Kaiser N. 1984. *Ap. J. Lett.* 284:L9
Kajino T, Mathews GJ, Fuller GM. 1990. *Ap. J.* 364:7
Karimabadi H, Blitz L. 1984. *Ap. J.* 283:169
Kashlinsky A, Rees MJ. 1983. *MNRAS* 205:955
Kassiola A, Kovner I, Blandford RD. 1991. *Ap. J.* 381:6
Katz N, White SDM. 1993. *Ap. J.* 412:455
Kayser R, Refsdal S, Stabell R. 1986. *Astron. Astrophys.* 166:36
Kerins E. 1994. Preprint
Kerins E, Carr BJ. 1994. *MNRAS* 266:775
Knapp GR. 1988. In *The Mass of the Galaxy,* ed. M. Fich, p. 35. CITA Workshop
Kovner I. 1991. *Ap. J.* 376:70
Krauss LM, Small TA. 1991. *Ap. J.* 378:22
Kronberg PP, Dyer CC, Burbidge EM, Junkkarinen VT. 1991. *Ap. J. Lett.* 367:L1
Kroupa P, Tout CA, Gilmore G. 1993. *MNRAS* 262:545
Kuijken K, Gilmore G. 1989. *MNRAS* 239:571,605,651
Kuijken K, Gilmore G. 1991. *Ap. J. Lett.* 367:L9
Kurki-Suonio H, Matzner RA, Olive KA, Schramm DN. 1990. *Ap. J.* 353:406
Lacey CG. 1991. In *Dynamics of Disk Galaxies,* ed. B Sundelius, p 257. Göteborg
Lacey CG, Field G. 1988. *Ap. J. Lett.* 330:L1
Lacey CG, Ostriker JP. 1985. *Ap. J.* 299:633
Lake G. 1990. *Ap. J. Lett.* 356:L43
Lake G. 1992. In *Trends in Particle Astrophysics,* ed. D. Cline, R. Peccei. World Scientific
Lake G, Schommer RA, van Gorkom JH. 1990. *Astron. J.* 99:547
Lange AE, Richards PL, Hayakawa S, Matsumoto T, Matsuo H, et al. 1991. Preprint
Lanzetta KM, Wolf AM, Turnshek DA, Lu LM, McMahon RG, Hazard C. 1991. *Ap. J. Suppl.* 77:1
Larson R. 1986. *MNRAS* 218:409
Latham DW, Tonry J, Bahcall JN, Soneira RM, Schechter PS. 1984. *Ap. J. Lett.* 281:L41

Layzer D, Hively RM. 1973. *Ap. J.* 179:361
Lewis GF, Miralda-Escude J, Richardson DC, Wambsganss J. 1993. *MNRAS* 261:647
Lin DNC, Faber SM. 1983. *Ap. J. Lett.* 266:L21
Linder EV, Schneider P, Wagoner RV. 1988. *Ap. J.* 324:786
Loeb A. 1993. *Ap. J.* 403:542
Loeb A, Rasio F. 1993. Preprint
Low FJ. 1986. In *Astrophysics of Brown Dwarfs,* ed. MC Kafatos, RS Harrington, SP Maran, p. 66. Cambridge:Cambridge Univ. Press
Low C, Lynden-Bell D. 1976. *MNRAS* 176:367
Lynden-Bell D, Gilmore G. 1990. *Baryonic Dark Matter.* Dordrecht:Kluwer
Maeder A. 1992 *Astron. Astrophys.* 264:105
Malaney RA, Fowler WA. 1988. *Ap. J.* 333:14
Malaney P, Mathews G. 1993. *Phys. Rep.* 229:147
Mao S. 1992. *Ap. J. Lett.* 389:L41
Matsumoto T, Akiba M, Murakami H. 1988b. *Ap. J.* 332:575
Matsumoto T, Hayakawa S, Matsuo H, Murakami H, Sato S, et al. 1988a. *Ap. J.* 329:567
Mather JC, Cheng ES, Eplee RE, Isaacman RB, Meyer SS, et al. 1990. *Ap. J. Lett.* 354:L37
Mather JC, Cheng ES, Cottingham DA, Eplee RE, Fixen DJ, et al. 1994. *Ap. J.* 420:439
Mathews GJ Schramm DN, Meyer BS. 1993. *Ap. J.* 404:476
Mateo M. Olszewski E, Welch DL, Fischer P, Kunkel W. 1991. *Astron. J.* 102:914
McDonald JM, Clarke CJ. 1993. *MNRAS* 262:800
McDowell JC. 1986. *MNRAS* 223:763
McGaugh S. 1994. *Nature* 367:538
Meszaros P. 1975. *Astron. Astrophys.* 38:5
Moore B. 1993. *Ap. J. Lett.* 413:L93
Moore B, Silk J. 1994. Preprint
Mould JR, Oke JB, de Zeeuw PT, Nemec JM. 1990. *Astron. J.* 245:454
Mulchaey JS, Davis DS, Mushotzky RF, Burstein D. 1993. *Ap. J. Lett.* 404:L9
Najita J, Silk J, Wachter JW. 1990. *Ap. J.* 348:383
Narayan R, Wallington S. 1992. *Ap. J.* 399:368
Negroponte J. 1986. *MNRAS* 222:19
Nelson LA, Rappaport SA, Joss PC. 1993 *Ap. J.* 404:723
Nemiroff RJ. 1989. *Ap. J.* 341:579
Nemiroff RJ. 1991a. *Comm. Astrophys.* 15:139
Nemiroff RJ. 1991b. *Phys. Rev. Lett.* 66:538
Nemiroff RJ, Bistolas V. 1990. *Ap. J.* 358:5
Nemiroff RJ, Norris JP, Wickramasinghe WADT, Horack JM, Kouveliotou C, et al. 1993. *Ap. J.* 414:36
Noda M, Christov VV, Matsuhara H, Matsumoto T, Matsuura S, et al. 1992. *Ap. J.* 381:456
Nottale L. 1986. *Astron. Astrophys.* 157:383
Nusser A, Dekel A. 1993. *Ap. J.* 405:437
Ober WW, El Eid MF, Fricke KL. 1983. *Astron. Astrophys.* 119:61
Oliver S, Rowan-Robison M, Saunders W. 1992.

MNRAS 256:15P

Oort J. 1932. *Bull. Astron. Inst. Neth.* 6:249

Ostriker JP, Cowie LL. 1981. *Ap. J. Lett.* 243:L127

Ostriker JP, Heisler J. 1984. *Ap. J.* 278:1

Ostriker JP, Binney J, Saha P. 1989. *MNRAS* 241:849

Ostriker JP, Peebles PJE, Yahil A. 1974. *Ap. J. Lett.* 193:L1

Ostriker JP, Vogeley MS, York DG. 1990. *Ap. J.* 364:405

Paczynski B. 1986. *Ap. J.* 304:1; 308:L43

Paczynski B. 1987. *Ap. J. Lett.* 317:L51

Pagel BEJ. 1987. In *The Galaxy,* ed. G Gilmore, B Carswell, p. 341. Dordrecht:Reidel

Pagel BEJ. 1990. See Lynden-Bell & Gilmore 1990, p. 237

Pagel BEJ. 1992. In *The Stellar Populations of Galaxies,* ed. B Barbuy, A Renzini. Dordrecht:Kluwer

Palla F, Salpeter EE, Stahler SW. 1983. *Ap. J.* 271:632

Patnaik AR, Browne IWA, Wilkinson PA, Wrobel JM. 1992. *MNRAS* 254:655

Peacock J. 1982. *MNRAS* 199:987

Peacock J. 1986. *MNRAS* 223:113

Peacock J, Dodds SJ. 1994. *MNRAS.* In Press

Peebles PJE. 1987a. *Nature* 237:210

Peebles PJE. 1987b. *Ap. J. Lett.* 315:L73

Peebles PJE, Dicke RH. 1968. *Ap. J.* 154:891

Peebles PJE, Partridge RB. 1967. *Ap. J.* 148:713

Persic M, Salucci P. 1990. *MNRAS* 245:257

Persic M, Salucci P. 1992. *MNRAS* 258:14P

Pfenniger D, Combes F, Martinet L. 1994. *Astron. Astrophys.* 285:79

Phinney ES. 1985. Preprint

Polnarev AG, Zemboricz R. 1988. *Phys. Rev. D* 43:1106

Ponman TJ, Bertram D. 1993. *Nature* 363:51

Press WH, Gunn JE. 1973. *Ap. J.* 185:397

Ramadurai S, Rees MJ. 1985. *MNRAS* 215:53P

Rana NC. 1991. *Annu. Rev. Astron. Astrophys.* 29:129

Rauch KP. 1991. *Ap. J.* 374:83

Rees MJ. 1976. *MNRAS* 176:483

Rees MJ. 1978. *Nature* 275:35

Rees MJ. 1984. In *Formation and Evolution of Galaxies and Large Structures in the Universe,* ed. J Audouze, J Tran Thanh Van, p. 271. Dordrecht:Reidel

Rees MJ. 1986. *MNRAS* 218:25P

Rees MJ, Ostriker JP. 1977. *MNRAS* 179:541

Refsdal S, Stabell R. 1991. *Astron. Astrophys.* 250:62

Reid N, Gilmore G. 1982. *MNRAS* 201:73

Richer HB, Fahlman GG. 1992. *Nature* 358:383

Richer HB, Fahlman GG, Buonanno R, Pecci FF, Searle L, Thompson IB. 1991. *Ap. J.* 381:147

Richstone D, Gould A, Guhathakurta P, Flynn C. 1992. *Ap. J.* 388:354

Rix H, Hogan CJ. 1988. *Ap. J.* 332:108

Rix H, Lake G. 1993. *Ap. J.* 417:1

Rodrigues-Williams LL, Hogan CJ. 1994. *Astron. J.* 107:451

Rowan-Robinson M. 1986. *MNRAS* 219:737

Rowan-Robinson M, Lawrence A, Saunders W, Crawford J, Ellis R, et al. 1990. *MNRAS* 247:1

Rubin VC, Ford WK, Thonnard N. 1980. *Ap. J.* 238:471

Ryu D, Olive KA, Silk J. 1990. *Ap. J.* 353:81

Sackett PD, Sparke LS. 1990. *Ap. J.* 361:408

Sakellariadou M. 1984. MS thesis. Cambridge Univ.

Salpeter EE. 1993. *Phys. Rep.* 227:309

Salpeter EE, Wasserman IM. 1993. Astron. Soc. Pac. 36:345.

Sancisi R, van Albada TS. 1987. In *Dark Matter in the Universe,* ed. J Kormendy, G Knapp, p. 67. Dordrecht:Reidel

Sarazin C. 1986. *Rev. Mod. Phys.* 58:1

Sargent WLW, Steidel CC. 1990. See Lynden-Bell & Gilmore 1990, p. 223

Saslaw WC, Narashima D, Chitre SM. 1985. *Ap. J.* 392:348

Saslaw WC, Valtonen MJ, Aarseth SJ. 1974. *Ap. J.* 190:253

Scaramella R, Bettolani G, Zamorani G. 1991. *Ap. J. Lett.* 376:L1

Scherrer RJ. 1993. *Ap. J.* 384:391

Schild H, Maeder A. 1985. *Astron. Astrophys. Lett.* 143:L7

Schneider DP, Schmidt M, Gunn JE. 1991. *Astron. J.* 102:837

Schneider P. 1993. *Astron. Astrophys.* 279:1

Schneider P, Ehlers J, Falco EE. 1992. In *Gravitational Lensing,* ed. R Kayser, T Schramm, L Nieser. Berlin:Springer-Verlag

Schneider P, Wagoner RV. 1987. *Ap. J.* 314:154

Schneider P, Weiss A. 1987. *Astron. Astrophys.* 171:49

Sellgren K, McGinn MT, Becklin EE, Hall DNB. 1990. *Ap. J.* 359:112

Shanks T. 1985. *Vistas Astron.* 28:595

Shu F, Adams F, Lizano S. 1987. *Annu. Rev. Astron. Astrophys.* 25:23

Silk J. 1968. *Ap. J.* 151:459

Silk J. 1977. *Ap. J.* 211:638

Silk J. 1991. *Science* 251:317

Silk J. 1992. In *The Stellar Populations of Galaxies,* ed. B Barbuy, A Renzini, p. 15. Dordrecht:Kluwer

Silk J. 1993. *Phys. Rep.* 227:143

Silk J, Stebbins A. 1993. *Ap. J.* 411:439

Simons DA, Becklin EE. 1992. *Ap. J.* 390:431.

Skrutskie MF, Shure M, Beckwith S. 1985. *Ap. J.* 299:303

Smecker TA, Wyse RFG. 1991. *Ap. J.* 372:448

Smoot GF, Bennett CL, Kogut A, Wright EL, Aymon J, et al. 1992. *Ap. J. Lett.* 396:L1

Speanhauer A, Jones BF, Whitford AE. 1992. *Astron. J.* 103:297

Spitzer L. 1958. *Ap. J.* 127:17

Steele IA, Jameson RF, Hambly NC. 1993. *MNRAS* 263:467

Steidel CC, Sargent WLW. 1988. *Ap. J. Lett.*

333:L5

Stevenson DJ. 1986. In *Astrophysics of Brown Dwarfs,* ed. MC Kafatos, RS Harrington, SP Maran, p. 66. Cambridge:Cambridge Univ. Press

Stevenson DJ. 1991. *Annu. Rev. Astron. Astrophys.* 29:163

Stromgren B. 1987. In *The Galaxy,* ed. G Gilmore, R Carswell, p. 229. Dordrecht:Reidel

Subramanian K, Chitre SM. 1987. *Ap. J.* 313:13

Suntzeff NB. 1992. In *The Stellar Populations of Galaxies,* ed. B Barbuy, A Renzini, p. 15. Dordrecht:Kluwer

Surdej J, Claeskens JF, Crampton D, Filippenko AV, Hutsemekers D, et al. 1993. *Astron. J.* 105:2064

Tamanaha CM, Silk J, Wood MA, Winget DE. 1990. *Ap. J.* 358:164

Taylor AN, Rowan-Robinson M. 1992. *Nature* 359:396

Terlevich RJ. 1985. In *Star Forming Dwarf Galaxies,* ed. D Kunth, TX Thuan, J Tran Thanh Van, p. 395. Paris:Editions Frontieres

Teuben PJ. 1991. In *Warped Disks and Inclined Rings Around Galaxies,* ed. S Casertano, PD Sackett, F Briggs, p. 40. Provo:Brigham Young Univ. Press

Thomas P, Fabian AC. 1990. *MNRAS* 246:156

Thorstensen JR, Partridge RB. 1975. *Ap. J.* 200:527

Tinney CG. 1993. *Ap. J.* 414:379

Tinney CG, Mould JR, Reid IN. 1992. *Ap. J.* 396:173

Tinney CG, Reid IN, Mould JR. 1993. *Ap. J.* 414:254

Tohline JE. 1980. *Ap. J.* 239:417

Trimble V. 1987. *Annu. Rev. Astron. Astrophys.* 25:425

Truran JW. 1984. *Annu. Rev. Nucl. Part. Phys.* 34:53

Truran JW, Cameron AGW. 1971. *Astrophys. Space Sci.* 14:179

Turner EL. 1980. *Ap. J. Lett.* 242:L135

Turner EL, Wardle MJ, Schneider DP. 1990. *Astron. J.* 100:146

Turner MS. 1991. *Phys. Scr.* T36:167

Udalski A, Szymanski M, Kaluzny J, Kubiak M, Krzeminski W, et al. 1993. *Acta Astron.* 43:289

Umemura M, Loeb A, Turner EL. 1993. *Ap. J.* 419:459

Valentijn EA. 1990. *Nature* 346:153

Van den Bergh S. 1969. *Nature* 224:891

Van der Kruit PC. 1987. In *Dark Matter in the Universe,* ed. GR Knapp, J Kormendy, p. 415. Dordrecht:Reidel

Vietri M, Ostriker JP. 1983. *Ap. J.* 267:488

Walker T, Steigman G Schramm DN, Olive KA, Kang HS. 1991. *Ap. J.* 376:51

Wambsganss J, Paczynski B. 1992. *Ap. J. Lett.* 397:L1

Wambsganss J, Paczynski B, Schneider P. 1990. *Ap. J. Lett.* 358:L33

Wandel A. 1985. *Ap. J.* 294:385

Wasserman I, Weinberg MD. 1987. *Ap. J.* 312:390

Wasserman I, Salpeter EE. 1993. Preprint

Webster RL, Fitchett M. 1986. *Nature* 324:617

Webster RL, Hewett PC, Harding ME, Wegner GA. 1988. *Nature* 336:358

Webster RL, Ferguson AMN, Corrigan RT, Irwin MJ. 1991. *Astron. J.* 102:1939

Wheeler JC, Sneden. C, Truran J. 1989. *Annu. Rev. Astron. Astrophys.* 27:279

White DA, Fabian AC, Johnstone RM, Mushotzky RF, Arnaud KA. 1991. *MNRAS* 252:72

White SDM. 1976. *MNRAS* 174:19

White SDM, Frenk CS. 1991. *Ap. J.* 379:52

White SDM, Navarro JF, Evrard AE, Frenk CS. 1993. *Nature* 366:429

White SDM, Rees MJ. 1978. *MNRAS* 183:341

Whitmore BC, McElroy DB, Schweizer F. 1987. *Ap. J.* 314:439

Wielen R. 1985. In *Dynamics of Star Clusters,* ed. J Goodwin, P Hut, p. 449. Dordrecht:Reidel

Wollmann ER. 1992. *Ap. J.* 392:80

Woosley SE, Weaver TA. 1982. In *Supernovae: A Survey of Current Research,* ed. MJ Rees, RJ Stoneham, p. 79. Dordrecht:Reidel

Wright EL. 1982. *Ap. J.* 255:401

Wright EL, Malkan MA. 1987. *Bull. Am. Astron. Soc.* 19:699

Wright EL, Mather JC, Fixen DJ, Kogut A, Schafer RA. 1994. *Ap. J.* 420:450

Yang J, Turner MS, Steigman G, Schramm DN, Olive K. 1984. *Ap. J.* 281:493

Yoshii Y, Saio H. 1986. *Ap. J.* 301:587

Zaritsky D, Smith R, Frenk CS, White SDM. 1993. *Ap. J.* 405:464

Zeldovich YaB. 1970. *Astron. Astrophys.* 5:84

Annu. Rev. Astron. Astrophys. 1994. 32: 591–639
Copyright © 1994 by Annual Reviews Inc. All rights reserved

BINARY AND MILLISECOND PULSARS

E. S. Phinney

Theoretical Astrophysics, 130-33, California Institute of Technology, Pasadena, California 91125

S. R. Kulkarni

Department of Astronomy, 105-24, California Institute of Technology, Pasadena, California 91125

KEY WORDS: radio pulsars, X-ray binaries, binary stars, neutron stars, white dwarfs

1. INTRODUCTION

Most of the ~600 known pulsars are single and located in the disk of our Galaxy. There is circumstantial evidence that the pulsars in this majority are created in supernova (SN) explosions, by the collapse of the cores of massive stars (initial mass $M_i \gtrsim M_{cr} \simeq 8\,M_\odot$). One is created roughly every 100 y in the Galaxy.

Figure 1 is a plot of the pulse period (P) versus the dipole field strength (B, inferred from the observed P and $\dot{P}$, and assuming a vacuum dipole model) for the 545 Galactic pulsars for which such measurements are available (cf Taylor et al 1993). Like the color-magnitude diagram for stars, this B-P diagram offers a convenient graphic representation on which to trace the evolution of pulsars. Young pulsars—those associated with supernova remnants (SNRs)—appear to be born with reasonably small periods, $P \lesssim 0.1$ s, and strong magnetic field strengths, $1 \lesssim B_{12} \lesssim 10$ where $B = 10^{12} B_{12}$ G is the inferred dipole field strength. Pulsars slow down as they age and thus move to the right in this diagram and cease emitting in the radio band as they approach the so-called death line (Figure 1).

The scale height of pulsars is much larger than that of their progenitors, the massive stars. Direct interferometric measurements have established that young

591

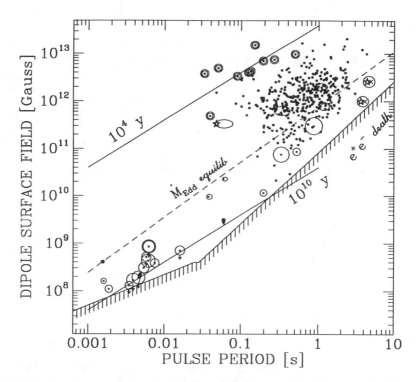

Figure 1 Plot of pulse period vs dipole field strength, for Galactic and Magellanic Cloud radio pulsars. Dipole field estimated by assuming energy loss from a vacuum magnetic dipole, $B^2 = 10^{39} P \dot{P}$ (P in s, B in G). Small points are single disk pulsars. Those surrounded by double circles are in supernova remnants. Binary pulsars lie at the left focus of ellipses with the orbital eccentricity, and semimajor axis proportional to $\log(P_b/0.01 \text{ d})$. The left focus is marked with a dot when the pulsar's companion is a white dwarf or neutron star, and by a 5-pointed star when it is an optically detected B-type star. The solid lines show where the characteristic age $\tau_c = P/2\dot{P}$ has the indicated values; the lower one shows an age equal to that of the Galaxy. Pulsars born with short period and evolving with constant dipole field must lie to the left of the line, and if there were no luminosity evolution, a majority would lie close to the line. The dashed line is the standard Eddington "Rebirth" line (cf Ghosh & Lamb 1992), specified by Equation (2.1). The shaded boundary line is the "death line" discussed in Section 6.2. Note the absence of radio pulsars to the right of the death line. The two binaries near the death line, with × s at the right focus, are not radio pulsars, but the two accreting X-ray pulsars with known orbital periods and magnetic fields determined from X-ray cyclotron harmonics, X0331+53 and X0115+634. If their B-star companions had lower rates of mass loss, these objects would not have spun down, and would have been radio pulsars. The location of the rebirth and death lines depends on the assumed magnetic topology, and is also subject to some physical uncertainty (see text).

single pulsars have large spatial motion with a median 3-dimensional velocity of ~ 400 km s^{-1}. The large velocity combined with the finite lifetime set by the death line (Figure 1), offer a first-order explanation for the scale height of pulsars. Considerable circumstantial evidence indicates that pulsars acquire a velocity kick of order ~ 100–600 km s^{-1} at their birth (Section 4).

Pulsars with short spin periods and low magnetic field strengths form a distinct group (Figure 1 and Table 1). The high abundance of binaries in this group indicates that binarity has played an important role in its formation. It is now being appreciated that this group of pulsars has a steady state population approximately *equal* to that of active ordinary radio pulsars (Section 3). The precision of their rotational clocks, and their close companions, allow them to be used for many remarkable experiments ranging from fundamental physics (nuclear equations of state, general relativity), to applied physics (Raman scattering of high-power microwaves in plasma), to astrophysics (planet formation, neutron star magnetospheres and winds, dynamical evolution of globular clusters).

Recently, a large number of pulsars with characteristics similar to the Galactic millisecond and binary pulsars have been discovered in globular clusters. Space limitations prevent us from including here any discussion of these objects, their formation mechanisms, and the remarkable inferences about globular cluster dynamics and evolution that they have made possible. The reader is referred to Phinney (1992, 1993), Manchester (1992), and Phinney & Kulkarni (1994) for a review.

We therefore concentrate on the Galactic millisecond and binary pulsars. Our division of labor assigned Sections 2–5 to SRK, and Sections 6–11, tables, and figures to ESP. Other useful review articles include those by Srinivasan (1989), Verbunt (1990), Bhattacharya & van den Heuvel (1991), and Lamb (1992), and those collected in Lewin et al (1994). Related recent Annual Review articles are by Verbunt (1993), Chanmugam (1992), and Canal et al (1990). Useful conference proceedings have been edited by Ögelman & van den Heuvel (1989) and van den Heuvel & Rappaport (1992).

2.   BACKGROUND AND FRAMEWORK

The binary pulsars in Table 1 are broadly divided into two categories: high mass binary pulsars (HMBPs; the upper two groups in the table) and low mass binary pulsars (LMBPs; the lower four groups). The few isolated pulsars present in Table 1 have been classified as follows: $P < 30$ ms, LMBPs; HMBPs, otherwise. The rationale for this is explained below. We now summarize our current understanding of the origin of these systems. The reader is referred to Bhattacharya & van den Heuvel (1991), Verbunt (1993), and van den Heuvel & Rappaport (1992) for more extensive reviews.

Table 1   Binary and millisecond pulsars in the Galaxy[†]

Pulsar	P (ms)	P_b (d)	e^{a}	$f(M)^{b}$ ($M_\odot$)	$M_2{}^{c}$ ($M_\odot$)	$\log(B)^{d}$ (G)	$P/(2\dot{P})^{e}$ (y)	Ref
J0045−7319	926.3	51	0.808	2.169	~10	12.3	3×10^6	1
1259−63	47.8	1237[E]	0.870	1.53	~10	11.5	3×10^5	2
1820−11	279.8	358	0.794	0.068	(0.8)	11.8	3×10^6	3
1534+12	37.9	0.42	0.274	0.315	1.34	10.0	2×10^8	4
1913+16	59.0	0.32	0.617	0.132	1.39	10.4	1×10^8	5
2303+46	1066.4	12.3	0.658	0.246	1.4	11.9	3×10^7	6
J2145−0750	16.0	6.8	0.000021	0.0241	$(0.51)^{O}$	<8.9	$> 8 \times 10^9$	7
0655+64	195.7	1.03	7×10^{-6}	0.071	$(0.8)^{O}$	10.1	5×10^9	8
0820+02	864.8	1232	0.0119	0.0030	$(0.23)^{O}$	11.5	1×10^8	9
J1803−2712	334	407	0.00051	0.0013	(0.17)	10.9	3×10^8	10
1953+29	6.1	117	0.00033	0.0024	(0.21)	8.6	3×10^9	11
J2019+2425	3.9	76.5	0.000111	0.0107	(0.37)	8.3	(1×10^{10})	12
J1713+0747	4.6	67.8	0.000075	0.0079	$(0.33)^{O}$	8.3	(9×10^9)	13
1855+09	5.4	12.3	0.000022	0.0056	0.26^{O}	8.5	5×10^9	14
J0437−4715	5.8	5.7	0.000018	0.0012	$(0.17)^{O}$	8.7	(2×10^9)	15
J1045−4509	7.5	4.1	0.000019	0.00177	(0.19)	8.6	6×10^9	7
J2317+1439	3.4	2.46	<0.000002	0.0022	(0.21)	8.1	(1×10^{10})	16
J0034−0534	1.9	1.6	<0.0001	0.0012	(0.17)	8.0	4×10^9	7
J0751+18	3.5	0.26	<0.01	(0.15)				17
1718−19	1004	0.26[E]	<0.005	0.00071	(0.14)	12.2	1×10^7	18
1831−00	520.9	1.8	<0.004	0.00012	(0.07)	10.9	6×10^8	9
1957+20	1.6	0.38[E]	$<4 \times 10^{-5}$	5×10^{-6}	0.02^{O}	8.1	2×10^9	19
1257+12	6.2	67,98	0.02,0.02	$5, 3 \times 10^{-16}$	$4,3 M_\oplus$	8.9	(8×10^8)	20
1937+21	1.6	single				8.6	2×10^8	21
J2235+1506	59.8	single				9.5	(6×10^9)	16
J2322+2057	4.8	single				8.3	(1×10^{10})	12

[a]Orbital eccentricity.

[b]Mass function $f(M_{psr}, M_2) = (M_2 \sin i)^3 (M_2 + M_{psr})^{-2}$.

[c]Mass of pulsar's companion (when in parentheses, tabulated value of M_2 is estimated from $f(M)$, assuming a pulsar mass of $1.4\,M_\odot$ and inclination $i = 60°$, the median for randomly oriented binaries).

[d]The value of B given is the dipole surface field, calculated as if the pulsar were an orthogonal vacuum rotator. Higher multipoles could be much stronger.

[e]Characteristic age enclosed in () when $\dot{P}/P$ is dominated by $V_\perp^2/(cD)$ centrifugal acceleration contribution (see Equation 8.1).

[E]Pulsar eclipsed when behind companion (or its wind).

[O]Optical radiation from the white dwarf companion is observed; its temperature combined with the theory of white dwarf cooling roughly confirm the age estimated from $P/(2\dot{P})$ (except for 1957+20's companion, which is heated by the pulsar's relativistic wind).

References: 1. Kaspi et al 1994a, 2. Johnston et al 1994, 3. Lyne & McKenna 1989, 4. Wolszczan 1991, 5. Damour & Taylor 1991, 6. Thorsett et al 1993b, 7. Bailes et al 1994,

2.1 *HMBPs*

HMBPs are believed to originate from massive binary systems, in which the primary first forms a neutron star. As the secondary evolves, its stellar wind or Roche lobe overflow feeds matter to the neutron star. The strong magnetic field of the young neutron star funnels the accreted matter to the polar cap, which gives rise to pulsed X-ray emission. This phase is identified with Massive X-ray Binaries (MXRBs). If the mass of the secondary is above M_{cr}, it will explode as a supernova. This will generally unbind the system, but a few combinations of recoil velocities and masses can leave two neutron stars bound in an eccentric orbit. Lower-mass secondaries evolve to a white dwarf in a circular orbit around a spun-up pulsar.

Matter that flows to the neutron star settles into an accretion disk, owing to its angular momentum. The accretion disk is terminated at the Alfvén radius— the radius at which the inward accretion ram pressure is balanced by magnetic pressure $B^2/8\pi$. The neutron star gets spun up until its rotation rate equals that of the Keplerian rotation rate at the inner edge of the accretion disk. This

---

Table 1 (*continued*)

8. Jones & Lyne 1988, 9. Taylor & Dewey 1988, 10. Taylor et al 1993, 11. Rawley et al 1988, 12. Nice et al 1993, 13. Foster et al 1993, 14. Kaspi et al 1994, 15. Johnston et al 1992, 16. Camilo et al 1993, 17. Lundgren 1994, 18. Lyne et al 1993, 19. Ryba & Taylor 1991, 20. Wolszczan & Frail 1992, 21. Thorsett & Phillips 1992.

†The horizontal lines divide the pulsars into six groups, depending on their probable evolutionary history. The *first* group of pulsars are antediluvian in the sense defined by Phinney & Verbunt (1991): The neutron stars have probably not yet accreted from their companions, and the orbits are eccentric. PSRs 1259–63 and J0045–7319 (in the Small Magellanic Cloud) have visible B stars as companions; 1820–11's companion may be lower main-sequence (Phinney & Verbunt 1991). The remaining pulsars are postdiluvian. For the *second* group, limits on the masses and sizes of the companions suggest that the companions are neutron stars. The short orbital periods and high eccentricities suggest that the pulsar and companion spiraled together in a common envelope, after which a second supernova created the second neutron star. Both stars were initially massive $>8\ M_\odot$. Pulsars in the *third* group have high-mass white dwarf companions, which must have formed in a massive red giant much larger than the current orbit, which implies that the neutron star spiraled into, and ejected the giant's envelope during unstable mass transfer. The remnant core was not massive enough to create a neutron star (as in the second group of pulsars), so a massive white dwarf was left. The *fourth* group has low-mass white dwarf companions in circular orbits, which could have formed in stable accretion from low-mass companion stars filling their Roche lobes. The residual masses and eccentricities are discussed in Section 10. The *fifth* group consists of systems with companions less massive than the core mass ($\sim 0.16 M_\odot$) of the least massive star to have evolved off the main sequence in the age of the universe. Their mass transfer must have been driven by something other than nuclear evolution of the companion, e.g. gravitational radiation, or loss of angular momentum in a magnetic wind. PSR $1718 - 19$ is located 2.3′ (about 1/4 of the tidal radius) from the core collapsed globular cluster NGC 6342, and might be a cluster pulsar, not a Galactic pulsar. In the *sixth* group, 1257+12 seems to have a planetary system; the other pulsars are single. These pulsars seem to have destroyed the stellar companions that provided the angular momentum to spin them up.

gives rise to the rebirth or "spin-up" line (Figure 1; see Ghosh & Lamb 1992 for subtleties):

$$P_{eq} = 1.3 \left(\frac{B}{10^{12}\,\text{G}} \right)^{6/7} \left(\frac{\dot{M}}{\dot{M}_{Edd}} \right)^{-3/7} \text{s}, \tag{2.1}$$

where $\dot{M}$ is the accretion rate and $\dot{M}_{Edd} \sim 2 \times 10^{-8} M_\odot\,\text{y}^{-1}$ is the Eddington accretion rate. The pulse periods of X-ray pulsars range from ~ 1 s to 10^3 s (White et al 1994). Detection of cyclotron line features in several of these sources directly confirm the existence of high magnetic field strengths, $0.5 \times 10^{12} \lesssim B \lesssim 5 \times 10^{12}\,\text{G}$ (see Figure 1).

2.2 LMBPs

There is considerable similarity between LMBPs and cataclysmic variables (CVs). Both contain a degenerate object whose progenitor at one time occupied a large volume, yet many CVs and LMBPs have orbital separations that are less than a fraction of the radius of the progenitor giant star. As with CVs, we believe that LMBPs underwent a common envelope (CE) phase during which the secondary was engulfed by the bloated primary and the two stars quickly spiraled closer, ejecting the primary's envelope in the process.

A stellar system forming an LMBP must overcome two obstacles that are not faced by systems forming CVs. First, the formation of a neutron star is accompanied by copious mass loss, potentially from $\gtrsim M_{cr}$ down to $M_n \sim 1.4 M_\odot$, the measured mass of neutron stars. Thus the system can be expected to unbind if the mass lost suddenly exceeds half the total pre-SN mass of the system. Second, natal velocity kicks (Section 4.2) add a further tendency to disrupt the binary. The magnetic field strengths and spin periods of the neutron stars in LMBPs are anomalously low compared to ordinary pulsars and HMBPs. Some mechanism is needed to reduce the magnetic field strength. Three scenarios which have been suggested (see Webbink 1992) are summarized below.

2.2.1 THE RECYCLED PULSAR MODEL   In this model (see Verbunt 1993, Bailes 1989), the progenitors of LMBPs are binaries with a primary mass (M_1) above M_{cr} and a secondary mass below M_{cr}, $M_2 \lesssim 1 M_\odot$. The primary evolves and expands; unstable mass transfer then leads to spiral-in of the secondary. A compact binary is thus formed, consisting of the secondary and the evolved He core of the primary which eventually explodes and leaves behind a neutron star. In some cases, a second spiral-in may occur during the late stages of the secondary's evolution, thus producing even tighter binaries. During the evolution of the secondary, mass is transferred from the secondary to the primary (a neutron star), which is spun up to millisecond periods. Mass transfer from the companion may be driven either by nuclear evolution (for orbital period,

$P_b \gtrsim 1$ d) or gravitational radiation aided by magnetic braking (for $P_b \lesssim 1$ d). This transfer phase circularizes the orbit, and is identified with the Low Mass X-ray Binaries (LMXBs), of which there are $\lesssim 10^2$ in the Galaxy. Unlike the MXRBs, LMXBs do not exhibit pulsed X-ray emission. This suggests that their neutron stars are weakly magnetized, and makes them plausible progenitors of the millisecond pulsars [cf Equation (2.1) and the "rebirth line" in Figure 1].

This scenario can explain the Population I origin of two compact X-ray pulsars with very low mass companions: 1E 2259 + 586 and 4U 1626 − 67. In particular, the model gives a satisfactory explanation for the location of 1E 2259 + 586 in an SNR located in a star forming region (Iwasawa et al 1992). In binaries, the primary star could have lost its H envelope to the companion; we are then left with a He core of much smaller mass. The minimum mass of a He core that will still undergo collapse is still controversial but is expected to be in the range 2.2–4 $M_\odot$ (Habets 1985, Bhattacharya & van den Heuvel 1991), precariously close to our limit. The SN mass loss is symmetric in the frame of the exploding star but not in the center-of-mass frame. Thus assuming that the system survives the SN explosion, the binary system acquires a systemic speed that is a sizable fraction of the orbital speed. As pointed out by Bailes (1989), suitably directed natal kicks can actually *stabilize* the system even if the mass loss exceeds half the total initial mass. A consequence is that this model predicts significant velocities for both LMXBs and LMBPs. This effect may have already been observed (Section 4.4).

There are two potential problems with the recycling model. 1. One must assume that magnetic fields do not decay unless there is accretion (see Section 5). This hypothesis remains to be proven. 2. The expected millisecond pulsations from LMXBs have not been detected. Statistical studies of the birthrates of LMXBs and LMBPs suggest a possible discrepancy (Section 3), which suggests that the observed LMXBs are not the only progenitors of LMBPs.

2.2.2   TRIPLE STAR EVOLUTION   In spirit, triple star evolution is similar to the recycled model except it does not appeal to velocity kicks to stabilize the system. Instead, a massive compact binary with a distant low mass tertiary is invoked (Eggleton & Verbunt 1986). Following the first SN and the evolution of the secondary, the neutron star spirals into the evolved secondary and sinks to the center. The binary is transformed into a red supergiant with a neutron star at the core—a Thorne-Żytkow object (Thorne & Żytkow 1977, Biehle 1994). The tertiary then undergoes a common envelope evolution to form an LMXB. From this point on, the evolutionary path is similar to the recycled model.

2.2.3   ACCRETION INDUCED COLLAPSE (AIC)   In the accretion induced collapse model, one assumes that the progenitor is an accreting white dwarf which is transmuted to a neutron star once its mass exceeds the Chandrashekar limit (see Canal et al 1990). There are a number of variants: neutron stars may be assumed to be born with or without natal kicks, and with or without low magnetic field strengths. One advantage of AIC without natal kicks is that very little mass is lost during the SNe (essentially the binding energy of the neutron star, $\sim 0.2\, M_{\odot}$). Thus all such systems survive the SNe. By the same token, we do not expect to see substantial systemic motion. However, the velocity data (Section 4) indicate that LMBPs have substantial motions, suggesting that neutron stars in LMBPs suffer velocity kicks for one reason or the other, removing the survival advantage of AIC. This leaves us two variants: The neutron star is born as a millisecond pulsar (Michel 1987) or as an ordinary high field pulsar. The latter model suffers from the same two problems as the recycled model and hence we ignore it.

The physics of the AIC mechanism is not well understood. Does the white dwarf explode (a popular model for Type Ia SNe), or implode to form a millisecond pulsar? Assuming implosion, why are the initial B and P values so much smaller than those of neutron stars formed via Type II SNe? Under what conditions does a white dwarf accrete matter as opposed to ejecting it via thermonuclear flashes as appears to be the case with CVs? In view of these rather fundamental uncertainties, our prejudice is to adopt as the standard model the recycling model, whose physics is relatively better understood. Throughout the article we compare the observations with the standard model. Only when the standard model fails should a search for an alternative be seriously considered.

3.   SEARCHES AND DEMOGRAPHY

The demography—distribution and birthrates—of millisecond pulsars offers a valuable clue to their origin. In particular, comparison of the birthrates of LMBPs and LMXBs is a particularly important exercise (Kulkarni & Narayan 1988). A number of searches are currently in progress and the essential results are summarized below.

3.1   *Pulsar Searches*

Pulsar searches are usually conducted at meter wavelengths since pulsars are steep-spectrum objects. Millisecond pulsars appear to have especially steep spectral indices: $\alpha \gtrsim 2$ (Foster et al 1991), where the flux at frequency ν, $S_{\nu} \propto \nu^{-\alpha}$. The conceptual basis of a pulsar search is quite simple. A spectrum with n channels spread over bandwidth B is recorded every Δt. The pulsar signal, owing to dispersive transmission through the interstellar medium, arrives earlier at the higher frequency channels compared to the lower frequency channels. The first step is to undo this dispersion. If the amount of dispersion is not known, as is the case in pulsar searches, one tries a variety of such shifts. Next, a search for a pulsed train is carried out in each of the dedispersed time series.

This is most easily implemented by a Fourier transform followed by a search for a pattern of evenly spaced peaks (the fundamental and harmonics).

Following the discovery of the first millisecond pulsar 1937 + 21 (Backer et al 1982), a number of searches were launched at Arecibo, Jodrell Bank, and Parkes. The search for millisecond pulsars requires enormous computing capacity since the memory and CPU requirements are $\propto P_{min}^{-2}$, where P_{min} is the minimum period to which the search is sensitive. For the succeeding eight years, searches proceeded rather slowly, primarily limited by both the recording and computing capacity. Most of these searches, motivated by the low latitude of the first millisecond pulsar 1937 + 21, were directed towards the Galactic plane.

The association of LMXBs with LMBPs meant that LMBPs should also be found away from the plane. This rationale led to a few searches at intermediate Galactic latitudes. However, it was the successful detection of a millisecond pulsar at high latitude by Wolszczan (1991) that demonstrated the importance of all-sky searches. At about the same time, two technological revolutions made it feasible to conduct large searches: the introduction of inexpensive recording media (8-mm Exabyte tapes), and the vast and relatively inexpensive computing power of workstations and supercomputers. The fully completed Parkes search (Bailes et al 1994; see Section 3.2) will have recorded a total of nearly a terabyte of data and used 20 Sparc-II-years or 2×10^{15} flop to fully reduce the data!

Each pulsar search is limited to pulsars with $P > 2\Delta t$ (from sampling theorem considerations). More importantly, the dispersion of the pulsar signal in the interstellar plasma causes smearing of the pulse signal in the time domain across the finite width B/n of the n frequency channels. This greatly reduces the sensitivity to pulsars with $P \lesssim 2\Delta t_{\rm D}$, where

$$\Delta t_{\rm D} = 0.5 \left(\frac{B/n}{250\,{\rm kHz}}\right)\left(\frac{DM}{20\,{\rm cm}^{-3}{\rm pc}}\right)\left(\frac{\nu}{430\,{\rm MHz}}\right)^{-3} {\rm ms}. \tag{3.1}$$

Here DM is the dispersion measure, the integrated electron density along the line of sight to the pulsar. Full sensitivity is obtained for pulsars with $P \gtrsim 10\max(\Delta t, \Delta t_D)$. Searches for HMBPs are not particularly taxing. Since these are relatively slow pulsars, the sky has been better searched for them. There is tremendous interest in obtaining the deathrate of double-neutron star systems like 1913+16 which coalesce in the Hubble time. However, the rapidly changing (Doppler-shifted) apparent periods of pulsars in such close binaries means that simple Fourier transforms smear the signal over a range of frequency bins, and lose sensitivity. More computationally intensive acceleration searches are needed to discover close neutron star binaries (Johnston & Kulkarni 1991; see Anderson 1992 for an implementation and the discovery of PSR2127 + 11C in this way). The gravitational waves generated just before coalescence of these systems are the dominant known source that defines sensitivity goals for on-going gravitational wave observatory efforts (Abramovici et al 1992). In addition, coalescing binary neutron stars are a leading candidate for models of gamma-ray bursts at cosmological distances (Narayan, Paczyński & Piran 1992). From the survey volumes that led to the detection of 1913 + 16 and 1534 + 12, a Galactic coalescence rate of $10^{-6}\,\mathrm{y}^{-1}$ to $10^{-7}\,\mathrm{y}^{-1}$ and a Galactic population $\gtrsim 3 \times 10^4$, comparable to the LMBP population, have been estimated (Narayan, Piran & Shemi 1991; Phinney 1991).

3.2   Summary of Searches for LMBPs

Three large searches are now in progress: at Parkes Observatory (all sky, $\delta <$ $0°$), at Arecibo (a variety of searches by several groups), and at Jodrell Bank ($\delta > 35°$). All these searches are being conducted at a frequency around 430 MHz.

The Arecibo surveys were done with either an $n = 128, B = 10$ MHz system or one with $n = 32, B = 8$ MHz. One sequence of surveys discovered three LMBPs and one HMBP in 800 square degrees of high latitude sky (R. S. Foster and A. Wolszczan, personal communication). Another sequence of surveys carried out by Princeton astronomers discovered three new LMBPs over 235 square degrees of Galactic plane (Fruchter 1989, Nice et al 1993) and three LMBPs and one HMBP over 464 square degrees of high latitude sky (F. Camilo, D. Nice, J. Taylor, personal communication). After accounting for lack of detections in other Arecibo searches, we conclude that the Arecibo high latitude success rate is one LMBP per 200 square degrees. Because of the small number n of frequency channels, these searches were sensitive only to millisecond pulsars with small dispersion measure, i.e. nearby pulsars.

The Parkes survey was done with an $n = 256, B = 32$ MHz system. Nine millisecond pulsars have been discovered in the 10^4 square degree area searched so far (M. Bailes, A. G. Lyne & R. N. Manchester, personal communication). The survey uncovered the nearest millisecond pulsar, J0437 − 4715 (Johnston et al 1993), located a mere 150 pc from Earth. As with the Arecibo sample, most of these pulsars also appear to be nearby ones.

3.3   Population and Birthrates

Because of their small n, the Arecibo searches are approximately volume-limited. They are sensitive to pulsars within a distance $\lesssim 1$ kpc, comparable to the scale height of LMBPs (Section 4). This, and their high sensitivity, make the Arecibo surveys an ideal tool to estimate the local surface density of LMBPs. If we then assume that LMBPs are distributed like other stars, with a surface density $n(R) \propto \exp(-R/R_\mathrm{d})$ kpc^{-2} (where R is the Galactocentric radius and $R_\mathrm{d} = 3.6$ kpc is the exponential scale length of the Galactic disk), and take the solar Galactocentric distance to be $R_0 = 8.5$ kpc, we can extrapolate the inferred local surface density to obtain the total number of LMBPs in the Galaxy, $N_{LMBP} = 5 \times 10^4 / \bar{f} \chi$. Here, $\bar{f}$ is the beaming fraction for millisecond pulsars, usually assumed to be close to unity, and $\chi \geq 1$ is a factor to account for faint pulsars that could have been missed by the Arecibo survey. Lorimer (1994) obtains a similar number for $N_{LMBP}(L > 2.5 \text{ mJy kpc}^2) \sim 5 \times 10^4$; here, L is the luminosity at 400 MHz defined as the product of the 400 MHz flux (Jy or mJy) and the square of the distance (kpc^2). This estimate was obtained by laying down pulsars according to a model Galactic distribution; calculating, considering the parameters of all the surveys, the probability p_i (roughly the ratio of the detection volume $V_{i, max}$ to the Galaxy's volume) that a pulsar of luminosity and period equal to that of a detected pulsar i would have been detected in one of the surveys; and estimating the total Galactic luminosity function as $\Sigma_i (p_i)^{-1} \delta (L - L_i)$.

In our crude estimate, we assume a mean age of $\tau_D / 2$, where $\tau_D \sim 10^{10}$y is the age of the disk. Thus the LMBP birthrate in the Galaxy is $B_{LMBP} = 10^{-5} \text{y}^{-1}$. Lorimer (1994) obtains a similar birthrate using the precise estimator $\Sigma_i (2t_p p_i)^{-1}$, where $t_p \sim \tau_c$ is the true age of each pulsar. This should be compared to the birthrate of LMXBs, $B_{LMXB} = N_{LMXB} / \tau_X$ where $N_{LMXB} \lesssim 10^2$ is the number of LMXBs in the Galaxy (van Paradijs 1994). Equality of these two rates is a fundamental expectation of the recycling model and would require $\tau_X \sim 10^7$y. This is a constraint on the progenitor systems and it needs to be demonstrated that a sufficient number of progenitors with suitable mass transfer histories exist.

An independent constraint can be obtained by noting that the mass that must be accreted to spin up a neutron star to period P_i is

$$\Delta M = f \, 0.08 \left(\frac{P_i}{2 \text{ ms}} \right)^{-4/3} I_{45} M_\odot, \tag{3.2}$$

where $f \gtrsim 1$ is a factor that depends on the details of the accretion; $f \sim 1$ if the star has a constant magnetic moment low enough that accretion just spins the star to its equilibrium period [Ghosh & Lamb (1992) favor $f \simeq 3$]. Further accretion ($f \gg 1$) will not change P_i unless the magnetic moment decays (as in Shibazaki et al 1989). The disk LMXBs (we exclude the bulge population since they are not relevant to the local population) have X-ray luminosities L_x ranging from $0.1 \, L_{Edd}$ to $0.01 \, L_{Edd}$ (Verbunt et al 1984, Naylor & Podsiadlowski 1993), with $(dN)/(d \ln L_X) \sim L_X^\alpha$ with $\alpha \sim 0$ to -0.5. Combining these luminosities with accretion efficiencies (Table 3) and Equation (3.2), implies that the observed X-ray sources would require τ_X ranging from $> 7 \times 10^7 (P_i / 2 \text{ ms})^{-4/3}$ y to $> 7 \times 10^8 (P_i / 2 \text{ ms})^{-4/3}$ y to produce pulsars with initial period P_i.

The birthrate of LMXBs must be estimated from stellar and binary evolution models. Wide-orbital period ($\gtrsim 1$ d) systems such as Cyg X-2 evolve by nuclear evolution of the secondary, whereas accretion in systems with short orbital periods is believed to be driven by angular momentum losses (magnetic stellar winds and gravitational radiation; Verbunt 1993). As discussed in Section 10.3, the inferred LMBP and LMXB birthrates are in rough agreement for systems of long orbital periods, but the single short-orbital period LMBPs have a birthrate $\gtrsim 10$ larger than the LMXBs believed to be appropriate progenitors. To bring the rates into agreement would require $\tau_X \sim 10^7$ y.

Our model-independent estimates of τ_X for observed X-ray sources exceed 10^7 y by a factor between 7 and $70 \times (P_i/2\mathrm{ms})^{-4/3}$. This highlights the importance of determining the typical intial period of millisecond pulsars. Camilo et al (1994) have argued that some millisecond pulsars are born with periods not very different from their observed periods, $P \sim 4$ ms. This is certainly the case for some pulsars (e.g. PSR J2145−0750; Bailes et al 1994). An independent reason to suspect that P_i is not small ($\ll 2$ ms) is that to spin many of the observed pulsars up to <1 ms would in most models require $\dot{M} \sim \dot{M}_{\mathrm{Edd}}$ [see Equation (2.1)] which, as discussed above, does not agree with the observed luminosity function of the LMXBs. Smaller $\dot{M}$ results in a larger equilibrium period.

We conclude that the LMBP birthrate is a factor of $\gtrsim 10$ higher than the (model-dependent) inferred birthrate of observed LMXBs. The birthrates of LMBPs with long orbital period systems agree fairly well with those inferred from well-established models of the X-ray binaries. As discussed in Section 10.3, the discrepancy in the total birthrate arises from "black-widow" pulsars and those in systems of short orbital period, for which models of the accretion phase are most controversial. It appears, as in the globular cluster system (Phinney & Kulkarni 1994), that the majority of these ill-understood LMBPs are produced in very short-lived events (not represented among observed LMXBs) of high $\dot{M}$, or that the mass transfer produces little hard X-ray emission. In the former case, the high $\dot{M}$ would remove the difficulty (see above) in producing LMBPs with $P_i \lesssim 2$ ms.

4.   VELOCITY AND KINEMATICS

4.1   *Observations*

The origin of pulsar velocities is a topic of much interest (see Radhakrishnan 1992, Bailes 1989). The ordinary pulsars, as a group, show large spatial motions, as has been demonstrated via interferometric proper motion observations (Lyne et al 1982). Another technique, less reliable but which can be applied to faint pulsars, is observation of interstellar scintillations (ISS) (Cordes 1986). Motion of the pulsar with respect to the intervening screen of interstellar plasma results in temporal changes of the speckle pattern as viewed on Earth. Recognizing that the scale height of the scattering screen is smaller than that of the interstellar electron layer, by almost an order of magnitude, Harrison & Lyne (1993) present a new calibration curve to convert timescales of changes to transverse speed and conclude that with due care, the ISS velocities are probably accurate to a factor of two. Proper motion measurements are also obtainable from timing observations for a few, mainly millisecond, pulsars.

There now exist almost 100 determinations and significant upper limits for pulsar proper motions (Harrison et al 1993). The mean transverse speed is 217 km s^{-1} using the DM-distance conversion of Lyne et al (1985), and 310 km s^{-1} using the DM-distance conversion of Taylor & Cordes (1993). Both these mean speeds include old pulsars which would have left the Galaxy if they had been born with high velocities. The mean transverse velocity of pulsars young enough not to suffer such velocity selection is (using the Taylor & Cordes scale) an astonishing 360 ± 70 km s^{-1} (Lyne & Lorimer 1994), giving a mean space velocity of 460 ± 90 km s^{-1}. These velocities are much larger than the birth velocities of massive stars, the progenitors of pulsars. Thus pulsars acquire such large velocities at or after their birth. The distribution of velocities, however, has not been well determined.

4.2 Natal Velocity Kicks

Pulsars born in binary systems can acquire velocity, immediately following the SN explosion. In the simplest, unavoidable model, the SN mass loss is symmetric in the frame of the exploding star. The resulting asymmetry in the frame of the center of mass of the binary system means that the binary center of mass must recoil. If more than half the total mass is lost, then the binary system is unbound and the neutron star will escape with a velocity of order its orbital velocity in the pre-supernova binary. However, this mechanism alone is insufficient to explain the observed velocity spectrum—in particular, pulsars with speeds approaching 10^3 km s^{-1} e.g. PSR 2224 + 65 (Harrison et al 1993, Cordes et al 1993). Several SNRs have been claimed to be associated with even faster moving pulsars (see Caraveo 1993). In addition, Cir X-1, a 16-day orbital period MXRB has a systemic velocity of ∼200 km s^{-1} (Duncan et al 1993). All of these results indicate that pulsars must acquire additional velocities (in the frame of the exploding star) at, or shortly after birth by some other mechanism.

Such velocity kicks have also been invoked to explain the paucity of binary pulsars. Most massive stars are thought to arise in binaries, yet most pulsars are single. Velocity kicks can disrupt the binary system during the second explosion in the system, releasing an old neutron star and a new pulsar (Radhakrishnan & Srinivasan 1982, Backus et al 1982, Bailes 1989). Dewey & Cordes (1987) simulated the binary population and found that velocity kicks of ∼100 to 200 km s^{-1} were sufficient to explain the observed binary and single pulsar population (though it is unclear if such low velocities would fit the new data discussed above).

The observed velocities require that kicks sometimes exceed 10^3 km s^{-1}. It is intriguing to note that large kicks appear to be associated with pulsars in or close to SNRs (Caraveo 1993). As many as a third of pulsars may be born with such high speeds. This is an astonishing conclusion because we do not find a large number of middle-aged ($\sim 10^6$y) fast moving pulsars. A related and more understandable observation is that the velocity spectrum of pulsars with $\tau_c < 10^7$ y is peaked at 300 km s^{-1}, a factor of 3 larger than that of older pulsars (Cordes 1986, Harrison et al 1993). This is usually explained as a selection effect by which such pulsars drift away from the Galactic plane in 10^7y. It is interesting that the proper motion of PSR 2224 + 65 ($\tau_c \sim 10^6$ y), the pulsar with the highest measured velocity, is along the Galactic plane—a confirmation of the above selection effect. The on-going all-sky and high latitude searches can significantly constrain the birthrate of fast moving pulsars.

4.3   *Kinematics of LMBPs*

In the recycling model, we expect LMBPs to have large systemic motions (Section 2.2.1), whether or not pulsars receive natal kicks. The systemic speed should be larger for systems of shorter orbital period (Bailes 1989).

Despite the fact that the current sample is heterogeneous, two simple conclusions can be drawn: 1. The scale height of LMBPs is at least about 0.6 kpc, and 2. the median transverse speed is 75 km s^{-1}, significantly smaller than that of the single pulsars. (In arriving at this latter inference we have used the measured proper motion, if available; otherwise, we derived the minimum z-velocity needed to attain their present z-height above the Galactic plane.) These results are in accord with the model discussed above. Velocity kicks between 100 and 200 km s^{-1} are sufficient to explain these trends. With a much larger data set, it should be possible to see a trend of larger systemic motions with smaller orbital periods. Conversely, long orbital period LMBPs should have smaller scale height than the short-period systems (Bailes et al 1994).

Johnston (1992), using the radial velocity data from optical observations of LMXBs, has carried out a similar exercise for LMXBs and arrives at a similar conclusion. Based on a kinematic analysis, Cowley et al (1988) argue that LMXBs have a tangential motion with respect to the Sun of $U \sim -68 \pm 46$ km s^{-1}, and that therefore they are kinematically intermediate between Population I and II objects. Given the large errors, one could also conclude that $U = 0$ km s^{-1} i.e. LMXBs are Population I objects, albeit with large random velocity. Naylor & Podsiadlowski (1993) also argue for a Population I origin based on the observed spatial distribution.

5.   MAGNETIC FIELD EVOLUTION

The magnetic field strength is the key parameter governing the evolution of neutron stars. High field strength is necessary for radio and X-ray pulsar activity. Low field strength is invoked to explain the absence of pulsations in LMXBs. The real mystery about millisecond pulsars is not the origin of their spin periods, but the origin of their distinctly lower magnetic field strengths. Unfortunately, we have no comprehensive theoretical model for either the origin or the evolution of magnetic fields in neutron stars. In view of this, our focus is primarily phenomenological. For reviews from a theoretical point of view, see Chanmugam (1992) and Bhattacharya & Srinivasan (1994).

5.1 *Early Ideas*

The earliest statistical study of pulsars (Ostriker & Gunn 1969) suggested that pulsars are born with large field strengths, $B_i \sim 10^{12}$ G, and that the field decays exponentially with a characteristic timescale, $\tau_B \sim 4 \times 10^6$y. Most but not all subsequent statistical studies appeared to essentially confirm this picture (e.g. Narayan & Ostriker 1990). A model-independent "proof" that was quoted in defense of field decay emerged from observations of pulsar proper motions (Lyne et al 1982). Comparison of the characteristic age τ_c with the kinematic age $\tau_z = |z|/v_z$, where z is the vertical distance from the Galactic plane and v_z is the vertical speed, showed that τ_c is systematically larger than τ_z, especially for large τ_c. Assuming an exponential field decay, $B(t) = B_i \exp(-t/\tau_B)$, one obtains

$$t_p = \frac{\tau_B}{2} \ln \left[1 + \frac{2\tau_c}{\tau_B} \left(1 - \frac{P_i^2}{P^2} \right) \right], \tag{5.1}$$

where P is the current period, t_p the true age of the pulsar, and P_i the initial period. These timescales need to be compared to the typical time taken by a pulsar, assumed to be born in the Galactic plane, to reach the highest $|z|$, the "turning point." For v_z between 100 km s^{-1} and 200 km s^{-1}, this is $\sim$50 My in the solar neighborhood. Since we believe that most pulsars are born in the Galactic plane and are rarely still luminous when they return to the plane, $\tau_z \sim t_p$. Interpreting the observations in light of Equation 4, Lyne et al (1982) inferred $\tau_B \lesssim 10$ My. This effect persists in the newer and larger proper motion data (Harrison et al 1993) with $\tau_B \lesssim 10$ My.

5.2 *Residual Fields*

A particularly powerful constraint on the secular evolution of the field strength can be obtained by optical observations of binary pulsars containing a white dwarf system. Since the white dwarf is born after the neutron star (Section 10.1), it is clear that the age of the white dwarf τ_{wd}, as deduced from comparison of the white dwarf luminosity with cooling models, must be a lower limit to t_p. White dwarfs have been detected in 0655+64 (Kulkarni 1986),

0820+02 (Kulkarni 1986, Koester et al 1992), and in the nearby millisecond pulsar J0437−4715 (Bailyn 1993, Bell et al 1993, Danziger et al 1993). Deep searches toward several millisecond pulsars, notably 1855+09 (Callanan et al 1989, Kulkarni et al 1991), have not detected the candidate companion; the white dwarf companion is presumed to be too cool to be seen by modern optical detectors. In all these cases, observations are consistent with $\tau_{wd} \sim \tau_c$. In the context of the ideas discussed in the previous section, Kulkarni (1986) proposed that magnetic fields stop decaying once they reach a "residual" value. The strength of this field appears to be in the range 3×10^8 G (PSR 1855+09) to 3×10^{11} G (PSR 0820 + 02).

The concept of a residual field has also arisen from other lines of reasoning, albeit more model dependent. Exponential decay of the field would result in radio lifetimes comparable to τ_B. The LMBP birthrate would then far exceed the birthrate of LMXBs (Bhattacharya & Srinivasan 1986, van den Heuvel et al 1986). From a stellar evolutionary point of view, Verbunt et al (1990) have argued that the LMXB 4U 1626−67 and Her X-1 (located at $z = 3$ kpc) are $\gtrsim 10^8$ y old; yet both are X-ray pulsars and have strong magnetic field strengths. It therefore appears that the residual field strength encompasses the entire range of pulsar magnetic field strengths: 10^8G to $\gtrsim 10^{12}$G. The simplest conclusion (ignoring the proper motion data), is that magnetic fields do not decay *at any field strength*.

5.3   No Field Decay?

The realization that fields may not decay has prompted reanalyses of the pulsar population. Monte Carlo comparison of the data with simulated pulsar populations, subjected to a detection process with appropriate selection effects, have led to confusing results. Narayan & Ostriker (1990) find evidence for field decay; Bhattacharya et al (1992) and Wakatsuki et al (1992) find evidence for little if any field decay ($\tau_B > 10^8$y). It is disturbing that these three analyses use essentially the same data but arrive at divergent conclusions. This simply may illustrate the dangers of trying to determine a $\gtrsim 7$ dimensional distribution function (in P, $\dot{P}$, S_{400}, DM, b, μ_b, μ_ℓ) from only a few hundred points. Perhaps the greatest systematic uncertainties in such studies are the treatment of selection effects and kinematics, and the unknown luminosity evolution law, an issue that bedevils such parametric pulsar population studies—see especially the last two paragraphs in Narayan & Ostriker (1990), the introduction in the Lorimer et al (1993) paper, and Michel (1991). Bhattacharya & Srinivasan (1994) in their review article refer to unpublished work which attempts to circumvent the parametric approaches taken in the above studies by analyzing the pulsar birth current as a function of P and B. They conclude that there is little need for field decay.

Despite the above discussion, there are still two observations that are hard to square with the assumption of *no* field decay: 1. the proper motion data (τ_z vs t_c) discussed above, and 2. the origin of pulsars with field strengths between 10^{11} and 10^{12}G, a group which we will henceforth refer to as the intermediate strength pulsars (see Figure 1). Bailes (1989) addressed these two issues, first arguing that the proper motion data were noisy and consistent with no field decay. However, in our opinion, the effect discussed above persists in the latest proper motion data (Harrison et al 1993) which has four times as many measurements as the Lyne et al (1982) sample. Next, even for $\tau_B = \infty$, τ_c is an overestimate of t_p if $P \sim P_i$ [see Equation (2.1)]. Thus Bailes argues, in the context of the recycling model, that the intermediate strength pulsars are the first-born pulsars, mildly spun up prior to the disruption of the system at the time of the second supernova. If so, then τ_c could be significantly overestimated, especially for those pulsars close to the rebirth line. Inspection of the five pulsars with largest τ_c in Figure 7 of Harrison et al (1993) shows that these pulsars have, for their observed period, an inferred magnetic field strength within a factor of two of the equilibrium value. Given the uncertainties in the constants in Equation (2.1), we conclude that most of these could well be recently recycled pulsars.

5.4   Tests for Field Constancy

The least model-dependent constraints on τ_B can be obtained by determining the fraction of pulsars returning to the Galactic plane and the fraction of pulsars above the interstellar electron layer, whose half thickness is estimated to be $h_e \lesssim 1$ kpc (Bhattacharya & Verbunt 1991). Pulsars will reach the death line shown in Figure 1 in time $t_D \sim 80/B_{12}$ My, where B_{12} is the field strength in units of 10^{12} G. Unless their luminosity drops on a timescale much shorter than t_D, pulsars with $B_{12} \lesssim 1$ will be able to reach the turning point ($|z| \sim 2$ kpc, for $v_z \sim 100$ km s^{-1}) and return to the Galactic plane within their radio lifetime. Bhattacharya et al (1992) present some evidence indicating that a small fraction of old pulsars lie above the electron layer. Harrison et al (1993) find 8 pulsars (out of $\lesssim 10^2$) returning to the Galactic plane. More high latitude pulsar searches and proper motion observations, especially of pulsars with $B \sim 10^{11}$–10^{12} G are needed to firmly constrain τ_B for single pulsars. Recently, Chen & Ruderman (1993) have argued that the location of the death line depends on the magnetic field topology. Thus pair production (and hence presumably radio emission) in pulsars with the same dipole field $P\dot{P}$ may cease at a range of periods, whose values depend on the pulsars' individual field topologies. They dub this range, the "Death Valley." Statistical analyses of the ongoing high latitude surveys will be most useful in demarcating the Death Valley in the B-P diagram.

5.5   *Bimodal Field Distribution*

In Section 3.1, we remarked that the binary pulsar systems with massive companions, the HMBPs 1913 + 16, 1534 + 12, and 0655 + 64, are supposed to emerge from massive binaries. These three pulsars have $B \sim 10^{10}$ G. In contrast, the magnetic field strengths of pulsars with low mass binary companions, the LMBPs such as 1855 + 09, etc have field strengths $\lesssim 10^9$ G. This led Kulkarni (1992) to propose that there is a "field gap" between $B = 10^9$ to $B = 10^{10}$ G since there is no known selection effect preventing the discovery of pulsars with field strengths in this range. Indeed, the selection effects favor the detection of pulsars with larger field strength.

Camilo et al (1994) argue that the field gap appears to persist in a larger sample. Correcting for the kinematic contributions to $\dot{P}$ (see Section 8), they find that for the LMBPs, $10^8 \lesssim B \lesssim 5 \times 10^8$ G. The discovery of an isolated pulsar, $J2235 + 1506$ with $B = 2.7 \times 10^9$ G (Camilo et al 1993), presumed to be a member of an HMBP system that was disrupted on the second explosion, suggests that the field gap is not due to some fundamental physics. Most likely, there are two distributions with two different mean values: one of intermediate field strength pulsars with B centered on $10^{11.5}$ G of which the HMBPs are the tail end, and one of low field strength with $B \sim 10^{8.5}$ G, constituting the rapidly rotating LMBPs. Note that LMBPs with slow rotation rates (e.g. PSR 0820+02) nominally belong to the intermediate group.

Note that there are no disk LMBPs with $B < 10^8$ G. However, the histogram of B for the rapidly rotating LMBPs shows that there are as many pulsars (six) with $B < 2 \times 10^8$ G as above this value, i.e. there is a steepening towards lower B. It is possible that the Galaxy contains a new class of neutron stars—low magnetic field strength neutron stars. It is unclear whether such objects would shine in the radio window at all (Section 6.2, Phinney 1994).

5.6   *A Phenomenological Model*

Following Bailes (1989), we find that the available data are consistent with the following picture:

1. Pulsars are born with Crab-pulsar-like properties: $B \sim 10^{12}$ G, $P \sim 20$ ms. The first half of the statement is motivated by the high field strengths of young pulsars in SNRs (Figure 1).

2. The magnetic field does not decay unless there is accretion. The age of the neutron star in Her X-1 is $\gtrsim 10^7$ y, given $z = 3$ kpc. The white dwarf in PSR 0820+02 is 2×10^8 y (Koester et al 1992). Thus, if there is any decay at all, $\tau_B \gtrsim 40$ My. A very gradual decay of the field strength over the age of the Galaxy, $\tau_D \sim 10^{10}$ y, cannot be ruled out. The highest B field of a cluster pulsar (PSR 1820−30B; Biggs et al 1993) is 10^{11} G (see also

Figure 1). Assuming that this pulsar has not been recycled, we conclude that neutron star magnetic fields cannot decay by a factor greater than $\sim$30 over a timescale τ_D.

5.7 *Models for Field Reduction*

Above, we have interpreted the observations as if neutron star magnetic fields do not evolve unless there is accretion. The first part of the statement may have theoretical blessings. In a recent review, Goldreich & Reisenegger (1992) find no physical mechanism for fast field decay. They speculate that magnetic fields buoyed by Hall drift may undergo a turbulent cascade terminated by ohmic dissipation on a timescale $\sim$5 $\times$ $10^8/B_{12}$ y.

The second part of the above statement, the origin of accretion induced field decay, is a mystery and without a proper physical model. There are two qualitatively different models: 1. crustal models where ΔM—the amount of matter accreted (Taam & van den Heuvel 1986, Shibazaki et al 1989, Romani 1990)— or an inverse battery effect (Blondin & Freese 1986) leads to field reduction, and 2. interior models, where interaction of the magnetic flux tubes with angular momentum vortices in the interior leads to field expulsion (Srinivasan et al 1990).

The crustal models are attractive from the point of view of the bimodal field distribution. HMBPs with their smaller inferred ΔM have higher field strengths, whereas those LMBPs with large inferred ΔM have smaller Bs. However, observational data do not support a simple relation between accreted matter and field strength (Verbunt et al 1990). In addition, crustal models need to be fine tuned since it is the accretion process that causes both field reduction and spin-up [see also the discussion following Equation (3.2)].

In the interior model, the angular momentum vortices are assumed to be strongly coupled to the magnetic flux lines in the interior. Thus as the young pulsar slows down, magnetic flux lines are brought up to the crust where they are assumed to decay. In this model, low field pulsars, at an earlier phase in their life, were spun down to very long periods before being spun up by accretion. The major difficulty in this model is that it requires rapid decay in the crust (e.g. 1E2259+586, discussed above) which appears to be ruled out by theoretical considerations. In addition, the model provides no natural explanation for the bimodal field distribution discussed above.

In a series of papers, Ruderman (1991) has developed a variant of an interior model. Here, the rotational-magnetic-field coupling results in rearranging the magnetic flux lines, thus effecting changes in the dipole field strength. As the pulsar slows down, the dipole moment diminishes, explaining the intermediate strength field pulsars. There is one basic difficulty with this model: While magnetic pole migration can reduce the dipole field strength, it does not reduce strengths of individual field lines. However, during the last stages of the spin-up process, when the pulsar is spinning at its equilibrium period, the inferred Alfvén radius must equal the Keplerian corotation radius, $R_{eq}/R_n \simeq 2(P/\mathrm{ms})^{2/3}$, where R_n is the radius of the neutron star. Strong multipole field strength ($\sim 10^{12}$ G) would result in large R_{eq} (Arons 1993), and it may be difficult to obtain millisecond rotation periods. Furthermore, if the magnetic geometries of millisecond pulsars were very nondipolar, one would expect the pulse and polarization profiles to differ from those of ordinary pulsars. However, the pulse shapes of millisecond pulsars now appear to be no different from those of ordinary pulsars (Bailes et al 1994); the same appears to be true of the meager polarization data (Thorsett & Stinebring 1990). Further polarimetric observations of millisecond pulsars can potentially refute or confirm Ruderman's model.

Romani's (1993) model combines aspects of Ruderman's model and crustal models by having the overburden of the accreted matter push magnetic field lines below the neutron star surface and thereby reduce the external dipole. The model claims to predict a floor value, $B \sim 10^8$ G by having advection cease when the field goes below 10^8 G. The model is phenomenologically appealing in that it explains the field gap (a consequence of different ΔM in the HMBPs versus LMBPs) and the near constancy of field strength of millisecond pulsars. In contrast to the disk LMBPs, cluster pulsars have field strengths ranging from 10^8 to 10^{11} G; this is explained in this model by appealing to the diversity of the tidal products, from long-lived LMXBs to short-lived accretion tori resulting from star-destroying encounters. However, it is fair to say that we still have no satisfactory physical model that can explain the magnetic field spectrum of binary and millisecond pulsars.

6. PULSAR AND NEUTRON STAR PHYSICS

The ultimate importance of millisecond and binary pulsars lies not so much in their origin and evolution, but rather in the fundamental experiments in physics and astrophysics which nature performs for us with them. Some of these experiments involve the emissions and environmental impact of the pulsars themselves. The thermal and crustal histories of their neutron stars differ from those of ordinary pulsars, and their magnetospheres are much smaller. The differences between the emissions and relativistic winds of pulsars with spin periods and dipole fields differing by four orders of magnitude might be expected to provide clues to the physics of neutron star interiors, crusts, and magnetospheres. The energy stored in the rotational energy of a millisecond pulsar can be comparable to the nuclear energy released by its main sequence progenitor, and to the energy released in the accretion required to spin up the neutron star. The release of this energy in the pulsar's relativistic wind can have striking effects on the pulsar's stellar companions, and on its interstellar environment.

Other experiments make use of two properties of millisecond pulsars: 1. their excellence as clocks, accurate to about 1 μs over decades, which allows changes in their distance from Earth to be determined to a precision of c (1 μs) = 30 m, and 2. the brevity and extremely high brightness temperature of the broad band radar pulses they emit. These powerful radar pulses act as both passive and active probes of the media through which they travel. We now review the many diverse experiments.

6.1 *Calorimetry*

One of the first compelling arguments that ordinary pulsars were neutron stars was the identification of the Crab nebula as a calorimeter for the Crab pulsar (cf Manchester & Taylor 1977). This showed that the Crab pulsar must have (and have had) a relativistic wind of luminosity $d(I\Omega^2)/dt$, where $I \simeq 10^{45}$ g cm^2 is the moment of inertia of the neutron star and Ω its angular spin frequency. Millisecond pulsars have several other types of calorimeters.

6.1.1 WIND NEBULAE   As a pulsar moves through the interstellar medium (ISM), its relativistic wind will form a bow shock around the pulsar. Ahead of the pulsar the contact discontinuity between the pulsar wind and the interstellar medium will lie at the point where the ram pressure (in the pulsar's rest frame) of the incoming ISM equals that of the pulsar's wind. Such a wind nebula is observed around PSR 1957+20. The ISM side of the contact discontinuity is an Hα-emitting nebula of cometary form (Kulkarni & Hester 1988, Aldcroft et al 1992) aligned along the direction of the pulsar's proper motion μ (Ryba & Taylor 1991). The head of the nebula is projected $\theta_s = 4''$ ahead of the pulsar. A parabolic model of the bow shock predicts that independent of the (unknown) line-of-sight velocity,

$$\theta_s = \left(\frac{I\Omega\dot{\Omega}}{4\pi C^2 \rho_i c} \right)^{1/2} \frac{1}{D^2 \mu} = \frac{2\overset{''}{.}4\, I_{45}}{C D_{\mathrm{kpc}}^2 n_0}, \tag{6.1}$$

where $C \sim 0.7$ if the ISM shock is adiabatic, $C \rightarrow 1$ if it is radiative, and the pre-shock ISM has hydrogen number density $1\, n_0$ cm^{-3}. Observations of the companion star to 1957+20 (Djorgovski & Evans 1988) suggest that $0.7 < D_{\mathrm{kpc}} < 1.3$ (see below). The dispersion measure distance (Taylor & Cordes 1993) is 1.5 kpc. Reasonable equations of state have $1 < I_{45} < 2$ (see Table 3). The fact that there is a Balmer-dominated shock in the ISM suggests that the pre-shock ISM is mostly neutral, so $n_0 \gtrsim 0.3$. The rough agreement of Equation 6.1 with the observed $\theta_s = 4''$ thus implies that the ISM encounters a (*a*) roughly isotropic pulsar wind, which (*b*) carries most of the spin-down luminosity, and (*c*) acts fluid-like on the $\sim 10^{17}$cm scale of the nebula. If most of the wind energy is carried in ions, then (*c*) requires that the gyro radii of the ions be smaller than the nebula. This is true as long as the ions were accelerated across no more than the voltage across a simple dipole polar cap (cf Arons & Tavani 1993).

6.1.2 COMPANION HEATING   The companions of millisecond pulsars are bombarded by relativistic particles and electromagnetic waves from the pulsar. The constituents of the pulsar wind deposit their energy at energy-dependent column densities in the companion and its atmosphere: hundredths of g cm^{-2} for X rays to hundreds of g cm^{-2} for ultrarelativistic ions and gamma rays. The energy deposited below the photosphere will be thermalized and reradiated within a few minutes. The side of the companion illuminated by the pulsar wind will therefore be hotter than the "dark" side of the companion. The flux at the surface of a companion with orbital period P_b from an isotropic pulsar wind is

$$F = \frac{I\Omega\dot{\Omega}}{4\pi a^2}$$
$$= 4.2 \times 10^{20} I_{45} \left(\frac{\tau_c}{\text{yr}}\right)^{-1} \left(\frac{P_b}{\text{d}}\right)^{-4/3} \left(\frac{P}{\text{ms}}\right)^{-2} \text{erg cm}^{-2} \text{ s}^{-1}. \quad (6.2)$$

For pulsar 1957+20, this, combined with models for the cooling of the companion, predicts (Phinney et al 1988) that the sub-pulsar part of the companion should have an effective temperature of $\sim$8000 K, while the dark side should be cooler than 2000 K. These predictions have been confirmed by observations (Djorgovski & Evans 1988, Eales et al 1990, Callanan 1992). This does not directly demonstrate that the bulk of the pulsar's spin-down energy is carried off in a form capable of penetrating below the $\sim$1 g cm^{-2} photosphere of the companion. This is because in PSR 1957+20, the pulsar wind must be shocked far above the companion's surface by the dense plasma in the companion's wind or magnetosphere (Ryba & Taylor 1991; see Figure 1 of Phinney et al 1988). It could be that radiation from the shock, or particles accelerated there—and not the pulsar wind itself—impinge on the companion.

Cleaner diagnostics of the penetrating power of pulsar winds may be provided by observations of pulsars with more massive white dwarf companions. Applying Equation (6.2) to Table 1 reveals that besides 1957+20, the strongest companion heating should occur in PSRs J0034 − 0534, J2317 + 1439, and J0437 − 4715, the illuminated sides of whose companions should be respectively about 25%, 2%, and 1% hotter than the $\sim$4000 K of their unirradiated sides. In addition to photometry, line spectra of these objects and 1957+20's companion will prove interesting. Spectral line profiles, heights, and depths are sensitive to the temperature profile above the photosphere, and thus can be used to determine the heat deposition as a function of column density. The limits on emission lines from 1957+20's companion (Aldcroft et al 1992) already rule out an X-ray heated wind from a compact companion (Levinson & Eichler 1991, Tavani & London 1993) as a significant source of mass loss in that system.

Heating of a convective companion can have a dramatic indirect effect on its structure. When one irradiates a star, which in isolation would have a low effective temperature, it initially develops a temperature inversion in its envelope. Heat is conducted inwards until, at the effective temperature set by the irradiation, an isothermal zone develops which extends inwards to the radius in the convective envelope at the same temperature. If this isothermal zone is at $\gtrsim 10^4$ K, its (hydrogen or helium ionization) opacity will be 1–2 orders of magnitude higher than it was at the unirradiated photosphere. In stars with deep convection zones, the isothermal layer can present the dominant bottleneck to the escape of radiation from the stellar interior and can therefore trap heat trying to escape from the stellar interior, bloating the star and causing it to expand on its thermal timescale. This effect, pointed out by Phinney et al (1988), has been studied in detail for isotropically irradiated stars by Podsiadlowski (1991), Harpaz & Rappaport (1991), and d'Antona & Ergma (1993). The star swells, and therefore transfers mass, on the thermal timescale, which is much shorter than the nuclear timescale. This results in short X-ray lifetimes, and rapid orbital evolution of low mass X-ray binaries, and expansion and perhaps destruction of close pulsar companions (Ruderman et al 1989, van den Heuvel & van Paradijs 1988, Phinney et al 1988).

In nature, however, the stars are irradiated on one side only. An isothermal zone with high opacity on one side of the star would simply force more of the star's internal luminosity to emerge on the unirradiated side—which would not necessitate much change in the convective envelope. Dramatic changes like those in isotropically irradiated stars would occur only if fast azimuthal winds, differential rotation, or asynchronous rotation carried the heat of irradiation to the unirradiated side of the companion in less than the cooling time of the isothermal zone. None of this is occurring, at least in PSR 1957+20's companion, otherwise the large temperature difference between its pulsar-illuminated face and its "dark" face would not be observed.

Furthermore, much of the observational evidence for irradiation-induced large $\dot{P}_b$ in X-ray binaries and pulsar companions has recently evaporated. Of the four cases adduced by Tavani (1991) as evidence, Cyg X-3's companion has been shown to be a luminous Wolf-Rayet star whose spontaneous wind naturally explains the magnitude and sign of the systems $\dot{P}_b$ (van Kerkwijk et al 1992, van Kerkwijk 1993); 4U1820 − 30's apparently large negative $\dot{P}_b$ (van der Klis et al 1993a) has recently been shown to be an artifact of changes in the shape of the X-ray light curve (van der Klis et al 1993b); and in both of the remaining systems, X1822 − 371 (Hellier et al 1990) and EX00748 − 676 (Parmar et al 1991), the total shift in the light curve on which the $\dot{P}_b$ is based is still a tiny fraction of an orbital period. Finally, what was once a large negative $\dot{P}_b$ of the eclipsing pulsar 1957+20 (Ryba & Taylor 1991) has changed sign since 1992 (Arzoumanian et al 1994), and is therefore probably a random walk in orbital phase like those commonly observed in cataclysmic variables (Warner 1988), which are produced by magnetic cycles (Applegate 1992) or other torques from the companion star.

6.2   Diagnostics of Pulsar Magnetospheres

A pulsar of spin period P_{-3} ms has its light cylinder at $c/\Omega \sim 5P_{-3}$ neutron star radii, and the radius where the orbital frequency equals the spin frequency is $R_{eq} = (GM/\Omega^2)^{1/3} \sim 2P_{-3}^{2/3}$ neutron star radii. The spin-down torques (Krolik 1991) and equilibrium spin-up line (see Equation 2.1) in millisecond pulsars are thus much more sensitive to nondipolar magnetic multipoles than they are in slower pulsars. This has been used to show, for example, that the millisecond pulsars' surface magnetic fields are not dominated by a quadrupole, nor concentrated in a single polar clump (see Section 5.7; Arons 1993).

The above arguments suggest that millisecond pulsars have surface fields of roughly dipole strengths. It seems not to be generally appreciated that millisecond neutron stars with $\dot{P}$s lower than the lowest observed $\dot{P}$s would not then be able to initiate traditional magnetic e^+e^- pair cascades, and consequently might not be radio luminous (Phinney 1994; but note that other types of pair cascades may occur—see below). The predominance of 10^8 G fields in millisecond pulsars may thus not be a consequence of magnetic field decay, but simply a radio selection effect. The high brightness temperature of pulsar radio emission is hard to explain except as coherent emission from a highly relativistic electron-positron plasma. [Melrose (1992) gives a critical review of the models.] Such a plasma can be produced by pair cascades above a vacuum gap in the magnetosphere of a rotating neutron star (Ruderman & Sutherland 1975). A cascade requires two ingredients: particles that radiate photons, and conversion of the photons to more particle pairs. A gap of thickness h has an electric field $E \simeq \Omega Bh/c$, and potential drop $\Delta V = Eh/2$. This can accelerate electrons to a maximum Lorentz factor

$$\gamma_{max} = \min[e\Delta V/(m_e c^2), (E\rho^2/e)^{1/4}], \tag{6.3}$$

where the second limit is imposed by curvature radiation reaction when the radius of field line curvature is $\rho \equiv 10^6 \rho_6$ cm. The electrons radiate curvature photons of frequency $\omega_c \sim \gamma_{max}^3 c/\rho$, which can pair create on the magnetic field if (cf Ruderman & Sutherland 1975)

$$\frac{\hbar\omega_c}{2m_e c^2}\frac{B_\perp}{B_q} > \frac{1}{14}, \tag{6.4}$$

where $B_q = m_e^2 c^3/e\hbar = 4.4\times10^{13}$ G. Except for a prescient paper by Goldreich & Keeley (1972), previous discussions of the pair creation limit, or "death line" (Chen & Ruderman 1993), have ignored the radiation reaction limit, and insert the first term on the right of (6.3) into (6.4), to derive death lines. These depend on ρ, and therefore on the assumed magnetic field structure. Lines with small ρ are consistent with the boundary of high field radio pulsars in the B-P plane (see Figure 1, Chen & Ruderman 1993). But for millisecond pulsars, the appropriate limit is the second, radiation reaction term in (6.3). Inserting this into (6.4) with the maximum vacuum E, gives pair creation only if

$$B > 5 \times 10^7 P_{\text{ms}}^{9/14} \rho_6^{-2/7} \text{ G} \tag{6.5}$$

(Phinney 1994; this conservative limit assumes a field line topology such that $B_{\perp} \sim B$), which for $\rho_6 \sim 1$ is plotted as the lower part of the "death boundary" in Figure 1. However, X rays from a heated polar cap (cf J0437$-$4715 discussed below) can pair create with curvature photons ($\gamma\gamma \rightarrow e^+e^-$) even for magnetic fields well below (6.5) (Phinney 1994); so it is unclear if this is a hard boundary. If magnetic pair creation were the only mechanism, pulsars with lower magnetic fields would be sources of GeV pulsed curvature gamma rays, and might be surrounded by bow shocks in the interstellar medium (Arons 1983), but would be very difficult to detect as pulsed sources unless they had outer gaps (Chen & Ruderman 1993, Halpern & Ruderman 1993) to produce X-ray or optical synchrotron emission.

Limits on gamma rays (Fichtel et al 1993) for many millisecond and binary pulsars from the EGRET sky survey (luminosity per octave of gamma ray energy E, $EL_E \lesssim 4 \times 10^{32}$erg s^{-1} sr$^{-1} D_{\text{kpc}}^2$ for 100 Mev $< E < 3$ GeV at distance D_{kpc} kpc) are already of order, or below the (crude) predictions of outer magnetosphere models ($L_\gamma \sim 10^{-2} I\Omega\dot{\Omega}$, Chen & Ruderman 1993), but do not constrain most polar cap models.

Kilovolt X rays have been detected from two millisecond pulsars: J0437$-$4715 (Becker & Trumper 1993) and 1957+20 (Kulkarni et al 1992b). In the former, the X rays are pulsed at the 5.75 ms radio period, and have $L_X \sim 4 \times 10^{-4} I\Omega\dot{\Omega}$. There is a hint that the spectrum may have both a thermal ($kT \sim 0.14$ keV) and a power-law component. In the latter ($L_X \sim 5 \times 10^{-5} I\Omega\dot{\Omega}$), too few X-ray photons were detected with high time resolution (Fruchter et al 1992) to tell if the X rays are pulsed or modulated at the orbital frequency (Kulkarni et al 1992b). They might therefore be generated at the shock around the eclipsing companion, or at the reverse shock in the wind nebula (Kulkarni et al 1992b, Arons & Tavani 1993). The soft X-ray emission from these pulsars carries a fraction of $I\Omega\dot{\Omega}$ similar to that observed from some nearby ordinary pulsars (Ögelman 1993, Yancopoulos et al 1994). This might suggest that all pulsars have heated polar caps, though the ratio of X-ray to spin-down luminosity varies over several orders of magnitude.

6.3 Neutron Star Physics

Binary and millisecond pulsars provide some of the most rigorous constraints on nuclear equations of state. Neutron stars constructed with very soft equations of

Table 2   Pulsars with measured masses

Pulsar	M_p^a	M_c	$M_p + M_c$	Reference
1913+16	1.4411 (7)	1.3874 (7)	2.82843 (2)	Taylor 92
1534+12	1.34 (5)	1.34 (5)	2.6781 (7)	Taylor 92
2127+11C	—	—	2.712 (5)	Anderson 92
2303+46	1.2 (3)	1.4 (2)	2.53 (8)	Thorsett et al 93b
1855+09	1.5 (2)	0.26 (2)	—	Kaspi et al 94b
1802−07	1.4 (3)	0.33 (1)	1.7 (4)	Thorsett et al 93b

[a] All masses in units of $M_\odot$.
Notes: Pulsars in the top group have eccentric orbits and high-mass (probably neutron star) companions. PSR 2127+11C is in the globular cluster M15, and probably formed by exchange (Phinney & Sigurdsson 1991). In the bottom group, PSR 1855 + 09 has a low-mass helium white dwarf companion in a circular orbit; PSR 1802 − 07 is in the globular cluster NGC 6539, and its companion, in an eccentric orbit, probably results from exchange or a tidal 3-body encounter (see Phinney 1992).

state, including pion-condensate and the recently popularized Kaon-condensate models (Brown & Bethe 1994) give maximum stable neutron star masses of $\simeq 1.4$–$1.5\, M_\odot$. This is because they have small radii, so the general relativistic contributions ($\sim GM/Rc^2$) to the radial stability criterion are large even for low masses. [Kunihiro et al 1993 have written a comprehensive monograph on recent work on pion and Kaon condensates; at sufficiently high densities in nuclear matter, the effective mass of the strange K^- meson can fall below that of the the π^- meson and become an energetically favored constituent of neutrons star cores, reducing the pressure support and the neutron star radii (Kaplan & Nelson 1986, Politzer & Wise 1991, Brown et al 1992).] General relativistic effects in timing binary pulsars (see below) allow gravitational masses of the component stars to be measured, sometimes with high precision (see Table 2).

6.3.1 MASSES AND SOFT EQUATIONS OF STATE   The high mass of PSR 1913+16 rules out, for example, the Reid soft-core equation of state of Pandharipande (1971) and that of Canuto & Chitre (1974), and is severely constraining to modern Kaon-condensate models. Even more severe constraints may well emerge in the future. The pulsars with the best determined masses so far are slowly rotating and in double neutron star binaries, for which standard evolutionary models (cf Verbunt 1993) do not require much accretion onto the pulsar. Nonstandard models with much higher accretion rates during common envelope evolution (Chevalier 1993) would turn most high mass X-ray binaries into black hole binaries, particularly if the equation of state is soft. However, even standard models for the evolution of many other binary pulsars predict large accreted masses, unfortunately all in systems where the neutron star masses are still ill-determined. Equation (3.2) shows that for a pulsar like 1957+20 to reach its

1.6 ms spin period, it must have accreted $\gtrsim 0.1\,M_\odot$. Models of the formation of pulsars with low mass white dwarf companions (the large fourth group in Table 1) suggest that much of the $\sim 0.7\,M_\odot$ lost by the progenitor of the white dwarf should have been accreted by the neutron star. Mass-conservative models are most compelling for the long-period binaries ($P_b > 50$ d) for which the predictions of mass and angular-momentum conserving models give reasonable accord between the birthrates of such pulsars (Section 3.3; Lorimer 1994) and their low-mass X-ray binary progenitors (Sections 10.1, 10.3, Verbunt & van den Heuvel 1994). Accretion of mass ΔM_0 onto an initially 1.3–1.4 $M_\odot$ neutron star increases its gravitational mass by $\Delta M \simeq 0.74\Delta M_0$. [See Woosley & Weaver's table 1 (1986), Woosley et al's table 2 (1993), and Weaver & Woosley (1993) for discussion of the iron core masses at the end of massive star evolution, and the neutron star masses expected to result after supernova explosion.] Thus conservative models for the formation of the neutron stars with white dwarf binaries predict that they should have gravitational masses of $\simeq 1.8$–$1.9\,M_\odot$. These greatly exceed the maximum stable masses of equations of state with Kaon or pion condensates, and are in the midst of the range of maximum stable masses for modern equations of state without Kaon condensation (cf Table 3, and Cook et al 1994). It may prove possible to measure accurate masses for neutron stars in some of these binaries. This can be done in fortuitously edge-on systems such as 1855+09 (see Table 2), where relativistic time delays can be measured, and more generally by optical spectroscopy of the white dwarf companions, whose orbital velocities, combined with the mass function derived from pulsar timing, would determine a strict lower limit to M_p (e.g. the preliminary results on PSR1957+20 by Aldcroft et al 1992). Such measurements will allow us do decide which is wrong: conservative evolution models, or soft nuclear equations of state.

6.3.2 MINIMUM STABLE PERIODS AND EQUATIONS OF STATE   Just as soft equations of state give small neutron stars with low maximum masses, hard equations of state give large neutron stars with high maximum masses. Large neutron stars cannot spin rapidly, or the speed ΩR of matter at their surface would exceed the escape speed $[(2GM)/R]^{1/2}$. If one spins up from rest a neutron star of mass M_s and areal radius R_s, the minimum spin period at which the star begins to shed mass from its equator is, to a good approximation (cf Figure 27 of Cook et al 1994, and Haensel & Zdunik 1989)

$$P_{min} = 0.77 \left(\frac{1.4\,M_\odot}{M_s}\right)^{1/2} \left(\frac{R_s}{10\,\text{km}}\right)^{3/2} \text{ ms.} \tag{6.6}$$

If the star were a rigid Newtonian sphere, the coefficient of 0.77 would be 0.46—the difference is due to centrifugal distortion, which makes the equatorial radius of a spinning star larger than that of a static one, and to relativistic effects. Assuming the masses of the two fastest known pulsars, 1937+21 and 1957+20

Table 3   Neutron star properties for three nuclear equations of state[a]

EOS	Model	M ($M_\odot$)	P_{min} (ms)	I (10^{45} g cm^{-2})	R_e (km)	ϵ	ρ_c (10^{15} g cm^{-3})
F	canonical	1.4	0.94	0.86	9.1	0.26	3.2
(soft)	M_{max}	1.46	—	0.74	7.9	0.33	5.1
	Ω_{max}	1.67	0.50	—	10.9	—	4.4
FPS	canonical	1.4	0.88	1.2	10.8	0.21	1.3
(med.)	M_{max}	1.80	—	1.4	9.3	0.35	3.4
	Ω_{max}	2.1	0.53	—	12.4	—	3.0
L	canonical	1.4	1.4	2.1	15.0	0.15	0.43
(hard)	M_{max}	2.7	—	4.7	13.7	0.35	1.5
	Ω_{max}	3.3	0.76	—	18.2	—	1.3

[a]EOS = equation of state; P_{min} is the minimum rotation period for a model of mass M with the EOS; I is the moment of inertia (for slowly rotating stars); R_e is the equatorial circumferential radius (for slowly rotating stars, it is the same as the Schwarzschild radius); $\epsilon = z/(1 + z)$ is the fraction of rest mass energy released in dropping a particle onto the neutron star from infinity, where z is the surface redshift for slowly rotating stars; ρ_c is the central energy density/c^2, to be compared to the baryon density at nuclear saturation ~ 0.17 fm^{-3} $\simeq 2.8 \times 10^{14}$ g cm^{-3}. Three models are given for each equation of state: one for a canonical neutron star of gravitational mass 1.4 $M_\odot$, one for the slowly rotating star of the maximum stable mass M_{max}, and one for the star with maximum spin frequency Ω_{max} (nearly the same as the rotating star of maximum mass). Data are from Friedman & Ipser (1992), Cook et al (1994), and Arnett & Bowers (1977).

Equation of state F (Arponen 1972) is one of the softest not yet conclusively ruled out.

Equation of state FPS (Lorenz, Ravenhall & Pethick 1993) is a recent model without a pion or kaon condensate phase. Adding such a phase transition at a few times nuclear density would reduce M_{max} by $\simeq 0.2\,M_\odot$, R_e by $\simeq 1$ km, and P_{min} by $\sim 30\%$ (cf pp. 266–70 of Kunihiro et al 1993).

Equation of state L (Pandharipande & Smith 1975) is one of the hardest not yet conclusively ruled out.

to be $\sim 1.4\,M_\odot$, Equation (6.6) would require the radii of static neutron stars of that mass to be $\lesssim 16$ km. A precise mass for PSR 1957+20 may eventually be determined by optical spectroscopy of its companion (Aldcroft et al 1992).

Since reasonable equations of state give $P_{min} \simeq 0.7$ ms, and since pulsars with commonly observed dipole fields of $1–2 \times 10^8$ G have even shorter equilibrium periods at Eddington accretion rates in simple models (Figure 1), we are encouraged to hope that pulsars with much shorter spin periods may yet be found.

If such discoveries are not made, the result in isolation could be interpreted in several ways. An absence of shorter periods would naturally occur if neutron stars have a hard equation of state, and P_{min} is indeed ~ 1.5 ms (Friedman et al 1988). It might also occur if they have a very soft equation of state, and all neutron stars that accrete enough to spin them to $\lesssim 1.5$ ms exceed M_{max} and collapse to black holes. It might also occur if accretion cannot spin a neutron star up to breakup. Newtonian stars without magnetospheres can be spun nearly to breakup by accretion (Popham & Narayan 1991, Paczyński 1991). However, it has been suggested that gravitational radiation from unstable nonaxisymmetric modes would remove angular momentum from rapidly spinning neutron stars so quickly that they could never be spun to periods less than 1.5 ms (Papaloizou & Pringle 1978, Wagoner 1984). These high-m modes have negative energy and in any dissipationless rotating star grow unstably through gravitational radiation reaction (Friedman & Schutz 1978). (In a frame rotating with the neutron star these modes are retrograde, and when excited, reduce the star's angular momentum; as measured by an inertial observer at infinity, they are prograde and thus carry postive angular momentum away, increasing their amplitude.). More recent calculations suggest that these modes are strongly damped: by bulk viscosity for $T > 10^{10}$ K (Cutler et al 1990, Friedman & Ipser 1992), and by scattering of normal particles on superfluid vortices for $T < 10^9$ K (Mendell 1991, Lindblom & Mendell 1992). If correct, and neutron stars are superfluid, this extinguishes hope for gravitational radiation reaction as a period-limiting device.

Another possibility is that the accretion disk itself removes more angular momentum, and therefore exerts less torque, than in simple models (Ghosh & Lamb 1992). It might appear that escaping photons could remove angular momentum from the disk effectively (Miller & Lamb 1993). This is not correct, however, since the maximum photon torque from a luminosity L_γ escaping from radius r, where the matter azimuthal velocity is v, is

$$\max \dot{J}_\gamma = \left(\frac{L_\gamma}{c}\right) r \left(\frac{v}{c}\right) = \left(\frac{L_\gamma}{\dot{M}c^2}\right) \dot{M} v r \equiv \epsilon \dot{J}_m, \tag{6.7}$$

where $\dot{J}_m = \dot{M} v r$ is the rate of advection of angular momentum by matter, and $\epsilon = L_\gamma/\dot{M}c^2 \sim 0.2$ (see Table 3) is the efficiency of accretion. A more physically acceptable way to remove angular momentum is via a magnetic wind from the inner accretion disk or neutron star surface. This could set a lower limit to the spin period if its relative importance increased as the spin period decreased.

If, in future, a pulsar is found with $P \lesssim 0.8$ ms, there are two ways it could be used to put strong constraints on the equation of state. First, if its mass could be measured (e.g. by optical spectroscopy of its companion), and was found to be $\sim 1.4 M_\odot$, it would suggest a pion or kaon condensed phase, because equations of state without those cannot reach such P at that mass (Table 3, and Figure 6-4 of Kunihiro et al 1993). If the equation of state were hard, the star's large moment of inertia would require it to have accreted enormously, and have $M \gtrsim 2 M_\odot$; such a measured mass would rule out both soft, condensed, and even "medium" equations of state. Second, if it had a large $\dot{P}$ and were in a short-period binary, it might prove possible to determine accurately the pulsar's moment of inertia (or more precisely, $dJ/d\Omega$), giving a quantitative constraint on the equation of state. The moment of inertia would be determined as follows. As the pulsar spins down, it loses mass-energy at a rate $\dot{M} = I\Omega\dot{\Omega}/c^2$. This loss of gravitating mass causes the orbit to expand, and the orbital period to increase at a rate

$$
\frac{\dot{P_b}}{P_b} = \frac{-2\dot{M_p}}{M_p + M_c} = \frac{-2I\Omega\dot{\Omega}}{M_p c^2 + M_c c^2}
$$

$$
\simeq \frac{I\Omega^2}{Mc^2}\frac{1}{\tau_c} \simeq 0.016 I_{45} M_{1.4}^{-1}\tau_c^{-1} P_{ms}^{-2}, \tag{6.8}
$$

where $M_{1.4} = M/(1.4\,M_\odot)$. To avoid confusion of the measured $\dot{P_b}/P_b$ by Galactic acceleration (see Section 8) would require $\tau_c P_{ms}^2 \leq 10^{8.5}$ y, hence the need for large $\dot{P}$ and short period. A pulsar similar to 1937+20 in a short-period binary would do.

There is also an interesting possibility that very massive pulsars with periods near P_{min} may have $\dot{P} < 0$, even though they are losing angular momentum. This is because although $\dot{J} = (dJ/d\Omega)\dot{\Omega} < 0$, and usually $dJ/d\Omega \simeq I$, so $\dot{\Omega} < 0$ and $\dot{P} > 0$, relativistic stars supported against collapse by rotation can have $dJ/d\Omega < 0$, so $\dot{\Omega} > 0$ when the star loses angular momentum (Cook et al 1992, 1994). The stars with $dJ/d\Omega < 0$ are those which are so massive that they are just barely supported by rotation against collapse to black holes. As such a neutron star loses angular momentum, its J approaches a few percent of the minimum value at which it can be supported at a large (quasi-Newtonian) radius. It then begins to evolve towards a second (unstable) equilibrium state of the same angular momentum, but with a smaller radius (confined by the enhanced gravitational attraction of strong-field relativity). As the star contracts at nearly constant J, its spin period must decrease by a small but finite amount, just before the pulsar collapses to a black hole. Negative observed $\dot{P}$s can be produced by accelerating pulsars with positive $\dot{P}$s (see Section 8), however, so caution would be required in interpreting any observed negative $\dot{P}$ as due to this manifestation of the effect of general relativity on neutron stars.

7.   TESTING GENERAL RELATIVITY

Millisecond and binary pulsars have been used to perform several high-precision tests of general relativity, its axioms, and predictions. Because these have been the subject of several recent reviews (Backer & Hellings 1986, Damour & Taylor 1992, Taylor et al 1992, Taylor 1992, Damour 1992; the book by Will 1993 gives a pedagogical introduction), we confine ourselves to a brief summary.

A pulsar has a much larger gravitational (binding energy) contribution to its mass-energy than a white dwarf or main sequence star. In most theories of gravity other than Einstein's, a pulsar and its white dwarf companion would therefore fall through the Galactic gravitational field with different accelerations, so their relative orbit would be eccentric (Nordvedt 1968, Damour & Schäfer 1991). The low eccentricity of the binary pulsars 0820+02 and J1803 − 2712 (see Table 1) already allow one to deduce that the mass-equivalent of gravitational binding energy feels the same acceleration as do baryons to within 0.2%. This limit is comparable to the best Solar System limits for violations linear in M/R, and much better for nonlinear violations. The best similar test of the weak equivalence principle for weak interaction energy was provided by the simultaneous arrival of photons and neutrinos from supernova 1987A (Krauss & Tremaine 1988).

The weak-field effects of general relativity (which have been verified in Solar System experiments) are all measured in binary pulsars: gravitational redshift, Shapiro delay, and the advance of periastron. In some cases the amplitudes are spectacular: The periastron of the orbit of PSR 1913+16 has advanced by nearly 90 degrees since its discovery in 1974 (Hulse & Taylor 1975). By contrast, relativity has contributed only 10^{-2} degrees to the advance of the perihelion of Mercury since the non-Newtonian portion was identified by LeVerrier and Newcomb. If one assumes the correctness of relativity, the parameters of these weak-field effects can be used to solve for the Newtonian parameters— individual stellar masses and orbital inclination, which are ordinarily unmeasurable in a Newtonian single-line spectroscopic binary. In some cases, the Newtonian quantities are overdetermined by the relativistic parameters, and the agreement between different determinations can be used to provide precision tests of relativity, and to constrain the post-Newtonian parameters of alternative theories of gravity (Damour & Taylor 1992, Taylor et al 1992, Will 1993).

With the Newtonian parameters determined, the rates of more exotic relativistic effects can be predicted and compared with observation. In PSR 1913+16, the rate of decrease of the orbital period due to the emission of gravitatonal radiation is predicted by the quadrupole formula of general relativity (Peters 1964, Thorne 1980), $\dot{P}_b/P_b = -8.60924\,(14) \times 10^{-17}\,\mathrm{s}^{-1}$, and observed to be $\dot{P}_b/P_b = -8.63\,(4) \times 10^{-17}\,\mathrm{s}^{-1}$ (after correction for the Galactic acceleration: see Section 8; Taylor 1992, Damour & Taylor 1991). This agreement to within 0.5% is a stunning confirmation of Einstein's prediction of waves in a dynamical space-time. It will soon be possible to repeat this test in pulsars 1534+12 and 2127+11C, with respectively greater and lesser precision.

Another relativistic effect, which ought to be measurable in one of these pulsars, is geodetic precession. The spin axis of a pulsar is parallel transported around the curved spacetime of its companion (torques on its quadrupole moment by companions are negligible), and in relativity is therefore predicted to precess at a rate $d\widehat{S}/dt = \Omega_p|\widehat{L} \times \widehat{S}|$, where $\widehat{S}$ is the unit vector along the pulsar spin, $\widehat{L}$ that along the normal to the orbit plane, and $\Omega_p \sim GM_c/(ac^2)$ is 1.2 degrees per year for PSR 1913+16, and 0.4 degrees per year for PSR 1534+12. Despite careful measurements, this has not been convincingly seen in PSR 1913+16 (Cordes et al 1990), but the pulse properties of PSR 1534+12 are more favorable, and its $|\widehat{L} \times \widehat{S}|$ appears larger.

Additional, and even more precise tests of general relativity would be possible in a binary system in which both components were millisecond pulsars. Such systems could form in a globular cluster by exchange, or by spin up from tidal debris in an encounter of two (neutron star–main sequence star) binaries (Sigurdsson & Hernquist 1992). Another type of binary system that ought to exist is a (young) pulsar orbiting a stellar mass black hole. Such black holes could either have formed directly in a supernova explosion, or could be neutron stars pushed beyond the maximum stable mass by accretion from the massive progenitor of the pulsar. If the orbital period were short, the black hole mass could be determined accurately by means of weak-field relativistic effects. However, unless the orbit happens to be improbably close to edge on, strong-field effects of the metric near the black hole horizon (e.g. faint "glory" pulses) are unlikely to be detectable.

8.   LARGE SCALE ACCELERATIONS

8.1   *Newtonian Accelerations*

The spin and orbital periods we measure for pulsars are shifted from their values in a local Lorentz frame moving with the pulsar or its binary center mass, respectively. The largest effect is the first-order Doppler shift. A pulse period P measured at the Solar System barycenter (velocity V_b) is thus related to a pulsar's rest-frame pulse period P_0 by $P = P_0[1 + (V_p - V_b) \cdot n/c]$, where V_p is the pulsar velocity and n is the unit vector pointing from the barycenter to the pulsar. If the Doppler shift were constant, the observed period would just differ from the true one by a small, uninteresting, and unmeasurable factor. However, the Doppler shift is never truly constant, because pulsars move (n varies) and are accelerated by gravititational forces (V_p varies). Differentiating the Doppler equation once more gives, to lowest order in v/c (Phinney 1992, 1993)

$$\frac{\dot{P}}{P} = \frac{\dot{P}_0}{P_0} + \frac{V_\perp^2}{cD} + \frac{(a_p - a_b) \cdot n}{c} = \frac{1}{2\tau_{c0}} + \frac{1}{10^{10}\,\text{y}} \left(\frac{V_\perp}{150\,\text{km s}^{-1}} \right)^2 D_{\text{kpc}}^{-1}$$

$$+ \frac{1}{4 \times 10^{10}\,\text{y}} \frac{(a_p - a_b) \cdot n}{A_\odot}, \qquad (8.1)$$

where $a = \dot{V}$ is acceleration, D is the distance to the pulsar, $V_\perp$ is the transverse relative velocity μD, μ is the pulsar proper motion, and $A_\odot$ is the acceleration of the solar local standard of rest about the center of the Galaxy.

These Doppler corrections to (8.1) are not important for the young majority of pulsars ($\tau_{c0} \simeq 10^7$ y), but they dominate the apparent $\dot{P}$ for many millisecond and binary pulsars, complicating determination of their true characteristic ages and magnetic field strengths. The second (centrifugal) term in (8.1) is, like the intrinsic $\dot{P}_0$, always positive, and is important for old, nearby pulsars with large space velocities (see Table 1 and Camilo et al 1994). The third (acceleration) term, which can have any sign, is not important to the $\dot{P}$s of Galactic pulsars (though it is important for their $\dot{P}_b$s, and uncertainty in estimating it limits the precision of measurement of the rate of decay of orbits by gravitational radiation reaction). Equations for estimating the term for Galactic pulsars can be found in Damour & Taylor (1991) and Phinney (1992). The third term is, however, extremely important for pulsars in globular clusters, and in several cases makes negative the apparent period derivative. Methods for predicting this term are discussed at length by Phinney (1992, 1993), as are the use of pulsar positions and negative $\dot{P}$ to determine model-independent mass-to-light ratios and other constraints on the structure of globular clusters.

8.2 *Gravitational Waves*

Pulsars can be accelerated by non-Newtonian forces as well. If the Earth and the millisecond pulsars surrounding it were bobbing in a sea of gravitational waves of periods ~years, the waves would introduce irregularities in the timing of the pulsars (Detweiler 1979). As it passes the Earth (or the pulsar), a gravitational wave of frequency f and dimensionless amplitude h of wavelength c/f shorter than the distance between the pulsar and Earth will periodically modulate the spacetime, and thus the arrival times of pulses, by of order $h/(2\pi f)$. The energy density in the waves, of order $f^2 h^2 c^2/G$, would be a fraction $\Omega_g \sim f^2 h^2/H_0^2$ of the closure density of the universe, with Hubble constant H_0. Thus the timing residuals produced by a gravitational wave background with Ω_g are of order $100\,\Omega_g^{1/2}(f/1\,\mathrm{y}^{-1})^{-2}\,\mu\mathrm{s}$ (the numerical coefficient depends somewhat on the number of pulsar parameters fitted by timing, the wave spectrum assumed, and the duration of the observations, assumed longer than $1/f$; Blandford et al 1984). Limits on these, most strongly from PSRs 1937+21 and 1855+09, constrain at 95% confidence the energy density in such waves to a fraction $\Omega_g < 6 \times 10^{-8} h_0^{-2}$ of the closure density of the universe (Kaspi et al 1994b; here h_0 is the Hubble constant in units of $100\,\mathrm{km\,s^{-1}\,Mpc^{-1}}$). This is a severe constraint on the presence of cosmic strings; in the scale-free model of Bennett & Bouchet (1991), the string mass-energy per unit length μ is constrained to $G\mu/c^2 < 71\,\Omega_g h^2 = 4 \times 10^{-6}$, while in the models of Caldwell & Allen (1992), which include additional modes of string excitation and emission by strings in the matter-dominated era and thus have more power at long periods, $G\mu/c^2 \lesssim 10^{-6}$. Pulsar timing is thus close to detecting, or ruling out, the presence of strings interesting for galaxy formation.

9. PLANETARY COMPANIONS

A planet of mass m orbiting a pulsar of mass M_p in an orbit of semimajor axis a and period P_y years moves the pulsar back and forth relative to Earth, and thus produces a periodic timing residual of semiamplitude

$$\frac{a_p \sin i}{c} = \frac{m}{M_p + m} \frac{a \sin i}{c} \simeq 1200 \frac{m}{M_\oplus} \left(\frac{1.4 \, M_\odot}{M_p + m} \right)^{2/3} P_y^{2/3} \sin i \, \mu s. \quad (9.1)$$

For example, in orbits of one year period around a $1.4 \, M_\odot$ neutron star at $i = 60°$, a Jupiter would produce a residual of semiamplitude 0.33 s, Earth of 1.0 ms, the Moon of 13 μs, and the asteroid Ceres 0.2 μs. Some millisecond pulsars can be timed to accuracies $\lesssim 1 \, \mu$s, so if they were surrounded by planetary systems, even bodies as small as the moon would be potentially detectable. The precision of such measurements is some 2000 times higher than radial velocity measurements with precisions of $10 \, \mathrm{m \, s^{-1}}$ (about the best obtained so far in searches for planetary systems about ordinary stars using stabilized I_2 absorption cells as wavelength references—cf Cochran et al 1991; Jupiter moves the Sun at $10 \, \mathrm{m \, s^{-1}}$ through a displacement of 3 light seconds, which would require an angular resolution of 6 μarcsec to detect at 1 kpc).

Before a neutron star forms, its planetary environment is subjected to the sublimating heat of a red giant star and a supernova blast, and whatever is left at large radii will generally be left behind as the newborn neutron star recoils at the $\gtrsim 100 \, \mathrm{km \, s^{-1}}$ speed typical of pulsars. Only in carefully contrived circumstances (Wijers et al 1992, Thorsett & Dewey 1993)—"Salamander scenarios" in the notation of Phinney & Hansen (1993)—might primordial planets survive around pulsars. Many pulsars clearly do not have any planetary companions (Thorsett & Phillips 1992, Thorsett et al 1993c).

It was therefore something of a surprise when the first unambiguous evidence for a planetary system outside our Sun's was found around the 6 ms pulsar 1257+12 (Wolszczan & Frail 1992). Two planets circle this pulsar. Both planets are a few times the mass of the Earth, and have nearly circular ($e = 0.02$) orbits of period $P_{in} = 66.6$ d and $P_{out} = 98.2$ d (Table 1). The timing evidence for the planets of PSR1257+12 seems secure, since its position determined by interferometry agrees with the position derived from timing, and the timing has by now been confirmed on several telescopes, with different timing hardware, and analyzed with independent software and ephemerides (Backer et al 1992).

PSR1257 + 12's high spatial velocity (290 km s^{-1}, Wolszczan 1993) combined with the circularity of its planets' orbits, make it very unlikely that its planets are survivors from an era when the neutron star had not yet been formed ("Salamanders"). The planets must instead have formed after the neutron star's progenitor metamorphosed into a neutron star ("Memnonides scenarios," in the notation of Phinney & Hansen 1993). Since the pulsar has the short period and small dipole field characteristic of recycled pulsars, the most natural scenarios for planet formation are those in which the planets formed from the debris of a companion star. This may have been destroyed gradually (Banit et al 1993) or catastrophically (e.g. Phinney & Hansen 1993). Most of the mass would accrete, but to conserve angular momentum, a small fraction must be excreted outwards; planets could form in this cool excretion disk (Ruden 1993, Phinney & Hansen 1993). For a review of these, and many other proposed scenarios, the reader is referred to the proceedings *Planets Around Pulsars* (Phillips et al 1993), particularly the summary by Podsiadlowski (1993).

Like the Solar System, the planetary system of PSR1257+12 will provide entertainment for generations of dynamicists. The two planets are close to a 3:2 resonance: The inner planet revolves $2.95 \simeq 3$ times for every two revolutions (196.4 d) of the outer planet. Since the relative configurations of the planets repeat almost exactly every 196.4 days, their mutual effects on each others' orbits (e.g. their longitudes of periastron ω and orbital eccentricities e) can build up over many orbital periods. The planets' masses were initially known only from the Keplerian mass functon—i.e. only the $m \sin i$ were known. If we had been viewing the planetary system close to face-on ($\sin i < 0.1$), the masses of the planets would have had to have been so large that their mutual interactions would have been strong enough to make their angles librate, locking them in resonance (Malhotra et al 1992). This would already have been noticed in the pulsar timing (Wolszczan 1993), so $\sin i > 0.1$ and the planets are not locked in the 3:2 resonance. Consequently, the difference of the planets' phases with respect to the resonance rotates cyclically with period $P_{\rm d}$ given by $1/P_{\rm d} = 2/P_{\rm in} - 3/P_{\rm out} = 1/(5.3\,{\rm y})$. Thus the planets' ω, e, and a should vary almost periodically with period $P_{\rm d}$, with its amplitude $\propto m \propto 1/\sin i$ (Rasio et al 1992). Timing residuals of about the amplitude expected (Malhotra 1993, Peale 1993) for $\sin i \sim 1$ are already detected at about the 2σ level (Wolszczan 1993).

PSR1257 + 12 may not be the only millisecond pulsar with planetary companions. The 11 ms pulsar $1620 - 26$ in the globular cluster M4 has a $\ddot{P}$ (Backer et al 1993a, Thorsett et al 1993a) much larger than could be produced by the jerking of randomly passing cluster stars (Phinney 1993). One plausible interpretation is that the jerk is produced by a companion with a mass 0.1–10 times that of Jupiter in a very long period orbit (Backer et al 1993a, Phinney 1993, Sigurdsson 1993), a hypothesis readily testable by further timing. Unlike PSR 1257 + 12, PSR 1620 − 26 also has a stellar mass companion with orbital period $P_{\rm b} = 191.4\,{\rm d}$. The mass function, period, and circular orbit are all consistent with this companion being the $\sim 0.4 M_\odot$ helium white dwarf remains of the red giant, the accretion of whose envelope spun up the pulsar (see Phinney 1992, and Section 10). Though some material might have been excreted, the low metallicity of stars in M4 makes it difficult to form such a massive planet in such a wide orbit (Sigurdsson 1992). It is more likely that the planet is a primordial planet which originally orbited a main sequence star, and was retained as a bystander during the exchange of partners which put the pulsar in orbit about the main sequence star (Sigurdsson 1993). This star then evolved into the current white dwarf companion, spinning up the pulsar in the process. In this model, the planet is predicted to have an eccentric orbit. If further observations reveal the orbit to be circular, the first ("Memnonides") scenario would be implicated, though in this case too the orbit is more likely to be eccentric, due to perturbations of passing cluster stars.

10. DYNAMICAL FOSSILS OF THE SPIN-UP ERA

As discussed in Section 2 and reviewed at length by Verbunt (1990, 1993) and van den Heuvel (1992), most millisecond and binary pulsars are believed to be products of mass-transfer, and their predecessors should have been X-ray binaries. But direct evidence for a connection between millisecond pulsars and low-mass X-ray binaries is still tenuous, given the apparent discrepancy between their birthrates (Section 3.3) and that searches have failed to reveal millisecond pulsations in LMXB (Wood et al 1991, Kulkarni et al 1992a). For evidence of the connection, one must search for dynamical fossils in the orbits of binary pulsars. Such fossil evidence is clearest among the binary pulsars with low mass companions and orbital periods $\gtrsim 1$ d. The two fossils are the $P_b(M_c)$ relation (predicted by Refsdal & Weigert 1971), and the $e(P_b)$ relation (predicted by Phinney 1992), illustrated respectively in Figures 2 and 3.

10.1 *Core Mass–Period Relation*

The $P_b(M_c)$ relation arises when a neutron star is in a wide orbit with a companion star of $\lesssim 2\,M_\odot$. Such a binary begins mass transfer when the $\lesssim 2\,M_\odot$ companion, with its electron-degenerate core surrounded by a thin hydrogen burning shell, evolves off the main sequence. The high gravity of the compact core requires a steep pressure gradient across the burning shell. Thus the properties of the shell are almost independent of the envelope above it, and are little affected by the mass loss that begins far above. As the shell burns outwards, its helium ash joins the degenerate core, increasing its mass and decreasing its radius. The temperature of the shell increases to remain in hydrostatic equilibrium; because of the steep temperature dependence of nuclear reaction rates, the shell's luminosity is thus a rapidly increasing function of core mass (Refsdal & Weigert 1970). For giants of solar metallicity, $L = [M_c/(0.16\,M_\odot)]^8 L_\odot$. Because the decreasing thermal conductivity of the increasingly large and opaque envelope ultimately forces the envelope to be convective (Renzini et al 1992), the giant lies on the Hayashi track $T_{\text{eff}}(L, M)$. As $L = 4\pi R^2 \sigma T_{\text{eff}}^4$, these relations give the radius R of the giant as a function of M_c:

$$R \simeq 1.3\left(\frac{M_c}{0.16\,M_\odot}\right)^5 R_\odot \qquad (10.1)$$

for solar metallicity. During mass transfer, the giant of mass M must fill its Roche lobe, whose volume-equivalent radius R_{L1} is given for a giant less

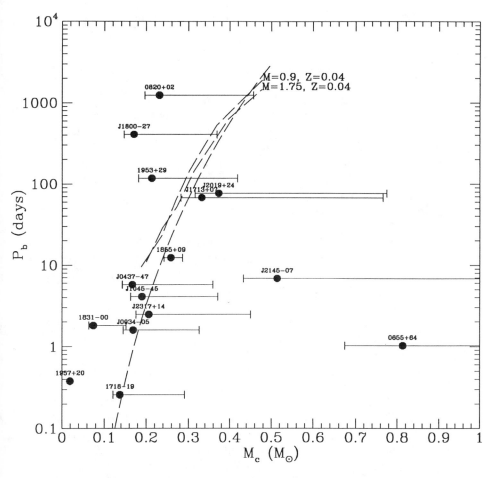

Figure 2   The predicted relation between core mass (i.e. final white dwarf mass) and orbital period for binaries in which a $\lesssim 1\,M_\odot$ giant star fills its Roche lobe with a $1.4\,M_\odot$ neutron star (*solid curve*, Refsdal & Weigert 1971; *dashed curves:* data from Sweigart & Gross 1978). The filled circles with error bars give most probable, and 90% confidence estimates of the masses of the white dwarf companions of the indicated pulsars (assumed, when not known, to be $1.4\,M_\odot$). All Galactic binary pulsars with orbital eccentricites <0.1 are shown. Note that hydrogen shell burning giants exist only for $0.16 < M_c < 0.45\,M_\odot$. White dwarf companions of mass outside this range (e.g. those of PSRs $0655 + 64$, $J2145 - 0750$, $1957 + 20$, etc) had different evolutionary histories (see notes to Table 1 and text).

massive than the neutron star, within a few percent, by

$$\frac{GM}{R_{L1}^3} = 10\,\Omega_b^2,$$

(10.2)

where $\Omega_b = 2\pi / P_b$ is the orbital frequency. Towards the end of transfer, when $M \simeq M_c$, (10.1) and (10.2) therefore give

$$P_b \simeq 1.3 \left(\frac{M_c}{0.16 \, M_\odot} \right)^7 \, \text{d}. \tag{10.3}$$

This is valid for $0.16 \, M_\odot < M_c < 0.45 \, M_\odot$, where the lower limit is the helium core mass at the end of main sequence evolution, and the upper limit is the core mass at the helium flash, for which $P_b \simeq 2000 \, \text{d}$. A more accurate version of Equation (10.3), derived from computed stellar evolutionary models, is plotted in Figure 2. Other approximations are given in Refsdal & Weigert (1971) (see also Joss et al 1987 and Verbunt 1993).

During stable mass transfer, the expanding giant will always just fill its Roche lobe, with radius given by (10.1), until the envelope has been reduced to a few times the mass of the burning shell ($\lesssim 10^{-2} M_\odot$). At that point, the envelope will become radiative (cf Renzini et al 1992) and begin to shrink on its thermal timescale (Refsdal & Weigert 1970, Taam 1983), and mass transfer will cease. As the burning shell consumes the shrinking envelope, it eventually has too little overburden to confine it, hydrogen burning ceases, and the core cools, becoming a helium white dwarf companion to the spun-up neutron star. The orbit should be nearly circular because tidal dissipation in the giant would have rapidly circularized the orbit (see Section 10.2).

The fossil evidence for stable mass transfer from a Roche-lobe filling giant star is therefore: a helium white dwarf companion of mass M_c between $0.16 \, M_\odot$ and $0.45 \, M_\odot$, in a circular orbit with period (between 1 and 2000 days) given by the P_b–M_c relation [Equation (10.3) or the more accurate versions plotted in Figure 2]. It is evident from Figure 2 that all 10 of the binary pulsars in the fourth group of Table 1 are consistent with this prediction. The other binary pulsars in the table, with companion masses less than $0.16 \, M_\odot$, or greater than $0.45 \, M_\odot$, cannot have had this evolutionary history, and therefore are not expected to obey the P_b–M_c relation (10.3). PSRs $0655 + 64$ and J2145 − 0750 have white dwarf companions with masses greater than the $\sim 0.45 \, M_\odot$ core mass at helium flash, and therefore cannot be the result of mass transfer from a red giant. Their properties are, however, just what would be expected from the evolution of a wide binary containing a neutron star and a main sequence companion more massive than $2 \, M_\odot$ (see note to Table 1 and Table 4; Iben & Livio 1993 review the physics of the common envelope phase expected in these circumstances, cf also van den Heuvel 1992).

10.2   *Eccentricity-Period Relation*

Another fossil of the mass transfer phase was recently pointed out by Phinney (1992). As a function of orbital phase, the difference in pulse arrival times between a pulsar in a circular orbit and one with a small eccentricity e will be

$$\delta t = \Delta t(t) - \Delta t_{\text{circ}}(t) = \frac{ea_p \sin i}{2c}[\sin \omega + \sin(2\pi t/P_b - \omega)]. \qquad (10.4)$$

Thus for a typical binary pulsar with $a_p \sin i/c = 10$ s, timing accuracy of 1 μs allows us to detect e as small as 1 μs/10 s $= 10^{-7}$.

After the supernova (or accretion-induced collapse) which creates a neutron star in a low mass binary, the orbital eccentricity will be high (>0.1). When the companion evolves to become a red giant, the time-dependent tides induced in the giant by the neutron star will be exponentially damped by convective eddy viscosity (Zahn 1966, 1977), on a timescale $\sim 10^4$ y, much shorter than the lifetime of the giant $\sim 10^7$–10^8 y. If this were the entire story, the orbital eccentricities would be predicted to be of order $\exp(-1000)$. As is clear from Table 1 and Figure 3, the observed eccentricities are small, but measurably nonzero, in the range 10^{-6}–10^{-2}. Perturbations by passing stars are inadequate to induce such eccentricities (Phinney 1992).

However, the fluctuation-dissipation theorem reminds us that the tidal dissipation cannot be the whole story. The fluctuating density of the convection cells in the convective giant star produces time-dependent moments of quadrupole and higher order. The neutron star in its orbit feels the noncentral forces produced by these fluctuating multipoles, and these randomly pump the eccentricity of the orbital motion. Dissipation of the time-dependent tide always tends to damp the eccentricity of the orbit. The resulting epicyclic motion is thus like that of a pendulum in air, which is continually bumped by air molecules. The pendulum is excited into Brownian motion—but if the amplitude becomes large, air drag slows it down. In equilibrium, a pendulum in air has energy that random walks on the drag timescale around a mean value of kT.

Phinney (1992) proves an analogous theorem for the binary star system: If convection is confined to a single thin layer, a (statistical) equilibrium eccentricity is reached when the energy of the orbital epicyclic motion is equal to the energy in a single convective eddy. The driving and damping forces can be summed over the convective layers of model red giants as a function of time during the mass transfer. The squared eccentricity random walks about its equilibrium value until the end of mass transfer, when the giant envelope begins to shrink as it becomes radiative (see Section 10.1). At that point, the tidal damping time rapidly increases, becoming longer than the evolution time, and the current (random) value of the eccentricity is frozen into perpetuity (Phinney 1992) as the star becomes a white dwarf. The predictions of this theory (Figure 3) are in good agreement with the measured eccentricities of the orbits of pulsars with white-dwarf companions—a remarkable result, since the theory has no adjustable parameters! It is curious that the orbit of PSR J2145 $-$ 0750, which clearly must have undergone spiral-in during unstable mass transfer (see Figure 2), lies in the predicted band along with all the other binaries. This

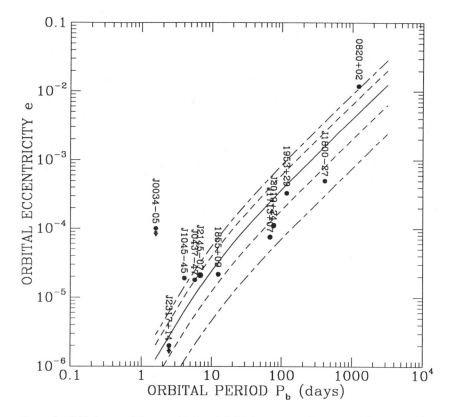

Figure 3 Orbital eccentricity vs orbital period for the pulsars whose companion masses and periods are consistent with their having formed by stable mass transfer from a Roche-lobe filling red giant (i.e. along the line of Figure 2). The solid line is the median eccentricity predicted by the convective fluctuation-dissipation theory of Phinney (1992; see also text). The inner and outer dashed lines are predicted to contain respectively 68% and 95% of the final eccentricities.

suggests that the entire red giant envelope was not ejected during the common-envelope phase, and the companion, of reduced mass, subsequently filled, or nearly filled its Roche lobe for $\gtrsim 10^5$ y.

10.3   *Puzzles in Mass Transfer*

As we have seen, conservative Roche lobe transfer models of binary evolution are successful at explaining the periods, masses, eccentricities, and, to some extent birthrates (Section 3.3) of LMBPs. However, some puzzles and issues of self-consistency remain.

The Eddington rate for accretion of a cosmic hydrogen/helium mixture onto a 1.4 $M_\odot$ neutron star is $1.9 \times 10^{-8}(0.2/\epsilon)M_\odot\,y^{-1}$, where ϵ is the surface binding energy given in column 7 of Table 3. Because large giant stars have nuclear evolution times shorter than 10^8 y, mass transfer on their nuclear timescale can lead to super-Eddington accretion rates. For standard conservative models of transfer, this occurs (see e.g. Figure 8 of Verbunt 1993) in binaries with initial orbital periods exceeding $\sim$10 d (hence final periods $\gtrsim$100 d). The evolution of such systems is therefore unlikely to be conservative of mass. If matter is ejected only from the regions of the accretion disk near the neutron star, tidal torques (Priedhorsky & Verbunt 1988 and references therein) on the outer part of the disk may still keep the evolution conservative of angular momentum. The evolution will then have the same stability properties as conservative transfer (see Table 4, case of a convective mass loser), and the evolution of the binary will not be qualitatively affected. If, however, angular momentum were lost (e.g. by ablation of the giant star), the evolution would be dramatically different, leading to unstable transfer on a dynamical timescale. The existence of pulsars with white dwarf companions near the predicted M_c-P_b relation for $P_b > 100$ d, and the rough agreement for long-period systems between the pulsar and the X-ray binary birthrates (inferred assuming evolution on the nuclear timescale), suggests that such angular momentum loss is *not* common. For pulsars with initial orbital periods between 2 and 10 d (final periods between 20 and 100 d), simple conservative models do not have any obvious physical contradictions, and the final period is almost linearly proportional to the initial period.

Table 4 Stability of mass loss in binary stars

CONVECTIVE LOSER	conservative transfer	nonconservative (of M, J) transfer	partly noncons. (conserve J, not M)
$M_{loser} \gg M_{gainer}$	unstable	stable	unstable
	$M_{loser} = \frac{2}{3}M_{gainer}$	$M_{loser} = M_{gainer}$	
$M_{loser} \ll M_{gainer}$	stable	unstable	stable

RADIATIVE LOSER	conservative transfer	nonconservative (of M, J) transfer	partly noncons. (conserve J, not M)
$M_{loser} \gg M_{gainer}$	delayed dynamically unstable	stable	delayed weakly unstable
	$M_{loser} = 2 M_{gainer}$ thermally unstable		
	$M_{loser} = 1.3 M_{gainer}$		
$M_{loser} \ll M_{gainer}$	stable	stable	stable

A different problem exists for short-period systems (initial period $\sim$0.7 d, final periods $1 \lesssim P_b \lesssim 6$ d for a 1 $M_\odot$ donor). For these, the companion comes into contact with its Roche lobe just as it is evolving off the main sequence, while it is still mostly radiative. For such stars, mass loss causes the star to shrink within its Roche lobe (see Table 4, and discussion in Hjellming & Webbink 1987). Thus the mass transfer will cease unless there is an angular momentum loss mechanism (e.g. magnetic braking) to shrink the orbit. The models thus predict a bifurcation, wherein systems with initial periods slightly less than 0.7 d spiral together to very short orbital periods, while systems with initial periods slightly greater spiral out to final periods $\gtrsim$6 d (Pylyser & Savonije 1988). For pulsars with initial periods between $\sim$0.7 d and 2 d, these models (see also Coté & Pylyser 1989) predict a steep dependence of the final period (0.7 to 20 d) on the initial period. Intermediate final periods are therefore expected to be very rare, since they require a fine-tuning of the initial period. The range of rare final periods can be reduced somewhat by increasing the donor mass into the range where mass transfer is initially unstable (Pylyser & Savonije 1988). Examining the list of the pulsars in Table 1, we see that the pulsars listed between PSR J0437 $-$ 4715 and PSR 1831 $-$ 00, inclusive, fall within this difficult category of intermediate periods. Furthermore, the birthrate inferred from these pulsars' discovery (Lorimer 1994; see also Section 3.3) seems substantially larger than the birthrate of X-ray binaries inferred in these models (Coté & Pylyser 1989), which suggests that we do not fully understand the origin of these systems.

The difficulties with the birthrates of these short-period systems may be related to the even greater difficulties with the birthrates of the "black widow" pulsars (PSRs 1957+20 to J0322+2057 in Table 1). These have companions of zero or very low mass ($<0.02 M_\odot$), yet have clearly been spun up by accretion from a companion now digested or destroyed. Despite initial enthusiasm (van den Heuvel & van Paradijs 1988, Phinney et al 1988, Ruderman et al 1989) following the discovery of the wind from the companion of the eclipsing pulsar PSR 1957 $+$ 20 (Fruchter et al 1988), it now seems that on both theoretical (Levinson & Eichler 1991) and observational grounds (Fruchter & Goss 1992) that PSR 1957 $+$ 20 is not significantly ablating its companion. If so, it makes unlikely the otherwise attractive model of van den Heuvel & van Paradijs (1988), in which a low mass companion has its orbit shrunk by magnetic braking to the "period gap," and is then ablated. The model proposed by Phinney et al (1988), in which the nonconservative mass loss from a Roche-lobe filling companion becomes unstable as its mass drops to the point that it becomes convective (see Table 4), is perhaps still viable, though it is doubtful if the remnant of the unstable mass transfer would be as small as $0.02 M_\odot$. The X-ray lifetimes in these models may be long or short, depending on the transport of the pulsar's heating (see Section 6.1.2).

Another process is the recoil of a newly formed neutron star into, or nearly into, its main-sequence companion. This would result respectively in the disruption, or severe bloating and loss of mass of the companion. The former could result in single pulsars; planets may form from the debris as in PSR1257 + 12 (Phinney & Hansen 1993). The tidal heating of near-misses could result in systems like PSRs 1831 − 00 and 1718 − 19 [if the latter is, as seems likely, not in the globular cluster NGC 6342; Wijers & Paczyński (1993) present cluster models]. Better aimed neutron stars might sink into their companions and form a single pulsar via a Thorne-Żytkow object (Leonard et al 1994). All versions of disruptive recoil will certainly result in a very short-lived X-ray phase, and thus avoid difficulty with the large birthrate of these types of pulsars relative to the number of X-ray sources (Section 3.3).

11. PULSARS AS PLASMA PROBES

The radio emission from several binary pulsars (PSRs 1957 + 20, 1718 − 19, 1744 − 24A and 1259 − 63) is delayed, pulse-smeared, and even eclipsed by plasma emitted from their companion stars. Thompson et al (1994) give a comprehensive and critical analysis of mechanisms for the pulse-smearing and eclipses. Remarkably, it appears that because of its high brightness temperature, pulsar emission can actively modify the plasma through which it passes (as lasers do to hydrogen pellets in inertial-confinement fusion experiments), and the eclipses of PSR1744 − 24A are most likely due to stimulated Raman scattering (Thompson et al 1994).

Pulsar radio waves propagating through the interstellar medium are more passive. Fluctuations in the electron density on scales $\sim 10^9$ cm, smaller than the Fresnel scale $(\lambda D/2\pi)^{1/2} \sim 10^{11}$ cm, contribute to the (observationally irritating) diffractive scintillation of pulsars (reviewed by Rickett 1990, Narayan 1992), while fluctuations on larger scales ($\sim 10^{13}$ cm) are probed by the long-term intensity fluctuations associated with refractive scintillation. Density fluctuations on still larger scales ($\sim 10^{13}$–10^{15} cm) can be probed by timing millisecond pulsars at multiple frequencies, and measuring the variations in the pulsar's dispersion measure as its motion changes our line of sight to it. Such precision measurements (Backer et al 1993b) complement low-frequency measurements on ordinary pulsars (Phillips & Wolszczan 1991, 1992), and tell us about the spectrum of typical fluctuations on the same scales as the rare "extreme scattering events" discovered by Fieldler et al (1987), and recently seen in both flux and delay in the millisecond pulsar 1937+21 (Cognard et al 1993). Differences in the dispersion measure of pulsars in globular clusters probe fluctuations on still larger scales ($\sim 10^{17}$ cm) (Anderson 1992), and the cluster pulsars also tell us that the scale height of ionized gas in the Galaxy is ~ 0.6–0.8 kpc (Bhattacharya & Verbunt 1991, Nordgren et al 1992).

12.   WHAT NEXT?

The discovery of millisecond and binary pulsars has revitalized the pulsar field. We expect that searches for millisecond pulsars will continue into the next millennium. Undoubtedly there will be some delights (a nearby long-period pulsar with a low-mass main sequence companion; a millisecond pulsar with an infrared-luminous asteroid belt and nine planets; a pulsar interacting with a magnetic white dwarf companion; a millisecond pulsar in a triple star system; a source that switches between an LMBP and an LMXB; and perhaps even a pulsar in a black hole binary) and surprises. The searches will improve our knowledge of the variety and demography of millisecond pulsars and their binary companions. Our present knowledge of the luminosity function, crucial to a quantitative understanding of the birthrate problem and estimation of the pulsar population in clusters, is in a state of some confusion. The Arecibo data suggest that unlike ordinary pulsars, millisecond pulsars may have a high minimum luminosity, ~ 10 mJy kpc^2. However, the detection efficiency at Arecibo is surprisingly higher than that of the Parkes survey, inconsistent with the previous conclusion.

As discussed in Section 5, the origin of the low magnetic field strengths of millisecond pulsars is an outstanding mystery. The bimodal field strength distribution might be interpreted to mean that LMBPs are a new class of neutron stars, which are born as millisecond pulsars (Arons 1983, Pacini 1983), e.g. born via the AIC mechanism. However, the standard recycling model is still more attractive. The high incidence of binarity, the P_b-M_c relation (Section 10.1) and the P_b-e relation (Section 10.2) clearly indicate that millisecond pulsars emerge from binary systems that had substantial mass transfer after the neutron star formed—consistent with the standard model, but inexplicable in AIC models if accretion ceased after formation of the neutron star. In models in which mass-accretion reduces the dipole field (Section 5.7), the observed bimodal distribution of magnetic field strengths would result from the very different histories of mass transfer of LMBPs and HMBPs. It will clearly be important to identify physical mechanisms by which accretion could lead to field reduction.

Accurate measurement of white dwarf masses, and the pulsar masses and kinematics offer the best quantitative checks of the formation scenarios. The present sample of LMBPs has $v_{rms} \sim 75$ km s^{-1}, which suggests that LMBPs suffered natal velocity kicks. It is important to measure velocities for as large a sample as possible and see if the predicted anti-correlation between P_b and v is seen (Bailes 1989, Johnston 1992). Finally, detection of millisecond pulsations (in radio or X rays) from LMXBs would provide direct evidence for the standard recycling model, in which millisecond pulsars were spun up by accretion.

It is of some importance to see if the histogram of magnetic field strength towards small B continues as the present (meager) data suggest. If so, there could be genuinely a large number of neutron stars with field strengths smaller than the weakest field LMBP, $B < 5 \times 10^7$ G. Such a hypothetical class would worsen the birthrate problem. A better understanding of the death line of such objects is needed as well as searches for sub-millisecond pulsars using coherent dedispersion techniques. X-ray and γ-ray detections will give insights into magnetospheric physics.

Finally, we predict that millisecond and binary pulsars will continue to perform serendipitous and exciting physics experiments. Just as commercial applications of the terrestrial Global Positioning System (GPS) are expanding exponentially, so applications of the pulsars' equally precise Galactic Positioning System are likely to expand, beyond gravitational wave detection, globular cluster dynamics, planetary dynamics, and convection in red giants. More systems in which the pulsar beams pass through companions and their winds will doubtless be found, as will systems in which a pulsar's wind had, and continues to have effects on its companion and environment. We have learned to expect unexpectedly exotic physics.

ACKNOWLEDGMENTS

ESP was partly supported by NASA Astrophysics Theory grant NAGW-2394 and the Alfred P. Sloan Foundation. SRK's work was supported by NASA, NSF, and the Packard Foundation.

Literature Cited

CALTECH—*Planets Around Pulsars,* ed. JA Phillips, SE Thorsett, SR Kulkarni. ASP Conf. Ser. Vol 36. San Francisco: Astron. Soc. Pac. (1993)
SANTA BARBARA—*X-Ray Binaries and Formation of Binary and Millisecond Pulsars,* ed. E van den Heuvel, S Rappaport. Dordrecht: Kluwer (1992)
XRAYBIN—*X-Ray Binaries,* ed. WHG Lewin, J van Paradijs, EPJ van den Heuvel. Cambridge: Cambridge Univ. Press. In press (1994)
Abramovici A, Althouse WE, Drever RWP, Gursel Y, Kawamura S, et al. 1992. *Science* 256:325–33
Aldcroft TL, Romani RW, Cordes JM. 1992. *Ap. J.* 400:638–46
Anderson SB. 1992. *A study of recycled pulsars in globular clusters.* PhD thesis. Calif. Inst. Technol.
Applegate JH. 1992. *Ap. J.* 385:621–29
Arnett WD, Bowers RL. 1977. *Ap. J. Suppl.* 33:415–36
Arons J. 1983. *Nature* 302:301–5
Arons J. 1993. *Ap. J.* 408:160–66
Arons J, Tavani M. 1993. *Ap. J.* 403:249–55
Arponen J. 1972. *Nucl. Phys.* A191:257–82
Arzoumanian Z, Fruchter AS, Taylor JH. 1994. *Ap. J.* 426:L85–L88
Backer D, Sallmen S, Foster R. 1992. *Nature* 358:24–25
Backer DC, Foster RS, Sallmen S. 1993a. *Nature* 365:817–19
Backer DC, Hama S, Van Hook S, Foster RS. 1993b. *Ap. J.* 404:636–42
Backer DC, Hellings RW. 1986. *Annu. Rev. Astron. Astrophys.* 24:537–75
Backer DC, Kulkarni SR, Heiles C, Davis MM, Goss WM. 1982. *Nature* 300:615–18
Backus PR, Taylor JH, Damashek M. 1982. *Ap. J. Lett.* 255:L63–67
Bailes M. 1989. *Ap. J.* 342:917–27
Bailes M, Harrison PA, Lorimer DR, Johnston S, Lyne AG, et al. 1994. Preprint

Bailyn CD. 1993. *Ap. J. Lett.* 411:L83–85
Banit M, Ruderman MA, Shaham J, Applegate JH. 1993. *Ap. J.* 415:779–96
Becker W, Trümper J. 1993. *Nature* 365:528–30
Bell JF, Bailes M, Bessell MS. 1993. *Nature* 364:603–5
Bennett DP, Bouchet FR. 1991. *Phys. Rev. D* 43:2733–35
Bhattacharya D, Srinivasan G. 1986. *Curr. Sci.* 55:327–30
Bhattacharya D, Srinivasan G. 1994. In XRAYBIN
Bhattacharya D, van den Heuvel EPJ. 1991. *Phys. Rep.* 203:1–124
Bhattacharya D, Verbunt F. 1991. *Astron. Astrophys.* 242:128–32
Bhattacharya D, Wijers RAMJ, Hartman JW, Verbunt F. 1992. *Astron. Astrophys.* 254:198–212
Biehle GT. 1994. *Ap. J.* 420:364–72
Biggs JD, Bailes M, Lyne AG, Goss WM, Fruchter AS. 1993. Preprint
Blandford RD, Narayan R, Romani RW. 1984. *J. Astrophys. Astron.* 5:369–88
Blondin JM, Freese K. 1986. *Nature* 323:786–88
Brown GE, Kubodera K, Rho M, Thorsson V. 1992. *Phys. Lett. B* 291:355–62
Brown GE, Bethe HA. 1994. *Ap. J.* 423:659–64
Caldwell RR, Allen B. 1992. *Phys. Rev. D* 45:3447–68
Callanan PJ. 1992. *Publ. Astron. Soc. Pac.* 104:775–79
Callanan PJ, Charles PA, Hassal BJM, Machin G, Mason KO, et al. 1989. *MNRAS* 238:25P–28P
Camilo F, Nice DJ, Taylor JH. 1993. *Ap. J. Lett.* 412:L37–40
Camilo F, Thorsett SE, Kulkarni SR. 1994. *Ap. J. Lett.* 421:L15–18
Canal R, Isern J, Labay J. 1990. *Annu. Rev. Astron. Astrophys.* 28:183–214
Canuto V, Chitre SM. 1974. *Phys. Rev. D* 9:1587–613
Caraveo PA. 1993. *Ap. J. Lett.* 415:L111–14
Chanmugam G. 1992. *Annu. Rev. Astron. Astrophys.* 30:143–84
Chen K, Ruderman M. 1993. *Ap. J.* 402:264–70
Chevalier RA. 1993. *Ap. J. Lett.* 411:L33–36
Cochran WD, Hatzes AP, Hancock TJ. 1991. *Ap. J. Lett.* 380:L35–38
Cognard I, Bourgois G, Lestrade J-F, Biraud F, Aubry D, et al. 1993. *Nature* 366:320–22
Cook GB, Shapiro SL, Teukolsky SA. 1992. *Ap. J.* 398:203–23
Cook GB, Shapiro SL, Teukolsky SA. 1994. *Ap. J.* 424:823–45
Cordes JM. 1986. *Ap. J.* 311:183–96
Cordes J, Romani R, Lundgren S. 1993. *Nature* 362:133–35
Cordes JM, Wasserman I, Blaskiewicz M. 1990. *Ap. J.* 349:546–52
Coté J, Pylyser EHP. 1989. *Astron. Astrophys.* 218:131–36
Cowley AP, Hutchings JB, Crampton D. 1988. *Ap. J.* 333:906–16
Cutler C, Lindblom L, Splinter RJ. 1990. *Ap. J.* 363:603–11
Damour T. 1992. *Philos. Trans. R. Soc. London Ser. A* 341:135–49
Damour T, Schäfer G. 1991. *Phys. Rev. Lett.* 66:2549–52
Damour T, Taylor JH. 1991. *Ap. J.* 366:501–11
Damour T, Taylor JH. 1992. *Phys. Rev. D* 45:1840–68
D'Antona F, Ergma E. 1993. *Astron. Astrophys.* 269:219–30
Danziger IJ, Baade D, Della Valle M. 1993. *Astron. Astrophys.* 276:382–88
Detweiler S. 1979. *Ap. J.* 234:1100–4
Dewey RJ, Cordes JM. 1987. *Ap. J.* 321:780–98
Djorgovski SJ, Evans CR. 1988. *Ap. J. Lett.* 335:L61–65
Duncan AR, Stewart RT, Haynes RF. 1993. *MNRAS* 265:157–60
Eales SA, Becklin EE, Zuckerman B, McLean IS. 1990. *MNRAS* 242:17P–19P
Eggelton PP, Verbunt F. 1986. *MNRAS* 220:13P–18P
Fichtel CE, Bertsch DL, Hartman RC, Hunter SD, Kanbach G, et al. 1993. *Astron. Astrophys. Suppl. Ser.* 97:13–16
Fiedler RL, Dennison B, Johnston KJ, Hewish A. 1987. *Nature* 326:675–78
Foster RS, Fairhead L, Backer DC. 1991. *Ap. J.* 378:687–95
Foster RS, Wolszczan A, Camilo F. 1993. *Ap. J.* 410:L91–94
Friedman JL, Imamura JN, Durisen RH, Parker L. 1988. *Nature* 336:560–62
Friedman JL, Ipser JR. 1992. *Philos. Trans. R. Soc. London Ser. A* 340:391–422
Friedman JL, Schutz BF. 1978. *Ap. J.* 222:281–96
Fruchter AS. 1989. *Pulsars lost and found: the second Princeton-Arecibo millisecond pulsar search.* PhD thesis. Princeton Univ.
Fruchter AS, Bookbinder J, Garcia MR, Bailyn CD. 1992. *Nature* 359:303–4
Fruchter AS, Goss WM. 1992. *Ap. J. Lett.* 384:L47–51
Fruchter AS, Stinebring DR, Taylor JH. 1988. *Nature* 333:237–39
Ghosh P, Lamb FK. 1992. In SANTA BARBARA, pp. 487–510
Goldreich P, Keeley DA. 1972. *Radiosorgenti Pulsate e Attività di Alta Energia nei Resti di Supernovae,* Ser. Problemi Attuali di Scienza e di Cultura, Quaderno 162, pp. 167–74. Rome:Accademia Nazionale dei Lincei
Goldreich P, Reisenegger A. 1992. *Ap. J.* 395:250–58
Habets GMHJ. 1985. PhD thesis. Univ. Amsterdam
Haensel P, Zdunik JL. 1989. *Nature* 340:617–19
Halpern JP, Ruderman M. 1993. *Ap. J.* 415:286–

97
Harrison PA, Lyne AG. 1993. *MNRAS* 265:778–80
Harrison PA, Lyne AG, Anderson B. 1993. *MNRAS* 261:113–24
Harpaz A, Rappaport S. 1991. *Ap. J.* 383:739–44
Hellier C, Mason KO, Smale AP, Kilkenny D. 1990. *MNRAS* 244:39P–43P
Hjellming MS, Webbink RF. 1987. *Ap. J.* 318:794–808
Hulse RA, Taylor JH. 1975. *Ap. J. Lett.* 195:L51–54
Iben I, Livio M. 1993. *Publ. Astron. Soc. Pac.* 105:1373–406
Iwasawa K, Koyama K, Halpern JP. 1992. *Publ. Astron. Soc. J.* 44:9–14
Johnston HM. 1992. *Compact objects in the disk and globular clusters.* PhD thesis. Calif. Inst. Technol.
Johnston HM, Kulkarni SR. 1991. *Ap. J.* 368:504–14
Johnston S, Lorimer DR, Harrison PA, Bailes M, Lyne AG, et al. 1993. *Nature* 361:613–15
Johnston S, Manchester RN, Lyne AG, Nicastro L, Spyromilio J. 1994. *MINRAS* 268:430–36
Jones AW, Lyne AG. 1988. *MNRAS* 232:473–80 [Note that $f(M)$ is incorrectly stated in this ref.]
Joss PC, Rappaport S, Lewis W. 1987. *Ap. J.* 319:180–87
Kaplan DB, Nelson AE. 1986. *Phys. Lett.* B175:57–63
Kaspi VM, Johnston S, Bell JF, Manchester RN, Bailes M, et al. 1994a. *Ap. J. Lett.* 423:L43–45
Kaspi VM, Taylor JH, Ryba MF. 1994b. *Ap. J.* 428:713–28
Koester D, Chanmugam G, Reimers D. 1992. *Ap. J. Lett.* 395:L107–10
Krauss LM, Tremaine S. 1988. *Phys. Rev. Lett.* 60:176–77
Krolik JH. 1991. *Ap. J. Lett.* 373:L69–72
Kulkarni SR. 1986. *Ap. J. Lett.* 306:L85–89
Kulkarni SR. 1992. *Philos. Trans. R. Soc. London Ser. A* 341:77–92
Kulkarni SR, Djorgovski S, Klemola AR. 1991. *Ap. J.* 367:221–27
Kulkarni SR, Hester JJ. 1988. *Nature* 335:801–3
Kulkarni SR, Narayan R. 1988. *Ap. J.* 335:755–68
Kulkarni SR, Navarro J, Vasisht G, Tanaka Y, Nagase F. 1992a. In SANTA BARBARA, pp. 99–104
Kulkarni SR, Phinney ES, Evans CR, Hasinger G. 1992b. *Nature* 359:300–2
Kunihiro T, Muto T, Takatsuka T, Tamagaki R, Tatsumi T. 1993. *Prog. Theor. Phys. Suppl.,* Vol. 112, *Various Phases in High-Density Nuclear Matter and Neutron Stars.* 315 pp.
Lamb DQ. 1992. In *Frontiers of X-Ray Astronomy,* ed. Y Tanaka, K Koyama, pp. 33–48. Tokyo: Universal Acad.

Leonard PJT, Hills JG, Dewey RJ. 1994. *Ap. J. Lett.* 423:L19–22
Levinson A, Eichler D. 1991. *Ap. J.* 379:359–65
Lewin WHG, van Paradijs J, van den Heuvel EPJ, eds. 1994. *X-Ray Binaries.* Cambridge:Cambridge Univ. Press. In press (XRAYBIN)
Lindblom L, Mendell G. 1992. In *The Structure and Evolution of Neutron Stars,* ed. D Pines, R Tamagaki, S Tsuruta, pp. 227–29. Redwood City, CA: Addison-Wesley
Lorenz CP, Ravenhall DG, Pethick CJ. 1993. *Phys. Rev. Lett.* 70:379–82
Lorimer DR. 1994. Preprint
Lorimer DR, Bailes M, Dewey RJ, Harrison PA. 1993. *MNRAS* 263:403–15
Lundgren SC. 1994. *A multi-wavelength study of rotation-driven pulsars.* PhD thesis. Cornell Univ.
Lyne AG, Anderson B, Salter MJ. 1982. *MNRAS* 201:503–20
Lyne AG, Biggs JD, Harrison PA, Bailes M. 1993. *Nature* 361:47–49
Lyne AG, Lorimer DR. 1994. *Nature* 369:127–29
Lyne AG, Manchester RN, Taylor JH. 1985. *MNRAS* 213:613–39
Lyne AG, McKenna J. 1989. *Nature* 340:367–69
Malhotra R. 1993. In CALTECH, pp. 89–106
Malhotra R, Black D, Eck A, Jackson A. 1992. *Nature* 356:583–85
Manchester RN. 1992. In *Back to the Galaxy,* ed. SS Holt, F Verter, pp 514–23. New York: Am. Inst. Phys.
Manchester RN, Taylor JH. 1977. *Pulsars* pp. 67–68. San Francisco:Freeman. 281pp.
Melrose DB. 1992. *Philos. Trans. R. Soc. London Ser. A* 341:105–15
Mendell G. 1991. *Ap. J.* 380:530–40
Michel FC. 1987. *Nature* 329:310–12
Michel FC. 1991. *Theory of Neutron Star Magnetospheres.* Chicago: Univ. Chicago Press
Miller MC, Lamb FK. 1993. *Ap. J. Lett.* 413:L43–46
Narayan R. 1992. *Philos. Trans. R. Soc. London Ser. A* 341:151–65
Narayan R, Ostriker JP. 1990. *Ap. J.* 352:222–46
Narayan R, Paczyński B, Piran T. 1992. *Ap. J. Lett.* 395:L83–86
Narayan R, Piran T, Shemi A. 1991. *Ap. J. Lett.* 379:L17–20
Naylor T, Podsiadlowski P. 1993. *MNRAS* 262:929–35
Nice DJ, Taylor JH, Fruchter AS. 1993. *Ap. J.* 402:L49–52
Nordgren TE, Cordes JM, Terzian Y. 1992. *Astron. J.* 104:1465–71
Nordvedt K. 1968. *Phys. Rev.* 169:1017–25
Ögelman H. 1993. In *Isolated Pulsars,* ed. KA van Riper, R Epstein, C Ho, pp. 96–109. Cambridge:Cambridge Univ. Press
Ögelman H, van den Heuvel EPJ, eds. 1989. *Timing Neutron Stars.* Dordrecht: Kluwer

638   PHINNEY & KULKARNI

Ostriker JP, Gunn JE. 1969. *Ap. J.* 157:1395–417
Paczyński B. 1991. *Ap. J.* 370:597–603
Pacini F. 1983. *Astron. Astrophys.* 126:L11–12
Pandharipande VR. 1971. *Nucl. Phys.* A178:123–44
Pandharipande VR, Smith RA. 1975. *Phys. Lett.* 59B:15–18
Papaloizou J, Pringle JE. 1978. *MNRAS* 184:501–8
Parmar AN, Smale AP, Verbunt F, Corbet RHD. 1991. *Ap. J.* 366:253–60
Peale SJ. 1993. *Astron. J.* 105:1562–570
Peters PC. 1964. *Phys. Rev.* B136:1224–32
Phillips JA, Thorsett SE, Kulkarni SR, eds. 1993. *Planets Around Pulsars.* San Francisco: Astron. Soc. Pac. (CALTECH)
Phillips JA, Wolszczan A. 1991. *Ap. J. Lett.* 382:L27–30
Phillips JA, Wolszczan A. 1992. *Ap. J.* 385:273–81
Phinney ES. 1991. *Ap. J. Lett.* 380:L17–21
Phinney ES. 1992. *Philos. Trans. R. Soc. London Ser. A* 341:39–75
Phinney ES. 1993. In *Structure and Dynamics of Globular Clusters,* ed. SG Djorgovski, G Meylan, ASP Conf. Ser. Vol. 50, pp. 141–69. San Francisco:Astron. Soc. Pac.
Phinney ES. 1994. *Ap. J.* Submitted
Phinney ES, Evans CR, Blandford RD, Kulkarni SR. 1988. *Nature* 333:832–34
Phinney ES, Hansen BMS. 1993. In CALTECH, pp. 371–90
Phinney ES, Kulkarni SR. 1994. *Nature* Submitted
Phinney ES, Sigurdsson S. 1991. *Nature* 349:220–23
Phinney ES, Verbunt F. 1991. *MNRAS* 248:21P–23P
Podsiadlowski P. 1991. *Nature* 350:136–38
Podsiadlowski P. 1993. In CALTECH, pp. 149–65
Politzer HD, Wise MB. 1991. *Phys. Lett. B* 273:156–62
Popham R, Narayan R. 1991. *Ap. J.* 370:604–14
Priedhorsky WC, Verbunt F. 1988. *Ap. J.* 333:895–905
Pylyser E, Savonije GJ. 1988. *Astron. Astrophys.* 191:57–70
Radhakrishnan V. 1992. In SANTA BARBARA, pp. 445–52
Radhakrishnan V, Srinivasan G. 1982. *Curr. Sci.* 51:1096–99
Rasio FA, Nicholson PD, Shapiro SL, Teukolsky SA. 1992. *Nature* 355:325–26
Rawley LA, Taylor JH, Davis MM. 1988. *Ap. J.* 326:947–53
Refsdal S, Weigert A. 1970. *Astron. Astrophys.* 6:426–40
Refsdal S, Weigert A. 1971. *Astron. Astrophys.* 13:367–73
Renzini A, Greggio L, Ritossa C, Ferrario L. 1992. *Ap. J.* 400:280–303
Rickett BJ. 1990. *Annu. Rev. Astron. Astrophys.* 28:561–605
Romani RW. 1990. *Nature* 347:741–43
Romani RW. 1993. In *Isolated Pulsars,* ed. KA van Riper, R Epstein, C Ho, pp. 75–83. Cambridge:Cambridge Univ. Press
Ruden SP. 1993. In CALTECH, pp. 197–215
Ruderman M. 1991. *Ap. J.* 382:576–86
Ruderman M, Shaham J, Tavani M, Eichler D. 1989. *Ap. J.* 343:292–312
Ruderman M, Sutherland P. 1975. *Ap. J.* 196:51–72
Ryba MF, Taylor JH. 1991. *Ap. J.* 380:557–63
Shibazaki N, Murakami T, Shaham J, Nomoto K. 1989. *Nature* 342:656–58
Sigurdsson S. 1992. *Ap. J. Lett.* 399:L95–97
Sigurdsson S. 1993. *Ap. J. Lett.* 415:L43–46
Sigurdsson S, Hernquist L. 1992. *Ap. J. Lett.* 401:L93–96
Srinivasan G. 1989. *Astron. Astrophys. Rev.* 1:209–60
Srinivasan G, Bhattacharya D, Muslimov AG, Tsygan AI. 1990. *Curr. Sci.* 59:31–38
Sweigart AV, Gross PG. 1978. *Ap. J. Suppl.* 36:405–37
Taam RE. 1983. *Ap. J.* 270:694–99
Taam RE, van den Heuvel EPJ. 1986. *Ap. J.* 305:235–45
Tavani M. 1991. *Nature* 351:39–41
Tavani M, London RA. 1993. *Ap. J.* 410:281–94
Taylor JH. 1992. *Philos. Trans. R. Soc. London Ser. A* 341:117–34
Taylor JH, Cordes JM. 1993. *Ap. J.* 411:674–84
Taylor JH, Dewey RJ. 1988. *Ap. J.* 332:770–76
Taylor JH, Wolszczan A, Damour T, Weisberg JM. 1992. *Nature* 355:132–36
Taylor JH, Manchester RN, Lyne AG 1993. *Ap. J. Suppl.* 88:529–68
Thompson C, Blandford RD, Evans CR, Phinney ES. 1994. *Ap. J.* 422:304–35
Thorne KS. 1980. *Rev. Mod. Phys.* 52:299–339
Thorne KS, Żytkow AN 1977. *Ap. J.* 212:832–58
Thorsett SE, Arzoumanian Z, McKinnon MM, Taylor JH. 1993b. *Ap. J.* 405:L29–32
Thorsett SE, Arzoumanian Z, Taylor JH. 1993a. *Ap. J. Lett.* 412:L33–36
Thorsett SE, Dewey RH. 1993. *Ap. J. Lett.* 419:L65–L68
Thorsett SE, Phillips JA. 1992. *Ap. J. Lett.* 387:L69–71
Thorsett SE, Phillips JA, Cordes JM. 1993c. In CALTECH, pp. 31–39
Thorsett SE, Stinebring DR. 1990. *Ap. J.* 361:644–49
van den Heuvel EPJ 1992. In SANTA BARBARA, pp. 233–56
van den Heuvel EPJ, Rappaport SA, eds. 1992. *X-Ray Binaries and Formation of Binary and Millisecond Pulsars.* Dordrecht: Kluwer (SANTA BARBARA)
van den Heuvel EPJ, van Paradijs JA. 1988. *Nature* 334:227–28

van den Heuvel EPJ, van Paradijs JA, Taam RE. 1986. *Nature* 322:153–55
van der Klis M, Hasinger G, Dotani T, Mitsuda K, Verbunt F, et al. 1993a. *MNRAS* 260:686–92
van der Klis M, Hasinger G, Verbunt F, van Paradijs J, Belloni T. 1993b. *Astron. Astrophys.* 279: L21–24
van Kerkwijk MH. 1993. *Astron. Astrophys.* 276:L9–12
van Kerkwijk MH, Charles PA, Geballe TR, King DL, Miley GK, et al. 1992. *Nature* 355:703–5
van Paradijs J. 1994. In XRAYBIN
Verbunt F. 1990. In *Neutron Stars and their Birth Events,* ed. W Kundt, pp. 179–218. Dordrecht:Kluwer
Verbunt F. 1993. *Annu. Rev. Astron. Astrophys.* 31:93–127
Verbunt F, van den Heuvel EPJ. 1994. In XRAYBIN
Verbunt F, van Paradijs J, Elson R. 1984. *MNRAS* 210:899–914
Verbunt F, Wijers RAMJ, Burm HMG. 1990. *Astron. Astrophys.* 234:195–202
Wagoner RV. 1984. *Ap. J.* 278:345–48
Wakatsuki S, Hikita A, Sato N, Itoh N. 1992. *Ap. J.* 392:628–36
Warner B. 1988. *Nature* 336:129–34
Weaver TA, Woosley SE. 1993. *Phys. Rep.* 227:65–96
Webbink RF. 1992. In SANTA BARBARA, pp. 269–80
White NE, Nagase F, Parmar AN. 1994. In XRAYBIN
Wijers RAMJ, Paczyński B. 1993. *Ap. J. Lett.* 415:L115–18
Wijers RAMJ, van den Heuvel EPJ, van Kerkwijk M.H, Bhattacharya D. 1992. *Nature* 355:593
Will CM. 1993. *Theory and Experiment in Gravitational Physics.* Cambridge: Cambridge Univ. Press. 380pp. revised edition.
Wolszczan A. 1991. *Nature* 350:688–90
Wolszczan A. 1994. In *Planetary Systems: Formation, Evolution and Detection,* ed. BF Burke, JH Rahe, E Roettger. Dordrecht:Kluwer. In press
Wolszczan A, Frail DA. 1992. *Nature* 355:145–47
Wood KS, Norris JP, Hertz P, Vaughan BA, Michelson PF, et al. 1991. *Ap. J.* 379:295–309
Woosley SE, Langer N, Weaver TA. 1993. *Ap. J.* 411:823–39
Woosley SE, Weaver TA. 1986. *Annu. Rev. Astron. Astrophys.* 24:205–53
Yancopoulos S, Hamilton TT, Helfand DJ. 1994. *Ap. J.* In press
Zahn J-P. 1966. *Ann. Astrophys.* 29:313–30, 489–506, 565–91
Zahn J-P. 1977. *Astron. Astrophys.* 57:383–94. Erratum 67:162

SUBJECT INDEX

A

A stars
  asteroseismology and, 61
  in nearby galaxies, 242
  pre-main-sequence binaries
    and, 508
Abundances
  in active galatic nuclei, 228
  beryllium, 18
  boron, 18
  carbon, 201–3
  chromium, 60
  cosmic, 15, 17
  deuterium, 18, 203–5, 210
  europium, 15, 60
  heavy element, 45, 47
  helium, 45, 57, 203, 206–11,
    551, 580
  holmium, 60
  Hubble sequence and, 137–40
  hydrogen, 70, 137–39
  in interstellar medium, 191–222
  iron, 137, 139, 279
  light element, 213–15
  lithium, 18, 203, 210–13
  on main sequence, 237–38
  in massive stars, 241–42, 247–
    49
  nucleosynthesis and, 153–55,
    157, 160, 162–64, 168–
    71, 174–76, 178–80
  oxygen, 137–39, 279
  in quasi-stellar objects, 228
  rare earth, 60
  silicon, 15, 60, 279
  strontium, 60
  sulfur, 279
Accretion
  black hole remnants of popula-
    tion III stars and, 552
  young binary environments
    and, 477–89
Accretion disks, 487–89
Accretion-induced collapse, 597–
  98, 634
Acetylenes, 305
ACME experiment, 348
Acoustic cutoff frequency, asteroseismology and, 67
ACRIM radiometer, stellar oscillations and, 44
Active galactic nuclei (AGNs)
  abundances in, 228
  Hubble sequence and, 129
Adiabatic modes, cosmic wave
  background anisotropies
  and, 332–34

Advanced Fiber Optic Echelle, as-
  teroseismology and, 77
AG Car, 239
AICVn, 62
AK Sco, 479–80, 484, 486–88,
  525
Alfvén velocity, 300, 455
α-α interactions, 213
α-process, 18
Amalthea, 439
Ammonia, in pre-main-sequence
  binaries, 516
Andromeda galaxy, 22
Anglo-Australian Telescope, 29,
  31
Angular degree, stellar oscilla-
  tions and, 39
Anisotropies, 319–63, 538–39,
  544, 547, 559
AO 0235+164, 297
APM galaxy survey, 342–43
Ap stars
  history of, 14
  rapidly oscillating, as-
    teroseismology and, 50,
    60–61
Aquarius, 388
Arcturus, 76
Arecibo General Catalog (AGC),
  118–19
Arecibo Telescope, 599–601,
  634
ARGO experiment, 342–44, 347–
  49 ASCA satellite observations, 290–91
Asteroids, 20, 546, 624
Asteroseismology
  introduction to, 37–38
  stellar oscillations and
  basic physics, 38–42
  Doppler-shift techniques,
    72–77
  information content esti-
    mates, 45–49
  photometric techniques, 68–
    72
  pulsating white dwarfs, 50–
    60
  pulsations in Sun-like stars,
    66–77
  rapidly oscillating Ap stars,
    60–61
  δ Scuti stars, 61–66
  solar example, 42–45
  stars unlike Sun, 49–66
  terminology, 38–42
ASTRO-1 satellite observations,
  288
ATCA experiment, 347

Atmosphere
  early hot H_2O-rich, 93–94
  evolution of, 83–112
Auger effect, 8
Average universe, 348–49
AWM7, 282
Axions, baryonic dark matter
  and, 537
Azimuthal order, stellar oscilla-
  tions and, 39
A85, 282
A119, 282
A262, 283
A399, 282, 307
A401, 282, 307
A426, 282
A478, 282, 284–86
A496, 282
A576, 283
A644, 282
A754, 282
A1060, 282
A1367, 282
A1644, 282
A1650, 283
A1651, 282
A1689, 283
A1736, 283
A1795, 282, 284, 294–96, 298
A2029, 282, 284–85, 293
A2052, 282, 298
A2063, 282
A2065, 283
A2142, 282
A2147, 282
A2199, 282, 296
A2204, 283
A2244, 283
A2255, 283
A2256, 282, 297
A2319, 282
A2597, 283–84, 294
A3112, 283
A3158, 282
A3266, 282
A3391, 283
A3532, 283
A3558, 282
A3562, 283
A3571, 282
A3667, 282
A4059, 283

B

B Spectrograph, 20, 25
B stars
  history of, 12, 14
  in nearby galaxies, 237, 241–42
  pulsars and, 592

Ba II stars, history of, 15, 18
Balmer decrement, 14
Bare cores, 245
Be stars, history of, 6–7, 9, 14–15
Beta-decay rate, nucleosynthesis and, 164, 170–71, 178–79
β values, dynamics of cosmic flows and, 406–10
Big Bang nucleosynthesis, 26–27, 175, 203–4, 206, 212–13, 215, 221–22, 319, 321, 531, 535–37, 551, 563
Binary stars, artificial, measuring and computing orbits of, 5
Binary stars, low-mass X-ray, pulsars and, 597–98, 601–2, 604–6, 610, 613, 626
Binary stars, massive X-ray, pulsars and, 595–97, 603
Binary stars, pre-main-sequence
  ages of, 493–95
  disk accretion and, 487–89
  disk frequency and, 478–80
  disk masses and, 481
  disk structure and, 482–87
  dynamical evolution of disks, 499–503
  frequency of, 467–72
  future research on, 517–19
  implications for binary formation, 511–15
  infrared companions and, 489–506
  introduction to, 465–67, 477–78
  masses of, 495
  orbital eccentricity distribution of, 472–76
  orbital evolution of, 506–10
  population of, 496–99
  protobinaries and, 515–16
  secondary mass distribution of, 476–77
Binary stars, short-period, abundances and, 237
Binary stars, visual, asteroseismology and, 47–48
Binary stars, X-ray, Hubble sequence and, 131
BK Cet, 62
Blackbodies, 319, 568
Black holes, 167, 209, 532, 536–37, 542, 546, 548, 550–54, 558–59, 562, 566–72, 581–85, 616, 618, 620, 622, 634
Black-widow pulsars, 602
BL Lac objects, 295, 297
Blobs, 299–300
Blue Hertzsprung gap, 242–43
Blueshift, cosmic wave background anisotropies and, 352–53
BN cnc, 62, 65
Boesgaard gap, 211–12
Boltzmann equation, 339–40, 352

Bondi accretion, 552
κ² Boo, 65
Boundary layer emission, pre-main-sequence binaries and, 518
Bow shock, cometary, cosmic dusty plasma and, 447
Bremsstrahlung radiation, 278
Brunt-Väisälä frequency
  massive stars and, 234
  stellar oscillations and, 42
Bubbly inflation, 393
BU CnC, 65
Buoyancy, g-modes and, 39
Butcher-Oemler effect, 143
B²FH, 9, 17–19, 24

C

Ca²⁺, Goldilocks problem and, 108
α Cam, 241
Cambridge Observatories, 28
Campbell shift, 211
ζ Cap, 18
Capella, 204
Cape Rapidly Oscillating Ap Star Survey, 61
Car OB1, 230
Carbon
  in interstellar medium, 201–3, 215–18
  in massive stars, 237, 241, 247–48, 250
  in population III stars, 548
  in pulsating white dwarfs, 51
  s-process and, 177
Carbon dioxide, Goldilocks problem and, 84–88, 92–97, 99–104, 106–12
Carbon monoxide, Hubble sequence and, 133–35
Cas A, 205, 216
γ Cassiopeiae, 7
CCD photometry
  asteroseismology and, 50, 70–72
  direct star counts and, 231–32
α Cen, 47–48, 74
Cen A, 21, 72–73, 75
Cen B, 73
Centaurus cluster, 282
Cepheid stars
  asteroseismology and, 51, 61, 63
  dwarf, 63
  stellar oscillations and, 49
Ceres, 624
Cetus, 390
CfA redshift survey, 342–43
Chipping, electrostatic, 435
Chlorofluorocarbons (CFCs), terraforming and, 110
Chromium, in roAp stars, 60
Circularization cutoff period, 475
Circulation radius, 438
Circumbinary disks, 478, 480–

82, 484–85, 487, 489, 500–2, 506, 509, 513, 516–17
Circumbinary envelopes, 505, 517
Circumprimary disks, 478, 500, 503
Circumsecondary disks, 478, 505
Circumstellar disks, 478–80, 482, 485–87, 489, 499–504, 506, 515–16, 518
Cir X-1, 603
Climate, early, Goldilocks problem and, 108–10
Cloud albedo, Goldilocks problem and, 84–85, 93, 102, 105
Cloud condensation nuclei, Goldilocks problem and, 104
Cloud fragmentation, 511
CLOUDY code, 309
Cluster environment, Hubble sequence and, 142–143 α
CMi, 75
CNO cycle
  interstellar abundances and, 215–21
  massive stars and, 235, 237, 241, 247–48
  s-process and, 175, 178, 180
COBE satellite observations, 319–21, 324, 330– 33, 335–38, 341–44, 347–49, 352, 359, 395, 402, 404, 539, 569, 584
CoKu Tau/3, 523
Cold starts, 107
Cold trap, 90
Collisions, cosmic wave background anisotropies and, 339
Coma cluster, 282, 297, 388–89, 395, 537
Cometary dust tails, 435, 446
Comet Giacobini-Zinner, 444, 446–50
Comet Halley, 419–20, 435–36, 446–47, 453
Comet Ikeya-Seki, 446
Comets, 20, 88, 92, 419, 435–36, 443, 445–48, 456–57, 611
Compton cooling, 361
Compton drag, 340
Continuity equation, 374
Continuously habitable zones, 105–7
Conti scenario, 252
Convective mode excitation, asteroseismology and, 67
Cooling catastrophe, 575
Cooling flows
  baryonic dark matter and, 545–46, 572–75
  condition summary for, 306
  distant cooling flows, 306–10
  galaxy formation and, 310–13
  global structure, 299–302
  hot intracluster medium, 278–79

infrared emission, 297
introduction to, 277–79
local structure, 302–06
non-X-ray wavebands, 298–99
observational evidence for, 279–93
optical waveband, 293–97
radio emission, 297–98
theoretical issues on, 299–306
X-ray imaging evidence for, 281–88, 292–93
X-ray spectral evidence for, 288–93
Copernican hypothesis, 393
Copernicus satellite observations, 204, 281
Coriolis forces, stellar oscillations and, 44
Corona Australis, 469, 471
Cosmic chemical memory theory, 181
Cosmic dusty plasma
coagulation of charged dust, 430–33
disruption of charged dust, 433–35
dynamics of, 436–50
electrostatic charging of dust, 421–30
grain ensemble and, 426–30
instabilities and, 450, 456–59
introduction to, 419–21
isolated grain and, 422–26
levitation of charged dust, 435–36
waves and, 450–56
wave scattering and, 450, 459–60
Cosmic flows
dynamics of
back in time, 403–5
β values from distortions in redshift space, 408–10
β values from galaxies, 406–8
bulk velocity, 399–401
cosmic microwave background, 393–95, 406–7
dark matter, 399–403
discrepancy on very large scale, 402–3
distance indicators, 377–79
elliptical galaxies, 397–99
environmental effects, 397–99
galaxies vs. dynamical mass, 395–97
Gaussianization, 404–5
gravitational instability, 374–76
homogenized catalogs, 380–82
initial fluctuations, 399–405

introduction to, 371–73
justification for hypotheses, 412–15
least action, 405
Mach number, 400
Malmquist biases, 379–80
Malmquist-free analysis, 387–88
mass density field, 388–90
non-linear biasing, 392
Ω value, 405–12
peculiar velocities, 377–90
potential analysis, 383–86
power spectrum, 399–03
probability distribution functions, 403–4, 410–11
quasilinear correction, 392–93
random error estimation, 384–86
regularized multi-parameter models, 386–87
sampling-gradient bias, 384
selection function, 391
shot-noise estimation, 392
spiral galaxies, 397–99
tensor window, 384
testing basic hypotheses, 393–99
toy models, 382–83
triple-valued zones, 392
velocity field, 388–90
voids, 411–12
Wiener filter, 386–87
z-distribution, 390–93
Zel'dovich time machines, 403
zone of avoidance, 391–92
Cosmic loops, 583
Cosmic microwave background
anisotropies in
adiabatic modes, 332–34
baryonic dark matter and, 538–39, 542, 544, 547, 559
beyond linear theory, 350–54
bias, 338
correlation function, 331
cosmological constant effects, 358
data analysis, 344–50
$\Delta T/T$ sources, 324–26
fitting data, 349–50
fluctuations, 331–36
Gaussian autocorrelation function, 344–45
gravitational waves, 335–36
inflation, 326–27
introduction to, 319–26
isocurvature, 334–35
non-standard cosmologies, 355–59
open universes, 356–57
polarization, 355–56

power spectrum, 336–44
quadrupole normalization, 337
recombination, 322–24
reionization, 350–52
second-order anisotropies, 352–54
sine wave chop, 346
specific entropy, 332
square wave chop, 346
structure formation theories, 327–30
theory, 326–36
topology, 358–59
uncertainties, 354–55
window functions, 346–49
dynamics of cosmic flows and, 393–95, 406–7
Cosmic rays, 199, 213, 302
Cosmic strings, 330, 537, 623
Cosmic variance, 354
Cosmic virial theorem, 409
Cosmological constant, 358
Cosmologies
inflationary, 321
nonstandard, 355–59
Cosmos, distribution of life in, 112
Crab Nebula, 611
Crab pulsar, 608, 611
Craters, on Mars, 94
Creation events, 35
CW Tau, 487
Cyg A, 24, 29, 130, 282, 285, 297–98, 307, 309
Cyg OB2, 230
Cyg X-2, 601
Cyg X-3, 613
CZ Tau, 522
3C 48, 22
3C 129, 282
3C 191, 27
3C 273, 26
3C 295, 298, 306
3C 356, 310
α^2CVn, 15–16, 18
4 CVn, 65–66

D

Damping, 539
Dark holes, 419
Dark matter
baryonic
anisotropies in cosmic microwave background and, 330, 334–35, 340–41, 353, 357
arguments for, 539–41
background dark matter and, 536
background light constraints and, 566–68

background radiation and, 566–72

black holes and, 566–72, 581–82

brown dwarfs and, 575–78, 580

candidates for baryonic dark matter, 578–84

cooling flows and, 572–75

cosmological nucleosynthesis and, 537–38

dark clusters and, 557–58

dark objects and, 558–59

dwarf galaxies and, 534–35

dynamical constraints and, 552–59

elliptical galaxies and, 534

evidence for, 532–35

future research on, 584–85

gravitational lensing effects and, 559–66

groups and clusters of galaxies, 535

halo holes and, 553–54

halo objects and, 554–56, 565–66

introduction to, 531–32

local dark matter and, 532–33

low mass objects and, 572–78

macrolensing and, 560–62

M-dwarfs and, 575–76

microlensing and, 562–66

microwave anisotropies and, 538–39

neutron stars and, 581

nonbaryonic dark matter vs., 536–42

population I stars and, 576–77

population II stars and, 576–77

population III stars and, 542–52

snowballs and, 579–80

spiral galaxies and, 533–34

supermassive objects and, 581–82

variants of baryonic dark matter scenario, 541–42

very massive objects and, 568–70, 581

white dwarfs and, 580

cold, 327–30, 334, 338, 340–41, 351–53, 359–60, 373, 399- 401, 410, 531, 535–37, 541, 544, 574–75

cooling flows and, 278, 311–12

cosmic microwave background anisotropies and, 320, 327–30, 334–35, 338–41, 351–53, 355, 357, 359–60

dynamics of cosmic flows and, 373, 399–403

hot, 531, 536, 541, 544

Hubble sequence and, 141

interstellar abundances and, 215

mixed, 329

DAV stars, asteroseismology and, 51–54, 63

DBV stars, asteroseismology and, 50–53, 57, 59

DD Tau, 478, 522

DDO 154, 535, 554

DDO 170, 535

Death line, pulsars and, 591–593, 614

Death Valley, 607

Debris-rain sediments, 109

Debye length, plasma, 419–20, 426–27, 430, 441, 458–59

Defect models, 330–31

$\Delta T/T$, cosmic microwave background anisotropies and, 324–26

Dense molecular clouds, interstellar abundances and, 197- 203

Deuterium

baryonic dark matter and, 538

Goldilocks problem and, 91–92

interstellar medium and, 203–5, 210

population III stars and, 547

DF Tau, 480, 485, 488–89, 502–3, 522

Diffuse molecular clouds, interstellar abundances and, 197

Dimethylsulfide gas, Goldilocks problem and, 104

DIRBE observations, 567, 584

Dirty window effect, 359

Discrepant velocities, 35

Disk fragmentation, 511

DI Tau, 522

DK Tau, 523

DoAr 24E, 478, 491–92, 506

Doppler peaks, cosmic wave background anisotropies and, 340–42, 344–45, 357, 359

Doppler-shift techniques, pulsations in Sun-like stars and, 72–77

30 Dor, 229–31, 253, 255–56, 260, 262, 264–65, 268

Double emission lines, in γ Cas, 7

DOV stars, asteroseismology and, 50–53, 55, 57–58

DQ Tau, 487

Dressed test particles, 459

DS Tau, 523–24, 481, 492

Dust, 227–28, 238, 305, 355, 360, 419–61, 566–70

Dust-in-plasma, 420

Dust packets, 435

E

E + A population, Hubble sequence and, 143

Earth

biogeochemical cycle of carbon on, 96

cosmic dusty plasma and, 436, 442, 449, 460

cosmic wave background anisotropies and, 319

fate of, 106, 108

Goldilocks problem and, 83–89, 91, 93, 97–106, 108–12

magnetosphere of, 436, 442, 449

mesopause of, 460

nucleosynthesis and, 161

Echelle spectroscopy, in asteroseismology, 74

Eddington limit, 238–40

EGRET sky survey, 615

Einstein equations, perturbed, 327

Einstein Medium Source Survey, 563–64

Einstein Observatory satellite observations, 281, 288, 290, 299, 306, 310

Ejecta deposits, 109

EK Cep, 475, 495, 507–8, 520, 525

Electrostatic charging, cosmic dusty plasma and, 421–30

Electrostatic ion cyclotron (EIC) waves, 453

Enceladus, 445

Entropy

cooling flows and, 280, 299

cosmic wave background anisotropies and, 332, 357

nucleosynthesis and, 155–60, 164–67

Envelopes, 39, 51, 53, 57, 63, 67–68, 184, 220, 296, 447, 481, 504–5, 510, 516–17, 551, 625, 628, 630

Environment, effect of, Hubble sequence and, 142–43

e-process, nucleosynthesis and, 18–19

γ Equ, 61

Equilibrium, nucleosynthesis and, 155–61

o^1 Eri, 65

Euler equation of motion, 374

Europium

in Apm stars, 15

in roAp stars, 60

r-process and, 170

EVRIS experiment, asteroseismology and, 78

EXOSAT satellite observations, 281, 290, 572

Explosion model, cosmic microwave background anisotropies and, 325
Explosive seeds, 545
Extra-mixing, 211
Extreme scattering events, 633
Extreme ultraviolet (EUV) radiation, Goldilocks problem and, 90, 108
EX00748-676, 613
1E2259+586, 609

F

F stars
  asteroseismology and, 68, 72
  baryonic dark matter and, 532
  in nearby galaxies, 238, 240, 242
Faint Object Spectrograph, 34–35
Faint young Sun paradox, 98–99
Falling short of equilibrium scenario, nucleosynthesis and, 161, 163
Faraday depolarization/rotation, 309
Far infrared (FIR) emissions
  Hubble sequence and, 119–20, 128–30
  massive stars and, 266
FF Tau, 522
Filaments, 299
Filter functions, 331
Fingers of God, 408
FIRAS observations, 319, 545, 566–67, 569–70, 581, 584
Fireballs, 435
FIRS experiment, 342–44, 347–49
Fluorine, s-process and, 175
Fornax cluster, 144
Fossils, 303, 321, 626–33
FO Tau, 522
Fourier transforms, asteroseismology and, 65
FQ Tau, 523
Free-free radiation
  cosmic wave background anisotropies and, 355
  Hubble sequence and, 130
Freeze-out, ionization, 323, 539
Freeze-out from equilibrium scenario, 162
Freezing in, 326
FRESIP photometric search, 72
Friedman equation, 327
FS Tau, 522
FU Ori, 491, 504
FV Tau, 523
FV Tau/c, 523
FW Tau, 522
FX Tau, 523
F2 V, 70
F134, 71

G

G stars
  asteroseismology and, 72
  Goldilocks problem and, 106
  in nearby galaxies, 238
Gaia hypothesis, 84, 103–5
Gain radius, r-process and, 173
Galactic Bulge, microlensing and, 565–66
Galactic Center, interstellar abundances and, 215–20
Galactic cirrus cloud population, Hubble sequence and, 129
Galactic clusters
  baryonic dark matter and, 535
  cooling flows in, 277–313
Galactic halos, baryonic dark matter and, 533–36, 540–42, 549, 554–59, 565–66
Galactic signal, 355
Galactic winds, 539
Galaxies
  dynamics of cosmic flows and, 395–97, 406–8
  formation of, 277–313
  Hubble sequence and, 116–46
  integrated spectra of, 259–65
  interstellar abundances and, 216
  massive stars in, 227–69
  mass-to-light ratios in, 17
Galaxies, amorphous, massive stars in, 262
Galaxies, binary, velocities of, 20
Galaxies, blue compact, Wolf-Rayet stars in, 246
Galaxies, blue compact dwarf, massive stars in, 262
Galaxies, brightest cluster, dynamics of cosmic flows and, 402–3
Galaxies, classical, Hubble sequence and, 140, 144–46
Galaxies, dwarf
  Hubble sequence and, 140–41, 144–46
  local dark matter in, 534–35
Galaxies, dwarf-elliptical, Hubble sequence and, 137
Galaxies, elliptical
  dynamics of cosmic flows and, 397–99
  Hubble sequence and, 129–30, 134, 139, 142, 145–46
  local dark matter in, 534
  UV upturn in, 222
Galaxies, emission line, massive stars in, 260, 266–68
Galaxies, field, Hubble sequence and, 140
Galaxies, giant, Hubble sequence and, 140, 144

Galaxies, HII, Wolf-Rayet stars in, 246–47
Galaxies, high-redshift, Hubble sequence and, 143
Galaxies, irregular-type, Hubble sequence and, 136–40
Galaxies, Markarian, massive stars in, 262
Galaxies, radio, cooling flows and, 306, 310, 312–13
Galaxies, Seyfert, Hubble sequence and, 129
Galaxies, SO, Hubble sequence and, 144
Galaxies, spheroidal, Hubble sequence and, 137
Galaxies, spiral
  dynamics of cosmic flows and, 397–99
  Hubble sequence and, 120, 126–37, 139–46
  local dark matter in, 533–34
  massive stars in, 262
Galaxies, Wolf-Rayet, massive stars in, 246, 261–64, 26–68
Galaxies, Zwicky, massive stars in, 262
Galaxy
  cosmic dusty plasma in, 419
  cosmic wave background anisotropies and, 319–21, 325, 356, 359
  Hubble sequence and, 129, 131, 133
  mapping spiral structure of, 12, 14
  massive stars and, 229–30, 241, 244–46, 251–58
  metal abundances for, 197
  pulsars in, 591–94, 600–1, 603–5, 607–8, 620, 622–23, 627, 634
  r-process in, 169–71
  s-process in, 180
  z distribution in, 390–93
  zone of avoidance in, 383–84, 390–92, 402
Galaxy catalogs, limitations of, 120–25
Gamma-process, 185–87
Gamma-ray bursts, 559–60, 565
Gamma Ray Observatory observations, 565
Gamma rays, 249
Gaussian auto-correlation function, 344–45
GD 358, 57, 59
General interstellar radiation field, Hubble sequence and, 129
GG Tau, 478, 480–83, 501–3, 517
GG Tau Aa, 522
GG Tau Bb, 523
GG Tau/c, 501

GHII regions, massive stars and, 229, 231, 247, 259–60, 262, 264, 266
GH Tau, 522
Giant molecular clouds (GMCs), interstellar abundances and, 197–98
Giant Red Envelope Galaxy, 296
Ginga satellite observations, 288–90
GK Tau, 523
Glacial deposits, early climate and, 108–10
Glass I, 478, 491–92, 506
Global Climate model (GCM) experiments, 96
Globular clusters
    Hubble sequence and, 131
    pulsars in, 593, 622–23, 625, 634
Globular cluster-type objects, 545
Glory pulses, 622
g-modes
    asteroseismology and, 51–52, 54–55, 63–64
    stellar oscillations and, 39, 41
GN Tau, 487, 489, 522
Goldilocks problem
    atmospheres and, 93–94
    early climate and, 93–94, 108–10
    Earth and, 97–103, 106, 108
    Gaia hypothesis and, 103–5
    glacial deposits and, 108–10
    introduction to, 83–84
    Mars and, 94–97
    moist greenhouse and, 89–91
    planetary surface temperatures and, 84–86
    runaway greenhouse and, 86–89
    terraforming and, 110–11
    water loss from Venus and, 91–93
GONG network, asteroseismology and, 54
Gossamer ring, 438–39
GR8, 535, 554
Gravitational instability, 373–76
Gravitational lensing, 325, 559–66
Gravitational radiation, from black holes, 570–72
Gravitational waves, 335–36, 623
Great Attractor, 354, 372, 383, 388–90, 395, 412
Great Wall, 354, 388
Greenhouse efect
    Goldilocks problem and, 85–86, 92–94, 96–100, 102, 104, 106, 110–11
    moist greenhouse, 89–91
    runaway greenhouse, 86–89

Group environment, Hubble sequence and, 143
GW Ori, 473–74, 480–81, 483–86, 488–89, 502–3, 509–10, 513, 526
GW Vir, 50
GX Peg, 62, 65
Gyrophase drift, 438–39

H

HI regions
    baryonic dark matter and, 534, 540
    Hubble sequence and, 132–33
HII regions
    Hubble sequence and, 135–37
    interstellar abundances and, 195–97
    in M31, 20
    in M33, 20
    massive stars and, 229, 253, 255–56, 260–62, 266
Hale 5-m telescope, 77, 468
Hall drift, 609
Halley's Comet, 92
Halo holes, 553–54
Halo field, 475
Halo objects, 554–56, 565–66
Halos
    cosmic wave background anisotropies and, 360
    dark, 312
    galactic, 533–36, 540–42, 549, 553–59, 565–66
Hamilton's action principle, 405
Haro 1–14c, 526
Haro 6–10, 478, 480, 491–92, 523
Haro 6–28, 522
Haro 6–37, 478, 523
Harrison-Zel'dovich spectrum, 328–29, 333, 344, 357, 360
Harvard College Observatory, 10–14
Hayashi tracks, 495, 626
HCG 62, 535, 537
HD 16723, 62
HD 24712, 60–61
HD 46407, 18–19
HD 60435, 61
HD 101065, 60–61
HD 119027, 61
HD 140283, 213
HD 155543, 70
HD 155555, 520, 525
HD 201601, 60, 61
He 2-10, 266–67
He 3-519, 240
Head-tail radio sources, 278
Heavy elements
    in population III stars, 550
    r-process and, 170
    stellar oscillations and, 45, 47
    synthesis of, 154

Heavy nuclei, making, 161–63
Helium
    in baryonic dark matter, 537–38, 542, 579–80
    in Be stars, 9
    B^2FH and, 18
    in γ Cas, 7
    cosmic wave background anisotropies and, 323
    Goldilocks problem and, 89
    in interstellar medium, 203, 206–11
    in massive stars, 234–37, 241, 243–44, 247–51
    in neutron stars, 626, 628, 631
    in population III stars, 549–52
    in pulsating white dwarfs, 50–51, 53, 57, 63
    r-process and, 168, 170
    in δ Scuti stars, 63
    s-process and, 177, 180
    stellar oscillations and, 45
Henry's law, 108
HEOS-2 spacecraft observations, 435
63 Her, 62, 65–66
Herbig Ae/Be stars, 466
Herbig-Haro objects, 490, 516
Hertzsprung-Russell diagram
    asteroseismic, 46, 49, 60–61
    of pre-main-sequence binaries, 493–95, 517
    of massive stars, 231, 236–39, 243, 250
Her X-1, 606, 608
HG 92, 535
HH 1-2 source, 478, 513
HH 111 source, 478
Hipparcos satellite observations, 61
HIRES spectrograph, asteroseismology and, 77
HK Tau, 478, 523
HL Tau, 504
HN Tau, 523
Holmium, in roAp stars, 60
HO Tau, 523–24
Hot dark matter model, 328–30, 334, 359, 373, 399, 401
Hot intracluster medium, 278–79
Hoyle's process, 18
HP Tau, 486, 522, 524
HP Tau G2, 524
HP Tau G2/G3, 523
HP Tau G3, 524
HQ Tau, 488, 522, 524
HR 1217, 70
HR 3831, 61
Hubble-Sandage variables, 238
Hubble sequence
    galaxy properties and
        carbon monoxide, 133–35
        chemical abundances, 137–40

classical galaxies, 144
cluster environment, 142–43
dwarfs, 144
environmental effect, 142–43
far infrared emission, 128–30
global properties, 141–42
group environment, 143
HII regions, 135–37
high-redshift galaxies, 143
luminosity function, 127–28
mass surface density, 141
morphological segregation, 142
neutral hydrogen mass and content, 132–33
optical colors, 126–27
optical linear size, 127
optical luminosity, 127–28
optical surface brightness, 128
radio continuum emission, 130–31
rotation curves, 140–41
SO problem, 144
total mass-to-light ratio, 140
variations in, 141–44
X-ray emission, 131–32
morphological dependence of fundamental properties, 117–18
construction of samples for analysis, 118–20
far infrared luminosity, 119–20
limitations of galaxy catalogs, 120–25
luminosity, 119
Malmquist bias, 121–22
neutral hydrogen mass, 119
optical size, 119
spiral galaxies, 120
summary of results, 121, 126
surface density, 119–20
surface magnitude, 119
total mass, 120
Hubble Space Telescope, 32–35, 202, 204, 221, 266–67, 296, 378, 470, 497, 519, 560–61
HV Tau Aa, 522
HV Tau AB, 523
Hyades cluster, 66, 475
Hydra A, 282, 284, 297–98
Hydrogen
in baryonic dark matter, 533, 545, 576–77, 579
in Be stars, 9, 14
B^2FH and, 18
in γ Cas, 7
cosmic wave background anisotropies and, 323

Goldilocks problem and, 89–90, 92, 97, 105, 108
Hubble sequence and, 119, 132–33, 135–39
in massive stars, 235–36, 239, 244, 247, 250–51
in neutron stars, 626–28, 631
in population III stars, 547–51
in pulsating white dwarfs, 51, 53, 63
p-process and, 184–85
s-process and, 175, 177–78
in supernovae, 17
β Hyi, 73, 75–77
Hypersurfaces, flat spatial, 326

I

IC 10, 253
IC 342, 221
IC 1613, 253–54
IC 4182, 19
Ice, on Mars, 94–97, 110
ICE spacecraft observations, 449–50
IE 2259+586, 597
Infall, 187, 372, 383, 503–5, 511, 514, 516, 534, 546, 556
Inflation, 326–27, 331, 333, 336, 393
Infrared (IR) companions, 478–79, 489–93, 503–6
Infrared (IR) emissions
baryonic dark matter and, 566–72, 577–78
brown dwarfs and, 577–78
cooling flows and, 297
Goldilocks problem and, 85, 96, 110
Initial mass function (IMF)
cooling flows and, 296, 299, 312–13
massive stars and, 245–46, 251, 255–57, 259–60, 262–63, 267–68
population III stars and, 542–43, 545, 547
universal, 129
Inos, 536–37, 582
Instabilities, 450, 456–59
Intermediate mass objects, 547–48
Intershell region, 244
Interstellar medium (ISM)
abundances
carbon and, 215–18
carbon isotope ratios and, 201
chemical fractionation and, 199–201
CNO isotopes and, 215–21
current enrichment and, 221
dense molecular clouds and, 197–203

determinations of, 193–203
deuterium and, 203–5
diffuse clouds and, 197
excitation and, 198–99
general background on, 193
helium and, 203–4, 206–11
HII regions and, 195–97
introduction to, 191–93
isotope ratios and, 201–3
light elements and, 213–15
lithium and, 203–4, 211–13
nitrogen isotopes and, 218
oxygen and, 219–20
selective dissociation and, 201
silicon isotopes and, 220
sulfur isotopes and, 220
VSGs in, 419
Interstellar scintillations, pulsars and, 602–3
Io, 439, 445–46
Iodine vapor, in asteroseismology, 74
IPHIR instrument, 42–44, 71
IRAS 0914+4109, 310
IRAS 10214+4724, 310
IRAS 16293-2422, 515–16
IRAS 19156+1906, 516
IRAS (Infrared Astronomical Satellite) observations, 118–20, 128, 134, 233, 248, 297, 320, 342–43, 360, 392–96, 404, 406–11, 479, 489, 491–92, 524, 536, 558, 567, 578–79
IRC+10216, 218, 220
Iron
Goldilocks problem and, 97
in hot intracluster medium, 279
Hubble sequence and, 137, 139
nucleosynthesis and, 157, 161
Iron-group nuclei, nucleosynthesis and, 163–65, 174
Irrotationality, 414
Isaac Newton 100-inch telescope, 30–31
Isochrones, 243
Isocurvature, 334–35, 539, 543–44, 546
ISO (Infrared Satellite Observatory) satellite observations, 558, 577–80
Isotopes
in interstellar medium, 215–21
neutron-rich, 15
nucleosynthesis and, 167–68
IS Tau, 522
IT Tau, 523
IUE (International Ultraviolet Explorer) satellite observations, 204, 264–65, 294–95
IW Tau, 522
IZw 18, 209

J

Jeans mass, 296, 304–6, 544, 573
Jeans scale, 341
Jets, 130, 307, 478, 559, 561
Jodrell Bank, 599
Jonker's curve, 432
Jupiter
   cosmic dusty plasma and, 436,
     438–39, 445–46
   lenticular dust halo of, 445
   magnetosphere of, 436, 438,
     445, 449
   moons of, 438–39, 445–46
   rings of, 438–39, 445
J0034-0534, 594, 612
J0322+2057, 632
J0437-4715, 594, 600, 606, 612,
   615, 632
J0751+18, 594
J1045-4509, 594
J1713+0747, 594
J1803-2712, 594, 621
J2019+2425, 594
J2145-0750, 594, 602, 627–28, 630
J2235+1506, 594, 608
J2317+1439, 594, 612
J2322+2507, 594
J4872, 523

K

K stars
   asteroseismology and, 72
   baryonic dark matter and, 532–
     33
   Goldilocks problem and, 106
   in nearby galaxies, 238
Kaon condensation, 616–17
κ mechanism, 53, 60, 63
Keck 10-m telescope, 77
Kellogg Radiation Laboratory, 9,
   17, 19
Kelvin-Helmholtz instability, 457
Kelvin-Helmholtz timescale, 236
Kelvin's circulation theorem, 383
Kitt Peak Observatory, 21–22,
   26–27, 31, 34
Klemola 44 cluster, 283
Knots, 130, 267–68
Kurtosis, 332, 404
K1–16, 50

L

Lakes, on Mars, 94
Landau damping, 456–58
LAOL opacities, stellar oscilla-
   tions and, 48
Large Magellanic Cloud (LMC)
   HII regions in, 208
   Hubble sequence and, 137
   massive stars in, 232, 238,
     241, 243–44, 246, 253–
     58, 260, 262

   microlensing and, 565
   p-process and, 186
Large-scale dynamics, 371–72
Large separation, stellar oscilla-
   tions and, 40
La Silla 3.6 meter telescope, 74
Lava flows, on Mars, 95
LBV stars, in nearby galaxies,
   238–40
Least action, 405
Ledge, 243–44
Ledoux criterion, 233–34, 241,
   243–44
Leiden Observatory, 28
Levitation, of fine dust, 435–36
LH 9, 230
LH 10, 230
LH 58, 230
LH 117, 230
LH 118, 230
Lick Observatory, 14, 22, 25–27,
   31
Light elements, in interstellar me-
   dium, 213–15
L'Institute d'Astrophysique, 10
Liouville's theorem, 339
LISA/SAGITTARIUS interfer-
   ometer, 571–72
Lithium
   in baryonic dark matter, 537–38
   in interstellar medium, 203,
     210–13
   in pre-main-sequence binaries,
     520
LkCa3, 467, 522, 526
LkCa7, 523
LkHα 332/G1, 522
LkHα 332/G2, 522
L'Observatoire de Haute Prov-
   ence, 9–10, 566
Local Group galaxies, 119, 131,
   232, 253–54, 319, 382, 389–
   91, 405–6, 412, 533, 535
Local Supercluster, 118–19, 147–
   50, 388
Local thermodynamic equilib-
   rium, interstellar abundances
   and, 198
Lorentz force, 444–45
Low mass objects (LMOs), 532,
   547, 572–78
Lyman-α clouds, 539, 547
Lyman continuum photons, 259–
   62
Lyman Far Ultraviolet Explorer
   satellite observations, 221
L723, 516
L1551 IRS5 region, 478

M

M stars
   asteroseismology and, 68
   Goldilocks problem and, 106

   interstellar abundances and,
     213
   in nearby galaxies, 238
Mach number, cosmic micro-
   wave background and, 400
Macrolensing, 560–62
Magellanic Clouds, see also
   Large Magellanic Cloud;
   Small Magellanic Cloud;
   massive stars and, 227–30,
   232, 252
Magellanic Stream, 533
Magic numbers, 18
Magnetic braking, 632
Magnetic fields
   cooling flows and, 299–301,
     303–6, 313
   cosmic dusty plasma and, 419
   pulsars and, 605–10
   r-process and, 166–67, 172
   stellar oscillations and, 39–40,
     57, 60–61
Magnetite, in cosmic dusty
   plasma, 447
Magneto-gravitational resistance,
   437
Magnetospheres
   planetary, 436–37, 440, 442–
     45, 449, 459
   pulsar, 614–15
Malin 1, 120
Malmquist bias, 121–22, 379–80
Malmquist-free analysis, 387–88
Mariner spacecraft observations,
   94
Mars
   Goldilocks problem and, 83–
     87, 93–98, 100, 107, 110–
     11
   satellites of, 444
Masers, water, 515
Massive compact halo objects
   (MACHOs), baryonic dark
   matter and, 540–42, 583–85
Massive stars
   in nearby galaxies
     abundances, 237–38
     blue Hertzsprung gap, 242–
      43
     blue supergiant, 240–45
     census, 230–31
     chemical abundances, 241–
      42, 247–49
     direct star counts, 229–32
     Eddington limit, 238–40
     emission line galaxies, 260,
      266–68
     far-infrared luminosities, 266
     HR diagram, 236–37
     input physics, 232–35
     introduction to, 227–28
     LBV stars, 238–40
     lifetimes, 236–37
     luminosity functions, 232

main sequence evolution, 236–38
masses, 236–37
metallicity, 235
nebular line analyses, 259–60
OB star distribution, 228–32
red supergiant, 240–45
SN 1987A progenitor, 243–44
spectral synthesis, 264–65
starbursts, 259–68
stellar models and observations, 232–45
Wolf-Rayet galaxies, 261–68
Wolf-Rayet stars, 245–59
MAX experiment, 341, 346–49
Max-flow algorithm, 385
MAX-GUM experiment, 342–44
Maximal cooling flows, 574
Maximum-probability method, 387
MAX-MuP experiment, 342–44
Mayer-Teller hypothesis, 15
McDonald Observatory, 10–13, 15–16, 19–21, 25–26
Melnick 42, 237–38
Memnonides scenarios, 624, 626
Metallicity
of massive stars, 235
of population III stars, 543, 545, 548
Meteorites, 15, 92, 95, 169, 181, 443
Meteors, 579
MFPOT model, 387
Mg^{2+}, Goldilocks problem and, 108
Microlensing, 560, 562–66
Micrometeoroids, 434, 443–44, 449
Microparticle plasma, 450
Milky Way, see Galaxy
MKW3s, 283
M/L ratios, 17, 22
Mode trapping, 52–53, 55, 63
Molecular clouds, 197–205, 216, 220, 229, 236
Molecular cores, 506, 511
Molecular outflow, 515
1 Mon, 62
21 Mon, 62, 65
Monopoles, 537
Mooms, 439
Moon, Goldilocks problem and, 91, 97
Moons
of Jupiter, 438–39, 445–46
of Mars, 444
of Saturn, 443–45
Morphological segregation, 142
Mount Hamilton Observatory, 25
Mount Wilson Observatory, 7–8, 14, 17–18, 20–21, 23

MSAM2 experiment, 342–44, 348–49
MSAM3 experiment, 342–44, 347–49
M4, 625
M15, 616
M31, 20, 22, 133, 137–38, 232, 238, 245, 253–54, 258
M33, 20, 133, 138, 238, 240, 244, 253–68
M51, 21, 117
M67, 62, 70
M81, 138, 239
M82, 221
M83, 138
M87, 24, 288, 296–98, 576
M101, 138, 239

N

Natal velocity kicks, 603–04, 635
N-body capture, 514, 534
Nebulae, ring, massive stars and, 247, 251
Nebulae, planetary
"case B" decrement for, 14
interstellar abundances and, 214
Nebulae, protoplanetary, triboelectric emission and, 421
Nebulae, wind, pulsars and, 611–12
Neodymium, r-process and, 168
Neon, in massive stars, 248
Neutrinos, cosmic wave background anisotropies and, 320, 322, 329, 537
Neutron stars
baryonic dark matter and, 537, 546, 548, 581
nucleosynthesis and, 163
pulsars and, 592, 595, 597–98, 600–1, 603, 605, 608–20, 622, 626–29, 631, 633–34
r-process and, 163, 165–68
Neutron star winds, nascent, r-process in, 172–74
New General Catalog (NGC), 116
NGC 100, 576
NGC 253, 221
NGC 300, 254
NGC 330, 244
NGC 346, 230, 232
NGC 592, 256
NGC 595, 256
NGC 604, 256, 268
NGC 1275, 128, 285, 287, 290, 293–94, 296–98, 310
NGC 1316, 128
NGC 1333 IRAS 4, 516
NGC 1365, 264
NGC 1741, 265
NGC 1976, 195

NGC 2024, 197
NGC 2403, 239
NGC 2974, 134
NGC 3242, 207, 214
NGC 3256, 221
NGC 3928, 134
NGC 4565, 576
NGC 4650A, 534
NGC 4945, 221
NGC 5018, 139
NGC 5044, 285
NGC 5128, 21
NGC 5907, 576
NGC 6166, 296
NGC 6334, 215
NGC 6342, 633
NGC 6539, 616
NGC 6611, 231
NGC 6822, 253–54
NGC 7027, 220
NGC 7252, 267
Nitrogen
Goldilocks problem and, 110
interstellar abundances and, 218
in massive stars, 237, 241, 243, 247–48
Noctilucent clouds, 420
No-disk-dark-matter hypothesis, 533
Non-baryonic matter, 322
Non-linear biasing, 392
Non-nuclear radiation, Hubble sequence and, 131
Non-X-ray wavebands, cooling flows at, 298, 299
Novae, interstellar abundances and, 213, 216
Nuclear statistical equilibrium, nucleosynthesis and, 155–66, 185
Nucleon shells, closed, 18
Nucleosynthesis
cosmological, 537–38
processes in, 153–187

O

O stars, in nearby galaxies, 227–69
OB 48, 232
OB 78, 232
OB stars
distribution of, 228–32
pre-main-sequence binaries and, 498–99
Oblique rotator model, for Ap stars, 60
Occam's razor, principle of, 393
Oceans, Goldilocks problem and, 88, 91, 94, 96–97, 99–100, 104–6
Olivine, in cosmic dusty plasma, 447

Ω value, cosmic flow dynamics and, 405–12
OPAL opacities, 48–49
Open universes, 356–57
ζ Oph, 201–2, 211
ρ Oph, 515
ρ Oph B1, 516
Ophiuchus cluster, 282
Optical waveband, cooling flows at, 293–97
ORI429, 525
ORI569, 475, 507
OriNTT 569, 525
Orion A, 195, 197
Orion Hot Core, 200
Orion KL nebula, 205
OS Gem, 62
Outgassing, 86–88, 92–94, 96–97
Overshooting, 233, 234, 241, 244
OVRO experiment, 343, 347
Oxygen
    in cosmic dusty plasma, 434, 451
    Goldilocks problem and, 92
    in hot intracluster medium, 279
    Hubble sequence and, 137–39
    interstellar abundances and, 219–20
    ions, excitation and de-excitation of, 8
    in massive stars, 237, 248, 250
    in pulsating white dwarfs, 50–51
Ozone layer, 32

P

P Cyg, 239, 489
Palomar Observatory, 14, 17, 23, 25
Palomar Sky Survey, 134
Pancakes, 325, 544
Parametric resonance, asteroseismology and, 64
Parkes Observatory, 599–600, 634
Paunch, in main sequence evolution, 236
Pavo-Indus-Telescopium (PIT), 388
Peculiar velocities, 373, 377–90
Pederson current, 441
ε Per, 204
ζ Per, 202, 211
Period gap, 632
Permafrost, on Mars, 94
Perseus A, 297
Perseus cluster, 285, 288–90, 293, 296, 298–99, 310, 388, 395
Pervasive pregalactic cooling flows, 574–75
PG1159-035, 50–51, 55–58
PHL 938, 27

PHL 5200, 27 *Phobos* spacecraft observations, 42, 444
Photinos, 537
Photodissociation, 87, 92, 99
Photoemission, cosmic dusty plasma and, 421
Photolysis, 105
Photometric techniques
    asteroseismology and, 50, 53–54, 64
    direct star counts and, 229, 231–32
    Hubble sequence and, 140
    pulsations in Sun-like stars and, 68–72
Photon drag, 328
Photon diffusion, 544
Photon trapping, 198–99
Pion condensation, 616–17
Pisces, 390, 395
PKS 0237-23, 27
PKS 0745, 284
PKS 0745-191, 296–98
Planetary nebula nuclei (PNNs), asteroseismology and, 50-53
Planetesimals, 92–93
Planets
    Goldilocks problem and, 83–112
    pulsars and, 624–26
Planck relics, 537
Planck spectrum, 319
Plasma drag, 439
Plasma oscillations, 16
Plasma probes, 633–34
Plate tectonics, Goldilocks problem and, 88, 92, 95, 111
Pleiades cluster, 475
P-modes
    asteroseismology and, 52, 60, 62–64, 66–69, 72–73, 75–77
    stellar oscillations and, 39–40, 42–45, 47
Poisson equations
    cosmic dusty plasma and, 427, 429
    cosmic flows and, 374
Poisson noise, 69, 72
Polar holes, 505
Polarization, cosmic wave background anisotropies and, 355–56
Polar mesospheric summer echoes, 460
Polar rings, 143, 534
Polyneutrons, primordial, 15
Popcorn, 231
Population III stars
    background light constraints on, 549–50
    black hole accretion and, 552
    as dark matter producers, 543–45

enrichment constraints on, 550–51
    expected mass of, 545–47
    helium production and, 551–52
    intermediate mass objects and, 547–48
    introduction to, 542, 543
    low mass objects and, 547
    as metal producers, 543
    remnants of, 547–48
    supermassive objects and, 548
    very massive objects and, 548
Possible universes, 331
POTENT procedure, 379–80, 383–85, 390, 392, 395–98, 400–1, 407–8, 410–11
Power spectrum
    cosmic wave background anisotropies and, 336–44
    dynamics of cosmic flows and, 399–403
Poynting-Robertzon radiation drag, 439
p-process
    constraints on, 186–87
    general considerations, 154, 182–84
    entropy and equilibrium, 155–56
    equilibrium nucleosynthesis, 156–61
    how to make heavy nuclei, 161–63
    history of, 18
    introduction to, 153–54, 181
    sites for, 184–85
Present day mass function, 230
Primeval fireball, 319
Primeval isocurvature, 539
Primordial holes, 537
Primordial isocurvature baryonic (PIB) model, 399
PRISMA mission, 78
Procyon, 73, 75–76, 204
Procyon A, 75
Protobinaries, 515–16
Protogalaxies, 533
Protostellar clouds, 236, 454, 457
Przybylski's star, 60
Pseudosynchronic bands, 435
Pulsars
    baryonic dark matter and, 560
    binary and millisecond
        accretion-induced collapse and, 597–98
        background and framework, 593–98
        bimodal field distribution and, 608
        birthrates of, 600–2
        calorimetry and, 611–14
        companion heating and, 612–14

core mass-period relation and, 626–28
dynamical fossils of spin-up era, 626–33
eccentricity-period relation and, 629–30
equations of state and, 617–20
field constancy and, 607
field decay and, 606–7
field reduction models and, 609–10
future research on, 634–35
general relativity and, 620–22
gravitational waves and, 623
high-mass binary pulsars, 595–96
introduction to, 591–93
kinematics of, 602–4
large-scale accelerations and, 622–23
low-mass binary pulsars, 596–98, 604
magnetic field evolution and, 605–10
magnetosphere diagnostics and, 614–15
masses and, 616–17
mass transfer and, 631–33
minimum stable periods and, 617–20
natal velocity kicks and, 603–4
neutron stars and, 610–20
Newtonian accelerations and, 622–23
phenomenological model of, 608–9
physics of, 610–20
planetary companions and, 624–26
as plasma probes, 633–34
populations of, 600–2
recycled pulsar model, 597
recycled pulsar model and, 596
residual fields and, 605–6
searches for, 598–600
soft equations of state and, 616–17
triple star evolution and, 597
velocity of, 602–4
wind nebulae and, 611–12
ζ Pup, 237
Python experiment, 342–44, 347–49
P1540, 495, 525
P1771, 526
P1925, 526
P2486, 508, 525
P2494, 525

Q

Quadrupole normalization, 337
Quantum fields, cosmic wave background anisotropies and, 326–27
Quark-hadron phase transition, 538, 541
Quark nuggets, 537
Quasars, 22, 26–27, 35, 228, 309–310, 313, 548, 559–64
Quasi-linear relations, 392–93, 374–75
Quasi-Steady-State-Cosmology, 35

R

R Mon, 504
Radial order, stellar oscillations and, 39
Radial spokes, 420
Radiation drag, 323, 355, 439
Radioactive emission, cosmic dusty plasma and, 421
Radio emissions
baryonic dark matter and, 559
cooling flows and, 297–98
Hubble sequence and, 130–31
pulsars and, 591–93
Radio-loud objects, 309
Rainout, 90
Rayleigh fractionation, 91
Rayleigh-Jeans region, 324
Rayleigh scattering, 93
Rebirth line, pulsars and, 592, 596–97
Recombination epoch, 322–24
Recycled pulsar model, 596–97
Redshift
baryonic dark matter and, 535, 539–40, 544, 547, 550, 562- 65, 568–71, 574, 585
cosmic wave background anisotropies and, 323–24, 327, 330, 333, 335, 340, 342, 352–53, 361, 363
dynamics of cosmic flows and, 394, 408–10
Rees-Sciama effect, 354
Reference Catalogues (RC), 116
Reionization, cosmic wave background anisotropies and, 350–52
Ringing, 338
Ring-of-fire effect, 356
Rings, planetary, cosmic dusty plasma and, 420, 425–27, 429, 437–45, 453, 457–59, 461
Riverbeds, on Mars, 94
Roche lobe overflow, 245, 252–53, 257, 261, 595, 627–28, 630–32

Rocks, surface, Goldilocks problem and, 86, 94, 99–101, 103, 106, 108, 111–12
ROSAT satellite observations, 131, 281, 284–87, 289, 306–8, 310, 535, 537
Royal Greenwich Observatory, 29–31
r-process
general considerations, 154
entropy and equilibrium, 155–56
equilibrium nucleosynthesis, 156–61
how to make heavy nuclei, 161–63
history of, 18
introduction to, 153–54, 163–64
in nascent neutron star winds, 172–74
observational constraints and, 169–72
primary, 164–67
secondary, 167–69
RR Lyrae, 49, 61
Runaway glaciation model, of planetary development, 96
RW Aur, 467
RW Aur AB, 523
RW Aur BC, 522
R 71, 239
R 84, 240
R 127, 239
R 136, 260

S

S Dor, 238, 239
S stars, slow neutron capture in, 18
Sachs-Wolfe effect, 324, 333–35, 339, 345, 354, 357, 362- 63, 395
Saha equation, 322
Salamander scenarios, 624
Salpeter value, 230, 256–57
SAS 3 satellite observations, 281
Saskatoon experiment, 342–44, 347–49
Saturn
A-ring of, 444
atmosphere of, 441
B-ring of, 420, 440–44
C-ring of, 444
cosmic dusty plasma and, 437, 440–45, 453, 457–59
E-ring of, 420, 445
F-ring of, 420, 426–27, 437, 453
G-ring of, 457–58
ionosphere of, 442–43
magnetosphere of, 436, 449
moons of, 443–45

spokes of, 420, 426, 440–42
Schwarzschild's criterion, 24, 233–34, 241, 244
Scintillation noise, asteroseismology and, 68–70
Sco-Oph association, 468–69, 471, 487, 494, 496, 498
Sculptor void, 411–12
δ Scuti stars, 38–39, 50, 60–66
Search for Extraterrestrial Intelligence (SETI), 105–6
Sedimentation, 109
Seed power, large-scale, 321
Ser/G1, 491, 493
Ser OB1, 230–31
Sersic 159–03, 293
Seyfert galaxies, Hubble sequence and, 129, 143
Sgr B2, 205
Shadow matter, 537
Shapley-Ames Survey, with update, 116, 144
Shapley concentration, 388, 402
Shells, 7, 143, 168, 170–71, 174, 185, 239, 244, 246–47, 551, 626–28
Shot-noise, 391–92
Siding Springs telescope, 31
Silicates
    in cosmic dusty plasma, 447
    Goldilocks problem and, 99–101, 106, 108, 111–12
    in pre-main-sequence binaries, 485–86, 490–92
Silicon
    in Apm stars, 15
    in hot intracluster medium, 279
    interstellar abundances and, 220
    in roAp stars, 60
Sine wave chop, 346
SIRTF satellite observations, 577, 580
Skewness
    cosmic wave background anisotropies and, 332
    dynamics of cosmic flows and, 404
Slingshot mechanism, 556
Small Magellanic Cloud (SMC)
    HII regions in, 208
    massive stars in, 232, 241, 244–46, 253–55, 257
    microlensing and, 565
    pulsars in, 595
Small separation, stellar oscillations and, 41
SN 1987A, 185–86, 228, 233–35, 241, 243–44, 621
SN 1993J, 233
Snowballs, 537, 547, 579–80
SO problem, Hubble sequence and, 144

Sodium, in massive stars, 242
Solar flares, 436
Solar System
    abundances in, 153, 154
    Goldilocks problem and, 88, 92–93, 97, 105
    interstellar abundances and, 200, 211, 216, 218–20
    r-nuclei in, 163
    s-process and, 176
Solar wind plasma, 435, 446, 456
Sound waves, stellar oscillations and, 39, 41
South African Astronomical Observatory, 50
Southern Wall, 390, 411
SP91–13pt experiment, 342–44, 347–49
Space Shuttle, 32–33
Spectral energy distributions, infrared, 482–84, 488, 490–93, 501–6, 515–16, 518
Spectroscopy
    asteroseismology and, 64
    direct star counts and, 229
Spheroids, 22
Spin flip, 204
Spin-up, 596, 626–33
Spiral-in, 596
Spite plateau, 212
Spitzer value, 301
s-process
    general considerations, 154
    entropy and equilibrium, 155–56
    equilibrium nucleosynthesis, 156–61
    how to make heavy nuclei, 161–63
    history of, 18–19
    introduction to, 153–54, 174
    mechanism for, 174–76
    sites for, 176–81
Square wave chop, 346
SR24N, 486
SSV 63, 491
Starbursts, 143, 228, 231, 246–47, 259–69, 546
Star clusters, 243, 258
Stardust, 181
Starlight, grain-thermalized, 570
Stars
    asymptotic giant branch, 177–78, 180–81, 187, 213
    Apm stars, 15
    A-type, see A stars
    Ba II, see Ba II stars
    Be, see Be stars
    binary, see Binary stars, artificial; Binary stars, low-mass X-ray; Binary stars, massive X-ray; Binary stars, pre-main-sequence; Binary stars, short-period;

Binary stars, visual; Binary stars, X-ray
    blue, 62
    blue supergiant, 227, 234, 240–45
    brown dwarf, 537, 546–47, 551, 575–78, 580, 583, 585
    B-type, see B stars
    carbon, 18, 216, 218–19
    cataclysmic variables, 596, 598
    Cepheid, see Cepheid stars
    continuously habitable zones around, 105–7
    formation of, 141–42, 145, 228, 231, 245–46, 255–57, 261- 64, 268, 302
    F-type, see F stars
    giant, 70
    G-type, see G stars
    hypergiant, 238–39
    intermediate-mass, 216
    K-type, see K stars
    LBV, see LBV stars
    low-mass, 178, 180–81, 187, 216
    luminous, 68
    magnetic, 39–41, 57, 60–61
    main-sequence, 49
    massive, see Massive stars
    M-dwarf, 537, 547, 566, 575–76, 580, 584
    M-type, see M stars
    neutron, see Neutron stars
    nucleosynthesis and, 161
    OB, see OB stars
    OBC supergiant, 241
    OBN, 241
    ON, 237, 238
    O-type, see O stars
    population I, 211–13, 576–77, 604
    population II, 211–13, 576–77, 604
    population III, see Population III stars
    post-AGB progenitor, 57
    proper motion, 20
    protostar, 511
    pulsating, 37–39, 50–63
    red giant, 15, 18, 180, 213, 216, 220, 595, 624, 629–30, 635
    red supergiant, 227, 233, 238, 240–45, 247, 597
    spheroid, 556
    S-type, see S stars
    subgiant, 49, 68, 70, 77
    supergiant, 213, 240–41
    WC, 248–49, 258–59
    white dwarf, 14–15, 50–50, 185, 537, 551, 580, 583, 592, 594–95, 597–98,

605–6, 617, 620, 625–31, 634
WN, 237, 247–48
WNL, 248
WN/WC, 234, 248, 252, 255–58
W-R/O, 255–58
STARS mission, asteroseismology and, 78
Starspots, 65
Steady-state universe, 27
Stefan-Boltzmann constant, Goldilocks problem and, 84
Steidel-Sargent absorption systems, 564
Stellar birthline, 493–94, 496, 507–8, 512–13, 517
Stellar holes, 537
Stellar Photometry International, 65
Stellar winds, 172–74, 177–78, 207–9, 227, 232–33, 235, 237, 250, 515, 550, 593–95
Stephen's Quintet, 25
Sternglass formula, 432
Steward Observatory telescope, 26
Striae, 435
Strontium, in roAp stars, 60
Sulfur
  in hot intracluster medium, 279
  interstellar abundances and, 220
Sulfuric acid, Goldilocks problem and, 93
Sun
  asteroseismology and, 42–45
  chemical elements in, 15
  cosmic wave background anisotropies and, 319
  Goldilocks problem and, 83–85, 87–89, 96, 98–3, 106, 111–12
  interstellar abundances and, 217
Sunyaev-Zel'dovich fluctuations, 324–25
Superclusters, 320
Supergalactic plane, 381, 383, 388–89, 394–96, 398, 411–12
Supermassive objects (SMOs), 537, 548, 581–83
Supernovae
  baryonic dark matter and, 559
  hot intracluster medium and, 279
  interstellar abundances and, 209
  massive stars and, 227
  neutrons and, 17

nucleosynthesis and, 161
p-process and, 184–87
rapid neutron capture in, 18
r-process and, 163, 166–72, 174
Supernova remnants, pulsars and, 591–92, 597, 604, 608
SVS 20, 493, 516
Synchrotron radiation
  cosmic wave background anisotropies and, 355
  from extragalactic sources, 16
  from radio galaxies, 24
  Hubble sequence and, 130
S27 shepherding satellite, 437
S984, 71

T

θ^2 Tau, 62, 65–66
T Tau, 478, 480, 489–92, 502, 504–5, 517, 523
T Tau N, 486, 490–92, 504, 506
T Tau S, 490–92, 504, 506
T Tauri, 18, 465, 467, 471, 473–74, 483, 487–88, 490–91, 493–94, 499, 502–4, 506, 520
Tau-Aur association, 467–71, 476, 479, 481, 486–88, 494, 496, 498–99, 519–20, 522–24
Tenerife experiment, 342–44, 347–49
Tensor spectrum, 336
Terraforming, 110–111
Thebe, 438
Theoretical uncertainty, 354
Theory gap, 511
Thermoionic emission, cosmic dusty plasma and, 421
Third dredge up, 178
Third Reference Catalog of Bright Galaxies (RC3), 116, 118–119, 121–25, 127, 144
Thomson depth, 297
Thomson scattering, 323, 339, 355
Thorne-Żytkow objects, 187, 597, 633
Tidal circularization, 506–8
Tidal bridges, 143
Tidal distortion, 252
Tidal phenomenon, 143, 167, 499, 501
Tidal spin-up, 533, 546
Tidal truncation, 485
Tillites, 109–10
Ton 1530, 27
Topology, cosmic wave background anisotropies and, 358–59
Tr 14, 257

Trapezium cluster, 195, 470, 497, 499, 511
Tri. Australis, 282
Triboelectric emission, cosmic dusty plasma and, 421
Triple star evolution, 597
Troposphere, Goldilocks problem and, 90, 95
Tully-Fisher relations, 140

U

Uhuru observations, 280
ULISSE experiment, 343
Ulysses spacecraft observations, 445
Ultraviolet (UV) radiation, Goldilocks problem and, 90, 104, 111
Ultraviolet (UV) upturn, 222
Universe
  cold dark matter in, 311–12
  cosmic microwave background anisotropies and, 319–20, 322- 23, 326, 328–29, 340, 348–49, 353–54, 356–57, 359
  creation of, 26–27, 35
  low density, 539
  recollapsing of, 536
University of London Observatory, 5–7, 9, 11
Uppsala General Catalog (UGC), 116, 118, 121–25, 127, 147-50
Uranium, nucleosynthesis and, 161
Urey silicate-rock weathering reactions, 88
UU Ari, 62
UX Tau, 467, 478
UX Tau A, 487
UX Tau AB, 523
UX Tau A/C, 523
UY Aur, 467, 491, 523
UZ Tau, 467, 491, 493
UZ Tau E, 503
UZ Tau EW, 478, 523
UZ Tau W, 503, 522
4U 1626–67, 597, 606
4U 1820–30, 613 V
Vacuum, cosmic wave background anisotropies and, 322
Van Rhijn function, 514
Var A, 240
VEGA spacecraft observations, 435
γ Vel, 249
Venus, Goldilocks problem and, 83–93, 97, 100, 107–8, 111
Very massive objects (VMOs), 537, 548, 551, 568–70, 581, 583

Very small grains (VSGs), 419, 432, 434–35, 453, 461
Viking spacecraft observations, 94, 110
Virgo A, 130, 297
Virgo cluster, 119, 127, 139, 143–44, 282, 288, 372, 383, 387–88, 559, 572
Virgo core galaxies, 133
Virial theorem, stellar oscillations and, 40
Vishniac effect, 334, 352–53
VLA (Very Large Array) observations, 343, 347, 513, 560–61
VLBA (Very Long Baseline Array) observations, 560–61
VLBI (Very Long Baseline Interferometry) mapping, 560–61
Voids, 354, 411–12
Volcanic plumes, 421
Voyager spacecraft observations, 420, 438, 440- 41
v-process, nucleosynthesis and, 169
VSB111, 526
VSB126, 526
*v*UMa, 65
VV CrA, 491
VW Ari, 62
VY Tau, 522
V410 Tau, 522
V650 Tau, 62, 65–66
V710 Tau, 523
V773 Tau, 522
V807 Tau, 488
V807 Tau NS, 522
V807 Tau S, 522
V826 Tau, 467, 507, 525
V853 Oph, 486
V927 Tau, 522
V928 Tau, 522
V955 Tau, 522
V4046 Sgr, 480, 484, 486–88, 507, 520, 525

W

Waiting point nuclei, r-process and, 170
Walborn hypothesis, 241
Warm dark matter model, 359
Water
  Goldilocks problem and, 85–95, 97–98, 104–5, 107, 110–11
  masers and, 515
Wavelet analysis, 386
Waves, cosmic dusty plasma and, 450–56, 459–60

Weakly interacting massive particles (WIMPS)
  baryonic dark matter and, 536, 541–42, 583–84
  solar neutrino deficit and, 43
Wiener filter, 386–87
Wells, 310
White Dish experiment, 343, 347
White Earth catastrophe, 83–84, 100
Whole Earth Telescope, 50, 52, 54, 56, 59, 65, 71
Wien region, 323
Window functions
  asteroseismology and, 53–54
  cosmic wave background anisotropies and, 331, 346–49
WKB approximation, asteroseismology and, 67
Wolf-Rayet stars
  in nearby galaxies, 227–28, 230, 232, 239–40, 245–59
  pulsars and, 613
Wolf-Rayet winds, 207–8
WR 40, 250
WSB 4, 491
W3, 207–8
W3A, 208
W43, 207, 214
W49, 214
W51, 228
W134, 475, 479, 508, 525

X

x-process, 18
X-ray emissions
  cooling flows and, 281–93
  Hubble sequence and, 131–32
XSPEC spectral-fitting package, 290
XZ Tau, 478, 491–92, 522
X0115+634, 592
X0331+53, 592
X1822-371, 613

Y

Yerkes Observatory, 10–15, 19–20, 23, 25–26
Ylem theory, 15–16

Z

Z CMa, 478, 480–81, 491, 505
z-distribution, 390–93
Zel'dovich time machines, 403
Zero-age main sequence

(ZAMS), 236, 465–66, 475, 495, 497, 508, 520
Zero-lag autocorrelation, 353
ZZ Ceti, 51
ZZ Tau, 522
Z3146, 308

MISCELLANEOUS

3°K background, generation of, 570
4 kpc molecular ring, 215–16, 219, 221
0045-73, 594–95
0655+64, 594, 606, 608, 627–28
0745-191, 282
0820+02, 594, 606, 608, 621
0846+51, 562
0957+561, 310
1257+12, 594–95, 624–25, 633
1259-63, 594–95, 633
1534+12, 594, 600, 608, 616, 621–22
1620-26, 625
1718-19, 594–95, 633
1744-24A, 633
1802+07, 616
1820-11, 594–95
1820+30B, 608
1831-00, 594, 632–33
1855+09, 594, 606, 608, 616–17, 623
1913+16, 594, 600, 608, 616, 621–22
1937+21, 594, 599, 618, 620, 623
1953+29, 594
1957+20, 594, 611–13, 617–18, 627, 632–33
2016+112, 564
2127+11C, 600, 616, 621
2224+65, 603–4
2237+0305, 562
2303+46, 594, 616
04320+1746, 523
034903+2431, 522
035120+3154, 523
035135+2528, 523
040012+2545, 523
040047+2603W, 523
040142+2150, 523
045251+3016, 467, 526
155808-2219, 526
155913-2233, 507, 526
160814-1857, 526
160905-1859, 526
162814-2427, 479–80, 484, 502, 525
162819-2423S, 479–80, 484, 486, 526
2A0335+096, 282, 285

CUMULATIVE INDEXES

CONTRIBUTING AUTHORS, VOLUMES 22–32

A

Abbott, D. C., 25:113–50
Adams, F. C., 25:23–81
Antonucci, R. R., 31:473–521
Arendt, R. G., 30:11–50
Arnett, W. D., 27:629–700
Athanassoula, E., 23:147–68

B

Backer, D. C., 24:537–75
Bahcall, J. N., 24:577–611;
  27:629–700
Bahcall, N. A., 26:631–86
Bai, T., 27:421–67
Baliunas, S. L., 23:379–412
Barcons, X., 30:429–56
Barnes, J. E., 30:705–42
Beckers, J. M., 31:13–62
Beichman, C. A., 25:521–63
Bertelli, G., 30:235–85
Bertout, C., 27:351–95
Bessell, M. S., 31:433–71
Binggeli, B., 26:509–60
Binney, J. J., 30:51–74
Blandford, R. D., 30:311–58
Bloemen, H., 27:469–516
Bodenheimer, P., 26:145–97
Boesgaard, A. M., 23:319–78
Boggess, A., 27:397–420
Bosma, A., 23:147–68
Bowyer, S., 29:59–88
Bradt, H. V. D., 21:13–66;
  30:391–427
Branch, D. R., 30:359–89
Brault, J. W., 22:291–317
Bressan, A., 30:235–85
Bridle, A. H., 22:319–58
Brown, R. L., 22:223–65
Brown, T. M., 32:37–82
Burbidge, E. M., 32:1–36

C

Caldeira, K., 32:83–114
Cameron, A. G. W., 26:441–72
Canal, R., 28:183–214
Carr, B., 32:531–90
Carroll, S. M., 30:499–542
Chanmugam, G., 30:143–84
Chapman, G. A., 25:633–67
Chincarini, G. L., 22:445–70
Chiosi, C., 24:329–75; 30:235–85
Chupp, E. L., 22:359–87
Colavita, M. M., 30:457–98
Combes, F., 29:195–237

Condon, J. J., 30:575–611
Conti, P. S., 25:113–50; 32:227–
  75
Coulman, C. E., 23:19–57
Cowan, J. J., 29:447–97
Cowie, L. L., 24:499–535
Cowley, A. P., 30:287–310
Cowling, T. G., 19:115–35;
  23:1–18
Cox, D. P., 25:303–44

D

D'Antona, F., 28:139–81
Davidson, K., 23:119–46
de Pater, I., 28:347–99
de Zeeuw, T., 29:239–74
Dekel, A., 32:371–417
Deubner, F.-L., 22:593–619
Dickey, J. M., 28:215–61
Djorgovski, S., 27:235–77
Done, C., 31:716–61
Draine, B. T., 31:373–432
Dressler, A., 22:185–222
Dulk, G. A., 23:169–224
Duncan, M. J., 31:265–95
Dupree, A. K., 24:377–420
Dwek, E., 30:11–50

E

Eggen, O. J., 31:1–11
Elitzur, M., 30:75–112
Ellis, G. F. R., 22:157–84
Elson, R., 25:565–601

F

Fabbiano, G., 27:87–138
Fabian, A. C., 30:429–56;
  32:277–318
Feast, M. W., 25:345–75
Fesen, R. A., 23:119–46
Fich, M., 29:409–45
Fowler, W. A., 21:165–76;
  30:1–9
Franx, M., 29:239–74
Freeman, K. C., 19:319–56;
  25:603–32
Frogel, J. A., 26:51–92
Fujimoto, M., 24:459–97
Fusi Pecci, F., 26:199–244

G

Gallagher, J. S., 22:37–74
Garstang, R. H., 27:19–40

Gehrz, R. D., 26:377–412
Genzel, R., 25:377–423; 27:41–
  85
Gilliland, R. L., 32:37–82
Gilmore, G., 27:555–627
Ginzburg, V. L., 28:1–36
Giovanelli, R., 22:445–70;
  29:499–541
Golub, L., 23:413–52
Gough, D., 22:593–619; 29:627–
  84
Greenstein, J. L., 22:1–35
Gustafsson, B., 27:701–56

H

Haisch, B. M., 29:275–324
Harris, W. E., 29:543–79
Hartmann, L. W., 25:271–301
Hartwick, F. D. A., 28:437–89
Haynes, M. P., 22:445–70;
  29:499–541; 32:115–52
Hellings, R. W., 24:537–75
Henry, R. C., 29:89–127
Hernquist, L. E., 30:705–42
Higdon, J. C., 28:401–36
Hillas, A. M., 22:425–44
Hodge, P. W., 19:357–72;
  27:139–59
Holzer, T. E., 27:199–234
Houck, J. R., 25:187–230
Howard, R., 22:131–55
Hudson, H. S., 26:473–507
Hummer, D. G., 28:303–45
Hunter, D. A., 22:37–74
Hut, P., 25:565–601

I

Inagaki, S., 25:565–601
Isern, J., 28:183–214

J

Joss, P. C., 22:537–92

K

Kahler, S. W., 30:113–41
Kaler, J. B., 23:89–117
Kirshner, R. P., 27:629–700
Kondo, Y., 27:397–420
Koo, D. C., 30:613–52
Kormendy, J., 27:235–77
Kron, R. G., 30:613–52
Kudritzki, R. P., 28:303–45
Kuijken, K., 27:555–627
Kulkarni, S. R., 32:591–639

Kurtz, D. W., 28:607–55
Kwok, S., 31:63–92

L

Labay, J., 28:183–214
Lada, C. J., 23:267–317
Lawrence, C. R., 30:653–703
Léger, A., 27:161–98
Liebert, J., 25:473–519
Lingenfelter, R. E., 28:401–36
Lissauer, J. J., 31:129–74
Liszt, H. S., 22:223–65
Lizano, S., 25:23–81
Lockman, F. J., 28:215–61
Low, B. C., 28:491–524
Lunine, J. I., 31:217–63

M

Mackay, C. D., 24:255–83
Maeder, A., 24:329–75; 32:227–75
Majewski, S. R., 31:575–638
Maran, S. P., 27:397–420
Marcus, P. S., 31:523–73
Margon, B., 22:507–36
Mariska, J. T., 24:23–48
Mathews, W. G., 24:171–203
Mathieu, R. D., 32:465–530
Mathis, J. S., 28:37–70
Mazzitelli, I., 28:139–61
McAlister, H. A., 23:59–87
McCammon, D., 28:657–88
McCarthy, P. J., 31:639–88
McCray, R., 31:175–216
McCrea, W. H., 25:1–22
McKee, C. F., 31:373–432
Melrose, D. B., 29:31–57
Mendis, D. A., 26:11–49; 32:419–63
Meyer, B., 32:153–90
Mikkola, S., 29:9–29
Monaghan, J. J., 30:543–74
Monet, D. G., 26:413–40
Moore, R., 23:239–66
Morgan, W. W., 26:1–9
Mushotzky, R. F., 31:717–61

N

Narayan, R., 24:127–70; 30:311–58
Narlikar, J. V., 29:325–62
Neugebauer, G., 25:187–230
Nityananda, R., 24:127–70
Nordlund, Å., 28:263–301
Noyes, R. W., 25:271–301

O

Ohashi, T., 30:391–427
Oort, J. H., 19:1–5; 21:373–428
Osterbrock, D. E., 24:171–203
Ostriker, J. P., 31:689–716

P

Padmanabhan, T., 29:325–62
Pearson, T. J., 22:97–130
Perley, R. A., 22:319–58
Phinney, E. S., 32:591–639
Pollack, J. B., 22:389–424
Pounds, K. A., 30:391–427; 31:717–61
Press, W. H., 30:499–542
Probst, R. G., 25:473–519
Puget, J. L., 27:161–98

Q

Quinn, T., 31:265–95

R

Rabin, D., 23:239–66
Rampino, M. R., 32:83–114
Rana, N. C., 29:129–62
Rappaport, S. A., 22:537–92
Raymond, J. C., 22:75–95
Readhead, A. C. S., 22:97–130; 30:653–703
Rees, M. J., 22:471–506
Reid, M., 31:345–72
Renzini, A., 21:271–342; 26:199–244
Reynolds, R. J., 25:303–44
Rickett, B. J., 28:561–605
Ridgway, S. T., 22:291–317
Roberts, M. S., 32:115–52
Rodonò, M., 29:275–324
Rood, H. J., 26:245–94
Rood, R. T., 32:191–226
Rosenberg, M., 32:419–63
Rosner, R., 23:413–52
Rossi, B., 29:1–8

S

Saikia, D. J., 26:93–144
Salter, C. J., 26:93–144
Sandage, A., 24:421–58; 26:509–60, 561–630
Sanders, W. T., 28:657–88
Sargent, A. I., 31:297–343
Schade, D., 28:437–89
Scott, D., 32:319–70
Scoville, N. Z., 29:581–625
Sellwood, J. A., 25:151–86
Shao, M., 30:457–98
Shields, G. A., 28:525–60
Shu, F., 19:277–93; 25:23–81
Silk, J., 32:319–70
Sneden, C., 27:279–349
Sofue, Y., 24:459–97
Soifer, B. T., 21:177–207; 25:187–230
Songaila, A., 24:499–535
Spinrad, H., 25:231–69
Spite, F., 23:225–38
Spite, M., 23:225–38

Spitzer, L. Jr., 27:1–17; 28:71–101
Spruit, H. C., 28:263–301
Sramek, R. A., 26:295–341
Steigman, G., 23:319–78
Stern, S. A., 30:185–233
Stevenson, D. J., 29:163–93
Stinebring, D. R., 24:285–327
Stringfellow, G. S., 31:433–71
Strong, K. T., 29:275–324
Sturrock, P. A., 27:421–67
Stutzki, J., 27:41–85

T

Tammann, G. A., 26:509–60; 29:363–407; 30:359–89
Taylor, J. H., 24:285–327
Telesco, C. M., 26:343–76
Tenorio-Tagle, G., 26:145–97
Thielemann, F.-K., 29:447–97
Title, A. M., 28:263–301
Toomre, J., 29:627–84
Townes, C. H., 21:239–70; 25:377–423
Tremaine, S., 29:409–45
Trimble, V., 25:425–72
Truran, J. W. Jr., 27:279–349; 29:447–97
Tsuji, T., 24:89–125
Turner, E. L., 30:499–542

V

Vaiana, G. S., 23:413–52
Valtonen, M., 29:9–29
van den Bergh, S., 29:363–407
van der Klis, M., 27:517–53
Vaughan, A. H., 23:379–412
Verbunt, F., 31:93–127

W

Wagner, W. J., 22:267–89
Walker, A. R., 25:345–75
Weaver, T. A., 24:205–53
Weidemann, V., 28:103–37
Weiler, K. W., 26:295–341
Welch, W. J., 31:297–343
Wheeler, J. C., 27:279–349
White, M., 32:319–70
Whitford, A. E., 24:1–22
Wielebinski, R., 24:459–97
Wilson, T. L., 32:191–226
Woosley, S. E., 24:205–53; 27:629–700
Wyse, R. F. G., 27:555–627

Y

Yorke, H. W., 24:49–87
Young, J. S., 29:581–625

Z

Zwaan, C., 25:83–111

CHAPTER TITLES, VOLUMES 22–32

PREFATORY CHAPTER

An Astronomical Life	J. L. Greenstein	22:1–35
Astronomer by Accident	T. G. Cowling	23:1–18
A Half-Century of Astronomy	A. E. Whitford	24:1–22
Clustering of Astronomers	W. H. McCrea	25:1–22
A Morphological Life	W. W. Morgan	26:1–9
Dreams, Stars, and Electrons	L. Spitzer, Jr.	27:1–17
Notes of an Amateur Astrophysicist	V. L. Ginzburg	28:1–36
The Interplanetary Plasma	B. Rossi	29:1–8
From Steam to Stars to the Early Universe	W. A. Fowler	30:1–9
Notes from a Life in the Dark	O. J. Eggen	31:1–11
Watcher of the Skies	E. M. Burbidge	32:1–36

SOLAR SYSTEM ASTROPHYSICS

Origin and History of the Outer Planets: Theoretical Models and Observational Constraints	J. B. Pollack	22:389–424
Comets and Their Composition	H. Spinrad	25:231–69
A Postencounter View of Comets	D. A. Mendis	26:11–49
Origin of the Solar System	A. G. W. Cameron	26:441–72
Radio Images of the Planets	I. de Pater	28:347–99
The Pluto-Charon System	S. A. Stern	30:185–233
Planet Formation	J. J. Lissauer	31:129–74
The Atmospheres of Uranus and Neptune	J. I. Lunine	31:217–63
The Long-Term Dynamical Evolution of the Solar System	M. J. Duncan, T. Quinn	31:265–95
Jupiter's Great Red Spot and Other Vortices	P. S. Marcus	31:523–73
The Goldilocks Problem: Climatic Evolution and Long-Term Habitability of Terrestrial Planets	M. R. Rampino, K. Caldeira	32:83–114
Cosmic Dusty Plasmas	D. A. Mendis, M. Rosenberg	32:419–63

SOLAR PHYSICS

Solar Rotation	R. Howard	22:131–55
Coronal Mass Ejections	W. J. Wagner	22:267–89
High-Energy Neutral Radiations From the Sun	E. L. Chupp	22:359–87
Helioseismology: Oscillations as a Diagnostic of the Solar Interior	F.-L. Deubner, D. Gough	22:593–619
Radio Emission From the Sun and Stars	G. A. Dulk	23:169–224
Sunspots	R. Moore, D. Rabin	23:239–66
The Quiet Solar Transition Region	J. T. Mariska	24:23–48
Elements and Patterns in the Solar Magnetic Field	C. Zwaan	25:83–111
Variations of Solar Irradiance due to Magnetic Activity	G. A. Chapman	25:633–67
Observed Variability of the Solar Luminosity	H. S. Hudson	26:473–507
Interaction Between the Solar Wind and the Interstellar Medium	T. E. Holzer	27:199–234
Classification of Solar Flares	T. Bai, P. A. Sturrock	27:421–67
Solar Convection	H. C. Spruit, Å. Nordlund, A. M. Title	28:263–301
Equilibrium and Dynamics of Coronal Magnetic Fields	B. C. Low	28:491–524

Flares on the Sun and Other Stars | B. M. Haisch, K. T. Strong, M. Rodonò | 29:275–324
Seismic Observations of the Solar Interior | D. Gough, J. Toomre | 29:627–84
Solar Flares and Coronal Mass Ejections | S. W. Kahler | 30:113–41
Proto-Planetary Nebulae | S. Kwok | 31:63–92
Supernova 1987A Revisited | R. McCray | 31:175–216

STELLAR PHYSICS

Observations of Supernova Remnants | J. C. Raymond | 22:75–95
High Angular Resolution Measurements of Stellar Properties | H. A. McAlister | 23:59–87
Planetary Nebulae and Their Central Stars | J. B. Kaler | 23:89–117
Radio Emission From the Sun and Stars | G. A. Dulk | 23:169–224
The Composition of Field Halo Stars and the Chemical Evolution of the Halo | M. Spite, F. Spite | 23:225–38
Stellar Activity Cycles | S. L. Baliunas, A. H. Vaughan | 23:379–412
On Stellar X-Ray Emission | R. Rosner, L. Golub, G. S. Vaiana | 23:413–52
Molecules in Stars | T. Tsuji | 24:89–125
The Physics of Supernova Explosions | S. E. Woosley, T. A. Weaver | 24:205–53
Recent Progress in the Understanding of Pulsars | J. H. Taylor, D. R. Stinebring | 24:285–327
The Evolution of Massive Stars With Mass Loss | C. Chiosi, A. Maeder | 24:329–75
Mass Loss From Cool Stars | A. K. Dupree | 24:377–420
The Population Concept, Globular Clusters, Subdwarfs, Ages, and the Collapse of the Galaxy | A. Sandage | 24:421–58
Pulsar Timing and General Relativity | D. C. Backer, R. W. Hellings | 24:537–75
Star Formation in Molecular Clouds: Observation and Theory | F. H. Shu, F. C. Adams, S. Lizano | 25:23–81
Wolf-Rayet Stars | D. C. Abbott, P. S. Conti | 25:113–50
Rotation and Magnetic Activity in Main-Sequence Stars | L. W. Hartmann, R. W. Noyes | 25:271–301
Very Low Mass Stars | J. Liebert, R. G. Probst | 25:473–519
Tests of Evolutionary Sequences Using Color-Magnitude Diagrams of Globular Clusters | A. Renzini, F. Fusi Pecci | 26:199–244
Supernovae and Supernova Remnants | K. W. Weiler, R. A. Sramek | 26:295–341
The Infrared Temporal Development of Classsical Novae | R. D. Gehrz | 26:377–412
Abundance Ratios as a Function of Metallicity | J. C. Wheeler, C. Sneden, J. W. Truran, Jr. | 27:279–349
T Tauri Stars: Wild as Dust | C. Bertout | 27:351–95
Quasi-Periodic Oscillations and Noise in Low-Mass X-Ray Binaries | M. van der Klis | 27:517–53
Supernova 1987A | W. D. Arnett, J. N. Bahcall, R. P. Kirshner, S. E. Woosley | 27:629–700
Chemical Analyses of Cool Stars | B. Gustafsson | 27:701–56
Masses and Evolutionary Status of White Dwarfs and Their Progenitors | V. Weidemann | 28:103–37
Cooling of White Dwarfs | F. D'Antona, I. Mazzitelli | 28:139–81
The Origin of Neutron Stars in Binary Systems | R. Canal, J. Isern, J. Labay | 28:183–214
Quantitative Spectroscopy of Hot Stars | R. P. Kudritzki, D. G. Hummer | 28:303–45
Gamma-Ray Bursts | J. C. Higdon, R. E. Lingenfelter | 28:401–36
Rapidly Oscillating Ap Stars | D. W. Kurtz | 28:607–55
The Search for Brown Dwarfs | D. J. Stevenson | 29:163–93
Flares on the Sun and Other Stars | B. M. Haisch, K. T. Strong, M. Rodonò | 29:275–324
Magnetic Fields of Degenerate Stars | G. Chanmugam | 30:143–84
New Developments in Understanding the HR Diagram | C. Chiosi, G. Bertelli, A. Bressan | 30:235–85
Evidence for Black Holes in Stellar Binary Systems | A. P. Cowley | 30:287–310
Type Ia Supernovae as Standard Candles | D. R. Branch, G. A. Tammann | 30:359–89
Asteroseismology | T. M. Brown, R. L. Gilliland | 32:37–82

The r, s, and p Processes in Nucleosynthesis	B. Meyer	32:153–90
Massive Star Populations in Nearby Galaxies	A. Maeder, P. S. Conti	32:227–75
Pre-Main-Sequence Binary Stars	R. D. Mathieu	32:465–530

DYNAMICAL ASTRONOMY
Dynamical Evolution of Globular Clusters	R. Elson, P. Hut, S. Inagaki	25:565–601
The Galactic Spheroid and Old Disk	K. C. Freeman	25:603–32
Recent Advances in Optical Astrometry	D. G. Monet	26:413–40

INTERSTELLAR MEDIUM
Observations of Supernova Remnants	J. C. Raymond	22:75–95
The Influence of Environment on the H I Content of Galaxies	M. P. Haynes, R. Giovanelli, G. L. Chincarini	22:445–70
Planetary Nebulae and Their Central Stars	J. B. Kaler	23:89–117
Cold Outflows, Energetic Winds, and Enigmatic Jets Around Young Stellar Objects	C. J. Lada	23:267–317
The Dynamical Evolution of H II Regions—Recent Theoretical Developments	H. W. Yorke	24:49–87
High-Resolution Optical and Ultraviolet Absorption-Line Studies of Interstellar Gas	L. L. Cowie, A. Songaila	24:499–535
Star Formation in Molecular Clouds: Observation and Theory	F. H. Shu, F. C. Adams, S. Lizano	25:23–81
The Local Interstellar Medium	D. P. Cox, R. J. Reynolds	25:303–44
Large-Scale Expanding Superstructures in Galaxies	G. Tenorio-Tagle, P. Bodenheimer	26:145–97
Supernovae and Supernova Remnants	K. W. Weiler, R. A. Sramek	26:295–341
The Orion Molecular Cloud and Star-Forming Region	R. Genzel, J. Stutzki	27:41–85
A New Component of the Interstellar Matter: Small Grains and Large Aromatic Molecules	J. L. Puget, A. Léger	27:161–98
Interaction Between the Solar Wind and the Interstellar Medium	T. E. Holzer	27:199–234
Diffuse Galactic Gamma-Ray Emission	H. Bloemen	27:469–516
Interstellar Dust and Extinction	J. S. Mathis	28:37–70
Theories of the Hot Interstellar Gas	L. Spitzer, Jr.	28:71–101
Extragalactic H II Regions	G. A. Shields	28:525–60
Radio Propagation Through the Turbulent Interstellar Medium	B. J. Rickett	28:561–605
Distribution of CO in the Milky Way	F. Combes	29:195–237
Dust-Gas Interactions and the Infrared Emission from Hot Astrophysical Plasmas	E. Dwek, R. G. Arendt	30:11–50
Theory of Interstellar Shocks	B. T. Draine, C. F. McKee	31:373–432
Abundances in the Interstellar Medium	T. L. Wilson, R. T. Rood	32:191–226

SMALL STELLAR SYSTEMS
Dynamical Evolution of Globular Clusters	R. Elson, P. Hut, S. Inagaki	25:565–601
The Galactic Nuclear Bulge and the Stellar Content of Spheroidal Systems	J. A. Frogel	26:51–92
Tests of Evolutionary Sequences Using Color-Magnitude Diagrams of Globular Clusters	A. Renzini, F. Fusi Pecci	26:199–244
Quasi-Periodic Oscillations and Noise in Low-Mass X-Ray Binaries	M. van der Klis	27:517–53
The Few-Body Problem in Astrophysics	M. Valtonen, S. Mikkola	29:9–29
Evidence for Black Holes in Stellar Binary Systems	A. P. Cowley	30:287–310
Origin and Evolution of X-Ray Binaries and Binary Radio Pulsars	F. Verbunt	31:93–127
Binary and Millisecond Pulsars	E. S. Phinney, S. R. Kulkarni	32:591–639

THE GALAXY
Sagittarius A and Its Environment	R. L. Brown, H. S. Liszt	22:223–65

Neutron Stars in Interacting Binary Systems   P. C. Joss, S. A. Rappaport   22:537–92

Star Counts and Galactic Structure   J. N. Bahcall   24:577–611

Physical Conditions, Dynamics, and Mass Distribution in the Center of the Galaxy   R. Genzel, C. H. Townes   25:377–423

The IRAS View of the Galaxy and the Solar System   C. A. Beichman   25:521–63

The Galactic Spheroid and Old Disk   K. C. Freeman   25:603–32

The Galactic Nuclear Bulge and the Stellar Content of Spheroidal Systems   J. A. Frogel   26:51–92

Large-Scale Expanding Superstructures in Galaxies   G. Tenorio-Tagle, P. Bodenheimer   26:145–97

Diffuse Galactic Gamma-Ray Emission   H. Bloemen   27:469–516

Kinematics, Chemistry, and Structure of the Galaxy   G. Gilmore, R. F. G. Wyse, K. Kuijken   27:555–627

H I in the Galaxy   J. M. Dickey, F. J. Lockman   28:215–61

The Cosmic Far Ultraviolet Background   S. Bowyer   29:59–88

Ultraviolet Background Radiation   R. C. Henry   29:89–127

Chemical Evolution of the Galaxy   N. C. Rana   29:129–62

The Search for Brown Dwarfs   D. J. Stevenson   29:163–93

Distribution of CO in the Milky Way   F. Combes   29:195–237

The Mass of the Galaxy   M. Fich, S. Tremaine   29:409–45

Millimeter and Submillimeter Interferometry of Astronomical Sources   A. I. Sargent, W. J. Welch   31:297–343

The Distance to the Center of the Galaxy   M. Reid   31:345–72

The Faint End of the Stellar Luminosity Function   M. S. Bessell, G. S. Stringfellow   31:433–71

Galactic Structure Surveys and the Evolution of the Milky Way   S. R. Majewski   31:575–638

EXTRAGALACTIC ASTRONOMY

Structure and Evolution of Irregular Galaxies   J. S. Gallagher, III, D. A. Hunter   22:37–74

The Evolution of Galaxies in Clusters   A. Dressler   22:185–222

Extragalactic Radio Jets   A. H. Bridle, R. A. Perley   22:319–58

Black Hole Models for Active Galactic Nuclei   M. J. Rees   22:471–506

Shells and Rings Around Galaxies   E. Athanassoula, A. Bosma   23:147–68

Emission-Line Regions of Active Galaxies and QSOs   D. E. Osterbrock, W. G. Mathews   24:171–203

Global Structure of Magnetic Fields in Spiral Galaxies   Y. Sofue, M. Fujimoto, R. Wielebinski   24:459–97

The IRAS View of the Extragalactic Sky   B. T. Soifer, J. R. Houck, G. Neugebauer   25:187–230

Cepheids as Distance Indicators   M. W. Feast, A. R. Walker   25:345–75

Existence and Nature of Dark Matter in the Universe   V. Trimble   25:425–72

Polarization Properties of Extragalactic Radio Sources   D. J. Saikia, C. J. Salter   26:93–144

Voids   H. J. Rood   26:245–94

Enhanced Star Formation and Infrared Emission in the Centers of Galaxies   C. M. Telesco   26:343–76

The Luminosity Function of Galaxies   B. Binggeli, A. Sandage, G. A. Tammann   26:509–60

Observational Tests of World Models   A. Sandage   26:561–630

Large-Scale Structure in the Universe Indicated by Galaxy Clusters   N. A. Bahcall   26:631–86

X Rays From Normal Galaxies   G. Fabbiano   27:87–138

Populations in Local Group Galaxies   P. Hodge   27:139–59

Surface Photometry and the Structure of Elliptical Galaxies   J. Kormendy, S. Djorgovski   27:235–77

The Space Distribution of Quasars   F. D. A. Hartwick, D. Schade   28:437–89

Extragalactic H II Regions   G. A. Shields   28:525–60

The Few-Body Problem in Astrophysics   M. Valtonen, S. Mikkola   29:9–29

The Cosmic Far Ultraviolet Background   S. Bowyer   29:59–88
Ultraviolet Background Radiation   R. C. Henry   29:89–127
Structure and Dynamics of Elliptical Galaxies   T. de Zeeuw, M. Franx   29:239–74
Galactic and Extragalactic Supernova Rates   S. van den Bergh, G. A. Tammann   29:363–407
Redshift Surveys of Galaxies   R. Giovanelli, M. P. Haynes   29:499–541
Globular Cluster Systems in Galaxies beyond
   the Local Group   W. E. Harris   29:543–79
Molecular Gas in Galaxies   J. S. Young, N. Z. Scoville   29:581–625
Warps   J. J. Binney   30:51–74
Type Ia Supernovae as Standard Candles   D. R. Branch, G. A. Tammann   30:359–89
The Origin of the X-Ray Background   A. C. Fabian, X. Barcons   30:429–56
Radio Emission From Normal Galaxies   J. J. Condon   30:575–611
Evidence for Evolution in Faint Field Galaxy
   Samples   D. C. Koo, R. G. Kron   30:613–52
Dynamics of Interacting Galaxies   J. E. Barnes, L. E. Hernquist   30:705–42
Millimeter and Submillimeter Interferometry of
   Astronomical Sources   A. I. Sargent, W. J. Welch   31:297–343
Unified Models for Active Galactic Nuclei and
   Quasars   R. R. Antonucci   31:473–521
High Redshift Radio Galaxies   P. J. McCarthy   31:639–88
X-Ray Spectra and Time Variability of Active
   Galactic Nuclei   R. F. Mushotzky, C. Done, K. A.   31:717–61
                                                          Pounds
Physical Parameters along the Hubble Sequence   M. S. Roberts, M. P. Haynes   32:115–52
Cooling Flows in Clusters of Galaxies   A. C. Fabian   32:277–318
Dynamics of Cosmic Flows   A. Dekel   32:371–417

OBSERVATIONAL PHENOMENA
   The Evolution of Galaxies in Clusters   A. Dressler   22:185–222
   Observations of SS 433   B. Margon   22:507–36
   Recent Developments Concerning the Crab
      Nebula   K. Davidson, R. A. Fesen   23:119–46
   The IRAS View of the Extragalactic Sky   B. T. Soifer, J. R. Houck, G.   25:187–230
                                                          Neugebauer
   Existence and Nature of Dark Matter in the
      Universe   V. Trimble   25:425–72
   Polarization Properties of Extragalactic Radio
      Sources   D. J. Saikia, C. J. Salter   26:93–144
   X Rays From Normal Galaxies   G. Fabbiano   27:87–138
   Populations in Local Group Galaxies   P. Hodge   27:139–59
   Astrophysical Contributions of the International
      Ultraviolet Explorer   Y. Kondo, A. Boggess, S. P. Maran   27:397–420
   The Search for Brown Dwarfs   D. J. Stevenson   29:163–93
   X-Ray Astronomy Missions   H. V. D. Bradt, T. Ohashi, K. A.   30:391–427
                                                          Pounds
   Observations of the Isotropy of the Cosmic
      Microwave Background Radiation   A. C. S. Readhead, C. R. Lawrence   30:653–703
   The Faint End of the Stellar Luminosity
      Function   M. S. Bessell, G. S. Stringfellow   31:433–71
   Anisotropies in the Cosmic Microwave
      Background   M. White, D. Scott, J. Silk   32:319–70
   Binary and Millisecond Pulsars   E. S. Phinney, S. R. Kulkarni   32:591–639

GENERAL RELATIVITY AND COSMOLOGY
   Alternatives to the Big Bang   G. F. R. Ellis   22:157–84
   Big Bang Nucleosynthesis: Theories and
      Observations   A. M. Boesgaard, G. Steigman   23:319–78
   Pulsar Timing and General Relativity   D. C. Backer, R. W. Hellings   24:537–75
   Existence and Nature of Dark Matter in the
      Universe   V. Trimble   25:425–72
   Observational Tests of World Models   A. Sandage   26:561–630
   Large-Scale Structure in the Universe Indicated
      by Galaxy Clusters   N. A. Bahcall   26:631–86

The Space Distribution of Quasars — F. D. A. Hartwick, D. Schade — 28:437–89
Inflation for Astronomers — J. V. Narlikar, T. Padmanabhan — 29:325–62
Radioactive Dating of the Elements — J. J. Cowan, F.-K. Thielemann, J. W. Truran, Jr. — 29:447–97

Cosmologial Applications of Gravitational Lensing — R. D. Blandford, R. Narayan — 30:311–58
The Cosmological Constant — S. M. Carroll, W. H. Press, E. L. Turner — 30:499–542

Evidence for Evolution in Faint Field Galaxy Samples — D. C. Koo, R. G. Kron — 30:613–52
Observations of the Isotropy of the Cosmic Microwave Background Radiation — A. C. S. Readhead, C. R. Lawrence — 30:653–703
Astronomical Tests of the Cold Dark Matter Scenario — J. P. Ostriker — 31:689–716
Anisotropies in the Cosmic Microwave Background — M. White, D. Scott, J. Silk — 32:319–70
Baryonic Dark Matter — B. Carr — 32:531–90

INSTRUMENTATION AND TECHNIQUES
Image Formation by Self-Calibration in Radio Astronomy — T. J. Pearson, A. C. S. Readhead — 22:97–130
Astronomical Fourier Transform Spectroscopy Revisited — S. T. Ridgway, J. W. Brault — 22:291–317
Fundamental and Applied Aspects of Astronomical "Seeing" — C. E. Coulman — 23:19–57
High Angular Resolution Measurements of Stellar Properties — H. A. McAlister — 23:59–87
Maximum Entropy Image Restoration in Astronomy — R. Narayan, R. Nityananda — 24:127–70
Charge-Coupled Devices in Astronomy — C. D. Mackay — 24:255–83
The Art of N-Body Building — J. A. Sellwood — 25:151–86
Recent Advances in Optical Astrometry — D. G. Monet — 26:413–40
The Status and Prospects for Ground-Based Observatory Sites — R. H. Garstang — 27:19–40
Astrophysical Contributions of the International Ultraviolet Explorer — Y. Kondo, A. Boggess, S. P. Maran — 27:397–420
X-Ray Astronomy Missions — H. V. D. Bradt, T. Ohashi, K. A. Pounds — 30:391–427

Long Baseline Optical and Infrared Stellar Interferometry — M. Shao, M. M. Colavita — 30:457–98
Adaptive Optics for Astronomy: Principles, Performance, and Applicat — J. M. Beckers — 31:13–62
Millimeter and Submillimeter Interferometry of Astronomical Sources — A. I. Sargent, W. J. Welch — 31:297–343

PHYSICAL PROCESSES
The Origin of Ultra-High-Energy Cosmic Rays — A. M. Hillas — 22:425–44
Observations of SS 433 — B. Margon — 22:507–36
The Physics of Supernova Explosions — S. E. Woosley, T. A. Weaver — 24:205–53
Quasi-Periodic Oscillations and Noise in Low-Mass X-Ray Binaries — M. van der Klis — 27:517–53
The Origin of Neutron Stars in Binary Systems — R. Canal, J. Isern, J. Labay — 28:183–214
Gamma-Ray Bursts — J. C. Higdon, R. E. Lingenfelter — 28:401–36
Collective Plasma Radiation Processes — D. B. Melrose — 29:31–57
Flares on the Sun and Other Stars — B. M. Haisch, K. T. Strong, M. Rodonò — 29:275–324

Astronomical Masers — M. Elitzur — 30:75–112
Smoothed Particle Hydrodynamics — J. J. Monaghan — 30:543–74
Jupiter's Great Red Spot and Other Vortices — P. S. Marcus — 31:523–73
Binary and Millisecond Pulsars — E. S. Phinney, S. R. Kulkarni — 32:591–639

ANNUAL REVIEWS

a nonprofit scientific publisher
4139 El Camino Way
P.O. Box 10139
Palo Alto, CA 94303-0139 • USA

Annual Reviews publications may be ordered directly from our office; through booksellers and subscription agents, worldwide; and through participating professional societies. **Prices are subject to change without notice. We do not ship on approval.**

- **Individuals:** Prepayment required on new accounts. in US dollars, checks drawn on a US bank.
- **Institutional Buyers:** Include purchase order. Calif. Corp. #161041 • ARI Fed. I.D. #94-1156476
- **Students / Recent Graduates:** $10.00 discount from retail price, per volume. *Requirements:* 1. be a degree candidate at, or a graduate within the past three years from, an accredited institution; 2. present proof of status (photocopy of your student I.D. or proof of date of graduation); 3. Order direct from Annual Reviews; 4. prepay. This discount **does not** apply to standing orders, *Index on Diskette*, Special Publications, ARPR, or institutional buyers.
- **Professional Society Members:** Many Societies offer *Annual Reviews* to members at reduced rates. Check with your society or contact our office for a list of participating societies.
- **California orders** add applicable sales tax. • **Canadian orders** add 7% GST. Registration #R 121 449-029.
- **Postage paid** by Annual Reviews (4th class bookrate/surface mail). UPS ground service is available at S2.00 extra per book within the contiguous 48 states only. UPS air service or US airmail is available to any location at actual cost. UPS requires a street address. P.O. Box, APO, FPO, not acceptable.
- **Standing Orders:** Set up a standing order and the new volume in series is sent automatically each year upon publication. Each year you can save 10% by prepayment of prerelease invoices sent 90 days prior to the publication date. Cancellation may be made at any time.
- **Prepublication Orders:** Advance orders may be placed for any volume and will be charged to your account upon receipt. Volumes not yet published will be shipped during month of publication indicated.

N O T E	For copies of individual articles from any *Annual Review*, or copies of any article cited in an *Annual Review*, call **Annual Reviews Preprints and Reprints (ARPR)** toll free 1-800-347-8007 (fax toll free 1-800-347-8008) from the USA or Canada. From elsewhere call 1-415-259-5017.

ANNUAL REVIEWS SERIES *Volumes not listed are no longer in print*	**Prices, postpaid, per volume. USA/other countries**	Regular Order Please send Volume(s):	Standing Order Begin with Volume:
❑ *Annual Review of* ANTHROPOLOGY			
Vols. 1-20 (1972-91).................................$41 / $46			
Vols. 21-22 (1992-93).................................$44 / $49			
Vol. 23 (avail. Oct. 1994).... ....$47 / $52 Vol(s). _____ Vol. _____			
❑ *Annual Review of* ASTRONOMY AND ASTROPHYSICS			
Vols. 1, 5-14, 16-29 (1963, 67-76, 78-91) ....$53 / $58			
Vols. 30-31 (1992-93).................................$57 / $62			
Vol. 32 (avail. Sept. 1994).... ....$60 / $65 Vol(s). _____ Vol. _____			
❑ *Annual Review of* BIOCHEMISTRY			
Vols. 31-34, 36-60 (1962-65,67-91).....................$41 / $47			
Vols. 61-62 (1992-93).................................$46 / $52			
Vol. 63 (avail. July 1994).....................$49 / $55 Vol(s). _____ Vol. _____			
❑ *Annual Review of* BIOPHYSICS AND BIOMOLECULAR STRUCTURE			
Vols. 1-20 (1972-91).................................$55 / $60			
Vols. 21-22 (1992-93).................................$59 / $64			
Vol. 23 (avail. June 1994)....................$62 / $67 Vol(s). _____ Vol. _____			

ANNUAL REVIEWS SERIES *Volumes not listed are no longer in print*	**Prices, postpaid, per volume. USA/other countries**	Regular Order Please send Volume(s):	Standing Order Begin with Volume:
❏ *Annual Review of* **CELL BIOLOGY**			
Vols. 1-7 (1985-91)......................................$41 / $46			
Vols. 8-9 (1992-93)......................................$46 / $51			
Vol. 10 (avail. Nov. 1994)........................$49 / $54		Vol(s). _____	Vol. _____
❏ *Annual Review of* **COMPUTER SCIENCE** (Series suspended)			
Vols. 1-2 (1986-87)......................................$41 / $46			
Vols. 3-4 (1988-89/90)...............................$47 / $52		Vol(s). _____	
Special package price for			
Vols. 1-4 (if ordered together)...............$100 / $115 ❏ Send all four volumes.			
❏ *Annual Review of* **EARTH AND PLANETARY SCIENCES**			
Vols. 1-6, 8-19 (1973-78, 80-91)..............$55 / $60			
Vols. 20-21 (1992-93)..................................$59 / $64			
Vol. 22 (avail. May 1994).....................$62 / $67		Vol(s). _____	Vol. _____
❏ *Annual Review of* **ECOLOGY AND SYSTEMATICS**			
Vols. 2-12, 14-17, 19-22..(1971-81, 83-86, 88-91)...$40 / $45			
Vols. 23-24 (1992-93)..................................$44 / $49			
Vol. 25 (avail. Nov. 1994)......................$47 / $52		Vol(s). _____	Vol. _____
❏ *Annual Review of* **ENERGY AND THE ENVIRONMENT**			
Vols. 1-16 (1976-91)...................................$64 / $69			
Vols. 17-18 (1992-93)..................................$68 / $73			
Vol. 19 (avail. Oct. 1994)......................$71 / $76		Vol(s). _____	Vol. _____
❏ *Annual Review of* **ENTOMOLOGY**			
Vols. 10-16, 18, 20-36 (1965-71, 73, 75-91)...........$40 / $45			
Vols. 37-38 (1992-93)..................................$44 / $49			
Vol. 39 (avail. January 1994)................$47 / $52		Vol(s). _____	Vol. _____
❏ *Annual Review of* **FLUID MECHANICS**			
Vols. 2-4, 7 (1970-72, 75)			
9-11, 16-23 (1977-79, 84-91).....................$40 / $45			
Vols. 24-25 (1992-93)..................................$44 / $49			
Vol. 26 (avail. January 1994)................$47 / $52		Vol(s). _____	Vol. _____
❏ *Annual Review of* **GENETICS**			
Vols. 1-12, 14-25 (1967-78, 80-91)............$40 / $45			
Vols. 26-27 (1992-93)..................................$44 / $49			
Vol. 28 (avail. Dec. 1994)......................$47 / $52		Vol(s). _____	Vol. _____
❏ *Annual Review of* **IMMUNOLOGY**			
Vols. 1-9 (1983-91)......................................$41 / $46			
Vols. 10-11 (1992-93)..................................$45 / $50			
Vol. 12 (avail. April 1994).....................$48 / $53		Vol(s). _____	Vol. _____
❏ *Annual Review of* **MATERIALS SCIENCE**			
Vols. 1, 3-19 (1971, 73-89)........................$68 / $73			
Vols. 20-23 (1990-93)..................................$72 / $77			
Vol. 24 (avail. August 1994)..................$75 / $80		Vol(s). _____	Vol. _____
❏ *Annual Review of* **MEDICINE: Selected Topics in the Clinical Sciences**			
Vols. 9, 11-15, 17-42 (1958, 60-64, 66-42) ....$40 / $45			
Vols. 43-44 (1992-93)..................................$44 / $49			
Vol. 45 (avail. April 1994).....................$47 / $52		Vol(s). _____	Vol. _____